Cosmology

Steven Weinberg
University of Texas at Austin

OXFORD
UNIVERSITY PRESS

OXFORD
UNIVERSITY PRESS

Great Clarendon Street, Oxford OX2 6DP

Oxford University Press is a department of the University of Oxford.
It furthers the University's objective of excellence in research, scholarship,
and education by publishing worldwide in

Oxford New York

Auckland Cape Town Dar es Salaam Hong Kong Karachi
Kuala Lumpur Madrid Melbourne Mexico City Nairobi
New Delhi Shanghai Taipei Toronto

With offices in

Argentina Austria Brazil Chile Czech Republic France Greece
Guatemala Hungary Italy Japan Poland Portugal Singapore
South Korea Switzerland Thailand Turkey Ukraine Vietnam

Oxford is a registered trade mark of Oxford University Press
in the UK and in certain other countries

Published in the United States
by Oxford University Press Inc., New York

British Library Cataloguing in Publication Data

Data available

Library of Congress Cataloging in Publication Data

Data available

Typeset by Newgen Imaging Systems (P) Ltd., Chennai, India
Printed in Great Britain
on acid-free paper by
CPI Group (UK) Ltd, Croydon CR0 4YY

ISBN 978–0–19–852682–7

To Louise, Elizabeth, and Gabrielle

Preface

Research in cosmology has become extraordinarily lively in the past quarter century. In the early 1980s the proposal of the theory of inflation offered a solution to some outstanding cosmological puzzles and provided a mechanism for the origin of large-scale structure, which could be tested by observations of anisotropies in the cosmic microwave background. November 1989 saw the launch of the Cosmic Background Explorer Satellite. Measurements with its spectrophotometer soon established the thermal nature of the cosmic microwave background and determined its temperature to three decimal places, a precision unprecedented in cosmology. A little later the long-sought microwave background anisotropies were found in data taken by the satellite's radiometer. Subsequent observations by ground-based and balloon-borne instruments and eventually by the Wilkinson Microwave Anisotropy Probe showed that these anisotropies are pretty much what would be expected on the basis of inflationary theory. In the late 1990s the use of Type Ia supernovae as standard candles led to the discovery that the expansion of the universe is accelerating, implying that most of the energy of the universe is some sort of dark energy, with a ratio of pressure to density less than $-1/3$. This was confirmed by precise observations of the microwave background anisotropies, and by massive surveys of galaxies, which together provided increasingly accurate values for cosmological parameters.

Meanwhile, the classic methods of astronomy have provided steadily improving independent constraints on the same cosmological parameters. The spectroscopic discovery of thorium and then uranium in the atmospheres of old stars, together with continued study of the turn-off from the main sequence in globular clusters, has narrowed estimates of the age of the universe. The measurement of the deuterium to hydrogen ratio in interstellar absorption combined with calculations of cosmological nucleosynthesis has given a good value for the cosmic density of ordinary baryonic matter, and shown that it is only about a fifth of the density of some sort of mysterious non-baryonic cold dark matter. Observations with the Hubble Space Telescope as well as ground-based telescopes have given increasingly precise values for the Hubble constant. It is greatly reassuring that some of the parameters measured by these other means have values consistent with those found in studies of the cosmic microwave background and large scale structure.

Progress continues. In the years to come, we can expect definite information about whether the dark energy density is constant or evolving, and we hope for signs of gravitational radiation that would open the era of inflation to observation. We may discover the nature of dark matter, either by artificially producing dark matter particles at new large accelerators, or by direct observation of natural dark matter particles impinging on the earth. It remains to be seen if in our times fundamental physical theory can provide a specific theory of inflation or explain dark matter or dark energy.

This new excitement in cosmology came as if on cue for elementary particle physicists. By the 1980s the Standard Model of elementary particles and fields had become well established. Although significant theoretical and experimental work continued, there was now little contact between experiment and new theoretical ideas, and without this contact, particle physics lost much of its liveliness. Cosmology now offered the excitement that particle physicists had experienced in the 1960s and 1970s.

In 1999 I finished my three-volume book on the quantum theory of fields (cited here as "QTF"), and with unaccustomed time on my hands, I set myself the task of learning in detail the theory underlying the great progress in cosmology made in the previous two decades. Although I had done some research on cosmology in the past, getting up to date now turned out to take a fair amount of work. Review articles on cosmology gave good summaries of the data, but they often quoted formulas without giving the derivation, and sometimes even without giving a reference to the original derivation. Occasionally the formulas were wrong, and therefore extremely difficult for me to rederive. Where I could find the original references, the articles sometimes had gaps in their arguments, or relied on hidden assumptions, or used unexplained notation. Often massive computer programs had taken the place of analytic studies. In many cases I found that it was easiest to work out the relevant theory for myself.

This book is the result. Its aim is to give self-contained explanations of the ideas and formulas that are used and tested in modern cosmological observations. The book divides into two parts, each of which in my experience teaching the subject provides enough material for a one-semester graduate course. The first part, Chapters 1 through 4, deals chiefly with the isotropic and homogeneous average universe, with only a brief introduction to the anisotropies in the microwave background in Section 2.6. These chapters are more-or-less in reverse chronological order; Chapter 1 concentrates on the universe since the formation of galaxies, corresponding roughly to redshifts $z < 10$; Chapter 2 deals with the microwave background, emitted at a redshift $z \simeq 1,000$; Chapter 3 describes the early universe, from the beginning of the radiation-dominated expansion to a redshift $z \approx 10^4$ when the density of radiation fell below that of matter; and Chapter 4 takes up the period of inflation that is believed to have preceded the radiation-dominated era. The second part, Chapters 5 through 10, concentrates on the departures from the average universe. After some general formalism in Chapter 5 and its application to the evolution of inhomogeneities in Chapter 6, I return in Chapter 7 to the microwave background anisotropies, and take up the large scale structure of matter in Chapter 8. Gravitational lensing is discussed late, in Chapter 9, because its most important cosmological application may be in the use of weak lensing to study large scale structure. The treatment of inflation in Chapter 4 deals only with the average properties of the universe in the inflationary era; I return to inflation in Chapter 10, which discusses the growth of inhomogeneities from quantum fluctuations during inflation.

To the greatest extent possible, I have tried throughout this book to present analytic calculations of cosmological phenomena, and not just report results obtained elsewhere by numerical computation. The calculations that are used in the literature to compare observation with theory necessarily take many details into account, which either make an analytic treatment impossible, or obscure the main physical features of the calculation. Where this is the case, I have not hesitated to sacrifice some degree of accuracy for greater transparency. This is especially the case in the hydrodynamical treatment of cosmic fluctuations in Sections 6.2 through 6.5, and in the treatment of large scale structure in Chapter 8. But in Section 6.1 and Appendix H I also give an account of the more accurate kinetic theory on which the modern cosmological computer codes are based. Both approaches are applied to the cosmic microwave background anisotropies in Chapter 7.

So much has happened in cosmology since the 1960s that this book necessarily bears little resemblance to my 1972 treatise, *Gravitation and Cosmology*. On occasion I refer back to that book (cited here as "G&C") for material that does not seem worth repeating here. Classical general relativity has not changed much since 1972 (apart from a great strengthening of its experimental verification) so it did not seem necessary to cover gravitation as well as cosmology in the present book. However, as a convenience to readers who want to refresh their knowledge of general relativity, and to establish my notation, I provide a brief introduction to general relativity in Appendix B. Other appendices deal with technical material that is needed here and there in the book. I have also supplied at the back of this book a glossary of symbols that are used in more than one section and an assortment of problems.

In order to keep the book to manageable proportions, I decided to exclude material that was highly speculative. Thus this book does not go into cosmological theory in higher dimensions, or anthropic reasoning, or holographic cosmology, or conjectures about the details of inflation, or many other new ideas. I may perhaps include some of them in a follow-up volume. The present book is largely concerned with what has become mainstream cosmology: a scenario according to which inflation driven by one or more scalar fields is followed by a big bang dominated by radiation, cold dark matter, baryonic matter, and vacuum energy.

I believe that the discussion of topics that are treated in this book is up to date as of $200n$, where n is an integer that varies from 1 to 7 through different parts of the book. I have tried to give full references to the relevant astrophysical literature up to these dates, but I have doubtless missed some articles. The mere absence of a literature reference should not be interpreted as a claim that the work presented is original, though perhaps some of it is. Where I knew them, I included references to postings in the Cornell archive, `http://arxiv.org`, as well as to the published literature. In some cases I had to list only the Cornell archive number, where the article in question had not yet appeared in print, or where it had never been submitted to publication. I have quoted the latest measurements of cosmological parameters known to me, in part because I want to give the reader a sense of what is now observationally possible. But I have not tried to combine measurements from observations of different types, because I did not think that it would add any additional physical insight, and any such cosmological concordance would very soon be out of date.

I owe a great debt to my colleagues at the University of Texas, including Thomas Barnes, Fritz Benedict, Willy Fischler, Karl Gebhardt, Patrick Greene, Richard Matzner, Paul Shapiro, Craig Wheeler, and especially Duane Dicus, who did some of the numerical calculations and supplied many corrections. I am grateful above all among these colleagues to Eiichiro Komatsu, who read through a draft of the manuscript and was a never-failing source of insight and information about cosmological research. I received much help with figures and calculations from my research student Raphael Flauger, and I was warned of numerous errors by Flauger and other students: Yingyue Li Boretz, Kannokkuan Chaicherdsakul, Bo Li, Ian Roederer, and Yuki Watanabe. Matthew Anderson helped with numerical calculations of cosmological nucleosynthesis. I have also benefited much from correspondence on special topics with Ed Bertschinger, Dick Bond, Latham Boyle, Robert Cahn, Alan Guth, Robert Kirshner, Andrei Linde, Eric Linder, Viatcheslav Mukhanov, Saul Perlmutter, Jonathan Pritchard, Adam Riess, Uros Seljak, Paul Steinhardt, Edwin Turner, and Matias Zaldarriaga. Thanks are also due to Jan Duffy and Terry Riley for many helps. Of course, I alone am responsible for any errors that may remain in the book. I hope that readers will let me know of any mistakes they may notice; I will post them on a web page, `http://zippy.ph.utexas.edu/~weinberg/corrections.html`.

Austin, Texas
June 2007

Notation

Latin indices i, j, k, and so on generally run over the three spatial coordinate labels, usually taken as 1, 2, 3.

Greek indices μ, ν, etc. generally run over the four spacetime coordinate labels 1, 2, 3, 0, with x^0 the time coordinate.

Repeated indices are generally summed, unless otherwise indicated.

The flat spacetime metric $\eta_{\mu\nu}$ is diagonal, with elements $\eta_{11} = \eta_{22} = \eta_{33} = 1$, $\eta_{00} = -1$.

Spatial three-vectors are indicated by letters in boldface.

A hat over any vector indicates the corresponding unit vector: Thus, $\hat{v} \equiv \mathbf{v}/|\mathbf{v}|$.

A dot over any quantity denotes the time-derivative of that quantity.

∇^2 is the Laplacian, $\dfrac{\partial^2}{\partial(x^1)^2} + \dfrac{\partial^2}{\partial(x^2)^2} + \dfrac{\partial^2}{\partial(x^3)^2}$.

Except on vectors and tensors, a subscript 0 denotes the present time.

On densities, pressures, and velocities, the subscripts B, D, γ, and ν refer respectively to the baryonic plasma (nuclei plus electrons), cold dark matter, photons, and neutrinos, while the subscripts M and R refer respectively to non-relativistic matter (baryonic plasma plus cold dark matter) and radiation (photons plus neutrinos).

The complex conjugate, transpose, and Hermitian adjoint of a matrix or vector A are denoted A^*, A^{T}, and $A^{\dagger} = A^{*\mathrm{T}}$, respectively. +H.c. or +c.c. at the end of an equation indicates the addition of the Hermitian adjoint or complex conjugate of the foregoing terms.

Beginning in Chapter 5, a bar over any symbol denotes its unperturbed value.

In referring to wave numbers, q is used for co-moving wave numbers, with an arbitrary normalization of the Robertson–Walker scale factor $a(t)$, while k is the present value q/a_0 of the corresponding physical wave number $q/a(t)$. (N.B. This differs from the common practice of using k for the co-moving wave number, with varying conventions for the normalization of $a(t)$.)

Except where otherwise indicated, we use units with $\hbar$ and the speed of light taken to be unity. Throughout $-e$ is the rationalized charge of the electron, so that the fine structure constant is $\alpha = e^2/4\pi \simeq 1/137$.

Numbers in parenthesis at the end of quoted numerical data give the uncertainty in the last digits of the quoted figure.

For other symbols used in more than one section, see the Glossary of Symbols on page 565.

Contents

Contents

The Expansion of the Universe

The visible universe seems the same in all directions around us, at least if we look out to distances larger than about 300 million light years.[1] The isotropy is much more precise (to about one part in 10^{-5}) in the cosmic microwave background, to be discussed in Chapters 2 and 7. As we will see there, this radiation has been traveling to us for about 14 billion years, supporting the conclusion that the universe at sufficiently large distances is nearly the same in all directions.

It is difficult to imagine that we are in any special position in the universe, so we are led to conclude that the universe should appear isotropic to observers throughout the universe. But not to all observers. The universe does not seem at all isotropic to observers in a spacecraft whizzing through our galaxy at half the speed of light. Such observers will see starlight and the cosmic microwave radiation background coming toward them from the direction toward which they are moving with much higher intensity than from behind. In formulating the assumption of isotropy, one should specify that the universe seems the same in all directions to a family of "typical" freely falling observers: those that move with the average velocity of typical galaxies in their respective neighborhoods. That is, conditions must be the same at the same time (with a suitable definition of time) at any points that can be carried into each other by a rotation about any typical galaxy. But any point can be carried into any other by a sequence of such rotations about various typical galaxies, so the universe is then also homogeneous — observers in all typical galaxies at the same time see conditions pretty much the same.[2]

The assumption that the universe is isotropic and homogeneous will lead us in Section 1.1 to choose the spacetime coordinate system so that the metric takes a simple form, first worked out by Friedmann[3] as a solution of the Einstein field equations, and then derived on the basis of isotropy and homogeneity alone by Robertson[4] and Walker.[5] Almost all of modern cosmology is based on this Robertson–Walker metric, at least as a first

---

[1] K. K. S. Wu, O. Lahav, and M. J. Rees, *Nature* **397**, 225 (January 21, 1999). For a contrary view, see P. H. Coleman, L. Pietronero, and R. H. Sanders, *Astron. Astrophys.* **200**, L32 (1988): L. Pietronero, M. Montuori, and F. Sylos-Labini, in *Critical Dialogues in Cosmology*, (World Scientific, Singapore, 1997): 24; F. Sylos-Labini, F. Montuori, and L. Pietronero, *Phys. Rep.* **293**, 61 (1998).

[2] The Sloan Digital Sky Survey provides evidence that the distribution of galaxies is homogeneous on scales larger than about 300 light years; see J. Yadav, S. Bharadwaj, B. Pandey, and T. R. Seshadri, *Mon. Not. Roy. Astron. Soc.* **364**, 601 (2005) [astro-ph/0504315].

[3] A. Friedmann, *Z. Phys.* **10**, 377 (1922); *ibid.* **21**, 326 (1924).

[4] H. P. Robertson, *Astrophys. J.* **82**, 284 (1935); *ibid.*, **83**, 187, 257 (1936).

[5] A. G. Walker, *Proc. Lond. Math. Soc.* (2) **42**, 90 (1936).

approximation. The observational implications of these assumptions are discussed in Sections 1.2–1.4, without reference to any dynamical assumptions. The Einstein field equations are applied to the Robertson–Walker metric in Section 1.5, and their consequences are then explored in Sections 1.6–1.13.

1.1 Spacetime geometry

As preparation for working out the spacetime metric, we first consider the geometry of a three-dimensional homogeneous and isotropic space. As discussed in Appendix B, geometry is encoded in a *metric* $g_{ij}(\mathbf{x})$ (with i and j running over the three coordinate directions), or equivalently in a *line element* $ds^2 \equiv g_{ij}\, dx^i\, dx^j$, with summation over repeated indices understood. (We say that ds is the *proper distance* between $\mathbf{x}$ and $\mathbf{x} + d\mathbf{x}$, meaning that it is the distance measured by a surveyor who uses a coordinate system that is Cartesian in a small neighborhood of the point $\mathbf{x}$.) One obvious homogeneous isotropic three-dimensional space with positive definite lengths is flat space, with line element

$$ds^2 = d\mathbf{x}^2 \; . \tag{1.1.1}$$

The coordinate transformations that leave this invariant are here simply ordinary three-dimensional rotations and translations. Another fairly obvious possibility is a spherical surface in four-dimensional Euclidean space with some radius a, with line element

$$ds^2 = d\mathbf{x}^2 + dz^2 \; , \qquad z^2 + \mathbf{x}^2 = a^2 \; . \tag{1.1.2}$$

Here the transformations that leave the line element invariant are *four-dimensional rotations*; the direction of $\mathbf{x}$ can be changed to any other direction by a four-dimensional rotation that leaves z unchanged (that is, an ordinary three-dimensional rotation), while $\mathbf{x}$ can be carried into any other point by a four-dimensional rotation that does change z. It can be proved[6] that the only other possibility (up to a coordinate transformation) is a hyperspherical surface in four-dimensional pseudo-Euclidean space, with line element

$$ds^2 = d\mathbf{x}^2 - dz^2 \; , \qquad z^2 - \mathbf{x}^2 = a^2 \; , \tag{1.1.3}$$

where a^2 is (so far) an arbitrary positive constant. The coordinate transformations that leave this invariant are four-dimensional pseudo-rotations, just like Lorentz transformations, but with z instead of time.

---

[6]See S. Weinberg, *Gravitation and Cosmology* (John Wiley & Sons, New York, 1972) [quoted below as G&C], Sec. 13.2.

We can rescale coordinates

$$\mathbf{x}' \equiv a\mathbf{x} , \qquad z' \equiv az .\tag{1.1.4}$$

Dropping primes, the line elements in the spherical and hyperspherical cases are

$$ds^2 = a^2\left[d\mathbf{x}^2 \pm dz^2\right] , \qquad z^2 \pm \mathbf{x}^2 = 1 .\tag{1.1.5}$$

The differential of the equation $z^2 \pm \mathbf{x}^2 = 1$ gives $zdz = \mp\mathbf{x}\cdot d\mathbf{x}$ so

$$ds^2 = a^2\left[d\mathbf{x}^2 \pm \frac{(\mathbf{x}\cdot d\mathbf{x})^2}{1\mp\mathbf{x}^2}\right] .\tag{1.1.6}$$

We can extend this to the case of Euclidean space by writing it as

$$ds^2 = a^2\left[d\mathbf{x}^2 + K\frac{(\mathbf{x}\cdot d\mathbf{x})^2}{1-K\mathbf{x}^2}\right] ,\tag{1.1.7}$$

where

$$K = \begin{cases} +1 & \text{spherical} \\ -1 & \text{hyperspherical} \\ 0 & \text{Euclidean} \end{cases}\tag{1.1.8}$$

(The constant K is often written as k, but we will use upper case for this constant throughout this book to avoid confusion with the symbols for wave number or for a running spatial coordinate index.) Note that we must take $a^2 > 0$ in order to have ds^2 positive at $\mathbf{x} = 0$, and hence everywhere.

There is an obvious way to extend this to the geometry of spacetime: just include a term (1.1.7) in the spacetime line element, with a now an arbitrary function of time (known as the *Robertson–Walker scale factor*):

$$d\tau^2 \equiv -g_{\mu\nu}(x)dx^\mu dx^\nu = dt^2 - a^2(t)\left[d\mathbf{x}^2 + K\frac{(\mathbf{x}\cdot d\mathbf{x})^2}{1-K\mathbf{x}^2}\right] .\tag{1.1.9}$$

Another theorem[7] tells us that this is the unique metric (up to a coordinate transformation) if the universe appears spherically symmetric and isotropic to a set of freely falling observers, such as astronomers in typical galaxies. The components of the metric in these coordinates are:

$$g_{ij} = a^2(t)\left(\delta_{ij} + K\frac{x^i x^j}{1-K\mathbf{x}^2}\right) , \qquad g_{i0} = 0 , \qquad g_{00} = -1 ,\tag{1.1.10}$$

----

[7]G&C, Sec. 13.5.

with i and j running over the values 1, 2, and 3, and with $x^0 \equiv t$ the time coordinate in our units, with the speed of light equal to unity. Instead of the quasi-Cartesian coordinates x^i, we can use spherical polar coordinates, for which

$$d\mathbf{x}^2 = dr^2 + r^2 d\Omega , \quad d\Omega \equiv d\theta^2 + \sin^2\theta \, d\phi^2 .$$

so

$$d\tau^2 = dt^2 - a^2(t)\left[\frac{dr^2}{1-Kr^2} + r^2 d\Omega\right] . \tag{1.1.11}$$

in which case the metric becomes diagonal, with

$$g_{rr} = \frac{a^2(t)}{1-Kr^2} , \quad g_{\theta\theta} = a^2(t)r^2 , \quad g_{\phi\phi} = a^2(t)r^2 \sin^2\theta , \quad g_{00} = -1 . \tag{1.1.12}$$

We will see in Section 1.5 that the dynamical equations of cosmology depend on the overall normalization of the function $a(t)$ only through a term $K/a^2(t)$, so for $K = 0$ this normalization has no significance; all that matters are the ratios of the values of $a(t)$ at different times.

The equation of motion of freely falling particles is given in Appendix B by Eq. (B.12):

$$\frac{d^2 x^\mu}{du^2} + \Gamma^\mu_{\nu\kappa}\frac{dx^\nu}{du}\frac{dx^\kappa}{du} = 0 , \tag{1.1.13}$$

where $\Gamma^\mu_{\nu\kappa}$ is the affine connection, given in Appendix B by Eq. (B.13),

$$\Gamma^\mu_{\nu\kappa} = \frac{1}{2}g^{\mu\lambda}\left[\frac{\partial g_{\lambda\nu}}{\partial x^\kappa} + \frac{\partial g_{\lambda\kappa}}{\partial x^\nu} - \frac{\partial g_{\nu\kappa}}{\partial x^\lambda}\right] . \tag{1.1.14}$$

and u is a suitable variable parameterizing positions along the spacetime curve, proportional to τ for massive particles. (A spacetime path $x^\mu = x^\mu(u)$ satisfying Eq. (1.1.13) is said to be a *geodesic*, meaning that the integral $\int d\tau$ is stationary under any infinitesimal variation of the path that leaves the endpoints fixed.) Note in particular that the derivatives $\partial_i g_{00}$ and $\dot{g}_{0i}$ vanish, so $\Gamma^i_{00} = 0$. A particle at rest in these coordinates will therefore stay at rest, so *these are co-moving coordinates,* which follow the motion of typical observers. Because $g_{00} = -1$, the proper time interval $(-g_{\mu\nu}dx^\mu dx^\nu)^{1/2}$ for a co-moving clock is just dt, so t is the time measured in the rest frame of a co-moving clock.

The meaning of the Robertson–Walker scale factor $a(t)$ can be clarified by calculating the proper distance at time t from the origin to a co-moving object at radial coordinate r:

$$d(r,t) = a(t) \int_0^r \frac{dr}{\sqrt{1 - Kr^2}} = a(t) \times \begin{cases} \sin^{-1} r & K = +1 \\ \sinh^{-1} r & K = -1 \\ r & K = 0 \end{cases} \quad (1.1.15)$$

In this coordinate system a co-moving object has r time-independent, so the proper distance from us to a co-moving object increases (or decreases) with $a(t)$. Since there is nothing special about our own position, the proper distance between any two co-moving observers anywhere in the universe must also be proportional to $a(t)$. The rate of change of any such proper distance $d(t)$ is just

$$\dot{d} = d\,\dot{a}/a \,. \quad (1.1.16)$$

We will see in the following section that in fact $a(t)$ is increasing.

We also need the non-zero components of the affine connection, given by Eq. (1.1.14) as:

$$\Gamma^0_{ij} = -\frac{1}{2}\left(g_{0i,j} + g_{0j,i} - g_{ij,0}\right) = a\dot{a}\left(\delta_{ij} + K\frac{x^i x^j}{1 - K\mathbf{x}^2}\right)$$

$$= a\dot{a}\tilde{g}_{ij} \,, \quad (1.1.17)$$

$$\Gamma^i_{0j} = \frac{1}{2}g^{il}\left(g_{l0,j} + g_{lj,0} - g_{0j,l}\right) = \frac{\dot{a}}{a}\delta_{ij} \,, \quad (1.1.18)$$

$$\Gamma^i_{jl} = \frac{1}{2}\tilde{g}^{im}\left(\frac{\partial \tilde{g}_{jm}}{\partial x^l} + \frac{\partial \tilde{g}_{lm}}{\partial x^j} - \frac{\partial \tilde{g}_{jl}}{\partial x^m}\right) \equiv \tilde{\Gamma}^i_{jl} \,. \quad (1.1.19)$$

Here $\tilde{g}_{ij}$ and $\tilde{\Gamma}^i_{jl}$ are the purely spatial metric and affine connection, and $\tilde{g}^{ij}$ is the reciprocal of the 3×3 matrix $\tilde{g}_{ij}$, which in general is different from the ij component of the reciprocal of the 4×4 matrix $g_{\mu\nu}$. In quasi-Cartesian coordinates,

$$\tilde{g}_{ij} = \delta_{ij} + K\frac{x^i x^j}{1 - K\mathbf{x}^2} \,, \qquad \tilde{\Gamma}^i_{jl} = K\tilde{g}_{jl}x^i \,. \quad (1.1.20)$$

We can use these components of the affine connection to find the motion of a particle that is not at rest in the co-moving coordinate system. First, let's calculate the rate of change of the momentum of a particle of non-zero mass m_0. Consider the quantity

$$P \equiv m_0 \sqrt{g_{ij}\frac{dx^i}{d\tau}\frac{dx^j}{d\tau}} \quad (1.1.21)$$

where $d\tau^2 = dt^2 - g_{ij}dx^i dx^j$. In a locally inertial Cartesian coordinate system, for which $g_{ij} = \delta_{ij}$, we have $d\tau = dt\sqrt{1 - \mathbf{v}^2}$ where $v^i = dx^i/dt$, so Eq. (1.1.21) is the formula given by special relativity for the magnitude of the momentum. On the other hand, the quantity (1.1.21) is evidently invariant under arbitrary changes in the *spatial* coordinates, so we can evaluate it just as well in co-moving Robertson–Walker coordinates. This can be done directly, using Eq. (1.1.13), but to save work, suppose we adopt a spatial coordinate system in which the particle position is near the origin $x^i = 0$, where $\tilde{g}_{ij} = \delta_{ij} + O(\mathbf{x}^2)$, and we can therefore ignore the purely spatial components Γ^i_{jk} of the affine connection. General relativity gives the equation of motion

$$\frac{d^2 x^i}{d\tau^2} = -\Gamma^i_{\mu\nu} \frac{dx^\mu}{d\tau} \frac{dx^\nu}{d\tau} = -\frac{2}{a} \frac{da}{dt} \frac{dx^i}{d\tau} \frac{dt}{d\tau} .$$

Multiplying with $d\tau/dt$ gives

$$\frac{d}{dt} \left(\frac{dx^i}{d\tau} \right) = -\frac{2}{a} \frac{da}{dt} \frac{dx^i}{d\tau} ,$$

whose solution is

$$\frac{dx^i}{d\tau} \propto \frac{1}{a^2(t)} . \tag{1.1.22}$$

Using this in Eq. (1.1.21) with a metric $g_{ij} = a^2(t)\delta_{ij}$, we see that

$$P(t) \propto 1/a(t) . \tag{1.1.23}$$

This holds for any non-zero mass, however small it may be compared to the momentum. Hence, although for photons both m_0 and $d\tau$ vanish, Eq. (1.1.23) is still valid.

It is important to characterize the paths of photons and material particles in interpreting astronomical observations (especially of gravitational lenses, in Chapter 9). Photons and particles passing through the origin of our spatial coordinate system obviously travel on straight lines in this coordinate system, which are spatial geodesics, curves that satisfy the condition

$$\frac{d^2 x^i}{ds^2} + \tilde{\Gamma}^i_{jl} \frac{dx^j}{ds} \frac{dx^l}{ds} = 0 , \tag{1.1.24}$$

where ds is the three-dimensional proper length

$$ds^2 \equiv \tilde{g}_{ij} \, dx^i \, dx^j . \tag{1.1.25}$$

But the property of being a geodesic is invariant under coordinate transformations (since it states the vanishing of a vector), so the path of the photon or particle will also be a spatial geodesic in any spatial coordinate system, including those in which the photon or particle's path does *not* pass through the origin. (This can be seen in detail as follows. Using Eqs. (1.1.17) and (1.1.18), the equations of motion (1.1.13) of a photon or material particle are

$$0 = \frac{d^2x^i}{du^2} + \Gamma^i_{jl}\frac{dx^j}{du}\frac{dx^l}{du} + \frac{2\dot{a}}{a}\frac{dx^i}{du}\frac{dt}{du} \qquad (1.1.26)$$

$$0 = \frac{d^2t}{du^2} + a\dot{a}\tilde{g}_{ij}\frac{dx^i}{du}\frac{dx^j}{du} \ . \qquad (1.1.27)$$

Eq. (1.1.26) can be written

$$0 = \left(\frac{ds}{du}\right)^2\left[\frac{d^2x^i}{ds^2} + \Gamma^i_{jl}\frac{dx^j}{ds}\frac{dx^l}{ds}\right] + \left[\frac{d^2s}{du^2} + \frac{2\dot{a}}{a}\frac{dt}{du}\frac{ds}{du}\right]\frac{dx^i}{ds} \ , \qquad (1.1.28)$$

where s is so far arbitrary. If we take s to be the proper length (1.1.25) in the spatial geometry, then as we have seen

$$du^2 \propto d\tau^2 \propto dt^2 - a^2\,ds^2$$

Dividing by du^2, differentiating with respect to u, and using Eq. (1.1.27) shows that

$$\frac{d^2s}{du^2} + \frac{2\dot{a}}{a}\frac{dt}{du}\frac{ds}{du} = 0 \ ,$$

so that Eq. (1.1.28) gives Eq. (1.1.24).)

There are various smoothed-out vector and tensor fields, like the current of galaxies and the energy-momentum tensor, whose mean values satisfy the requirements of isotropy and homogeneity. Isotropy requires that the mean value of any three-vector v^i must vanish, and homogeneity requires the mean value of any three-scalar (that is, a quantity invariant under purely spatial coordinate transformations) to be a function only of time, so the current of galaxies, baryons, etc. has components

$$J^i = 0 \ , \quad J^0 = n(t) \ , \qquad (1.1.29)$$

with $n(t)$ the number of galaxies, baryons, etc. per proper volume in a co-moving frame of reference. If this is conserved, in the sense of Eq. (B.38), then

$$0 = J^\mu_{\ ;\mu} = \frac{\partial J^\mu}{\partial x^\mu} + \Gamma^\mu_{\mu\nu}J^\nu = \frac{dn}{dt} + \Gamma^i_{i0}n = \frac{dn}{dt} + 3\frac{\dot{a}}{a}n$$

so

$$n(t) = \frac{\text{constant}}{a^3(t)} \ . \qquad (1.1.30)$$

This shows the decrease of number densities due to the expansion of the co-moving coordinate mesh for increasing $a(t)$.

Likewise, isotropy requires the mean value of any three-tensor t^{ij} at $\mathbf{x} = 0$ to be proportional to δ_{ij} and hence to g^{ij}, which equals $a^{-2}\delta_{ij}$ at $\mathbf{x} = 0$. Homogeneity requires the proportionality coefficient to be some function only of time. Since this is a proportionality between two three-tensors t^{ij} and g^{ij} it must remain unaffected by an arbitrary transformation of space coordinates, including those transformations that preserve the form of g^{ij} while taking the origin into any other point. Hence homogeneity and isotropy require the components of the energy-momentum tensor *everywhere* to take the form

$$T^{00} = \rho(t) , \quad T^{0i} = 0 , \quad T^{ij} = \tilde{g}^{ij}(\mathbf{x}) \, a^{-2}(t) \, p(t) . \quad (1.1.31)$$

(These are the conventional definitions of proper energy density ρ and pressure p, as given by Eq. (B.43) in the case of a velocity four-vector with $u^i = 0$, $u^0 = 1$.) The momentum conservation law $T^{i\mu}{}_{;\mu} = 0$ is automatically satisfied for the Robertson–Walker metric and the energy-momentum tensor (1.1.31), but the energy conservation law gives the useful information

$$0 = T^{0\mu}{}_{;\mu} = \frac{\partial T^{0\mu}}{\partial x^\mu} + \Gamma^0_{\mu\nu} T^{\nu\mu} + \Gamma^\mu_{\mu\nu} T^{0\nu}$$

$$= \frac{\partial T^{00}}{\partial t} + \Gamma^0_{ij} T^{ij} + \Gamma^i_{i0} T^{00} = \frac{d\rho}{dt} + \frac{3\dot{a}}{a}\left(p + \rho\right) ,$$

so that

$$\frac{d\rho}{dt} + \frac{3\dot{a}}{a}\left(p + \rho\right) = 0 . \quad (1.1.32)$$

This can easily be solved for an equation of state of the form

$$p = w\rho \quad (1.1.33)$$

with w time-independent. In this case, Eq. (1.1.32) gives

$$\rho \propto a^{-3-3w} . \quad (1.1.34)$$

In particular, this applies in three frequently encountered extreme cases:

- **Cold Matter** (e.g. dust): $p = 0$

$$\rho \propto a^{-3} \quad (1.1.35)$$

- **Hot Matter** (e.g. radiation): $p = \rho/3$

$$\rho \propto a^{-4} \quad (1.1.36)$$

- **Vacuum energy**: As we will see in Section 1.5, there is another kind of energy-momentum tensor, for which $T^{\mu\nu} \propto g^{\mu\nu}$, so that $p = -\rho$, in which case the solution of Eq. (1.1.32) is that ρ is a constant, known (up to conventional numerical factors) either as the *cosmological constant* or the *vacuum energy*.

These results apply separately for coexisting cold matter, hot matter, and a cosmological constant, provided that there is no interchange of energy between the different components. They will be used together with the Einstein field equations to work out the dynamics of the cosmic expansion in Section 1.5.

So far, we have considered only local properties of the spacetime. Now let us look at it in the large. For $K = +1$ space is finite, though like any spherical surface it has no boundary. The coordinate system used to derive Eq. (1.1.7) with $K = +1$ only covers half the space, with $z > 0$, in the same way that a polar projection map of the earth can show only one hemisphere. Taking account of the fact that z can have either sign, the circumference of the space is $2\pi a$, and its volume is $2\pi^2 a^3$.

The spaces with $K = 0$ or $K = -1$ are usually taken to be infinite, but there are other possibilities. It is also possible to have finite spaces with the same local geometry, constructed by imposing suitable conditions of periodicity. For instance, in the case $K = 0$ we might identify the points $\mathbf{x}$ and $\mathbf{x} + n_1\mathbf{L}_1 + n_2\mathbf{L}_2 + n_3\mathbf{L}_3$, where n_1, n_2, n_3 run over all integers, and $\mathbf{L}_1$, $\mathbf{L}_2$, and $\mathbf{L}_3$ are fixed non-coplanar three-vectors that characterize the space. This space is then finite, with volume $a^3 \mathbf{L}_1 \cdot (\mathbf{L}_2 \times \mathbf{L}_3)$. Looking out far enough, we should see the same patterns of the distribution of matter and radiation in opposite directions. There is no sign of this in the observed distribution of galaxies or cosmic microwave background fluctuations, so any periodicity lengths such as $|\mathbf{L}_i|$ must be larger than about 10^{10} light years.[8]

There are an infinite number of possible periodicity conditions for $K = -1$ as well as for $K = +1$ and $K = 0$.[9] We will not consider these possibilities further here, because they seem ill-motivated. In imposing conditions of periodicity we give up the rotational (though not translational) symmetry that led to the Robertson–Walker metric in the first place, so there seems little reason to impose these periodicity conditions while limiting the local spacetime geometry to that described by the Robertson–Walker metric.

---

[8]N. J. Cornish *et al.*, *Phys. Rev. Lett.* **92**, 201302 (2004); N. G. Phillips & A. Kogut, *Astrophys. J.* **545**, 820 (2006) [astro-ph/0404400].

[9]For reviews of this subject, see G. F. R. Ellis, *Gen. Rel. & Grav.* **2**, 7 (1971); M. Lachièze-Rey and J.-P. Luminet, *Phys. Rept.* **254**, 135 (1995); M. J. Rebouças, in *Proceedings of the Xth Brazilian School of Cosmology and Gravitation*, eds. M. Novello and S. E. Perez Bergliaffa (American Institute of Physics Conference Proceedings, Vol. 782, New York, 2005): 188 [astro-ph/0504365].

1.2 The cosmological redshift

The general arguments of the previous section gave no indication whether the scale factor $a(t)$ in the Robertson–Walker metric (1.1.9) is increasing, decreasing, or constant. This information comes to us from the observation of a shift in the frequencies of spectral lines from distant galaxies as compared with their values observed in terrestrial laboratories.

To calculate these frequency shifts, let us adopt a Robertson–Walker coordinate system in which we are at the center of coordinates, and consider a light ray coming to us along the radial direction. A ray of light obeys the equation $d\tau^2 = 0$, so for such a light ray Eq. (1.1.11) gives

$$dt = \pm a(t)\frac{dr}{\sqrt{1 - Kr^2}} \qquad (1.2.1)$$

For a light ray coming toward the origin from a distant source, r decreases as t increases, so we must choose the minus sign in Eq. (1.2.1). Hence if light leaves a source at co-moving coordinate r_1 at time t_1, it arrives at the origin $r = 0$ at a later time t_0, given by

$$\int_{t_1}^{t_0} \frac{dt}{a(t)} = \int_0^{r_1} \frac{dr}{\sqrt{1 - Kr^2}} . \qquad (1.2.2)$$

Taking the differential of this relation, and recalling that the radial coordinate r_1 of co-moving sources is time-independent, we see that the interval δt_1 between departure of subsequent light signals is related to the interval δt_0 between arrivals of these light signals by

$$\frac{\delta t_1}{a(t_1)} = \frac{\delta t_0}{a(t_0)} \qquad (1.2.3)$$

If the "signals" are subsequent wave crests, the emitted frequency is $\nu_1 = 1/\delta t_1$, and the observed frequency is $\nu_0 = 1/\delta t_0$, so

$$\nu_0/\nu_1 = a(t_1)/a(t_0) . \qquad (1.2.4)$$

If $a(t)$ is increasing, then this is a redshift, a decrease in frequency by a factor $a(t_1)/a(t_0)$, equivalent to an increase in wavelength by a factor conventionally called $1 + z$:

$$1 + z = a(t_0)/a(t_1) . \qquad (1.2.5)$$

Alternatively, if $a(t)$ is decreasing then we have a blueshift, a decrease in wavelength given by the factor Eq. (1.2.5) with z negative. These results are frequently interpreted in terms of the familiar Doppler effect; Eq. (1.1.15) shows that for an increasing or decreasing $a(t)$, the proper distance to any co-moving source of light like a typical galaxy increases or decreases with time, so that such sources are receding from us or approaching us, which naturally produces a redshift or blueshift. For this reason, galaxies with redshift (or blueshift) z are often said to have a cosmological radial velocity cz. (The meaning of relative velocity is clear only for $z \ll 1$, so the existence of distant sources with $z > 1$ does not imply any violation of special relativity.) However, the interpretation of the cosmological redshift as a Doppler shift can only take us so far. In particular, the increase of wavelength from emission to absorption of light does not depend on the rate of change of $a(t)$ at the times of emission or absorption, but on the increase of $a(t)$ in the whole period from emission to absorption.

We can also understand the frequency shift (1.2.4) by reference to the quantum theory of light: The momentum of a photon of frequency v is hv/c (where h is Planck's constant), and we saw in the previous section that this momentum varies as $1/a(t)$.

For nearby sources, we may expand $a(t)$ in a power series, so

$$a(t) \simeq a(t_0)\,[1 + (t - t_0)H_0 + \ldots] \tag{1.2.6}$$

where H_0 is a coefficient known as the *Hubble constant*:

$$H_0 \equiv \dot{a}(t_0)/a(t_0) . \tag{1.2.7}$$

Eq. (1.2.5) then gives the fractional increase in wavelength as

$$z = H_0\ (t_0 - t_1) + \ldots . \tag{1.2.8}$$

Note that for close objects, $t_0 - t_1$ is the proper distance d (in units with $c = 1$). We therefore expect a redshift (for $H_0 > 0$) or blueshift (for $H_0 < 0$) that increases linearly with the proper distance d for galaxies close enough to use the approximation (1.2.6):

$$z = H_0 d + \ldots . \tag{1.2.9}$$

The redshift of light from other galaxies was first observed in the 1910s by Vesto Melvin Slipher at the Lowell Observatory in Flagstaff, Arizona. In 1922, he listed 41 spiral nebulae, of which 36 had positive z up to 0.006, and only 5 had negative z, the most negative being the Andromeda nebula M31, with $z = -0.001$.[1] From 1918 to 1925 C. Wirtz and K. Lundmark[2]

---

[1] V. M. Slipher, table prepared for A. S. Eddington, *The Mathematical Theory of Relativity*, 2nd ed. (Cambridge University Press, London, 1924): 162.

[2] C. Wirtz, *Astr. Nachr.* **206**, 109 (1918); *ibid.* **215**, 349 (1921); *ibid.* **216**, 451 (1922); *ibid.* **222**, 21 (1924); *Scientia* **38**, 303 (1925); K. Lundmark, *Stock. Hand.* **50**, No. 8 (1920); *Mon. Not. Roy. Astron. Soc.* **84**, 747 (1924); *ibid.* **85**, 865 (1925).

discovered a number of spiral nebulae with redshifts that seemed to increase with distance. But until 1923 it was only possible to infer the *relative* distances of the spiral nebulae, using observations of their apparent luminosity or angular diameter. With the absolute luminosity and physical dimensions unknown, it was even possible that the spiral nebulae were outlying parts of our own galaxy, as was in fact believed by many astronomers. Edwin Hubble's 1923 discovery of Cepheid variable stars in the Andromeda nebula M31 (discussed in the next section) allowed him to estimate its distance and size, and made it clear that the spiral nebulae are galaxies like our own, rather than objects in our own galaxy.

No clear linear relation between redshift and distance could be seen in the early data of Slipher, Wirtz, and Lundmark, because of a problem that has continued to bedevil measurements of the Hubble constant down to the present. Real galaxies generally do not move only with the general expansion or contraction of the universe; they typically have additional "peculiar" velocities of hundreds of kilometers per second, caused by gravitational fields of neighboring galaxies and intergalactic matter. To see a linear relation between redshift and distance, it is necessary to study galaxies with $|z| \gg 10^{-3}$, whose cosmological velocities zc are thousands of kilometers per second.

In 1929 Hubble[3] announced that he had found a "roughly linear" relation between redshift and distance. But at that time redshifts and distances had been measured only for galaxies out to the large cluster of galaxies in the constellation Virgo, whose redshift indicates a radial velocity of about 1,000 km/sec, not much larger than typical peculiar velocities. His data points were therefore spread out widely in a plot of redshift versus distance, and did not really support a linear relation. But by the early 1930s he had measured redshifts and distances out to the Coma cluster, with redshift $z \simeq 0.02$, corresponding to a recessional velocity of about 7,000 km/sec, and a linear relation between redshift and distance was evident. The conclusion was clear (at least, to some cosmologists): the universe really is expanding. The correctness of this interpretation of the redshift is supported by observations to be discussed in Section 1.7.

From Hubble's time to the present galaxies have been discovered with ever larger redshifts. Galaxies were found with redshifts of order unity, for which expansions such as Eq. (1.2.9) are useless, and we need formulas that take relativistic effects into account, as discussed in Sections 1.4 and 1.5. At the time of writing, the largest accurately measured redshift is for a galaxy observed with the Subaru telescope.[4] The Lyman alpha line from

[3]E. P. Hubble, *Proc. Nat. Acad. Sci.* **15**, 168 (1929).
[4]M. Iye *et al.*, *Nature* **443**, 186 (2006) [astro-ph/0609393].

this galaxy (emitted in the transition from the $2p$ to $1s$ levels of hydrogen), which is normally at an ultraviolet wavelength of 1,215 Å, is observed at the infrared wavelength of 9,682 Å, indicating a redshift $1 + z = 9682/1215$, or $z = 6.96$.

It may eventually become possible to measure the expansion rate $H(t) \equiv \dot{a}(t)/a(t)$ at times t earlier than the present, by observing the change in very accurately measured redshifts of individual galaxies over times as short as a decade.[5] By differentiating Eq. (1.2.5) we see that the rate of change of redshift with the time of observation is

$$\frac{dz}{dt_0} = \frac{\dot{a}(t_0)}{a(t_1)} - \frac{a(t_0)\,\dot{a}(t_1)}{a^2(t_1)} \frac{dt_1}{dt_0} = \left[H_0 - H(t_1)\frac{dt_1}{dt_0} \right](1 + z) \ .$$

From the same argument that led to Eq. (1.2.3) we have $dt_1/dt_0 = 1/(1+z)$, so if we measure dz/dt_0 we can find the expansion rate at the time of light emission from the formula

$$H(t_1) = H_0(1 + z) - \frac{dz}{dt_0} \ . \tag{1.2.10}$$

1.3 Distances at small redshift: The Hubble constant

We must now think about how astronomical distances are measured. In this section we will be considering objects that are relatively close, say with z not much greater than 0.1, so that effects of the spacetime curvature and cosmic expansion on distance determinations can be neglected. These measurements are of cosmological importance in themselves, as they are used to learn the value of the Hubble constant H_0. Also, distance measurements at larger redshift, which are used to find the shape of the function $a(t)$, rely on the observations of "standard candles," objects of known intrinsic luminosity, that must be identified and calibrated by studies at these relatively small redshifts. Distance determinations at larger redshift will be discussed in Section 1.6, after we have had a chance to lay a foundation in Sections 1.4 and 1.5 for an analysis of the effects of expansion and spacetime geometry on measurements of distances of very distant objects.

It is conventional these days to separate the objects used to measure distances in cosmology into primary and secondary distance indicators. The absolute luminosities of the primary distance indicators in our local group

---

[5]A. Loeb, *Astrophys. J.* **499**, L111 (1998) [astro-ph/9802122]; P-S. Corasaniti, D. Huterer, and A. Melchiorri, *Phys. Rev. D* **75**, 062001 (2007) [astro-ph/0701433]. For an earlier suggestion along this line, see A. Sandage, *Astrophys. J.* **139**, 319 (1962).

of galaxies are measured either directly, by kinematic methods that do not depend on an *a priori* knowledge of absolute luminosities, or indirectly, by observation of primary distance indicators in association with other primary distance indicators whose distance is measured by kinematic methods. The sample of these relatively close primary distance indicators is large enough to make it possible to work out empirical rules that give their absolute luminosities as functions of various observable properties. Unfortunately, the primary distance indicators are not bright enough for them to be studied at distances at which z is greater than about 0.01, redshifts at which cosmological velocities cz would be greater than typical random departures of galactic velocities from the cosmological expansion, a few hundred kilometers per second. Thus they cannot be used directly to learn about $a(t)$. For this purpose it is necessary to use secondary distance indicators, which are bright enough to be studied at these large distances, and whose absolute luminosities are known through the association of the closer ones with primary distance indicators.

A. Primary distance indicators[1]

Almost all distance measurements in astronomy are ultimately based on measurements of the distance of objects within our own galaxy, using one or the other of two classic kinematic methods.

1. Trigonometric parallax

The motion of the earth around the sun produces an annual motion of the apparent position of any star around an ellipse, whose maximum angular radius π is given in radians (for $\pi \ll 1$, which is the case for all stars) by

$$\pi = \frac{d_E}{d} \tag{1.3.1}$$

where d is the star's distance from the solar system, and d_E is the mean distance of the earth from the sun,[2] defined as the *astronomical unit*,

---

[1] For a survey, see M. Feast, in *Nearby Large-Scale Structures and the Zone of Avoidance*, eds. A. P. Fairall and P. Woudt (ASP Conference Series, San Francisco, 2005) [astro-ph/0405440].

[2] The history of measurements of distances in the solar system goes back to Aristarchus of Samos (circa 310 BC–230 BC). From the ratio of the breadth of the earth's shadow during a lunar eclipse to the angular diameter of the moon he estimated the ratio of the diameters of the moon and earth; from the angular diameter of the moon he estimated the ratio of the diameter of the moon to its distance from the earth; and from the angle between the lines of sight to the sun and moon when the moon is half full he estimated the ratio of the distances to the sun and moon; and in this way he was able to measure the distance to the sun in units of the diameter of the earth. Although the method of Aristarchus was correct, his observations were poor, and his result for the distance to the sun was far too low. [For an account of Greek astronomy before Aristarchus and a translation of his work, see T. L. Heath, *Aristarchus of*

$1\,\mathrm{AU} = 1.496 \times 10^8$ km. A parsec (pc) is defined as the distance at which $\pi = 1''$; there are 206,264.8 seconds of arc per radian so

$$1\,\mathrm{pc} = 206{,}264.8\,\mathrm{AU} = 3.0856 \times 10^{13}\,\mathrm{km} = 3.2616\,\text{light years}.$$

The parallax in seconds of arc is the reciprocal of the distance in parsecs.

The first stars to have their distances found by measurement of their trigonometric parallax were α Centauri, by Thomas Henderson in 1832, and 61 Cygni, by Friedrich Wilhelm Bessel in 1838. These stars are at distances 1.35 pc and 3.48 pc, respectively. The earth's atmosphere makes it very difficult to measure trigonometric parallaxes less than about 0.03″ from ground-based telescopes, so that for many years this method could be used to find the distances of stars only out to about 30 pc, and at these distances only for a few stars and with poor accuracy.

This situation has been improved by the launching of a European Space Agency satellite known as Hipparcos, used to measure the apparent positions and luminosities of large numbers of stars in our galaxy.[3] For stars of sufficient brightness, parallaxes could be measured with an accuracy (standard deviation) in the range of 7 to 9 $\times 10^{-4}$ arc seconds. Of the 118,000 stars in the Hipparcos Catalog, it was possible in this way to find distances with a claimed uncertainty of no more than 10% for about 20,000 stars, some at distances over 100 pc.

2. Proper motions
A light source at a distance d with velocity $v_\perp$ perpendicular to the line of sight will appear to move across the sky at a rate μ in radians/time given by

$$\mu = v_\perp/d . \tag{1.3.2}$$

This is known as its *proper motion*. Of course, astronomers generally have no way of directly measuring the transverse velocity $v_\perp$, but they can measure the component v_r of velocity along the line of sight from the Doppler shift of the source's spectral lines. The problem is to infer $v_\perp$ from the measured value of v_r. This can be done in a variety of special cases:

- Moving clusters are clusters of stars that were formed together and hence move on parallel tracks with equal speed. (These are *open*

Samos (Oxford University Press, Oxford, 1913).] The first reasonably accurate determination of the distance of the earth to the sun was made by the measurement of a parallax. In 1672 Jean Richer and Giovanni Domenico Cassini measured the distance from the earth to Mars, from which it was possible to infer the distance from the earth to the sun, by observing the difference in the apparent direction to Mars as seen from Paris and Cayenne, which are separated by a known distance of 6,000 miles. Today distances within the solar system are measured very accurately by measurement of the timing of radar echoes from planets and of radio signals from transponders carried by spacecraft.

[3]M. A. C. Perryman *et al.*, *Astron. Astrophys.* **323**, L49 (1997).

clusters, in the sense that they are not held together by gravitational attraction, in distinction to the much larger globular clusters whose spherical shape indicates a gravitationally bound system.) The most important such cluster is the Hyades (called by Tennyson's Ulysses the "rainy Hyades"), which contains over 100 stars. The velocities of these stars along the line of sight are measured from their Doppler shifts, and if we knew the distance to the cluster then the velocities of its stars at right angles to the line of sight could be measured from their proper motions. The distance to the cluster was determined long ago to be about 40 pc by imposing the constraint that all these velocities are parallel. Distances measured in this way are often expressed as *moving cluster parallaxes*. Since the advent of the Hipparcos satellite, the moving cluster method has been supplemented with a direct measurement of the trigonometric parallax of some of these clusters, including the Hyades.

• A second method is based on the statistical analysis of the Doppler shifts and proper motions of stars in a sample whose *relative* distances are all known, for instance because they all have the same (unknown) absolute luminosity, or because they all at the same (unknown) distance. The Doppler shifts give the velocities along the line of sight, and the proper motions and the relative distances give the velocities transverse to the line of sight, up to a single overall factor related to the unknown absolute luminosity or distance. This factor can be determined by requiring that the distribution of velocities transverse to the line of sight is the same as the distribution of velocities along the line of sight. Distances measured in this way are often called *statistical parallaxes*, or *dynamical distances*.

• The distance to the Cepheid variable star ζ Geminorum has been measured[4] by comparing the rate of change of its physical diameter, as found from the Doppler effect, with the rate of change of its angular diameter, measured using an optical interferometer. (About Cepheids, more below.) The distance was found to be 336 ± 44 pc, much greater than could have been found from a trigonometric parallax. This method has subsequently been extended to eight other Cepheids.[5]

• It is becoming possible to measure distances by measuring the proper motion of the material produced by supernovae, assuming a

---

[4]B. F. Lane, M. J. Kuchner, A. F. Boden, M. Creech-Eakman, and S. B. Kulkarni, *Nature* **407**, 485 (September 28, 2000).
[5]P. Kervella *et al.*, *Astron. Astrophys.* **423**, 327 (2004) [astro-ph/0404179].

more-or-less cylindrically symmetric explosion, so that the transverse velocity $v_\perp$ can be inferred from the radial velocity v_r measured by Doppler shifts. This method has been applied[6] to the ring around the supernova SN1987A, observed in 1987 in the Large Magellanic Cloud, with the result that its distance is 52 kpc (thousand parsecs).

- The measurement of the time-varying Doppler shift and proper motion of an object in orbit around a central mass can be used to find the distance to the object. For instance, if the line of sight happens to be in the plane of the orbit, and if the orbit is circular, then the Doppler shift is a maximum when the object is moving along the line of sight, and hence gives the orbital velocity v, while the proper motion μ is a maximum when the object is moving with the same velocity at right angles to the line of sight, and gives the distance as v/μ. This method can also be used for orbits that are inclined to the line of sight and not circular, by studying the time-variation of the Doppler shift and proper motion. The application of this method to the star S2, which orbits the massive black hole in the galactic center, gives what is now the best value for the distance of the solar system from the galactic center,[7] as 8.0 ± 0.4 kpc. This method also allows the measurement of some distances outside our galaxy, by using the motion of masers — point microwave sources — in the accretion disks of gas and dust in orbit around black holes at the centers of galaxies. The orbital velocity can be judged from the Doppler shifts of masers at the edge of the accretion disk, which are moving directly toward us or away from us, and if this is the same as the orbital velocities of masers moving transversely to the line of sight, then the ratio of this orbital velocity to their observed proper motion gives the distance to the galaxy. So far, this method has been used to measure the distance to the galaxy NGC 4258,[8] as 7.2 ± 0.5 Mpc (million parsecs), and to the galaxy M33,[9] as 0.730 ± 0.168 Mpc.

These kinematic methods have limited utility outside the solar neighborhood. We need a different method to measure larger distances.

3. Apparent luminosity

The most common method of determining distances in cosmology is based on the measurement of the apparent luminosity of objects of known (or

[6]N. Panagia, *Mem. Soc. Astron. Italiana* **69**, 225 (1998).

[7]F. Eisenhauer *et al.*, *Astrophys. J. Lett.* **597**, L121 (2003) [astro-ph/0306220].

[8]J. Herrnstein *et al.*, *Nature* **400**, 539 (3 August 1999).

[9]A. Brunthaler, M. J. Reid, H. Falcke, L. J. Greenhill, and C. Henkel, *Science* **307**, 1440 (2005) [astro-ph/0503058].

supposedly known) absolute luminosity. The absolute luminosity L is the energy emitted per second, and the apparent luminosity ℓ is the energy received per second per square centimeter of receiving area. If the energy is emitted isotropically, then we can find the relation between the absolute and apparent luminosity in Euclidean geometry by imagining the luminous object to be surrounded with a sphere whose radius is equal to the distance d between the object and the earth. The total energy per second passing through the sphere is $4\pi d^2\ell$, so

$$\ell = \frac{L}{4\pi d^2} \,. \tag{1.3.3}$$

This relation is subject to corrections due to interstellar and/or intergalactic absorption, as well as possible anisotropy of the source, which though important in practice involve too many technicalities to go into here.

Astronomers unfortunately use a traditional notation for apparent and absolute luminosity in terms of apparent and absolute *magnitude*.[10] In the second century A.D., the Alexandrian astronomer Claudius Ptolemy published a list of 1,022 stars, labeled by categories of apparent brightness, with bright stars classed as being of first magnitude, and stars just barely visible being of sixth magnitude.[11] This traditional brightness scale was made quantitative in 1856 by Norman Pogson, who decreed that a difference of five magnitudes should correspond to a ratio of a factor 100 in apparent luminosities, so that $\ell \propto 10^{-2m/5}$. With the advent of photocells at the beginning of the twentieth century, it became possible to fix the constant of proportionality: the apparent bolometric luminosity (that is, including all wavelengths) is given in terms of the apparent bolometric magnitude m by

$$\ell = 10^{-2m/5} \times 2.52 \times 10^{-5} \,\text{erg cm}^{-2}\,\text{s}^{-1} \,. \tag{1.3.4}$$

For orientation, Sirius has a visual magnitude $m_{\text{vis}} = -1.44$, the Andromeda nebula M31 has $m_{\text{vis}} = 0.1$, and the large galaxy M87 in the nearest large cluster of galaxies has $m_{\text{vis}} = 8.9$. The absolute magnitude in any wavelength band is defined as the apparent magnitude an object would have at a distance of 10 pc, so that the absolute bolometric luminosity is given in terms of the absolute bolometric magnitude M by

$$L = 10^{-2M/5} \times 3.02 \times 10^{35} \,\text{erg s}^{-1} \,. \tag{1.3.5}$$

---

[10]For the history of the apparent magnitude scale, see J. B. Hearnshaw, *The Measurement of Starlight: Two centuries of astronomical photometry* (Cambridge University Press, Cambridge, 1996); K. Krisciunas, astro-ph/0106313.

[11]For the star catalog of Ptolemy, see M. R. Cohen and I. E. Drabkin, *A Source Book in Greek Science* (Harvard University Press, Cambridge, MA, 1948): p. 131.

For comparison, in the visual wavelength band the absolute magnitude M_{vis} is +4.82 for the sun, +1.45 for Sirius, and −20.3 for our galaxy. Eq. (1.3.3) may be written as a formula for the distance in terms of the *distance-modulus* $m - M$:

$$d = 10^{1+(m-M)/5}\text{pc} .$$ (1.3.6)

There are several different kinds of star that have been used in measurements of distance through the observation of apparent luminosity:

- **Main Sequence**: Stars that are still burning hydrogen at their cores obey a characteristic relation between absolute luminosity and color, both depending on mass. This is known as the main sequence, discovered in the decade before the First World War by Ejnar Hertzsprung and Henry Norris Russell. The luminosity is greatest for blue-white stars, and then steadily decreases for colors tending toward yellow and red. The *shape* of the main sequence is found by observing the apparent luminosities and colors of large numbers of stars in clusters, all of which in each cluster may be assumed to be at the same distance from us, but we need to know the distances to the clusters to calibrate absolute luminosities on the main sequence. For many years the calibration of the main sequence absolute luminosities was based on observation of a hundred or so main sequence stars in the Hyades cluster, whose distance was measured by the moving cluster method described above. With the advent of the Hipparcos satellite, the calibration of the main sequence has been greatly improved through the observation of colors and apparent luminosities of nearly 100,000 main sequence stars whose distance is known through measurement of their trigonometric parallax. Including in this sample are stars in open clusters such as the Hyades, Praesepe, the Pleiades, and NGC 2516; these clusters yield consistent main sequence absolute magnitudes if care is taken to take proper account of the varying chemical compositions of the stars in different clusters.[12]   With the main sequence calibrated in this way, we can use Eq. (1.3.3) to measure the distance of any star cluster or galaxy in which it is possible to observe stars exhibiting the main sequence relation between apparent luminosity and color. Distances measured in this way are sometimes known as *photometric parallaxes.*

   The analysis of the Hipparcos parallax measurements revealed a discrepancy between the distances to the Pleiades star cluster measured by observations of main sequence stars and by measurements of

---

[12]S. M. Percival, M. Salaris, and D. Kilkenny, *Astron. Astrophys.* **400**, 541 (2003) [astro-ph/0301219].

trigonometric parallax.[13]   The traditional method, using a main
sequence calibration based on the application of the moving cluster
method to the closer Hyades cluster, gave a distance to the Pleiades[14]
of 132 ± 4 pc. Then trigonometric parallaxes of a number of stars in
the Pleiades measured by the Hipparcos satellite gave a distance[15] of
118 ± 4 pc, in contradiction with the results of main sequence fitting.
More recently, these Hipparcos parallaxes have been contradicted by
more accurate measurements of the parallaxes of three stars in the
Pleiades with the Fine Guidance Sensor of the Hubble Space Telescope,[16] which gave a distance of 133.5 ± 1.2 pc, in good agreement
with the main sequence results. At the time of writing, the balance
of astronomical opinion seems to be favoring the distances given by
main sequence photometry.[17]

- **Red Clump Stars**: The color–magnitude diagram of clusters in metal-
  rich[18] parts of the galaxy reveals distinct clumps of red giant stars
  in a small region of the diagram, with a spread of only about 0.2 in
  visual magnitude. These are stars that have exhausted the hydrogen
  at their cores, with helium taking the place of hydrogen as the fuel for
  nuclear reactions at the stars' cores. The absolute magnitude of the
  red clump stars in the infrared band (wavelengths around 800 nm) has
  been determined[19] to be $M_I = -0.28 \pm 0.2$ mag, using the distances
  and apparent magnitudes measured with the Hipparcos satellite and in
  an earlier survey.[20] In this band there is little dependence of absolute
  magnitude on color, but it has been argued that even the infrared
  magnitude may depend significantly on metallicity.[21]

- **RR Lyrae Stars**: These are variable stars that have been used as
  distance indicators for many decades.[22] They can be recognized by
  their periods, typically 0.2 to 0.8 days. The use of the statistical parallax, trigonometric parallax and moving cluster methods (with data

[13]B. Paczynski, *Nature* **227**, 299 (22 January, 2004).

[14]G. Meynet, J.-C. Mermilliod, and A. Maeder, *Astron. Astrophys. Suppl. Ser.* **98**, 477 (1993).

[15]J.-C. Mermilliod, C. Turon, N. Robichon, F. Arenouo, and Y. Lebreton, in *ESA SP-402 Hipparcos–Venice '97*, eds. M.A.C. Perryman and P. L. Bernacca (European Space Agency, Paris, 1997), 643; F. van Leeuwen and C. S. Hansen Ruiz, *ibid*, 689; F. van Leeuwen, *Astron. Astrophys.* **341**, L71 (1999).

[16]D. R. Soderblom *et al.*, *Astron. J.* **129**, 1616 (2005) [astro-ph/0412093].

[17]A new reduction of the raw Hipparcos data is given by F. van Leeuwen and E. Fantino, *Astron. Astrophys.* **439**, 791 (2005) [astro-ph/0505432].

[18]Astronomers use the word "metal" to refer to all elements heavier than helium.

[19]B. Paczyński and K. Z. Stanek, *Astrophys. J.* **494**, L219 (1998).

[20]A. Udalski *et al.*, *Acta. Astron.* **42**, 253 (1992).

[21]L. Girardi, M. A. T. Groenewegen, A. Weiss, and M. Salaris, astro-ph/9805127.

[22]For a review, see G. Bono, *Lect. Notes Phys.* **635**, 85 (2003) [astro-ph/0305102].

from both ground-based observatories and the Hipparcos satellite) give respectively[23] an absolute visual magnitude for RR Lyrae stars in our galaxy's halo of 0.77 ± 0.13, 0.71 ± 0.15, and 0.67 ± 0.10, in good agreement with an earlier result[24] $M_{\rm vis} = 0.71 \pm 0.12$ for halo RR Lyrae stars and 0.79 ± 0.30 for RR Lyrae stars in the thick disk of the galaxy. RR Lyrae stars are mostly too far for a measurement of their trigonometric parallax, but recently measurements[25] with the Hubble Space Telescope have given a value of 3.82×10^{-3} arcsec for the trigonometric parallax of the eponymous star RR Lyr itself, implying an absolute visual magnitude of $0.61^{-0.11}_{+0.10}$.

- **Eclipsing Binaries**: In favorable cases it is possible to estimate the intrinsic luminosity of a star that is periodically partially eclipsed by a smaller companion, without the use of any intermediate distance indicators. The velocity of the companion can be inferred from the Doppler shift of its spectral lines (with the ellipticity of the orbit inferred from the variation of the Doppler shift with time), and the radius of the primary star can then be calculated from the duration of the eclipse. The temperature of the primary can be found from measurement of its spectrum, typically from its apparent luminosity in various wavelength bands. Knowing the radius, and hence the area, and the temperature of the primary, its absolute luminosity can then be calculated from the Stefan–Boltzmann law for black body radiation. This method has been applied to measure distances to two neighboring dwarf galaxies, the Large Magellanic Cloud (LMC)[26] and the Small Magellanic Cloud (SMC),[27] and to the Andromeda galaxy M31[28] and its satellite M33.[29]

- **Cepheid variables**: Because they are so bright, these are by far the most important stars used to measure distances outside our galaxy. Named after the first such star observed, δ Cephei, they can be

---

[23]P. Popowski and A. Gould, *Astrophys. J.* **506**, 259, 271 (1998); also astro-ph/9703140, astro-ph/9802168; and in *Post-Hipparcos Cosmic Candles*, eds. A. Heck and F. Caputo (Kluwer Academic Publisher, Dordrecht) [astro-ph/9808006]; A. Gould and P. Popowski, *Astrophys. J.* **568**, 544 (1998) [astro-ph/9805176]; and references cited therein.

[24]A. Layden, R. B. Hanson, S. L. Hawley, A. R. Klemola, and C. J. Hanley, *Astron. J.* **112**, 2110 (1996).

[25]G. F. Benedict *et al.*, *Astrophys. J.* **123**, 473 (2001) [astro-ph/0110271]

[26]E. F. Guinan *et al.*, *Astrophys. J.* **509**, L21 (1998); E. L. Fitzpatrick *et al.* *Astrophys. J.* **587**, 685 (2003).

[27]T. J. Harries, R. W. Hilditch, and I. D. Howarth, *Mon. Not. Roy. Astron. Soc.* **339**, 157 (2003); R. W. Hilditch, I. D. Howarth, and T. J. Harries, *Mon. Not. Roy. Astron. Soc.* **357**, 304 (2005).

[28]I. Ribas *et al.*, *Astrophys. J.* **635**, L37 (2005).

[29]A. Z. Bonanos *et al.*, *Astrophys. Space Sci.* **304**, 207 (2006) [astro-ph/0606279].

recognized from the characteristic time dependence of their luminosity, with periods ranging from 2 to 45 days. (Cepheids in other galaxies have been observed with periods extending up to 100 days.) In 1912 Henrietta Swan Leavitt discovered that the Cepheid variables in the Small Magellanic Cloud (SMC) have apparent luminosities given by a smooth function of the period of the variation in luminosity, but the distance to the SMC was not known. Having measured the distances and apparent luminosities of several Cepheids in open clusters, and hence their absolute luminosities, it became possible to calibrate the relation between period and luminosity. Cepheid variables thus became a "standard candle" that could be used to measure the distance to any galaxy close enough for Cepheids to be seen. It was the discovery of Cepheids in M31, together with Leavitt's calibration of the Cepheid period–luminosity relation, that allowed Edwin Hubble in 1923 to measure the distance of M31, and show that it was far outside our own galaxy, and hence a galaxy in its own right.

Today the form of the Cepheid period–luminosity relation is derived more from the Large Magellanic Cloud (LMC), where there are many Cepheids, and the dependence of the absolute luminosity on color is also taken into account. The calibration of Cepheid absolute luminosities can therefore be expressed as (and often in fact amounts to) a measurement of the distance to the LMC. Main sequence photometry and other methods gave what for some years was a generally accepted LMC distance modulus of 18.5 mag, corresponding according to Eq. (1.3.6) to a distance of 5.0×10^4 pc. The use of red clump stars[30] has given a distance modulus of 18.47, with a random error ± 0.01, and a systematic error $^{+0.05}_{-0.06}$. A large catalog[31] of Cepheids in the LMC has been interpreted by the members of the Hubble Space Telescope Key Project[32] to give the Cepheid visual and infrared absolute magnitudes as functions of the period P in days:

$$M_V = -2.760 \log_{10} P - 1.458 , \qquad M_I = -2.962 \log_{10} P - 1.942 ,$$
$$(1.3.7)$$

under the assumption that the LMC distance modulus is 18.5.

This result was challenged in two distinct ways, which illustrate the difficulty of this sort of distance measurement:

First, there have been discordant measurements of the distance to the LMC. Under the assumption that red clump stars in the LMC

---

[30] M. Salaris, S. Percival, and L. Girardi, *Mon. Not. Roy. Astron. Soc.* **345**, 1030 (2003) [astro-ph/0307329].

[31] A. Udalski *et al.*, *Acta Astr.* **49**, 201 (1999): Table 1.

[32] W. L. Freedman *et al.*, *Astrophys. J.* **553**, 47 (2001).

have the same infrared luminosity as those in the local galactic disk, a distance modulus was found[33] that was 0.45 magnitudes smaller, giving a distance to the LMC that is smaller by a factor 0.8. This has in turn been challenged on the grounds that the stars in the LMC have distinctly lower metallicity than in the local disk; two groups taking this into account[34] have given LMC distance moduli of 18.36 ± 0.17 mag and 18.28 ± 0.18 mag, in fair agreement with the previously accepted value. This also agrees with the measurement of the distance to the LMC inferred[35] from observations of RR Lyrae stars, which gives a distance modulus of 18.33 ± 0.06 mag. This distance modulus for the LMC is further confirmed by the measurement of the distance of the eclipsing binary HV2274; taking account of its distance from the center of the LMC gives[36] a distance modulus for the LMC of 18.30 ± 0.07.

Second, there have been new calibrations of the Cepheid period–luminosity relation, that do not rely on Cepheids in the LMC, which together with observations of Cepheids in the LMC can be used to give an independent estimate of the LMC distance.[37] In recent years the satellite Hipparcos[38] has measured trigonometric parallaxes for 223 Cepheid variables in our galaxy, of which almost 200 can be used to calibrate the period–luminosity relation, without relying on main sequence photometry, red clump stars, or RR Lyrae stars. The nearest Cepheids are more than 100 pc away from us (the distance to Polaris is about 130 pc), so the parallaxes are just a few milliarcseconds, and individual measurements are not very accurate, but with about 200 Cepheids measured it has been possible to get pretty good accuracy. One early result[39] gave the relation between the absolute visual magnitude M_V and the period P (in days) as

$$M_V = -2.81 \log_{10} P - 1.43 \pm 0.10 .$$

This was a decrease of about 0.2 magnitudes from previous results, i.e., an increase of the intrinsic luminosity of Cepheids by a factor $10^{0.2 \times 2/5} = 1.20$ leading to a 10% increase in all cosmic distances based

[33] K. Z. Stanek, D. Zaritsky, and J. Harris, *Astrophys. J.* **500**, L141 (1998) [astro-ph/9803181].

[34] A. A. Cole, Astrophys. J. **500**, L137 (1998) [astro-ph/9804110]; L. Girardi *et al., op. cit.*.

[35] P. Popowski and A. Gould, *op. cit.*.

[36] E. F. Guinan *et al., op. cit.*.

[37] For a review, see M. Feast, *Odessa Astron. Publ.* **14** [astro-ph/0110360].

[38] M. A. C. Perryman, *Astron. Astrophys.* **323**, L49 (1997).

[39] M. W. Feast and R. M. Catchpole, *Mon. Not. Roy. Astron. Soc.* **286**, L1 (1997); also see F. Pont, in *Harmonizing Cosmic Distances in a Post-Hipparcos Era*, eds. D. Egret and A. Heck (ASP Conference Series, San Francisco, 1998) [astro-ph/9812074]; H. Baumgardt, C. Dettbarn, B. Fuchs, J. Rockmann, and R. Wielen, in *Harmonizing Cosmic Distance Scales in a Post-Hipparcos Era, ibid* [astro-ph/9812437].

directly or indirectly on the Cepheid period–luminosity relation. With this value of Cepheid absolute luminosity, the LMC distance modulus would be 18.66, or slightly less with corrections for the metallicity of the LMC (though with the absolute luminosity of Cepheids calibrated by Hipparcos observations, the only relevance of the LMC for the Cepheid period–luminosity relation is to determine its shape.) This result for the Cepheid absolute luminosities has in turn been contradicted.[40]

These uncertainties may now have been resolved by measurements of the trigonometric parallax of Cepheids in our galaxy with the Fine Guidance Sensor of the Hubble Space Telescope. First, the trigonometric parallax of δ Cephei[41] gave a distance of 273 ± 11 pc, corresponding to an LMC distance modulus of 18.50 ± 0.13. More recently, trigonometric parallaxes have been measured for nine Galactic Cepheids, giving an LMC distance modulus of 18.50 ± 0.03, or with metallicity corrections, 18.40 ± 0.05.[42]

There has also been an independent calibration of the Cepheid period–luminosity relation through observations[43] of Cepheids in the galaxy NGC 4258, whose distance 7.2 ± 0.5 Mpc has been measured using the observations of proper motions of masers in this galaxy mentioned above. This distance is in satisfactory agreement with the distance 7.6 ± 0.3 Mpc obtained from the Cepheids in NGC 4258 under the assumption that these Cepheids have the period–luminosity relation (1.3.7) obtained under the assumption that the LMC distance modulus is 18.5, which tends to confirm this period–luminosity relation. But there are differences in the metallicity of the Cepheids in NGC 4258 and in the LMC, which makes this conclusion somewhat controversial.[44] A 2006 calibration of the Cepheid period–luminosity relation based on the study of 281 Cepheids in NGC 4258[45] (whose distance, as we have seen, is known from observations of maser Doppler shifts and proper motions) gave an LMC distance modulus 18.41 ± 0.10 (stat.) ± 0.13 (syst.). This study includes both a field that is metal rich, like our Galaxy, and a field that is metal poor, like the LMC, so

---

[40] See, e.g., B. F. Madore and W. L. Freedman, *Astrophys. J.* **492**, 110 (1998) For a recent survey of the theory underlying the Cepheid period–luminosity relation, see A. Gautschy, in *Recent Results on H_0–19th Texas Symposium on Relativistic Astrophysics* [astro-ph/9901021].

[41] G. F. Benedict *et al.*, *Astrophys. J.* **124**, 1695 (2002).

[42] G. F. Benedict *et al.*, *Astron. J.* **133**, 1810 (2007), [astro-ph/0612465].

[43] J. A. Newman *et al.*, *Astrophys. J.* **553**, 562 (2001) [astro-ph-0012377].

[44] For instance, see B. Paczynski, *Nature* **401**, 331 (1999); F. Caputo, M. Marconi, and I. Musella, *Astrophys. J.* [astro-ph/0110526].

[45] L. M. Macri *et al.*, *Astrophys. J.* **652**, 1133 (2006) [astro-ph/0608211].

it provides a calibration of the metallicity dependence of the Cepheid period–luminosity relation.

In a 2003 survey[46] the LMC distance modulus measured using a variety of distance indicators *other* than Cepheid variables (including RR Lyrae stars, red clump stars, etc.) was found to be 18.48 ± 0.04, in very good agreement with the earlier value 18.52 ± 0.05 found by observation of Cepheids, with the corrections adopted by the Hubble Space Telescope group.

B. Secondary distance indicators

None of the above distance indicators are bright enough to be used to measure distances at redshifts large enough so that peculiar velocities can be neglected compared with the expansion velocity, say, $z > 0.03$. For this we need what are called *secondary distance indicators* that are brighter than Cepheids, such as whole galaxies, or supernovae, which can be as bright as whole galaxies.

For many years Cepheids could be used as distance indicators only out to a few million parsecs (Mpc), which limited their use to the Local Group (which consists of our galaxy and the Andromeda nebula M31, and a dozen or so smaller galaxies like M33 and the LMC and SMC) and some other nearby groups (the M81, M101, and Sculptor groups). This was not enough to calibrate distances to an adequate population of galaxies or supernovae, and so it was necessary to use a variety of intermediate distance indicators: globular clusters, HII regions, brightest stars in galaxies, etc. Now the Hubble Space Telescope allows us to observe Cepheids in a great many galaxies at much greater distances, out to about 30 Mpc, and so the secondary distance indicators can now be calibrated directly, without the use of intermediate distance indicators. Four chief secondary distance indicators have been developed:

1. The Tully–Fisher relation
Although whole galaxies can be seen out to very large distances, it has not been possible to identify any class of galaxies with the same absolute luminosity. However, in 1977 Tully and Fisher[47] developed a method for estimating the absolute luminosity of suitable spiral galaxies. The 21 cm absorption line in these galaxies (arising in transitions of hydrogen atoms from lower to the higher of their two hyperfine states) is widened by the

---

[46]M. Feast, *Lect. Notes Phys.* **635**, 45 (2003) [astro-ph/0301100].

[47]R. B. Tully and J. R. Fisher, *Astron. Astrophys.* **54**, 661 (1977)

Doppler effect, caused by the rotation of the galaxy. The line width W gives an indication of the maximum speed of rotation of the galaxy, which is correlated with the mass of the galaxy, which in turn is correlated with the galaxy's absolute luminosity.[48] (It is also possible to apply the Tully–Fisher relation using the width of other lines, such as a radio frequency transition in the carbon monoxide molecule.[49])

In one application of this approach[50] the *shape* of the function $L_I(W)$ that gives the infrared band absolute luminosity as a function of 21 cm line width (that is, the absolute luminosity up to a common constant factor) was found from a sample of 555 spiral galaxies in 24 clusters, many with redshifts less than 0.01. (The relative distances to these galaxies were found from the ratios of their redshifts, using Eq. (1.2.9), so that the peculiar velocities of these galaxies introduced considerable errors into the estimated ratios of absolute luminosities of individual pairs of galaxies, but with 555 galaxies in the sample it could be assumed that these errors would cancel in a least-squares fit of the measured relative values of absolute luminosity to a smooth curve.) Roughly speaking, $L_I(W)$ turned out to be proportional to W^3. The overall scale of the function $L_I(W)$ was then found by fitting it to the absolute luminosities of 15 spiral galaxies whose distances were accurately known from observations of Cepheid variables they contain. (These 15 galaxies extended out only to 25 Mpc, not far enough for them to be used to measure the Hubble constant directly.) The Hubble constant could then be found by using the function $L_I(W)$ calibrated in this way to find the distances to galaxies in 14 clusters with redshifts ranging from 0.013 to 0.03, and comparing the results obtained with Eq. (1.2.8). (These redshifts may not be large enough to ignore peculiar velocities altogether, but again, this problem is mitigated by the use of a fairly large number of galaxies.) The Hubble constant found in this way was 70 ± 5 km s^{-1} Mpc^{-1}. More recently, the Hubble Space Telescope Key Project to Measure the Hubble Constant has used Cepheid variables to recalibrate the Tully–Fisher relation (assuming an LMC distance of 50 kpc) and then found H_0 by plotting distances found from the Tully–Fisher relation against redshift for a sample of 19 clusters with redshifts from 0.007 to 0.03,[51] taken from the G97 survey of Giovanelli et al.[52] The result was $H_0 = 71 \pm 3 \pm 7$ km s^{-1} Mpc^{-1}, with the first quoted uncertainty statistical and the second systematic.

---

[48] M. Aaronson, J. R. Mould, and J. Huchra, *Astrophys. J.* **229**, 1 (1979).

[49] Y. Tutui et al., *Publ. Astron. Soc. Japan* **53**, 701 (2001) [astro-ph/0108462].

[50] R. Giovanelli, in *The Extragalactic Distance Scale - Proceedings of the Space Telescope Science Institute Symposium held in Baltimore, MD, May, 1996* (Cambridge University Press, Cambridge, 1997): 113; R. Giovanelli et al., *Astron. J.* **113**, 22 (1997).

[51] S. Sakai et al., *Astrophys. J.* **529**, 698 (2000); W. L. Freedman et al., *Astrophys. J.* **553**, 47 (2001); and references cited therein.

[52] Giovanelli et al., op. cit.

2. Faber–Jackson relation

Just as the Tully–Fisher method is based on a correlation of orbital velocities with absolute luminosities in spiral galaxies, the Faber–Jackson method is based on a correlation of random velocities with absolute luminosities in elliptical galaxies.[53]   An advantage of this method over the Tully–Fisher method is that it has a firmer theoretical foundation, provided by the virial theorem to be discussed in Section 1.9, which directly relates the mean square random velocity to the galaxy mass.

3. Fundamental plane

The Faber–Jackson method was improved by the recognition that the correlation between orbital velocity and absolute luminosity depends also on the surface brightness of the cluster, and hence on its area.[54] (The term "fundamental plane" refers to the way that data on elliptical galaxies are displayed graphically.) This method has been used[55] to estimate that $H_0 = 78 \pm 5$ (stat.) ± 9 (syst.) km sec^{-1} Mpc^{-1}.

4. Type Ia supernovae

Supernovae of Type Ia are believed to occur when a white dwarf star in a binary system accretes sufficient matter from its partner to push its mass close to the Chandrasekhar limit, the maximum possible mass that can be supported by electron degeneracy pressure.[56] When this happens the white dwarf becomes unstable, and the increase in temperature and density allows the conversion of carbon and oxygen into ^{56}Ni, triggering a thermonuclear explosion that can be seen at distances of several thousand megaparsecs. The exploding star always has a mass close to the Chandrasekhar limit, so there is little variation in the absolute luminosity of these explosions, making them nearly ideal distance indicators.[57] What variation there is seems

---

[53] S. M. Faber and R. E. Jackson, *Astrophys. J.* **204**, 668 (1976).

[54] A. Dressler *et al.*, Astrophys. J. **313**, 42 (1987).

[55] D. D. Kelson *et al.*, *Astrophys. J.* **529**, 768 (2000) [astro-ph/9909222]; J. P. Blakeslee, J. R. Lucey, J. L. Tonry, M. J. Hudson, V. K. Nararyan, and B. J. Barris, *Mon. Not. Roy. Astron. Soc.* **330**, 443 (2002) [astro-ph/011183].

[56] W. A. Fowler and F. Hoyle, *Astrophys. J.* **132**, 565 (1960). Calling a supernova Type I simply means that hydrogen lines are not observed in its spectrum. In addition to Type Ia supernovae, there are other Type I supernovae that occur in the collapse of the cores of stars much more massive than white dwarfs, whose outer layer of hydrogen has been lost in stellar winds, as well as Type II supernovae, produced by core collapse in massive stars that have not lost their outer layer of hydrogen. For a discussion of the Chandrasekhar limit, see G&C, Section 11.3.

[57] The use of Type Ia supernovae as distance indicators was pioneered by A. Sandage and G. A. Tammann, *Astrophys. J.* **256**, 339 (1982), following an earlier observation that they had fairly uniform luminosity by C. T. Kowal, *Astron. J.* **73**, 1021 (1968). In 1982 it was necessary to use brightest supergiant stars as intermediate distance indicators, to bridge the gap between the distances that could then be measured using Cepheids and the distances at which the Type Ia supernova could be found. For reviews of the use of type Ia supernovae as standard candles, see D. Branch, *Ann. Rev. Astron. & Astrophys.* **36**, 17 (1998); P. Höflich, C. Gerardy, E. Linder, and H. Marion, in *Stellar Candles*, eds. W. Gieren *et al.* (*Lecture Notes in Physics*) [astro-ph/0301334].

to be well correlated with the rise time and decline time of the supernova light: the slower the decline, the higher the absolute luminosity.[58]

This relation has been calibrated by measurements of Type Ia super-novae in several galaxies of known distance. From 1937 to 1999 there were ten supernovae in galaxies whose distance had been measured by observation of Cepheid variables they contain.[59] Of these, six Type Ia supernovae were used by the HST Key H_0 Group[60] to calibrate the relation between absolute luminosity and decline time. This relation was then used to calculate distances to a sample of 29 Type Ia supernovae in galaxies with redshifts extending from 0.01 to 0.1, observed at the Cerro Tololo Inter-American Observatory.[61] Plotting these distances against measured redshifts gave a Hubble constant[62] of 71 ± 2(statistical) ± 6(systematic) km s^{-1} Mpc^{-1}. This agrees well with an older determination using Type Ia supernovae by a Harvard group,[63] which found $H_0 = 67 \pm 7$ km s^{-1} Mpc^{-1}. Members of this group have superceded this result,[64] now giving a Hubble constant $H_0 = 73 \pm 4$(stat.) ± 5(syst.) km s^{-1} Mpc^{-1}. On the other hand, a group headed by Sandage using Type Ia supernovae and the Tully–Fisher relation has consistently found lower values of H_0.[65] The gap seems to be narrowing; in 2006, this group quoted[66] a value $H_0 = 62.3 \pm 1.3$(stat.) ± 5.0(syst.) km s^{-1} Mpc^{-1}. (According to Sandage *et al.*, the difference between these results is due to a difference in the Cepheid period–luminosity relation used to measure distances to the galaxies that host the supernovae that are used to calibrate the relation between supernova absolute luminosity and decline time. Sandage *et al.* use a metallicity-dependent period–luminosity relation. However Macri *et al.*[45] subsequently reported no difference in the period-luminosity relation for Cepheids in a metal-rich and a metal-poor region of NGC 4258.)

It is an old hope that with a sufficient theoretical understanding of supernova explosions, it might be possible to measure their distance

---

[58]M. Phillips, *Astrophys. J.* **413**, L105 (1993); M. Hamuy *et al.*, *Astron. J.* **109**, 1 (1995); A. Reiss, W. Press, and R. Kirshner, *Astrophys. J.* **438**, L17 (1996); S. Jha, A. Riess, & R. P. Kirshner, Astrophys. J. **659**, 122 (20007). A dependence of absolute luminosity on color as well as decline time has been considered by R. Tripp and D. Branch, *Astrophys. J.* [astro-ph/9904347].

[59]For a list, see Tripp and Branch, *op. cit.*.

[60]B. Gibson *et al.*, *Astrophys. J.* **529**, 723 (2000) [astro-ph/9908149].

[61]M. Hamuy *et al.*, *Astron. J.* **112**, 2398 (1996).

[62]L. Ferrarese *et al.*, in *Proceedings of the Cosmic Flows Workshop*, eds. S. Courteau *et al.* (ASP Conference Series) [astro-ph/9909134]; W. L. Freedman *et al.*, *Astrophys. J.* **553**, 47 (2001).

[63]A. G. Riess, W. H. Press, And R. P. Kirshner, *Astrophys. J.* **438**, L17 (1995)

[64]A. Riess *et al.*, *Astrophys. J.* **627**, 579 (2005) [astro-ph/0503159].

[65]For a 1996 summary, see G. A. Tammann and M. Federspeil, in *The Extragalactic Distance Scale*, eds. M. Livio, M. Donahue, and N. Panagia (Cambridge University Press, 1997): 137.

[66]A. Sandage *et al.*, *Astrophys. J.* **653**, 843 (2006) [astro-ph/0603647].

without use of primary distance indicators. A 2003 comparison[67] of observed light curves (apparent magnitude as a function of time) and spectra with theory for 26 Type Ia supernovae with redshifts extending up to 0.05, plus one with redshift 0.38, gave $H_0 = 67$ km sec^{-1}Mpc^{-1}, with a two standard deviation uncertainty of 8 km sec^{-1}Mpc^{-1}. It is too soon for this method to replace the older method based on the use of primary distance indicators to calibrate the supernova absolute luminosities, but the agreement between the values of H_0 found in these two ways provides some reassurance that no large error is being made with the older method.

It is instructive to consider a fifth secondary distance indicator that is also used to measure the Hubble constant:

5. Surface brightness fluctuations

In 1988 Tonry and Schneider[68] suggested using the fluctuations in the observed surface brightness of a galaxy from one part of the image to another as a measure of the galaxy's distance. Suppose that the stars in a galaxy can be classified in luminosity classes, all the stars in a luminosity class i having the same absolute luminosity L_i. The rate of receiving energy per unit area of telescope aperture in a small part of the galactic image (as for instance, a single pixel in a charge-coupled device) is

$$\ell = \sum_i \frac{N_i L_i}{4\pi d^2} \tag{1.3.8}$$

where N_i the number of stars of class i in this part of the galaxy's image, and d is the distance of the galaxy. Usually only the brightest stars can be resolved, so it is not possible to measure all the N_i directly, but one can measure the fluctuations in ℓ from one part of the image to another due to the finite values of the N_i. Suppose that the different N_i fluctuate independently from one small part of the galaxy's image to another, and obey the rules of Poisson statistics, so that

$$\left\langle (N_i - \langle N_i \rangle)(N_j - \langle N_j \rangle) \right\rangle = \delta_{ij} \langle N_i \rangle , \tag{1.3.9}$$

with brackets denoting an average over small parts of the central portion of the galaxy's image. It follows then that

$$\frac{\langle (\ell - \langle \ell \rangle)^2 \rangle}{\langle \ell \rangle} = \frac{\bar{L}}{4\pi d^2} , \tag{1.3.10}$$

---

[67]P. Höflich, C. Gerardy, E. Linder, and H. Marion, *op. cit.*
[68]J. Tonry and D. P. Schneider, *Astron. J.* **96**, 807 (1988).

where $\bar{L}$ is a luminosity-weighted mean *stellar* luminosity

$$\bar{L} \equiv \frac{\sum_i \langle N_i \rangle L_i^2}{\sum_i \langle N_i \rangle L_i} \tag{1.3.11}$$

which is expected to vary much less from one galaxy to another than the luminosities of the galaxies themselves. Eq. (1.3.10) can be used to measure distances once this relation is calibrated by measuring $\bar{L}$. By studying surface brightness fluctuations in a survey of galaxies whose distances were found by observations of Cepheids they contain, Tonry *et al.*[69] found an absolute magnitude $\bar{M}_I$ that in the infrared band is equivalent to the absolute luminosity $\bar{L}$:

$$\bar{M}_I = (-1.74 \pm 0.07) + (4.5 \pm 0.25) [m_V - m_I - 1.15] \tag{1.3.12}$$

where $m_V - m_I$ is a parameter characterizing the color of the galaxy, equal to the difference of its apparent magnitudes in the infrared and visual bands, assumed here to lie between 1.0 and 1.5. Using Eq. (1.3.10) to find distances of galaxies of higher redshift, they obtained a Hubble constant $81 \pm 6 \, \text{km} \, \text{s}^{-1} \, \text{Mpc}^{-1}$.

There are other phenomena that are used to measure the Hubble constant, including the comparison of apparent and absolute luminosity of supernovae of other types, novae, globular clusters, and planetary nebulae, the diameter–velocity dispersion relation for elliptical galaxies, gravitational lenses (discussed in Section 1.12), the Sunyaev–Zel'dovich effect (discussed in Section 2.3), etc.[70] The HST Key H_0 Group have put together their results of measurements of the Hubble constant using the Tully–Fisher relation, Type Ia supernovae, and several of these other secondary distance indicators, and conclude that[71]

$$H_0 = 71 \pm 6 \, \text{km} \, \text{s}^{-1} \, \text{Mpc}^{-1} \, .$$

As we will see in Section 7.2, the study of anisotropies in the cosmic microwave background has given a value $H_0 = 73 \pm 3 \, \text{km} \, \text{s}^{-1} \, \text{Mpc}^{-1}$. This does not depend on any of the tools discussed in this section, but it does depend on some far-reaching cosmological assumptions: including flat spatial geometry, time-independent vacuum energy, and cold dark matter. For this reason, the increasingly precise measurement of H_0 provided by the

---

[69] J. L. Tonry, J. P. Blakeslee, E. A. Ajhar, and A. Dressler, *Astrophys. J.* **473**, 399 (1997). For a more recent survey, see J. L. Tonry *et al.*, *Astrophys. J.* **546**, 681 (2001) [astro-ph/0011223].

[70] For a survey of most of these methods, with references, see G. H. Jacoby, D. Branch, R. Ciardullo, R. L. Davies, W. E. Harris, M. J. Pierce, C. J. Pritchet, J. L. Tonry, and D. L. Welch, *Publ. Astron. Soc. Pacific* **104**, 599 (1992).

[71] L. Ferrarese *et al.*, *op. cit.*; W. L. Freedman *et al.*, *Astrophys. J.* **553**, 47 (2001).

cosmic microwave background will not supplant the older measurements discussed in this section — rather, the agreement (or possible future disagreement) between the values of H_0 provided by these very different methods will serve to validate (or possibly invalidate) the cosmological assumptions made in the analysis of the cosmic microwave background.

To take account of the remaining uncertainty in the Hubble constant, it is usual these days to take

$$H_0 = 100\,h \text{ km s}^{-1} \text{ Mpc}^{-1}, \tag{1.3.13}$$

with the dimensionless parameter h assumed to be in the neighborhood of 0.7. This corresponds to a Hubble time

$$1/H_0 = 9.778 \times 10^9 \, h^{-1} \text{ years}. \tag{1.3.14}$$

1.4 Luminosity distances and angular diameter distances

We must now consider the measurement of distances at large redshifts, say $z > 0.1$, where the effects of cosmological expansion on the determination of distance can no longer be neglected. It is these measurements that can tell us whether the expansion of the universe is accelerating or decelerating, and how fast. Before we can interpret these measurements, we will need to consider in this section how to define distance at large redshifts, and we will have to apply Einstein's field equations to the Robertson–Walker metric in the following section. After that, we will return in Section 1.6 to the measurements of distances for large redshift, and their interpretation.

In the previous section we derived the familiar relation $\ell = L/4\pi d^2$ for the apparent luminosity ℓ of a source of absolute luminosity L at a distance d. At large distances this derivation needs modification for three reasons:

1. At the time t_0 that the light reaches earth, the proper area of a sphere drawn around the luminous object and passing through the earth is given by the metric (1.1.10) as $4\pi r_1^2 a^2(t_0)$, where r_1 is the coordinate distance of the earth as seen from the luminous object, which is just the same as the coordinate distance of the luminous object as seen from the earth. The fraction of the light received in a telescope of aperture A on earth is therefore $A/4\pi r_1^2 a^2(t_0)$, and so the factor $1/d^2$ in the formula for ℓ must be replaced with $1/r_1^2 a^2(t_0)$.

2. The rate of arrival of individual photons is lower than the rate at which they are emitted by the redshift factor $a(t_1)/a(t_0) = 1/(1+z)$.

3. The energy $h\nu_0$ of the individual photons received on earth is less than the energy $h\nu_1$ with which they were emitted by the same redshift factor $1/(1+z)$.

31

Putting this together gives the correct formula for apparent luminosity of a source at radial coordinate r_1 with a redshift z of any size:

$$\ell = \frac{L}{4\pi r_1^2 a^2(t_0)(1+z)^2} . \tag{1.4.1}$$

It is convenient to introduce a "luminosity distance" d_L, which is defined so that the relation between apparent and absolute luminosity and luminosity distance is the same as Eq. (1.3.3):

$$\ell = \frac{L}{4\pi d_L^2} . \tag{1.4.2}$$

Eq. (1.4.1) can then be expressed as

$$d_L = a(t_0)r_1(1+z) . \tag{1.4.3}$$

For objects with $z \ll 1$, we can usefully write the relation between luminosity distance and redshift as a power series. The redshift $1 + z \equiv a(t_0)/a(t_1)$ is related to the "look-back time" $t_0 - t_1$ by

$$z = H_0(t_0 - t_1) + \frac{1}{2}(q_0 + 2)H_0^2(t_0 - t_1)^2 + \dots \tag{1.4.4}$$

where H_0 is the Hubble constant (1.2.7) and q_0 is the *deceleration parameter*

$$q_0 \equiv \frac{-1}{H_0^2 a(t_0)} \left. \frac{d^2 a(t)}{dt^2} \right|_{t=t_0} . \tag{1.4.5}$$

This can be inverted, to give the look-back time as a power series in the redshift

$$H_0(t_0 - t_1) = z - \frac{1}{2}(q_0 + 2)z^2 + \dots . \tag{1.4.6}$$

The coordinate distance r_1 of the luminous object is given by Eq. (1.2.2) as

$$\frac{t_0 - t_1}{a(t_0)} + \frac{H_0(t_0 - t_1)^2}{2a(t_0)} + \dots = r_1 + \dots , \tag{1.4.7}$$

with the dots on the right-hand side denoting terms of *third* and higher order in r_1. Using Eq. (1.4.6), the solution is

$$r_1 a(t_0) H_0 = z - \frac{1}{2}(1 + q_0)z^2 + \dots . \tag{1.4.8}$$

This gives the luminosity distance (1.4.3) as a power series

$$d_L = H_0^{-1}\left[z + \frac{1}{2}(1 - q_0)z^2 + \cdots\right]. \tag{1.4.9}$$

We can therefore measure q_0 as well as H_0 by measuring the luminosity distance as a function of redshift to terms of order z^2. The same reasoning has been used to extend the expression (1.4.9) to fourth order in z:[1]

$$d_L(z) = H_0^{-1}\left[z + \frac{1}{2}(1 - q_0)z^2 - \frac{1}{6}\left(1 - q_0 - 3q_0^2 + j_0 + \frac{K}{H_0^2 a_0^2}\right)z^3\right.$$

$$+ \frac{1}{24}\left(2 - 2q_0 - 15q_0^2 - 15q_0^3 + 5j_0 + 10q_0 j_0\right.$$

$$\left.\left. + s_0 + \frac{2K(1 + 3q_0)}{H_0^2 a_0^2}\right)z^4 + \cdots\right],$$

where j_0 and s_0 are parameters known as the *jerk* and *snap*:

$$j_0 \equiv \frac{1}{H_0^3 \, a(t_0)} \left.\frac{d^3 a(t)}{dt^3}\right|_{t=t_0}, \qquad s_0 \equiv \frac{1}{H_0^4 \, a(t_0)} \left.\frac{d^4 a(t)}{dt^4}\right|_{t=t_0}.$$

Years ago cosmology was called "a search for two numbers," H_0 and q_0. The determination of H_0 is still a major goal of astronomy, as discussed in the previous section. On the other hand, there is less interest now in q_0. Instead of high-precision distance determinations at moderate redshifts, of order 0.1 to 0.2, which would give an accurate value of q_0, we now have distance determinations of only moderate precision at high redshifts, of order unity, which depend on the whole form of the function $a(t)$ over the past few billion years. For redshifts of order unity, it is not very useful to expand in powers of redshift. In order to interpret these measurements, we will need a dynamical theory of the expansion, to be developed in the next section. As we will see there, modern observations suggest strongly that there are not two but at least three parameters that need to be measured to calculate $a(t)$.

Before turning to this dynamical theory, let's pause a moment to clarify the distinction between different measures of distance. So far, we have encountered the proper distance (1.1.15) and the luminosity distance (1.4.3). There is another sort of distance, which is what we measure when we compare angular sizes with physical dimensions. Inspection of the metric

---

[1] M. Visser, *Class. Quant. Grav.* **21**, 2603 (2004) [gr-qc/0309109]. The term of third order in z was previously calculated by T. Chiba and T. Nakamura, *Prog. Theor. Phys.* **100**, 1077 (1998).

(1.1.12) shows that a source at co-moving radial coordinate r_1 that emits light at time t_1 and is observed at present to subtend a small angle θ will extend over a proper distance s (normal to the line of sight) equal to $a(t_1)r_1\theta$. The *angular diameter distance* d_A is defined so that θ is given by the usual relation of Euclidean geometry

$$\theta = s/d_A \qquad (1.4.10)$$

and we see that

$$d_A = a(t_1)r_1 . \qquad (1.4.11)$$

Comparison of this result with Eq. (1.4.3) shows that the ratio of the luminosity and angular-diameter distances is simply a function of redshift:

$$d_A/d_L = (1+z)^{-2} . \qquad (1.4.12)$$

Therefore if we have measured the luminosity distance at a given redshift (and if we are convinced of the correctness of the Robertson–Walker metric), then we learn nothing additional about $a(t)$ if we also measure the angular diameter distance at that redshift. Neither galaxies nor supernovas have well-defined edges, so angular diameter distances are much less useful in studying the cosmological expansion than are luminosity distances. However, as we shall see, they play an important role in the theoretical analysis of both gravitational lenses in Chapter 9 and of the fluctuations in the cosmic microwave radiation background in Chapters 2 and 7. We will see in Section 8.1 that the observation of acoustic oscillations in the matter density may allow a measurement of yet another distance, a *structure distance*, equal to $a(t_0)r_1 = (1+z)d_A$.

1.5 Dynamics of expansion

All our results up to now have been very general, not depending on assumptions about the dynamics of the cosmological expansion. To go further we will need now to apply the gravitational field equations of Einstein, with various tentative assumptions about the cosmic energy density and pressure.

The expansion of the universe is governed by the Einstein field equations (B.71), which can be put in the convenient form

$$R_{\mu\nu} = -8\pi GS_{\mu\nu} , \qquad (1.5.1)$$

where $R_{\mu\nu}$ is the *Ricci tensor*:

$$R_{\mu\nu} = \frac{\partial \Gamma^\lambda_{\lambda\mu}}{\partial x^\nu} - \frac{\partial \Gamma^\lambda_{\mu\nu}}{\partial x^\lambda} + \Gamma^\lambda_{\mu\sigma}\Gamma^\sigma_{\nu\lambda} - \Gamma^\lambda_{\mu\nu}\Gamma^\sigma_{\lambda\sigma} , \qquad (1.5.2)$$

and $S_{\mu\nu}$ is given in terms of the energy-momentum tensor $T_{\mu\nu}$ by

$$S_{\mu\nu} \equiv T_{\mu\nu} - \tfrac{1}{2}g_{\mu\nu}T^{\lambda}{}_{\lambda} \ . \tag{1.5.3}$$

As we saw in Section 1.1, for the Robertson–Walker metric the components of the affine connection with two or three time indices all vanish, so

$$
\begin{aligned}
R_{ij} = & \frac{\partial \Gamma^{k}_{ki}}{\partial x^{j}} - \left[\frac{\partial \Gamma^{k}_{ij}}{\partial x^{k}} + \frac{\partial \Gamma^{0}_{ij}}{\partial t} \right] \\
& + \left[\Gamma^{0}_{ik}\Gamma^{k}_{j0} + \Gamma^{k}_{i0}\Gamma^{0}_{jk} + \Gamma^{l}_{ik}\Gamma^{k}_{jl} \right] \\
& - \left[\Gamma^{k}_{ij}\Gamma^{l}_{kl} + \Gamma^{0}_{ij}\Gamma^{l}_{0l} \right]
\end{aligned}
\tag{1.5.4}
$$

$$R_{00} = \frac{\partial \Gamma^{i}_{i0}}{\partial t} + \Gamma^{i}_{0j}\Gamma^{j}_{0i} \tag{1.5.5}$$

We don't need to calculate $R_{i0} = R_{0i}$, because it is a three-vector, and therefore must vanish due to the isotropy of the Robertson–Walker metric. Using the formulas (1.1.17)–(1.1.19) for the non-vanishing components of the affine connection gives

$$
\frac{\partial \Gamma^{0}_{ij}}{\partial t} = \tilde{g}_{ij}\frac{d}{dt}(a\dot{a}) \ , \quad \Gamma^{0}_{ik}\Gamma^{k}_{j0} = \tilde{g}_{ij}\dot{a}^2 \ , \quad \Gamma^{0}_{ij}\Gamma^{l}_{0l} = 3\tilde{g}_{ij}\dot{a}^2 \ ,
$$

$$
\frac{\partial \Gamma^{i}_{i0}}{\partial t} = 3\frac{d}{dt}\left(\frac{\dot{a}}{a}\right) \ , \quad \Gamma^{i}_{0j}\Gamma^{j}_{i0} = 3\left(\frac{\dot{a}}{a}\right)^2 \ ,
\tag{1.5.6}
$$

where dots denote time derivatives. Using this in Eqs. (1.5.4) and (1.5.5), we find that the non-vanishing components of the Ricci tensor are

$$R_{ij} = \tilde{R}_{ij} - 2\dot{a}^2\tilde{g}_{ij} - a\ddot{a}\tilde{g}_{ij} \ , \tag{1.5.7}$$

$$R_{00} = 3\frac{d}{dt}\left(\frac{\dot{a}}{a}\right) + 3\left(\frac{\dot{a}}{a}\right)^2 = 3\frac{\ddot{a}}{a} \ , \tag{1.5.8}$$

where $\tilde{R}_{ij}$ is the purely spatial Ricci tensor

$$\tilde{R}_{ij} = \frac{\partial \Gamma^{k}_{ki}}{\partial x^{j}} - \frac{\partial \Gamma^{k}_{ij}}{\partial x^{k}} + \Gamma^{l}_{ik}\Gamma^{k}_{jl} - \Gamma^{l}_{ij}\Gamma^{k}_{kl} \ . \tag{1.5.9}$$

According to Eq. (1.1.19), the spatial components Γ^{i}_{jk} of the four-dimensional affine connection are here the same as those of the affine connection that would be calculated in three dimensions from the three-metric $\tilde{g}_{ij}$:

$$\Gamma^{k}_{ij} = Kx^{k}\tilde{g}_{ij} \ . \tag{1.5.10}$$

To calculate $\tilde{R}_{ij}$, we use a trick used earlier in calculating particle trajectories: we calculate $\tilde{R}_{ij}$ where the calculation is simplest, at $\mathbf{x} = 0$, and express the result as a relation that is invariant under all transformations of the spatial coordinates, so that the homogeneity of the three-dimensional metric insures that this relation is valid everywhere. The spatial Ricci tensor at $\mathbf{x} = 0$ is

$$\tilde{R}_{ij} = \frac{\partial \Gamma^l_{li}}{\partial x^j} - \frac{\partial \Gamma^l_{ji}}{\partial x^l} = K\delta_{ij} - 3K\delta_{ij} = -2K\delta_{ij} \ . \tag{1.5.11}$$

At $\mathbf{x} = 0$ the spatial metric $\tilde{g}_{ij}$ is just δ_{ij}, so this can be rewritten as

$$\tilde{R}_{ij} = -2K\tilde{g}_{ij} \ , \tag{1.5.12}$$

which, since it is an equality between two three-tensors, is then true in all spatial coordinate systems, including systems in which the point $\mathbf{x} = 0$ is transformed into any other point. Hence Eq. (1.5.12) is true everywhere, and together with Eq. (1.5.7) gives

$$R_{ij} = -\left[2K + 2\dot{a}^2 + a\ddot{a}\right]\tilde{g}_{ij} \ . \tag{1.5.13}$$

We also need the values of S_{ij} and S_{00}. For this, we use Eq. (1.1.31) in the form

$$T_{00} = \rho \ , \qquad T_{i0} = 0 \ , \qquad T_{ij} = a^2 p \, \tilde{g}_{ij} \ , \tag{1.5.14}$$

where $\rho(t)$ and $p(t)$ are the proper energy density and pressure. Eq. (1.5.3) gives then

$$S_{ij} = T_{ij} - \frac{1}{2}\tilde{g}_{ij}a^2 \left(T^k{}_k + T^0{}_0\right) = a^2 p \tilde{g}_{ij} - \frac{1}{2}a^2 \tilde{g}_{ij}(3p - \rho) = \frac{1}{2}(\rho - p) \, a^2 \, \tilde{g}_{ij} \ , \tag{1.5.15}$$

$$S_{00} = T_{00} + \frac{1}{2}\left(T^k{}_k + T^0{}_0\right) = \rho + \frac{1}{2}(3p - \rho) = \frac{1}{2}(\rho + 3p) \ , \tag{1.5.16}$$

and $S_{i0} = 0$. The Einstein equations are therefore

$$-\frac{2K}{a^2} - \frac{2\dot{a}^2}{a^2} - \frac{\ddot{a}}{a} = -4\pi \, G(\rho - p) \ , \tag{1.5.17}$$

$$\frac{3\ddot{a}}{a} = -4\pi \, G(3p + \rho) \ . \tag{1.5.18}$$

We can eliminate the second derivative terms by adding three times the first equation to the second, and find

$$\dot{a}^2 + K = \frac{8\pi G \rho a^2}{3} . \tag{1.5.19}$$

This is the fundamental Friedmann equation[1] governing the expansion of the universe.

The remaining information in Eqs. (1.5.17) and (1.5.18) just reproduces the conservation law (1.1.32):

$$\dot{\rho} = -\frac{3\dot{a}}{a}(\rho + p) . \tag{1.5.20}$$

(This should come as no surprise. Under all circumstances, the energy-momentum conservation law may be derived as a consequence of the Einstein field equations.) Given p as a function of ρ, we can solve Eq. (1.5.20) to find ρ as a function of a, and then use this in Eq. (1.5.19) to find a as a function of t.

There is another way of deriving Eq. (1.5.19), at least for the case of non-relativistic matter. Imagine a co-moving ball cut out from the expanding universe, with some typical galaxy at its center, and suppose it then emptied of the matter it contains. According to Birkhoff's theorem,[2] in any system that is spherically symmetric around some point, the metric in an empty ball centered on this point must be that of flat space. This holds whatever is happening outside the empty ball, as long as it is spherically symmetric. Now imagine putting the matter back in the ball, with a velocity proportional to distance from the center of symmetry, taken as $\mathbf{X} = 0$:

$$\dot{\mathbf{X}} = H(t)\mathbf{X} . \tag{1.5.21}$$

(Here the components X^i of $\mathbf{X}$ are ordinary Cartesian coordinates, not the co-moving coordinates x^i used in the Robertson–Walker metric. Note that this is the one pattern of velocities consistent with the principle of homogeneity: The velocity of a co-moving particle at $\mathbf{X}_1$ relative to a co-moving particle at $\mathbf{X}_2$ is $\dot{\mathbf{X}}_1 - \dot{\mathbf{X}}_2 = H(t)(\mathbf{X}_1 - \mathbf{X}_2)$.) The solution of Eq. (1.5.21) is

$$\mathbf{X}(t) = \left(\frac{a(t)}{a(t_0)}\right)\mathbf{X}(t_0) , \tag{1.5.22}$$

where $a(t)$ is the solution of the equation

$$\dot{a}(t)/a(t) = H(t) . \tag{1.5.23}$$

---

[1] A. Friedmann, *Z. Phys.* **16**, 377 (1922); *ibid* **21**, 326 (1924).
[2] G&C, Section 11.7.

As long as the radius of the ball is chosen to be not too large, the expansion velocity (1.5.21) of the matter we put into it will be non-relativistic, and the gravitational field will be weak, so that we can follow its motion using Newtonian mechanics. The kinetic energy of a co-moving particle of mass m at $\mathbf{X}$ is

$$K.E. = \frac{1}{2}m\dot{\mathbf{X}}^2 = \frac{m\dot{a}^2\mathbf{X}^2}{2\,a^2} . \qquad (1.5.24)$$

The mass interior to the position of the particle is $M(\mathbf{X}) = 4\pi\rho|\mathbf{X}|^3/3$, so the potential energy of the particle is

$$P.E. = -\frac{G\,m\,M(\mathbf{X})}{|\mathbf{X}|} = -\frac{4\pi\,G\,m\,\rho\,|\mathbf{X}|^2}{3} . \qquad (1.5.25)$$

The condition of energy conservation thus tells us that

$$E = K.E. + P.E. = \frac{m\,|\mathbf{X}(t_0)|^2}{a^2(t_0)}\left[\frac{\dot{a}^2}{2} - \frac{4\pi\,G\,\rho\,a^2}{3}\right] = \text{constant} . \quad (1.5.26)$$

This is the same as Eq. (1.5.19), providing we identify the particle energy as

$$E = -\frac{K\,m\,|\mathbf{X}(t_0)|^2}{2\,a^2(t_0)} . \qquad (1.5.27)$$

Particles will be able to escape to infinity if and only if $E \geq 0$, which requires $K = 0$ or $K = -1$. For $K = +1$ they have less than escape velocity, so the expansion eventually stops, and particles fall back toward each other.

Returning now to the relativistic formalism and an arbitrary dependence of ρ on a, even without knowing this dependence we can use Eq. (1.5.19) to draw important consequences about the general features of the expansion. First, as long as ρ remains positive, it is only possible for the expansion of the universe to stop if $K = +1$, the case of spherical geometry. Also, for any value of the Hubble constant $H_0 \equiv \dot{a}(t_0)/a(t_0)$, we may define a critical present density

$$\rho_{0,\text{crit}} \equiv \frac{3H_0^2}{8\pi\,G} = 1.878 \times 10^{-29}\,h^2\;g/cm^3 , \qquad (1.5.28)$$

where h is the Hubble constant in units of $100\ \text{km s}^{-1}\ \text{Mpc}^{-1}$. According to Eq. (1.5.19), whatever we assume about the constituents of the universe, the curvature constant K will be $+1$ or 0 or -1 according to whether the present density ρ_0 is greater than, equal to, or less than $\rho_{0,\text{crit}}$. *If* the quantity $3p+\rho$ is positive (as it is for any mixture of matter and radiation, in the absence of a vacuum energy density) then Eq. (1.5.18) shows that $\ddot{a}/a \leq 0$, so the expansion must have started with $a = 0$ at some moment in the past; the present age of the universe t_0 is *less* than the Hubble time

$$t_0 < H_0^{-1} . \tag{1.5.29}$$

Also, if $K = +1$ and the expansion stops, then with $\ddot{a}/a \leq 0$ the universe will again contract to a singularity at which $a = 0$.

We can use Eq. (1.5.18) to give a general formula for the deceleration parameter $q_0 \equiv -\ddot{a}(t_0)a(t_0)/\dot{a}^2(t_0)$:

$$q_0 = \frac{4\pi G(\rho_0 + 3p_0)}{3H_0^2} = \frac{\rho_0 + 3p_0}{2\rho_{0,\text{crit}}} , \tag{1.5.30}$$

with a subscript 0 denoting a present value. If the present density of the universe were dominated by non-relativistic matter then $p_0 \ll \rho_0$, and the curvature constant K would be $+1$ or 0 or -1 according to whether $q_0 > \frac{1}{2}$ or $q_0 = \frac{1}{2}$ or $q_0 < \frac{1}{2}$. On the other hand, if the present density of the universe were dominated by relativistic matter then $p_0 = \rho_0/3$, and the critical value of the deceleration parameter at which $K = 0$ would be $q_0 = 1$. Finally, if the present density of the universe were dominated by vacuum energy then $p_0 = -\rho_0$, and the value of the deceleration parameter at which $K = 0$ would be $q_0 = -1$.

There is a peculiar aspect to these results. The contribution of non-relativistic and relativistic matter to the quantity ρa^2 in Eq. (1.5.19) grows as a^{-1} and a^{-2}, respectively, as $a \to 0$, so at sufficiently early times in the expansion we may certainly neglect the constant K, and Eq. (1.5.19) gives

$$\frac{\dot{a}^2}{a^2} \to \frac{8\pi G\rho}{3} . \tag{1.5.31}$$

That is, at these early times the density becomes essentially equal to the critical density $3H^2/8\pi G$, where $H \equiv \dot{a}/a$ is the value of the Hubble "constant" at those times. On the other hand, we will see later that the total energy density of the present universe is still a fair fraction of the critical density. How is it that after billions of years, ρ is still not very different from ρ_{crit}? This is sometimes called the *flatness problem*.

The simplest solution to the flatness problem is just that we are in a spatially flat universe, in which $K = 0$ and ρ is always precisely equal to ρ_{crit}. A more popular solution is provided by the inflationary theories discussed in Chapter 4. In these theories K may not vanish, and ρ may not start out close to ρ_{crit}, but there is an early period of rapid growth in which ρ/ρ_{crit} rapidly approaches unity. In inflationary theories it is expected though not required that ρ should now be very close to ρ_{crit}, in which case it is a good approximation to take $K = 0$.

For $K = 0$ we get very simple solutions to Eq. (1.5.19) in the three special cases listed in Section 1.1:

Non-relativistic matter: Here $\rho = \rho_0(a/a_0)^{-3}$, and the solution of Eq. (1.5.19) with $K = 0$ is

$$a(t) \propto t^{2/3} . \tag{1.5.32}$$

This gives $q_0 \equiv -a\ddot{a}/\dot{a}^2 = 1/2$, and a simple relation between the age of the universe and the Hubble constant

$$t_0 = \frac{2}{3H_0} = 6.52 \times 10^9 \, h^{-1} \, \text{yr} . \tag{1.5.33}$$

Eqs. (1.5.32) and (1.5.18) show that for $K = 0$, the energy density at time t is $\rho = 1/6\pi \, Gt^2$. This is known as the *Einstein–de Sitter model.* It was for many years the most popular cosmological model, though as we shall see, the age (1.5.33) is uncomfortably short compared with the ages of certain stars.

Relativistic matter: Here $\rho = \rho_0(a/a_0)^{-4}$, and the solution of Eq. (1.5.19) with $K = 0$ is

$$a(t) \propto \sqrt{t} . \tag{1.5.34}$$

This gives $q_0 = +1$, while the age of the universe and the Hubble constant are related by

$$t_0 = \frac{1}{2H_0} . \tag{1.5.35}$$

The energy density at time t is $\rho = 3/32\pi \, Gt^2$.

Vacuum energy: Lorentz invariance requires that in locally inertial coordinate systems the energy-momentum tensor $T_V^{\mu\nu}$ of the vacuum must be proportional to the Minkowski metric $\eta^{\mu\nu}$ (for which $\eta^{ij} = \eta_{ij} = \delta_{ij}$, $\eta^{i0} = \eta_{i0} = \eta^{0i} = \eta_{0i} = 0$, $\eta^{00} = \eta_{00} = -1$), and so in general coordinate systems $T_V^{\mu\nu}$ must be proportional to $g^{\mu\nu}$. Comparing this with Eq. (B.43) shows that the vacuum has $p_V = -\rho_V$, so that $T_V^{\mu\nu} = -\rho_V \, g^{\mu\nu}$. In the absence of any other form of energy this would satisfy the conservation law $0 = T_V^{\mu\nu}{}_{;\mu} = g^{\mu\nu}\partial\rho_V/\partial x^\mu$, so that ρ_V would be a constant, independent of spacetime position. Eq. (1.5.19) for $K = 0$ requires that $\rho_V > 0$, and has the solutions

$$a(t) \propto \exp(Ht) \tag{1.5.36}$$

where H is the Hubble constant, now really a constant, given by

$$H = \sqrt{\frac{8\pi \, G\rho_V}{3}} . \tag{1.5.37}$$

Here $q_0 = -1$, and the age of the universe in this case is infinite. This is known as the *de Sitter model*.[3] Of course, there is *some* matter in the universe, so even if the energy density of the universe is now dominated by a constant vacuum energy, there was a time in the past when matter and/or radiation were more important, and so the expansion has a finite age, although greater than it would be without a vacuum energy.

More generally, for arbitrary K and a mixture of vacuum energy and relativistic and non-relativistic matter, making up fractions Ω_Λ, Ω_M, and Ω_R of the critical energy density,[4] we have

$$\rho = \frac{3 H_0^2}{8 \pi G} \left[\Omega_\Lambda + \Omega_M \left(\frac{a_0}{a} \right)^3 + \Omega_R \left(\frac{a_0}{a} \right)^4 \right], \qquad (1.5.38)$$

where the present energy densities in the vacuum, non-relativistic matter, and and relativistic matter (i.e., radiation) are, respectively,

$$\rho_{V0} \equiv \frac{3 H_0^2 \Omega_\Lambda}{8 \pi G}, \quad \rho_{M0} \equiv \frac{3 H_0^2 \Omega_M}{8 \pi G}, \quad \rho_{R0} \equiv \frac{3 H_0^2 \Omega_R}{8 \pi G}, \quad (1.5.39)$$

and, according to Eq. (1.5.19),

$$\Omega_\Lambda + \Omega_M + \Omega_R + \Omega_K = 1, \quad \Omega_K \equiv - \frac{K}{a_0^2 H_0^2}. \qquad (1.5.40)$$

Using this in Eq. (1.5.19) gives

$$dt = \frac{dx}{H_0 x \sqrt{\Omega_\Lambda + \Omega_K x^{-2} + \Omega_M x^{-3} + \Omega_R x^{-4}}}$$

$$= \frac{-dz}{H_0 (1 + z) \sqrt{\Omega_\Lambda + \Omega_K (1 + z)^2 + \Omega_M (1 + z)^3 + \Omega_R (1 + z)^4}}, \quad (1.5.41)$$

where $x \equiv a/a_0 = 1/(1 + z)$. Therefore, if we define the zero of time as corresponding to an infinite redshift, then the time at which light was emitted that reaches us with redshift z is given by

$$t(z) = \frac{1}{H_0} \int_0^{1/(1+z)} \frac{dx}{x \sqrt{\Omega_\Lambda + \Omega_K x^{-2} + \Omega_M x^{-3} + \Omega_R x^{-4}}}. \qquad (1.5.42)$$

---

[3]W. de Sitter, *Proc. Roy. Acad. Sci.* (Amsterdam), **19**, 1217 (1917); *ibid.* **20**, 229 (1917); *ibid.* **20**, 1309 (1917); *Mon. Not. Roy. Astron. Soc.*, **78**, 2 (1917).

[4]The use of the symbol Ω_Λ instead of Ω_V for the ratio of the vacuum energy density to the critical energy density has become standard, because of a connection with the cosmological constant discussed in a historical note below.

In particular, by setting $z = 0$, we find the present age of the universe:

$$t_0 = \frac{1}{H_0} \int_0^1 \frac{dx}{x\sqrt{\Omega_\Lambda + \Omega_K x^{-2} + \Omega_M x^{-3} + \Omega_R x^{-4}}} \,. \qquad (1.5.43)$$

In order to calculate luminosity or angular diameter distances, we also need to know the radial coordinate $r(z)$ of a source that is observed now with redshift z. According to Eqs. (1.2.2) and (1.5.41), this is

$$r(z) = S\left[\int_{t(z)}^{t_0} \frac{dt}{a(t)}\right]$$

$$= S\left[\frac{1}{a_0 H_0} \int_{1/(1+z)}^1 \frac{dx}{x^2\sqrt{\Omega_\Lambda + \Omega_K x^{-2} + \Omega_M x^{-3} + \Omega_R x^{-4}}}\right],$$

where

$$S[y] \equiv \begin{cases} \sin y & K = +1 \\ y & K = 0 \\ \sinh y & K = -1 \,. \end{cases}$$

This can be written more conveniently by using Eq. (1.5.40) to express $a_0 H_0$ in terms of Ω_K. We then have a single formula

$$a_0 r(z) = \frac{1}{H_0 \Omega_K^{1/2}}$$

$$\times \sinh\left[\Omega_K^{1/2} \int_{1/(1+z)}^1 \frac{dx}{x^2\sqrt{\Omega_\Lambda + \Omega_K x^{-2} + \Omega_M x^{-3} + \Omega_R x^{-4}}}\right],$$

$$(1.5.44)$$

which can be used for any curvature. (Eq. (1.5.43) has a smooth limit for $\Omega_K \to 0$, which gives the result for zero curvature. Also, for $\Omega_K < 0$, the argument of the hyperbolic sine is imaginary, and we can use $\sinh ix = i \sin x$.) Using Eq. (1.5.44) in Eq. (1.4.3) gives the luminosity distance of a source observed with redshift z as

$$d_L(z) = a_0 r(z)(1 + z) = \frac{1+z}{H_0 \Omega_K^{1/2}}$$

$$\times \sinh\left[\Omega_K^{1/2} \int_{1/(1+z)}^1 \frac{dx}{x^2\sqrt{\Omega_\Lambda + \Omega_K x^{-2} + \Omega_M x^{-3} + \Omega_R x^{-4}}}\right].$$

$$(1.5.45)$$

For $K = 0$ we have $\Omega_K = 0$ and Eq. (1.5.45) becomes

$$d_L(z) = a_0 r_1(1+z) = \frac{1+z}{H_0} \int_{1/(1+z)}^{1} \frac{dx}{x^2 \sqrt{\Omega_\Lambda + \Omega_M x^{-3} + \Omega_R x^{-4}}} .$$

$$(1.5.46)$$

As we will see in Section 2.1, Ω_R is much less than Ω_M, and the integral (1.5.46) converges at its lower bound for $z \to \infty$ whether or not Ω_R vanishes, so it is a good approximation to take $\Omega_R = 0$ here.

It is of some interest to express the deceleration parameter q_0 in terms of the Ωs. The p/ρ ratio w for vacuum, matter, and radiation is -1, 0, and $1/3$, respectively, so Eq. (1.5.39) gives the present pressure as

$$p_0 = \frac{3H_0^2}{8\pi G} \left(-\Omega_\Lambda + \frac{1}{3}\Omega_R \right) .$$

$$(1.5.47)$$

Eq. (1.5.30) then gives

$$q_0 = \frac{4\pi G(3p_0 + \rho_0)}{3H_0^2} = \frac{1}{2} (\Omega_M - 2\Omega_\Lambda + 2\Omega_R) .$$

$$(1.5.48)$$

One of the reasons for our interest in the values of Ω_K, Ω_M, etc. is that they tell us whether the present expansion of the universe will ever stop. According to Eq. (1.5.38), the expansion can only stop if there is a real root of the cubic equation

$$\Omega_\Lambda u^3 + \Omega_K u + \Omega_M = 0 ,$$

$$(1.5.49)$$

where $u \equiv a(t)/a(t_0)$ is greater than one. (We are ignoring radiation here, since it will become even less important as the universe expands.) This expression has the value $+1$ for $u = 1$. If $\Omega_\Lambda < 0$ then the left-hand side of Eq. (1.5.49) becomes negative for sufficiently large u, so it must take the value zero at some intermediate value of u, and the expansion will stop when this value of u is reached. Even for $\Omega_\Lambda \geq 0$ it is still possible for the expansion to stop, provided $\Omega_K = 1 - \Omega_\Lambda - \Omega_M$ is sufficiently negative (which, among other things, requires that $K = +1$).

Historical Note 1: If we express the total energy momentum tensor $T_{\mu\nu}$ as the sum of a possible vacuum term $-\rho_V g_{\mu\nu}$ and a term $T_{\mu\nu}^M$ arising from matter (including radiation), then the Einstein equations take the form

$$R_{\mu\nu} - \frac{1}{2}g_{\mu\nu}R^\lambda{}_\lambda = -8\pi G T_{\mu\nu}^M + 8\pi G \rho_V g_{\mu\nu} .$$

$$(1.5.50)$$

Thus the effect of a vacuum energy is equivalent to modifying the Einstein field equations to read

$$R_{\mu\nu} - \frac{1}{2}g_{\mu\nu}R - \Lambda g_{\mu\nu} = -8\pi G T^M_{\mu\nu} , \qquad (1.5.51)$$

where

$$\Lambda = 8\pi G \rho_V . \qquad (1.5.52)$$

The quantity Λ is known as the *cosmological constant*. It was introduced into the field equation by Einstein in 1917 in order to satisfy a condition that at the time was generally regarded as essential, that the universe should be static.[5] According to Eqs. (1.5.18) and (1.5.19), a static universe is only possible if $3p + \rho = 0$ and $K = 8\pi G\rho a^2/3$. If the contents of the universe are limited to vacuum energy and non-relativistic matter, then $\rho = \rho_M + \rho_V$, $p = -\rho_V$, and $\rho_M \geq 0$. It follows that $\rho_M = 2\rho_V \geq 0$, so $K > 0$, which by convention means $K = +1$, so that a takes the value $a_E = 1/\sqrt{8\pi G\rho_V} = 1/\sqrt{\Lambda}$. This is known as the *Einstein model*.

Einstein did not realize it, but his cosmology was unstable: If a is a little less than a_E then ρ_M is a little larger than $2\rho_V$, so Eq. (1.5.18) shows that $\ddot{a}/a < 0$, and a thus begins to decrease. Likewise, if a is a little greater than a_E then it begins to increase. The models with $K = +1$ and $\Lambda > 0$ in which a starts at the Einstein radius $a = a_E$ with $\rho_M = 2\rho_V$ and then expands to infinity (or starts at $a = 0$ and approaches a_E as $t \to \infty$ with just enough matter so that $\rho_M = 2\rho_V$ at the Einstein radius), are known as *Eddington–Lemaître models*.[6] There are also models with $K = +1$ and a little more matter, that start at $a = 0$, spend a long time near the Einstein radius, and then expand again to infinity, approaching a de Sitter model. These are known as *Lemaître models*.[7]

Oddly, de Sitter also invented his cosmological model (with $a \propto \exp(Ht)$) in order to satisfy a supposed need for a static universe. He originally proposed a time-independent metric, given by

$$d\tau^2 = (1 - r^2/R^2)dt^2 - \frac{dr^2}{1 - r^2/R^2} - r^2 d\theta^2 - r^2 \sin^2\theta \, d\phi^2 ,$$
$$(1.5.53)$$

[5] A. Einstein, *Sitz. Preuss. Akad. Wiss.* 142 (1917). For an English translation, see *The Principle of Relativity* (Methuen, 1923; reprinted by Dover Publications, New York, 1952), p. 35.

[6] A. S. Eddington, *Mon. Not. Roy. Astron. Soc.* **90**, 668 (1930); G. Lemaître, *Ann. Soc. Sci. Brux.* **A47**, 49 (1927); *Mon. Not. Roy. Astron. Soc.* **91**, 483 (1931). The interpretation of the cosmological constant in terms of vacuum energy was stated by Lemaître in *Proc. Nat. Acad. Sci.* **20**, 12L (1934).

[7] G. Lemaître, *op. cit.*

or equivalently, setting $r = R \sin \chi$,

$$d\tau^2 = \cos^2 \chi \, dt^2 - R^2 \left(d\chi^2 + \sin^2 \chi \, d\theta^2 + \sin^2 \chi \, \sin^2 \theta \, d\phi^2 \right), \quad (1.5.54)$$

with $R = \sqrt{3/\Lambda}$ constant. De Sitter did not realize at first that this metric has $\Gamma^i_{00} \neq 0$, so that his coordinate system was not co-moving.[8] Only later was it noticed that, using co-moving spatial coordinates and cosmological standard time, de Sitter's model is equivalent to a Robertson–Walker metric with $K = 0$ and $a \propto \exp(t/R)$.

After the discovery of the expansion of the universe, cosmologists lost interest in a static universe, and Einstein came to regret his introduction of a cosmological constant, calling it his greatest mistake. But as we shall see in the next section, there are theoretical reasons to expect a non-vanishing vacuum energy, and there is observational evidence that in fact it does not vanish. Einstein's mistake was not that he introduced the cosmological constant — it was that he thought it was a mistake.

Historical Note 2: There is a cosmological model due to Bondi and Gold[9] and in a somewhat different version to Hoyle,[10] known as the *steady state theory*. In this model nothing physical changes with time, so the Hubble constant really is constant, and hence $a(t) \propto \exp(Ht)$, just as in the de Sitter model. To keep the curvature constant, it is necessary to take $K = 0$. In this model new matter must be continually created to keep ρ constant as the universe expands. Since the discovery of the cosmic microwave background (discussed in Chapter 2) the steady state theory in its original form has been pretty well abandoned.

1.6 Distances at large redshift: Accelerated expansion

We now return to our account of the measurement of distances as a function of redshift, considering now redshifts $z > 0.1$, which are large enough so that we can ignore the peculiar motions of the light sources, and also large enough so that we need to take into account the effects of cosmological expansion on distance determination.

For many years, the chief "standard candles" used at large redshift were the brightest galaxies in rich clusters. It is now well established that the

---

[8] A. S. Eddington, *The Mathematical Theory of Relativity*, 2nd ed. (Cambridge University Press, Cambridge, 1924), Section 70. It is interesting that Eddington interpreted Slipher's observation that most spiral nebulae exhibit redshifts rather than blueshifts in terms of the de Sitter model, rather than Friedmann's models.

[9] H. Bondi and T. Gold, *Mon. Not. Roy. Astron. Soc.* **108**, 252 (1948).

[10] F. Hoyle, *Mon. Not. Roy. Astron. Soc.* **108**, 372 (1948), *ibid.* **109**, 365 (1949).

absolute luminosity of these brightest galaxies evolves significantly over cosmological time scales. There are also severe selection effects: there is a tendency to pick out larger clusters with brightest galaxies of higher absolute luminosity at large distances. The evolution of brightest galaxies is of interest in itself, and continues to be the object of astronomical study,[1] but the use of these galaxies as distance indicators has been pretty well abandoned. Similarly, although the Tully–Fisher relation discussed in Section 1.3 has been applied to galaxies with redshifts of order unity, at these redshifts it is used to study galactic evolution, rather than to measure cosmological parameters.[2]

Fortunately, the Type Ia supernovae discussed in Section 1.3 provide an excellent replacement as standard candles.[3] They are very bright; the peak blue absolute magnitude averages about -19.2, which compares well with the absolute magnitude -20.3 estimated for our own galaxy. Also, as described in Section 1.3, a Type Ia supernova typically occurs when a white dwarf member of a binary pair has accreted just enough mass to push it over the Chandrasekhar limit, so that the nature of the explosion does not depend much on when in the history of the universe this happens, or on the mass with which the white dwarf started or the nature of the companion star. But it might depend somewhat on the metallicity (the proportion of elements heavier than helium) of the white dwarf, which can depend on the epoch of the explosion. The absolute luminosity of Type Ia supernovae is observed to vary with environmental conditions, but fortunately in the use of supernovae as distance indicators the bulk of this variation is correctable empirically.

Observations of Type Ia supernovae have been compared with theoretical predictions (equivalent to Eq. (1.5.45)) for luminosity distance as a function of redshift at about the same time by two groups: The Supernova Cosmology Project[4] and the High-z Supernova Search Team.[5]

---

[1] See, e.g., D. Zaritsky *et al.*, in *Proceedings of the Sesto 2001 Conference on Tracing Cosmic Evolution with Galaxy Clusters* [astro-ph/0108152]; S. Brough *et al.*, in *Proceedings of the Sesto 2001 Conference on Tracing Cosmic Evolution with Galaxy Clusters* [astro-ph/0108186].

[2] N. P. Vogt *et al.*, *Astrophys. J.* **465**, 115 (1996). For a review and more recent references, see A. Aragón-Salmanca, in *Galaxy Evolution Across the Hubble Time – Proceedings of I.A.U. Symposium 235*, eds. F. Combes and J. Palous [astro-ph/0610587].

[3] For reviews, see S. Perlmutter and B. P. Schmidt, in *Supernovae & Gamma Ray Bursts*, ed. K. Weiler (Springer, 2003) [astro-ph/0303428]; P. Ruiz-Lapuente, *Astrophys. Space Sci.* **290**, 43 (2004) [astro-ph/0304108]; A. V. Filippenko, in *Measuring and Modeling of the Universe* (Carnegie Observatories Astrophysics Series, Vol 2., Cambridge University Press) [astro-ph/0307139]; Lect. Notes Phys. **645**, 191 (2004) [astro-ph/0309739]; N. Panagia, *Nuovo Cimento* **B 210**, 667 (2005) [astro-ph/0502247].

[4] S. Perlmutter *et al.*, *Astrophys. J.* **517**, 565 (1999) [astro-ph/9812133]. Also see S. Perlmutter *et al.*, *Nature* **391**, 51 (1998) [astro-ph/9712212].

[5] A. G. Riess *et al.*, *Astron. J.* **116**, 1009 (1998) [astro-ph/9805201]. Also see B. Schmidt *et al.*, *Astrophys. J.* **507**, 46 (1998) [astro-ph/9805200].

The Supernova Cosmology Project analyzed the relation between apparent luminosity and redshift for 42 Type Ia supernovae, with redshifts z ranging from 0.18 to 0.83, together with a set of closer supernovae from another supernova survey, at redshifts below 0.1. Their original results are shown in Figure 1.1.

With a confidence level of 99%, the data rule out the case $\Omega_\Lambda = 0$ (or $\Omega_\Lambda < 0$). For a flat cosmology with $\Omega_K = \Omega_R = 0$, so that $\Omega_\Lambda + \Omega_M = 1$, the data indicate a value

$$\Omega_M = 0.28^{+0.09}_{-0.08} \text{ (1}\sigma \text{ statistical)}^{+0.05}_{-0.04} \text{ (identified systematics)}$$

(These results are independent of the Hubble constant or the absolute calibration of the relation between supernova absolute luminosity and time scale, though they do depend on the shape of this relation.) This gives the age (1.5.43) as

$$t_0 = 13.4^{+1.3}_{-1.0} \times 10^9 \left(\frac{70 \text{ km s}^{-1} \text{ Mpc}^{-1}}{H_0} \right) \text{ yr} .$$

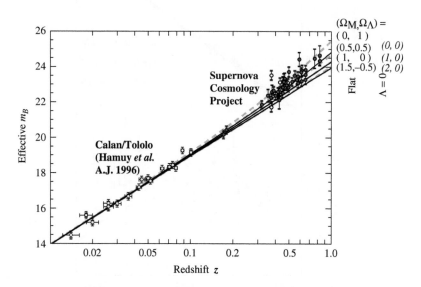

Figure 1.1: Evidence for dark energy, found in 1998 by the Supernova Cosmology Project, from S. Perlmutter *et al.*, *Astrophys. J.* **517**, 565 (1999) [astro-ph/9812133]. Here the effective blue apparent magnitude (corrected for variations in absolute magnitude, as indicated by supernova light curves) are plotted versus redshift for 42 high redshift Type Ia supernovae observed by the Supernova Cosmology Project, along with 18 lower redshift Type Ia supernovae from the Calán–Tololo Supernovae Survey. Horizontal bars indicate the uncertainty in cosmological redshift due to an assumed peculiar velocity uncertainty of 300 km sec^{-1}. Dashed and solid curves give the theoretical effective apparent luminosities for cosmological models with $\Omega_K = 0$ or $\Omega_\Lambda = 0$, respectively, and various possible values of Ω_M.

For $\Omega_M = 0.28$ and $\Omega_\Lambda = 1 - \Omega_M$, Eq. (1.5.48) gives a negative deceleration parameter, $q_0 = -0.58$, indicating that *the expansion of the universe is accelerating*.

The High-*z* Supernova Search Team originally studied 16 Type Ia supernovae of high redshift (with redshifts ranging from 0.16 to 0.97), including 2 from the Supernova Cosmology Project, together with 34 nearby supernovae, and conclude that $\Omega_\Lambda > 0$ at the 99.7% confidence level, with no assumptions about spatial curvature. Their original results are shown in Figure 1.2.

Their best fit for a flat cosmology is $\Omega_M = 0.28 \pm 0.10$ and $\Omega_\Lambda = 1 - \Omega_M$, giving an age of about $(14.2 \pm 1.5) \times 10^9$ years, including uncertainties in the Cepheid distance scale. Assuming only $\Omega_M \geq 0$, and with a conservative

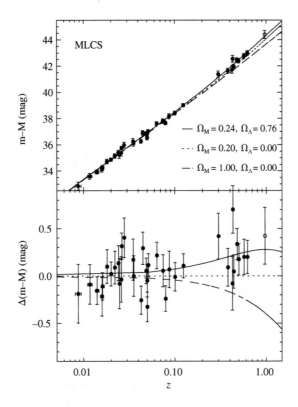

Figure 1.2: Evidence for dark energy, found in 1998 by the High-*z* Supernova Search Team, from A. G. Riess *et al.*, *Astron. J.* **116**, 1009 (1998) [astro-ph/9805201]. In the upper panel distance modulus is plotted against redshift for a sample of Type Ia supernovae. The curves give the theoretical results for two cosmologies with $\Omega_\Lambda = 0$ and a good-fit flat cosmology with $\Omega_M = 0.24$ and $\Omega_\Lambda = 0.76$. The bottom panel shows the difference between data and a formerly popular Einstein–de Sitter model with $\Omega_M = 0.2$ and $\Omega_\Lambda = 0$, represented by the horizontal dotted line.

fitting method, with 99.5% confidence they conclude that $q_0 < 0$, again strongly indicating an accelerated expansion. Including 8 new supernovae in a sample of 230 supernovae of Type Ia gave[6] $1.4\,\Omega_M - \Omega_\Lambda = -0.35 \pm 0.14$, providing further evidence that $\Omega_\Lambda > 0$. The case for vacuum energy was then strenghtened when the Supernova Cosmology Project,[7] including a new set of supernova, found for a flat universe that $\Omega_\Lambda = 0.75^{+0.06}_{-0.07}(\text{stat.}) \pm 0.032(\text{syst.})$.

Both groups agree that their results are chiefly sensitive to a linear combination of Ω_Λ and Ω_M, given as $0.8\Omega_M - 0.6\Omega_\Lambda$ by the Supernova Cosmology Project and $\Omega_M - \Omega_\Lambda$ or $1.4\Omega_M - \Omega_\Lambda$ by the High-z Supernova Search Team. The minus sign in these linear combinations, as in Eq. (1.5.48), reflects the fact that matter and vacuum energy have opposite effects on the cosmological acceleration: Matter causes it to slow down, while a positive vacuum energy causes it to accelerate. The negative values found for these linear combinations shows the presence of a component of energy something like vacuum energy, with $p \simeq -\rho$. This is often called *dark energy*.

Incidentally, these linear combinations of Ω_Λ and Ω_M are quite different from the expression $\Omega_M/2 - \Omega_\Lambda$, which according to Eq. (1.5.48) gives the deceleration parameter q_0 that was the target of much cosmological work of the past. Thus the observations of Type Ia supernovae at cosmological distances should not be regarded as simply measurements of q_0.

The High-z Supernova Search Team subsequently began to use the same survey observations to follow the time development of supernovae that were used to find them.[8] They discovered 23 new high redshift supernovae of Type Ia, including 15 with $z > 0.7$. Using these new supernovae along with the 230 used earlier by Tonry *et al.*, and with the assumption that $\Omega_M + \Omega_\Lambda = 1$, they found the best-fit values $\Omega_M = 0.33$ and $\Omega_\Lambda = 0.67$.

The crucial feature of the supernova data that indicates that $\Omega_\Lambda > \Omega_M$ is that the apparent luminosity of Type Ia supernovae falls off more rapidly with redshift than would be expected in an Einstein–de Sitter cosmology with $\Omega_M = 1$ and $\Omega_\Lambda = 0$. We can see the effect of vacuum energy on apparent luminosity by comparing the luminosity distance calculated in two extreme cases, both with no matter or radiation. For a vacuum-dominated flat model with $\Omega_\Lambda = 1$ and $\Omega_K = \Omega_M = \Omega_R = 0$, Eq. (1.5.46) gives

$$d_L(z) = \frac{z + z^2}{H_0} \quad \text{(vacuum dominated)}, \tag{1.6.1}$$

---

[6]J. L. Tonry *et al.*, *Astrophys. J.* **594**, 1 (2003) [astro-ph/0305008].
[7]R. Knop *et al.*, *Astrophys. J.* **598**, 102 (2003) [astro-ph/0309368].
[8]B. J. Barris *et al.*, *Astrophys. J.* **502**, 571 (2004) [astro-ph/0310843].

while for an empty model with $\Omega_K = 1$ and $\Omega_\Lambda = \Omega_M = \Omega_R = 0$, Eq, (1.5.46) gives

$$d_L(z) = \frac{z + z^2/2}{H_0} \quad \text{(empty)} , \qquad (1.6.2)$$

Evidently for all z, vacuum energy increases the luminosity distance. The same increase is seen if we compare the more realistic case $\Omega_\Lambda = 0.7$, $\Omega_M = 0.3$, $\Omega_K = \Omega_R = 0$ with the corresponding case without vacuum energy and $\Omega_K = 0.7$, $\Omega_M = 0.3$, $\Omega_\Lambda = \Omega_R = 0$, as can be seen in Figure 1.3.

Both the Supernova Cosmology Project and the High-z Supernova Search Team found that curve of measured luminosity distances *vs.* redshift of Type Ia supernovae was closer to the upper than the lower curve in Figure 1.3. Indeed, according to Eq. (1.4.9), the negative value of q_0 found by all groups corresponds to the fact that the apparent luminosity of the type Ia super-novae seen at moderate redshifts is *less* than in the empty model, for which $q_0 = 0$, in contrast with what had been expected, that the expansion is dom-inated by matter, in which case we would have had $q_0 > 0$, and the apparent luminosities at moderate redshifts would have been larger than for $q_0 = 0$.

The connection between an accelerating expansion and a reduced apparent luminosity can be understood on the basis of the naive Newtonian cosmological model discussed in Section 1.5. In this model, the redshift we observe from a distant galaxy depends on the speed the galaxy had when the light we observe was emitted, but the apparent luminosity is inversely proportional to the square of the distance of this galaxy *now*, because the galaxy's light is now spread over an area equal to 4π times this squared dis-tance. If the galaxies we observe have been traveling at constant speed since the beginning, as in the empty model, then the distance of any galaxy from

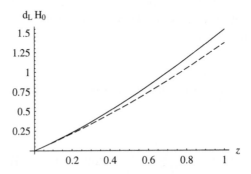

Figure 1.3: Luminosity distance versus redshift for two cosmological models. The upper solid curve is for the case $\Omega_\Lambda = 0.7$, $\Omega_M = 0.3$, $\Omega_K = \Omega_R = 0$; the lower dashed curve is for an empty model, with $\Omega_K = 1$, $\Omega_\Lambda = \Omega_M = \Omega_R = 0$. The vertical axis gives the luminosity distance times the Hubble constant.

us now would be proportional to its speed when the light was emitted. In the absence of a vacuum energy, we would expect the galaxies to be slowing down under the influence of their mutual gravitational attraction, so that the speed we observe would be greater than the speed they have had since the light was emitted, and their distances now would therefore be smaller than they would be if the speeds were constant. Thus in the absence of vacuum energy we would expect an enhanced apparent luminosity of the supernovae in these galaxies. In fact, it seems that the luminosity distances of supernovae are *larger* than they would be if the speeds of their host galaxies were constant, indicating that these galaxies have not been slowing down, but speeding up. This is just the effect that would be expected from a positive vacuum energy.

Of course, it is also possible that the reduction in apparent luminosity is due to absorption or scattering of light by intervening material rather than an accelerated expansion. It is possible to distinguish such effects from a true increase in luminosity distance by the change in the apparent color produced by such absorption or scattering, but this is a complicated business.[9] This concern has been allayed by careful color measurements.[10] But it is still possible to invent intergalactic media (so-called gray dust) that would reduce the apparent luminosity while leaving the color unchanged.

This concern has been largely put to rest, first by the study[11] of the supernovae SN1997ff found in the Hubble Deep Field[12] in a galaxy with a redshift $z = 1.7 \pm 0.1$, the greatest yet found for any supernova, and then by the discovery and analysis by a new team, the Higher-z Supernova Team,[13] of 16 new Type Ia supernovae, of which six have $z > 1.25$. These redshifts are so large that during a good part of the time that the light from these supernovae has been on its way to us, the energy density of the universe would have been dominated by matter rather than by a cosmological constant, and so the expansion of the universe would have been decelerating rather than accelerating as at present. Thus if the interpretation of the results of the two groups at smaller redshifts in terms of Ω_M and Ω_Λ is correct, then the apparent luminosity of these supernovae should be *larger* than would be given by a linear relation between luminosity distance and redshift, a result that could not be produced by absorption or scattering of light. We see this in Figure 1.4, which shows the difference between the luminosity distance (in units H_0^{-1}) for the realistic case with $\Omega_\Lambda = 0.7$, $\Omega_M = 0.3$, $\Omega_K = \Omega_R = 0$ and for

---

[9]See e.g., A. Aguirre, *Astrophys. J.* **525**, 583 (1999) [astro-ph/9904319].

[10]R. Knop *et al.*, ref. 7; also see M. Sullivan *et al.*, *Mon. Not. Roy. Astron. Soc.* **340**, 1057 (2003) [astro-ph/0211444].

[11]A. G. Riess *et al.*, *Astrophys. J.* **560**, 49 (2001) [astro-ph/0104455].

[12]R. L. Gilliland, P. E. Nugent, and M. M. Phillips, *Astrophys. J.* **521**, 30 (1999).

[13]A. G. Riess *et al.*, *Astrophys. J.* **607**, 665 (2004) [astro-ph/0402512].

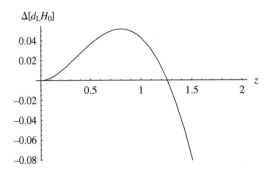

Figure 1.4: The luminosity distance times H_0 for the realistic case $\Omega_\Lambda = 0.7$, $\Omega_M = 0.3$, $\Omega_K = \Omega_R = 0$, minus its value for the empty case $\Omega_K = 1$, $\Omega_\Lambda = \Omega_M = \Omega_R = 0$, plotted against redshift.

the empty model with $\Omega_K = 1$, $\Omega_M = \Omega_K = \Omega_R = 0$. We see that luminosity distances for the realistic model are greater than for the empty model for moderate redshift, but become less than for the empty model for $z > 1.25$. This is just what is seen. The apparent luminosity of all supernovae is consistent with the parameters $\Omega_M \approx 0.3$, $\Omega_\Lambda \approx 0.7$ found in the 1998 studies, but not consistent with what would be expected for gray dust and $\Omega_\Lambda = 0$. These conclusions have subsequently been strengthened by the measurement of luminosity distances of additional Type Ia supernovae with redshifts near 0.5.[14] In 2006 Riess *et al.*[15] announced the discovery with the Hubble Space Telescope of 21 new Type Ia supernovae, which included 13 supernovae with redshifts $z \geq 1$ measured spectroscopically (not just photometrically). Their measured luminosity distances and redshifts, together with data on previously discovered Type Ia supernovae, gave further evidence of a transition from a matter-dominated to a vacuum energy-dominated expansion, and showed that the pressure/density ratio of the vacuum energy for $z > 1$ is consistent with $w = -1$, and not rapidly evolving.

Another serious concern arises from the possibility that the absolute luminosity of Type Ia supernovae may depend on when the supernovae occur. Because Type Ia supernovae occur at a characteristic moment in the history of a star, evolution effects on the luminosities of these supernovae are not expected to be as important as for whole galaxies, which at great distances are seen at an earlier stage in their history.[16] Even so, the absolute luminosity of a Type Ia supernova is affected by the chemical composition

[14] A. Clocchiatti *et al.*, *Astrophys. J.* **642**, 1 (2006) [astro-ph/0510155].

[15] A. Riess *et al.*, *Astrophys. J.* **659**, 98 (2007) [astro-ph/0611572].

[16] D. Branch, S. Perlmutter, E. Baron, and P. Nugent, contribution to the *Supernova Acceleration Probe Yellow Book* (Snowmass, 2001) [astro-ph/0109070].

of the two progenitor stars of the supernova, which is in turn affected by the evolution of the host galaxy.[17] Such effects are mitigated by taking account of the correlation of supernova absolute luminosity with decay time and with intrinsic color, both of which presumably depend on the progenitor's chemical composition. Also, evidence for dark energy has been found in studies of subsets of Type Ia supernovae found in very different environments with very different histories.[18] The study of the seven supernovae with $z > 1.25$ mentioned above rules out models with $\Omega_\Lambda = 0$ and any sort of dramatic monotonic evolution of supernovae absolute luminosities that would mimic the effects of dark energy.

There are other effects that might possibly impact the observed relation between supernova apparent luminosities and redshifts:

1. The effect of weak gravitational lensing on the implications of the supernova observations is expected to be small,[19] except perhaps for small area surveys.[20] (Gravitational lensing is discussed in Chapter 9.) It had been thought that the apparent luminosity of the most distant observed supernova, SN1997ff, may be enhanced by gravitational lensing,[21] conceivably reopening the possibility that the reduction of the apparent luminosity of the nearer supernovae *is* due to gray dust. However, a subsequent analysis by the same group[22] reported that the magnification of this supernova due to gravitational lensing is less than had been thought, and that the effect of the corrections due to gravitational lensing on current cosmological studies is small. Members of the High-z Supernova project have reported that instead this effect is likely to improve agreement with the estimate that $\Omega_M = 0.35$ and $\Omega_\Lambda = 0.65$.[23]

2. It has been argued that inhomogeneities in the cosmic distribution of matter could produce an accelerating expansion, without the need for any sort of exotic vacuum energy.[24] Given the high degree of

---

[17]P. Podsiadlowski *et al.*, astro-ph/0608324. Evolution may also affect the extinction of light by dust in the host galaxy; see T. Totani and C. Kobayashi, *Astrophys. J.* **526**, 65 (1999).

[18]M. Sullivan *et al.*, ref. 10.

[19]A. J. Barber, *Astron. Soc. Pacific Conf. Ser.* **237**, 363 (2001) [astro-ph/0109043].

[20]A. Cooray, D. Huterer, and D. E. Holz, *Phys. Rev. Lett.* **96**, 021301 (2006).

[21]E. Mörtstell, C. unnarsson, and A. Goobar, *Astrophys. J.* **561**, 106 (2001); C. Gunnarsson, in *Proceedings of a Conference on New Trends in Theoretical and Observational Cosmology – Tokyo, 2001* [astro-ph/0112340].

[22]J. Jönsson *et al.*, *Astrophys. J.* **639**, 991 (2006) [astro-ph/0506765].

[23]N. Benítez *et al.*, *Astrophys. J.* **577**, L1 (2002) [astro-ph/0207097].

[24]E. W. Kolb, S. Matarrese, A. Notari, and A. Riotto, *Astrophys. J.* **626**, 195 (2005) [hep-th/0503117]; E. W. Kolb, S. Matarrese, and A. Riotto, *New J. Phys.* **8**, 322 (2006) [astro-ph/0506534]; E. Barausse, S. Matarrese, and A. Riotto, *Phys. Rev. D* **71**, 063537 (2005).

homogeneity of the universe when averaged over sufficiently large scales, this seems unlikely.[25]

3. There is some evidence for two classes of type Ia supernovae,[26] with the minority associated perhaps with merging white dwarfs, or with a variation in explosion physics. The effect on cosmological studies remains to be evaluated.

4. Other uncertainties that can degrade the accuracy of measurements of dark energy (without casting doubt on its existence) arise from the circumstance that the shape of the curve of luminosity distance versus redshift is found by numerous observatories, both ground-based and space-based, and there are various flux calibration errors that can arise between these different observatories.

5. The measurement of luminosity distance of any source of light at large redshift has historically been plagued by the fact that measurements are not "bolometric," that is, equally sensitive to all wavelengths, but are rather chiefly sensitive to wavelengths in a limited range. The cosmological redshift changes the apparent colors of sources, and thereby changes the sensitivity with which apparent luminosity is measured. To take this into account, the observed apparent magnitude is corrected with a so-called *K-correction*.[27] The K-correction for supernovae were worked out before the discovery of dark energy,[28] and has been refined subsequently.[29] As the precision of supernovae observations improves, further improvements may also be needed in the K-correction.

These observations of an accelerated expansion are consistent with the existence of a constant vacuum energy, but do not prove that this energy density really is constant. According to Eq. (1.5.18), the existence of an accelerated expansion does however require that a large part of the energy density of the universe is in a form that has $\rho + 3p < 0$, unlike ordinary matter or radiation. This has come to be called *dark energy*.[30]

---

[25]É. É. Flanagan, *Phys. Rev. D* **71**, 103521 (2005) [hep-th/0503202]; G. G. Geshnizjani, D. J. H. Chung, and N. Afshordi, *Phys. Rev. D* **72**, 023517 (2005) [astro-ph/0503553]; C. M. Hirata and U. Seljak, *Phys. Rev. D* **72**, 083501 (2005) [astro-ph/0503582]; A. Ishibashi and R. M. Wald, Class. Quant. Grav. **23**, 235 (2006) [gr-qc/0509108].

[26]D. Howell *et al.*, *Nature* **443**, 308 (2006); S. Jha, A. Riess, & R. P. Kirshner, Astrophys. J. **654**, 122 (2007); R. Quimby, P. Höflich, and J. C. Wheeler, 0705.4467.

[27]For a discussion of the K-correction applied to observations of whole galaxies, and original references, see G&C, p. 443.

[28]A. Kim, A. Goobar, and S. Perlmutter, *Proc. Astron. Soc. Pacific* **108**, 190 (1995) [astro-ph/9505024].

[29]P. Nugent, A. Kim, and S. Perlmutter, *Proc. Astron. Soc. Pacific* **114**, 803 (2002) [astro-ph/0205351].

[30]For a general review, see P. J. E. Peebles and B. Ratra, *Rev. Mod. Phys.* **75**, 559 (2003).

To take into account the possibility that the dark energy density is not constant, it has become conventional to analyze observations in terms of its pressure/density ratio $p_{D.E.}/\rho_{D.E.} \equiv w$. Except in the case of a constant vacuum energy density, for which $w = -1$, there is no special reason why w should be time-independent. (A different, more physical, possibility is explored at the end of Section 1.12.) Still, it is popular to explore cosmological models with w constant but not necessarily equal to -1. As long as the dark energy density and Ω_K are non-negative, the expansion of the universe will continue, with $\dot{a}$ always positive. As shown in Eq. (1.1.34), the dark energy density in this case goes as a^{-3-3w}, so if w is negative (as indicated by the supernova observations) the energy density of radiation and matter must eventually become negligible compared with the dark energy density. For $w < -1/3$, the effect of a possible curvature in the Friedmann equation (1.5.19) also eventually becomes negligible. The solution of this equation for $w > -1$ with $\dot{a} > 0$ then becomes $t - t_1 \rightarrow Ca^{(3+3w)/2}$, with $C > 0$, and t_1 an integration constant. This is a continued expansion, with a decreasing expansion rate. But for $w < -1$, sometimes known as the case of *phantom energy*, the solution with $\dot{a} > 0$ is instead $t_1 - t \rightarrow Ca^{(3+3w)/2}$, again with $C > 0$. This solution has the remarkable feature that $a(t)$ becomes infinite at the time t_1. In contrast with the case $w \geq -1$, for $w < -1$ all structures — galaxy clusters, galaxy clusters, stars, atoms, atomic nuclei, protons and neutrons — eventually would be ripped apart by the repulsive forces associated with dark energy.[31]

To further study the time dependence of the dark energy, a five year supernova survey, the Supernova Legacy Survey,[32] was begun in 2003 at the Canada–France–Hawaii telescope on Mauna Kea. At the end of the first year, 71 high redshift Type Ia supernovae had been discovered and studied, with the result that $\Omega_M = 0.263 \pm 0.042(\text{stat}) \pm 0.032(\text{sys})$. Combining this supernova data with data from the Sloan Digital Sky Survey (discussed in Chapter 8), and assuming that the dark energy has $w \equiv p/\rho$ time-independent, it is found that if w is constant then $w = -1.023 \pm 0.09(\text{stat}) \pm 0.054(\text{sys})$, consistent with the value $w = -1$ for a constant vacuum energy. At the time of writing, results have just become available for 60 Type Ia supernovae from another supernova survey, ESSENCE.[33] (The acronym is for Equation of State: Supernovae trace Cosmic Expansion). Combining these with the results of the Supernova Legacy Survey, the ESSENCE group found that if w is constant then $w = -1.07 \pm 0.09(\text{stat}, 1\sigma) \pm 0.13(\text{syst})$, and $\Omega_M = 0.267^{-0.028}_{-0.018}(\text{stat}, 1\sigma)$.

[31] R. R. Caldwell, M. Kamionkowski, and N. N. Weinberg, *Phys. Rev. Lett.* **91**, 071301 (2003) [astro-ph/0302506].

[32] P. Astier, *et al.*, *Astron. Astrophys.* **447**, 31 (2006) [astro-ph/0510447].

[33] M. Wood-Vesey *et al.*, astro-ph/0701041

The conclusion that dark energy makes up a large fraction of the energy of the universe has been confirmed by observations of the cosmic microwave background, as discussed in Section 7.2. This conclusion has also received support from the use of a different sort of secondary distance indicator, the emission of X-rays from hot gas in galaxy clusters. In Section 1.9 we will see that the measurement of redshift, temperature, apparent X-ray luminosity and angular diameter of a cluster allows a determination of the ratio of hot gas ("baryons") to all matter in the cluster, with this ratio proportional to $d_A^{-3/2}$, where d_A is the assumed angular-diameter distance of the cluster. This can be turned around: under the assumption that the ratio of hot gas to all matter is the same for all clusters in a sample, X-ray observations can be used to find the dependence of the cluster angular diameter distances on redshift.[34] In this way, observations by the Chandra satellite of X-rays from 26 galaxy clusters with redshifts in the range $0.07 < z < 0.9$ have been used to determine that in a cosmology with a constant vacuum energy and cold dark matter, $\Omega_\Lambda = 0.94^{+0.21}_{-0.25}$, within 68% confidence limits.[35] Relaxing the assumption that the cosmological dark energy density is constant, but assuming $\Omega_K = 0$ and a constant w, and taking the baryon density to have the value indicated by cosmological nucleosynthesis (discussed in Section 3.2), this analysis of the Chandra data yields $1 - \Omega_M = 0.76 \pm 0.04$ and a dark energy pressure/density ratio $w = -1.20^{+0.24}_{-0.28}$.

It is possible that measurements of luminosity distance can be pushed to much larger redshifts by the use of long gamma ray bursts as secondary distance indicators. These bursts definitely do not have uniform absolute luminosity, but there are indications that the absolute gamma ray luminosity is correlated with the peak gamma ray energy and a characteristic time scale.[36]

The discovery of dark energy is of great importance, both in interpreting other observations and as a challenge to fundamental theory. It is profoundly puzzling why the dark energy density is so small. The contribution of quantum fluctuations in known fields up to 300 GeV, roughly the highest energy at which current theories have been verified, gives a vacuum energy

---

[34]S. Sasaki, *Publ. Astron. Soc. Japan* **48**, 119 (1996) [astro-ph/9611033]; U.-L. Pen, *New Astron.* **2**, 309 (1997) [astro-ph/9610147].

[35]S. W. Allen, R. W. Schmidt, H. Ebeling, A. C. Fabian, and L. van Speybroeck, *Mon. Not. Roy. Astron. Soc.* **353**, 457 (2004) [astro-ph/0405340]. For earlier applications of this technique, see K. Rines *et al.*, *Astrophys. J.* **517**, 70 (1999); S. Ettori and A. Fabian, *Astron. Soc. Pac. Conf. Ser.* **200**, 369 (2000); S. W. Allen, R. W. Schmidt, and A. C. Fabian, *Mon. Not. Roy. Astron. Soc.* **334**, L11 (2002); S. Ettori, P, Tozzi, and P. Rosati, *Astron. & Astrophys.* **398**, 879 (2003). The possibility of a variable ratio of hot gas to all matter is explored by R. Sadat *et al.*, *Astron. & Astrophys.* **437**, 310 (2005); L. D. Ferramacho and A. Blanchard, *Astron. & Astrophys.* **463**, 423 (2007) [astro-ph/0609822].

[36]C. Firmani, V. Avila-Reese, G. Ghisellini, and G. Ghirlanda, *Mon. Not. Roy. Astron. Soc.* **372**, 28 (2006) [astro-ph/0605430]; G. Ghirlanda, G. Ghisellini, and C. Firmani, *New J. Phys.* **8**, 123 (2006) [astro-ph/0610248].

density of order $(300 \, \text{GeV})$,[4] or about $10^{27} \, \text{g/cm}^3$. This of course is vastly larger than the observed dark energy density, $\Omega_V \rho_{0,\text{crit}} \simeq 10^{-29} \, \text{g/cm}^3$, by a factor of order 10^{56}. There are other unknown contributions to the vacuum energy that might cancel this contribution, coming from fluctuations in fields at higher energies or from the field equations themselves, but this cancelation would have to be precise to about 56 decimal places. There is no known reason for this remarkable cancelation.[37] The discovery of dark energy now adds a second problem: why is the dark energy density comparable to the matter energy density at this particular moment in the history of the universe?

In thinking about these problems, it is crucial to know whether the vacuum energy is really time-independent, or varies with time, a question that may be answered by future studies of distant Type Ia supernovae or other measurements at large redshift. The possibility of a varying dark energy (known as *quintessence*) will be considered further in Section 1.12.

1.7 Cosmic expansion or tired light?

In comparing observations of redshifts and luminosity distances with theory, we rely on the general understanding of redshifts and luminosities outlined in Sections 1.2 and 1.4. One thing that might invalidate this understanding is absorption or scattering, which reduces the number of photons reaching us from distant sources. This possibility is usually taken into account by measuring the color of the source, which would be affected by absorption or scattering, though as mentioned in the previous section there is a possibility of gray dust, which could not be detected in this way. Another possible way that apparent luminosities could be reduced is through the conversion of photons into particles called axions by intergalactic magnetic fields. There is also a more radical possibility. Ever since the discovery of the cosmological redshift, there has been a nagging doubt about its interpretation as evidence of an expanding universe. Is it possible that the universe is really static, and that photons simply suffer a loss of energy and hence a decrease in frequency as they travel to us, the loss of energy and hence the redshift naturally increasing with the distance that the photons have to travel?

It is possible to rule out all these possibilities by comparing the luminosity distance $d_L(z)$ with the angular diameter distance $d_A(z)$ of the same distant source. None of the possibilities mentioned above can affect the angular diameter distance, while the conventional interpretation of

---

[37] For a survey of efforts to answer this question, see S. Weinberg, *Rev. Mod. Phys.* **61**, 1 (1989).

redshifts and luminosities provides the model-independent result (1.4.12), that $d_L(z)/d_A(z) = (1+z)^2$, so a verification of this ratio can confirm the conventional understanding of $d_L(z)$.

We can check this formula for $d_L(z)/d_A(z)$ by a "surface brightness test" suggested long ago by Tolman.[1] If a light source has an absolute luminosity per proper area $\mathcal{L}$, then the apparent luminosity of a patch of area A will be $\ell = \mathcal{L}A/4\pi d_L^2$. This patch will subtend a solid angle $\Omega = A/d_A^2$. The *surface brightness B* is defined as the apparent luminosity per solid angle, so

$$B \equiv \frac{\ell}{\Omega} = \frac{\mathcal{L}\, d_A^2}{4\pi d_L^2} . \tag{1.7.1}$$

In the conventional big bang cosmology the ratio d_A/d_L is given by Eq. (1.4.12), so

$$B = (1+z)^{-4} \left(\frac{\mathcal{L}}{4\pi} \right) . \tag{1.7.2}$$

If we can find a class of light sources with a common value for the absolute luminosity per proper area $\mathcal{L}$, then their surface brightness should be found to decrease with redshift precisely as $(1+z)^{-4}$.

For instance, one important difference between "tired light" theories and the conventional big bang theory is that in the conventional theory all rates at the source are decreased by a factor $(1+z)^{-1}$, while in tired light theories there is no such slowing down. One rate that is slowed down at large redshifts in the conventional theory is the rate at which photons are emitted by the source. This is responsible for one of the two factors of $(1+z)^{-1}$ in formula (1.4.1) for apparent luminosity, the other factor being due to the reduction of energy of individual photons. On the other hand, if the rate of photon emission is not affected by the redshift, then in a static Euclidean universe in which photons lose energy as they travel to us, the apparent luminosity of a distant source L at a distance d will be given by $L/4\pi(1+z)d^2$, with only a single factor $1+z$ in the denominator to take account of the photon energy loss. That is, the luminosity distance will be $(1+z)^{1/2}d$, while the angular diameter distance in a Euclidean universe is just d, so here $d_L/d_A = (1+z)^{1/2}$, and the surface brightness of distant galaxies should decrease as $(1+z)^{-1}$.

Lubin and Sandage[2] have used the Hubble Space Telescope to compare the surface brightness of galaxies in three distant clusters with redshifts

---

[1] R. C. Tolman, *Proc. Nat. Acad. Sci* **16**, 5111 (1930); *Relativity, Thermodynamics, and Cosmology* (Oxford Press, Oxford, 1934): 467.

[2] L. M. Lubin and A. Sandage, Astron. J. **122**, 1084 (2001) [astro-ph/0106566]. Their earlier work is described in A. Sandage and L. M. Lubin, Astron. J. **121**, 2271 (2001); L. M. Lubin and A. Sandage, *ibid*, 2289 (2001) and Astron. J. **122**, 1071 (2001) [astro-ph/0106563.]

0.76, 0.90, and 0.92 with the surface brightness measured in closer galaxies. They detect a decrease of B with increasing z that is consistent with Eq. (1.7.2) with reasonable corrections for the effects of galaxy evolution, and is quite inconsistent with the behavior $B \propto (1+z)^{-1}$ expected in a static universe with "tired light."

In the standard big bang cosmology all rates observed from a distant source are slowed by a factor $1/(1 + z)$, not just the rate at which photons are emitted. This slowing has been confirmed[3] for the rate of decline of light from some of the Type Ia supernovae used by the Supernova Cosmology Project discussed in the previous section. The hypothesis that the absolute luminosity is simply correlated with the intrinsic decline time is found to work much better if the observed decline time is taken to be the intrinsic decline time stretched out by a factor $1 + z$. Nothing like this would be expected in a static Euclidean universe with redshifts attributed to tired light.

1.8 Ages

As we have seen, a knowledge of the Hubble constant and of the matter and vacuum density parameters Ω_M and Ω_Λ allows us to estimate the age of the universe. In this section we will discuss independent estimates of the age of the universe, based on calculations of the ages of some of the oldest things it contains.

Since metals (elements heavier than helium) found in the outer parts of stars arise chiefly from earlier generations of stars, the oldest stars are generally those whose spectra show small abundances of metals. These are the so-called Population II stars, found in the halo of our galaxy, and in particular in globular clusters. There are two main ways of estimating ages of old stars:

A. Heavy element abundances

If a nucleus decays with decay rate λ, and has an initial abundance A_{init}, then the abundance A after a time T is $A = A_{init} \exp(-\lambda T)$. Hence if we knew A_{init} and could measure A, we could determine T from $T = \lambda^{-1} \ln(A_{init}/A)$. Unfortunately neither condition is likely to be satisfied. On the other hand, it is often possible to calculate the *ratio* of the initial abundances $A_{1\,init}$ and $A_{2\,init}$ of two nuclei, and to measure their *relative* present abundance A_1/A_2.

---

[3]B. Leibundgut *et al.*, Astrophys. J. *466*, L21 (1996); G. Goldhaber *et al.* (Supernova Cosmology Project), *Astrophys. J.* **558**, 359 (2001) [astro-ph/0104382].

This relative abundance is given by

$$\frac{A_1}{A_2} = \left(\frac{A_{1\,\text{init}}}{A_{2\,\text{init}}}\right) \exp\left((\lambda_2 - \lambda_1)T\right),$$

so

$$T = \frac{1}{\lambda_2 - \lambda_1}\left[\ln\left(\frac{A_1}{A_2}\right) - \ln\left(\frac{A_{1\,\text{init}}}{A_{2\,\text{init}}}\right)\right] \tag{1.8.1}$$

If the initial abundances are similar and the observed abundances are very different, then the estimated value of T will be insensitive to the precise value of the initial abundance ratio.

The initial relative abundances of heavy, radioactive elements are estimated on the well-founded assumption that these elements are made in the so-called r-process, the rapid addition of neutrons to lighter elements such as iron in core-collapse supernova explosions, after which the neutron-rich isotopes formed in this way undergo multiple beta decays, transforming them to the most deeply bound nuclei with the same number of nucleons. This method has been used to put a lower bound on the age of our galaxy from the terrestrial abundance of ^{235}U, which has a decay rate of 0.971×10^{-9}/yr. To avoid uncertainties in the distribution of ^{235}U in earth, its abundance is measured relative to the isotope ^{238}U, which has a slower decay rate of 0.154×10^{-9}/yr, but behaves the same chemically and is presumably distributed in the same way. The initial abundance ratio of ^{235}U to ^{238}U is estimated to be 1.65 ± 0.15; it is larger than one because three additional neutrons must be added to the progenitor of ^{235}U to form the progenitor of ^{238}U. On the other hand, the larger decay rate of ^{235}U makes it (fortunately) less abundant than ^{238}U now. The present abundance ratio of uranium isotopes on earth is 0.00723, so this uranium has been decaying for a time

$$\frac{\ln(1.65) - \ln(.00723)}{0.971 \times 10^{-9}/\text{yr} - 0.154 \times 10^{-9}/\text{yr}} = 6.6 \text{ Gyr} \qquad [1 \text{ Gyr} = 10^9 \text{ yr}].$$

But the sun is a second (or perhaps third) generation (called "Population I") star, and presumably its uranium was being produced over a long time interval before the formation of the solar system. The uranium abundance ratio has been supplemented with measurements of other abundance ratios on the earth and meteorites, such as ^{232}Th/^{238}U and ^{187}Re/^{187}Os ratios, and analyzed with the length of the period of heavy element formation left as a free parameter. This gives a more stringent (but less certain) lower bound of 9.6 Gyr[1] on the age of the heavy elements in the neighborhood of the solar system.

---

[1] B. S. Meyer and D. N. Schramm, *Astrophys. J.* **311**, 406 (1986).

A much more stringent lower bound on the age of the galaxy is given by applying these methods to heavy elements in metal-poor stars beyond the solar system. First thorium was observed spectroscopically in a very metal-poor star (and hence presumably old) K giant star, CS 22892-052.[2] The relative abundances in this star of the more stable elements produced in the r-process, as measured from the intensity of absorption lines in the star's spectrum, matches those of the same elements in the solar system, except for a much lower abundance of the heaviest detected element thorium, which (for ^{232}Th) has a half life 14 Gyr. Attributing the decrease in thorium to its radioactive decay, the age of the thorium in this star is estimated as 14.1 ± 3 Gyr. Other estimates of the ages of CS 22892-052 and other metal-poor stars have been made using the measured abundance ratios of thorium to europium and lanthanum.[3]

Uranium-238 decays more rapidly than ^{232}Th, so we can get a more sensitive estimate of the age of a star by using both its uranium and its thorium abundances, providing of course that uranium as well as thorium lines can be observed in the star's spectrum. No uranium absorption lines were observed in the spectrum of CS 22892-052, but absorption lines from singly ionized uranium were subsequently observed in two other metal-poor star with an abundance of r-process elements, CS31082-001 and BD+17°3248. The observed abundance ratio of uranium to thorium in CS31082-001 is $10^{-0.74\pm0.15}$, while the initial abundance ratio has been variously estimated as $10^{-0.255}$ or $10^{-0.10}$. Using these numbers in Eq. (1.8.1) gives this star an age of 12.5 ± 3 Gyr.[4] Subsequent observations indicated ages of 14 ± 2 Gyr,[5] 15.5 ± 3.2 Gyr,[6] and 14.1 ± 2.5 Gyr.[7] In a similar way, the age of BD+17°3248 has been calculated as 13.8 ± 4 Gyr.[8] (See Figure 1.5.) More recently, both uranium and thorium lines have been found in the spectrum of the newly discovered metal-poor star HE 1523-0903; the ratio of thorium and uranium abundance to the abundances of other r-process elements, and to each other, was used to give an age of the star as 13.2 Gyr.[9]

---

[2]C. Sneden *et al.*, *Astrophys. J.* **467**, 819 (1996); *Astrophys. J.* **591**, 936 (2003) [astro-ph/0303542]. A review with references to earlier work on thorium abundances was given by C. Sneden and J. J. Cowan, *Astronomia y Astrofísica (Serie de Conferencia)* **10**, 221 (2001) [astro-ph/0008185].

[3]I. I Ivans *et al.*, *Astrophys. J.* **645**, 613 (2006) [astro-ph/0604180], and earlier references cited therein.

[4]R. Cayrel *et al.*, *Nature* **409**, 691 (2001).

[5]V. Hill *et al.*, *Astron. Astrophys.* **387**, 580 (2002).

[6]H. Schatz *et al.*, *Astrophys. J.* **579**, 626 (2002).

[7]S. Wanajo, *Astrophys. J.* **577**, 853 (2002).

[8]J. J. Cowan, *et al.*, *Astrophys. J.* **572**, 861 (2002) [astro-ph/0202429].

[9]A. Frebel *et al.*, *Astrophys.* **660**, L117 (2007). [astro-ph/0703414].

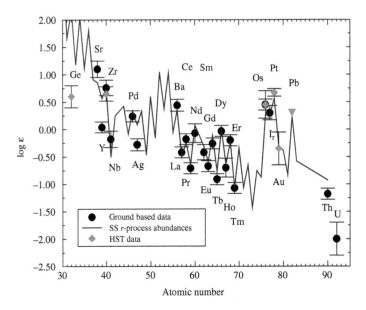

Figure 1.5: Abundances of elements produced by the *r*-process in the star BD+17°3248, obtained by ground-based and Hubble Space Telescope spectroscopic observations. For comparison, the solid curve gives theoretical initial abundances, based on solar system data. Note the low observed abundances of thorium and uranium, compared with the theoretical initial abundances, which indicate an age for BD+17°3248 of 13.8±4 Gyr. From J. J. Cowan *et al.*, *Astrophys. J.* **572**, 861 (2002) [astro-ph/0202429].

B. Main sequence turn-off

The stars that satisfy the main sequence relation between absolute luminosity and surface temperature are still burning hydrogen at their core. When the hydrogen is exhausted at the core, hydrogen-burning continues in a shell around a (temporarily) inert helium core. The star then moves off the main sequence, toward higher luminosity and lower surface temperature. The heavier a star is, the more luminous it is on the main sequence, and the faster it evolves. Thus as time passes, the main sequence of a cluster of stars of different masses but the same age shows a turn-off that moves to lower and lower luminosities. (See Figure 1.6). Roughly, the absolute luminosity of stars at the turn-off point is inversely proportional to the age of the cluster. In particular, observations of the main sequences of a number of globular clusters gave ages variously calculated[10] as 11.5±1.3 Gyr, 12±1 Gyr, 11.8±1.2 Gyr, 14.0±1.2 Gyr, 12±1 Gyr, and 12.2±1.8 Gyr. A summary by Schramm[11] found that most of the discrepancies disappear when

---

[10]For references, see B. Chaboyer, *Phys. Rep.* **307**, 23 (1998) [astro-ph/9808200].

[11]D. Schramm, in *Critical Dialogues in Cosmology*, N. Turok, ed. (World Scientific, Singapore, 1997): 81

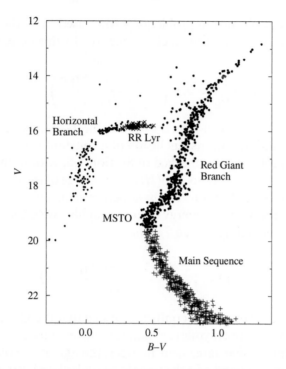

Figure 1.6: Color–magnitude diagram for the globular cluster M15. Visual apparent magnitudes of M15 stars are plotted on the vertical axis. Since all stars in M15 are at about the same distance from earth, the apparent visual magnitude differs from the absolute visual magnitude by a constant term, with absolute luminosities increasing upwards. The difference of apparent blue and visual magnitudes is plotted on the horizontal axis. This is a measure of surface temperature, with temperature decreasing to the right. If M15 were young, the main sequence would continue upwards and to the left; the position of the main sequence turn-off (MSTO) and other features of the diagram indicate that the age of the cluster is 15 ± 3 Gyr. Diagram from B. Chaboyer, *Phys. Rep.* **307**, 23 (1998), based on data of P. R. Durrell and W. E. Harris, *Astron. J.* **105**, 1420 (1993) [astro-ph/9808200].

the various calculations are done with the same input values for parameters like the initial abundance of helium, oxygen, and iron, and gave a consensus value as 14 ± 2(statistical) ± 2(systematic) Gyr. Note that all these ages are sensitive to the distance scale; a fractional change $\delta d/d$ in estimates of distances would produce a fractional change $\delta L/L = -2\delta d/d$ in estimates of absolute luminosities, and hence a fractional change $\delta t/t \approx +2\delta d/d$ in estimates of ages. Using measurements of distances to nine globular clusters with the Hipparcos satellite yields an estimated galactic age[12] of 13.2 ± 2.0 Gyr.

---

[12]E. Carretta, R. G. Gratton, G. Clementini, and F. F. Pecci, *Astrophys. J.* **533**, 215 (2000) [astro-ph/9902086].

These ages would pose a problem for what used to be the most popular model, with $\Omega_M = 1$ and no vacuum energy. In this case, the age of the universe is

$$t_0 = \frac{2}{3H_0} = 9.3 \left(\frac{70 \text{ km/sec/Mpc}}{H_0}\right) \text{ Gyr} ,$$

which is somewhat younger than the oldest objects in the galaxy, though not by many standard deviations. Inclusion of a constant vacuum energy helps to avoid this problem; as remarked in Section 1.5, with nothing else in the universe we would have $a(t) \propto \exp(Ht)$, and the age of the universe would be infinite. As we saw in Section 1.6, the supernovae distance–redshift relation indicates that the vacuum energy is now roughly twice the matter energy, giving an age much longer than $2/3H_0$:

$$t_0 = 13.4^{+1.3}_{-1.0} \left(\frac{70 \text{ km/sec/Mpc}}{H_0}\right) \text{ Gyr} ,$$

This removes the danger of a conflict, *provided* that the globular clusters in our galaxy are not much younger than the universe itself. In fact, there is now a truly impressive agreement between the age of the oldest stars and star clusters on one hand and the cosmic age calculated using values of H_0, Ω_M, and Ω_Λ found from the redshift–distance relation. As we will see in Section 7.2, there is also an excellent agreement between these ages and the age calculated using parameters measured in observations of anisotropies in the cosmic microwave background.

So far in this section, we have considered only the present age of our own galaxy. It is also possible to estimate the ages of other galaxies at high redshift, at the time far in the past when the light we now observe left these galaxies. Of course, it is not possible to distinguish individual stars or globular clusters in these galaxies, but the spectrum of the galaxy gives a good idea of the age. We need the whole spectrum to separate the effects of metallicity, scattering, etc., but roughly speaking, the redder the galaxy, the more of its bright bluer stars have left the main sequence, and hence the older it is. In this way, it has been found[13] that the radio galaxies 53W091($z = 1.55$) and 53W069($z = 1.43$) have ages $\simeq$ 3.5 Gyr and 3 to 4 Gyr, respectively. This sets useful lower bounds on the vacuum energy. In a model with non-relativistic matter and a constant vacuum energy, the age of the universe at the time of emission of light that is seen at present with

---

[13] J. S. Dunlop *et al.*, *Nature* **381**, 581 (1996); J. S. Dunlop, in *The Most Distant Radio Galaxies - KNAW Colloquium, Amsterdam, October 1997*, eds. Best *et al.* [astro-ph/9801114].

redshift z is given by Eq. (1.5.42) as

$$t(z) = \frac{1}{H_0} \int_0^{\frac{1}{1+z}} \frac{dx}{x\sqrt{\Omega_\Lambda + \Omega_K x^{-2} + \Omega_M x^{-3} + \Omega_R x^{-4}}} . \quad (1.8.2)$$

Any galaxy observed with redshift z must have been younger than this at the time that its light was emitted. For instance, for a flat universe with $\Omega_K = \Omega_R = 0$, so that $\Omega_M = 1 - \Omega_\Lambda$, the existence of a galaxy at $z = 1.55$ with age $\simeq 3.5$ Gyr sets a lower bound[14] on Ω_Λ of about 0.6 for $H_0 = 70$ km s^{-1} Mpc^{-1}.

Eventually the accuracy of these age determinations may become good enough to allow us to measure at least the dependence of redshift on the cosmic age. Of course, galaxies form at various times in the history of the universe, so the age of any one galaxy does not allow us to infer the age of the universe at the time light we now see left that galaxy. However, the homogeneity of the universe implies that the *distribution* of cosmic times of formation for any one variety of galaxy is the same anywhere in the universe. From differences in the distributions of ages of a suitable species of galaxy at different redshifts, we can then infer the difference of cosmic age t at these redshifts. The Robertson–Walker scale factor $a(t)$ is related to the redshift $z(t)$ observed now of objects that emitted light when the cosmic age was t by $1 + z(t) = a(t_0)/a(t)$, so $\dot{z} = -H(t)(1+z)$. To calculate $\ddot{z}$, we note that for $K = 0$, $H^2(t) = 8\pi\, G\rho(t)/3$, and $\dot{\rho} = -3H(\rho + p)$,

$$\dot{H}(t) = -4\pi\, G\Big(\rho(t) + p(t)\Big) . \quad (1.8.3)$$

Then for $K = 0$

$$\ddot{z} = \frac{\dot{z}^2}{1+z}\left(\frac{5}{2} + \frac{3p}{2\rho}\right) . \quad (1.8.4)$$

Thus measurements of differences in t for various differences in redshift may allow a measurement of the ratio p/ρ at various times in the recent history of the universe.[15]

1.9 Masses

We saw in Section 1.6 that the observed dependence of luminosity distance on redshift suggests that the fraction Ω_M of the critical density provided by

---

[14]L. M. Krauss, *Astrophys. J.* **489**, 486 (1997); J. S. Alcaniz and J. A. S. Lima, *Astrophys. J.* **521**, L87 (1999) [astro-ph/9902298].

[15]R. Jiminez and A. Loeb, *Astrophys. J.* **573**, 37 (2002) [astro-ph/0106145].

non-relativistic matter is roughly 30%. In this section we will consider other independent ways that Ω_M is measured.

A. Virialized clusters of galaxies

The classic approach[1] to the measurement of Ω_M is to use the virial theorem to estimate the masses of various clusters of galaxies, calculate a mean ratio of mass to absolute luminosity, and then use observations of the total luminosity of the sky to estimate the total mass density, under the assumption that the mass-to-light ratio of clusters of galaxies is typical of the universe as a whole.

To derive the virial theorem, consider a non-relativistic gravitationally bound system of point masses m_n (either galaxies, or stars, or single particles) with positions relative to the center of mass (in an ordinary Cartesian coordinate system) $\mathbf{X}_n$. The equations of motion are

$$m_n \ddot{X}_n^i = -\frac{\partial V}{\partial X_n^i} , \tag{1.9.1}$$

where the potential energy V is

$$V = -\frac{1}{2} \sum_{n \neq \ell} \frac{G m_n m_\ell}{|\mathbf{X}_n - \mathbf{X}_\ell|} . \tag{1.9.2}$$

Multiplying Eq. (1.9.1) with X_n^i and summing over n and i gives

$$-\sum_n X_n^i \frac{\partial V}{\partial X_n^i} = \sum_n m_n \mathbf{X}_n \cdot \ddot{\mathbf{X}}_n = \frac{1}{2} \frac{d^2}{dt^2} \sum_n m_n \mathbf{X}_n^2 - 2T , \tag{1.9.3}$$

where T is the internal kinetic energy (not counting any motion of the center of mass)

$$T = \frac{1}{2} \sum_n m_n \dot{\mathbf{X}}_n^2 . \tag{1.9.4}$$

Let us assume that the system has reached a state of equilibrium ("become virialized"), so that although the individual masses are moving there is no further statistical evolution, and in particular that

$$0 = \frac{d^2}{dt^2} \sum_n m_n \mathbf{X}_n^2 \tag{1.9.5}$$

[1] F. Zwicky, *Astrophys. J.* **86**, 217 (1937); J. H. Oort, in *La Structure et l'Evolution de l'Universe* (Institut International de Physique Solvay, R. Stoops, Brussels, 1958): 163.

(This is why it was important to specify that X_n is measured relative to the center of mass; otherwise a motion of the whole cluster would give the sum a term proportional to t^2, invalidating Eq. (1.9.5).) But V is of order -1 in the coordinates, so the left-hand side of Eq. (1.9.3) is just V, giving the *virial theorem*:

$$2T + V = 0 . \tag{1.9.6}$$

We may express T and V as

$$T = \frac{1}{2} M \langle v^2 \rangle , \qquad V = -\frac{1}{2} GM^2 \langle \frac{1}{r} \rangle , \tag{1.9.7}$$

where $\langle v^2 \rangle$ is the mean (mass weighted) square velocity relative to the center of mass, $\langle 1/r \rangle$ is the mean inverse separation, and $M = \sum_n m_n$ is the total mass. Eq. (1.9.6) thus gives the virial formula for M:

$$M = \frac{2 \langle v^2 \rangle}{G \langle 1/r \rangle} . \tag{1.9.8}$$

This derivation does not apply to irregular clusters of galaxies, like the nearby one in Virgo. Clusters like this do not seem to have settled into a configuration in which the condition (1.9.5) is satisfied, and therefore probably do not satisfy the virial theorem. On the other hand, the virial theorem probably does apply at least approximately to other clusters of galaxies, like the one in Coma, which appear more or less spherical. According to general ideas of statistical equilibrium, we may expect the rms velocity dispersion $\sqrt{\langle v^2 \rangle}$ of the dominant masses in such clusters to equal the velocity dispersion of the visible galaxies in the cluster, which can be measured from the spread of their Doppler shifts, and also to equal the velocity dispersion of the ionized intergalactic gas in the cluster, which since the advent of X-ray astronomy can be measured from the X-ray spectrum of the gas. The values obtained in these ways for $\langle v^2 \rangle$ are independent of the distance scale. On the other hand, values for $\langle 1/r \rangle$ are obtained from angular separations: the true transverse proper distance d is given in terms of the angular separation θ by $d = \theta d_A$, where d_A is the angular diameter distance (1.4.11). For clusters with $z \ll 1$, Eqs. (1.4.9) and (1.4.11) give $d_A \simeq z/H_0$, so $d \simeq \theta z/H_0$. Thus the estimated values of $\langle 1/r \rangle$ for galaxy clusters with $z \ll 1$ scale as H_0, and the values of M inferred from Eq. (1.9.8) scale as $1/H_0$. The absolute luminosity L of a cluster of galaxies with redshift z and apparent luminosity ℓ is given for $z \ll 1$ by Eqs. (1.4.2) and (1.4.9) as $L = 4\pi z^2 \ell / H_0^2$, so the values of L scale as H_0^{-2}, and the mass-to-light ratios obtained in this way therefore scale as $H_0^{-1}/H_0^{-2} = H_0$.

Estimates of M/L for rich clusters have generally given results of order 200 to 300 $h\ M_\odot/L_\odot$, where h is the Hubble constant in units of 100 km s^{-1} Mpc^{-1}, and $M_\odot$ and $L_\odot$ are the mass and absolute luminosity of the sun. For instance, a 1996 study[2] of 16 clusters of galaxies with redshifts between 0.17 and 0.55 gave $M/L = (295 \pm 53)\, h\, M_\odot/L_\odot$. Some of the same group[3] have corrected this result for various biases, and now find $M/L = (213 \pm 59)\, h\, M_\odot/L_\odot$. A more recent application[4] of the virial theorem to 459 clusters has found a value $M/L \simeq 348\, h\, M_\odot/L_\odot$.

All these values of M/L for clusters of galaxies are very much larger than the mass-to-light ratios of the visible regions of individual galaxies.[5] The mass-to-light ratios of individual elliptical galaxies can be measured using the virial theorem, with $\sqrt{<v^2>}$ taken as the velocity dispersion of stars contained in the galaxy; this gives mass-to-light ratios generally in the range of 10 to 20 $h\, M_\odot/L_\odot$.[6] All of the visible light from clusters comes from their galaxies, so we must conclude that most of the mass in clusters of galaxies is in some non-luminous form, either in the outer non-luminous parts of galaxies or in intergalactic space. It has been argued that this mass is in large dark halos surrounding galaxies, extending to 200 kpc for bright galaxies.[7] The nature of this dark matter is an outstanding problem of cosmology, to which we will frequently return.

Incidentally, the large value of M/L given by the virial theorem for elliptical galaxies shows that most of the mass of these galaxies is not in the form of stars as bright as the sun. It is harder to estimate M/L for spiral galaxies, but since the work of Vera Rubin[8] it has been known that most of their mass is also not in luminous stars.[9] If most of the mass of a spiral galaxy were in the luminous central regions of the galaxy, then the rotational speeds of stars outside this region would follow the Kepler law, $v \propto r^{-1/2}$. Instead, it is observed that v outside the central region is roughly constant, even beyond the visible disk of the galaxy, which is what would be expected for a spherical halo with a mass density that decreases only as $1/r^2$, in which case most of the mass of the galaxy would be in the dark outer

[2]R. G. Carlberg *et al.*, *Astrophys. J.* **462**, 32 (1996).

[3]R. G. Carlberg, H. K. C. Yee, and E. Ellingson, *Astrophys. J.* **478**, 462 (1997).

[4]H. Andernach, M. Plionis, O. López-Cruz, E. Tago, and S. Basilakos, *Astron. Soc. Pacific Conf. Ser.* **329**, 289 (2005) [astro-ph/0407098].

[5]This conclusion was first reached in a study of the Coma cluster by F. Zwicky, *Helv. Phys. Acta* **6**, 110 (1933).

[6]T. R. Lauer, *Astrophys. J.* **292**, 104 (1985); J. Binney and S. Tremaine, *Galactic Dynamics* (Princeton University Press, Princeton, 1987).

[7]N. A. Bahcall, L. M. Lubin, and V. Dorman, *Astrophys. J.* **447**, L81 (1995).

[8]V. C. Rubin, W. K. Ford, and N. Thonnard, *Astrophys. J.* **225**, L107; **238**, 471 (1980).

[9]M. Persic and P. Salucci, *Astrophys. J. Supp.* **99**, 501 (1995); M. Persic, P. Salucci, and F. Stel, *Mon. Not. Roy. Astron. Soc.* **281**, 27P (1996).

parts of the halo. There is some evidence from the absence of gravitational microlensing by the halo (discussed in Section 9.2) that this mass is not in the form of dark stars either, but it is still possible that most of the matter in galaxies is baryonic. We will not go into this in detail, because the formation of galaxies involves cooling processes that requires baryonic matter of the same sort as in stars, so that we would only expect the value of M/L for galaxies to be similar to the value for the universe as a whole if the matter of the universe were mostly baryonic.

In using the value of M/L derived from the virial theorem for clusters of galaxies to find the mass density of the universe, we cannot just add up the luminosity per volume of clusters, because most of the light of the universe comes from "field" galaxies that are not in clusters. Instead, if we assume that the field galaxies are accompanied by the same amount of dark matter as the galaxies in clusters, as argued in ref. 7, then we can find Ω_M by using the value of M/L for clusters together with an estimate of the total luminosity density $\mathcal{L}$ to estimate the total mass density as

$$\rho_M = (M/L)\mathcal{L} . \tag{1.9.9}$$

Since values of absolute luminosities inferred from apparent luminosities and redshifts scale as H_0^{-2}, and distances inferred from redshifts scale as H_0^{-1}, the total luminosity density of the universe calculated by adding up the absolute luminosities of galaxies per volume scales as $H_0^{-2}/(H_0^{-1})^3 = H_0$. For example, a 1999 estimate[10] gave $\mathcal{L} = 2 \pm 0.2 \times 10^8\, h\, L_\odot\, \mathrm{Mpc}^{-3}$. For the purpose of calculating Ω_M it is more convenient to write this as a ratio of the critical mass density to the luminosity density:

$$\rho_{0,\mathrm{crit}}/\mathcal{L} = (1390 \pm 140)\, h\, M_\odot/L_\odot .$$

(Here we use $M_\odot = 1.989 \times 10^{33}$ g, 1 Mpc $= 3.0857 \times 10^{24}$ cm, and $\rho_{0,\mathrm{crit}} = 1.878 \times 10^{-29}\, h^2$ g/cm^3.) Taking $M/L = (213 \pm 53) h M_\odot/L_\odot$ gives then

$$\Omega_M = \frac{M/L}{\rho_{0,\mathrm{crit}}/\mathcal{L}} = 0.15 \pm 0.02 \pm .04 ,$$

with the first uncertainty arising from $\mathcal{L}$ and the second from M/L. It is important to note that this is independent of the Hubble constant, as both $\mathcal{L}M/L$ and $\rho_{0,\mathrm{crit}}$ scale as H_0^2.

This estimate of Ω_M is somewhat lower than those derived from the redshift–luminosity relation of supernovae and from the anisotropies in the

---

[10]S. Folkes *et al.*, *Mon. Not. Roy. Astron. Soc.* **308**, 459 (1999); M. L. Blanton *et al.*, *Astron. J.* **121**, 2358 (2001).

cosmic microwave background, to be discussed in Section 2.6 and Chapter 7. But all these estimates agree that Ω_M is distinctly less than unity.

B. X-ray luminosity of clusters of galaxies

Does the dark matter in clusters of galaxies consist of ordinary nuclei and electrons? We can find the ratio of the fraction Ω_B of the critical density provided by baryonic matter (nuclei and electrons) to the fraction Ω_M provided by all forms of non-relativistic matter by studying the X-rays from clusters of galaxies, for it is only the collisions of ordinary baryonic particles that produces these X-rays. Because these collision processes involve pairs of particles of baryonic matter, the absolute X-ray luminosity per unit proper volume takes the form

$$\mathcal{L}_X = \Lambda\left(T_B\right)\rho_B^2 \,, \tag{1.9.10}$$

where T_B and ρ_B are the temperature and density of the baryonic matter, and $\Lambda(T)$ is a known function of temperature and fundamental constants. The baryonic density satisfies the equation of hydrostatic equilibrium, which (assuming spherical symmetry) follows from the balance of pressure and gravitational forces acting on the baryons in a small area A and between radii r and $r + \delta r$:

$$A\left(p_B(r + \delta r) - p_B(r)\right) = -\frac{A\delta r \, \rho_B(r) \, G}{r^2} \int_0^r 4\pi r^2 \, \rho_M(r) \, dr \,,$$

or, canceling factors of A and δr and using the ideal gas law $p_B = k_B T_B \rho_B / m_B$,

$$\frac{d}{dr}\left(\frac{k_B T_B(r)\rho_B(r)}{m_B}\right) = -\frac{G\rho_B(r)}{r^2} \int_0^r 4\pi r^2 \rho_M(r) \, dr \,,$$

where $\rho_M(r)$ is the total mass density, k_B is Boltzmann's constant, m_B is a characteristic mass of the baryonic gas particles, and r is here the proper distance to the center of the cluster. Multiplying by $r^2/\rho_B(r)$ and differentiating with respect to r yields

$$\frac{d}{dr}\left[\frac{r^2}{\rho_B(r)}\frac{d}{dr}\left(\frac{k_B T_B(r)\rho_B(r)}{m_B}\right)\right] = -4\pi \, G r^2 \rho_M(r) \,. \tag{1.9.11}$$

If we make the assumption that cold dark matter particles, or whatever particles dominate the dark intergalactic matter, have an isotropic velocity distribution (which is not very well motivated), then the same derivation applies to these particles, and their density $\rho_D = \rho_M - \rho_B$ satisfies the non-linear differential equation

$$\frac{d}{dr}\left[\frac{r^2}{\rho_D(r)}\frac{d}{dr}\left(\frac{k_B\, T_D(r)\rho_D(r)}{m_D} \right) \right] = -4\pi\, Gr^2\rho_M(r)\,. \qquad (1.9.12)$$

where $T_D(r)$ and m_D are the temperature and mass of the dark matter particles. With perfect X-ray data and a knowledge of the distance of the source, one could measure the X-ray luminosity density $\mathcal{L}_X(r)$ and (using the X-ray spectrum) the baryon temperature $T_B(r)$ at each point in the cluster, then use Eq. (1.9.10) to find the baryon density $\rho_B(r)$ at each point, and then use Eq. (1.9.11) to find the total mass density at each point. We could then calculate the fractional baryon density ρ_B/ρ_M, and if we were interested we could also use Eq. (1.9.12) to find the velocity dispersion $k_B T_D(r)/m_D$ of the dark matter.

In practice, it is usually necessary to use some sort of cluster model. In the simplest sort of model, one assumes an *isothermal sphere*: the temperatures T_B and T_D are taken to be independent of position, at least near the center of the cluster where most of the X-rays come from. It is also often assumed that the same gravitational effects that causes the concentration of the hot intergalactic gas in the cluster is also responsible for the concentration of the dark matter, so that the densities $\rho_B(r)$ and $\rho_M(r)$ are the same, up to a constant factor, which represents the cosmic ratio Ω_B/Ω_M of baryons to all non-relativistic matter. (These gravitational effects are believed to be a so-called "violent relaxation,"[11] caused by close encounters of clumps of matter whose gravitational attraction cannot be represented as an interaction with a smoothed average gravitational field. The condensation of galaxies out of this mixture requires quite different cooling processes that can affect only the baryonic gas, which is why galaxies have a lower proportion of dark matter and a lower mass-to-light ratio.) Comparison of Eqs. (1.9.11) and (1.9.12) shows that $\rho_B(r)$ and $\rho_D(r)$ will be proportional to each other, and hence also to $\rho_M(r)$ if the velocity dispersions of the dark matter and hot baryonic gas are the same:

$$k_B T_M/m_M = k_B T_D/m_D \equiv \sigma^2\,. \qquad (1.9.13)$$

Equations (1.9.11) and (1.9.12) both then tell us that

$$\rho_M(r) = \rho_M(0)\, F(r/r_0) \qquad (1.9.14)$$

---

[11] D. Lynden-Bell, *Mon. Not. Roy. Astron. Soc.* **136**, 101 (1967).

where $F(0) \equiv 1$; r_0 is a *core radius*, defined conventionally by

$$r_0 \equiv \sqrt{\frac{9\sigma^2}{4\pi G \rho_M(0)}} \; ; \qquad (1.9.15)$$

and $F(u)$ is a function satisfying the differential equation

$$\frac{d}{du}\left(\frac{u^2}{F(u)}\frac{dF(u)}{du}\right) = -9u^2 F(u) . \qquad (1.9.16)$$

We must also impose the boundary condition that ρ_M is analytic in the coordinate $\mathbf{X}$ at $\mathbf{X} = 0$, which for a function only of r means that it is given near $r = 0$ by a power series in r^2, so that $F(u)$ is given near $u = 0$ by a power series in u^2, $F(u) = 1 + O(u^2)$. Together with this boundary condition, Eq. (1.9.16) defines a unique function[12] that for small u has the approximate behavior[13]

$$F(u) \simeq (1 + u^2)^{-3/2} . \qquad (1.9.17)$$

The solution to Eq. (1.9.16) is shown together with the approximation (1.9.17) in Figure 1.7.

For large u, $F(u)$ approaches the exact solution $2/9u^2$. Taken literally, this would make the integral for the total mass diverge at large r, which shows that the assumption of constant σ^2 must break down at some large radius. Often the function $F(u)$ is taken simply as[14]

$$F(u) = (1 + u^2)^{-3\beta/2} ,$$

where β is an exponent of order unity.

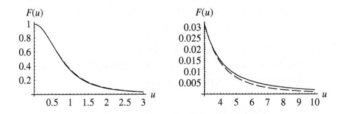

Figure 1.7: The solution to Eq. (1.9.16) (solid line) and the approximation (1.9.17) (dashed line). For the lower values of u in the figure at the left, the two curves are indistinguishable.

---

[12] For a tabulation of values of $F(u)$, see e.g. J. Binney and S. Tremaine, *Galactic Dynamics* (Princeton University, Princeton, 1987): Table 4.1.

[13] I. R. King, *Astron. J.* **67**, 471 (1962).

[14] A. Cavaliere and R. Fusco-Fermiano, *Astron. Astrophys.* **49**, 137 (1976).

Also, Eq. (1.9.11) has the solution

$$\rho_B(r) = \rho_B(0)F(r/r_0) , \qquad (1.9.18)$$

with the same function $F(u)$ and the same core radius r_0. We can measure the core radius from the X-ray image of the cluster, and measure σ^2 from the X-ray spectrum, so that Eq. (1.9.15) can be used to find the central density $\rho_M(0)$ of all non-relativistic matter. The central density $\rho_B(0)$ of the baryonic matter can then be found from the total X-ray luminosity, which with these approximations (and using Eq. (1.9.10)) is

$$L_X \equiv \int d^3x \, \mathcal{L}_X = 4\pi \Lambda(T_B) \, r_0^3 \, \rho_B(0)^2 \mathcal{I} , \qquad (1.9.19)$$

where

$$\mathcal{I} \equiv \int_0^\infty u^2 F^2(u) \, du . \qquad (1.9.20)$$

Even though the solution of Eq. (1.9.16) gives an infinite mass, it gives a finite total X-ray luminosity, with $\mathcal{I} = 0.1961$. (The approximation (1.9.17) would give $\mathcal{I} = \pi/16 = 0.1963$.)

For a cluster at redshift z, the core radius r_0 inferred from observation of the angular size of the cluster will be proportional to the angular diameter distance $d_A(z)$, while the temperature and velocity dispersion found from the X-ray spectrum will not depend on the assumed distance. Thus the value of the central total matter density $\rho_M(0)$ given by Eq. (1.9.15) will be proportional to $1/d_A^2(z)$. On the other hand, the absolute X-ray luminosity L_X inferred from the apparent X-ray luminosity will (like all absolute luminosities) be proportional to the value assumed for $d_L^2(z)$, the square of the luminosity distance, so with $r_0 \propto d_A$, the central baryon density $\rho_B(0)$ given by Eq. (1.9.19) will be proportional to $[d_L^2(z)/d_A^3(z)]^{1/2}$. The value of the ratio of central densities inferred from observations of a given cluster at redshift z will therefore have a dependence on the distance assumed for the cluster given by

$$\frac{\rho_B(0)}{\rho_M(0)} \propto d_L(z)d_A^{1/2}(z) = (1+z)^2 d_A^{3/2}(z) , \qquad (1.9.21)$$

in which we have used the relation (1.4.12) between luminosity and angular diameter distances.

For $z \ll 1$, we have $d_A(z) \simeq d_L(z) \simeq z/H_0$, and so according to Eq. (1.9.21) the value of $\rho_B(0)/\rho_M(0)$ obtained from observations of clusters of small redshift will be proportional to the assumed value of $H_0^{-3/2}$. It is believed that most of the baryonic mass in a cluster of galaxies is in the hot gas outside the galaxies, and if we suppose that this mass is the same fraction of the total mass as in the universe as a whole,[15] then we should get the same value of $\rho_B(0)/\rho_M(0)$, equal to Ω_B/Ω_M, for all clusters, whatever value we assume for H_0, but this value of Ω_B/Ω_M will be proportional to the assumed value of $H_0^{-3/2}$. For example, Schindler[16] quotes various studies that give $\rho_B(0)/\rho_M(0)$ as 0.14, 0.11, 0.12, and 0.12 for $H_0 = 65$ km s^{-1} Mpc^{-1}, so if we take the average 0.12 of these values as the cosmic value of Ω_B/Ω_M for $H_0 = 65$ km s^{-1} Mpc^{-1}, then for a general Hubble constant we find

$$\Omega_B/\Omega_M \simeq 0.06\, h^{-3/2}\,, \tag{1.9.22}$$

where h as usual is Hubble's constant in units of 100 km s^{-1} Mpc^{-1}. We can thus conclude pretty definitely that *only a small fraction of the mass in clusters of galaxies is in a baryonic form that can emit X-rays.*

On the other hand, when we study clusters with a range of redshifts that are not all small, we will not get a uniform value of $\rho_B(0)/\rho_M(0)$ unless we use values of $d_A(z)$ with the correct dependence on z. As remarked in Section 1.6, observations of clusters have been used in this way to learn about the z-dependence of $d_A(z)$.

It should be mentioned that computer simulations that treat galaxy clusters as assemblages of collisionless particles do not show evidence for a central core,[17] but instead indicate that the dark matter density at small distances r from the center should diverge as r^{-1} to $r^{-3/2}$. On the other hand, it has been shown[18] that the density of a baryonic gas in hydrostatic equilibrium in the gravitational field of such a distribution of dark matter does exhibit the core expected from Eq. (1.9.18). In any case, the dark matter and baryonic gas densities do have the same distributions at distances from the center that are larger than r_0.

As we will see in Section 3.2, it is possible to infer a value for $\Omega_B h^2$ from the abundances of deuterium and other light isotopes, which together with Eq. (1.9.22) can be used to derive a value for $\Omega_M h^{1/2}$. There are several other methods for estimating Ω_M or $\Omega_M h^2$ that will be discussed elsewhere

---

[15]This is argued by S. D. M. White, J. F. Navarro, A. E. Evrard, and C. S. Frenk, *Nature* **366**, 429 (1993). Calculations supporting this assumption are described in Section 8.3.

[16]S. Schindler, in *Space Science Reviews* **100**, 299 (2002), ed. P. Jetzer, K. Pretzl, and R. von Steiger (Kluwer) [astro-ph/0107028].

[17]J. F. Navarro, C. S. Frenk, and S. D. M. White, *Astrophys. J.* **462**, 563 (1996) [astro-ph/9508025]; **490**, 493 (1997) [astro-ph/9610188]; T. Fukushige and J. Makino, *Astrophys. J.* **477**, L9 (1997) [astro-ph/9610005]; B. Moore *et al.*, *Mon. Not. Roy. Astron. Soc.* **499**, L5 (1998).

[18]N. Makino, S. Sasaki, and Y. Suto, *Astrophys. J.* **497**, 555 (1998). Also see Y. Suto, S. Sasaki, and M. Makino, *Astrophys. J.* **509**, 544 (1998); E. Komatsu and U. Seljak, *Mon. Not. Roy. Astron. Soc.* **327**, 1353 (2001).

in this book, using gravitational lenses (Section 9.3), the Sunyaev–Zel'dovich effect (Section 2.5), and anisotropies in the cosmic microwave background (Sections 2.6 and 7.2), the last of which also gives a value for $\Omega_B h^2$. In addition to these, there are methods[19] based on the evolution of clusters of galaxies, cosmic flows, cluster correlations, etc., that depend on detailed dynamical theories of structure formation.

1.10 Intergalactic absorption

Some of the cosmic gas of nuclei and electrons from which the first galaxies and clusters of galaxies condensed must be still out there in intergalactic space. Atoms or molecules in this gas could be observed through the resonant absorption of light or radio waves from more distant galaxies or quasars, but it is believed that most of the gas was ionized by light from a first generation of hot massive stars, now long gone, that are sometimes called stars of Population III. It now appears that some quasars formed before this ionization was complete, giving us the opportunity to observe the intergalactic gas through resonant absorption of the light from these very distant quasars.

Let us suppose that an atomic transition in a distant source produces a ray of light of frequency v_1 that leaves the source at time t_1 and arrives at the Earth with frequency v_0 at time t_0. At time t along its journey the light will have frequency redshifted to $v_1 a(t_1)/a(t)$, so if the intergalactic medium absorbs light of frequency v at a rate (per proper time) $\Lambda(v, t)$, and does not emit light, then the intensity $I(t)$ of the light ray will decrease according to the equation

$$\dot{I}(t) = -\Lambda\Big(v_1 a(t_1)/a(t), t\Big) I(t) \ .$$

But if the intergalactic gas is at a non-zero temperature $T(t)$, then photons will also be added to the light ray through the process of stimulated emission, as a rate per photon given by the Einstein formula[1] $\exp(-hv/k_B T)\,\Lambda(v, t)$, so the intensity of the light ray will satisfy

$$\dot{I}(t) = -\left[1 - \exp\left(-\frac{hv_1 a(t_1)}{k_B T(t) a(t)}\right)\right] \Lambda\Big(v_1 a(t_1)/a(t), t\Big) I(t) \quad (1.10.1)$$

The intensity observed at the earth will then be

$$I(t_0) = \exp(-\tau) I(t_1) \ , \qquad (1.10.2)$$

---

[19]For surveys, see N. A. Bahcall, *Astrophys. J.* **535**, 593 (2000) [astro-ph/0001076]; M. Turner, *Astrophys. J.* **576**, L101 (2002) [astro-ph/0106035]; S. Schindler, *op. cit.*; K. A. Olive, lectures given at Theoretical Advanced Study Institute on Elementary Particle Physics, Boulder, June 2002 [astro-ph/0301505].
[1]A. Einstein, *Phys. Z.* **18**, 121 (1917).

where τ is the *optical depth*:

$$\tau = \int_{t_1}^{t_0} \left[1 - \exp\left(-\frac{h\nu_1 a(t_1)}{k_B T(t)a(t)} \right) \right] \Lambda(\nu_1 a(t_1)/a(t), t)\, dt \ . \quad (1.10.3)$$

The absorption rate is given by

$$\Lambda(\nu, t) = n(t)\,\sigma(\nu)\ , \quad (1.10.4)$$

where $\sigma(\nu)$ is the absorption cross section at frequency ν, and $n(t)$ is the number density (per proper volume) of absorbing atoms. Often the absorption cross section is sharply peaked at some frequency ν_R, so the absorption takes place only close to a time t_R, given by

$$a(t_R) = \nu_1 a(t_1)/\nu_R \ . \quad (1.10.5)$$

Therefore the optical depth can be approximated as

$$\tau \simeq n(t_R) \left[1 - \exp\left(-h\nu_R/k_B T(t_R) \right) \right] \int \sigma\left(\nu_1 a(t_1)/a(t) \right) dt \ .$$

By changing the variable of integration from time to frequency, we can write this as

$$\tau \simeq n(t_R) \left[1 - \exp\left(-h\nu_R/k_B T(t_R) \right) \right] [a(t_R)/\dot{a}(t_R)] \mathcal{I}_R \ , \quad (1.10.6)$$

where

$$\mathcal{I}_R \equiv \frac{1}{\nu_R} \int \sigma(\nu)\, d\nu \ , \quad (1.10.7)$$

the integral being taken over a small range of frequencies containing the absorption line. The only thing in the formula for τ that depends on a cosmological model is the Hubble expansion rate $\dot{a}(t_R)/a(t_R)$ at the time of absorption, given by Eq. (1.5.19) and (1.5.38) as

$$\frac{\dot{a}(t_R)}{a(t_R)} = H_0 \sqrt{\Omega_\Lambda + \Omega_K(1 + z_R)^2 + \Omega_M(1 + z_R)^3 + \Omega_R(1 + z_R)^4} \ ,$$

$$(1.10.8)$$

where $z_R = a(t_0)/a(t_R) - 1 = \nu_R/\nu_0 - 1$ is the redshift of the location of the resonant absorption. For a source at redshift z, the absorption takes place over a range of *observed* frequencies $\nu_0 = \nu_1/(1 + z)$ given by the condition that the time t_R defined by Eq. (1.10.5) should be between t_1 and t_0:

$$\nu_R/(1 + z) \leq \nu_0 \leq \nu_R \ . \quad (1.10.9)$$

For example, in 1959 Field[2] suggested looking for the effects of absorption of radio frequencies in the 21 cm transition in hydrogen atoms, caused in transitions from the spin zero to spin one hyperfine states in the $1s$ state of intergalactic hydrogen. Here $\nu_R = 1420$ MHz, so the radio spectrum of the galaxy Cygnus A at a redshift $z = 0.056$ should show an absorption trough (1.10.9) from 1342 MHz to 1420 MHz. Unfortunately, the temperature of neutral hydrogen in intergalactic space is much larger than $h\nu_R/k_B = 0.068$K, so the optical depth (1.10.6) is suppressed by a factor $\simeq 0.068$ K$/T(t_R)$. No sign of this absorption trough has been discovered. It is hoped that in the future a new generation of low frequency radio telescopes with good angular resolution may be able to use the emission and absorption of 21 cm radiation at large redshifts to study both the growth of structure and primordial density perturbations from which they grew.[3] For instance, by 2010 the Low Frequency Array (LOFAR) should be able to study 21 cm radiation from sources at redshift between 5 and 15 with good sensitivity and high angular resolution.[4]

For the present, a much better probe of intergalactic hydrogen atoms is provided by absorption of photons in the Lyman α transition from the $1s$ ground state to the $2p$ excited state, known as the *Gunn–Peterson effect*.[5] This has a resonant frequency in the ultraviolet, $\nu_R = 2.47 \times 10^{15}$ Hz, corresponding to a wavelength 1,215 Å, but for a source of redshift $z > 1.5$ the lower part or the absorption trough (1.10.9) will be observable on the Earth's surface at wavelengths greater than 3,000 Å, in the visible or infrared part of the spectrum. Here $h\nu_R/k_B = 118,000$ K, which is likely to be larger than the temperature of the intergalactic medium, in which case the factor $1 - \exp\left(-h\nu_R/k_B T(t_R)\right)$ in Eq. (1.10.6) can be set equal to unity. The integral (1.10.7) here has the value 4.5×10^{-18} cm^2, so Eq. (1.10.6) gives the optical depth just above the lower end of the absorption trough (1.10.9) as

$$
\tau_{\nu_0 = \nu_R/(1+z)+} = \left(\frac{n(t_R)}{2.4h \times 10^{-11} \text{ cm}^{-3}}\right)\left(\Omega_\Lambda + \Omega_K(1+z)^2\right.
$$
$$
\left. + \Omega_M(1+z)^3 + \Omega_R(1+z)^4\right)^{-1/2}, \qquad (1.10.10)
$$

where again h is Hubble's constant in units of 100 km s^{-1} Mpc^{-1}. For instance, if a fraction f of the baryons of the universe at a time corresponding

---

[2]G. Field, *Astrophys. J.* **129**, 525 (1959).

[3]A. Loeb and M. Zaldarriaga, *Phys. Rev. Lett.* **92**, 211301 (2004) [astro-ph/0312134]; S. Furlanetto, S. P. Oh, and F. Briggs, *Phys. Rep.* **433**, 181 (2006) [astro-ph/0608032].

[4]H. J. A. Röttgering *et al.*, in *Cosmology, Galaxy Formation, and Astroparticle Physics on the Pathway to the SKA*, eds. H.-R. Klöckner *et al.* [astro-ph/0610596].

[5]J. E. Gunn and B. A. Peterson, *Astrophys. J.* **142**, 1633 (1965). Also see I. S. Shklovsky, *Astron. Zh.* **41**, 408 (1964); P. A. G. Scheuer, *Nature* **207**, 963 (1965).

to $z = 5$ were in the form of neutral intergalactic hydrogen atoms, and $\Omega_B h^2 = 0.02$, then the number density of hydrogen atoms at $z = 5$ would be $4.8f \times 10^{-5}$ cm^{-3}. Taking $h = 0.65$, $\Omega_\Lambda = 0.7$, $\Omega_M = 0.3$, and $\Omega_K = \Omega_R = 0$, the optical depth (1.10.10) would be $3.8f \times 10^5$. Thus with these parameters intergalactic neutral hydrogen that makes up a fraction of baryonic matter $f \gg 2.6 \times 10^{-6}$ would completely block any light with a frequency above the redshifted Lyman α line from sources beyond $z = 5$. Evidently the Gunn–Peterson effect provides a very sensitive probe of even a small proportion of neutral hydrogen atoms.

For many years the search for the Lyman α absorption trough was unsuccessful. Quasar spectra show numerous Lyman α absorption lines, forming what are sometimes called "Lyman α forests," which are believed to arise from clouds of neutral hydrogen atoms along the line of sight, but for quasars out to $z \approx 5$ there was no sign of a general suppression of frequencies above the redshifted Lyman α frequency,[6] that would be produced by even a small fraction f of the baryons in the universe in the form of neutral intergalactic hydrogen atoms. Then in 2001 the spectrum of the quasar SDSSp J103027.10+052455.0 with redshift $z = 6.28$ discovered by the Sloan Digital Sky Survey was found to show clear signs of a complete suppression of light in the wavelength range from just below the redshifted Lyman α wavelength at 8,845 Å down to 8,450 Å, indicating a significant fraction f of baryons in the form of neutral intergalactic hydrogen atoms at redshifts greater than $8,450/1,215 - 1 = 5.95$.[7] (See Figure 1.8.) Thus a redshift of order 6 may mark the end of a "dark age," in which the absorption of light by neutral hydrogen atoms made the universe opaque to light with frequencies above the redshifted Lyman α frequency. Further evidence for this conclusion is supplied by the spectrum of intense gamma ray sources, known as *gamma ray bursters*, at large redshifts.[8]

This does not mean that all or even most of the hydrogen in the universe was in the form of neutral atoms at $z > 6$. As we have seen, even small concentrations of neutral hydrogen could have produced an absorption trough in the spectrum of distant quasars. In fact, we shall see in Chapter 7 that there is now some evidence from the study of the cosmic microwave background that hydrogen became mostly ionized at redshifts considerably larger than $z \approx 6$, perhaps around $z \approx 10$.

---

[6]A. Songalia, E. Hu, L. Cowie, and R. McMahon, *Astrophys. J.* **525**, L5 (1999).

[7]R. H. Becker *et al.*, *Astron. J.* **122**, 2850 (2001) [astro-ph/0108097]. See S. G. Djorgovski *et al.*, *Astrophys. J.* **560**, L5 (2001) [astro-ph/0108069], for a hint of absorption by neutral hydrogen at slightly smaller redshifts. Also see X. Fan *et al.*, *Astrophys. J.* **123**, 1247 (2002) [astro-ph/0111184].

[8]T. Totani *et al.*, *Publ. Astron. Soc. Pacific* **58**, 485 (2006) [astro-ph/0512154].

The clouds of neutral hydrogen at redshifts $z < 6$ which produce the
Lyman α forest can provide an independent means of measuring Ω_M and
Ω_Λ. The idea goes back to a 1979 paper of Alcock and Paczyński.[9] Suppose
we observe a luminous object at a redshift z that extends a proper distance
$D_\perp$ perpendicular to the line of sight and a proper distance $D_\parallel$ along the
line of sight. According to the definition of the angular diameter distance,
the object will subtend an angle

$$\Delta\theta = D_\perp/d_A(z) . \tag{1.10.11}$$

Also, when we observe light from the whole object at the same time t_0, the
difference in the time t_1 that the light was emitted from the far and near
points of the object will be $\Delta t_1 = D_\parallel$. The redshift is $a(t_0)/a(t_1) - 1$, so the
absolute value of the difference of redshift from the far and near points of
the object will be

$$\Delta z = \frac{a(t_0)}{a^2(t_1)}\dot{a}(t_1)\Delta t_1 = (1+z)H(z)D_\parallel , \tag{1.10.12}$$

where $H(z) \equiv \dot{a}(t_1)/a(t_1)$ is the Hubble constant at the time of emission.
Taking the ratio, we have

$$\frac{\Delta z}{\Delta\theta} = (1+z)\,H(z)\,d_A(z)\left(D_\parallel/D_\perp\right) \tag{1.10.13}$$

It is then only necessary to use Eq. (1.5.19) to write $H(z)$ as

$$H(z) = \sqrt{\left(\frac{8\pi G}{3}\right)\left(\rho_{M0}(1+z)^3 + \rho_V + \rho_{R0}(1+z)^4\right) - \frac{K}{a_0^2}(1+z)^2}$$

$$= H_0\sqrt{\Omega_M(1+z)^3 + \Omega_\Lambda + \Omega_R(1+z)^4 + \Omega_K(1+z)^2} , \tag{1.10.14}$$

and use Eqs. (1.4.12) and (1.5.45) to write $d_A(z)$ as

$$d_A(z) = \frac{1}{(1+z)H_0\Omega_K^{1/2}}$$

$$\times \sinh\left[\Omega_K^{1/2}\int_{1/(1+z)}^1 \frac{dx}{x^2\sqrt{\Omega_\Lambda + \Omega_K x^{-2} + \Omega_M x^{-3} + \Omega_R x^{-4}}}\right]. \tag{1.10.15}$$

---

[9]C. Alcock and B. Paczyński, *Nature* **281**, 358 (1979).

The Hubble constant H_0 cancels in the product, and we find a result that depends only on z, $D_\parallel/D_\perp$, and the Ωs:

$$\frac{\Delta z}{\Delta \theta} = \left(D_\parallel/D_\perp\right)\Omega_K^{-1/2}\sqrt{\Omega_M(1+z)^3 + \Omega_\Lambda + \Omega_R(1+z)^4 + \Omega_K(1+z)^2}$$

$$\times \sinh\left[\Omega_K^{1/2}\int_{1/(1+z)}^1 \frac{dx}{x^2\sqrt{\Omega_\Lambda + \Omega_K x^{-2} + \Omega_M x^{-3} + \Omega_R x^{-4}}}\right].$$

$$(1.10.16)$$

For instance, if the object is known to be a sphere, such as a spherical cluster of galaxies, then $D_\parallel/D_\perp = 1$, and we can use a measurement of Δz and $\Delta \theta$ to set a model-independent constraint on the Ωs, with no need to worry about effects of evolution or intergalactic absorption.

Unfortunately, it is not so easy to find spherical objects at large redshift. But there are various objects whose distribution functions *are* spherically symmetric. For instance, the distribution of field galaxies is presumably spherically symmetric about any point in space, and it has been proposed that the application of the Alcock–Paczyński method to galaxies might allow a determination of the cosmological constant.[10] This method has been applied[11] instead to the distribution of quasars measured in the 2dF QSO Redshift Survey.[12] Assuming $K = 0$, this analysis gives $\Omega_\Lambda = 0.71_{-0.17}^{0.09}$.

Recently the Alcock–Paczyński idea has been applied to the distribution function of Lyman α clouds.[13] As already mentioned, these are intergalactic clouds containing neutral hydrogen atoms, which absorb light from more distant quasars along the line of sight in $1s \to 2p$ transitions, showing up as dark lines in the spectrum of the quasar at wavelengths $1215(1+z)$ Å for clouds at redshift z. Suppose we measure the number density $N(z,\hat{n})$ of Lyman α clouds at various redshifts z in various directions $\hat{n}$. Assuming a spherically symmetric distribution of Lyman α clouds, the mean value of the product of the number densities of these clouds at two nearby points with redshifts z and $z+\Delta z$ (with $\Delta z \ll 1$) and directions $\hat{n}$ and $\hat{n}+\Delta\hat{n}$ separated by a small angle $\Delta\theta$ will be a function only of z and the proper distance between the points, and will be analytic in the components of the vector

[10]W. E. Ballinger, J. A. Peacock, and A. F. Heavens, *Mon. Not. Roy. Astron. Soc.* **281**, 877 (1996).

[11]P. J. Outram *et al.*, *Mon. Not. Roy. Astron. Soc.* **348**, 745 (2004) [astro-ph/0310873].

[12]S. M. Croom *et al.*, *Mon. Not. Roy. Astron. Soc.* **349**, 1397 (2004); available at www.2df quasar.org.

[13]L. Hui, A. Stebbins, and S. Burles, *Astrophys. J.* **511**, L5 (1999); P. McDonald and J. Miralda-Escudeé, *Astrophys. J.* **518**, 24 (1999); W-C. Lin and M. L. Norman, talk at the Theoretical Astrophysics in Southern California meeting, Santa Barbara, October 2002 [astro-ph/0211177]; P. McDonald, *Astrophys. J.* **585**, 34 (2003).

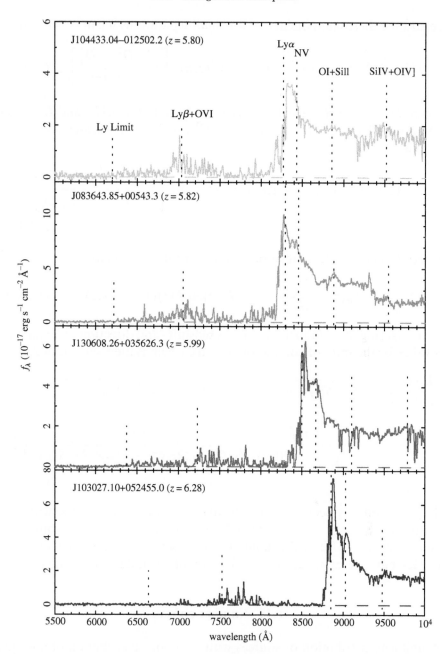

Figure 1.8: Observed intensity versus wavelength for four high-redshift quasars, from R. H. Becker *et al.*, *Astron. J.* **122**, 2850 (2001) [astro-ph/0108097]. Vertical dashed lines indicate the redshifted wavelengths for various spectral lines. In the direction of the quasar with $z = 6.28$ the intensity drops to zero within experimental accuracy just to the left of the Lyman α line at 8845 Å, a feature not seen for the quasar with $z = 5.99$, indicating the onset of patches of nearly complete ionization at a redshift between 5.99 and 6.28.

separating these components, so for small separations it can be written

$$\langle N(z,\hat{n})\,N(z+\Delta z,\hat{n}+\Delta\hat{n})\rangle \simeq \langle N^2(z,\hat{n})\rangle\left[1-\frac{D_\perp^2+D_\parallel^2}{L^2(z)}\right], \quad (1.10.17)$$

where $D_\perp$ and $D_\parallel$ are given by Eqs. (1.10.11) and (1.10.12), and L is some correlation length. This can be written in terms of the observed Δz and $\Delta\theta$, as

$$\langle N(z,\hat{n})\,N(z+\Delta z,\hat{n}+\Delta\hat{n})\rangle \simeq \langle N^2(z,\hat{n})\rangle\left[1-\frac{\Delta z^2}{L_z^2(z)}-\frac{\Delta\theta^2}{L_\theta^2(z)}\right],$$

$$(1.10.18)$$

where L_z and L_θ are correlation lengths for redshift and angle

$$L_\theta(z)=\frac{L(z)}{d_A(z)}\,, \qquad L_z(z)=L(z)(1+z)H(z)\,. \quad (1.10.19)$$

By measuring this product for various redshifts and directions, we can infer a value for the ratio of correlation lengths, which is independent of L:

$$\frac{L_z(z)}{L_\theta(z)}=\Omega_K^{-1/2}\sqrt{\Omega_M(1+z)^3+\Omega_\Lambda+\Omega_R(1+z)^4+\Omega_K(1+z)^2}$$

$$\times\sinh\left[\Omega_K^{1/2}\int_{1/(1+z)}^{1}\frac{dx}{x^2\sqrt{\Omega_\Lambda+\Omega_K x^{-2}+\Omega_M x^{-3}+\Omega_R x^{-4}}}\right].$$

$$(1.10.20)$$

This method has been applied[14] to five pairs of close quasars, with redshifts in the range from 2.5 to 3.5 and separations ranging from 33 to 180 arcseconds. Use of this limited sample sets only weak constraints on the Ωs, but it rules out $\Omega_\Lambda=0$ at the level of 2 standard deviations.

1.11 Number counts

A uniform distribution of sources with a smooth distribution of absolute luminosity leads in ordinary Euclidean space to a unique distribution in apparent luminosity. If there are $N(L)dL$ sources per unit volume with absolute luminosity between L and $L+dL$, then the number $n(>\ell)$ of

---

[14]A. Lidz, L. Hui, A. P. S. Crotts, and M. Zaldarriaga, astro-ph/0309204 (unpublished).

sources observed with apparent luminosity greater than ℓ is given by

$$n(> \ell) = \int_0^\infty N(L)\, dL \int_0^{\sqrt{L/4\pi\ell}} 4\pi r^2\, dr$$

$$= \frac{1}{3\sqrt{4\pi}\,\ell^{3/2}} \int_0^\infty L^{3/2}\, N(L)\, dL \qquad (1.11.1)$$

Thus whatever the distribution in absolute luminosity, we expect that $n(> \ell) \propto \ell^{-3/2}$.

This analysis needs several changes in a cosmological setting:

1. Instead of the volume element $r^2 \sin\theta\, dr\, d\theta\, d\phi$, the proper volume element here is $(\text{Det}\, g^{(3)})^{1/2} dr\, d\theta\, d\phi$, where $g_{ij}^{(3)} \equiv a^2 \tilde{g}_{ij}$ is the three-dimensional metric, with non-vanishing components $g_{rr}^{(3)} = a^2/(1 - Kr^2)$, $g_{\theta\theta}^{(3)} = a^2 r^2$, $g_{\phi\phi}^{(3)} = a^2 r^2 \sin^2\theta$, so

$$dV = \frac{a^3(t)\, r^2\, \sin\theta\, dr\, d\theta\, d\phi}{\sqrt{1 - Kr^2}}. \qquad (1.11.2)$$

2. The apparent luminosity is related to the absolute luminosity by

$$\ell = \frac{L}{4\pi d_L^2(z)}, \qquad (1.11.3)$$

where $d_L(z)$ is the luminosity distance (1.4.3).

3. Except in the steady state cosmology, the number density of sources changes with time, even if only through the cosmic expansion.

4. We can often measure the redshift z as well as the apparent luminosity.

Eq. (1.11.2) gives the number of sources with redshift between z and $z + dz$ and apparent luminosity between ℓ and $\ell + d\ell$ as

$$n(z, \ell)\, dz\, d\ell = 4\pi \mathcal{N}(t, L) dL \frac{a^3(t)\, r^2\, dr}{\sqrt{1 - Kr^2}}, \qquad (1.11.4)$$

where $\mathcal{N}(t, L)\, dL$ is the number of sources per proper volume at time t with absolute luminosity between L and $L + dL$; t and z are related by $1 + z = a(t_0)/a(t)$, and t and r are related by Eq. (1.2.2):

$$\int_t^{t_0} \frac{dt'}{a(t')} = \int_0^r \frac{dr'}{\sqrt{1 - Kr'^2}}. \qquad (1.11.5)$$

We use (1.11.5) to express the differential dr in terms of dt, and then express dt in terms of dz:

$$\frac{dr}{\sqrt{1-Kr^2}} = -\frac{dt}{a(t)} = \frac{dz}{H(z)\,a_0} \,,$$

where $H(z) \equiv \dot{a}(t)/a(t)$ and $a_0 \equiv a(t_0)$. As a reminder, for a universe containing radiation, matter, and a constant vacuum energy, Eq. (1.5.41) gives

$$H(z) = H_0\sqrt{\Omega_\Lambda + \Omega_K(1+z)^2 + \Omega_M(1+z)^3 + \Omega_R(1+z)^4}\,.$$

Canceling dz in Eq. (1.11.4), we then have

$$n(z,\ell)\,d\ell = \frac{4\pi\,\mathcal{N}\!\left(t(z),L\right)r^2(z)a_0^2\,dL}{(1+z)^3\,H(z)}\,,$$

We next use Eq. (1.11.3) to write (with z now held fixed):

$$dL = 4\pi d_L^2(z)\,d\ell\,,$$

so that canceling $d\ell$ gives

$$n(z,\ell) = \frac{16\pi^2\,\mathcal{N}\!\left(t(z),4\pi d_L^2(z)\ell\right)d_L^4(z)}{H(z)\,(1+z)^5}\,, \qquad (1.11.6)$$

in which we have used Eq. (1.4.3) to express $a_0 r$ in terms of d_L.

In particular, for a sample of sources that are not evolving at a time $t(z)$, the time dependence of the number density $\mathcal{N}$ is just proportional to $a^{-3} \propto (1+z)^3$:

$$\mathcal{N}\!\left(t(z),L\right) = (1+z)^3 \mathcal{N}_0(L)\,. \qquad (1.11.7)$$

If all members of this sample are bright enough to be visible at a redshift z, then the total number of sources observed with redshifts between z and $z+dz$ will be $n(z)\,dz$, where

$$n(z) \equiv \int_0^\infty n(z,\ell)\,d\ell = \frac{4\pi\,\mathcal{N}_0\,d_L^2(z)}{H(z)\,(1+z)^2} \qquad (1.11.8)$$

where $d_L(z)$ is given by Eq. (1.5.45), and

$$\mathcal{N}_0 \equiv \int_0^\infty \mathcal{N}_0(L)\,dL\,. \qquad (1.11.9)$$

In principle, even without knowing $\mathcal{N}_0$ or H_0, if $n(z)$ were accurately measured we could compare the observed *shape* of this function with Eq. (1.11.8) to find the Ωs.

There are several obvious dangers in using Eq. (1.11.8) in this way. For one thing, it is necessary to avoid missing sources that have high redshift and hence low apparent luminosity. Also, evolution in the number of sources can introduce an additional dependence on the light emission time t, and hence on z. In 1986 Loh and Spillar[1] carried out a survey of galaxy numbers as a function of redshift. The redshifts were measured photometrically (i. e., from their luminosities at various colors rather than by the shift of specific spectral lines), which generally gives less reliable results. Comparing their results with Eq. (1.11.8) in the case $\Omega_K = \Omega_R = 0$ (so that $\Omega_\Lambda + \Omega_M = 1$), they found that $\Omega_\Lambda / \Omega_M = 0.1^{-0.4}_{+0.2}$. By now it has been realized that the evolution of sources cannot be neglected at redshifts large enough for $n(z)$ to be sensitive to cosmological parameters, and this result for $\Omega_\Lambda / \Omega_M$ has been abandoned.[2]

Useful results can be obtained when evolution is taken into account. One group[3] used number counts of very faint galaxies[4] as a function of apparent luminosity to estimate the free parameters in a model of galactic luminosity evolution (assuming the number of galaxies per coordinate volume to be constant), and then used this model together with a redshift survey[5] extending to $z \simeq 0.47$ to conclude that Ω_M is small and that Ω_Λ is in the range of 0.5 to 1. More recently, several surveys[6] of numbers of galaxies at different redshifts that yield important results about galactic evolution, and with the use of dynamical models they can yield information about Ω_M and Ω_Λ.[7] But it appears that number counts of galaxies will be more useful in learning about galactic evolution than in making precise determinations of cosmological parameters. In a dramatic application of this approach,[8] a

---

[1] E. D. Loh, *Phys. Rev. Lett.* **57**, 2865 (1986); E. D. Loh and E. J. Spillar, *Astrophys. J.* **284**, 439 (1986).

[2] For a discussion of future prospects for measuring Ω_Λ in redshift surveys, see W. E. Ballinger, J. A. Peacock, and A. F. Heavens, *Mon. Not. Roy. Astron. Soc.* **282**, 877 (1996).

[3] M. Fukugita, F. Takahara, K. Yamashita, and Y. Yoshii, *Astrophys. J.* **361**, L1 (1990).

[4] J. A. Tyson, *Astron. J.* **96**, 1 (1988).

[5] T. J. Broadhurst, R. S. Ellis, and T. Shanks, *Mon. Not. Roy. Astron. Soc.* **235**, 827 (1988).

[6] G. Efstathiou, R. S. Ellis, B. A. Peterson, *Mon. Not. Roy. Astron. Soc.* **232**, 431 (1988); J. Loveday, B. A. Peterson, G. Efstathiou, and S. J. Maddox, *Astrophys. J.* **390**, 338 (1992); L. da Costa, in *Proceedings of the Conference on Evolution of Large Scale Structure*, Garching, August 1998 [astro-ph/9812258]; S. Borgani, P. Rosati, P. Tozzi, and C. Norman, *Astrophys. J.* **517**, 40 (1999) [astro-ph/9901017]; S. J. Oliver, in *Highlights of the ISO Mission: Special Scientific Session of the IAU General Assembly*. eds. D. Lemke *et al.* (Kluwer) [astro-ph/9901272]; M. Colless, in Publ. Astron. Soc. Australia [astro-ph/9911326]; S. Rawlings, astro-ph/0008067.

[7] W. J. Percival *et al.*, *Mon. Not. Roy. Astron. Soc.* **327**, 1297 (2001) [astro-ph/0105252]; S. Borgnani *et al.*, *Astrophys. J.* **561**, 13 (2001) [astro-ph/0106428].

[8] R. J. Bouwens and G. D. Illingworth, *Nature* **443**, 189 (2006).

search at the Lick Observatory for galaxies with redshifts in the range $z \approx$ 7 to 8 found at most just one galaxy, while it is estimated that if Eq. (1.11.7) were valid then, on the basis of the number of galaxies observed (with the same conservative selection criteria) at redshifts $z \approx 6$, ten galaxies should have been found with $z \approx 7$ to 8. The implication is that there must have been a spurt in the formation of luminous galaxies at a redshift in the range 6 to 7. This fits in well with the conclusion discussed in Section 1.10, that the ionization of intergalactic hydrogen became essentially complete at a redshift of order 6, presumably due to ultraviolet radiation from massive stars formed around that time.

* * *

Historically the first important application of number counts was in radio source surveys, where redshifts are not generally available. These surveys take place at a fixed receiving frequency v, corresponding to a variable emitted frequency $v(1+z)$, so the source counts are affected by the frequency dependence of the distribution of intrinsic source powers.

If a source with a redshift z emits a power[9] $P(v)dv$ between frequencies v and $v + dv$, then the power received at the origin per unit antenna area between frequencies v and $v + dv$ is

$$S(v)dv = \frac{P\big(v(1+z)\big)\, dv\,(1+z)}{4\pi d_L^2(z)}.$$ (1.11.10)

Many radio sources have a "straight" spectrum, i.e.

$$P(v) \propto v^{-\alpha}$$ (1.11.11)

with the spectral index α typically about 0.7 to 0.8. This allows a great simplification in Eq. (1.11.10):

$$S(v)dv = \frac{P(v)\, dv}{4\pi\, d_L^2(z)(1+z)^{\alpha-1}}.$$ (1.11.12)

From now on we will take the observed frequency v as fixed, and write $S(v) = S$ and $P(v) = P$. Canceling dv, Eq. (1.11.12) then reads

$$S = \frac{P}{4\pi\, d_L^2(z)\,(1+z)^{\alpha-1}}.$$ (1.11.13)

---

[9]In G&C, P was defined as the power emitted per solid angle, while here it is the power emitted in all directions.

If at time t there are $N(P, t)\, dP$ sources per proper volume with power between P and $P + dP$, then the number of sources observed with power per antenna area greater than S is

$$n(> S) = \int_0^\infty dP \int N(P, t) \frac{4\pi r^2 a^3(t)\, dr}{\sqrt{1 - Kr^2}} , \qquad (1.11.14)$$

with the upper limit on the integral over r set by the condition that

$$a_0^2 r^2 (1 + z)^{1+\alpha} < \frac{P}{4\pi S} . \qquad (1.11.15)$$

Of course, r, z, and t are related by the familiar formulas

$$\int_0^r \frac{dr'}{\sqrt{1 - Kr'^2}} = \int_t^{t_0} \frac{dt'}{a(t')} , \qquad 1 + z = a(t_0)/a(t) . \quad (1.11.16)$$

This becomes much simpler if we assume that the time-dependence of the source number density can be parameterized as

$$N(P, t) = N(P) \left(\frac{a(t)}{a_0} \right)^\beta . \qquad (1.11.17)$$

For instance, if sources do not evolve and are neither created nor destroyed, then $\beta = -3$, while in the steady-state model $\beta = 0$. Eq. (1.11.14) now reads

$$n(> S) = a_0^3 \int_0^\infty N(P)\, dP \int \frac{4\pi r^2 (1 + z)^{-\beta-3}\, dr}{\sqrt{1 - Kr^2}} , \qquad (1.11.18)$$

with the same P/S-dependent upper limit (1.11.15) on r.

The coordinate r is given in terms of z by the power series (1.4.8)

$$a_0 H_0 r = z - \tfrac{1}{2}(1 + q_0)z^2 + \dots . \qquad (1.11.19)$$

We can then convert the integral over r to one over z, with

$$a_0 H_0\, dr = dz\, [1 - (1 + q_0)z + \dots] , \qquad (1.11.20)$$

and the upper limit on z is given by

$$z^2 [1 + z(\alpha - q_0) + \dots] < \frac{PH_0^2}{4\pi S} ,$$

or, in other words,

$$z < \sqrt{\frac{PH_0^2}{4\pi S}} \left[1 - \frac{1}{2}(\alpha - q_0)\sqrt{\frac{PH_0^2}{4\pi S}} + \dots \right] . \qquad (1.11.21)$$

Then Eq. (1.11.18) becomes

$$n(> S) = \frac{4\pi}{H_0^3} \int_0^\infty N(P)\, dP$$

$$\times \left[\frac{1}{3} \left(\frac{PH_0^2}{4\pi\, S} \right)^{3/2} \left(1 - \frac{3}{2}(\alpha - q_0) \left(\frac{PH_0^2}{4\pi\, S} \right)^{1/2} + \cdots \right) \right.$$

$$\left. - \frac{1}{4} \left(\frac{PH_0^2}{4\pi\, S} \right)^2 (\beta + 5 + 2q_0) + \cdots \right],$$

or, collecting terms,

$$n(> S) = \frac{1}{3\sqrt{4\pi}\, S^{3/2}} \int_0^\infty P^{3/2} N(P)\, dP$$

$$\times \left[1 - \frac{3}{4}(5 + \beta + 2\alpha) \left(\frac{PH_0^2}{4\pi\, S} \right)^{1/2} + \cdots \right]. \quad (1.11.22)$$

We see that $n(> S)$ has a term with the familiar $S^{-3/2}$ dependence, plus a correction proportional to S^{-2} with a coefficient proportional to $5 + \beta + 2\alpha$. It is noteworthy that this coefficient is independent of q_0 or K. For the standard cosmology with no evolution of sources $\beta = -3$, and we have mentioned that $\alpha \approx 0.75$, so $5 + \beta + 2\alpha = 3.5$. Although the precise value is uncertain, this coefficient is definitely positive, which means that for faint sources $n(> S)$ should fall off more *slowly* than $S^{-3/2}$. This is definitely not what is observed.[10] It has been known for many years that for $S > 5 \times 10^{-26} \text{Wm}^{-2}/\text{Hz}$, the source count function $N(> S)$ falls off more rapidly than $S^{-3/2}$. The conclusion is inevitable that the number of radio sources per co-moving volume is decreasing, with $\beta < -6.5$. Radio source counts are useful in studying this evolution, but not for measuring cosmological parameters.

On the other hand, for the steady state cosmology (discussed in Section 1.5) we have $\beta = 0$, so the coefficient $5 + \beta + 2\alpha \approx 6.5$, and the predicted number count $N(> S)$ decreases even more slowly with S, making the disagreement with experiment even worse than for the standard cosmology with no evolution of sources. Here it is not possible to save the situation by appealing to evolution, because the essence of the steady state model is that on the average there is no evolution. This observation

---

[10] For a list of major radio source surveys, and references to the original literature, see G&C, Sec. 14.8.

discredited the steady state model even before the discovery of the cosmic microwave radiation background.

1.12 Quintessence

So far, we have taken into account only non-relativistic matter, radiation, and a constant vacuum energy in calculating the rate of expansion of the universe. It appears that the vacuum energy is not only much smaller than would be expected from order-of-magnitude estimates based on the quantum theory of fields, but is only a few times greater than the present matter density. This has led to a widespread speculation that the vacuum energy is not in fact constant; it may now be small because the universe is old. A time-varying vacuum energy is sometimes called *quintessence*.[1]

The natural way to introduce a varying vacuum energy is to assume the existence of one or more scalar fields, on which the vacuum energy depends, and whose cosmic expectation values change with time. Scalar fields of this sort play a crucial part in the modern theory of weak and electromagnetic interactions, and are also introduced in theories of inflation, as discussed in Chapters 4 and 10.

For simplicity, let us consider a single real scalar field $\varphi(\mathbf{x}, t)$. We will be concerned here with fields that are vary little on elementary particle spacetime scales, so the action of these field is taken to have a minimum number of spacetime derivatives:

$$I_\varphi = -\int d^4x \, \sqrt{-\mathrm{Det}g} \left[\frac{1}{2} g^{\lambda\kappa} \frac{\partial\varphi}{\partial x^\lambda} \frac{\partial\varphi}{\partial x^\kappa} + V(\varphi) \right], \qquad (1.12.1)$$

with an unspecified potential function $V(\varphi)$. We are interested here in the case of a Robertson–Walker metric, and a scalar field that depends only on time, not position. In this case the formulas (B.66) and (B.67) for the scalar field energy density and pressure become

$$\rho_\varphi = \frac{1}{2}\dot{\varphi}^2 + V(\varphi) \qquad (1.12.2)$$

$$p_\varphi = \frac{1}{2}\dot{\varphi}^2 - V(\varphi) . \qquad (1.12.3)$$

It follows immediately that $(1 + w)\rho_\varphi \geq 0$, where $w \equiv p_\varphi/\rho_\varphi$, so as long as $\rho_\varphi \geq 0$ this model has $w \geq -1$, and the phantom energy disaster discussed in Section 1.6 does not occur.

---

[1] For reviews with references to the original literature, see B. Ratra and P. J. E. Peebles, *Rev. Mod. Phys.* **75**, 559 (2003); E. V. Linder, 0704.2064.

The equation (1.1.32) of energy conservation here reads

$$\ddot{\varphi} + 3H\dot{\varphi} + V'(\varphi) = 0 , \qquad (1.12.4)$$

(where as usual $H(t) \equiv \dot{a}(t)/a(t)$), which is the same as the field equation derived from the action (1.12.1). This is the equation of a particle of unit mass with one-dimensional coordinate φ, moving in a potential $V(\varphi)$ with a frictional force $-3H\dot{\varphi}$. The field will run toward lower values of $V(\varphi)$, finally coming to rest if it can reach any field value where $V(\varphi)$ is at least a local minimum. Unfortunately, we do not know any reason why the value of $V(\varphi)$ where it is stationary should be small.

Nevertheless, there are potentials that have some attractive properties once we adjust an additive constant in the potential to make them vanish at their stationary point. The original and simplest example is provided by a potential[2]

$$V(\varphi) = M^{4+\alpha}\varphi^{-\alpha} , \qquad (1.12.5)$$

where α is positive but otherwise arbitrary, and M is a constant with the units of mass (taking $\hbar = c = 1$), which gives $V(\varphi)$ the dimensions of an energy density. There is no special reason to believe that the potential has this form, and in particular there is no known reason for excluding an additive constant (including effects of quantum fluctuations in all other fields), which would give the potential a non-zero value at its stationary point, at $\varphi = \infty$. Nevertheless, it may be illuminating to work out the consequences of this one specific model of quintessence.

For any potential it is necessary to assume that at sufficiently early times ρ_φ was much less than the energy density ρ_R of radiation because, as we will see in Section 3.2, any appreciable increase in the energy density at the time of cosmological nucleosynthesis would lead to a helium abundance exceeding what is observed. At these early times the energy density of radiation (including particles like neutrinos with masses less than $k_B T$) is also greater than that of non-relativistic matter, so Eq. (1.5.34) gives $a(t) \propto t^{1/2}$, and therefore $H = 1/2t$. The field equation (1.12.4) with potential (1.12.5) then reads

$$\ddot{\varphi} + \frac{3}{2t}\dot{\varphi} - \alpha M^{4+\alpha}\varphi^{-\alpha-1} = 0 . \qquad (1.12.6)$$

[2]P. J. E. Peebles and B. Ratra, *Astrophys. J.* **325**, L17 (1988); B. Ratra and P. J. E. Peebles, *Phys. Rev.* **D 37**, 3406 (1988); C. Wetterich, *Nucl. Phys.* **B302**, 668 (1988). Quintessence models with this potential were intensively studied by I. Zlatev, L. Wang, and P. J. Steinhardt, *Phys. Rev. Lett.* **82**, 896 (1999); P. J. Steinhardt, L. Wang, and I. Zlatev, *Phys. Rev.* **D 59**, 123504 (1999).

This has a solution

$$\varphi = \left(\frac{\alpha(2+\alpha)^2 M^{4+\alpha} t^2}{6+\alpha} \right)^{\frac{1}{2+\alpha}} . \tag{1.12.7}$$

Both $\dot{\varphi}^2$ and $V(\varphi)$ then go as $t^{-2\alpha/(2+\alpha)}$, and therefore at very early times ρ_φ must have been less than ρ_R, which goes as t^{-2}. This solution is not unique, but it is an attractor, in the sense that any other solution that comes close to it will approach it as t increases. (To see this, note that a small perturbation $\delta\varphi$ of the solution (1.12.7) will satisfy

$$0 = \delta\ddot{\varphi} + \frac{3}{2t}\delta\dot{\varphi} + \alpha(1+\alpha) M^{4+\alpha} \varphi^{-\alpha-2}\delta\varphi = \delta\ddot{\varphi} + \frac{3}{2t}\delta\dot{\varphi} + \frac{(6+\alpha)(1+\alpha)}{(2+\alpha)^2 t^2}\delta\varphi .$$

This has two independent solutions of the form

$$\delta\varphi \propto t^\gamma , \quad \gamma = -\frac{1}{4} \pm \sqrt{\frac{1}{16} - \frac{(6+\alpha)(1+\alpha)}{(2+\alpha)^2}} .$$

The square root is imaginary for $\alpha > 0$, so both solutions for $\delta\varphi$ decay as $t^{-1/4}$ for increasing t, while φ itself is increasing.) For this reason, the particular solution of Eq. (1.12.6) that goes as Eq. (1.12.7) for $t \to 0$ is known as the *tracker solution*. There is no particular physical reason to require that the initial conditions for the scalar field are such that the scalar field has approached the tracker solution by the present moment (the set of such initial conditions is called the "basin of attraction"), but since this requirement would make the present evolution of the scalar field insensitive to the initial conditions, it has the practical advantage of providing a model of quintessence with just two free parameters: M and α.

Nothing much changes when the radiation energy density drops below the energy density of non-relativistic matter. The tracker solution for the scalar field continues to grow as $t^{2/(2+\alpha)}$ (though with a different constant factor), so $\dot{\varphi}^2$ and $V(\varphi)$ continue to fall as $t^{-2\alpha/(2+\alpha)}$. But ρ_M and ρ_R are decreasing faster, like t^{-2} and $t^{-8/3}$, respectively, so eventually ρ_M and ρ_R will fall below ρ_φ. It is interesting that the value of φ where ρ_φ becomes equal to ρ_M is independent of the unknown constant M. When the expansion is dominated by matter ρ_M is given by Eq. (1.5.31) as $1/6\pi G t^2$, while (1.1.2), (1.12.5) and (1.12.7) give $\rho_\varphi \approx M^{2(4+\alpha)/(2+\alpha)} t^{-2\alpha/(2+\alpha)}$, so the time t_c at which $\rho_\varphi = \rho_M$ is of order

$$t_c \approx M^{-(4+\alpha)/2} G^{-(2+\alpha)/4} . \tag{1.12.8}$$

Using this in Eq. (1.12.7) then gives

$$\varphi(t_c) \approx G^{-1/2} . \tag{1.12.9}$$

Once ρ_M falls well below ρ_φ, the equation of motion of $\varphi(t)$ becomes

$$\ddot{\varphi} + \sqrt{24\pi G \rho_\varphi}\, \dot{\varphi} - \alpha M^{4+\alpha} \varphi^{-\alpha-1} = 0 , \tag{1.12.10}$$

with ρ_φ given by Eq. (1.12.2). The tracker solution in this era has a complicated time dependence, but it becomes simple again at sufficiently late times, times that may be later than the present. We can guess that the damping term proportional to $\dot{\varphi}$ in this equation will eventually slow the growth of φ, so that $\dot{\varphi}^2$ will become less than $V(\varphi)$, and also guess that the inertial term proportional to $\ddot{\varphi}$ will become negligible compared to the damping and potential terms. (Similar "slow roll" conditions will play an important role in the theory of inflation, described in Chapters 4 and 10.) Equation (1.12.10) then becomes

$$\sqrt{24\pi G M^{4+\alpha} \varphi^{-\alpha}}\, \dot{\varphi} = \alpha M^{4+\alpha} \varphi^{-\alpha-1} ,$$

and so

$$\dot{\varphi} = \frac{\alpha M^{2+\alpha/2} \varphi^{-\alpha/2-1}}{\sqrt{24\pi G}} . \tag{1.12.11}$$

The solution is

$$\varphi = M \left(\frac{\alpha(2+\alpha/2)\, t}{\sqrt{24\pi G}} \right)^{1/(2+\alpha/2)} . \tag{1.12.12}$$

(In general this involves a redefinition of the zero of time, to avoid a possible integration constant that might be added to t.) We can now check the approximations used in deriving Eq. (1.12.11), of which this is the solution. From Eq. (1.12.12) we see that $\dot{\varphi}^2 \propto t^{-(2+\alpha)/(2+\alpha/2)}$ while $V(\varphi) \propto t^{-\alpha/(2+\alpha/2)}$, so the kinetic energy term in Eq. (1.12.2) does become small compared with the potential term at late times. Also, $\ddot{\varphi} \propto t^{-(3+\alpha)/(2+\alpha/2)}$ while $V'(\varphi) \propto t^{-(1+\alpha)/(2+\alpha/2)}$, so the inertial term in Eq. (1.12.10) does become small compared with the potential term at late times. Eq. (1.12.12) is therefore a valid asymptotic solution of Eq. (1.12.10) for $t \to \infty$. Numerical calculations show that it is not only a solution for $t \to \infty$; it is the asymptotic form approached for $t \to \infty$ by the tracker solution.

With $\rho_\varphi \propto t^{-\alpha/(2+\alpha/2)}$ dominating the expansion rate at late times, we have $\dot{a}/a \propto t^{-\alpha/2(2+\alpha/2)}$, so

$$\ln a \propto t^{2/(2+\alpha/2)} . \tag{1.12.13}$$

This is a similar but less rapid growth of a than would be produced by a cosmological constant, for which $\ln a \propto t$. The difference between the deceleration parameter q_0 and the value -1 for an expansion dominated by a cosmological constant vanishes as $t^{-(2+\alpha)/(2+\alpha/2)}$. Note that the radiation and matter densities decrease as $1/a^4$ and $1/a^3$ respectively, and the curvature decreases as $1/a^2$, all of which have a much faster rate of decrease with time than the power-law decrease of ρ_φ, so the expansion rate is indeed dominated by ρ_φ at late times, justifying the derivation of Eq. (1.12.10).

We have found that, at least for a range of initial conditions, the potential (1.12.5) leads to an expansion that is dominated by radiation and then matter at early times, but becomes dominated by the scalar field energy at late times. But to get agreement with observation it is necessary arbitrarily to exclude a large constant term that might be added to (1.12.5), and also to adjust the value of M to make the critical time (1.12.8) at which the values of ρ_φ and ρ_M cross be close to the present moment $t_0 \approx 1/H_0$. Specifically, Eq. (1.12.8) shows that we need the constant factor in $V(\varphi)$ to take the value

$$M^{4+\alpha} \approx G^{-1-\alpha/2}H_0^2 . \tag{1.12.14}$$

There is no known reason why this should be the case.

Several groups of observers are now planning programs to discover whether the vacuum energy density is constant, as in the case of a cosmological constant, or changing with time. In such programs, one would compare the observed luminosity distance (or angular diameter distance) with a formula obtained by replacing the term Ω_Λ in the argument of the square root in Eq. (1.5.45) with a time-varying dark energy term. These observations will not actually measure the value w_0 of w at the present time, much less the present time derivatives $\dot{w}_0$, $\ddot{w}_0$, etc., because for that purpose it would be necessary to have extremely precise measurements of the luminosity distance or angular-diameter distance for small redshifts. Instead, measurements will be made with only moderate precision, but over a fairly large range of redshifts. To compare such measurements with theory, one needs a model of the time-variation of the dark energy. One model is simply to assume that w is constant, or perhaps varying linearly with time or redshift, but there is no physical model that entails such behavior.[3] It seems preferable to compare observation with the model of a scalar field rolling down a potential, which (whatever reservations may have about its naturalness) at least provides a physically possible model of varying dark

---

[3]Other assumptions about the form of w as a function of redshift that can mimic scalar field models have been considered by J. Weller and A. Albrecht, *Phys. Rev. D* **65**, 103512 (2002); E. V. Linder, *Phys. Rev. Lett.* **90**, 091301 (2003) [astro-ph/0208512].

energy.[4] Because these observations are difficult, it pays to adopt scalar field models with just two parameters, which can if we like be expressed in terms of $\Omega_\Lambda = 1 - \Omega_M$ (assuming flatness and neglecting the radiation energy density) and w_0.

One possibility is to suppose that over the latest *e*-folding of cosmic expansion the scalar field φ has taken values for which $V(\varphi)$ is only slowly varying. If $V(\varphi)$ were constant, we would have a constant vacuum energy, with $w = -1$, and the only parameter to measure would be Ω_V. For a two-parameter fit, we can take $V(\varphi)$ to vary linearly with φ:

$$V(\varphi) = V_0 + \left(\varphi - \varphi_0\right)V_0' \,. \tag{1.12.15}$$

This is valid if the fractional change in $V'(\varphi)$ in a time interval of order $1/H_0$ is small; that is, if $|V_0''\dot\varphi_0| \ll H_0|V_0'|$.

The field equation (1.12.4) for $\varphi(t)$ can be put in a convenient dimensionless form by replacing the dependent variable t and independent variable φ with dimensionless variables x and ω, defined by

$$x \equiv H_0\sqrt{\Omega_M}\,t \,, \qquad \omega \equiv \frac{8\pi\,G V(\varphi)}{3\Omega_M H_0^2} \,. \tag{1.12.16}$$

Because V is linear in φ, we have

$$\dot\varphi = \frac{3\Omega_M H_0^2 \dot\omega}{8\pi\,G V_0'} = \frac{3\Omega_M^{3/2} H_0^3}{8\pi\,G V_0'}\frac{d\omega}{dx} \,.$$

Then Eq. (1.12.4) becomes

$$\frac{d^2\omega}{dx^2} + 3\mathcal{H}\frac{d\omega}{dx} + \lambda = 0 \,, \tag{1.12.17}$$

where λ is the dimensionless parameter

$$\lambda \equiv \frac{8\pi\,G V_0'^2}{3H_0^4\Omega_M^2} \,, \tag{1.12.18}$$

and $\mathcal{H}$ is a function of ω and $d\omega/dx$:

$$\mathcal{H} \equiv \frac{H}{H_0\sqrt{\Omega_M}} = \sqrt{(1+z)^3 + \omega + \frac{1}{2\lambda}\left(\frac{d\omega}{dx}\right)^2} \,. \tag{1.12.19}$$

---

[4]This approach is followed by D. Huterer and H. V. Peiris, *Phys. Rev. D* **75**, 083502 (2007) [astro-ph/0610427]; R. Crittenden, E. Majerotto, and F. Piazza, astro-ph/0702003.

We will also need the differential equation for the redshift:

$$\frac{dz}{dx} = -\mathcal{H}(1+z) .$$

(1.12.20)

In general, even if we wrote all derivatives with respect to x in terms of derivatives with respect to z, to solve these equations we would need initial conditions for ω and $d\omega/dz$ at some initial z, which with λ would give a three-parameter set of solutions. However, assuming that for large redshift the energy density is dominated by matter rather than vacuum energy (which as we shall see is the case), the derivative $d\omega/dx$ sufficiently late in the matter-dominated era becomes quite insensitive to initial conditions.[5] For $z \gg 1$, Eq. (1.12.19) gives

$$\mathcal{H} \to (1+z)^{3/2} ,$$

(1.12.21)

and (1.12.17) and (1.12.20) then have the solution

$$1+z \to \left(\frac{3x}{2}\right)^{-2/3} , \qquad \frac{d\omega}{dx} \to -\frac{\lambda x}{3} .$$

(1.12.22)

(An integration constant in the solution for z has been absorbed into the definition of x, setting the zero of time. An integration constant in the solution for $d\omega/dx$ has been dropped, because it gives a term in $d\omega/dx$ that dies away with increasing time as $x^{-2} \propto t^{-2}$.) The free parameters in our solution are then λ, together with the value of ω at some arbitrary initial value x_1 of x, taken sufficiently small so that at x_1 the energy density is dominated by matter rather than vacuum energy. (Note that the constant V_0 appears nowhere in these equations; it contributes a term to $\omega(x_1)$, but there is no need to isolate this term.) One must adopt various trial values of λ and $\omega(x_1)$; use Eq. (1.12.22) to calculate $1 + z$ and $d\omega/dx$ at $x = x_1$; with these initial conditions, integrate the differential equations (1.12.17) and (1.12.20) numerically from x_1 to a value x_0 where $z = 0$; and then if we like calculate the values of $\Omega_V = 1 - \Omega_M$ and the present value w_0 of the ratio p_φ/ρ_φ for this particular solution,[6] using

$$\frac{\Omega_\Lambda}{\Omega_M} = \omega(x_0) + \frac{1}{2\lambda}\left(\frac{d\omega}{dx}\right)^2_{x=x_0} , \qquad w_0 = \frac{(d\omega/dx)^2_{x=x_0} - 2\lambda\omega(x_0)}{(d\omega/dx)^2_{x=x_0} + 2\lambda\omega(x_0)} .$$

(1.12.23)

---

[5] R. Cahn, private communication. Cahn has also shown that the approximation of neglecting the second derivative term in the field equation does not work well in this context.

[6] As already mentioned, with models of this sort one can only have $w_0 > -1$. To compare the case $w_0 < -1$ with observation, it is necessary to adopt a model with the opposite sign for the derivative term in the action (1.12.1). The analysis given here can then be applied, with only obvious sign changes here and there.

The ratio of the dark energy at a given time to its value at the present is

$$\xi \equiv \frac{\rho_V(t)}{\rho_V(t_0)} = \frac{(d\omega(x)/dx)^2 + 2\lambda\,\omega(x)}{(d\omega/dx)^2_{x=x_0} + 2\,\lambda\omega(x_0)} \tag{1.12.24}$$

For instance, if we take $\Omega_\Lambda = 1 - \Omega_M = 0.76$ and $w_0 = -0.777$, the ratio ξ of the dark energy density to its value at present rises to 1.273 at $z = 1$ and to 1.340 at infinite redshift.[7] The leveling off of $\xi(z)$ for large z occurs because the growth of the matter density for increasing redshift makes the expansion rate grow, so that the friction term $3H\dot\varphi$ in Eq. (1.12.4) freezes the value of the scalar field at early times.

It should not be thought that the leveling off of the dark energy for large z for the potential (1.12.15) means that in analyzing dark energy observations with this potential one must give up the idea motivating theories of quintessence, that the vacuum energy is now small because the universe is old. In fact, for the potential $V(\varphi) \propto \varphi^{-\alpha}$, for typical initial conditions the quintessence energy drops at first precipitously, and then levels off while the scalar field rolls slowly down the potential until the field approaches the tracker solution, with the tracker solution not reached by the present time if α is small.[8] The condition $|V_0''\dot\varphi_0| \ll H_0|V_0'|$ for treating this potential as linear over a time of order $1/H_0$ is satisfied if $\alpha(1+\alpha)\varphi_0^{-2} \ll 8\pi\,G$, which in light of Eq. (1.12.9) is likely to be satisfied if $\alpha < 1$.

Another possible two-parameter model is provided by the same potential, $V(\varphi) \propto \varphi^{-\alpha}$, but now under the assumption that the tracker solution is reached by some early time (say, for $z \leq 10$) in the matter-dominated era. With this assumption the observable history of dark energy is insensitive to initial conditions, so the model has just two parameters: M and α. The equations of this model can be put in dimensionless form by writing the coupling constant of this potential in terms of a dimensionless parameter β as

$$M^{4+\alpha} \equiv \beta\,\Omega_M\,H_0^2\,(8\pi\,G)^{-1-\alpha/2} \tag{1.12.25}$$

and replacing the dependent variable t and independent variable φ with dimensionless variables x and f, defined by

$$t \equiv x/H_0\sqrt{\Omega_M}\,, \qquad \varphi(t) \equiv f(x)/\sqrt{8\pi\,G}\,. \tag{1.12.26}$$

---

[7]Numerical results for various values of redshift are given in Table 1.1. These results for the linear potential were calculated by R. Cahn, private communication.

[8]Steinhardt, Wang, and Zlatev, ref. 2.

The field equation (1.12.4) (with no slow roll approximation) in the era dominated by matter and vacuum energy then takes the form

$$\frac{d^2f}{dx^2} + 3\mathcal{H}\frac{df}{dx} - \alpha\beta f^{-\alpha-1} = 0 , \tag{1.12.27}$$

where

$$\mathcal{H} \equiv H/\sqrt{\Omega_M}H_0 = \sqrt{\frac{1}{6}\left(\frac{df}{dx}\right)^2 + \frac{\beta}{3}f^{-\alpha} + (1+z)^3} \tag{1.12.28}$$

in which we also need

$$\frac{dz}{dx} = -\mathcal{H}(1+z) . \tag{1.12.29}$$

Because the large z solution (1.12.7) is an attractor, the initial conditions introduce no new free parameters; in terms of these dimensionless variables, the initial conditions are that, for $x \to 0$,

$$f \to \left[\frac{\alpha\beta(\alpha+2)^2x^2}{2(\alpha+4)}\right]^{1/(\alpha+2)} , \quad 1+z \to \left(\frac{2}{3x}\right)^{2/3} . \tag{1.12.30}$$

We need to integrate the equations (1.12.27) and (1.12.29) from some small x (say, $x = 0.01$) to a value x_0 at which $z = 0$, with the initial conditions (1.12.30), and then evaluate $\Omega_M = 1 - \Omega_\Lambda$ from the condition that $\mathcal{H}(x_0) = 1/\sqrt{\Omega_M}$. We can also evaluate the present value w_0 of $w \equiv p_\varphi/\rho_\varphi$ from the formula

$$w_0 = \frac{f'^2(x_0)f(x_0)^\alpha/2\beta - 1}{f'^2(x_0)f(x_0)^\alpha/2\beta + 1} , \tag{1.12.31}$$

and then replace the parameters α and β with Ω_M and w_0. For instance, if we arbitrarily take $\alpha = 1$, then to get the realistic value $\Omega_M = 0.24$ we must take $\beta = 9.93$, in which case $w_0 = -0.777$. Of course, we can get any other values of w_0 greater than -1 by choosing different values of α and re-adjusting β to give the same value of Ω_M (though for small α, the range of initial conditions that allow the tracker solution to be reached well before the present is relatively small.) For instance, for $\alpha = 1/2$ we must take $\beta = 7.82$ to have $\Omega_M = 0.24$, and in this case we calculate that $w_0 = -0.87$. (For the case $w < -1$, see footnote 6.) The ratios of dark energy to its value at present calculated in this way for $\Omega_M = 0.24$ and $w_0 = -0.777$ are shown in Table 1.1, along with the values calculated with the same choice of Ω_M and w_0 for both the case of constant w and for the linear potential (1.12.15). The tracker and linear models evidently represent opposite extreme assumptions about the time-dependence of dark

Table 1.1: Ratio of dark energy to its present value, for the tracker solution with the potential (1.12.5), and for the linear potential (1.12.15), calculated for $\Omega_M = 1 - \Omega_\Lambda = 0.24$ and $w_0 = -0.777$, compared with the results for a constant $w = -0.777$.

z	tracker	linear	constant w
0	1	1	1
0.1	1.067	1.062	1.066
0.5	1.347	1.200	1.312
1	1.712	1.273	1.590
2	2.469	1.318	2.086
3	3.224	1.331	2.528
$\gg 1$	$\gg 1$	1.340	$\gg 1$

energy, but both are better motivated physically than the assumption of a constant w.

1.13 Horizons

Modern cosmological theories can exhibit horizons of two different types, which limit the distances at which past events can be observed or at which it will ever be possible to observe future events. These are called by Rindler[1] *particle horizons* and *event horizons*, respectively.

According to Eq. (1.2.2), if the big bang started at a time $t = 0$, then the greatest value $r_{max}(t)$ of the Robertson–Walker radial coordinate from which an observer at time t will be able to receive signals traveling at the speed of light is given by the condition

$$\int_0^t \frac{dt'}{a(t')} = \int_0^{r_{max}(t)} \frac{dr}{\sqrt{1 - Kr^2}} \tag{1.13.1}$$

Thus there is a particle horizon unless the integral $\int dt/a(t)$ does not converge at $t = 0$. It does converge in conventional cosmological theories; whatever the contribution of matter or vacuum energy at the present, it is likely that the energy density will be dominated by radiation at early times, in which case $a(t) \propto t^{1/2}$, and the integral converges. The proper distance

---

[1] W. Rindler, *Mon. Not. Roy. Astron. Soc.* **116**, 663 (1956).

of the horizon is given by Eq. (1.1.15) and (1.13.1) as

$$d_{\max}(t) = a(t) \int_0^{r_{\max}(t)} \frac{dr}{\sqrt{1 - Kr^2}} = a(t) \int_0^t \frac{dt'}{a(t')} . \qquad (1.13.2)$$

For instance, during the radiation-dominated era $a(t) \propto t^{1/2}$, so $d_{\max}(t) = 2t = 1/H$. Well into the matter-dominated era most of the integral over time in Eq. (1.13.1) comes from a time when $a \propto t^{2/3}$, so that $d_{\max}(t) \simeq 3t = 2/H$. At present most of the integral over t' comes from a period when the expansion is dominated by matter and the vacuum energy, and perhaps curvature as well. According to Eq. (1.5.41), the particle horizon distance at present is

$$d_{\max}(t_0) = \frac{1}{H_0} \int_0^1 \frac{dx}{x^2 \sqrt{\Omega_\Lambda + \Omega_K x^{-2} + \Omega_M x^{-3}}} . \qquad (1.13.3)$$

We will see in Chapter 4 that there may have been a time before the radiation-dominated era in which there was nothing in the universe but vacuum energy, in which case the particle horizon distance would actually be infinite. But as far as telescopic observations are concerned, Eq. (1.13.3) gives the proper distance beyond which we cannot now see.

Just as there are past events that we cannot now see, there may be events that we never will see. Again returning to Eq. (1.2.2), if the universe re-collapses at a time T, then the greatest value r_{MAX} of r from which an observer will be able to receive signals traveling at the speed of light emitted at any time later than t is given by the condition

$$\int_t^T \frac{dt'}{a(t')} = \int_0^{r_{\mathrm{MAX}}(t)} \frac{dr}{\sqrt{1 - Kr^2}} \qquad (1.13.4)$$

Even if the future is infinite, if the integral $\int dt/a(t)$ converges at $t = \infty$ there will be an event horizon given by

$$\int_t^\infty \frac{dt'}{a(t')} = \int_0^{r_{\mathrm{MAX}}(t)} \frac{dr}{\sqrt{1 - Kr^2}} \qquad (1.13.5)$$

Since co-moving sources are labeled with a fixed value of r, the condition $r < r_{\mathrm{MAX}}$ limits the events occurring at time t that we can ever observe. In the case where the universe does not recollapse, the proper distance to the event horizon is given by

$$d_{\mathrm{MAX}}(t) = a(t) \int_0^{r_{\mathrm{MAX}}(t)} \frac{dr}{\sqrt{1 - Kr^2}} = a(t) \int_t^\infty \frac{dt'}{a(t')} . \qquad (1.13.6)$$

In the absence of a cosmological constant, $a(t)$ grows like $t^{2/3}$, and the integral diverges, so that there is no event horizon. But with a cosmological constant $a(t)$ will eventually grow as $\exp(Ht)$ with $H = H_0 \Omega_\Lambda^{1/2}$ constant, and there really is an event horizon, which approaches the value $d_{\text{MAX}}(\infty) = 1/H$. As time passes all sources of light outside our gravitationally bound Local Group will move beyond this distance, and become unobservable. The same is true for the quintessence theory described in the previous section. In that case $a(t)$ eventually grows as $\exp(\text{constant} \times t^{2/(2+\alpha/2)})$, so for any $\alpha \geq 0$ the integral (1.13.6) again converges.

   If a source is at a radial coordinate r in a Robertson–Walker coordinate system based on us, then we are at a radial coordinate r in a Robertson–Walker coordinate system based on the source. Hence Eq. (1.13.4) or (1.13.5) also gives the greatest radial coordinate to which, starting at time t, we will ever be able to travel.

2

The Cosmic Microwave Radiation Background

Before the mid-1960s by far the greatest part of our information about the structure and evolution of the universe came from observations of the redshifts and distances of distant galaxies, discussed in the previous chapter. In 1965 a nearly isotropic background of microwave radiation was discovered, which has provided a wealth of new cosmological data. After reviewing the expectations and discovery of this radiation, this chapter will explore some of its implications. We will only be able to give a first look at the anisotropies in this radiation in this chapter. In Chapter 7 we will return to this very important topic, applying the analysis of the evolution of cosmological perturbations presented in Chapters 5 and 6, and in Chapter 10 we will consider the origin of these perturbations in the very early universe.

2.1 Expectations and discovery of the microwave background

The work done by pressure in an expanding fluid uses heat energy drawn from the fluid. The universe is expanding, so we expect that in the past matter was hotter as well as denser than at present. If we look far enough backward in time we come to an era when it was too hot for electrons to be bound into atoms. At sufficiently early times the rapid collisions of photons with free electrons would have kept radiation in thermal equilibrium with the hot dense matter. The number density of photons in equilibrium with matter at temperature T at photon frequency between ν and $\nu + d\nu$ is given by the *black-body spectrum*:

$$n_T(\nu)d\nu = \frac{8\pi \nu^2 \, d\nu}{\exp(h\nu/k_B T) - 1},$$

(2.1.1)

where h is the original Planck's constant (which first made its appearance in a formula equivalent to this one), and k_B is Boltzmann's constant. (Recall that we are using units with $c = 1$.)

As time passed, the matter became cooler and less dense, and eventually the radiation began a free expansion, but *its spectrum has kept the same form*. We can see this most easily under an extreme assumption, that there was a time t_L when radiation suddenly went from being in thermal equilibrium with matter to a free expansion. (The subscript L stands for "last scattering.") Under this assumption, a photon that has frequency ν at some later time t when photons are traveling freely would have had frequency

101

$va(t)/a(t_L)$ at the time the radiation went out of equilibrium with matter, and so the number density at time t of photons with frequency between v and $v + dv$ would be

$$n(v,t)\,dv = \Big(a(t_L)/a(t)\Big)^3 n_{T(t_L)}\Big(va(t)/a(t_L)\Big)\,d(va(t)/a(t_L)) , \quad (2.1.2)$$

with the factor $\Big(a(t_L)/a(t)\Big)^3$ arising from the dilution of photons due to the cosmic expansion. Using Eq. (2.1.1) in (2.1.2), we see that the redshift factors $a(t)/a(t_L)$ all cancel except in the exponential, so that the number density at time t is given by

$$n(v,t)dv = \frac{8\pi v^2\,dv}{\exp\left(hv/k_B T(t)\right) - 1} = n_{T(t)}(v)\,dv , \quad (2.1.3)$$

where

$$T(t) = T(t_L)a(t_L)/a(t) . \quad (2.1.4)$$

Thus the photon density has been given by the black-body form even after the photons went out of equilibrium with matter, but with a redshifted temperature (2.1.4).

This conclusion is obviously unchanged if the transition from opacity to transparency occupied a finite time interval, as long as the interactions of photons with matter during this interval are limited to elastic scattering processes in which photon frequencies are not changed. This is a very good approximation. We will see in Section 2.3 that the last interaction of photons with matter (until near the present) took place at a time when the cosmic temperature T was of order 3,000 K, when by far the most important interaction was the elastic scattering of photons with electrons, in which the fractional shift of photon frequency was of order $k_B T/m_e c^2 \approx 3 \times 10^{-7}$. In the following section we shall show that, because of the large photon entropy, even the small shift of photon frequency in elastic scattering and the relatively infrequent inelastic interactions of photons with hydrogen atoms had almost no effect on the photon spectrum.

It was George Gamow and his collaborators who first recognized in the late 1940s that the universe should now be filled with black-body radiation.[1] The first plausible estimate of the present temperature of this radiation was

[1]G. Gamow, *Phys. Rev.* **70**, 572 (1946); R. A. Alpher, H. A. Bethe, and G. Gamow, *Phys. Rev.* **73**, 803 (1948); G. Gamow, *Phys. Rev.* **74**, 505 (1948); R. A. Alpher and R. C. Herman, *Nature* **162**, 774 (1948); R. A. Alpher, R. C. Herman, and G. Gamow, *Phys. Rev.* **74**, 1198 (1948); *ibid* **75**, 332A (1949); *ibid* **75**, 701 (1949); G. Gamow, *Rev. Mod. Phys.* **21**, 367 (1949); R. A. Alpher, *Phys. Rev.* **74**, 1577 (1948); R. A. Alpher and R. C. Herman, *Phys. Rev.* **75**, 1089 (1949).

made in 1950 by Ralph Alpher and Robert Herman.[2]   On the basis of considerations of cosmological nucleosynthesis, to be discussed in Section 3.2, they found a present temperature of 5 K. This work was largely forgotten in subsequent decades, until in 1965 a group at Princeton started to search for a cosmic radiation background left over from the early universe. They had only a rough idea of the temperature to be expected, based on a nucleosynthesis calculation of P. J. E. Peebles, which suggested a value of 10 K.[3] Before they could complete their experiment the radiation was discovered in a study of noise backgrounds in a radio telescope by Arno Penzias and Robert Wilson,[4] who published their work along with a companion article[5] by the Princeton group explaining its possible cosmological significance.[6]

Originally Penzias and Wilson could only report that the antenna temperature at a wavelength 7.5 cm was 3.5±1.0 K, meaning that the intensity of the radiation *at this one wavelength* agreed with Eq. (2.1.1) for this temperature. This of course did not show that they were observing black-body radiation. Then Roll and Wilkinson[7] measured the radiation intensity at a wavelength of 3.2 cm, finding an antenna temperature of 3.0 ± 0.5 K, in agreement with what would be expected for black-body radiation at the temperature measured by Penzias and Wilson. In the following few years a large number of measurements were made by other radio astronomers at other wavelengths. These measurements also gave antenna temperatures at the wavelengths being studied around 3 K, with uncertainties that gradually improved to of order 0.2 K. But this also did not establish the black-body nature of the radiation, because these measurements were all at wavelengths greater than about 0.3 cm, where the black-body energy distribution $h\nu n_T(\nu)$ with $T \approx 3$ K has its maximum. For these long wavelengths the argument of the exponential is small, and Eq. (2.1.1) gives

$$h\nu\,n_T(\nu) \simeq 8\pi\nu^2 k_B T \; , \qquad (2.1.5)$$

This is the *Rayleigh–Jeans formula* of classical statistical mechanics, but it describes the long-wavelength distribution of radiant energy under vari-

---

[2]R. A. Alpher and R. C. Herman, *Rev. Mod. Phys.* **22**, 153 (1950)

[3]This work was never published. According to A. Guth, *The Inflationary Universe* (Perseus Books, Reading, MA, 1997), Peebles' paper was rejected by The Physical Review, apparently because of the issue of the credit to be given to earlier work by R. Alpher, G. Gamow, and R. Herman. This earlier work and the subsequent work of Peebles and others is briefly described here in Section 3.2.

[4]A. A. Penzias and R. W. Wilson, *Astrophys. J.* **142**, 419 (1965).

[5]R. H. Dicke, P. J. E. Peebles, P. G. Roll, and D. T. Wilkinson, *Astrophys. J.* **142**, 414 (1965).

[6]For a more detailed history of these developments, see A. Guth, *op. cit.*, and S. Weinberg, *The First Three Minutes* (Basic Books, New York, 1977; second edition 1993).

[7]P. G. Roll and D. T. Wilkinson, *Phys. Rev. Lett.* **16**, 405 (1966).

ous circumstances more general than black-body radiation. For instance, black-body radiation diluted by an expansion that preserves the energy of individual photons also has $h\nu\, n(\nu) \propto \nu^2$ for low frequencies.[8] To confirm that the cosmic microwave background radiation is really described by the black-body formula, it was necessary to see at least the beginning of the exponential fall-off of $n_T(\nu)$ for wavelengths shorter than about 0.3 cm.

This was difficult because the earth's atmosphere becomes increasingly opaque at wavelengths shorter than about 0.3 cm. However, there had been a measurement of the radiation temperature at a wavelength 0.264 cm in 1941, long before the discovery by Penzias and Wilson. In between the star ζ Oph and the earth there is a cloud of cold molecular gas, whose absorption of light produces dark lines in the spectrum of the star. In 1941 W. S. Adams,[9] following a suggestion of Andrew McKellar, found two dark lines in the spectrum of ζ Oph that could be identified as due to absorption of light by cyanogen (CN) in the molecular cloud. The first line, observed at a wavelength of 3,874.62 Å, could be attributed to absorption of light from the CN ground state, with rotational angular momentum $J = 0$, leading to the component of the first vibrationally excited state with $J = 1$. But the second line, at 3,874.00 Å, represented absorption from the $J = 1$ rotationally excited vibrational ground state, leading to the $J = 2$ component of the first vibrationally excited state.[10] From this, McKellar concluded[11] that a fraction of the CN molecules in the cloud were in the first excited rotational component of the vibrational ground state, which is above the true $J = 0$ ground state by an energy $hc/(0.264\ \text{cm})$. and from this fraction he estimated an equivalent molecular temperature of 2.3 K. Of course, he did not know that the CN molecules were being excited by radiation, much less by black-body radiation. After the discovery by Penzias and Wilson several astrophysicists[12] independently noted that the old Adams–McKellar result could be explained by radiation with a

---

[8]The sunlight falling on the earth's surface provides a pretty good example of dilute black-body radiation; it is described by the Planck formula (2.1.1), with $T \approx 6,000$ K the temperature of the sun's surface, but with the right-hand side of Eq. (2.1.1) multiplied by a factor $(R_\odot/r)^2$, where $R_\odot$ is the radius of the sun and r is the distance from the sun to the earth.

[9]W. S. Adams, *Astrophys. J.* **93**, 11 (1941)

[10]Today the wavelengths of these two lines are more accurately known to be 3,874.608 and 3,873.998 Å. There is another line at 3875.763 Å, produced by transitions from the $J = 1$ rotationally excited vibrational ground state to the $J = 0$ component of the first vibrationally excited state.

[11]A. McKellar, *Publs. Dominion Astrophys. Observatory (Victoria, B.C.)* **7**, 251 (1941).

[12]G. Field, G. H. Herbig, and J. L. Hitchcock, *Astron. J.* **71**, 161 (1966); G. Field and J. L. Hitchcock, *Phys. Rev. Lett.* **16**. 817 (1966); *Astrophys. J.* **146**, 1 (1966); N. J. Woolf, quoted by P. Thaddeus and J. F. Clauser, *Phys. Rev. Lett.* **16**, 819 (1966); I. S. Shklovsky, Astronomicheskii Tsircular No. **364** (1966).

black-body temperature at wavelength 0.264 cm in the neighborhood of 3 K. Theoretical analysis showed that nothing else could explain the excitation of this rotational state.[13] This interpretation was then borne out by continuing observations on this and other absorption lines in CN as well as CH and CH$^+$ in the spectrum of ζ Oph and other stars.[14]

The black-body spectrum of the cosmic microwave radiation background began to be established by balloon-borne and rocket-borne observations above the earth's atmosphere at wavelengths below 0.3 cm. For some years there were indications of an excess over the black-body formula at these short wavelengths. It was clearly necessary to do these observations from space, but this is difficult; to measure the absolute value of the microwave radiation intensity it is necessary to compare the radiation received from space with that emitted by a "cold load" of liquid helium, which rapidly evaporates. Finally, the Planck spectrum of the microwave background was settled in the 1990s by observations with the FIRAS radiometer carried by the Cosmic Background Explorer Satellite (COBE), launched in November 1989.[15] When a slide showing the agreement of the observed spectrum with the Planck black-body spectrum was shown by J. C. Mather at a meeting of the American Astronomical Society in January 1990, it received a standing ovation. It was found that the background radiation has a nearly exact black-body spectrum in the wavelength range from 0.5 cm to 0.05 cm.[16] The comparison of observation with the black-body spectrum is shown in Figure 2.1. After six years of further analysis, the temperature was given as 2.725 ± 0.002 K (95% confidence).[17] Other observations at wavelengths between 3 cm and 75 cm and at 0.03 cm are all consistent with a Planck distribution at this temperature.[18]

The energy density in this radiation is given by

$$\int_0^\infty h\nu\, n(\nu)\, d\nu = a_B T^4 \tag{2.1.6}$$

where a_B is the radiation energy constant; in c.g.s. units,

$$a_B = \frac{8\pi^5 k_B^4}{15 h^3 c^3} = 7.56577(5) \times 10^{-15} \text{ erg cm}^{-3} \text{ deg}^{-4} \tag{2.1.7}$$

---

[13] Field *et al.*, ref. 12; Thaddeus and Clauser, ref. 12.

[14] For a summary of this early work with references to the original literature, see G&C, Table 15.1.

[15] J. C. Mather *et al.*, *Astrophys. J.* **354**, 237 (1990).

[16] J. C. Mather *et al.*, *Astrophys. J.* **420**, 439 (1994).

[17] J. C. Mather, D. J. Fixsen, R. A. Shafer, C. Mosier, and D. T. Wilkinson, *Astrophys. J.* **512**, 511 (1999). A 1999 review by G. F. Smoot, in *Proc. 3K Cosmology Conf.*, eds. A. Melchiorri *et al.* [astro-ph/9902027], gave a temperature 2.7377 ± 0.0038 K (95% confidence), but the result of Mather *et al.* seems to be the one usually quoted.

[18] For a review, see G. Sironi *et al.*, in *Proc. Third Sakharov Conf. – Moscow 2002* [astro-ph/0301354].

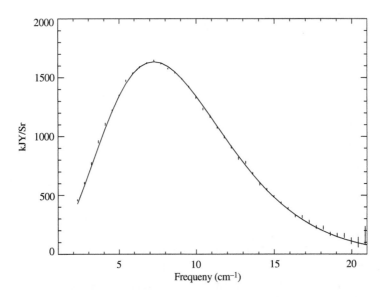

Figure 2.1: Comparison of the intensity of radiation observed with the FIRAS radiometer carried by COBE with a black-body spectrum with temperature 2.728 K, from D. J. Fixsen *et al.*, *Astrophys. J.* **473**, 576 (1996) [astro-ph/9605054]. The vertical axis gives the intensity in kiloJansky per steradian (one Jansky equals 10^{-23} erg cm^{-2} s^{-1} Hz^{-1}); the horizontal axis gives the reciprocal wavelength in cm^{-1}. The 1σ experimental uncertainty in intensity is indicated by the tiny vertical bars; the uncertainty in wavelength is negligible.

Using $T = 2.725$ K, this gives an equivalent mass density (reverting to $c = 1$)

$$\rho_{\gamma 0} = a_B T_{\gamma 0}^4 = 4.64 \times 10^{-34} \text{g cm}^{-3} .$$

Taking the ratio of this with the critical density (1.5.28) gives

$$\Omega_\gamma \equiv \frac{\rho_{\gamma 0}}{\rho_{0\text{crit}}} = 2.47 \times 10^{-5} h^{-2} \tag{2.1.8}$$

We will see in Section 3.1 that the photons are accompanied with neutrinos and antineutrinos of three different types, giving a total energy density in radiation (that is, in massless or nearly massless particles):

$$\rho_{R0} = \left[1 + 3 \left(\frac{7}{8} \right) \left(\frac{4}{11} \right)^{4/3} \right] \rho_{\gamma 0} = 7.80 \times 10^{-34} \text{g cm}^{-3} , \tag{2.1.9}$$

or in other words, using Eq. (1.5.28),

$$\Omega_R \equiv \frac{\rho_{R0}}{\rho_{0,\text{crit}}} = 4.15 \times 10^{-5} h^{-2} . \tag{2.1.10}$$

106

We see that ρ_{R0} is much less than the critical mass density needed to give $K = 0$, and much less even than the mass density of ordinary matter seen in stars. It is for this reason that we have generally neglected Ω_R in calculating luminosity distances as a function of redshift.

On the other hand, even at present the *number* density of photons is relatively very large. Eq. (2.1.1) gives

$$n_{\gamma 0} = \int_0^\infty \frac{8\pi v^2 \, dv}{\exp\left(hv/k_B T\right) - 1} = \frac{30 \, \zeta(3)}{\pi^4} \frac{a_B T^3}{k_B}$$

$$= 0.3702 \frac{a_B T^3}{k_B} = 20.28 \left[T(\deg K)\right]^3 \text{ cm}^{-3}, \qquad (2.1.11)$$

where $\zeta(3) = 1.202057\ldots$ For $T = 2.725$ K this gives a present number density

$$n_{\gamma 0} = 410 \text{ photons/cm}^3 . \qquad (2.1.12)$$

This is much larger than the present number density n_{B0} of nucleons, given by

$$n_{B0} = \frac{3\Omega_B H_0^2}{8\pi \, Gm_N} = 1.123 \times 10^{-5} \, \Omega_B \, h^2 \text{ nucleons/cm}^3 . \qquad (2.1.13)$$

Both n_γ and n_B vary with time as $a^{-3}(t)$, so the ratio n_γ/n_B has been the same at least during the whole period that photons have been traveling freely.

* * *

There is an effect of the cosmic microwave background that has long been expected but has been difficult to observe. A cosmic ray proton of moderate energy striking a photon in the cosmic microwave background can only scatter the photon, a process whose rate is proportional to the square of the fine structure constant $\alpha \simeq 1/137$. However, if the proton has sufficiently high energy then it is also possible for the photon to be converted into a π meson in the reactions $\gamma + p \rightarrow \pi^0 + p$ or $\gamma + p \rightarrow \pi^+ + n$, processes whose rate is proportional to α, not α^2. Assuming that high energy cosmic rays come to us from outside our galaxy, we therefore expect a dip in the spectrum of cosmic ray protons at an energy where the cross section for these processes becomes appreciable.[19] Although some pions can be produced at lower energy, the effective threshold is at a value of the total energy W of the

---

[19]K. Greisen, *Phys. Rev. Lett.* **16**, 748 (1966); G. T. Zatsepin and V. A. Kuzmin, *Pis'ma Sh. Exsp. Teor. Fiz.* **4**, 114 (1966) [transl. *Sov. Phys. JETP Lett.* **4**, 78 (1966)]; F. W. Stecker, *Phys. Rev. Lett.* **21**, 1016 (1968).

initial proton and photon in the center-of-mass system equal to $m_\Delta = 1232$ MeV, the mass of the pion–nucleon resonance with spin-parity $3/2^+$ and isospin $3/2$. The center of mass energy is

$$W = \left(\left(q + \sqrt{p^2 + m_p^2} \right)^2 - \left| \mathbf{q} + \mathbf{p} \right|^2 \right)^{1/2}$$

$$\simeq \left(2qp\left(1 - \cos\theta \right) + m_p^2 \right)^{1/2} , \tag{2.1.14}$$

where $\mathbf{q}$ and $\mathbf{p}$ are the initial photon and proton momenta (with $p \gg m_p \gg q$) and θ is the angle between them. The threshold condition that $W > m_\Delta$ thus requires that

$$qp(1 - \cos\theta) > m_\Delta^2 - m_p^2 . \tag{2.1.15}$$

The typical energy of photons in black-body radiation at temperature $T_{\gamma 0} = 2.725$K is $\rho_{\gamma 0}/n_{\gamma 0} \simeq 6 \times 10^{-4}$ eV, while the largest value for $1 - \cos\theta$ is 2, so the effective threshold is roughly at a proton energy

$$p_{\text{threshold}} \approx \frac{m_\Delta^2 - m_p^2}{2\rho_{\gamma 0}/n_{\gamma 0}} \simeq 10^{20} \text{ eV} . \tag{2.1.16}$$

This effect is not easy to see. The flux of cosmic ray protons with energies between E and $E + dE$ goes roughly as $E^{-3} \, dE$, so there are few protons at these very high energies, roughly one per square kilometer per year above 10^{19} eV and 0.01 per square kilometer per year above 10^{20} eV. At these rates, direct observation is clearly impossible, and the cosmic rays have had to be studied indirectly by observation at ground level of the large showers of photons and charged particles that they produce. Also, there is a smooth distribution of photon energies and directions, so one is not looking for a sharp cut-off at 10^{20} eV, but rather for a dip below the E^{-3} curve at around this energy.[20] No such effect was observed by the Akeno Giant Air Shower Array,[21] but a subsequent analysis of this and other observations showed the effect.[22] More recently signs of a drop appeared in the High Resolution Fly's Eye experiment.[23] In 2006 this group announced the observation of a "sharp suppression" of the primary cosmic ray spectrum at an energy of 6×10^{19} GeV, just about where expected.[24]

---

[20]For instance, see I. F. M. Albuquerque and G. F. Smoot, *Astroparticle Phys.* **25**, 375 (2006) [astro-ph/0504088].

[21]M. Takeda *et al. Phys. Rev. Lett.* **81**, 1163 (1998).

[22]J. N. Bahcall and E. Waxman, *Phys. Lett.* **B556**, 1 (2003) [hep-ph/0206217].

[23]R. U. Abbasi *et al.*, *Phys. Rev. Lett.* **92**, 151101 (2004) [astro-ph/0208243]; *Phys. Lett.* **B619**, 271 (2005) [astro-ph/0501317].

[24]G. B. Thompson, for HiRes Collaboration, in *Proc. Quarks '06 Conf.* [astro-ph/0609403]; R. U. Abbasi *et al.*, astro-ph/0703099.

2.2 The equilibrium era

As already remarked in the previous section, if we look back in time sufficiently far, we surely must come to an era when the temperature and density were sufficiently high so that radiation and matter were in thermal
equilibrium. We will now consider this era, jumping over the intermediate
time when radiation was going out of equilibrium with matter, which will
be the subject of the next section.

As we saw in the previous section, in a free expansion of photons, the
frequency distribution preserves the Planck black-body form (2.1.1), but
with a temperature that falls as $1/a(t)$. On the other hand, in a free expansion of non-relativistic particles such as electrons or nuclei, the momentum
distribution preserves the Maxwell–Boltzmann form, $n(p)dp \propto$
$\exp(-p^2/2mk_BT)$, but since (as shown in Eq. (1.1.23)) the momentum of
any particle decreases as the universe expands with $p \propto 1/a$, the temperature of the Maxwell–Boltzmann distribution decreases as $1/a^2(t)$. So if
radiation is in equilibrium with matter, who wins? Does the temperature
decrease as $1/a(t)$, or $1/a^2(t)$, or something more complicated?

The issue is settled democratically, on the basis of one particle, one vote.
Since there are so many more photons than electrons or nucleons, the photons win, and the temperature decreases almost exactly as $1/a(t)$.

We can see this in more detail by applying the second law of thermodynamics. In equilibrium both the entropy and the baryon number (that
is, at temperatures $\ll 10^{13}$ K, the number of protons and neutrons) in any
co-moving volume were constant, and so their ratio, the entropy per baryon,
was also constant. It is convenient to write the entropy per baryon as $k_B\sigma$,
with k_B the Boltzmann constant and σ dimensionless. The second law of
thermodynamics tells us that this satisfies

$$d\left(k_B\sigma\right) = \frac{d(\epsilon/n_B) + p\,d(1/n_B)}{T} , \qquad (2.2.1)$$

where n_B is the baryon number density (so that $1/n_B$ is the volume per
baryon), ϵ is the thermal energy density and p is the pressure. For simplicity,
let us consider an ideal gas of photons and non-relativistic particles (mostly
protons, helium nuclei, and electrons), with a fixed number (of order unity)
$\mathcal{N}$ of the non-relativistic particles per baryon. Then

$$\epsilon = a_B T^4 + \frac{3}{2}n_B \mathcal{N} k_B T , \qquad p = \frac{1}{3}a_B T^4 + n_B \mathcal{N} k_B T , \qquad (2.2.2)$$

with a_B the radiation energy constant (2.1.7). The solution of Eq. (2.2.1) is
here

$$\sigma = \frac{4a_B T^3}{3n_B k_B} + \mathcal{N}\ln\left(\frac{T^{3/2}}{n_B C}\right) , \qquad (2.2.3)$$

where C is an arbitrary constant of integration. It is the quantity σ that remains constant in thermal equilibrium.

We saw in the previous section that the first term in Eq. (2.2.3) is larger than 10^8 at present. (Compare Eqs. (2.1.12) and (2.1.13).) Let us tentatively assume that this quantity was also much larger than unity when photons were in equilibrium with matter. Since σ was constant in this era, this means that the ratio T^3/n_B would also have been very close to constant during the era of equilibrium, unless the quantity $T^{3/2}/n_B$ changed by an enormous amount. For instance, if at some time in the equilibrium era the first term in Eq. (2.2.3) were of order 10^8, then in order for the first term in Eq. (2.2.3) to have changed by even 0.01%, to keep σ constant the value of $T^{3/2}/n_B$ would have had to change by a factor $e^{10000/\mathcal{N}}$. We are not going to be considering such enormous density or temperature ratios here, so we can conclude that T^3/n_B was essentially constant in thermal equilibrium. We saw in the previous section that this ratio was also constant when the photons were traveling freely, and also when photons were interacting only by purely elastic Thomson scattering, so it has been close to constant from the beginning of the era considered here to the present. If we define the constant C in Eq. (2.2.3) to equal the value of $T^{3/2}/n_B$ at some typical time during the era of equilibrium, then the entropy per baryon throughout this era can be taken as

$$\sigma = \frac{4a_B T^3}{3n_B k_B} = \frac{3.60\, n_{\gamma 0}}{n_{B0}} = 1.31 \times 10^8 h^{-2}\Omega_B^{-1}. \tag{2.2.4}$$

The conservation of baryon number tells us that $n_B a^3$ is constant, so the constancy of (2.2.3) has the consequence that $T \propto 1/a$. (Note incidentally that if the first term in Eq. (2.2.3) were of order 10^{-8} instead of greater than 10^8, then it would be the logarithm in the second term that would have to remain constant during thermal equilibrium, in which case $T \propto n_B^{2/3} \propto a^{-2}$, as expected if non-relativistic particles dominate the thermal evolution.)

We can now see why the black-body spectrum with $T \propto 1/a$ is preserved as photons go out of equilibrium with matter, even when we take into account small inelastic effects, like the loss of energy to electron recoil in photon–electron scattering. Photons effectively stop gaining or losing energy to matter when $\Gamma_\gamma < H$, where $H \equiv \dot{a}/a$ and Γ_γ is the rate at which an individual photon loses or gains an energy $k_B T$ through scattering on electrons (to be calculated later). But the rate Γ_e at which an individual electron would gain or lose an energy $k_B T$ by scattering on photons is greater than Γ_γ by a factor $n_\gamma/n_e > 10^8$, so at the time that Γ_γ drops below H we still have $\Gamma_e \gg H$. Thus instead of the electron kinetic energies decreasing like $1/a^2$, as they would in a free expansion, they continue to remain in thermal equilibrium with the photons. While in equilibrium with matter the photon temperature goes as $1/a$, and we saw in the previous section that when they stop interacting with matter the temperature continues to drop like $1/a$, so through this whole period the electron temperature also drops as $1/a$, and the last few exchanges of energy that photons have with electrons do not affect the photon energy distribution.

According to results quoted in the previous section, the present ratio of the equivalent photon and neutrino mass density and the total matter mass density $\rho_{M0} = \Omega_M \, \rho_{0,\text{crit}}$ is

$$\frac{\rho_{R0}}{\rho_{M0}} = \frac{\Omega_R}{\Omega_M} = 4.15 \times 10^{-5} \Omega_M^{-1} h^{-2} \; . \tag{2.2.5}$$

As we have seen, the photon and neutrino energy density varied as $T^4 \propto a^{-4}$ even before the photons began their free expansion, while the density of pressureless matter varied as a^{-3}, so ρ_R/ρ_M varied as $1/a \propto T$. Therefore the energy density of photons and neutrinos was equal to that of matter when the temperature was

$$T_{EQ} = \frac{T_{\gamma 0} \Omega_M}{\Omega_R} = 6.56 \times 10^4 \, \text{K} \times \Omega_M h^2 \; , \tag{2.2.6}$$

and ρ_R was greater than ρ_M at earlier times. For $\Omega_M h^2 \simeq 0.15$, T_{EQ} is about 10^4 K.

Although collisions cannot change the *distribution* of photon energies as long as the photon number is much greater than the number of charged particles, at sufficiently high temperatures collisions can drastically change the energy of an individual photon. It is of some interest to work out when photons stopped exchanging energies of order $k_B T$ with electrons. The rate at which any individual photon is scattered by electrons is $\Lambda_\gamma = \sigma_T n_e c$, where n_e is the number density of electrons, and $\sigma_T = 0.66525 \times 10^{-24}$ cm^2 is the cross section for Thomson scattering, the elastic scattering of photons by non-relativistic electrons. As we will see in Section 3.2, about 76% of the matter of the universe in this era was ionized hydrogen, with the rest helium, completely ionized at temperatures above $20,000$ K, so with one electron per nucleon for hydrogen and half an electron per nucleon for helium, the net number of electrons per nucleon is $0.76 + (1/2)0.24 = 0.88$, and the number density of electrons at temperature T is $n_e \simeq 0.88 n_B = 0.88 n_{B0}(T/T_{\gamma 0})^3$, with a subscript zero as usual indicating the present moment. Using Eq. (2.1.13), the rate at which a photon is scattered by electrons is

$$\Lambda_\gamma = 0.88 n_{B0}\left(\frac{T}{T_{\gamma 0}}\right)^3 \sigma_T c = 1.97 \times 10^{-19} \, \text{s}^{-1} \times \Omega_B h^2 \left(\frac{T}{T_{\gamma 0}}\right)^3 . \tag{2.2.7}$$

But this is not the rate that governs the effectiveness of the energy exchange between matter and radiation. A photon with energy much less than $m_e c^2$ that strikes a non-relativistic electron will transfer a momentum to the electron of the order of its own momentum, typically about $k_B T$ (in units with $c = 1$), and so it will gain or lose an energy of order $(k_B T)^2/m_e$. (This agrees with Eq. (C.20) for $\omega \approx k_B T \ll m_e$.) Hence the rate for energy transfer of order $k_B T$ between a given photon and the electrons equals the rate of collisions times the fraction of the energy $k_B T$ that is transferred per collision:

$$\Gamma_\gamma \simeq \left(\frac{k_B T}{m_e}\right) \Lambda_\gamma \approx 9.0 \times 10^{-29} \, \mathrm{s}^{-1} \, \Omega_B h^2 \left(\frac{T}{T_{\gamma 0}}\right)^4 . \tag{2.2.8}$$

We have to compare this with the cosmic expansion rate. Let's tentatively assume that at the time that concerns us here the density of the universe was dominated by photons and neutrinos, not matter, in which case the cosmic expansion rate was

$$H \equiv \frac{\dot a}{a} = H_0 \sqrt{\Omega_R T^4/T_{\gamma 0}^4} = 2.1 \times 10^{-20} \, \mathrm{s}^{-1} \left(\frac{T}{T_{\gamma 0}}\right)^2 . \tag{2.2.9}$$

Therefore Γ_γ was greater than H until the temperature dropped to a value of order

$$T_{\text{freeze}} = 1.5 \times 10^4 \, \mathrm{K} \left(\Omega_B h^2\right)^{-1/2} . \tag{2.2.10}$$

For $\Omega_B h^2 \simeq 0.02$, this is about 10^5 K. (Comparison with Eq. (2.2.6) shows that for plausible values of $\Omega_M h^2$ and $\Omega_B h^2$, as for instance $\Omega_M h^2 \simeq 0.15$ and $\Omega_B h^2 \simeq 0.02$, we have $T_{\text{freeze}} > T_{\text{EQ}}$, so the temperature (2.2.10) was reached while the expansion rate was still dominated by radiation, as we have been assuming.)

After the temperature dropped below about 10^5 K photons no longer exchanged appreciable energy with electrons, but at this temperature the rate (2.2.7) of elastic scattering was still roughly 10^5 times greater than H. If Eq. (2.2.7) remained valid then Λ_γ would remain larger than H until the temperature dropped to much lower temperature. For $3 \, \mathrm{K} \ll T \ll 10^4 \, \mathrm{K}$ the universe was matter-dominated, with

$$H = H_0 \sqrt{\Omega_M T^3/T_{\gamma 0}^3} = 3.3 \times 10^{-18} \, \mathrm{s}^{-1} \sqrt{\Omega_M h^2} \left(\frac{T}{T_{\gamma 0}}\right)^{3/2} .$$

This became equal to the rate given by Eq. (2.2.7) at a temperature $T \simeq 18 \, \mathrm{K} \, (\Omega_M h^2)^{1/3}/(\Omega_B h^2)^{2/3}$, or about 130 K for $\Omega_M h^2 = 0.15$ and $\Omega_B h^2 =$

0.02, and until then each photon would be scattered many times by electrons in each doubling of $a(t)$. This is not what actually happens, because when the temperature became low enough for electrons and nuclei to hold together as neutral atoms, the elastic scattering rate Λ_γ dropped sharply below the value (2.2.7). As we shall see in the next section, this was at a temperature of about 3,000 K, which marked the end of the era of rapid scattering of photons by electrons.

2.3 Recombination and last scattering

We saw in the previous section that photons stopped exchanging energy effectively with electrons when the temperature of the expanding universe dropped to about 10^5 K. After that, photons continued to be scattered by free electrons, but without appreciable gain or loss of energy. This terminated when the free electrons became bound into hydrogen and helium atoms, ending the scattering of photons. This is called *recombination*.[1] Let's consider when this happened.

We start our calculation at a time early enough so that protons, electrons, and hydrogen and helium atoms were in thermal equilibrium at the temperature of the radiation. In a gas in equilibrium at temperature T, the number density of any non-relativistic non-degenerate particle of type i is given by the Maxwell–Boltzmann formula:

$$n_i = g_i \, (2\pi\hbar)^{-3} e^{\mu_i/k_B T} \int d^3q \, \exp\left[-\left(m_i + \frac{q^2}{2m_i} \right) \Big/ k_B T \right], \quad (2.3.1)$$

where m_i is the particle mass, g_i is the number of its spin states, and μ_i is a characteristic of the gas known as the *chemical potential* of particles of type i. (The generalization of Eq. (2.3.1) to include the effects of relativity and/or degeneracy is given in the following chapter.) The property of the chemical potentials that make this a useful formula is that they are conserved in any reaction that is occurring rapidly in the gas. In our case, the particles are protons, electrons, and hydrogen atoms in any bound state, for which we take i as p, e, $1s$, $2s$, $2p$, etc. (As already mentioned, about 24% of the mass of the early universe was in the form of helium nuclei, but helium atoms are more tightly bound than hydrogen atoms, so that at the time that concerns us now, say for $T < 4{,}400$ K, almost all the helium was locked up in the form of neutral atoms, and therefore played no role here.) The electron and proton

---

[1]The "re" in "recombination" may be misleading; before this time electrons and protons had never been combined into atoms. This has become the standard term throughout astrophysics for the capture of electrons into atoms.

have spin one-half, so $g_p = g_e = 2$, while the $1s$ ground state of the hydrogen atom has two hyperfine states with spins 0 and 1, so $g_{1s} = 1+3 = 4$. At first the recombination and ionization reactions $p + e \rightleftharpoons H_{1s}$ occurred rapidly by cascades of radiative transitions through excited states, so the chemical potentials satisfied

$$\mu_p + \mu_e = \mu_{1s} . \tag{2.3.2}$$

(Photons can be freely created and destroyed in reactions like $e + p \rightleftharpoons e + p + \gamma$, so their chemical potential vanishes.) The integrals are

$$(2\pi\hbar)^{-3} \int d^3p \, \exp\left(-\frac{p^2}{2mk_BT}\right) = \left(\frac{mk_BT}{2\pi\hbar^2}\right)^{3/2},$$

so the conservation law (2.3.2) gives

$$\frac{n_{1s}}{n_p n_e} = \left(\frac{m_e k_B T}{2\pi\hbar^2}\right)^{-3/2} \exp(B_1/k_BT) , \tag{2.3.3}$$

where $B_1 \equiv m_p + m_e - m_H = 13.6$ eV is the binding energy of the $1s$ ground state of hydrogen. (Here we ignore the difference between the hydrogen mass and the proton mass, except in the exponential.) Also, the charge neutrality of cosmic matter requires that

$$n_e = n_p . \tag{2.3.4}$$

Further, in equilibrium the number density of hydrogen atoms in any one excited state is less than the number density in the ground states by a factor of order $\exp(-\Delta E/k_BT)$, where ΔE is the excitation energy, which is necessarily not less than the difference 10.6 eV in the binding energies of the $n = 1$ and $n = 2$ states of hydrogen. (Excited states had the same chemical potential as the ground state, since in equilibrium the atom could go rapidly from one to the other by emitting or absorbing photons.) For temperatures below 4,200 K this exponential factor is less than 6×10^{-13}, so that to a good approximation we can neglect the presence of excited hydrogen atoms as long as thermal equilibrium persisted. As we will see in Section 3.2, the matter at the time of recombination was about 76% by weight neutral or ionized hydrogen, so we can take

$$n_p + n_{1s} = 0.76 \, n_B , \tag{2.3.5}$$

where n_B is the number density of baryons (i.e. at the temperatures of interest here, of neutrons and protons.) The fractional hydrogen ionization $X \equiv n_p/(n_p + n_{1s})$ therefore satisfies the *Saha equation*:

$$X(1 + SX) = 1 \tag{2.3.6}$$

where

$$S \equiv \frac{(n_p + n_{1s})n_{1s}}{n_p^2} = 0.76\, n_B \left(\frac{m_e k_B T}{2\pi \hbar^2} \right)^{-3/2} \exp(B_1/k_B T) . \qquad (2.3.7)$$

One might think that the fractional ionization in equilibrium would become small when the temperature drops below the value $B_1/k_B = 157{,}894$ K, but even in equilibrium the recombination is considerably delayed, because of the small value of the coefficient of the exponential in Eq. (2.3.7). With $n_B = n_{B0}(T/T_{\gamma 0})^3$ and n_{B0} given by Eq. (2.1.13), we have

$$S = 1.747 \times 10^{-22}\, e^{157894/T}\, T^{3/2}\, \Omega_B\, h^2 , \qquad (2.3.8)$$

where h is here again the Hubble constant in units of 100 km s^{-1} Mpc^{-1}, and T is the temperature in degrees Kelvin. This function of temperature is extremely rapidly varying where it is of order unity, so the Saha equation gives a quite sharp temperature of recombination, as shown in Table 2.1. We see from Table 2.1 that the equilibrium value of the ionization dropped from over 97% for $T = 4{,}200$ K to less than 1% for $T = 3{,}000$ K.

This gives the correct order of magnitude of the temperature of the steep decline in fractional ionization, but it is not correct in detail, because

Table 2.1: Equilibrium hydrogen ionization X for various values of the temperature T and $\Omega_B h^2$.

T (K)	$\Omega_B h^2 = 0.01$	$\Omega_B h^2 = 0.02$	$\Omega_B h^2 = 0.03$
4,500	0.999	0.998	0.997
4,200	0.990	0.981	0.971
4,000	0.945	0.900	0.863
3,800	0.747	0.634	0.565
3,600	0.383	0.290	0.244
3,400	0.131	0.094	0.078
3,200	0.0337	0.0240	0.0196
3,000	0.00693	0.00491	0.00401
2,800	0.00112	0.00079	0.00065
2.725	2.8×10^{-12571}	2.0×10^{-12571}	1.6×10^{-12571}

equilibrium was not actually maintained for low ionization levels. The photon that is emitted when a free electron is captured by a proton into the ground state has more than enough energy to ionize another hydrogen atom, so this process produces no net decrease in ionization. Similarly, the photon emitted when an electron in a high orbit of the hydrogen atom with principal quantum number $n \geq 3$ falls into the ground state has more than enough energy to lift an electron in the ground state of some other hydrogen atom to the $n = 2$ excited state, so this process also produces no net increase in the number of electrons in the ground state. The ground state of hydrogen is typically reached by formation of excited states H* in the reaction $e + p \rightarrow$ H* $+ \gamma$, followed by a cascade of radiative decays down to the $n = 2$ excited state. The final transition from the $2p$ excited state to the $1s$ ground state by emission of a single Lyman α photon is impeded by the same effect that impedes the transition of free electrons or electrons in higher excited states to the ground state: the Lyman α photon does not simply merge into the thermal radiation background; it has a large resonant cross section for exciting another hydrogen atom from the ground state to the first excited state, from which it is most often reionized (as shown by the small equilibrium numbers of hydrogen atoms in excited states). But in cosmology this process is not entirely ineffective, because the Lyman α photon emitted by one atom will have just barely enough energy to excite another hydrogen atom in its ground state to the $2p$ state, so if it does not interact very soon with another atom then the cosmological redshift will take its energy outside the resonant line, after which it no longer has enough energy to excite a hydrogen atom in its ground state. Even so, the formation of the ground state by radiative decay from the $2p$ state is so inefficient that we also have to consider slower pathways to the ground state, such as the formation of the $2s$ excited state, which can only decay into the ground state by emitting two photons, neither of which has enough energy to re-excite a hydrogen atom in its ground state. When the ionization became small the rate of these reactions could no longer compete with the cosmic expansion rate, and the ionization no longer fell as fast as it would in thermal equilibrium.[2]

---

[2] The classic paper on this subject is by P. J. E. Peebles, *Astrophys. J.* **153**, 1 (1968). Also see Ya. B. Zel'dovich, V. G. Kurt, and R. A. Sunyaev, *Soviet Physics JETP* **28**, 146 (1969). At the time of writing the most thorough analysis known to me is that of S. Seager, D. D. Sasselov, and D. Scott , *Astrophys. J.* **523**, L1 (1999); *Astrophys. J. Suppl. Ser.* **128**, 407 (2000) [astro-ph/9909275]. Corrections of the order of a percent of the calculated value due to additional transitions and stimulated emission were subsequently found by V. K. Dubrovich and S. I. Grachev, *Astron. Lett.* **31**, 359 (2005) [astro-ph/051672]; J. Chluba and R. A. Sunyaev, *Astron. Astrophys.* **446**, 39 (2006) [astro-ph/0508144]; W. Y. Wong and D. Scott, *Mon. Not. Roy. Astron. Soc.* **375**, 1441 (2007) [astro-ph/0610691]. The possible use of observations of the cosmic microwave background to resolve these uncertainties is discussed by A. Lewis, J. Weller, and R. Battye, *Mon. Not. Roy. Astron. Soc.* **373**, 561 (2006) [astro-ph/0606552]. For the sake of simplicity here I will make the same approximations as Peebles, but will use the "escape probability"

This is a complicated business, but it is not hard to see the main outlines of the recombination process. We make the following approximations:

1. Collisions between hydrogen atoms and radiative transitions between the states of these atoms are sufficiently rapid so that all states of the atoms are in equilibrium with each other at the temperature T of the radiation, except for the 1s ground state, which as already mentioned is reached only through slow or inefficient processes. This has the consequence that the number density of states $n\ell$ of hydrogen having principal quantum number n (with $n > 1$) and orbital angular momentum ℓ can be expressed in terms of the number density of any one state, say the 2s state:[3]

$$n_{n\ell} = (2\ell + 1)\, n_{2s} \exp\left((B_2 - B_n)/k_B T\right), \qquad (2.3.9)$$

where B_n is the binding energy[4] of the state with principal quantum number n.

2. The net rate of change in the population of hydrogen atoms in their 1s state is given by the rate of radiative decay from the 2s and 2p states, minus the rate of excitation of these states from the 1s state. In accordance with the discussion above, all other processes leading to the ground state are assumed to be canceled by the reionization or re-excitation of other atoms by the emitted photon. Because recombination occurs in collisions of electrons and protons, it decreases the number $n_e a^3$ of free electrons in a co-moving volume a^3 at a rate $\alpha(T)n_p n_e a^3$ that is proportional both to $n_p = n_e$ and $n_e a^3$, with the coefficient $\alpha(T)$ depending only on temperature, not on n_e or a. (We do not include recombination directly to the ground state here, since it is canceled by the ionization of other atoms by the emitted photon, so $\alpha(t)$ is what in astrophysics is called the "case B recombination coefficient.") Also, leaving aside ionization from the ground state which just cancels recombination to the ground state, the ionization from excited states of hydrogen increases $n_e a^3$ at a rate given by a sum of terms proportional to the $n_{n\ell} a^3$ with $n > 1$, with coefficients depending only on temperature. Eq. (2.3.9) gives all the $n_{n\ell}$ with $n > 1$ as proportional to n_{2s}, with coefficients that also depend only on temperature, so ionization increases the number of electrons in a co-moving volume a^3 at a rate that can be written as $\beta(T)n_{2s}a^3$, with $\beta(T)$

---

method discussed by Seager *et al.*, which seems to me more direct and easier to justify.

[3]This formula does not apply to hydrogen atoms in very high excited states, which have such large radii and small binding energies that they cannot be treated as free particles. See, e.g., D. Mihalas, *Stellar Atmospheres*, 2nd edition (Freeman, San Francisco, 1978).

[4]We are here neglecting the fine structure, hyperfine structure, and Lamb energy shifts, which give the binding energies a very small dependence on ℓ and on the total (including spin) angular momentum j.

a function only of temperature, not of a or n_e or any of the $n_{n\ell}$. Putting these rates together, we have

$$\frac{d}{dt}\left(n_e a^3\right) = -\alpha n_e^2 a^3 + \beta \, n_{2s} a^3 \,.$$

Dividing by the constant na^3 (where $n \equiv n_p + n_H = n_p + \sum_{n\ell} n_{n\ell} = 0.76 n_B$), this gives

$$\frac{d}{dt}\left(\frac{n_e}{n}\right) = -\frac{\alpha \, n_e^2}{n} + \frac{\beta n_{2s}}{n} \,. \tag{2.3.10}$$

Further, this must vanish under conditions of equilibrium, where (since the transitions $e + p \rightleftharpoons 2s$ occur rapidly) instead of Eq. (2.3.3) we would have

$$\frac{n_{2s}}{n_e^2} = \left(\frac{m_e k_B T}{2\pi \hbar^2}\right)^{-3/2} \exp(B_2/k_B T) \,,$$

so the coefficients in Eq. (2.3.10) are related by

$$\beta/\alpha = \left(\frac{n_e^2}{n_{2s}}\right)_{\text{equilibrium}} = \left(\frac{m_e k_B T}{2\pi \hbar^2}\right)^{3/2} \exp(-B_2/k_B T) \,. \tag{2.3.11}$$

In order for Eq. (2.3.10) to be useful, we need to relate n_{2s} to the total number density n of protons and hydrogen atoms.

3.   *The total number of excited hydrogen atoms in a co-moving volume $1/n$ changes so much more slowly than individual radiative processes that the net increase in this number due to recombination and reionization of hydrogen is balanced by the net decrease in this number by transitions to and from the $1s$ state.* That is,

$$\alpha \, n_e^2 - \beta \, n_{2s} = (\Gamma_{2s} + 3P\Gamma_{2p}) n_{2s} - \mathcal{E} n_{1s} \tag{2.3.12}$$

where Γ_{2s} and Γ_{2p} are the rates for the radiative decay processes $2s \rightarrow 1s + \gamma + \gamma$ and $2p \rightarrow 1s + \gamma$, respectively;[5] P is the probability that the Lyman α photon emitted in the decay $2p \rightarrow 1s + \gamma$ will escape to infinity without exciting some other hydrogen atom in the $1s$ state back to the $2p$ state (to be calculated below); and $\mathcal{E}$ is the rate at which hydrogen atoms in the $1s$ state are excited to the $2s$ or $2p$ state, not including those that are excited to the $2p$ state by Lyman α photons from the decays $2p \rightarrow 1s + \gamma$, which we take into account with the factor P. (The factor 3 in Eq. (2.3.12)

---

[5] We neglect the process $2p \rightarrow 1s + \gamma + \gamma$, which is much slower than $2s \rightarrow 1s + \gamma + \gamma$.

enters because Eq. (2.3.9) gives $n_{2p} = 3n_{2s}$.) From this, we obtain the needed formula for n_{2s}:

$$n_{2s} = \frac{\alpha\, n_e^2 + \mathcal{E}\, n_{1s}}{\Gamma_{2s} + 3P\Gamma_{2p} + \beta} \qquad (2.3.13)$$

We will be concerned here with temperatures $T \ll (B_2-B_3)/k_B = 21{,}900\,\mathrm{K}$, so according to Eq. (2.3.9) all $n_{n\ell}$ with $n > 2$ are much less than n_{2s}, and hence to a very good approximation the total number of hydrogen atoms is

$$n_H = n_{1s} + n_{2s} + n_{2p} = n_{1s} + 4n_{2s} . \qquad (2.3.14)$$

We can therefore eliminate n_{1s} in favor of n_H in Eq. (2.3.13), so

$$n_{2s} = \frac{\alpha\, n_e^2 + \mathcal{E}\, n_H}{\Gamma_{2s} + 3P\Gamma_{2p} + \beta + 4\mathcal{E}} \qquad (2.3.15)$$

Further, in equilibrium the number of hydrogen atoms in the $1s$ state would be constant, so the coefficients on the right-hand side of Eq. (2.3.12) (which gives the net rate of increase of the number density of hydrogen atoms in the ground state) must have the ratio

$$\frac{\mathcal{E}}{\Gamma_{2s} + 3P\Gamma_{2p}} = \left(\frac{n_{2s}}{n_{1s}}\right)_{\text{equilibrium}} = \exp\left(-(B_1 - B_2)/k_B T\right) . \qquad (2.3.16)$$

Using Eqs. (2.3.15) and (2.3.16) in Eq. (2.3.10) now gives the rate equation in a more useful form

$$\frac{d}{dt}\left(\frac{n_e}{n}\right) = \frac{\Gamma_{2s} + 3P\Gamma_{2p}}{(\Gamma_{2s} + 3P\Gamma_{2p})\left[1 + 4\exp\left(-(B_1 - B_2)/k_B T\right)\right] + \beta}$$

$$\times \left(-\left[1 + 4\exp\left(-(B_1 - B_2)/k_B T\right)\right]\frac{\alpha n_e^2}{n}\right.$$

$$\left. + \exp\left(-(B_1 - B_2)/k_B T\right)\frac{\beta n_H}{n}\right) .$$

At the temperatures of interest the factor $1 + 4\exp(-(B_1 - B_2)/k_B T)$ can be replaced with unity. Using Eq. (2.3.11) and the definition of fractional ionization $X \equiv n_e/n = n_p/n = 1 - n_H/n$, we have at last

$$\frac{dX}{dt} = \left(\frac{\Gamma_{2s} + 3P\Gamma_{2p}}{\Gamma_{2s} + 3P\Gamma_{2p} + \beta}\right) \alpha\, n \left(-X^2 + S^{-1}(1 - X)\right) , \qquad (2.3.17)$$

where $S(T)$ is the function (2.3.7) appearing in the Saha equation. Note that if the temperature were constant then this would be satisfied by any solution of the Saha equation (2.3.6). In fact, X is always larger than the value given by the Saha equation, so Eq. (2.3.17) gives a monotonically decreasing fractional ionization. The first factor in Eq. (2.3.17) represents the suppression of recombination that occurs when the transitions of the $2s$ and $2p$ states to the ground state are slower than the reionization of the atom.

It remains to calculate the photon survival probability P. This is given in general by

$$P(t) = \int_{-\infty}^{+\infty} d\omega \, \mathcal{P}(\omega) \exp\left[-\int_t^\infty dt' \, n_{1s}(t') \, c \, \sigma\Big(\omega a(t)/a(t')\Big)\right], \quad (2.3.18)$$

where $\mathcal{P}(\omega) \, d\omega$ is the probability that a photon emitted in the transition $2p \to 1s$ has energy between $\hbar\omega$ and $\hbar(\omega + d\omega)$, normalized so that $\int \mathcal{P}(\omega) \, d\omega = 1$, and $\sigma(\omega)$ is the cross section for the excitation $1s \to 2p$ by a photon of energy $\hbar\omega$. The factor $a(t)/a(t')$ arises from the cosmological redshift of the photon. The cross section is given by the Breit–Wigner formula[6]

$$\sigma(\omega) = \left(\frac{3}{2}\right) \left(\frac{2\pi^2 \Gamma_{2p}}{k_\alpha^2}\right) \mathcal{P}(\omega) \,, \quad (2.3.19)$$

where k_α is the mean wave number $(B_1 - B_2)/\hbar c$ of the Lyman α photon emitted in the transition $2p \to 1s$. The probability density here is

$$\mathcal{P}(\omega) = \frac{\Gamma_{2p}}{2\pi} \frac{1}{(\omega - \omega_\alpha)^2 + \Gamma_{2p}^2/4} \,, \quad (2.3.20)$$

where ω_α is the circular frequency $c k_\alpha$ corresponding to the wave number k_α.

Now, if a photon is captured at all then it is captured in a time much less than the characteristic expansion time of the universe, so we can set the density $n_{1s}(t')$ in Eq. (2.3.18) equal to $n_{1s}(t)$. The cross section $\sigma(\omega)$ varies very rapidly with ω, so we cannot neglect the time-dependence of $a(t)$, but we can take the expansion rate to be constant in the integral over t', so that $a(t)/a(t') = 1 - H(t)(t' - t)$, where $H(t) = \dot{a}(t)/a(t)$. It is convenient then

---

[6]See, e.g., S. Weinberg, *The Quantum Theory of Fields*, (Cambridge University Press, Cambridge, UK, 1995, 1996, 2000) [referred to henceforth as QTF]: Vol. I, Eq. (3.8.16). The factor 3/2 arises from the number $2\ell + 1 = 3$ of $2p$ angular momentum states, and the number 2 of photon helicity states. If the spin of the electron and proton are taken into account then the numerator is 12 and the denominator is 8, but the ratio is the same.

to change the variable of integration in the exponential in Eq. (2.3.18) from t' to $\omega' = (1 - H(t)(t' - t))\omega$, so that Eq. (2.3.18) reads

$$P(t) = \int_{-\infty}^{+\infty} d\omega \, \mathcal{P}(\omega) \exp\left[-\frac{3\pi^2 \Gamma_{2p} n_{1s}(t) \, c}{\omega \, H(t) \, k_\alpha^2} \int_{-\infty}^{\omega} d\omega' \, \mathcal{P}(\omega') \right] , \quad (2.3.21)$$

The function $\mathcal{P}(\omega)$ is negligible except for ω very near ω_α, so we can replace ω with ω_α in the factor $1/\omega$ in the argument of the exponential (but not in the upper limit of the integral over ω', which rises steeply from zero to one for ω in the neighborhood of ω_α.) The integral over ω is now trivial, and gives

$$P(t) = F\left(\frac{3\pi^2 \Gamma_{2p} n_{1s}(t) \, c}{\omega_\alpha \, H(t) \, k_\alpha^2} \right) , \quad (2.3.22)$$

where

$$F(x) \equiv \left(1 - e^{-x}\right) / x . \quad (2.3.23)$$

The Lyman α transition rate[7] is $\Gamma_{2p} = 4.699 \times 10^8 \text{ s}^{-1}$, so that the argument of the function $F(x)$ in Eq. (2.3.22) is very large. For instance, if we take $\Omega_M h^2 = 0.15$ and $\Omega_B h^2 = 0.01$, then for $T < 6{,}000$ K the argument x is greater than 10. Hence for the temperatures of interest here we can drop the exponential in Eq. (2.3.23), and find

$$P = \frac{\omega_\alpha \, H \, k_\alpha^2}{3\pi^2 \Gamma_{2p} \, n_{1s} \, c} = \frac{8\pi \, H}{3\lambda_\alpha^3 \Gamma_{2p} \, n_{1s}} , \quad (2.3.24)$$

where $\lambda_\alpha = 1215.682 \times 10^{-8}$ cm is the wavelength of Lyman α photons, and the argument t is now understood. We see that the quantity $3P\Gamma_{2p}$ in the rate equation (2.3.17) is independent of Γ_{2p}:

$$3P\Gamma_{2p} = \frac{8\pi \, H}{\lambda_\alpha^3 \, n_{1s}} , \quad (2.3.25)$$

With this result for P, equation (2.3.17) is the same as the rate equation found in a different way by Peebles.[2] In this equation we may use the approximation that at the temperatures of interest, which are much less than $(B_1 - B_2)/k_B = 118{,}420$ K, even though n_{1s}/n_{2s} is less than it would be in thermal equilibrium it is still much larger than unity, so that we can replace n_{1s} with $n_H = (1 - X)n$. We also replace time with temperature as the independent variable, using

$$\frac{dt}{dT} = -\frac{1}{HT} . \quad (2.3.26)$$

---

[7]W. L. Wiese, M. W. Smith, and R. M. Glennon, *Atomic Transition Probabilities*, Vol. I, National Standard Reference Data Series NBS 4 (1966).

The rate equation (2.3.17) then becomes

$$\frac{dX}{dT} = \frac{\alpha n}{HT}\left(1 + \frac{\beta}{\Gamma_{2s} + 8\pi H/\lambda_\alpha^3 n(1-X)}\right)^{-1}\left[X^2 - (1-X)/S\right]. \quad (2.3.27)$$

Now we must put in some numbers. In calculating the expansion rate we need to include both the energy density of non-relativistic matter and that of neutrinos and photons, but at temperatures where neutrinos are important their mass is negligible, and for temperatures $T > 30$ K we may neglect vacuum energy, so

$$H = H_0\left[\Omega_M\left(\frac{T}{T_{\gamma 0}}\right)^3 + \Omega_R\left(\frac{T}{T_{\gamma 0}}\right)^4\right]^{1/2}$$

$$= 7.204 \times 10^{-19}\, T^{3/2}\sqrt{\Omega_M h^2 + 1.523 \times 10^{-5} T}\, \text{s}^{-1}\,, \quad (2.3.28)$$

where again in all numerical expressions T is the temperature in degrees Kelvin. The number density of ionized and un-ionized hydrogen is

$$n = 0.76 \times \frac{3H_0^2\Omega_B}{8\pi Gm_p}\left(\frac{T}{T_{\gamma 0}}\right)^3 = 4.218 \times 10^{-7}\,\Omega_B h^2\, T^3\,\text{cm}^{-3}. \quad (2.3.29)$$

The two-photon decay rate of the $2s$ state is[8]

$$\Gamma_{2s} = 8.22458\,\text{s}^{-1}\,. \quad (2.3.30)$$

The coefficient of the proton number density in the electron recombination rate used by Peebles is[9]

$$\alpha = 2.84 \times 10^{-11} T^{-1/2}\,\text{cm}^3\,\text{s}^{-1}\,.$$

The factor $T^{-1/2}$ here is what would be expected from a factor velocity^{-1} that generally appears in the cross sections for exothermic reactions like $e + p \to H + \gamma$.[10] Actually α represents a chain of reactions more complicated than just the radiative capture of an electron, including the cascade of radiative decays down to the $2s$ and $2p$ states, and so it has a more complicated temperature dependence. A variety of detailed numerical calculations of the effective rate of recombination to the $2s$ and $2p$ states can be fit

---

[8] S. P. Goldman, *Phys. Rev.* **A40**, 1185 (1989). Peebles used an earlier value, 8.227 s^{-1}, given by L. Spitzer and J. L. Greenstein, *Astrophys. J.* **114**, 407 (1951).

[9] W. J. Boardman, *Astrophys. J. Suppl.* **9**, 185 (1964).

[10] See, e.g., QTF, Vol. I, p. 157.

with the simple formula:[11]

$$\alpha = \frac{1.4377 \times 10^{-10} \, T^{-0.6166} \, \text{cm}^3 \, \text{s}^{-1}}{1 + 5.085 \times 10^{-3} \, T^{0.5300}}.$$ (2.3.31)

The value of β is given in terms of α by Eq. (2.3.11):

$$\beta = \left(\frac{m_e k_B T}{2\pi \hbar^2}\right)^{3/2} \exp(-B_2/k_B T) \, \alpha = 2.4147 \times 10^{15} \, \text{cm}^{-3} \, T^{3/2} \, e^{-39474/T} \alpha.$$ (2.3.32)

Finally, the function $S[T]$ is given by Eq. (2.3.8).

   The values of $X[T]$ calculated[12] from the differential equation (2.3.27) with the inputs (2.3.28)–(2.3.32) (and Eq. (2.3.8) for $S[T]$) are given in Table 2.2 for $\Omega_M h^2 = 0.15$ and a range of values of $\Omega_B h^2$. The initial condition here is taken to be that $X[T]$ is given by the solution of the Saha equation (2.3.6) for thermal equilibrium at the highest temperature considered, which we take to be $T = 4,226$ K (i.e. $z = 1,550$), high enough so that thermal equilibrium should be a good approximation (because the equilibrium value of X is very close to one, while the true value must be between one and the equilibrium value), but low enough so that hardly any of the helium is still ionized. (Almost all helium was doubly ionized until the temperature dropped below $20,000$ K, and there was still appreciable singly ionized helium until the temperature dropped below about $4,400$ K.) The actual value of $X[4226]$ is slightly higher than the equilibrium value, but to three significant figures there would be no effect on the results at lower temperatures if we increased the assumed value of $X[4226]$ by a small amount, as say from 0.984 to 0.99 for $\Omega_B h^2 = 0.02$.

   Comparison of Tables 2.1 and 2.2 shows that for the small values of $\Omega_B h^2$ considered here, the Saha equation stopped giving a good approximation to the fractional ionization as soon as the equilibrium ionization dropped appreciably below unity. In particular, although in equilibrium the fractional ionization would have dropped to vanishingly small values for temperatures below $2,000$ K, the ionization calculated from Eq. (2.3.27) leveled off at low temperatures to a small but non-zero asymptotic value,

---

[11]D. Péquignot, P. Petijean, and C. Boisson, *Astron. Astrophys.* **251**, 680 (1991). This includes an over-all "fudge factor" of 1.14 recommended by S. Seager, D. D. Sasselov, and D. Scott, *Astrophys. J.* **523**, L1 (1999). Eq. (2.3.31) agrees well with subsequent calculations of D. G. Hummer, *Mon. Not. Roy. Astron. Soc.* **268**, 109 (1994).

[12]I thank D. Dicus for the calculation of these numerical results. The calculation is stopped at $z = 10$, because at later times hydrogen is reionized by the first generation of stars, and also the rate of expansion is affected by the vacuum energy.

Table 2.2: Hydrogen ionization X calculated from Eq. (2.3.27) and time t for the temperature to drop to T from $10^6\,^\circ$K calculated from Eq. (2.3.26). The fourth through sixth
columns give results for various values of $\Omega_B h^2$, with $\Omega_M h^2 = 0.15$. The last column gives
results for parameters $\Omega_B h^2 = 0.02238$, $\Omega_M h^2 = 0.13229$ used in Section 7.2 to compare
the analytic calculation of microwave background anisotropies with a numerical calculation.

z	T(K)	t(yrs)	$X_{\Omega_B h^2=0.01}$	$X_{\Omega_B h^2=0.02}$	$X_{\Omega_B h^2=0.03}$	$X_{\text{Sec.7.2}}$
1550	4226	202,600	0.992	0.984	0.977	0.982
1500	4090	213,200	0.976	0.958	0.943	0.954
1450	3954	225,900	0.935	0.902	0.878	0.895
1400	3818	239,800	0.861	0.815	0.780	0.805
1350	3681	255,200	0.759	0.703	0.659	0.690
1300	3545	272,000	0.645	0.580	0.529	0.564
1250	3409	290,600	0.526	0.456	0.402	0.437
1200	3273	311,300	0.409	0.339	0.289	0.321
1150	3136	334,600	0.299	0.236	0.194	0.220
1100	3000	360,400	0.205	0.154	0.122	0.142
1050	2864	389,600	0.129	0.0928	0.0721	0.0846
1000	2728	422,600	0.0752	0.0520	0.0396	0.0470
950	2591	460,500	0.0405	0.0270	0.0203	0.0243
900	2455	503,600	0.0210	0.0136	0.0101	0.0121
800	2183	611,400	0.00662	0.00387	0.00276	0.00339
700	1910	761,300	0.00319	0.00174	0.00120	0.00150
600	1638	977,700	0.00203	0.00107	0.000731	0.00920
500	1365	1.312×10^6	0.00147	0.000762	0.000517	0.000653
400	1093	1.872×10^6	0.00114	0.000585	0.000395	0.000499
250	684	3.922×10^6	0.000829	0.000423	0.000285	0.000361
100	275	1.604×10^7	0.000632	0.000321	0.000216	0.000273
50	139	4.535×10^7	0.000579	0.000294	0.000197	0.000250
10	30.0	4.568×10^8	0.000537	0.000272	0.000183	0.000231

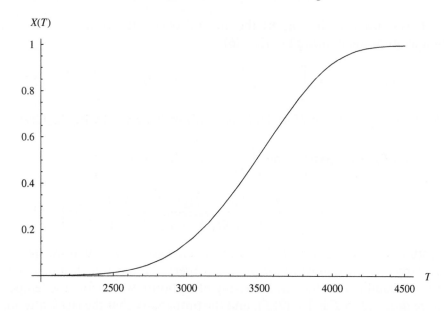

Figure 2.2: The fractional ionization given by the rate equation (2.3.17) as a function of temperature in degrees Kelvin, for $\Omega_M h^2 = 0.132$, $\Omega_B h^2 = 0.0224$.

due to the increasing rarity of encounters of the few remaining free protons and electrons. This residual ionization played an important role in the formation of the first stars.

The fractional ionization given by Eq. (2.3.17) for a currently favored set of cosmological parameters is also shown in Figure 2.2. On the scale of this figure, the results are indistinguishable from those given by the more elaborate calculations of Seager, Sasselov, and Scott,[2] except for temperatures above $T > 4,300$ K, where the contribution of electrons from the ionization of helium (ignored in Eq. (2.3.17)) was still significant.

The most important application of these results for the fractional ionization is the calculation of the *opacity* $\mathcal{O}(T)$, the probability that a photon present at a time $t(T)$ when the temperature is T will undergo at least one more scattering before the present, given by

$$\mathcal{O}(T) = 1 - \exp\left[-\int_{t(T)}^{t_0} c\,\sigma_T\,n_e(t)\,dt\right]. \qquad (2.3.33)$$

This rises from near zero at low temperature, where the integral in the exponent is small, to near one at high temperature, where the integral becomes large.

It is convenient to convert the integral over time to an integral over temperature, again using Eq. (2.3.26):

$$\mathcal{O}(T) = 1 - \exp\left[-c\,\sigma_T \int_{2.725}^{T} n_e(T')\,dT'/H(T')T'\right].\qquad(2.3.34)$$

We use Eq. (2.3.28) for H and take $n_e = Xn$ with n given by Eq. (2.3.29), so

$$\mathcal{O}(T) = 1 - \exp\left[-c\,\sigma_T \int_{2.725}^{T} n_e(T')\,dT'/H(T')T'\right]$$

$$= 1 - \exp\left[-\int_{2.725}^{T} \frac{0.01168\,\Omega_B h^2\,T'^{1/2}\,X(T')\,dT'}{\sqrt{\Omega_M h^2 + 1.523\times 10^{-5}\,T'}}\right].\ (2.3.35)$$

In studies of anisotropies in the cosmic microwave background we are particularly interested in when the photons observed today were last scattered. The probability that the last scattering of a photon was before the temperature dropped to T is $1 - \mathcal{O}(T)$, and the probability that the last scattering was after the temperature dropped further to $T - dT$ is $\mathcal{O}(T - dT)$, so the probability that the last scattering of a photon was at a temperature between T and $T - dT$ is

$$1 - (1 - \mathcal{O}(T)) - \mathcal{O}(T - dT) = \mathcal{O}'(T)\,dT.\qquad(2.3.36)$$

The opacity function $\mathcal{O}(T)$ increases monotonically with temperature from $\mathcal{O} = 0$ at $T = T_0$ to $\mathcal{O} \to 1$ for $T \to \infty$, so $\mathcal{O}'(T)$ is a positive normalized probability distribution, with unit integral. Values[12] of $\mathcal{O}'(T)$ for $\Omega_M h^2 = 0.15$ and a range of values of $\Omega_B h^2$ are given in Table 2.3. The distribution $\mathcal{O}'(T)$ is peaked at a temperature T_L, with a standard deviation σ, given in the two bottom rows of Table 2.3.

* * *

It may be useful to note an approximation[13] for the fractional ionization, even though it was not used in the calculations of Tables 2.2 or 2.3. In equilibrium the fractional ionization depends only on the temperature and $\Omega_B h^2$. But as soon as the Saha equation stopped giving a good approximation to the fractional ionization, the ionization then depended not only on $\Omega_B h^2$, but also on $\Omega_M h^2$. The approximation derived below shows that at sufficiently low temperatures the fractional ionization depends on $\Omega_B h^2$ and $\Omega_M h^2$ chiefly through a multiplicative factor $(\Omega_M h^2)^{1/2}/\Omega_B h^2$, while the opacity $\mathcal{O}'(T)$ is nearly independent of these parameters.

---

[13]B. J. T. Jones and R. F. Wyse, *Astron. Astrophys.* **149**, 144 (1985).

Table 2.3: The normalized probability distribution $\mathcal{O}'(T)$ of the temperature of last scattering, as a function of the temperature T (in degrees Kelvin), calculated from Eq. (2.3.35). The second through fourth columns give results for various values of $\Omega_B h^2$, with $\Omega_M h^2 = 0.15$. The last column gives results for parameters $\Omega_B h^2 = 0.02238$, $\Omega_M h^2 = 0.13229$ used in Section 7.2 to compare the analytic calculation of microwave background anisotropies with a numerical calculation. The bottom two rows give the parameters for a fit of $\mathcal{O}'(T)$ to the Gaussian $\exp[-(T - T_L)^2/2\sigma^2]/\sigma\sqrt{2\pi}$, found by setting T_L equal to temperature at which $\mathcal{O}'$ is a maximum, and $1/\sigma\sqrt{2\pi}$ equal to the value of $\mathcal{O}'$ at that maximum.

$T(\text{K})$	$\mathcal{O}'_{\Omega_B h^2 = 0.01}$	$\mathcal{O}'_{\Omega_B h^2 = 0.02}$	$\mathcal{O}'_{\Omega_B h^2 = 0.03}$	$\mathcal{O}'_{\text{Sec.7.2}}$
4000	6.80×10^{-7}	5.39×10^{-10}	1.08×10^{-12}	5.75×10^{-11}
3500	2.26×10^{-4}	4.12×10^{-5}	1.15×10^{-5}	2.53×10^{-5}
3400	0.000451	0.000152	0.000069	0.000112
3300	0.000759	0.000412	0.000262	0.000345
3200	0.00109	0.000826	0.000664	0.000759
3100	0.00132	0.00127	0.00118	0.00124
3000	0.00139	0.00155	0.00157	0.00156
2900	0.00127	0.00154	0.00164	0.00158
2800	0.00102	0.00130	0.00142	0.00135
2700	0.000746	0.000965	0.00107	0.00101
2600	0.000502	0.000650	0.000721	0.000680
2500	0.000320	0.000411	0.000455	0.000429
2000	4.66×10^{-5}	5.16×10^{-5}	5.39×10^{-5}	5.25×10^{-5}
1000	9.50×10^{-6}	9.76×10^{-6}	9.87×10^{-6}	9.84×10^{-6}
$T_L(\text{K})$	3017	2954	2930	2941
$\sigma(\text{K})$	287	253	241	248

To derive this approximation, note first that as soon as the fractional ionization dropped well below its equilibrium value, the term $(1 - X)/S$ in the square brackets in Eq. (2.3.27) (which would equal X^2 in equilibrium) became much less than the term X^2. Comparison of Tables 2.1 and 2.2 shows that for a plausible range of cosmological parameters, this is the case for $T < 3,400$ K. At such temperatures we can also neglect the radiation energy density compared with the mass density, so that the expansion rate (2.3.28) becomes simply $7.2 \times 10^{-19} \sqrt{\Omega_M h^2}\, \text{s}^{-1}$. Also, for temperatures in the range from 3,400 K down to 2000 K the effective $2p$ decay rate $3P\Gamma_{2p}$ was less than the $2s$ decay rate, while for $T < 2,000$ K the reionization rate β was less than either $3P\Gamma_{2p}$ or Γ_{2s}, so for all temperatures below $T < 3,400$ K, Eq. (2.3.27) took the form

$$\frac{dX}{dT} = f(T)X^2 \,, \tag{2.3.37}$$

where

$$f(T) = \frac{\alpha n}{HT}\left(1 + \frac{\beta}{\Gamma_{2s}}\right)^{-1} = \frac{\Omega_B h^2}{(\Omega_M h^2)^{1/2}}\, g(T) \,, \tag{2.3.38}$$

with $g(T)$ a function of temperature alone

$$g(T) = \frac{84.2\, T^{-0.1166}}{1 + 5.085 \times 10^{-3}\, T^{0.53} + 4.22 \times 10^4\, T^{.8834} e^{-39474/T}} \,. \tag{2.3.39}$$

The solution is

$$X(T) \simeq \left[X(3400)^{-1} + \frac{\Omega_B h^2}{(\Omega_M h^2)^{1/2}} \int_T^{3400} g(T')\, dT' \right]^{-1} \,. \tag{2.3.40}$$

Comparison of this formula with the results shown in Table 2.2 for $\Omega_M = 0.15$ and $\Omega_B = 0.02$ shows that the error introduced by these approximations rises to about 25% as the temperature drops from $3,400$ K to $\simeq 2,500$ K, and then drops to less than 10% for $T < 1,400$ K. For temperatures less than about 2,600 K the fractional ionization is so much less than at 3,400 K that we can neglect the term $X(3400)^{-1}$ in the denominator of Eq. (2.3.36) without introducing any appreciable additional error, giving

$$X(T) \simeq \frac{(\Omega_M h^2)^{1/2}}{\Omega_B h^2} \left[\int_T^{3400} g(T')\, dT' \right]^{-1} \,. \tag{2.3.41}$$

so that $X(T)$ is proportional at sufficiently low temperature to $(\Omega_M h^2)^{1/2}/\Omega_B h^2$, as was to be shown. Inspection of Table 2.2 confirms that for $T < 1,700$ K and with $\Omega_M h^2 = 0.15$ fixed, the fractional ionization is indeed inversely proportional to $\Omega_B h^2$ with fair accuracy, but this is not a good approximation in the neighborhood of 3,000 K.

It is striking that the values of $\mathcal{O}'(T)$ depend only weakly on $\Omega_B h^2$ for $T < 3,200$ K. The reason, as noted by Jones and Wyse,[13] is that at temperatures low enough to neglect the radiation energy density, Eq. (2.3.35) gives

In $\mathcal{O}(T)$ as proportional to $\Omega_B h^2/\sqrt{\Omega_M h^2}$ times an integral of $T^{1/2}X(T)$, while the approximate formula (2.3.41) gives $X(T)$ proportional at low temperature to $\sqrt{\Omega_M h^2}/\Omega_B h^2$, so at sufficiently low temperature $\mathcal{O}(T)$ and hence $\mathcal{O}'(T)$ are independent of $\Omega_B h^2$ and also of $\Omega_M h^2$. Table 2.3 shows that for plausible values of cosmological parameters, the temperature T_L of last scattering is always close to 3,000 K, with a spread (in the sense of standard deviation) of about 10%.

2.4 The dipole anisotropy

In the previous sections of this chapter we have treated the cosmic microwave background as perfectly isotropic and homogeneous. This is certainly a good approximation. Indeed, in the discovery of the cosmic microwave background in 1965, the one thing that enabled Penzias and Wilson to distinguish the background radiation from radiation emitted by earth's atmosphere was that the microwave background did not seem to vary with direction in the sky.

Of course, the cosmic microwave background does have small variations in direction that are too small to have been detected by Penzias and Wilson. The departures from perfect isotropy provides some of the most important information we have about the evolution of the universe. This section deals with the simplest and earliest detected departure from isotropy of the *observed* cosmic microwave background, arising from the earth's motion. The following two sections deal with anisotropies due to scattering of photons by intergalactic electrons in clusters of galaxies, and with the primary anisotropies left over from the early universe.

To the extent that the cosmic microwave background is itself perfectly homogeneous and isotropic, it provides a frame of reference for the whole universe, with respect to which we can measure the peculiar velocities of individual galaxies. To analyze the measurement of our own galaxy through the cosmic microwave background, it is useful to consider the density $N_\gamma(\mathbf{p})$ of photons in phase space, defined by specifying that there are $N_\gamma(\mathbf{p})\,d^3p$ photons of *each* polarization (right or left circularly polarized) per unit spatial volume in a momentum-space volume d^3p centered at $\mathbf{p}$. Since $|\mathbf{p}| = h\nu/c$ and the momentum-space volume between frequencies ν and $d\nu$ is $4\pi\,h^3\nu^2 d\nu/c^3$, Eq. (2.1.1) gives

$$N_\gamma(\mathbf{p}) = \frac{1}{2}\frac{n_T(c|\mathbf{p}|/h)}{4\pi\,h^3\nu^2/c^3} = \frac{1}{h^3}\frac{1}{\exp\left(|\mathbf{p}|c/kT\right) - 1}\,, \qquad (2.4.1)$$

(The factor $1/2$ takes account of the fact that n_T includes *both* possible photon polarization states.) This is of course the density that would be measured by an observer at rest in the radiation background. The phase space volume is Lorentz invariant, and the number of photons is also Lorentz invariant, so N_γ is a scalar, in the sense that a Lorentz transformation to a coordinate system moving with respect to the radiation background that takes $\mathbf{p}$ to $\mathbf{p}'$ also takes N_γ to N_γ', where

$$N_\gamma'(\mathbf{p}') = N_\gamma(\mathbf{p}) . \tag{2.4.2}$$

If the earth is moving in the three-direction with a velocity (in units of c) of β, and we take $\mathbf{p}$ to be the photon momentum in the frame at rest in the cosmic microwave background and $\mathbf{p}'$ to be the photon momentum measured on the earth, then

$$\begin{pmatrix} p_1 \\ p_2 \\ p_3 \\ |\mathbf{p}| \end{pmatrix} = \begin{pmatrix} 1 & 0 & 0 & 0 \\ 0 & 1 & 0 & 0 \\ 0 & 0 & \gamma & \beta\gamma \\ 0 & 0 & \beta\gamma & \gamma \end{pmatrix} \begin{pmatrix} p_1' \\ p_2' \\ p_3' \\ |\mathbf{p}'| \end{pmatrix}, \tag{2.4.3}$$

where as usual $\gamma \equiv (1 - \beta^2)^{-1/2}$. In particular

$$|\mathbf{p}| = \gamma\left(1 + \beta\cos\theta\right)|\mathbf{p}'| \tag{2.4.4}$$

where θ is the angle between $\mathbf{p}'$ and the three-axis. Thus

$$N_\gamma'(\mathbf{p}') = \frac{1}{h^3} \frac{1}{\exp\left(|\mathbf{p}'|c/kT'\right) - 1}, \tag{2.4.5}$$

where the temperature is a function of the angle between the direction of the photon and the earth's velocity

$$T' = \frac{T}{\gamma\left(1 + \beta\cos\theta\right)} . \tag{2.4.6}$$

Since the galaxy can be expected to be moving at a velocity of several hundred kilometers per second, comparable to the peculiar velocities observed for other galaxies relative to the mean Hubble flow, and the solar system is moving with a similar velocity within the galaxy, we expect β to be roughly of order 10^{-3}, in which case γ is essentially unity. The apparent temperature is greatest if we observe photons coming from the direction toward which the earth is moving, for which $\cos\theta = -1$, where it is greater than the intrinsic temperature by a fractional amount β_{earth}. It is least if we observe photons moving in the same direction as the earth, for which $\cos\theta = +1$, and the temperature is decreased by the same fractional amount.

This effect was first observed in 1969 with a ground-based radiometer, but at the time it was only possible to measure the component of the earth's velocity in the earth's equatorial plane, found to be 350 km/sec in a direction corresponding to right ascension 11 h 20 m.[1] The full velocity vector of the earth was measured in 1977 by a Berkeley group,[2] using measurements from a U2 aircraft flying above most of the earth's atmosphere. Our knowledge of this effect has been greatly improved by measurements from the COBE satellite. The Far Infrared Absolute Spectrophotometer group[3] found a maximum temperature increase ΔT of 3.372 ± 0.014 mK (95% confidence level) in a direction with galactic coordinates[4] $\ell = 264°.14 \pm 0°.30$, $b = 48°.26 \pm 0°.30$; the Differential Microwave Interferometer group[5] found $\Delta T = 3.353 \pm 0.024$ mK (95% confidence level) in a direction with galactic coordinates $\ell = 264°.26 \pm 0°.33$, $b = 48°.22 \pm 0°.13$, corresponding to right ascension $11^h 12^m.2 \pm 0^m.8$, declination $-7°.06 \pm 0°.16$.

More recently, the WMAP satellite experiment[6] (discussed in detail in Chapter 7) has given a maximum temperature increase of 3.346 ± 0.017 mK in a direction $\ell = 263°.85 \pm 0°.1$, $b = 48°.25 \pm 0°.04$. These results indicate a motion of the solar system with a velocity $(0.00335)c/(2.725) = 370$ km/sec, not quite in the direction of the Virgo cluster, which has $\ell \approx 284$ and $b \approx 74$. For comparison, the rotation of the galaxy gives the earth a velocity relative to the center of the galaxy of about 215 km/sec, more or less in the opposite direction. Taking this into account gives a net velocity of the local group of galaxies[7] relative to the microwave background of 627 ± 22 km/sec in a direction ($\ell = 276° \pm 3°$, $b = 30° \pm 3°$) between the Hydra and Centaurus clusters of galaxies.

Expanding Eq. (2.4.6) in powers of β, the temperature shift can be expressed as a sum of Legendre polynomials

$$\Delta T \equiv T' - T = T\left[-\frac{\beta^2}{6} - \beta P_1(\cos\theta) + \frac{2\beta^2}{3}P_2(\cos\theta) + \ldots\right]. \quad (2.4.7)$$

Because $\beta = 370$ km/sec$/c = 0.0013$ is small, the temperature shift is primarily a dipole, but Eq. (2.4.7) also exhibits a "kinematic quadrupole"

---

[1] E. K. Conklin, *Nature* **222**, 971 (1969).

[2] G. F. Smoot, M. V. Gorenstein, and R. A. Muller, *Phys. Rev. Lett.* **39**, 898 (1977).

[3] D. J. Fixsen *et al.*, *Astrophys. J.* **473**, 576 (1996).

[4] The galactic coordinate b is the angle between the line of sight and the plane of our galaxy, so that the north galactic pole is at $b = 90°$; the galactic coordinate ℓ is the azimuthal angle around the axis of rotation of our galaxy, with the center of the galaxy at $\ell = 0°$.

[5] C. L. Bennett *et al.*, *Astrophys. J.* **464**, L1 (1996).

[6] C. L. Bennett *et al.*, *Astrophys. J. Suppl.* **148**, 1 (2003).

[7] G. F. Smoot, C. L. Bennett, A. Kogut, J. Aymon, C. Backus *et al.*, *Astrophys. J.* **371**, L1 (1991); A. Kogut, C. Lineweaver, G. F. Smoot, C. L. Bennett, A. Banday *et al.*, *Astrophys. J.* **419**, 1 (1993).

term that is not much smaller than the intrinsic quadrupole term in the temperature, to be discussed in Section 2.6.

2.5 The Sunyaev–Zel'dovich effect

There is another contribution to the anisotropy of the cosmic microwave radiation background, due to the scattering of this radiation by electrons in intergalactic space within clusters of galaxies along the line of sight. This is known as *the Sunyaev–Zel'dovich effect.*[1] Eq. (C.26) (the Kompaneets equation) of Appendix C shows that scattering of the cosmic microwave background by a non-relativistic[2] electron gas changes the observed photon occupation number $N(\omega)$ (defined so that $4\pi\omega^2 N(\omega) d\omega$ is the number of photons of each of the two polarization states with energy between $\hbar\omega$ and $\hbar(\omega + d\omega)$) at photon energy $\hbar\omega \ll m_e c^2$ at a rate (here in cgs units)

$$
\dot{N}(\omega) = \frac{n_e \sigma_T}{m_e\, c\, \omega^2} \frac{\partial}{\partial\omega}\left[k_B T_e \omega^4 \frac{\partial N(\omega)}{\partial\omega} + \hbar\omega^4 N(\omega)\left(1 + N(\omega)\right)\right],
$$

where σ_T is the Thomson scattering cross section, n_e is the electron number density, and T_e is the electron temperature. This formula can be used directly in calculating the rate of change of the occupation number $N(\omega)$ of a homogeneous isotropic photon gas through interactions with a Maxwell–Boltzmann distribution of electrons at temperature T_e. In the application that concerns us now, we are interested instead in the change of the appearance to us of the cosmic microwave background due to scattering by a cloud of electrons along the line of sight. In this context, we rewrite the Kompaneets equation as

$$
\frac{\partial}{\partial\ell} N(\omega, \ell) = \frac{n_e(\ell)\sigma_T}{m_e\, c^2\, \omega^2} \frac{\partial}{\partial\omega}
$$
$$
\times \left[k_B T_e(\ell)\omega^4 \frac{\partial N(\omega, \ell)}{\partial\omega} + \hbar\omega^4 N(\omega, \ell)\left(1 + N(\omega, \ell)\right)\right],
$$

$$(2.5.1)$$

where ℓ is the proper distance coordinate along the line of sight through the cloud.

The ionized plasma in clusters of galaxies is typically at temperatures greater than 10^6 degrees, so that $k_B T_e$ is very much greater than the typical

---

[1] Ya. B. Zel'dovich and R. A. Sunyaev, *Astrophys. Space Sci.* **4**, 301 (1969); R. A. Sunyaev and Ya. B. Zel'dovich, *Comments Astrophys. and Space Physics* **2**, 66 (1970); **4**, 173 (1972).

[2] The ionized gas in galaxy clusters is hot, so that relativistic corrections, though small, are not negligible; see Y. Rephaeli, *Astrophys. J.* **445**, 33 (1995); A. Challinor and A. Lasenby, *Astrophys. J.* **499**, 1 (1998); N. Itoh, U. Kohyama, and S. Nozawa, *Astrophys. J.* **502**, 7 (1998).

value of the photon energy $\hbar\omega$, which for the cosmic microwave background is 10^{-4} eV to 10^{-3} eV. In this case Eq. (2.5.1) simplifies to

$$\frac{\partial}{\partial\ell}N(\omega,\ell) = \frac{n_e(\ell)\sigma_T k_B T_e(\ell)}{m_e c^2 \omega^2}\frac{\partial}{\partial\omega}\left[\omega^4\frac{\partial N(\omega,\ell)}{\partial\omega}\right]. \tag{2.5.2}$$

This is now a linear differential equation, so it can be solved exactly,[3] but in the usual case this is unnecessary, because the cloud of electrons is optically thin. Eq. (2.5.2) then tells us that the change in the occupation number due to passage of the radiation through the cloud is simply

$$\Delta N(\omega) = \frac{y}{\omega^2}\frac{\partial}{\partial\omega}\left[\omega^4\frac{\partial N(\omega)}{\partial\omega}\right], \tag{2.5.3}$$

where y is the dimensionless parameter

$$y \equiv \frac{\sigma_T}{m_e c^2}\int d\ell\, n_e(\ell)k_B T_e(\ell)\,, \tag{2.5.4}$$

the integral being taken along the line of sight through the cloud. For black-body radiation at temperature T_γ we have

$$N(\omega) = \frac{1}{\exp\left(\hbar\omega/k_B T_\gamma\right) - 1}. \tag{2.5.5}$$

Using this in Eq. (2.5.3) gives

$$\Delta N = y\left(\frac{-x + (x^2/4)\coth(x/2)}{\sinh^2(x/2)}\right), \tag{2.5.6}$$

where $x \equiv \hbar\omega/k_B T_\gamma$. We see that in general the scattering changes the shape of the photon energy distribution, not just the effective temperature. The characteristic dependence on ω of Eq. (2.5.6) makes it possible in principle for radio astronomers to distinguish between anisotropies due to the Sunyaev–Zel'dovich effect and the primary anisotropies discussed in Section 2.6 and Chapter 7.

But in the Rayleigh–Jeans part of the spectrum, where $x \ll 1$, Eq. (2.5.6) gives $\Delta N \to -2y/x$, while Eq. (2.5.5) gives $N \to 1/x$, so the shape of the spectrum is preserved, with a fractional change in photon temperature equal to

$$\frac{\Delta T_\gamma}{T_\gamma} = \frac{\Delta N}{N} = -2y\,. \tag{2.5.7}$$

---

[3] Ya. B. Zel'dovich and I. D. Novikov, *Relativistic Astrophysics, Volume 2: The Structure and Evolution of the Universe*, transl. L. Fishbone (University of Chicago Press, Chicago, 1983): Eq. (8.7.3).

It may seem surprising that scattering of photons by a cloud of electrons with much higher temperature would *lower* the photon temperature, but this is only in the Rayleigh–Jeans part of the spectrum. Far beyond the peak in the black-body spectrum, where $x \gg 1$, Eqs. (2.5.5) and (2.5.6) give $\Delta N / N$ the *positive* value $y x^2$; the cooling observed in the Rayleigh–Jeans region is due to the transfer of photons from low to high energy. Indeed, Eq. (2.5.3) shows that $\int \Delta N(\omega) \omega^2 d\omega$ vanishes, so scattering by the electron cloud preserves the total number of photons we receive.

Eq. (2.5.7) gives the fractional change in photon temperature observed at low frequency due to scattering in a cloud of electrons, typically associated with a cluster of galaxies. If the cluster is at a redshift z, then both T_γ and ΔT_γ are reduced by a factor $(1+z)^{-1}$, but the ratio is independent of z, and independent also of the Hubble constant and other cosmological parameters, except in so far as these quantities are needed in estimating n_e and proper lengths. The electron temperature T_e can be measured from observations of the luminosity of the cluster as a function of photon wavelength, with a result that is independent of the cluster redshift or H_0, but the proper length along the line of sight has to be inferred from the angular size of the cluster and the angular diameter distance d_A, which is inversely proportional to H_0, so the electron density calculated from measurements of the Sunyaev–Zel'dovich parameter y is proportional to the value assumed for H_0.

Measurements of the Sunyaev–Zel'dovich effect can usefully be combined with measurements of the X-ray luminosity of the cluster of galaxies.[4] The cross section[5] $d\sigma$ for the bremsstrahlung of photons of energy between $\hbar\omega$ and $\hbar(\omega + d\omega)$ (with $\hbar\omega$ comparable to but not very close to $m_e v^2/2$) in collisions between non-relativistic electrons of velocity v (with $1/137 \ll v \ll 1$) and atomic nuclei of small velocity is proportional to $v^{-2} d\omega/\omega$; the density of protons equals the density of electrons; and v is typically of order $(k_B T / m_e)^{1/2}$; so the rate at which an antenna of radius R receives bremsstrahlung photons of wavelength λ in this energy range and in the resolution solid angle $(\lambda/R)^2$ from an optically thin plasma cloud with electron temperature T_e and number density n_e is

$$[R^2/d_A^2][d_A^2 (\lambda/R)^2] \int d\ell \, \langle v \, d\sigma \rangle n_e^2 \propto \lambda^2 \left(\frac{d\omega}{\omega} \right) \int d\ell \, T_e^{-1/2} n_e^2 .$$

$$(2.5.8)$$

Hence the value of n_e inferred from measurements of the X-ray luminosity and the angular size of the cluster is proportional to the assumed value of $H_0^{1/2}$. The requirement that the value of n_e inferred from the

[4]J. Silk and S. D. M. White, *Astrophys. J.* **226**, L103 (1978); A. Cavaliere, L. Danese, and G. De Zotti, *Astron. Astrophys.* **75**, 322 (1979). For a review, see M. Birkinshaw, *Phys. Rep.* **310**, 97 (1999).

[5]See, *e.g.*, V. B. Berestetskii, E. M. Lifshitz, and L. P. Pitaevskii, *Quantum Electrodynamics*, 2nd edition, transl. by J. B. Sykes and J. S. Bell (Pergamon, Oxford, 1982).

X-ray luminosity should agree with the value obtained from the Sunyaev–Zel'dovich effect thus allows an estimate of H_0. For example, study of two galaxy clusters with $z \simeq 0.55$ (with d_A calculated assuming $\Omega_M = 0.3$ and $\Omega_\Lambda = 0.7$) has given a Hubble constant of $63^{+12}_{-9} \pm 21$ km/sec/Mpc, with the first uncertainty statistical and the second systematic, both with 68% confidence.[6] Any such measurement depends so much on assumptions about the shape of the galaxy cluster and about the clumpiness of the electron distribution, that it seems likely that measurements of the Sunyaev–Zel'dovich effect and X-ray luminosities will be more useful in providing information about galaxy clusters than in fixing the Hubble constant. In one respect, the Sunyaev–Zel'dovich effect is a nuisance — it will interfere with future measurements of the correlation between the primary temperature fluctuations at very small angular separation.[7] But if combined with a model of structure formation, such as discussed in Chapter 8, observations of the Sunyaev–Zel'dovich effect in small angle correlations of the microwave background temperature fluctuations can provide useful cosmological information.[8] Observations of such small-angle correlations[9] have been interpreted as due to the Sunyaev–Zel'dovich effect.[10]

2.6 Primary fluctuations in the microwave background: A first look

In the two previous sections we have considered anisotropies in the cosmic microwave background that arise from effects in the recent universe: the motion of the earth relative to the cosmic microwave background, and the scattering of light by intergalactic electrons in clusters of galaxies along the line of sight. Now we turn to general anisotropies, including the highly revealing *primary* anisotropies that have their origin in the early universe.[1]

It is convenient to expand the difference $\Delta T(\hat{n})$ between the microwave radiation temperature observed in a direction given by the unit vector $\hat{n}$ and the present mean value T_0 of the temperature in spherical harmonics $Y_\ell^m(\hat{n})$:

$$\Delta T(\hat{n}) \equiv T(\hat{n}) - T_0 = \sum_{\ell m} a_{\ell m} Y_\ell^m(\hat{n}) , \qquad T_0 \equiv \frac{1}{4\pi} \int d^2\hat{n}\, T(\hat{n}) , \quad (2.6.1)$$

---

[6] E. D. Reese *et al.*, *Astrophys. J.* **533**, 38 (2000).

[7] Y. Rephaeli, *Astrophys. J.* **245**, 351 (1981); S. Cole and N. Kaiser, *Mon. Not. Roy. Astron. Soc.* **233**, 637 (1988).

[8] E. Komatsu and T. Kitayam, *Astrophys. J.* **526**, L1 (1999).

[9] B. S. Mason *et al.*, *Astrophys. J.* **591**, 540 (2003).

[10] E. Komatsu and U. Seljak, *Mon. Not. Roy. Astron. Soc.* **336**, 1256 (2002); J. R. Bond *et al.*, *Astrophys. J.* **626**, 12 (2005).

[1] P. J. E. Peebles and J. T. Yu, *Astrophys. J.* **162**, 815 (1970); R. A. Sunyaev and Ya. B. Zel'dovich, *Astrophys. Space Sci.* **7**, 20 (1970); Ya. B. Zel'dovich, *Mon. Not. Roy. Astron. Soc.* **160**, 1 (1972).

the sum over ℓ running over all positive-definite integers, and the sum over m running over integers from $-\ell$ to ℓ. Since $\Delta T(\hat{n})$ is real, the coefficients $a_{\ell m}$ must satisfy the reality condition

$$a_{\ell m}^* = a_{\ell\,-m} \,. \tag{2.6.2}$$

(We are defining the spherical harmonics so that $Y_\ell^m(\hat{n})^* = Y_\ell^{-m}(\hat{n})$.) As we saw in Section 2.4, the earth's motion contributes to $\Delta T(\hat{n})$ an anisotropy that to a good approximation is proportional to $P_1(\cos\theta) \propto Y_1^0(\theta,\phi)$ (with the z-axis taken in the direction of the earth's motion), so the main $a_{\ell m}$ produced by this effect is that with $\ell = 1$ and $m = 0$.

The coefficients $a_{\ell m}$ reflect not only what was happening at the time of last scattering, but also the particular position of the earth in the universe. No cosmological theory can tell us this. The quantities of greatest cosmological interest are *averages*, which may be regarded either as averages over the possible positions from which the radiation could be observed, or averages over the historical accidents that produced a particular pattern of fluctuations. The ergodic theorem described in Appendix D shows that, under reasonable assumptions, these two kinds of average are the same. These averages will be denoted $\langle\cdots\rangle$. As discussed in Chapter 10, for anisotropies that arise from quantum fluctuations during inflation, it is these averages over historical accidents that are related to quantum mechanical expectation values. We will return shortly to the question of how to use observations from *one* position in a universe produced by *one* specific sequence of accidents to learn about these averages, but first we must introduce some notation.

We assume that the universe is rotationally invariant on the average, so all averages $\langle\Delta T(\hat{n}_1)\Delta T(\hat{n}_2)\Delta T(\hat{n}_3)\cdots\rangle$ are rotationally invariant functions of the directions $\hat{n}_1$, $\hat{n}_2$, etc. In particular, $\langle\Delta T(\hat{n})\rangle$ is independent of $\hat{n}$, Since $\Delta T(\hat{n})$ is defined as the departure of the temperature from its angular average, its angular average $\int \Delta T(\hat{n})d^2\hat{n}/4\pi$ vanishes. Averaging over the position of the observer, we have $\int\langle\Delta T(\hat{n})\rangle d^2\hat{n} = 0$, so since $\langle\Delta T(\hat{n})\rangle$ is independent of $\hat{n}$, it too vanishes.

The simplest non-trivial quantity characterizing the anisotropies in the microwave background is the average of a product of *two* ΔTs. Rotational invariance requires that the product of two as takes the form

$$\langle a_{\ell m} a_{\ell' m'}\rangle = \delta_{\ell\ell'}\delta_{m\,-m'} C_\ell \,, \tag{2.6.3}$$

for in this case the average of the product of two ΔTs is rotationally invariant:

$$\langle\Delta T(\hat{n})\Delta T(\hat{n}')\rangle = \sum_{\ell m} C_\ell Y_\ell^m(\hat{n}) Y_\ell^{-m}(\hat{n}') = \sum_\ell C_\ell \left(\frac{2\ell+1}{4\pi}\right) P_\ell(\hat{n}\cdot\hat{n}') \,,$$

$$\tag{2.6.4}$$

where P_ℓ is the usual Legendre polynomial. Given the left-hand side, we can find C_ℓ by inverting the Legendre transformation

$$C_\ell = \frac{1}{4\pi} \int d^2\hat{n}\, d^2\hat{n}'\, P_\ell(\hat{n} \cdot \hat{n}')\langle \Delta T(\hat{n}) \Delta T(\hat{n}') \rangle \ . \qquad (2.6.5)$$

Instead of Eq. (2.6.3), we could equivalently define the multipole coefficients C_ℓ by

$$\langle a_{\ell m} a^*_{\ell' m'} \rangle = \delta_{\ell\ell'} \delta_{mm'} C_\ell \ ,$$

which shows that the C_ℓ are real and positive. For perturbations ΔT that are Gaussian in the sense described in Appendix E, a knowledge of the C_ℓ tells us all we need to know about averages of all products of ΔTs.

Of course, we cannot average over positions from which to view the microwave background. What is actually observed is a quantity averaged over m but not position:

$$C^{obs}_\ell \equiv \frac{1}{2\ell + 1} \sum_m a_{\ell m}\, a_{\ell\,-m} = \frac{1}{4\pi} \int d^2\hat{n}\, d^2\hat{n}'\, P_\ell(\hat{n} \cdot \hat{n}') \Delta T(\hat{n}) \Delta T(\hat{n}') \ .$$

$$(2.6.6)$$

The fractional difference between the cosmologically interesting C_ℓ and the observed C^{obs}_ℓ is known as the *cosmic variance*. Fortunately, for Gaussian perturbations, the mean square cosmic variance decreases with ℓ. The mean square fractional difference is

$$\left\langle \left(\frac{C_\ell - C^{obs}_\ell}{C_\ell} \right)^2 \right\rangle = 1 - 2 + \frac{1}{(2\ell+1)^2 C_\ell^2} \sum_{mm'} \langle a_{\ell m} a_{\ell\,-m} a_{\ell m'} a_{\ell\,-m'} \rangle \ .$$

$$(2.6.7)$$

If ΔT is governed by a Gaussian distribution, then so are its multipole coefficients $a_{\ell m}$ (but *not* quantities quadratic in the $a_{\ell m}$, such as C_ℓ.) It follows then that[2]

$$\langle a_{\ell m} a_{\ell\,-m} a_{\ell m'} a_{\ell\,-m'} \rangle = \langle a_{\ell m} a_{\ell\,-m} \rangle \langle a_{\ell m'} a_{\ell\,-m'} \rangle$$

$$+ \langle a_{\ell m} a_{\ell m'} \rangle \langle a_{\ell\,-m} a_{\ell\,-m'} \rangle$$

$$+ \langle a_{\ell m} a_{\ell\,-m'} \rangle \langle a_{\ell\,-m} a_{\ell m'} \rangle \ . \qquad (2.6.8)$$

---

[2] Non-Gaussian terms in the probability distribution of anisotropies would show up as non-vanishing averages of products of odd numbers of the $a_{\ell m}$, as well as corrections to formulas like Eq. (2.6.8) for the averages of products of even numbers of the $a_{\ell m}$. Such non-Gaussian terms are produced both in the early universe and at relatively late times. For a review, see N. Bartolo, E. Komatsu, S. Matarrese, and A. Riotto, *Phys. Rep.* **402**, 103 (2004). Non-Gaussian terms produced by quantum fluctuations during inflation were calculated in the tree graph approximation by J. Maldacena, *J. High Energy Phys.*

Using Eq. (2.6.3), we find that the first term on the right-hand side of Eq. (2.6.8) contributes $(2\ell + 1)^2 C_\ell^2$ to the sum in Eq. (2.6.7), while the second and third terms each contribute $(2\ell + 1)C_\ell^2$ to the sum, so that

$$
\left\langle \left(\frac{C_\ell - C_\ell^{\text{obs}}}{C_\ell} \right)^2 \right\rangle = \frac{2}{2\ell + 1} \,. \tag{2.6.9}
$$

This sets a limit on the accuracy with which we can measure C_ℓ for small values of ℓ. On the other hand, the same analysis shows that for $\ell \neq \ell'$,

$$
\left\langle \left(\frac{C_\ell - C_\ell^{\text{obs}}}{C_\ell} \right) \left(\frac{C_{\ell'} - C_{\ell'}^{\text{obs}}}{C_{\ell'}} \right) \right\rangle = 0 \,, \tag{2.6.10}
$$

so the fluctuations of C_ℓ^{obs} away from the smoothly varying quantity C_ℓ are uncorrelated for different values of ℓ. This means that when C_ℓ^{obs} is measured for all ℓ in some range $\Delta\ell$ in which C_ℓ actually varies little, the uncertainty due to cosmic variance in the value of C_ℓ obtained in this range is reduced by a factor $1/\sqrt{\Delta\ell}$. Even so, measurements of C_ℓ for $\ell < 5$ probably tell us little about cosmology. Also, measurements for $\ell > 2,000$ are corrupted by foreground effects, such as the Sunyaev–Zel'dovich effect discussed in the previous section. Fortunately there is lots of structure in C_ℓ at values of ℓ between 5 and 2,000 that provides invaluable cosmological information.

The primary anisotropies in the cosmic microwave background arise from several sources:

1. Intrinsic temperature fluctuations in the electron–nucleon–photon plasma at the time of last scattering,[3] at a redshift of about 1,090.

---

05 (2003) 013. The effect of loop graphs is considered by S. Weinberg, *Phys. Rev. D* **72**, 043514 (2005) [hep-ph/0506236]; *Phys. Rev. D.* **74**, 023508 (2006) [hep-ph/0605244]; K. Chaicherdsakul, *Phys. Rev. D* **75**, 063522 (2007) [hep-th/0611352]. For late-time contributions, see M. Liguori, F. K. Hansen, E. Komatsu, S. Matarrese, and A. Riotto, *Phys. Rev. D* **73**, 043505 (2006) [astro-ph/0509098]. The weakness of microwave background anisotropies indicates that any non-Gaussian terms are likely to be quite small. So far, there is no observational evidence of such terms.

[3]Strictly speaking, in the approximation of a sudden drop in opacity at a fixed temperature $T_L \simeq 3,000$ K, it is not the intrinsic fluctuation in temperature we observe, but the consequent fluctuation in the redshift z_L of last scattering. Since the unperturbed temperature $\bar{T}(t)$ goes as $1/a(t)$, the value a_L of $a(t)$ at which the total temperature $\bar{T}(t) + \delta T(t)$ reaches a fixed value T_L is shifted by an amount δa_L such that $-(T_L/a_L)\delta a_L + \delta T(t_L) = 0$. The observed temperature (leaving aside other effects) is $T_L a(t_L)/a_0$, so to first order this is shifted by a fractional amount $T_L \delta a_L / a_0 T_0 = \delta T(t_L) a_L / a_0 T_0 = \delta T(t_L)/T_L$, just as if it were the intrinsic temperature fluctuation that we observe.

2. The Doppler effect due to velocity fluctuations in the plasma at last scattering.

3. The gravitational redshift or blueshift due to fluctuations in the gravitational potential at last scattering. This is known as the *Sachs–Wolfe effect*.[4]

4. Gravitational redshifts or blueshifts due to time-dependent fluctuations in the gravitational potential between the time of last scattering and the present. (It is necessary that the fluctuations be time-dependent; a photon falling into a time-independent potential well will lose the energy it gains when it climbs out of it.) This is known, somewhat confusingly, as the *integrated Sachs–Wolfe* effect.[4]

A proper treatment of these effects requires the use of general relativity. This will be the subject of Chapters 5–7. On the other hand, from the time the temperature dropped below 10^4 K until vacuum energy became important at a redshift of order unity, the gravitational field of the universe was dominated to a fair approximation by cold dark matter, which can be treated by the methods of Newtonian physics. Therefore in this introductory section we will concentrate on the Sachs–Wolfe and integrated Sachs–Wolfe effect, which turn out to dominate the multipole coefficients C_ℓ for relatively small ℓ, less than about 40. We will make only a few tentative remarks in this section about the contribution of intrinsic temperature fluctuations and of the Doppler effect.

In considering the Sachs–Wolfe and integrated Sachs–Wolfe effects, we return to the Newtonian approach to cosmology outlined in Eqs. (1.5.21)–(1.5.27). The treatment of perturbations to a homogeneous isotropic cosmology in this approach is presented in Appendix F. For the moment, we need only one result of this analysis, that the perturbation to the gravitational potential, when expressed as a function of the co-moving coordinate $\mathbf{x}$, is a *time-independent* function $\delta\phi(\mathbf{x})$. This perturbation has two effects. First, there is the usual gravitational redshift: A photon emitted at a point $\mathbf{x}$ at the time of last scattering will have its frequency and hence its energy shifted by a fractional amount $\delta\phi(\mathbf{x})$, so the temperature seen when we look in a direction $\hat{n}$ will be shifted from the average over the whole sky by an amount

$$\left(\frac{\Delta T(\hat{n})}{T_0} \right)_1 = \delta\phi(\hat{n}r_L) . \tag{2.6.11}$$

---

[4]R. K. Sachs and A. M. Wolfe, *Astrophys. J.* **147**, 73 (1967).

Here r_L is the radial coordinate of the surface of last scattering, given by Eq. (1.5.44) as

$$r_L = \frac{1}{\Omega_K^{1/2} H_0 a(t_0)} \sinh \left[\Omega_K^{1/2} \int_{1/(1+z_L)}^{1} \frac{dx}{\sqrt{\Omega_\Lambda x^4 + \Omega_K x^2 + \Omega_M x + \Omega_R}} \right] ,$$
(2.6.12)

where $\Omega_K = 1 - \Omega_\Lambda - \Omega_M - \Omega_R$; $z_L \simeq 1090$ is the redshift of last scattering; and t_0 is the present. The perturbation to the gravitational potential also has the effect of changing the rate at which the universe expands by a fractional amount $\delta\phi(\mathbf{x})$, and since the temperature in a matter-dominated universe is falling like $a^{-1} \propto t^{-2/3}$, this shifts the value of the redshift at which the universe reaches the temperature $\simeq 3,000$ K of last scattering in direction $\hat{n}$ by a fractional amount

$$\left(\frac{\delta z}{1+z} \right)_{T=3000 \text{ K}} = -\left(\frac{\delta a(t)}{a(t)} \right)_{T=3000 \text{ K}} = \left(\frac{\dot{a}}{a} \right)_{T=3000 \text{ K}}$$

$$\delta\phi(r_L \hat{n}) t_L = \frac{2}{3} \delta\phi(r_L \hat{n}) .$$

With all other effects neglected, the temperature observed in direction $\hat{n}$ would be 3,000 K divided by $1 + z$, so this fractional shift in $1 + z$ changes the observed temperature by a fractional amount

$$\left(\frac{\Delta T(\hat{n})}{T_0} \right)_2 = -\frac{2}{3} \delta\phi(\hat{n} r_L) .$$
(2.6.13)

The sum of the fractional shifts (2.6.11) and (2.6.13) gives the Sachs–Wolfe effect:

$$\left(\frac{\Delta T(\hat{n})}{T_0} \right)_{\text{SW}} = \frac{1}{3} \delta\phi(\hat{n} r_L)$$
(2.6.14)

The factor $1/3$ will be obtained in Chapter 7 as a result of a better-grounded relativistic treatment.

It is convenient to write $\delta\phi(\mathbf{x})$ as a Fourier transform

$$\delta\phi(\mathbf{x}) = \int d^3q \, e^{i\mathbf{q}\cdot\mathbf{x}} \delta\phi_\mathbf{q} .$$
(2.6.15)

We make use of the well-known Legendre expansion of the exponential

$$e^{i\mathbf{q}\cdot\mathbf{x}} = \sum_\ell (2\ell + 1) i^\ell P_\ell(\hat{q} \cdot \hat{n}) j_\ell(qr) ,$$
(2.6.16)

where j_ℓ is the spherical Bessel function, defined in terms of the usual Bessel function $J_\nu(z)$ by $j_\ell(z) \equiv (\pi/2z)^{1/2} J_{\ell+1/2}(z)$. Eq. (2.6.14) then gives

$$\left(\frac{\Delta T(\hat{n})}{T_0}\right)_{\text{SW}} = \frac{1}{3}\sum_{\ell=0}^{\infty}(2\ell+1)\,i^{\ell}\int d^3q\,\delta\phi_{\mathbf{q}}\,j_{\ell}\,(q\,r_L)\,P_{\ell}(\hat{q}\cdot\hat{n}) \quad (2.6.17)$$

Now we must consider how to calculate the average of a product of two of these fractional temperature shifts in two different directions. Although $\delta\phi(\mathbf{x})$ depends on position, the probability distribution of $\delta\phi(\mathbf{x})$ as seen by observers in different parts of the universe is invariant under spatial rotations and translations. This implies among other things that

$$\langle\delta\phi_{\mathbf{q}}\,\delta\phi_{\mathbf{q}'}\rangle = \mathcal{P}_{\phi}(q)\,\delta^3(\mathbf{q}+\mathbf{q}') . \quad (2.6.18)$$

where $\mathcal{P}_{\phi}(q)$ is a function only of the magnitude of $\mathbf{q}$. (The delta function is needed so that $\langle\delta\phi(\mathbf{x})\delta\phi(\mathbf{y})\rangle$ should be only a function of $\mathbf{x}-\mathbf{y}$.) Because $\delta\phi(\mathbf{x})$ is real, its Fourier transform satisfies the reality condition $\delta\phi_{\mathbf{q}}^* = \delta\phi_{-\mathbf{q}}$, which together with Eq. (2.6.18) tells us that $\mathcal{P}_{\phi}(q)$ is real and positive.

Now, using Eqs. (2.6.17) and (2.6.18), together with the reflection property $P_{\ell}(-z) = (-1)^{\ell}P_{\ell}(z)$ and the orthogonality property of Legendre polynomials

$$\int d\Omega_{\hat{q}}\,P_{\ell}\left(\hat{n}\cdot\hat{q}\right)\,P_{\ell'}\left(\hat{n}'\cdot\hat{q}\right) = \left(\frac{4\pi}{2\ell+1}\right)\delta_{\ell\ell'}P_{\ell}\left(\hat{n}\cdot\hat{n}'\right) , \quad (2.6.19)$$

we have

$$\langle\Delta T(\hat{n})\,\Delta T(\hat{n}')\rangle_{\text{SW}} = \frac{4\pi T_0^2}{9}\sum_{\ell}(2\ell+1)\,P_{\ell}(\hat{n}\cdot\hat{n}')\int_0^{\infty}q^2\,dq\,\mathcal{P}_{\phi}(q)j_{\ell}^2(qr_L) , \quad (2.6.20)$$

or, comparing with Eq. (2.6.4),

$$C_{\ell,\,\text{SW}} = \frac{16\pi^2 T_0^2}{9}\int_0^{\infty}q^2\,dq\,\mathcal{P}_{\phi}(q)j_{\ell}^2(qr_L) \quad (2.6.21)$$

To the extent that the gravitational potential is produced by pressureless cold dark matter, the differential equation for $\delta\phi$ does not involve gradients, and so the differential equation for its Fourier transform does not involve the wave vector $\mathbf{q}$. (See Eqs. (F.12) and (F.18).) The dependence of $\delta\phi_{\mathbf{q}}$ on $\mathbf{q}$ then can arise only from the initial conditions for these differential equations.[5] It is therefore natural to try the hypothesis that the function

---

[5]This is not true even for the Sachs–Wolfe effect if q is so large that q/a became greater than H before the density of the universe became dominated by cold matter. As we will see in Chapter 7, this qualification affects $C_{\ell\,\text{SW}}$ only for $\ell > 100$.

$\mathcal{P}_\phi(q)$ has a simple form, such as a power of q. This power is conventionally written as $n - 4$:

$$\mathcal{P}_\phi(q) = N_\phi^2 q^{n-4} \qquad (2.6.22)$$

where N_ϕ^2 is a positive constant. Then we can use a standard formula:

$$\int_0^\infty j_\ell^2(s)\, s^{n-2} ds = \frac{2^{n-4}\pi \Gamma(3-n)\, \Gamma\left(\ell + \frac{n-1}{2}\right)}{\Gamma^2\left(\frac{4-n}{2}\right) \Gamma\left(\ell + 2 - \frac{n-1}{2}\right)} , \qquad (2.6.23)$$

and find that for $\ell < 100$, Eq. (2.6.21) gives

$$C_{\ell \, \mathrm{SW}} \to \frac{16\pi^3 2^{n-4}\Gamma(3-n)r_L^{1-n} N_\phi^2 T_0^2}{9\,\Gamma^2\left(\frac{4-n}{2}\right)} \; \frac{\Gamma\left(\ell + \frac{n-1}{2}\right)}{\Gamma\left(\ell + 2 - \frac{n-1}{2}\right)} . \qquad (2.6.24)$$

In particular, even before the discovery of primary fluctuations in the cosmic microwave background there was a general expectation about the values of n and N, based not on the microwave background, but on the large scale structure of matter observed relatively close to the present. The perturbation $\delta\rho$ in the total mass density is related to the Sachs–Wolfe effect through the Poisson equation, which gives

$$a^{-2}\nabla^2\delta\phi = 4\pi G\delta\rho , \qquad (2.6.25)$$

with the factor $a^{-2}(t)$ inserted because it is $\mathbf{X} = a(t)\mathbf{x}$ that measures proper distances. (See Eq. (F.12).) Expressing the Fourier transform of $\delta\rho$ in terms of the Fourier transform of $\delta\phi$, we find the correlation function of the density fluctuations to be

$$\langle \delta\rho(\mathbf{x}, t)\delta\rho(\mathbf{x}', t') \rangle = (4\pi G a(t)a(t'))^{-2} \int d^3q\, q^4\, \mathcal{P}_\phi(q)\, e^{i\mathbf{q}\cdot(\mathbf{x}-\mathbf{x}')} . \qquad (2.6.26)$$

The use of this formula to measure $\mathcal{P}_\phi$ is discussed in Chapter 8. For the present, it is enough to note that observations of the density correlation function long ago led to the expectation that $\mathcal{P}_\phi$ takes the so-called Harrison–Zel'dovich form[6] with $n = 1$, and that $N_\phi \approx 10^{-5}$. As we will see in Chapter 10, inflationary theories generally predict that n is close but not

---

[6]E. R. Harrison, *Phys. Rev.* **D1**, 2726 (1970); Ya. B. Zel'dovich, *Mon. Not. Roy. Astron. Soc.* **160**, 1P (1972).

precisely equal to unity. For $n = 1$, we obtain a result that is scale-invariant, in the sense of being independent of r_L:

$$C_{\ell,\text{sw}} \rightarrow \frac{8\pi^2 N_\phi^2 T_0^2}{9\ell(\ell+1)}. \qquad (2.6.27)$$

This is why experimental data on the cosmic microwave background anisotropies is usually presented as a plot of $\ell(\ell+1)C_\ell$ versus ℓ.

What about the other contributions to C_ℓ? Pressure gradients are important in the dynamics of the photon–nucleon–electron plasma, so in estimating the contributions of the Doppler effect and intrinsic temperature fluctuations we must deal with differential equations in which the wave number q enters in an important way. Whatever sort of perturbation we are considering, we can always write it as a Fourier integral like Eq. (2.6.15) and use Eq. (2.6.16) to express $e^{i\mathbf{q}\cdot\hat{n}r_L}$ as a series of Legendre polynomials $P_\ell(\hat{q}\cdot\hat{n})$ with coefficients proportional to $j_\ell(qr_L)$. The integral over q is then dominated by values of q of order ℓ/r_L in the case $\ell \gg 1$. (This is the most interesting case because Eq. (2.6.9) shows that it is only for $\ell \gg 1$ that cosmic variance can be neglected in measurements of C_ℓ.) This is because for $\ell \gg 1$, the spherical Bessel function $j_\ell(z)$ is peaked at $z \simeq \ell$. Specifically, for $\nu \equiv \ell+1/2 \rightarrow \infty$ and $z \rightarrow \infty$ with ν/z fixed at a value other than unity, we have

$$j_\ell(z) \rightarrow \begin{cases} 0 & z < \nu \\ z^{-1/2}(z^2-\nu^2)^{-1/4}\cos\left(\sqrt{z^2-\nu^2} - \nu\arccos(\nu/z) - \frac{\pi}{4}\right) & z > \nu. \end{cases} \qquad (2.6.28)$$

Hence C_ℓ for large ℓ chiefly reflects the behavior of the Fourier components of perturbations for $q \approx \ell/r_L$. To put this another way, the physical wave number at time t_L is $\mathbf{k}_L \equiv \mathbf{q}/a(t_L)$ (because it is $a(t_L)\mathbf{x}$ that measures proper distances at this time) so C_ℓ for large ℓ reflects the behavior of the perturbations for $k_L \approx \ell/d_A$, where d_A is the *angular diameter distance* of the surface of last scattering

$$d_A \equiv r_L a(t_L) = \frac{1}{\Omega_K^{1/2} H_0 (1+z_L)}$$
$$\times \sinh\left[\Omega_K^{1/2}\int_{\frac{1}{1+z_L}}^1 \frac{dx}{\sqrt{\Omega_\Lambda x^4 + \Omega_K x^2 + \Omega_M x + \Omega_R}}\right],$$

$$(2.6.29)$$

with $1+z_L \equiv a(t_0)/a(t_L)$.

For physical wave numbers q/a that are much less than the expansion rate H the differential equation governing any one perturbation is essentially independent of the wave number q, so that the whole dependence of the perturbation on q comes from the initial conditions. (Ratios of different perturbations, such as $\delta\phi_q$ and $\delta\rho_q$, may of course depend on q.) Such perturbations are said to be "outside the horizon" because the wavelength $2\pi a/q$ is much larger than the horizon distance (strictly speaking, the "particle horizon" distance), which we saw in Section 1.13 is roughly of order $1/H$. During the radiation- or matter-dominated eras q/a decreased like $t^{-1/2}$ or $t^{-2/3}$, while H decreased faster, like $1/t$, so perturbations that were outside the horizon at early times subsequently came within the horizon, those with high wave number entering the horizon earlier than those with lower wave numbers.

We need to be a little more precise about the horizon distance. At the time of last scattering the universe was largely matter dominated, so as shown in Section 1.13 the horizon distance at that time was approximately $2/H(t_L)$. But this is the maximum proper distance that *light* could have traveled since the beginning of the present phase of the expansion of the universe. As we will see in Chapters 5 and 6, the dominant perturbations to the plasma of nucleons, electrons, and photons that are relevant to the anisotropies in the cosmic microwave background are *sound* waves, so we need to take into account the smaller speed of sound. During the era before recombination, when radiation and matter were in thermal equilibrium, the speed of sound was $v_s = (\delta p/\delta\rho)^{1/2}$, where δp and $\delta\rho$ are infinitesimal variations in the pressure and density in an *adiabatic* fluctuation. In such a perturbation, the entropy per baryon remains unperturbed:

$$0 = \delta\sigma \propto \delta\left(\frac{\epsilon}{n_B}\right) + p\,\delta\left(\frac{1}{n_B}\right),$$

where ϵ is the thermal energy density, n_B is the baryon number density, and p is the pressure. As we saw in Section 2.2, there are so many more photons than baryons that both ϵ and p are dominated by radiation:

$$\epsilon = a_B T^4, \qquad p = \frac{1}{3}a_B T^4,$$

and therefore for adiabatic perturbations

$$3\frac{\delta T}{T} = \frac{\delta n_B}{n_B} = \frac{\delta\rho_B}{\rho_B}, \tag{2.6.30}$$

where ρ_B is the baryonic mass density. This gives a sound speed

$$v_s = \left(\frac{4a_B T^3 \delta T/3}{\delta\rho_B + 4a_B T^3 \delta T}\right)^{1/2} = \frac{1}{\sqrt{3(1+R)}}, \tag{2.6.31}$$

where $R \equiv 3\rho_B/4a_B T^4$. Since $\rho_B \propto a^{-3}$ and $T \propto a^{-1}$, we have $R \propto a$, and hence $dt = dR/HR$, or in more detail,

$$dt = \frac{dR}{RH_0\sqrt{\Omega_M(R_0/R)^3 + \Omega_R(R_0/R)^4}} = \frac{R\,dR}{H_0\sqrt{\Omega_M}R_0^{3/2}\sqrt{R_{\rm EQ} + R}} ,$$

where $R_{\rm EQ} \equiv \Omega_R R_0/\Omega_M = 3\Omega_R\Omega_B/4\Omega_M\Omega_\gamma$ is the value of R at matter–radiation equality. The acoustic horizon distance is then

$$d_H \equiv a_L \int_0^{t_L} \frac{dt}{a\sqrt{3(1+R)}} = R_L \int_0^{t_L} \frac{dt}{R\sqrt{3(1+R)}}$$

$$= \frac{2}{H_0\sqrt{3R_L\Omega_M}(1+z_L)^{3/2}} \ln\left(\frac{\sqrt{1+R_L} + \sqrt{R_{\rm EQ}+R_L}}{1+\sqrt{R_{\rm EQ}}}\right) , \quad (2.6.32)$$

where $R_L = 3\Omega_B/4\Omega_\gamma(1+z_L)$ is the value of R at last scattering. Gradients become important when $k_L \approx 1/d_H$, and since in C_ℓ the integral over wave number is dominated by $k_L \approx \ell/d_A$, gradients become important when ℓ reaches the value ℓ_H, where $\ell_H \equiv d_A/d_H$. Both d_A and d_H are proportional to $1/H_0$, so H_0 cancels in the ratio.

For a crude estimate of ℓ_H, we note that d_A is proportional to $(1+z_L)^{-1}$, while d_H is proportional to $(1+z_L)^{-3/2}$, so ℓ_H is of the order of $(1+z_L)^{1/2} = 33$. To get a closer estimate, we will take cosmological parameters to have sample values suggested by supernova observations and cosmological nucleosynthesis (discussed in Section 3.2): $\Omega_M = 0.26$, $\Omega_\Lambda = 0.74$, $\Omega_B = 0.043$. This gives $R_L = 0.62$, $R_{\rm EQ} = 0.21$, $d_A = 3.38\,H_0^{-1}(1+z_L)^{-1}$ and $d_H = 1.16\,H_0^{-1}(1+z_L)^{-3/2}$, and hence $\ell_H = 2.9\sqrt{1+z_L} = 96$.

We can now estimate the relative magnitude of the contributions to C_ℓ other than the Sachs–Wolfe effect:

- **Doppler effect:** Like any vector field, the plasma velocity can be decomposed into a term given by the gradient of a scalar, plus a "vector" term whose divergence vanishes. Appendix F shows that the vector term decays as $1/a$, so the dominant perturbations are compressional modes, for which the velocity is the gradient of a scalar. We can therefore expect that in the integral over wave numbers q for $\Delta T/T_0$, the contribution of the Doppler effect will be suppressed for small wave numbers by a factor of order $\mathbf{k}_L d_H$. We will see in Chapter 7 that, because it is proportional to the vector $\mathbf{k}_L$, the Doppler effect contribution does not interfere with the contribution of the Sachs–Wolfe effect, so for a multipole order $\ell \ll \ell_H$, the contribution of the Doppler effect will be less than that of the Sachs–Wolfe effect by a factor of order $[(\ell/d_A)d_H]^2 = \ell^2/\ell_H^2$.

- **Intrinsic temperature fluctuations**: As we have seen in Eq. (2.6.30) the intrinsic fractional temperature fluctuation at the time of last scattering will be just one-third the intrinsic fractional perturbation in the plasma density. As discussed in Chapters 5 and 6, the particular perturbations that are believed to dominate outside the horizon are adiabatic in the further sense that the fractional perturbation in the plasma density is equal to the fractional perturbation $\delta\rho/\bar{\rho}$ in the total matter density. But the perturbation to the total matter density is related to the perturbation to the gravitational potential by Poisson's equation (2.6.25), which if evaluated at the time of last scattering gives for the Fourier transform:

$$\delta\rho_{\mathbf{q}}(t_L) = -\frac{q^2}{4\pi\,Ga^2(t_L)}\delta\phi_{\mathbf{q}} = -\frac{k_L^2}{4\pi\,G}\delta\phi_{\mathbf{q}}\,,$$

where, as before, $k = q/a(t_L)$. Also, the mean total mass density $\bar{\rho}(t_L)$ at last scattering is related to the horizon size d_H by

$$\bar{\rho}(t_L) \simeq \frac{3H^2(t_L)}{8\pi\,G} \approx \frac{1}{2\pi\,Gd_H^2}\,,$$

so the order of magnitude of the intrinsic fractional temperature perturbation is related to the perturbation to the gravitational potential by

$$\frac{\delta T(t_L)}{\bar{T}(t_L)} = \frac{\delta\rho(t_L)}{3\bar{\rho}(t_L)} \approx k_L^2 d_H^2 \delta\phi_q\,. \tag{2.6.33}$$

Thus we expect that in the integral over wave numbers q for $\Delta T/T_0$, the contribution of intrinsic temperature fluctuations will be suppressed for small wave numbers by a factor $\mathbf{k}_L^2 d_H^2$. The *interference* of this contribution with the Sachs–Wolfe term then makes a contribution to C_ℓ for $\ell \ll \ell_{\text{horizon}}$ that, like the contribution of the Doppler effect, is smaller than the Sachs–Wolfe contribution by a factor of order $[(\ell/d_A)d_H]^2 \approx \ell^2/\ell_{\text{horizon}}^2$. (It should be noted that the distinction between the Sachs–Wolfe effect and the effect of intrinsic temperature fluctuations depends on how the time coordinate is defined. The non-relativistic estimates made here correspond to what in the relativistic treatment of Chapters 5–7 would be the use of what are called Newtonian gauge coordinates.)

- **Integrated Sachs–Wolfe effect**: As already mentioned, to the extent that $\delta\phi$ is truly time independent, there is no integrated Sachs–Wolfe effect. The blueshift caused by a photon falling into a time-independent gravitational potential well along the line of sight would be canceled by the redshift caused when the photon climbs out of the well.[7] In fact, the perturbation to the gravitational field is not strictly time-independent, both because radiation continues to make a non-negligible contribution to the gravitational field for some time after last scattering, and also because at late times vacuum energy requires modifications to the Newtonian treatment presented in Appendix F. The late-time integrated Sachs–Wolfe effect chiefly affects C_ℓ for ℓ less than about 10.

It is reasonable then to assume that for $10 < \ell < 50$, the dominant contribution to C_ℓ is from the Sachs–Wolfe effect.

Apart from observation of the $\ell = 1$ anisotropy due to the earth's motion through the microwave background, the first detection of an anisotropy in the cosmic microwave radiation background was achieved by the COBE satellite in 1992.[8] This experiment scanned the sky with two microwave antennae separated by $60°$, at frequencies 31.5, 53, and 90 GHz near the minimum of emission from our galaxy and the maximum of the Planck distribution for 2.7 K. The 1992 data showed an rms fluctuation in the temperature with angle of 30 ± 5 μK, with an angular distribution consistent with $n = 1$. After four years, values of C_ℓ had been measured[9] with the same instruments at values of ℓ ranging from $\ell = 2$ to $\ell = 40$. For $\ell \geq 4$ the results were fit to the ℓ-dependence given by Eq. (2.6.24), with the result that $n = 1.13^{+0.3}_{-0.4}$, which is consistent with the value $n = 1$ for what is called a Harrison–Zel'dovich spectrum. This result is often written as

$$C_\ell = \frac{24\pi Q^2}{5\ell(\ell + 1)} , \tag{2.6.34}$$

with Q known as the *quadrupole moment*. Fitting the values of C_ℓ for $10 \leq \ell \leq 40$, the 1996 COBE results gave $Q = 18 \pm 1.4$ μK. Comparing Eq. (2.6.34) with (2.6.27), we see that

$$N_\phi = \sqrt{\frac{27}{5\pi} \frac{Q}{T_0}} = (8.7 \pm 0.7) \times 10^{-6} . \tag{2.6.35}$$

Surprisingly, the multipole coefficients for $\ell = 2$ and $\ell = 3$ were found to be much less than would be expected by extrapolation of Eq. (2.6.34) from

---

[7]There is also an anisotropy produced by time-dependent fluctuations in the gravitational potential of cosmological inhomogeneities (such as concentrations of cold dark matter), known as *the Rees–Sciama effect*; M. J. Rees and D. W. Sciama, *Nature* **217**, 511 (1968). This is expected to be quite small; see U. Seljak, *Astrophys. J.* **460**, 549 (1996). For possible larger effects due to local structures, see A. Rakić, Syksy Räsänen, and D. J. Schwarz, *Mon. Not. Roy. Astron. Soc.* **363**, L27 (2006) [astro-ph/0601445].

[8]G. F. Smoot *et al.*, *Astrophys. J.* **396**, L1 (1992)

[9]C. L. Bennett *et al.*, *Astrophys. J.* **464**, L1 (1996) [astro-ph/9601067]; A. Kogut *et al.*, *ibid*, L5 (1996) [astro-ph/9601066]; K. M. Górski *et al.*, *ibid*, L11 (1996) [astro-ph/9601063]; G. Hinshaw *et al.*, *ibid*, L17 (1996) [astro-ph/9601088]: E. L. Wright *et al.*, *ibid*, L21 (1996) [astro-ph/9601059].

the fit for $\ell \geq 4$,[10] as shown by the apparent absence of two-point correlations for angles greater than about 60°. The discrepancy with theory is even worse when the integrated Sachs–Wolfe effect is taken into account. This result has since been confirmed by observations with the **WMAP** satellite,[11] to be discussed in Chapter 7. It is quite possible that this discrepancy is due to a combination of foreground contamination and cosmic variance,[12] which according to Eq. (2.6.9) is 63% for $\ell = 2$.

We will have to wait to discuss calculations of C_ℓ for large ℓ, above the range of validity of Eq. (2.6.34), until Chapter 7, after we have developed the general relativistic theory of cosmological fluctuations in Chapters 5 and 6. In Chapter 7 we will also come back to the observations over the past decade that have refined the COBE measurements and extended them to higher ℓ.

[10]G. Hinshaw *et al.*, *Astrophys. J.* **464**, L25 (1996) [astro-ph/9601061].

[11]D. N. Spergel *et al.*, *Astrophys. J. Suppl.* **148**, 175 (2003).

[12]G. Efstathiou, *Mon. Not. Roy. Astron. Soc.* **346**, L26 (2003) [astro-ph/0306431]; A. Slosar, U. Seljak, and A. Makarov, *Phys. Rev.* D **69**, 123003 (2004) [astro-ph/0403073]; A. Slosar and U. Seljak, *Phys. Rev.* D **70**, 083002 (2004); G. Hinshaw *et al.*, *Astrophys. J. Suppl. Ser.* **170**, 288 (2007).

3

The Early Universe

We have been exploring the period in the history of the universe when the radiation temperature dropped from a little over 10^4 K down to its present value of 2.725 K. We now want to look back to the era when the temperature was greater than about 10^4 K, well before the energy density in radiation fell below that of baryons and cold dark matter. We will carry the story back to when the temperature was above 10^{10} K, when electron–positron pairs were abundant and neutrinos were in equilibrium with these pairs, and even farther back, as far as our current knowledge of the laws of physics will take us.

3.1 Thermal history

We first want to work out the history of the falling temperature of the early universe. In this section we will look back only to a time when the temperature was between 10^4 K and 10^{11} K, which is low enough so that muon–antimuon and hadron–antihadron pairs were no longer being produced in appreciable numbers.

There are two circumstances that greatly simplify this task. The first is that the collision rate of photons with electrons and other charged particles during this era was so much greater than the expansion rate of the universe that the photons and charged particles can be assumed to have been in thermal equilibrium, with a common falling temperature. At sufficiently early times even the neutrinos and perhaps the cold dark matter particles were also in thermal equilibrium with the photons and charged particles; later, when no longer colliding rapidly with other particles, they can be treated separately as free particles. The other circumstance is that the number density of baryons (or more strictly, the number density of baryons minus the number density of antibaryons) is so much less than the number density of photons that we can ignore the chemical potential associated with baryon number. Baryons will be put back into the picture in the following section. Also, because the electron/photon number ratio is so small now, it is reasonable to assume that the universe has always had a very small *net* lepton number density (the number density of leptons of all sorts minus that of antileptons) per photon. This means that even at temperatures of order 10^{10} K, when electron–positron pairs were abundant and the energy density and pressure were not simply proportional to T^4 and the entropy density was not simply proportional to T^3, the entropy density, energy density, and pressure were functions $s(T)$, $\rho(T)$, and $p(T)$ of the temperature alone.

149

(The possibility of a non-negligible net lepton number is discussed at the end of this section.)

Before studying the history of the universe during this era we will have to take a brief look at the thermodynamics and statistical mechanics of this sort of matter, in thermal equilibrium with negligible chemical potentials. The condition of thermal equilibrium tells us that the entropy in a co-moving volume is fixed

$$s(T)a^3 = \text{constant} . \tag{3.1.1}$$

The second law of thermodynamics says that any adiabatic change in a system of volume V produces a change in the entropy given by

$$d(s(T)V) = \frac{d(\rho(T)\,V) + p(T)\,dV}{T} . \tag{3.1.2}$$

Equating the coefficients of dV gives our formula for the entropy density

$$s(T) = \frac{\rho(T) + p(T)}{T} . \tag{3.1.3}$$

Also, equating the coefficients of $V\,dT$ and using Eq. (3.1.3) give the law of conservation of energy:

$$T\frac{dp(T)}{dT} = \rho(T) + p(T) . \tag{3.1.4}$$

(For instance, for radiation we have $p(T) = \rho(T)/3$, so Eq. (3.1.4) gives the Stefan–Boltzmann law $\rho = a_B T^4$ with a_B a constant that cannot be determined from thermodynamics alone; Eq. (3.1.3) then gives the entropy density for radiation as $s(T) = 4\,a_B\,T^3/3$. This is why in we said in Section 2.2 that the constant $\sigma k_B \equiv 4a_B T^3/3n_B$ may be interpreted as the radiation entropy per baryon.)

With equal numbers of particles and antiparticles, the number density $n(p)dp$ of a species of fermions (such as electrons) or bosons (like photons) of mass m and momentum between p and $p + dp$ is given by the Fermi–Dirac or Bose–Einstein distributions (with zero chemical potential)

$$n(p, T) = \frac{4\pi g p^2}{(2\pi\hbar)^3} \left(\frac{1}{\exp(\sqrt{p^2 + m^2}/k_B T) \pm 1} \right) , \tag{3.1.5}$$

where g is the number of spin states of the particle and antiparticle, and the sign is $+$ for fermions and $-$ for bosons. For instance, for photons $g = 2$ (and of course $m = 0$), because photons have two polarization states and they are their own antiparticles, while for electrons and positrons $g = 4$, because they have two spin states and electrons and positrons are distinct

150

particle species. The energy density and pressure of a particle of mass m are given by the integrals[1]

$$\rho(T) = \int_0^\infty n(p, T)\, dp\, \sqrt{p^2 + m^2}\,, \qquad (3.1.6)$$

$$p(T) = \int_0^\infty n(p, T)\, dp\, \frac{p^2}{3\sqrt{p^2 + m^2}}\,. \qquad (3.1.7)$$

The entropy density of this particle is then given by Eq. (3.1.3) as

$$s(T) = \frac{1}{T} \int_0^\infty n(p, T)\, dp\, \left[\sqrt{p^2 + m^2} + \frac{p^2}{3\sqrt{p^2 + m^2}} \right]. \qquad (3.1.8)$$

In particular, for massless particles Eq. (3.1.6) gives

$$\rho(T) = g \int_0^\infty \frac{4\pi p^3\, dp}{(2\pi\hbar)^3} \left(\frac{1}{\exp(p/k_B T) \pm 1} \right)$$

$$= \begin{cases} g a_B T^4 / 2 & \text{bosons} \\ 7 g a_B T^4 / 16 & \text{fermions} \end{cases}, \qquad (3.1.9)$$

and of course $p(T) = \rho(T)/3$ and $s(T) = 4\rho(T)/3T$. In other words, each species and spin of massless fermions makes a contribution to the energy density, pressure, and entropy density that is just the same as for each polarization state of photons, except for an additional factor 7/8.

During a period of thermal equilibrium, the variation with time of the temperature is governed by the equation (3.1.1) of entropy conservation and the Einstein field equation, with curvature neglected,

$$\frac{\dot{a}^2}{a^2} = \frac{8\pi G\rho(T)}{3}\,. \qquad (3.1.10)$$

Combining these gives

$$t = -\int \frac{s'(T)\, dT}{s(T)\sqrt{24\pi G\rho(T)}} + \text{constant}\,. \qquad (3.1.11)$$

(The minus sign is inserted in taking the square root of $\dot{T}^2$, to take account of the fact that the temperature decreases as time passes.) In particular, during any epoch in which the dominant constituent of the universe is a highly relativistic ideal gas, the entropy and energy densities are

---

[1] Eq. (3.1.6) follows directly from the definition of $n(p, T)$, and Eq. (3.1.7) can then be derived from Eq. (3.1.4). Or both Eqs. (3.1.6) and (3.1.7) can be obtained from Eqs. (B.41) and (B.43).

given by

$$s(T) = \frac{2\mathcal{N}a_B T^3}{3} \tag{3.1.12}$$

$$\rho(T) = \frac{\mathcal{N}a_B T^4}{2}, \tag{3.1.13}$$

where $\mathcal{N}$ is the number of particle types, counting particles and antiparticles and each spin state separately, and with an extra factor of 7/8 for fermions. Then Eq. (3.1.11) becomes

$$t = \sqrt{\frac{3}{16\pi G\mathcal{N}a_B}} \frac{1}{T^2} + \text{constant}. \tag{3.1.14}$$

With this background, let us now start our history at a time when the temperature was around 10^{11} K, which is in the range $m_\mu \gg k_B T \gg m_e$. Even though it was too cold at this time for reactions like $\nu_\mu + e \to \mu + \nu_e$ or $\nu_\tau + e \to \tau + \nu_e$, the μ and τ neutrinos and antineutrinos were kept in thermal equilibrium by neutral current reactions, like neutrino-electron scattering or $e^+ + e^- \rightleftharpoons \nu + \bar\nu$. Hence the constituents of the universe at this time were photons with two spin states, plus three species of neutrinos and three of antineutrinos, each with one spin state, plus electrons and positrons, each with two spin states, all in equilibrium and all highly relativistic, giving

$$\mathcal{N} = 2 + \frac{7}{8}(6 + 4) = \frac{43}{4}, \tag{3.1.15}$$

so that Eq. (3.1.14) gives, in cgs units:

$$t = \sqrt{\frac{3c^2}{172\pi G a_B}} \frac{1}{T^2} + \text{constant} = 0.994 \text{ sec} \left[\frac{T}{10^{10}\text{K}}\right]^{-2} + \text{constant}. \tag{3.1.16}$$

For instance, with muons ignored and the mass of the electron neglected, it took 0.0098 sec for the temperature to drop from a value 10^{12} K to 10^{11} K, and another 0.98 sec for the temperature to drop to 10^{10} K.

At a temperature of about 10^{10} K neutrinos were just going out of equilibrium and beginning a free expansion. The weak interaction cross section for neutrino-electron scattering is roughly $\sigma_{wk} \approx (\hbar G_{wk} k_B T)^2$, where $G_{wk} \simeq 1.16 \times 10^{-5} \text{GeV}^{-2}$ is the weak interaction coupling constant, and the factor $\hbar^2$ is included to convert a quantity with the units (energy)$^{-2}$ to a quantity with the units (length)^2 of a cross section. (Recall that we are using units with $c = 1$.) The number density of electrons at temperatures above 10^{10} K is roughly given by $n_e \approx (k_B T/\hbar)^3$, so the collision rate of a neutrino with electrons or positrons at such temperatures is

$$\Gamma_\nu = n_e \sigma_{wk} \approx G_{wk}^2 (k_B T)^5/\hbar.$$

This may be compared with the expansion rate, which during the radiation-dominated era is of the order

$$H \approx \sqrt{G(k_B T)^4/\hbar^3} \, ,$$

with the factor $\hbar^{-3}$ included to convert a quantity with the units (energy)^4 to a quantity with the units mass/length^3 of a mass density. The ratio of the collision rate to the expansion rate is thus

$$\frac{\Gamma_\nu}{H} \approx G_{\text{wk}}^2 (\hbar/G)^{1/2} (k_B T)^3 \simeq \left(\frac{T}{10^{10} \text{ K}}\right)^3$$

Hence neutrinos were scattered rapidly enough to remain in thermal equilibrium at temperatures above 10^{10} K. This is just a little greater than m_e/k_B, so for lower temperatures electrons and positrons rapidly disappeared from equilibrium, the collision rate dropped rapidly below G_{wk}^2 $(k_B T)^5/\hbar$, and hence the ratio Γ_ν/H dropped rapidly below unity. The neutrinos then began a free expansion, in which (as we saw in Section 2.1) the number density distribution n_ν continued to keep the form (3.1.5), with a temperature $T_\nu \propto 1/a$.

At lower temperatures we must take into consideration the finite mass of the electron, so the temperature T of the electrons, positrons, and photons (which were still in equilibrium with each other) no longer fell as $1/a$. On the other hand, the freely expanding massless neutrinos preserved a Fermi–Dirac momentum distribution,[2] with a temperature that continued to drop as $1/a$. We must therefore now distinguish between the photon temperature T, and the neutrino temperature T_ν.

The entropy density of the photons, electrons, and positrons is

$$s(T) = \frac{4 a_B T^3}{3} + \frac{4}{T} \int_0^\infty \frac{4\pi p^2 \, dp}{(2\pi\hbar)^3} \left(\sqrt{p^2 + m_e^2} + \frac{p^2}{3\sqrt{p^2 + m_e^2}} \right)$$

$$\times \frac{1}{\exp\left(\sqrt{p^2 + m_e^2}/k_B T\right) + 1}$$

$$= \frac{4 a_B T^3}{3} \mathcal{S}(m_e/k_B T) \, , \tag{3.1.17}$$

---

[2]This is not exact; even at temperatures under 10^{10} K, the neutrino distribution was slightly affected by weak interaction processes, such as $e^- + e^+ \to \nu + \bar\nu$. See A. D. Dolgov, S. H. Hansen, and D. V. Semikoz, *Nucl. Phys.* B **503**, 426 (1997); **543**, 269 (1999); G. Mangano, G. Miele, S. Pastor, and M. Peioso, *Phys. Lett.* B **534**, 8 (2002). For a review, see A. D. Dolgov, *Phys. Rep.* **370**, 333 (2002). The weak interactions provide some thermal contact between the neutrinos and the plasma, which is being heated by electron–positron annihilation, so the effect is to slightly increase the neutrino energy density, by an amount usually represented as an increase in the effective number of neutrino species, from 3 to 3.04. This effect is neglected in what follows.

where, recalling that $a_B = \pi^2 k_B^4 / 15 \hbar^3$,

$$S(x) \equiv 1 + \frac{45}{2\pi^4} \int_0^\infty y^2 \, dy \left(\sqrt{y^2 + x^2} + \frac{y^2}{3\sqrt{y^2 + x^2}} \right) \frac{1}{\exp \sqrt{y^2 + x^2} + 1} \, .$$

(3.1.18)

The entropy conservation law (3.1.1) gives $a^3 T^3 S(m_e/k_B T)$ constant, and since $T_\nu \propto 1/a$, this means that T_ν is proportional to $T S^{1/3}(m_e/k_B T)$. The temperatures were equal for $k_B T \gg m_e$, and

$$S(0) = 1 + 2\frac{7}{8} = 11/4 \, ,$$

(3.1.19)

so

$$T_\nu = \left(4/11 \right)^{1/3} T S^{1/3}(m_e/k_B T) \, .$$

(3.1.20)

The ratio T/T_ν rose from very close to unity for $T > 10^{11}$ K to 1.008 at $T = 10^{10}$ K and to 1.346 at $T = 10^9$ K. To find the asymptotic value of T/T_ν without a computer calculation, we note that $S(\infty) = 1$, so for $k_B T \ll m_e$, Eq. (3.1.20) gives

$$T/T_\nu \rightarrow \left(11/4 \right)^{1/3} = 1.401 \, .$$

(3.1.21)

In particular, at the present time, when $T = 2.725$ K, the neutrino temperature is 1.945 K. Unfortunately there does not seem to be any way of detecting such a neutrino background.

With three flavors of neutrinos and antineutrinos, the total energy density during this period is

$$\rho(T) = 6 \cdot \frac{7}{8} \cdot \frac{a_B T_\nu^4}{2} + a_B T^4 + 4 \int_0^\infty \frac{4\pi p^2 \, dp}{(2\pi \hbar)^3} \frac{\sqrt{p^2 + m_e^2}}{\exp\left(\sqrt{p^2 + m_e^2}/k_B T \right) + 1}$$

$$= a_B T^4 \mathcal{E}(m_e/k_B T) \, ,$$

(3.1.22)

where

$$\mathcal{E}(x) = 1 + \frac{21}{8} \left(\frac{4}{11} \right)^{4/3} S^{4/3}(x) + \frac{30}{\pi^4} \int_0^\infty \frac{y^2 \sqrt{y^2 + x^2} dy}{\exp \sqrt{y^2 + x^2} + 1} \, .$$

(3.1.23)

We insert Eqs. (3.1.17) and (3.1.22) in Eq. (3.1.11), and find

$$t = \int \left(\frac{(m_e/k_B T) S'(m_e/k_B T)}{S(m_e/k_B T)} - 3 \right) \frac{dT}{T \sqrt{24\pi \, G a_B T^4 \mathcal{E}(m_e/k_B T)}}$$

$$= t_e \int \left(3 - \frac{x \, S'(x)}{S(x)} \right) \mathcal{E}^{-1/2}(x) \, x \, dx \, ,$$

(3.1.24)

where $x \equiv m_e/k_B T$, and in cgs units

$$t_e \equiv \left(24\pi\, G\, c^6\, a_B\, m_e^4/k_B^4\right)^{-1/2} = 4.3694 \text{ sec} \,. \tag{3.1.25}$$

The values of T/T_ν and of the time required for the temperature to fall to T (calculated from Eq. (3.1.11)) are given for various values of T in Table 3.1.

After the era of electron–positron annihilation, the energy density of the universe was dominated for a long while by photons, neutrinos, and antineutrinos, all of them highly relativistic, so during this period we have $s(T) \propto T^3$, and

$$\rho(T) = a_B T^4 + \frac{7}{8} \cdot 3 \cdot a_B T_\nu^4 = a_B T^4 \left(1 + \frac{7}{8} \cdot 3 \cdot (4/11)^{4/3}\right) = 3.363\, a_B T^4/2 \,. \tag{3.1.26}$$

Table 3.1: Ratio of electron-photon temperature T to neutrino temperature T_ν and the time t required for the temperature to drop from 10^{11} K to T, for various values of T.

T (K)	T/T_ν	t(sec)
10^{11}	1.000	0
6×10^{10}	1.000	0.0177
3×10^{10}	1.001	0.101
2×10^{10}	1.002	0.239
10^{10}	1.008	0.998
6×10^{9}	1.022	2.86
3×10^{9}	1.080	12.66
2×10^{9}	1.159	33.1
10^{9}	1.345	168
3×10^{8}	1.401	1980
10^{8}	1.401	1.78×10^{4}
10^{7}	1.401	1.78×10^{6}
10^{6}	1.401	1.78×10^{8}

That is, during this period the effective number of particle species is $\mathcal{N} = 3.363$. Using Eq. (3.1.14) then gives, in cgs units:

$$t = \sqrt{\frac{3c^2}{3.363 \cdot 16 \cdot \pi a_B G}\frac{1}{T^2}} + \text{constant} = 1.78 \text{ sec} \left[\frac{T}{10^{10}\text{K}}\right]^{-2} + \text{constant}.$$

$$(3.1.27)$$

For instance, the time required for the universe to cool from a temperature of 10^9 K (where electrons and positrons have mostly annihilated) to a temperature of 10^8 K is 1.76×10^4 sec, or 4.9 hours.

According to Eq. (3.1.27), the time required for the temperature to drop to 10^6 K from much higher values is 1.78×10^8 sec, or 5.64 years. At lower temperatures we must take into account the energy density of non-relativistic matter, and Eq. (3.1.27) no longer applies. We saw in Section 2.3 that for $\Omega_M h^2 = 0.15$, it took an additional 360,000 years for the universe to cool to the temperature $3,000$ K of last scattering.

* * *

So far in this section we have been assuming that neutrinos are massless, and that the net neutrino number of each of the three types (that is, the number of neutrinos minus the number of antineutrinos) is much less than the number of photons. In the general case of an ideal gas of particles of mass m, the number $n(p)\, dp$ of particles of momentum between p and $p + dp$ is given by the Fermi–Dirac and Bose–Einstein distributions

$$n(p, T, \mu)\, dp = \frac{4\pi g p^2\, dp}{(2\pi\hbar)^3}\left(\frac{1}{\exp[(\sqrt{p^2 + m^2} - \mu)/k_B T] \pm 1}\right), \quad (3.1.28)$$

where μ is the chemical potential for the particle in question, a quantity that is conserved in any reaction occurring rapidly in thermal equilibrium, and again g is the number of spin states of the particle and antiparticle, and the sign is $+$ for fermions and $-$ for bosons. This reduces to Eq. (3.1.5) in the case of zero chemical potential, and it yields the number density (2.3.1) for non-relativistic particles with $p \ll m$ and $k_B T \ll m$. During the whole of the era of interest here, electrons and positrons rapidly annihilated into photons, so their chemical potentials were equal and opposite, and since we are assuming charge neutrality and neglecting the tiny number of baryons per photon, we can conclude that the chemical potentials of the electrons and photons were much less than $k_B T$. At temperatures at which neutrinos and antineutrinos were in thermal equilibrium with electrons, positrons, and photons, reactions like $e^+ + e^- \rightleftharpoons \nu_i + \bar{\nu}_i$ were occurring rapidly (where $i = e, \mu, \tau$ label the three types of neutrino), so the chemical potential μ_i of each type of neutrino was equal and opposite to the chemical potential of the corresponding antineutrino. But if we do not assume zero net neutrino (or lepton) number, then there is no *a priori* reason why the μ_i had to be less than $k_B T$.

If neutrino masses are less than about 1 eV, then they may be neglected at the temperatures of interest in this section. The observations of oscillations between different flavors of neutrinos from the sun, nuclear reactors, and cosmic rays shows[3] that the two *differences* in the squares of the masses of the three types of neutrinos of definite mass (which are mixtures of neutrinos of electron, muon, and tau flavor) are $8.0^{+0.4}_{-0.3} \times 10^{-5}$ eV^2 and between 1.9 and 3.0 times 10^{-3} eV^2. Thus the neutrino masses are all much less than 1 eV, unless they are highly degenerate, which there is no reason to expect. If degenerate, then from the absence of anomalies in the low-energy beta decay of tritium, their common mass must be less than 2 eV.[3] Whether degenerate or not, it is clear from this that all three neutrino types (if there are only three) have masses very much less than 1 MeV, and were therefore highly relativistic at the time that they went of thermal equilibrium with electrons and positrons, at a temperature of about 10^{10} K. Once out of equilibrium, their momentum simply decayed as $1/a$ (as shown in Section 1.1), so if their chemical potential was negligible their momentum distribution remained the same as that of photons, with a temperature less by a factor $(4/11)^{1/3}$. Thus once $k_B T$ dropped below the smallest neutrino mass, their energy density became just $n_\nu \sum_\nu m_\nu = (3/11) n_\gamma \sum_\nu m_\nu$. (For $k_B T$ much larger than the mass, the integral for the number density of each spin state of fermions is 3/4 the corresponding integral for bosons, and after neutrinos decouple $T_\nu^3 = (4/11) T_\gamma^3$.) With a non-zero chemical potential the energy density is larger. The agreement between theory (with massless neutrinos) and observation for the cosmic microwave background anisotropies discussed in Sections 2.6 and 7.2 and for the large scale structure discussed in Chapter 8 shows that the sum of the three neutrino masses is less than 0.68 eV (95% confidence level),[4] so if they are degenerate the common mass is less than 0.23 eV. This result has been contradicted by the observation of neutrinoless double beta decay in a Heidelberg–Moscow experiment,[5] which suggests a value greater than 1.2 eV for the sum of neutrino masses.[6] There has not yet been an opportunity to confirm the double beta decay results, and for the present it seems reasonable to continue to neglect neutrino masses.

---

[3] W.-M. Yao *et al.* (Particle Data Group), *J. Phys. G.* **33**, 1 (2006).

[4] D. N. Spergel *et al.*, *Astrophys. J. Suppl. Ser.* **170**, 377 (2007) [astro-ph/0603449].

[5] H. V. Klapdor-Kleingroth, I. V. Krivosheina, A. Dietz and O. Chkvorets, *Phys. Lett. B* **586**, 198 (2004).

[6] A. De La Macorra, A. Melchiorri, P. Serra, and R. Bean, *Astropart. Phys.* **27**, 406 (2007) [astro-ph/0608351].

With neutrino masses neglected, the energy density, pressure, and net lepton number density of neutrinos and antineutrinos of type i is

$$\rho_i = 3p_i$$

$$= \frac{4\pi}{(2\pi\hbar)^3} \int p^3 \, dp \left(\frac{1}{\exp[(p - \mu_i)/k_B T] + 1} + \frac{1}{\exp[(p + \mu_i)/k_B T] + 1} \right)$$

$$= 4\pi \frac{(k_B T)^4}{(2\pi\hbar)^3} \mathcal{P}(\mu_i/k_B T) , \tag{3.1.29}$$

$$n_i = \frac{4\pi}{(2\pi\hbar)^3} \int p^2 \, dp$$

$$\times \left(\frac{1}{\exp[(p - \mu_i)/k_B T] + 1} - \frac{1}{\exp[(p + \mu_i)/k_B T] + 1} \right)$$

$$= 4\pi \frac{(k_B T)^3}{(2\pi\hbar)^3} \mathcal{M}(\mu_i/k_B T) , \tag{3.1.30}$$

where

$$\mathcal{P}(x) \equiv \int_0^\infty \left[\left(e^{y-x} + 1 \right)^{-1} + \left(e^{y+x} + 1 \right)^{-1} \right] y^3 \, dy , \tag{3.1.31}$$

$$\mathcal{M}(x) \equiv \int_0^\infty \left[\left(e^{y-x} + 1 \right)^{-1} - \left(e^{y+x} + 1 \right)^{-1} \right] y^2 \, dy . \tag{3.1.32}$$

As we have seen, at temperatures above 10^{10} K the energy density of the photons and electron–positron pairs is $11 \, a_B T^4/4$, and the pressure is one-third as great, so the total energy density and pressure are given by

$$\rho = 3p = T^4 \left[\frac{11 a_B}{4} + \frac{4\pi \, k_B^4}{(2\pi\hbar)^3} \sum_{i=e,\mu,\tau} \mathcal{P}(\mu_i/k_B T) \right] . \tag{3.1.33}$$

The equation (1.5.20) of energy conservation tells us that under these circumstances ρa^4 is constant, while the conservation of each type of neutrino number also tells us that $n_i a^3$ is constant. Since ρ and n_i depend in different ways on the chemical potentials, this requires that as the universe expands in this era the $\mu_i/k_B T$ remain constant, and also $T \propto 1/a$, just as in the case of zero chemical potential.

As the temperature dropped below 10^{10} K the temperature of the photons and electron–positron pairs no longer varied as $1/a$, but as we have seen the neutrinos and antineutrinos entered on a free expansion. With each neutrino's momentum p varying as $1/a$, the form of the Fermi–Dirac distributions for each type of massless neutrino was preserved, with a temperature $T_\nu \propto 1/a$ and $\mu_i/k_B T$ constant, just as before decoupling. We conclude that to a good approximation $T_\nu \propto 1/a$ and each $\mu_i \propto 1/a$ throughout the whole era of interest here.

The only effect that a non-zero neutrino chemical potential would have on the calculations of this section is that it would increase the total energy density and hence shorten the time scale. For any non-zero μ the function $\mathcal{P}(\mu/k_B T)$ is greater than $\mathcal{P}(0) = 7\pi^4/60$, so if $\mu_i \neq 0$ then $\rho_i > 7\pi^2 (k_B T)^4/120(\hbar)^3 = (7/8)a_B T^4$. In particular, if $|\mu_i| \gg k_B T$ then $\mathcal{P}(\mu_i/k_B T) \simeq (\mu_i/k_B T)^4/4$, so $\rho_i \simeq \pi(\mu_i)^4/(2\pi\hbar)^3 \gg a_B T^4$, and these neutrinos (or antineutrinos, if $\exp(-\mu_i/k_B T)$ is large) dominated the energy density of the universe, at least until the cross-over of non-relativistic matter and radiation. Inspection of Eq. (3.1.28) shows that for a chemical potential $\mu \gg k_B T$, the fermion distribution function $n(p, T, \mu)dp$ is equal to $4\pi g p^2 dp/(2\pi\hbar)^3$ for particle energies less than μ, and then falls off rapidly for higher energies, indicating that all fermion energy levels up to energy μ are filled, while higher energy levels are empty. This is the case of complete neutrino degeneracy.[7] Experiments on the beta decay of tritium, $^3\text{H} \to {^3\text{He}} + e^- + \bar{\nu}_e$ show that $|\mu_e|$ is less than a few eV, because otherwise for $\mu_e > 0$ there would be a rise in the electron spectrum beyond the expected endpoint $m(^3\text{H}) - m(^3\text{He})$, due to absorption of degenerate cosmic neutrinos in the reaction $\nu_e + {^3\text{H}} \to {^3\text{He}} + e^-$, while for $\mu_e < 0$ the Pauli exclusion principle would produce a *dip* in the electron spectrum within a few eV of its expected endpoint, where antineutrino energies are less than a few eV, putting them within the degenerate antineutrino sea. (The absence of a dip within a few eV of the expected electron spectrum endpoint also sets a limit of a few eV on the electron neutrino mass.) This yields an upper bound on the time-independent quantity $|\mu_e|/k_B T$ of order 10^4, which is very much weaker than the upper bound that will be provided in the following section by considerations of cosmological nucleosynthesis.

3.2 Cosmological nucleosynthesis

We have worked out the thermal history of the universe from temperatures above 10^{10} K down to the crossover temperature $\approx 10^4$ K, ignoring the presence of a small number of nucleons (and a small excess of electrons over positrons). Now let us consider what happens to the nucleons during this era.[1]

---

[7]The possibility of cosmic neutrino degeneracy was raised by S. Weinberg, *Phys. Rev.* **128**, 1457 (1962).

[1]I outlined the history of these calculations in *The First Three Minutes* (Basic Books, 1977, 1988). Briefly, the first calculations of cosmological nucleosynthesis were undertaken by Ralph Alpher, George Gamow, and Robert Herman in the late 1940s; see G&C, Chapter 15, footnotes 51 and 52. In this

The weak interactions allow neutron–proton conversion through six processes:

$$n + \nu \rightleftharpoons p + e^- , \qquad n + e^+ \rightleftharpoons p + \bar{\nu} , \qquad n \rightleftharpoons p + e^- + \bar{\nu} . \qquad (3.2.1)$$

(Here ν is ν_e; the other neutrino flavors do not contribute to these reactions.) In this range of temperatures $k_B T \ll m_N$, so the nucleons can be treated as essentially at rest. The initial and final lepton energies are therefore simply related, by

$$
\begin{aligned}
E_e - E_\nu &= Q \quad \text{for } n + \nu \rightleftharpoons p + e^- , \\
E_\nu - E_e &= Q \quad \text{for } n + e^+ \rightleftharpoons p + \bar{\nu} \qquad (3.2.2)\\
E_\nu + E_e &= Q \quad \text{for } n \rightleftharpoons p + e^- + \bar{\nu} ,
\end{aligned}
$$

where

$$Q = m_n - m_p = 1.293 \text{ MeV} . \qquad (3.2.3)$$

The total rates at which an individual neutron is converted to a proton or a proton to a neutron take the form

$$\lambda(n \rightarrow p) = A \int \left(1 - \frac{m_e^2}{(Q+q)^2} \right)^{1/2} \frac{(Q+q)^2 q^2 dq}{\left(1 + e^{q/k_B T_\nu} \right)\left(1 + e^{-(Q+q)/k_B T} \right)} ,$$

$$(3.2.4)$$

---

work it was assumed that nucleons start as pure neutrons, which then convert to protons by the process of neutron beta decay. It was then pointed out that the conversion of neutrons into protons and *vice versa* occurs primarily through two-particle collisions, and that the rapid rate of these processes at very early times has the consequence that nucleons start as 50% neutrons and 50% protons, by C. Hayashi, *Prog. Theor. Phys. (Japan)* **5**, 224 (1950). Following this, a modern calculation of the evolution of the neutron/proton ratio was presented by R. A. Alpher, J. W. Follin, Jr., and R. C. Herman, *Phys. Rev.* **92**, 1347 (1953), but the results were not applied to the problem of cosmological nucleosynthesis. Several authors noted that the abundance of helium in the universe is too large to be accounted for by stellar nucleosynthesis; see G. Burbidge, *Pub. Astron. Soc. Pacific* **70**, 83 (1958); F. Hoyle and R. J. Tayler, *Nature* **203**, 1108 (1964). The modern theory of cosmological nucleosynthesis is due to P. J. E. Peebles, *Astron. J.* **146**, 542 (1966). (Related calculations done by Ya. B. Zel'dovich, *Adv. Astron. Astrophys.* **3**, 241 (1965) were not known in the West until much later.) Nucleosynthesis calculations were then extended to more nuclides and reactions by R. V. Wagoner, W. A. Fowler, and F. Hoyle, *Astrophys. J.* **148**, 3 (1967), and many small corrections were included by D. A. Dicus, E. W. Kolb, A. M. Gleeson, E. C. G. Sudarshan, V. L. Teplitz, and M. S. Turner, *Phys. Rev. D* **26**, 2694 (1982). Modern reviews are given by G. Steigman, in *Measuring and Modeling the Universe – Carnegie Observatories Astrophysics Series, Volume 2*, ed. W. Freedman (Cambridge University Press, Cambridge, UK) [astro-ph/0307244]; in *The Local Group as an Astrophysical Laboratory – Proceedings of the May 2003 STScI Symposium* [astro-ph/0308511]; in *Chemical Abundances and Mixing in Stars in the Milky Way and its Satellites – Proceedings of the ESO/Arcetrei Workshop*, eds. L. Pasquini and S. Randich (Springer–Verlag) [astro-ph/0501591]; and *Int. J. Mod. Phys.* **E15**, 1 (2006) [astro-ph/0511534]. For discussions emphasizing analytic calculations, see G&C, Section 15.7, and V. Mukhanov, *Int. J. Theor. Phys.* **43**, 669 (2004) [astro-ph/0303073].

$$\lambda(p \to n) = A \int \left(1 - \frac{m_e^2}{(Q+q)^2}\right)^{1/2} \frac{(Q+q)^2 q^2 dq}{\left(1 + e^{-q/k_B T_\nu}\right)\left(1 + e^{(Q+q)/k_B T}\right)},$$

(3.2.5)

where[2]

$$A \equiv \frac{G_{\mathrm{wk}}^2 \left(1 + 3g_A^2\right) \cos^2 \theta_C}{2\pi^3 \hbar},$$

(3.2.6)

and the integrals over q run from $-\infty$ to $+\infty$, leaving out a gap from $q = -Q - m_e$ to $q = -Q + m_e$, where the square root would be imaginary. These rates include the effect of the Pauli Principle in the presence of partly filled lepton seas. For instance, the cross section for the process $n + e^+ \to p + \bar{\nu}$ is $2\pi^2 \hbar^3 A E_\nu^2 / v_e$, the number density of positrons of each helicity with momentum between p_e and $p_e + dp_e$ is $4\pi p_e^2 dp_e (2\pi \hbar)^{-3} \big[\exp(E_e/k_B T) + 1\big]^{-1}$, and the fraction of unfilled antineutrino levels with energy E_ν is

$$1 - [\exp(E_\nu/k_B T_\nu) + 1]^{-1} = [\exp(-E_\nu/k_B T_\nu) + 1]^{-1}$$

so the total rate per neutron of the process $n + e^+ \to p + \bar{\nu}$ is

$$\lambda(n + e^+ \to p + \bar{\nu}) = A \int_0^\infty E_\nu^2 p_e^2 dp_e \, [\exp(E_e/k_B T) + 1]^{-1}$$

$$\times [\exp(-E_\nu/k_B T_\nu) + 1]^{-1}$$

Changing the variable of integration to $q \equiv -E_\nu = -Q - E_e$, we see that this partial rate is just the part of the integral (3.2.4) that runs from $q = -\infty$ to $q = -Q - m_e$. Likewise, the part of the integral that runs from $q = -Q + m_e$ to $q = 0$ is supplied by the neutron-decay process $n \to p + e^- + \bar{\nu}$, with $q = -E_\nu$, and the part of the integral that runs from $q = 0$ to $q = +\infty$ arises from the process $n + \nu \to p + e^-$, with $q = E_\nu$. Similar remarks apply to the integral (3.2.5).

With the rates (3.2.4)–(3.2.5) known in principle, we can calculate the change in the ratio X_n of neutrons to all nucleons from the differential equation

$$\frac{dX_n}{dt} = -\lambda(n \to p)X_n + \lambda(p \to n)(1 - X_n) .$$

(3.2.7)

---

[2]Here $G_{\mathrm{wk}} = 1.16637(1) \times 10^{-5} \mathrm{GeV}^{-2}$ is the weak coupling constant, measured from the rate of the decay process $\mu^+ \to e^+ + \nu_e + \bar{\nu}_\mu$; $g_A = 1.257$ is the axial vector coupling of beta decay, measured from the rate of neutron decay; and θ_C is the Cabibbo angle, with $\cos\theta_C = 0.9745(6)$, measured from the rate of O^{14} beta decay and other $0^+ \to 0^+$ transitions.

As a check, note that for a time-independent temperature T equal to T_ν, the two rates (3.2.4) and (3.2.5) would have the ratio

$$\frac{\lambda(p \to n)}{\lambda(n \to p)} = \exp\left(-Q/k_B T\right) \quad \text{for } T = T_\nu . \tag{3.2.8}$$

It follows then that in this case, Eq. (3.2.7) would have the time-independent solution expected in thermal equilibrium

$$\frac{X_n}{X_p} = \frac{X_n}{1 - X_n} = \exp\left(-Q/k_B T\right) . \tag{3.2.9}$$

It is the inequality of T and T_ν as well as the time-dependence of these temperatures that drives X_n/X_p away from its equilibrium value (3.2.9).

For $k_B T \gg Q$ we can evaluate the integrals (3.2.4)–(3.2.5) by setting $T_\nu = T$ and $Q = m_e = 0$, so in this case

$$\lambda(n \to p) = \lambda(p \to n) = A \int_{-\infty}^{+\infty} \frac{q^4 \, dq}{\left(1 + e^{q/k_B T}\right)\left(1 + e^{-q/k_B T}\right)}$$

$$= \frac{7}{15}\pi^4 A(k_B T)^5 = 0.400 \text{ sec}^{-1} \left(\frac{T}{10^{10}\text{K}}\right)^5 . \tag{3.2.10}$$

For comparison, the time t for the temperature to drop to T from much higher values is given by Eq. (3.1.16) as 0.99 sec $\left(T/10^{10}\text{K}\right)^{-2}$. Also, $H \simeq 1/2t$. The ratio λ/H is therefore

$$\lambda/H \simeq 0.8 \times \left(\frac{T}{10^{10}\text{K}}\right)^3 . \tag{3.2.11}$$

This ratio is larger than 1 for $T > 1.1 \times 10^{10}$ K. It is true that temperatures near this lower bound are not much larger than Q/k, and T_ν at this epoch is not precisely equal to T, so the rates $\lambda(n \to p)$ and $\lambda(p \to n)$ are not precisely equal, and neither is given precisely by Eq. (3.2.10). Nevertheless Eq. (3.2.10) gives the order of magnitude of these rates in this temperature range, so we can still rely on the conclusion that $\lambda(n \to p)/H$ and $\lambda(p \to n)/H$ are large down to these temperatures. This means that the initial value of X_n at temperatures larger than about 3×10^{10} K is given by the condition that the right-hand side of Eq. (3.2.7) should vanish:

$$X_n \to \frac{\lambda(p \to n)}{\lambda(p \to n) + \lambda(n \to p)} . \tag{3.2.12}$$

If X_n were larger or smaller than this, then the right-hand side of Eq. (3.2.7) would be large and respectively negative or positive, so X_n would be rapidly driven to the value (3.2.12). For $T > 10^{10}$ K the neutrino temperature was within 1% of the photon-electron-positron temperature, so at these temperatures the rates have the ratio (3.2.8). Thus at the temperatures $T > 3 \times 10^{10}$ K at which the neutron fraction is given by the equilibrium formula (3.2.12), this formula gives simply

$$X_n = \frac{1}{1 + \exp(Q/k_BT)} . \qquad (3.2.13)$$

It is crucially important that this ratio is fixed at high temperatures by the condition of zero lepton chemical potential, so that we do not need to make any *a priori* assumptions about the initial neutron/proton ratio. The results of a numerical integration of Eq. (3.2.7) with initial condition provided by Eq. (3.2.13) are presented in Table 3.2.[3]

We have seen that for $k_BT > m_e$ the ratios $\lambda(p \to n)/H$ and $\lambda(n \to p)/H$ varied roughly as T^3, so there was a rather sharp end to the equilibrium era, in which $\lambda \gg H$, at a temperature between 3×10^{10} K and 10^{10} K. A little later, at a temperature between 10^{10} K and 3×10^9 K the two-body and three-body neutron–proton conversion reaction rates became negligible, due in part to the disappearance of electron–positron pairs. Neutron–proton conversion continued mostly through the process of neutron decay, with a mean lifetime τ_n of 885.7±0.9 sec, so the neutron fraction became proportional to $\exp(-t/\tau_n)$. This is confirmed by fitting the numerical results presented in Table 3.2 with an exponential that decays with the observed rate of neutron decay, which gives

$$X_n \to 0.1609 \, \exp\left(\frac{-t}{885.7 \text{ sec}}\right) . \qquad (3.2.14)$$

The conversion of neutrons into protons was eventually stopped by the formation of complex nuclei, in which neutrons are stable. As a guide to this process, note that in thermal (and chemical) equilibrium the number density of a nuclear species i is given by formulas like Eqs. (2.3.1):

$$n_i = g_i e^{\mu_i/k_BT} \left(\frac{2\pi m_i k_BT}{h^2}\right)^{3/2} \exp(-m_i/k_BT) , \qquad (3.2.15)$$

where m_i is the mass of nucleus i and g_i is the number of its spin states. *If these nuclei can be built up rapidly out of Z_i protons and $A_i - Z_i$ neutrons,*

---

[3]These results are somewhat different from those given in Table 15.5 of G&C, because in 1972 only two types of neutrinos were known, and we are now assuming three types of massless neutrinos, which increases the expansion rate at a given temperature. We are also using the modern value 1.257 for the axial vector coupling constant of beta decay in Eq. (3.2.6), instead of the value 1.18 used in G&C.

Table 3.2: Neutron fraction X_n as a function of temperature or time (with neglect of the formation of complex nuclei).

T (K)	t(sec)	X_n
10^{12}	0.0001	0.4962
3×10^{11}	0.0011	0.4875
10^{11}	0.0099	0.4626
3×10^{10}	0.1106	0.3798
10^{10}	1.008	0.2386
3×10^{9}	12.67	0.1654
1.3×10^{9}	91.09	0.1458
1.2×10^{9}	110.2	0.1425
1.1×10^{9}	135.1	0.1385
10^{9}	168.1	0.1333
9×10^{8}	212.7	0.1268
8×10^{8}	274.3	0.1182
7×10^{8}	362.6	0.1070
6×10^{8}	496.3	0.0919
3×10^{8}	1980	0.0172
10^{8}	17780	3.07×10^{-10}

then the chemical potential of nuclei of type i is

$$\mu_i = Z_i\mu_p + (A_i - Z_i)\mu_n . \tag{3.2.16}$$

We can eliminate the unknown nucleon chemical potentials by forming the quantities

$$\frac{n_i}{n_p^{Z_i} n_n^{A_i-Z_i}} = \frac{g_i}{2^{A_i}} A_i^{3/2} \left(\frac{2\pi m_N k_B T}{h^2}\right)^{3(1-A_i)/2} e^{B_i/k_B T} , \tag{3.2.17}$$

where B_i is the binding energy, defined by

$$m_i = Z_i m_p + (A_i - Z_i)m_n - B_i . \tag{3.2.18}$$

(In deriving Eq. (3.2.17), we neglect the binding energy and the neutron–proton mass difference outside the exponential.) Eq. (3.2.17) can be expressed in terms of the dimensionless ratios

$$X_i \equiv n_i/n_N \ , \qquad X_p \equiv n_p/n_N \ , \qquad X_n \equiv n_n/n_N \ , \qquad (3.2.19)$$

where n_N is here the number density of *all* nucleons. In these terms,

$$X_i = \frac{g_i}{2} X_p^{Z_i} X_n^{A_i - Z_i} A_i^{3/2} \epsilon^{A_i - 1} e^{B_i/k_B T} \ , \qquad (3.2.20)$$

where ϵ is the dimensionless quantity

$$\epsilon \equiv \frac{1}{2} n_N h^3 (2\pi m_N k_B T)^{-3/2}$$

$$= 2.96 \times 10^{-11} \left(\frac{a}{10^{-10} a_0} \right)^{-3} \left(\frac{T}{10^{10} \mathrm{K}} \right)^{-3/2} \Omega_B h^2 \ , \quad (3.2.21)$$

in which we have used $n_N = 3\Omega_B H_0^2 (a_0/a)^3/8\pi G m_N$ for the number density of nucleons, bound or free. During the period of interest (after electron–positron annihilation) the temperature T goes as $1/a$, so Eq. (3.2.21) may be written

$$\epsilon = 1.46 \times 10^{-12} \left(\frac{T}{10^{10} \mathrm{K}} \right)^{+3/2} \Omega_B h^2 \ . \qquad (3.2.22)$$

The coefficient ϵ is very small for temperatures in the range of interest, so in equilibrium a nuclear species i is nearly absent until the temperature drops to the value

$$T_i \simeq \frac{B_i}{k(A_i - 1)|\ln \epsilon|} \ . \qquad (3.2.23)$$

For $\Omega_B h^2 \simeq 0.02$ this temperature is 0.75×10^9 K for deuterium, 1.4×10^9 K for H^3, 1.3×10^9 K for He^3, and 3.1×10^9 K for He^4, with only a very weak dependence on $\Omega_B h^2$. The binding energy per nucleon for heavier nuclei is similar to that of He^4, so they have similar values of T_i.

If thermal and chemical equilibrium were really maintained during the time that the temperature drops from around 10^{10} K to below 10^9 K, then during this time He^4 and heavier nuclei would appear first, followed by He^3 and H^3 (which would later beta decay to He^3), followed finally by H^2. But this is not what happens. The density at this time is too low for any but two-body reactions to compete with the expansion rate, so nuclei are built up by a chain of two-body processes: first $p + n \rightarrow d + \gamma$, then $d + d \rightarrow$ H$^3 + p$ and $d + d \rightarrow$ He$^3 + n$, and next $d +$ H$^3 \rightarrow$ He$^4 + n$ and $d +$ He$^3 \rightarrow$ He$^4 + p$, as well as slower processes involving photons.

There is no trouble with the first step. The rate of deuterium production per free neutron is

$$\lambda_d = 4.55 \times 10^{-20} \text{cm}^3/\text{sec} \times n_p = 511 \text{ sec}^{-1} \left(\frac{a}{10^{-9}a_0}\right)^{-3} X_p \Omega_B h^2$$

$$= 2.52 \times 10^4 \text{ sec}^{-1} \left(\frac{T}{10^{10}\text{K}}\right)^3 X_p \Omega_B h^2 . \qquad (3.2.24)$$

Multiplying by the time (3.1.27), this gives

$$\lambda_d t \simeq 4.5 \times 10^4 \left(\frac{T}{10^{10}\text{K}}\right) X_p \Omega_B h^2 , \qquad (3.2.25)$$

which remains substantially greater than unity until well after the temperature drops below 10^9 K. Therefore the deuterium abundance during the period of interest is given to a good approximation by its equilibrium value, which according to Eq. (3.2.20) is

$$X_d = 3\sqrt{2} X_p X_n \epsilon \exp\left(\frac{B_d}{k_B T}\right) . \qquad (3.2.26)$$

The trouble is that, because of the small binding energy of the deuteron, the temperature $T_d \approx 0.7 \times 10^9$ K is quite small, so deuterons remained rare until long after He^4 would have been abundant in thermal equilibrium. With deuterons rare, the two-deuteron processes $d + d \rightarrow H^3 + p$ and $d + d \rightarrow He^3 + n$ had small rates per deuteron, blocking further nucleosynthesis. (The rarity of deuterons had no effect on the rate *per deuteron* of radiative processes like $p + d \rightarrow He^3 + \gamma$ and $n + d \rightarrow H^3 + \gamma$, and these reactions are included in modern nucleosynthesis calculations, but they have intrinsically small cross sections.) When finally the temperature dropped below T_d the neutrons that were still extant were very rapidly assembled into the most deeply bound of the light elements, He^4. Further cosmological nucleosynthesis was blocked by the non-existence of stable nuclear species with atomic weight 5 or 8. (In stars this blockage is overcome[4] by the brief formation of the unstable Be^8 nucleus in collisions of two He^4 nuclei, followed by resonant capture of another He^4 nucleus to form an excited state of C^{12}, but the time available in the early universe was too short for this to have been effective then.) Thus, to a good approximation, the fraction by weight Y of He^4 formed in the early universe is just equal to twice[5] the fraction X_n of all nucleons that are neutrons (because in He^4 each neutron

---

[4] E. E. Salpeter, *Astrophys. J.* **115**, 326 (1952).

[5] In a little more detail, with n_{He} helium nuclei and n_H hydrogen nuclei per unit proper volume, the helium mass per unit proper volume in atomic mass units is $4n_{He}$, while the total mass per unit proper

is accompanied with a bound proton) at the time that deuterium becomes abundant enough to allow the build-up of heavier nuclei.

This actually occurs at a temperature somewhat above T_d. The first steps in building up heavier elements from deuterium are the reactions $d + d \to H^3 + p$ and $d + d \to He^3 + n$. These are exothermic processes, so for initial velocity $v \to 0$ these have cross sections proportional to $1/v$. At the temperatures of interest here, we have

$$\langle \sigma \left(d + d \to H^3 + p \right) v \rangle \simeq 1.8 \times 10^{-17} \, \text{cm}^3/\text{sec} \,,$$

$$\langle \sigma \left(d + d \to He^3 + n \right) v \rangle \simeq 1.6 \times 10^{-17} \, \text{cm}^3/\text{sec} \,,$$

so the total rate of these processes per deuteron is[6]

$$\Lambda = \left[\langle \sigma \left(d + d \to H^3 + p \right) v \rangle + \langle \sigma \left(d + d \to He^3 + n \right) v \rangle \right] X_d n_N$$

$$\simeq 1.9 \times 10^7 \left(T/10^{10} \text{K} \right)^3 \left(\Omega_B h^2 \right) X_d \, \text{sec}^{-1} \,.$$

This may be compared with the expansion rate, which after electron–positron annihilation is given by Eq. (3.1.27) as $H = 1/2t = 0.28(T/10^{10}\text{K})^2 \, \text{sec}^{-1}$, so for T in the neighborhood of 10^9 K, we have $\Lambda = H$ at $X_d \simeq 1.2 \times 10^{-7}/\Omega_B h^2$, which is 0.6×10^{-5} for $\Omega_B h^2 = 0.02$. This value is reached in thermal equilibrium at a temperature $\simeq 10^9$ K, with only a weak dependence on $\Omega_B h^2$, so nucleosynthesis began at around 10^9 K, not at 0.75×10^9 K. According to table 3.2, this happened when $t = 168$ seconds, so according to Eq. (3.2.14), the abundance by weight of helium formed at this time was about

$$Y_p \simeq 2 \times 0.1609 \times \exp(-168/885) \simeq 0.27 \,.$$

(The conventional subscript p stands for "primordial.") The larger the nucleon density, the higher the temperature at which nucleosynthesis began, and so the less time there was for neutron decay before nucleosynthesis, leading to a higher final He^4 abundance. The results of modern calculations are usually given for different values of η, the ratio of nucleons to photons, which according to Eqs. (2.1.12) and (2.1.13) is related to $\Omega_B h^2$ by

$$\Omega_B h^2 = 3.65 \, \eta \times 10^7 \,,$$

---

volume is $4n_{He} + n_H$, so the fractional abundance by weight of helium is $Y = 4n_{He}/(4n_{He} + n_H)$. But the numbers of protons and neutrons per unit proper volume are $n_p = 2n_{He} + n_H$ and $n_n = 2n_{He}$, so the fraction of nucleons that are neutrons is $X_n = 2n_{He}/(4n_{He} + n_H)$. Hence $Y = 2X_n$.

[6]V. Mukhanov, ref. 1.

so for instance $\Omega_B h^2 = 0.02$ corresponds to $\eta = 5.5 \times 10^{-10}$. (The nucleon density at a given temperature can be expressed in terms of η without knowing the present microwave background temperature, which made η a more convenient parameter than $\Omega_B h^2$ before the 1990s, when the present microwave background temperature was not yet accurately known.) Detailed calculations[7] give $Y_p = 0.232$ for $\eta = 2 \times 10^{-10}$ and $Y_p = 0.240$ for $\eta = 4 \times 10^{-10}$.

The primordial helium/hydrogen ratio is inferred from spectroscopic study of HII regions (regions of ionized hydrogen) containing low abundances of "metals" (elements other than hydrogen or helium), especially in blue compact galaxies which have not yet formed many stars. Observations in the 1990s were divided between those giving lower values[8] $Y_p = 0.234 \pm 0.002 \pm 0.005$, and higher values[9] $Y_p = 0.243 \pm 0.003$. Even with this division among the observers, the helium abundance clearly called for a hot universe, with η in the range of 10^{-10} to 5×10^{-10}. More recent observations of HII regions, combined with new atomic data used to interpret the spectroscopic observations, have led to a more precise determination:[10] $Y_p = 0.2477 \pm 0.0029$, corresponding to $\eta = (5.813 \pm 1.81) \times 10^{-10}$. The uncertainty in the value of η given by observations of helium abundance is still quite large, because Y_p is only weakly dependent on the baryon/photon ratio.

Helium abundance is more useful as a test of the expansion rate than of the value of η. For instance, if there were four flavors of massless neutrinos that went out of equilibrium at temperatures between 10^{11} K and 10^{10} K, then at temperatures below 3×10^9 K, the effective number of particle species would have been $2 + (7/8) \cdot 8 \cdot (4/11)^{4/3} = 3.817$ instead of 3.363, so the time required to drop from 3×10^9 K to any lower temperature would have been shortened by a factor $\sqrt{3.363/3.817} = 0.94$. Shortening the time to reach a given temperature increases the neutron fraction at that temperature, and hence increases the abundance of helium produced at the temperature $T \simeq 10^9$K of nucleosynthesis. It is actually the shortening of the time scale at temperatures between 3×10^{10} K and 3×10^9 K, when electron–positron pairs were disappearing, that would have the largest effect on the helium abundance. Detailed calculations show[11] that for each additional

[7]R. E. Lopez and M. S. Turner, *Phys. Rev. D* **59**, 103502 (1999) [astro-ph/9807279].

[8]K. A. Olive and G. Steigman, *Astrophys. J. Suppl.* **97**, 49 (1995); K. A. Olive, E. Skillman, and G. Steigman, *Astrophys. J.* **483**, 788 (1997).

[9]Y. I. Izotov, T. X. Thuan, and V. A. Lipovetsky, *Astrophys. J.* **435**, 647 (1994); *Astrophys. J. Suppl.* **108**, 1 (1997); Y. I. Izotov and T. X. Thuan, *Astrophys. J.* **497**, 227 (1998); **500**, 188 (1998).

[10]A. Peimbert, M. Peimbert, and V. Luridiana, *Astrophys. J.* **565**, 668 (2002); V. Luridiana, A. Peimbert, M. Peimbert, and M. Cerviño, *Astrophys. J.* **592**, 846 (2003); M. Peimbert, V. Luridiana, and A. Peimbert, *Astrophys. J.* **667**, (2007) [astro-ph/0701580].

[11]R. E. Lopez and M. S. Turner, ref. 7.

neutrino species this effect increases Y by an amount 0.01276 for $\eta = 10^{-10}$, and by 0.01369 for $\eta = 5 \times 10^{-10}$. The agreement between theory and observation for the helium abundance gave an upper bound of four light neutrino flavors before this was accurately measured to be just three flavors in measurements of the Z^0 decay width, part of which is due to the unobserved processes $Z^0 \rightarrow \nu + \bar{\nu}$. (Strictly speaking, the Z^0 width measures the number of neutrinos with masses less than $m_Z/2 = 45.6$ GeV, while the helium abundance measures the number of neutrinos with masses less than about 1 MeV.) Although the number of light neutrino flavors is now definitely known on the basis of Z^0 decay to be no greater than three, there may be other light particles left over from the very early universe that contribute to the expansion rate, and for these the helium abundance continues to provide useful upper bounds.

The nuclear reactions that built up helium from free neutrons at $T \approx 10^9$ K were not perfectly efficient, but left over a small residue,[12] like an unburned ash, of the light elements H^2, H^3, He^3, Li^7, and Be^7. The nuclei of H^3 decayed later by β^+ decay to He^3, and the nuclei of Be^7 decayed later by electron capture to Li^7, leaving us with H^2, He^3, and Li^7, as well as protons and He^4. The calculated abundances are shown in Figure 3.1.

The higher the baryon density, the more complete will be the incorporation of neutrons into He^4, and hence the smaller the resulting abundance of deuterium. We saw earlier that nucleosynthesis began when $X_d \simeq 1.2 \times 10^{-7}/\Omega_B h^2$, or $X_d \simeq 0.6 \times 10^{-5}$ for $\Omega_B h^2 = 0.02$. The deuterium fraction continued to rise for a while, as the temperature dropped and the exponential in Eq. (3.2.26) increased, but X_d then decreased again as the incorporation of free neutrons into deuterium reduced the factor X_n in Eq. (3.2.26), and deuterium was converted to H^3 and He^3, and thence to He^4. The final results, shown in Figure 3.1, are not very different from the deuterium fraction $X_d \simeq 1.2 \times 10^{-7}/\Omega_B h^2$ that we found at $T = 10^9$ K, and in particular exhibit a strong decline with increasing values of $\Omega_B h^2$.

The measurement of the deuterium abundance is complicated by the fact that deuterium has a small binding energy, and can readily be destroyed in stars. Any measurement of the deuterium abundance therefore gives a lower bound on the primordial deuterium abundance, and hence an upper bound on the baryon density.

In the past, the deuterium/hydrogen ratio was measured in various ways:

1. *Interstellar medium.* Spectroscopic studies of the interstellar medium gave a deuterium/hydrogen ratio[13] of $(1.60 \pm 0.09^{+0.05}_{-0.10}) \times 10^{-5}$.

---

[12]The calculated abundances cited below are given by S. Burles, K. M. Nollett, and M. S. Turner, *Astrophys. J.* **552**, L1 (2001).

[13]J. L. Linsky *et al.*, *Astrophys. J.* **402**, 694 (1993); **451**, 335 (1995).

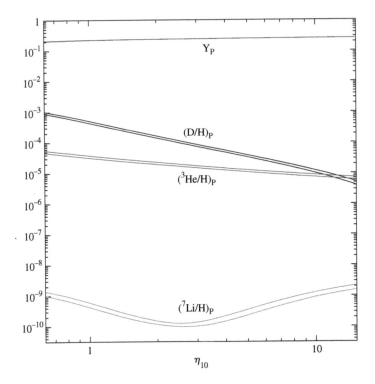

Figure 3.1: Calculated primordial abundances of deuterium, He^3, and Li^7 relative to hydrogen, and the fraction Y_P of the primordial mass of the universe in He^4, as functions of $\eta_{10} \equiv 10^{10}\eta$, where η is the ratio of nucleons to photons in the present universe. The widths of the bands indicate the effect of uncertainties in nuclear reaction rates. From G. Steigman, *Int. J. Mod. Phys.* **E15**, 1 (2006) [astro-ph/0511534].

2. *Solar wind.* Deuterium was converted to He^3 in the sun before it went onto the main sequence, so measurements of He^3 in the solar wind and meteorites is believed to give the total pre-solar value of He^3 and deuterium. Subtracting the abundance of He^3 obtained in other ways gave[14] a deuterium/hydrogen ratio of $(2.6 \pm 0.6 \pm 1.4) \times 10^{-5}$.

3. *Jovian atmosphere.* Spectroscopic studies of the Jovian atmosphere gave[15] a deuterium/hydrogen ratio of $(5 \pm 2) \times 10^{-5}$.

More recently, it has been possible to measure the deuterium/hydrogen ratio in very early intergalactic matter, by observing deuterium as well as hydrogen absorption lines in the spectra of quasistellar objects, due to

---

[14]K. A. Olive, lectures given at the Advanced School on Cosmology and Particle Physics, Peniscola, Spain, June 1998, and Theoretical and Observational Cosmology Summer School, Cargese, France, August 1998 [astro-ph/9901231], and references therein.
[15]H. B. Niemann *et al. Science* **272**, 846 (1996).

Table 3.3: Five quasi-stellar objects in whose spectra deuterium and hydrogen absorption lines are observed, together with the redshift of the intergalactic cloud responsible for the absorption, and the ratio of the numbers of deuterium and hydrogen atoms in the clouds inferred from the relative strength of the absorption lines, from ref. 17. (Observational uncertainties represent one standard deviation.)

QSO	z	D/H
PKS 1937-1009	3.572	$3.25 \pm 0.3 \times 10^{-5}$
Q1009+299	2.504	$3.98^{+0.59}_{-0.67} \times 10^{-5}$
HS 0105+1619	2.536	$2.54 \pm 0.23 \times 10^{-5}$
Q1243+3047	2.525675	$2.42^{+0.35}_{-0.25} \times 10^{-5}$
Q2206-199	2.0762	$1.65 \pm 0.35 \times 10^{-5}$

absorption in intervening intergalactic clouds of large redshift.[16] In 2003 the results of several years of observations of deuterium and hydrogen absorption lines in the spectra of four quasi-stellar objects were put together with results for one more QSO,[17] with the results shown in Table 3.3. The best value of the deuterium/hydrogen number ratio was found to be $2.78^{+0.44}_{-0.38} \times 10^{-5}$, from which is inferred a baryon/photon ratio $\eta = 5.9 \pm 0.5 \times 10^{-10}$, corresponding to

$$\Omega_B h^2 = 0.0214 \pm 0.0020 .$$

Even for H_0 as small as 50 km/sec/Mpc, it was clear that Ω_B is much less than the fraction that all non-relativistic matter contributes to the critical density, which we have seen had been given as $\Omega_M \simeq 0.2$ by studies of galaxy clusters and as $\Omega_M \simeq 0.3$ by the redshift–distance relation of type Ia supernovae. It is this discrepancy that provided the original evidence for non-baryonic dark matter in the universe.

The discrepancy has become sharper through measurements of anisotropies in the cosmic microwave background. As we will see in Section 7.2, from these measurements it is possible to infer that $\Omega_B h^2 = 0.0223^{+0.0007}_{-0.0009}$ and $\Omega_M h^2 = 0.127^{+0.007}_{-0.013}$, indicating a total mass density that is from 5 to 6 times larger than the density of ordinary baryonic matter. The nature of the missing matter is discussed in Section 3.4.

---

[16] For a summary of early results, see K. A. Olive, ref. 13; S. Sarker, talk at the Second International Workshop on Dark Matter in Astro- and Particle Physics, Heidelberg, July 1998 [astro-ph/9903183].

[17] D. Kirkman, D. Tytler, N. Suzuki, J. M. O'Meara, and D. Lubin, *Astrophys. J. Suppl.* **149**, 1 (2003) [astro-ph/0302006], and references cited therein.

The primordial abundance of He^3, like that of deuterium, is a monotonically decreasing function of the baryon/photon ratio. On the basis of a long term study of galactic HII regions and planetary nebula, a 2002 study[18] concluded that the He^3/H ratio in interstellar space is less than $1.1 \pm 0.2 \times 10^{-5}$, which is consistent with the value $1.04 \pm 0.06 \times 10^{-5}$ calculated for $\Omega_B h^2 = 0.0214$. But, unlike deuterium, He^3 is both produced and destroyed in stars, so it is not clear whether the observed interstellar abundance really represents the primordial abundance of He^3. Indeed, it had been thought that low-mass stars would inject a good deal of He^3 into the interstellar medium, in which case the apparent agreement between the amount observed in the interstellar medium and the amount expected from cosmological nucleosynthesis would actually represent a discrepancy.[19] This apparent discrepancy may have been removed by detailed calculations[20] of the movement of He^3 into the interior of these low-mass stars, which indicate that these stars do not in fact emit much He^3.

Some Li^6 is produced cosmologically, but in such small quantities (an abundance of about 10^{-13} to 10^{-14} that of hydrogen) that it has generally not been considered useful as a test of cosmological theories.[21] Much more attention has been given to Li^7. Its abundance has a more complicated dependence on the baryon/photon ratio, because Li^7 was formed in two different ways: directly, by the reactions $H^3 + He^4 \rightarrow Li^7 + \gamma$, and indirectly by $He^3 + He^4 \rightarrow Be^7 + \gamma$, followed much later by $e^- + Be^7 \rightarrow \nu + Li^7$. As we go to higher baryon densities, the amount of Li^7 produced directly increases at first, but then begins to decrease as Li^7 is destroyed in the reaction $p + Li^7 \rightarrow He^4 + He^4$. Eventually the indirect reaction takes over, and the Li^7 abundance rises again. The minimum Li^7/hydrogen ratio is calculated to be about 2×10^{-10}, and is reached at a baryon/photon ratio of about 3×10^{-10}, corresponding to $\Omega_B h^2 = 0.01$. The observed Li^7/H ratio was reported[22] in 2000 to be $2.07^{+0.15}_{-0.04} \times 10^{-10}$, close to this minimum. A subsequent study[23] of 63 dwarf stars in the galactic halo gave a Li^7/H ratio of $(2.37 \pm 0.05) \times 10^{-10}$. Either result is less than the value 3×10^{-10} predicted for $\Omega_B h^2 = 0.0214$, but it is plausible that although

[18]T. M. Bania, R. T. Rood, and D. S. Balser, *Nature* **415**, 54 (2002).

[19]N. Hata *et al.*, *Phys. Rev. Lett.* **75**, 3977 (1995); K. A. Olive *et al.*, *Astrophys. J.* **444**, 680 (1995).

[20]P. P. Eggleston, D. S. P. Dearborn, and J. C. Lattanzio, *Science* **314**, 1580 (2006).

[21]For a review, see E. Vangioni-Flam, M. Cassé, R. Cayrel J. Audouze, M. Spite, and F. Spite, *New Astron.* **4**, 245 (1999).

[22]T. K. Suzuki, Y. Yoshii, and T. C. Beers, *Astrophys. J.* **540**, 99 (2000) [astro-ph/0003164]. For earlier observations, see M. Spite and F. Spite, *Nature* **297**, 483 (1982); S. G. Ryan, J. E. Norris, and T. C. Beers, *Astrophys. J.* **523**, 654 (1999).

[23]J. Melendez and I. Ramirez, *Astrophys. J.* **615**, L33 (2004).

Li^7 is produced in stars and in the interaction of cosmic rays with matter, the Li^7 abundance has been depleted by convection in stellar atmospheres. Observation of Li^7 abundances in stars of varying temperature in the globular cluster NGC 6397 gave results in agreement with a theory of convective depletion of Li^7, with an assumed initial Li^7/H ratio equal to the expected value 3×10^{-10} (calculated taking $\Omega_B h^2$ to have the value estimated from deuterium abundance and microwave background anisotropies).[24] With the one possible exception of Li^7, there is now complete agreement between observations of light element abundances and calculations of cosmological nucleosynthesis, adopting the value $\Omega_B h^2$ provided by observations of anisotropies in the cosmic microwave background.

Although the baryon mass density inferred from cosmological nucleosynthesis and the cosmic microwave background is considerably less than the total mass density, it is also considerably greater than the density of baryonic matter observed in stars and luminous interstellar matter.[25] Some of this dark baryonic matter is in intergalactic space, but a fair fraction is believed to be present in galaxies, in the form of brown dwarf stars and clouds of hydrogen molecules.[26] But this is more a problem for the astrophysics of galaxies than for cosmology.

3.3 Baryonsynthesis and leptonsynthesis

We saw in the previous section that the ratio η of nucleons to photons at the time of nucleosynthesis had the tiny value $\simeq 5 \times 10^{-10}$. At earlier times, when the temperature was above 10^{13} K, nucleons would not yet have formed from their three constituent quarks, and there would have been roughly as many quark–antiquark pairs in thermal equilibrium as photons. But the conservation of baryon number (one-third the number of quarks minus the number of antiquarks) during the annihilation process tells us that before annihilation there must have been a slight excess, roughly of order η per photon, of quarks over antiquarks, so that some quarks would survive to form nucleons when all the antiquarks had annihilated with quarks. There was also a slight excess of electrons over positrons, to maintain the charge neutrality of the universe. It is conceivable that there is a compensating excess of antineutrinos over neutrinos, so that the total lepton number density (the number density of electrons, muons, tauons, and neutrinos, minus the number density of their antiparticles) vanishes, but it seems more natural to assume that before lepton–antilepton annihilation

---

[24]A. J. Korn *et al.*, *Nature* **442**, 657 (2006) [astro-ph/0608201].

[25]M. Fukugita, C. J. Hogan, and P. J. E. Peebles, *Astrophys. J.* **503**, 518 (1998).

[26]See e.g. M. Roncadelli, Recent Research Devel. *Astron. & Astrophys.* **1**, 407 (2003).

there was also a slight imbalance of leptons and antileptons, comparable to the excess of quarks over antiquarks.

These tiny imbalances in the numbers of quarks and antiquarks and of leptons and antileptons might be explained if the baryon and lepton number densities were generated by physical processes in a universe that at some early time had equal number of particles and antiparticles of all sorts. We could then hope to calculate η from first principles, and understand why it is so small. In 1967 Sakharov[1] outlined three conditions that must be met for this to be possible:

1. Obviously, in order for an excess of baryons over antibaryons or an inequality of leptons and antileptons to arise in a universe that begins with equal numbers of particles and antiparticles of each type, some physical processes must violate the conservation of baryon number or lepton number.

2. A universe with equal numbers of particles and antiparticles of each type (and each momentum and helicity) is invariant under the symmetry operators C (the exchange of particles with antiparticles) and CP (the exchange of particles with antiparticles, combined with a change of sign of all three-dimensional coordinate vectors), while a state with an excess of baryons over antibaryons or an imbalance of leptons and antileptons is clearly not invariant under either C or CP. Hence to produce such an state, some physical process must violate invariance under both C and CP. It is true that, whether or not C and/or CP are conserved, any relativistic quantum field theory will respect a symmetry[2] under CPT, the simultaneous exchange of particle with antiparticles, combined with a change of sign of all three-dimensional coordinate vectors, combined with a change in the direction of time's flow, but this does not prevent the production of baryon or lepton number, because the time-reversal symmetry T is violated by the expansion of the universe.

3. A little less obviously, in order to produce an excess of baryons over antibaryons or an imbalance of leptons and antileptons out of a state with equal numbers of particles and antiparticles, the universe must at some time depart from a state of thermal (including chemical) equilibrium. This is because in a state of thermal equilibrium, if baryon and/or lepton conservation are not respected, and all conserved quantities like electric charge vanish, then since chemical potentials must be conserved in all reactions, all chemical potentials must vanish. The CPT symmetry implies that even if C and CP are not conserved, the masses of particles and their

---

[1] A. D. Sakharov, *JETP Lett.* **5**, 24 (1967).
[2] See QTF, Vol. I, Sec. 5.8.

antiparticles are precisely equal, so with vanishing chemical potentials particles and antiparticles will have identical distribution functions, such as (3.1.28) with $\mu = 0$ for ideal gases. Thus, whatever the rates for various processes, no net baryon or lepton number will be produced.[3] An exception to this reasoning is presented at the end of this section.

All three conditions are now known to be satisfied:

(1) There is no direct experimental evidence for the nonconservation of baryon number, but a very weak baryon number nonconservation is expected according to modern views of the standard model of elementary particles. According to these views, the standard model is not a fundamental theory, which might be expected to be contain only interactions whose coupling parameters are either dimensionless or proportional to positive powers of mass, so that all infinities that arise in the standard model can be absorbed into a renormalization of these coupling parameters. Rather, we now think that the standard model is only an effective field theory, valid at energies much less than some fundamental mass scale M, which might be the Planck mass $M_P \equiv G^{-1/2} = 1.22 \times 10^{19}$ GeV, or perhaps the energy scale $\approx 10^{15}$ to 10^{16} GeV at which the three independent (suitably normalized) gauge coupling parameters of the standard model become equal. We would expect such an effective field theory to contain every possible interaction allowed by the gauge symmetries of the strong, weak, and electromagnetic interactions, but all but a finite number of these coupling parameters will have the dimensions of negative powers of mass. These "non-renormalizable" couplings are thus suppressed at energy $E \ll M$ by powers of E/M. Now, the gauge symmetries of the standard model do not allow any unsuppressed interactions among quarks and leptons that violate baryon or lepton number, so this picture makes it plausible that baryon and lepton number would be automatically conserved to a good approximation for energies $E \ll M$, even if baryon and lepton conservation are not respected by whatever fundamental theory describes physics at energies of order M. In other words, baryon and lepton number conservation may be mere "accidental" symmetries. In this case, there is no reason to exclude any suppressed interactions that violate conservation of baryon and lepton number. The least suppressed interactions of this sort are an interaction involving two lepton doublets and two scalar doublets, which is suppressed by a factor M^{-1} and violates lepton but not baryon conservation, and an interaction suppressed by a factor M^{-2} involving three quark fields and one lepton field, which violates both lepton and baryon

---

[3]For a more detailed argument and references to earlier discussions, see S. Weinberg, *Phys. Rev. Lett.* **42**, 850 (1978).

conservation.[4] (Also, as we shall see below, there is a quantum-mechanical violation of baryon-number and lepton-number conservation in the standard electroweak theory.) There is experimental evidence for the first but not the second interaction. With a coupling parameter of order $1/(10^{16} \text{ GeV})$, when the neutral scalar fields are replaced by their vacuum expectation values, this interaction provides a neutrino mass of order 10^{-2} eV, in good agreement with the results of neutrino oscillation experiments. (The existence of non-zero neutrino masses means that helicity $+1/2$ neutrinos, which are conventionally assigned lepton number $+1$, can be changed to helicity $-1/2$ neutrinos with lepton number -1 by a Lorentz transformation.) The second interaction would lead to decay processes like $p \rightarrow \pi^0 + e^+$, with decay rates of order $m_p^5/M^4\hbar$. Such events have not been seen, but could easily have escaped detection.

(2) The violation of invariance under C was discovered in 1957, while the violation of invariance under CP was discovered in 1964.[5]

(3) The expansion of the universe tends to pull states out of thermal equilibrium, either because the cooling temperature makes reaction rates decrease below the expansion rate, or because as it cools the universe goes through first-order phase transitions, similar to the condensation of water vapor or the freezing of liquid water. Although this in itself does not violate the conservation of baryon or lepton number, it opens the door for physical processes that do violate these conservation laws, as well as C and CP conservation, to create an imbalance between baryons and antibaryons and/or between leptons and antileptons.

All this just goes to show that it is possible for physical processes to produce a non-zero cosmological baryon and lepton number. It remains to find a specific theory in which the observed ratio of baryons to photons could be produced.[6] There are several theories of this type:

1. Delayed decay of heavy particles[7]

Suppose there is a species of heavy "X" particle, which decays into a pair of different channels, with baryon numbers B_1 and B_2 and lepton numbers L_1 and L_2, and branching ratios r and $1 - r$. (For instance, in some grand

---

[4]For a review, with references to the original literature, see QTF, Vol. II, Section 21.5.

[5]See QTF, Vol. I, Section 3.3.

[6]For the earliest attempts in this direction, see M. Yoshimura, *Phys. Rev. Lett.* **41**, 281 (1978); **42**, 746(E) (1979); S. Dimopoulos and L. Susskind, *Phys. Rev. D* **18**, 4500 (1979); *Phys. Lett.* **81B**, 416 (1979); A. Yu. Ignatiev, N. V. Krosnikov, V. A. Kuzmin, and A. N. Tavkhelidze, *Phys. Lett.* **76B**, 436 (1978); B. Toussaint, S. B. Treiman, F. Wilczek, and A. Zee, *Phys. Rev. D* **19**, 1036 (1978); J. Ellis, M. K. Gaillard, and D. V. Nanopoulos, *Phys. Lett.* **80B**, 360 (1979); **82B**, 464(E) (1979).

[7]S. Weinberg, *Phys. Rev. Lett.* **42**, 850 (1978).

unified theories there are "leptoquarks" with mass of order 10^{15} GeV, that decay into either two quarks, with baryon number 2/3 and lepton number zero, or into a lepton and antiquark, with baryon number $-1/3$ and lepton number $+1$.) The antiparticle will then decay into channels with baryon numbers $-B_1$ and $-B_2$ and lepton numbers $-L_1$ and $-L_2$, with branching ratios $\bar{r}$ and $1 - \bar{r}$. (Invariance under CPT tells us that any particle has the same *total* decay rate as its antiparticle, but as long as C and CP conservation are violated, it is possible for particles and antiparticles to have different branching ratios for their different decay channels.) The average total baryon number produced in the decay of one X particle and the decay of one of the corresponding antiparticles is then

$$\Delta B = rB_1 + (1 - r)B_2 - \bar{r}B_1 - (1 - \bar{r})B_2 = (r - \bar{r})(B_1 - B_2) , \quad (3.3.1)$$

and likewise for lepton number.

Similarly, there may be some heavy "N" particle that is its own antiparticle, and that decays with branching ratio r into one channel with baryon number B_1 and lepton number L_1, and with branching ratio $1 - r$ into the antichannel with baryon number $-B_1$ and lepton number $-L_1$. On the average, each decay produces a baryon number

$$\Delta B = rB_1 - (1 - r)B_1 , \quad (3.3.2)$$

and likewise for lepton number. If C and CP conservation are violated, then it is possible to have $r \neq 1 - r$, so that a net baryon number and lepton number may be produced. For instance, in some grand unified theories there are neutral fermions that decay both into scalar particles and leptons and into their antiparticles, producing a net lepton number if the branching ratios for these channels are unequal.

Such processes do not produce any net baryon or lepton number in equilibrium, because the inverse to the decay processes will destroy precisely as much baryon and lepton number as the decay processes create. The conditions for thermal equilibrium will be violated if H falls below the X-particle decay rate, but as long as $k_B T$ remained above all particle masses, whatever the rates of various processes, the expansion preserved the equilibrium form for all particle distributions, with a redshifted temperature $T \propto 1/a$. Specifically, if $k_B T$ was still above all particle masses when $H \equiv \dot{a}/a$ fell below the decay rate Γ_X, the inverse decay process would re-create as many X particles as had decayed. On the other hand, if at the time that $H \approx \Gamma_X$, $k_B T$ was less than the mass m_X of the X particles, then the Boltzmann factor $e^{-m_X/k_B T}$ would have blocked the inverse decay, and the X particles and antiparticles would have disappeared, yielding a net baryon number $(r - \bar{r})(B_1 - B_2)$ for each original heavy particle–antiparticle pair.

The condition for this to work is then that $k_B T \leq m_X$ at the temperature when $H = \Gamma_X$. We can estimate that (here taking $\hbar = c = 1$)

$$H = \sqrt{\frac{8\pi G a_B T^4 (\mathcal{N}/2)}{3}} = 1.66(k_B T)^2 \mathcal{N}^{1/2}/m_P \qquad (3.3.3)$$

where $m_P \equiv G^{-1/2} = 1.22 \times 10^{19}$ GeV, and $\mathcal{N}$ is the total number of helicity states of all elementary particles and antiparticles, with an extra factor 7/8 for fermions. (We assume here that the main contribution to the energy density of the universe at these temperatures comes from the large number $\mathcal{N}$ of highly relativistic particle types, rather than from the X particles themselves.) The decay rate of the X particle will be m_X times some dimensionless parameter α_X, which characterizes the strength of the interactions responsible for the decay and the number of decay channels. Hence decays start to be significant at a temperature T_X, given by

$$k_B T_X \approx \sqrt{\alpha_X m_X m_P/\mathcal{N}^{1/2}} \qquad (3.3.4)$$

For this to be less than m_X, we must have

$$m_X \geq \mathcal{N}^{-1/2} \alpha_X m_P . \qquad (3.3.5)$$

This is a fairly severe lower bound on m_X. For instance, if the X particles decay through ordinary electroweak interactions, then $\alpha_X \approx 10^{-2}$, so if $\mathcal{N} \approx 100$ we must have m_X greater than about 10^{16} GeV. From the point of view of theories that unify the strong and electroweak interactions, this is not an unreasonable value for the mass.

Assuming that this condition is satisfied, we can easily use Eq. (3.3.1) to make an estimate of η. At temperatures far above the heavy particle mass the number density of pairs of X particles and antiparticles is of the same order as the number density of photons. The entropy density (using energy units for temperature, with $k_B = 1$) at this time is of the order of $\mathcal{N}$ times the number density of photons, so the ratio of the number densities of X and $\bar{X}$ pairs to the entropy density is of the order of $1/\mathcal{N}$. With Eq. (3.3.4) satisfied, after the disappearance of these pairs the ratio of the baryon number to the entropy densities will be of order $(r - \bar{r})(B_1 - B_2)/\mathcal{N}$. The entropy density varies as a^{-3}, and *provided that baryon number is subsequently conserved* so does the baryon number density, and so the baryon number to entropy ratio will remain unchanged. The present entropy density is of the order of the photon number density, and at present the only baryons are nucleons, with no antinucleons, so we expect a nucleon to photon ratio

$$\eta \approx (r - \bar{r})(B_1 - B_2)/\mathcal{N} . \qquad (3.3.6)$$

Typically $B_1 - B_2$ is of order unity (in the leptoquark example mentioned above, it is equal to unity) but $r - \bar{r}$ is generally very small, both because CP conservation is weakly violated, and because the CPT theorem tells us that $r = \bar{r}$ in the lowest order of perturbation theory. The precise value of $r - \bar{r}$ is very model-dependent, but values of η of the desired order, 5×10^{-10}, appear quite natural.[8] This idea runs into difficulty in inflationary theories, which as we will see in Chapter 10 generally require that the temperature was never high enough to produce particles with masses satisfying Eq. (3.3.5).

2. Nonperturbative electroweak baryon and lepton number nonconservation

In the standard model of weak, electromagnetic, and strong interactions baryon and lepton number are automatically conserved to all orders of perturbation theory, but not when certain non-perturbative effects are taken into account.[9] This produces reactions that violate baryon and lepton number conservation, but such reactions are suppressed at low temperatures by a factor $\exp(-8\pi^2/g^2) \simeq 10^{-162}$, where g is the $SU(2)$ electroweak coupling constant. This tiny exponential is actually a barrier penetration factor, which accompanies the quantum mechanical tunneling transition through the barrier between topologically different configurations of the gauge fields. At high temperatures, above about 300 GeV, thermal fluctuations allow passage *over* this barrier, and the exponential suppression disappears.[10] Nevertheless, by themselves these reactions do not produce an appreciable net baryon or lepton number, both because they take place at a time of nearly perfect thermal equilibrium,[11] and because they are suppressed by small parameters associated with the need to violate CP conservation as well as the conservation of baryon and lepton number.[12]

3. Leptogenesis[13]

Although the non-perturbative effects of electroweak interactions described in the previous paragraph do not by themselves provide a way of accounting

---

[8]D. V. Nanopoulos and S. Weinberg, *Phys. Rev. D* **20**, 2484 (1979).

[9]G. 't Hooft, *Phys. Rev. Lett.* **37**, 8 (1976). Also see QTF, Vol. II, Section 23.5.

[10]V. A. Kuzmin, V. A. Rubakov, and M. E. Shaposhnikov, *Phys. Lett.* **155B**, 36 (1985). The transition between field configurations is dominated by intermediate field configurations known as sphalerons; see N. S. Manton, *Phys. Rev. D* **28**, 2019 (1983); F. R. Klinkhammer and N. S. Manton, *Phys. Rev. D* **30**, 2212 (1984); R. F. Dashen, B. Hasslacher, and A. Neveu, *Phys. Rev. D* **10**, 4138 (1974).

[11]The absence of a first-order phase transition in the electroweak theory is shown by K. Kajantie, M. Laine, K. Rummukainen, and M. Shaposhnikov, *Nucl. Phys. B* **466**, 189 (1996); K. Rummukainen, M. Tsypin, K. Kajantie, and M. Shaposhnikov, *Nucl. Phys. B* **532** (1998); F. Csikor, Z. Fodor, and J. Heitger, *Phys. Rev. Lett.* **82**, 21 (1999); and earlier references cited therein.

[12]M. E. Shaposhnikov, *JETP Lett.* **44**, 465 (1986); *Nucl. Phys. B* **287**, 757 (1987).

[13]M. Fukugita and T. Yanagida, *Phys. Lett. B* **174**, 45 (1986). For a review, see W. Buchmüller, R. D. Peccei, and T. Yanagida, *Ann. Rev. Nucl. Part. Sci.* **55**, 311 (2005) [hep-ph/0502169].

for the observed baryon/photon ratio, they can convert a cosmological lepton number density into a baryon number density, or vice versa. The only truly conserved quantum numbers in the $SU(3) \times SU(2) \times U(1)$ standard model are those associated with gauge symmetries — the electroweak isospin generator T_3, the electroweak hypercharge Y (defined so that the electric charge in units of e is $T_3 - Y$), and a pair of generators of the $SU(3)$ gauge group of quantum chromodynamics — together with $B-L$, the total baryon number minus the total lepton number, which is conserved because of a cancelation between Feynman diagrams containing loops of quarks or leptons. This creates a further problem for the proposal that the observed baryon number density of the universe is created in the decay of a leptoquark into both two quarks and into a lepton and antiquark. Although these channels have different values for B and L, they both have the same value (equal to 2/3) for $B - L$. Hence if the universe starts with equal numbers of particles and antiparticles of all types, then even if leptoquark decay produces equal non-zero baryon and lepton number densities, all truly conserved quantities will remain zero, so when thermal equilibrium is established at lower temperatures it will be with zero values for all chemical potentials. (This is demonstrated below.) Such a state has equal numbers of particles and antiparticles, and hence zero densities of B and L as well as $B - L$.

On the other hand, if some heavy particle (such as the N particle mentioned above) in the early universe decays in such a way as to produce a non-zero density of $B - L$ this will persist through the period of thermal equilibrium, though the relative densities of B and L may change. In general, whatever mixture of baryon and lepton number is produced when the heavy particle decay, *and even if only lepton number is produced*, we would expect the densities of baryon and lepton number to be comparable in a subsequent period of thermal equilibrium.

This can be made quantitative.[14] Suppose in thermal equilibrium there are a set of conserved quantum numbers Q_a, such as T_3, Y, and $B - L$. Suppose also that there are several species i of particles in equilibrium, such as quarks, leptons, etc., each carrying a value q_{ai} for the quantum number Q_a. The chemical potentials μ_i for these particles must be conserved for all reactions in thermal equilibrium, which requires that they be linear combinations of the conserved quantum numbers:

$$\mu_i = \sum_a q_{ai}\mu_a ,\qquad\qquad (3.3.7)$$

with coefficients μ_a that can be regarded as chemical potentials for the different conserved quantities. The densities of the different particle species

[14]J. A. Harvey and M. S. Turner, *Phys. Rev. D* **42**, 3344 (1990).

can then be expressed as functions of the μ_a and the temperature, and these relations can be used to calculate the densities n_a of the different conserved quantum numbers as functions of the μ_a and the temperature. But there are just as many n_a as there are μ_a, so these relations can be inverted to give the μ_a in terms of the n_a and the temperature, from which we can calculate the density of anything else as functions of the n_a and the temperature, including the density of a nonconserved quantity like baryon or lepton number that is not among the Q_a.

In the case that interests us here, at temperatures above about 10^{16} K, all particles of the Standard Model are highly relativistic, so that their masses can be neglected. The number density of particle species i is then

$$
\begin{aligned}
n_i &= \frac{g_i}{(2\pi\hbar)^3} \int \frac{d^3p}{e^{(p-\mu_i)/k_BT} \mp 1} \\
&= 4\pi g_i \left(\frac{k_BT}{2\pi\hbar}\right)^3 \int_0^\infty \frac{x^2\, dx}{e^{x-\mu_i/k_BT} \mp 1},
\end{aligned}
\tag{3.3.8}
$$

where g_i is the number of helicity (and other sources of multiplicity) states for each species, and the $\mp$ sign is $-$ for bosons and $+$ for fermions. The antiparticle density $\bar{n}_i$ will be given by the same formula, but with μ_i replaced with $-\mu_i$, so the difference is

$$
n_i - \bar{n}_i = 8\pi g_i \left(\frac{k_BT}{2\pi\hbar}\right)^3 \sinh\left(\frac{\mu_i}{k_BT}\right) \int_0^\infty \frac{x^2 e^x\, dx}{e^{2x} \mp 2e^x \cosh(\mu_i/k_BT) + 1}.
\tag{3.3.9}
$$

In the situation that concerns us here, the imbalance between particles and antiparticles of all sorts is small, so $|\mu_i| \ll 1$ for all particle species. In this case,

$$
n_i - \bar{n}_i = 8\pi g_i \left(\frac{k_BT}{2\pi\hbar}\right)^3 \frac{\mu_i}{k_BT} \int_0^\infty \frac{x^2 e^x\, dx}{(e^x \mp 1)^2}.
\tag{3.3.10}
$$

The integral over x has the value $\pi^2/3$ for bosons and $\pi^2/6$ for fermions, so we can write this as

$$
n_i - \bar{n}_i = f(T)\,\tilde{g}_i\,\mu_i,
\tag{3.3.11}
$$

where

$$
f(T) \equiv \frac{4\pi^3}{3} \frac{(k_BT)^2}{(2\pi\hbar)^3},
\tag{3.3.12}
$$

and $\tilde{g}_i$ is the number of spin states, but with an extra factor of 2 for bosons. Using Eq. (3.3.7), Eq. (3.3.11) becomes

$$
n_i - \bar{n}_i = f(T)\,\tilde{g}_i \sum_a q_{ai}\mu_a,
\tag{3.3.13}
$$

In particular, the density of the conserved quantum number Q_a is

$$n_a = \sum_i q_{ai}\left(n_i - \bar{n}_i\right) = f(T) \sum_b M_{ab}\, \mu_b \,, \qquad (3.3.14)$$

where M is the matrix

$$M_{ab} \equiv \sum_i \tilde{g}_i\, q_{ai}\, q_{bi} \,. \qquad (3.3.15)$$

This matrix is positive-definite (in the sense that, for any set of real numbers ξ_a, we have $\sum_{ab} M_{ab}\xi_a\xi_b > 0$, unless all ξ_a vanish), and therefore it has an inverse M^{-1}. We can thus invert the relation (3.3.14), and find $\mu_a = \sum_b M_{ab}^{-1} n_b / f(T)$. Using this in Eq. (3.3.13) gives

$$n_i - \bar{n}_i = \sum_{ab} \tilde{g}_i\, q_{ai}\, M_{ab}^{-1}\, n_b \qquad (3.3.16)$$

for any particle species i. Note in particular that if the densities n_b of all the conserved quantum numbers vanish, then there is an equal number of particles and antiparticles of every kind, as mentioned above.

In order to deal with the case where some conserved quantities such as $B - L$ have non-zero densities, we need to calculate the matrix M_{ab}. The particles of the Standard Model are listed in Table 3.4. For N_g generations of quarks and leptons and N_d scalar doublets, the independent elements of the matrix M_{ab} are

$$M_{B-L\,B-L} = \frac{13\,N_g}{3}\,, \quad M_{B-L\,Y} = -\frac{8\,N_g}{3}\,, \quad M_{Y\,Y} = \frac{10\,N_g}{3} + N_d \,. \qquad (3.3.17)$$

We don't need any of the matrix elements involving T_3, because the sum of the T_3 values vanishes for all the particles with any given values of $B - L$ and Y, so that $M_{B-L\,T_3} = M_{Y\,T_3} = 0$. That is, the matrix M_{ab} is block-diagonal, with a 2×2 block having a and b running over $B - L$ and Y. This has the consequence that without bothering to calculate $M_{T_3\,T_3}$, we can calculate that

$$M_{B-L\,B-L}^{-1} = \frac{10\,N_g}{3D} + \frac{N_d}{D}\,, \quad M_{B-L\,Y}^{-1} = \frac{8\,N_g}{3D}\,, \quad M_{Y\,Y}^{-1} = \frac{13\,N_g}{3D}\,, \qquad (3.3.18)$$

where D is the determinant

$$D = \frac{22N_g^2}{3} + \frac{13 N_g N_d}{3} \,. \qquad (3.3.19)$$

(Similarly, because the sum of the color quantum numbers vanishes for all the particles with any given values of $B - L$ and Y, we do not need to take

Table 3.4: Particles of the Standard Model, together with the number $\tilde{g}$ of their helicity and color states (with an extra factor 2 for bosons), and the values of their baryon number, lepton number, and gauge quantum numbers. Only one "generation" of quarks and leptons and only one doublet of scalar fields are shown. The subscripts L and R denote the helicity states of quarks u and d and leptons ν and e. Antiparticles are not shown separately, and the photon and Z^0 are not shown because they are their own antiparticles, and so do not contribute to the densities of any quantum numbers. Color quantum numbers are not shown, for reasons given in the text.

Particle	$\tilde{g}$	B	L	T_3	Y
u_L	3	1/3	0	1/2	−1/6
d_L	3	1/3	0	−1/2	−1/6
u_R	3	1/3	0	0	−2/3
d_R	3	1/3	0	0	1/3
ν_L	1	0	1	1/2	1/2
e_L	1	0	1	−1/2	1/2
e_R	1	0	1	0	1
W^+	4	0	0	1	0
φ^+	2	0	0	1/2	−1/2
φ^0	2	0	0	−1/2	−1/2
gluons	4	0	0	0	0

the color quantum numbers into account here.) Thus if $B - L$ is the only conserved quantum number with a non-vanishing number density, then Eqs. (3.3.16) and (3.3.18) tell us that the baryon number density in thermal equilibrium is

$$n_B \equiv \sum_i B_i(n_i - \bar{n}_i) = \sum_i \tilde{g}_i B_i \left((B-L)_i M^{-1}_{B-L\ B-L} + Y_i M^{-1}_{Y\ B-L} \right) n_{B-L}$$

$$= \left(\frac{4}{3} M^{-1}_{B-L\ B-L} - \frac{2}{3} M^{-1}_{Y\ B-L} \right) N_g\, n_{B-L}$$

$$= \left(\frac{8\,N_g + 4\,N_d}{22\,N_g + 13\,N_d} \right) n_{B-L} \tag{3.3.20}$$

For instance, in the minimal experimentally allowed case, with $N_g = 3$ and $N_d = 1$, this gives $n_B = (28/79)n_{B-L}$. In any case, n_B turns out to be of the same order of magnitude as n_{B-L}, as anticipated above. The reason for the smallness of n_B/n_γ in this scenario would be traced to the smallness of CP violation in the out-of-equilibrium heavy particle decay that produces a non-vanishing density of $B - L$.

4. Affleck–Dine mechanism[15]
It is possible for baryon number nonconservation to occur in the nonequilibrium dynamics of a scalar field that carries a non-zero baryon number. Both baryon-number conservation and CP-invariance need to be violated in the Lagrangian of the scalar field. Such theories find their motivation in supersymmetry, which lies outside the scope of this book.

5. Equilibrium baryon synthesis
The violation of CPT by the expansion of the universe means that the part of the Hamiltonian that is odd in C and CP can have a non-zero expectation value ΔE_i in the state of a single baryon of type i, and of course opposite expectation value $-\Delta E_i$ in the state of the corresponding antibaryon. Then even the universe starts in a state of zero baryon number, if it enters a state of thermal equilibrium in which baryon-number nonconserving processes occur rapidly, although there will be no chemical potential associated with baryon number, the difference in energy of baryons and antibaryons will lead to a net baryon number. When baryon-number nonconserving processes become ineffective the resulting baryon number density will survive, simply decreasing as a^{-3}. As long as baryon-conserving collisions remain sufficiently rapid, the one-particle distribution will have the form appropriate for thermal equilibrium, but now with a non-vanishing baryonic chemical potential.

As a general class of theories of this sort, suppose that in the expanding universe, there is a term in the Lagrangian density of the form

$$\Delta \mathcal{L}(x) = -\sqrt{-\mathrm{Det}g}\, V_\mu(x)\, J_B^\mu(x) , \qquad (3.3.21)$$

where $V_\mu(x)$ is a classical vector field and $J_B^\mu(x)$ is the current associated with baryon number (for which $J_B^0(x)$ is the baryon density, which is odd under C and CP). Two proposals of this sort have been made: the vector field could be $V_\mu = M^{-1}\partial_\mu \varphi$, where φ is some scalar field[16] and M is some large mass; or the vector field could be $V_\mu = M^{-2}\partial_\mu R$, where R is the curvature scalar[17] and again M is some large mass. In any case, the isotropy and homogeneity of the Robertson–Walker metric requires that

---

[15] L. Affleck and M. Dine, *Nucl. Phys.* B **249**, 361 (1985); M. Dine, L. Randall, and S. Thomas, *Phys. Rev. Lett.* **75**, 398 (1995); *Nucl. Phys.* B **458**, 291 (1996).

[16] A. G. Cohen and D. B. Kaplan, *Phys. Lett.* B **199**, 251 (1987). The production of baryon number is suppressed when the scalar field oscillates rapidly; see A. Dolgov, K. Freese, R. Rangarajan, and M. Srednicki, *Phys. Rev.* D **56**, 6155 (1997) [hep-ph/9610405].

[17] H. Davoudiasl, R. Kitano, G. D. Kribs, H. Murayama, and P. J. Steinhardt, *Phys. Rev. Lett.* **93**, 201301 (2004).

V_i vanishes, while V_0 is a function only of time. This interaction then shifts the energy of the state of a single particle of type i with baryon number b_i by an amount $\Delta E_i = V_0(t) b_i$. In thermal (including chemical) equilibrium, the baryon number density will be

$$n_B(t) = \sum_i b_i g_i \int \frac{4\pi p^2 \, dp}{(2\pi\hbar)^3} \left[e^{(\sqrt{p^2+m_i^2}+b_i V_0(t))/k_B T(t)} \mp 1 \right]^{-1}, \quad (3.3.22)$$

where the sum over i runs over all particle (and antiparticle) types; g_i and b_i are the number of spin states and the baryon number of a particle of type i; and the upper and lower signs again apply to bosons and fermions, respectively. If baryon non-conserving collisions shut off suddenly at time t_1, then subsequently the baryon number density will be

$$n_B(t) = \left(\frac{a(t_1)}{a(t)} \right)^3 \sum_i b_i g_i \int \frac{4\pi p^2 \, dp}{(2\pi\hbar)^3} \left[e^{(\sqrt{p^2+m_i^2}+b_i V_0(t_1))/k_B T(t_1)} \mp 1 \right]^{-1}.$$

$$(3.3.23)$$

For $V_0 = 0$ the cancelation between baryons and antibaryons of course makes this vanish. Since the baryon/entropy ratio is known to be small, we expect $V_0(t_1)$ to be small; to first order in $V_0(t_1)$, the baryon number density for $t > t_1$ will be

$$n_B(t) = \left(\frac{a(t_1)}{a(t)} \right)^3 \frac{V_0(t_1)}{k_B T(t_1)} \sum_i b_i^2 g_i$$

$$\times \int \frac{4\pi p^2 \, dp}{(2\pi\hbar)^3} e^{\sqrt{p^2+m_i^2}/k_B T(t_1)} \left[e^{\sqrt{p^2+m_i^2}/k_B T(t_1)} \mp 1 \right]^{-2}. \quad (3.3.24)$$

A similar mechanism could also be responsible for lepton synthesis.

The crucial confirmation of any theory of baryon synthesis would be a successful prediction of the present baryon/photon ratio. So far, none of the proposals discussed here are anywhere near this goal.

3.4 Cold dark matter

We saw in Section 3.2 that considerations of cosmological nucleosynthesis lead to the conclusion that most of the mass in the universe is *not* in the form of ordinary baryonic matter, i.e. atomic nuclei and electrons. We will see in Chapter 7 that this conclusion is powerfully reinforced by observations of anisotropies in the cosmic microwave background. So we face the question, if the particles that make up most of the mass of the universe are not baryons, then what are they?

We know that this matter is *dark*, in the sense that it does not interact significantly with radiation, both because we don't see it, and also because it has not lost its kinetic energy sufficiently to relax into the disks of galaxies, as has baryonic matter. This means in particular that these particles must be electrically neutral.[1] Detailed studies of the dynamics of galaxy clusters indicate that the dark matter particles must also be *cold*, in the sense that their velocities are highly non-relativistic.[2]

The study of a double galaxy cluster 1E0657-558 (the "bullet cluster," with $z = 0.296$) has provided vivid direct evidence of the existence of dark matter, which does not have non-gravitational interactions with itself or with ordinary baryonic matter.[3] The galaxies in this cluster are mostly grouped into two distinct subclusters, while hot gas (observed through its emission of X-rays) is concentrated between these subclusters. The interpretation is that two clusters of galaxies have collided; the galaxies which have little chance of close encounters have mostly continued on their original paths, while the two clouds of hot gas that previously accompanied them have collided and remained closer to the center of the double cluster. The total matter density in 1E0657-558 is mapped out through its effect in gravitationally deflecting light from more distant galaxies along the same line of sight. (Gravitational lensing is discussed in Chapter 9.) In this way, it is found that most of the matter in 1E0657-558 is not associated with the hot gas, but like the galaxies forms two subclusters that have evidently passed through each other without appreciable interaction. The ratio of the mass in hot gas to the mass in all matter is estimated to be about 1/6, in line with the value of Ω_B/Ω_M previously inferred from measurements of deuterium abundance and luminosity distance as a function of redshift, or from anisotropies in the cosmic microwave background.

Elementary particle theory offers several candidates for the particles making up the cold dark matter.

A. Weakly interacting massive particles (WIMPs)

Massive particles may survive to the present if they carry some sort of conserved additive or multiplicative quantum number. If there is a non-zero

---

[1] There are particularly strong limitations on the number density of any sort of charged stable particles that might be left over from the big bang, which are set by mass spectroscopy, the analysis of the charged particles contained in samples of matter according to their ratio of mass to charge. The number of electrically charged exotic particles with masses in the range of 6 GeV to 330 GeV has been found to be less than 10^{-21} of the number of nucleons, by P. F. Smith and J. R. J. Bennett, *Nucl. Phys.* **B 149**, 525 (1979).

[2] P. J. E. Peebles, *Astrophys. J.* **263**, L1 (1983); G. R. Blumenthal, S. M. Faber, J. R. Primack, and M. J. Rees, *Nature* **311**, 517 (1984).

[3] D. Clowe *et al.*, *Astrophys. J.* **648**, L109 (2006) [astro-ph/0608407].

chemical potential associated with this quantum number, then of course some particles (or antiparticles) must be left over after all the antiparticles (or particles) have annihilated. But even if there are no non-zero chemical potentials for these particles, so that the initial number densities of particles and antiparticles are equal, if they can only annihilate with their antiparticles then once their number density becomes sufficiently low the collision rate eventually becomes too small to reduce the density further.[4] We will call these particles L-particles (for "left-over"), to distinguish them from the other particles into which they may annihilate, which we will assume are all approximately in thermal and chemical equilibrium during the period of annihilation. The annihilation rate per particle of the L particles and antiparticles is $n\langle\sigma v\rangle$, where n is their number density, and $\langle\sigma v\rangle$ is the average value of the product of annihilation cross section and relative velocity. The rate of decrease in the number of L particles in a co-moving volume a^3 is then $na^3 \times n\langle\sigma v\rangle$. There is also an n-independent rate of creation of these particle–antiparticle pairs from the thermal background. Since this must balance the annihilation rate when everything is in equilibrium, in general the creation rate per volume a^3 must equal $n_{eq}^2 a^3 \langle\sigma v\rangle$, where n_{eq} is the number density of L particles and of antiparticles in equilibrium. The number na^3 of L particles and of antiparticles in a co-moving volume a^3 is therefore governed by a Boltzmann equation

$$\frac{d(na^3)}{dt} = -\left(n^2 - n_{eq}^2\right)a^3 \langle\sigma v\rangle . \tag{3.4.1}$$

For very high temperatures with $k_B T \gg m_L$ the equilibrium density n_{eq} varies as T^3, and T varies as $1/a$, so Eq. (3.4.1) has a solution $n = n_{eq}$. Eventually, with the decrease in temperature below the L-particle mass, the equilibrium density drops so low that the creation term in Eq. (3.4.1) becomes negligible, and we have

$$\frac{d(na^3)}{dt} = -n^2 a^3 \langle\sigma v\rangle . \tag{3.4.2}$$

The solution of Eq. (3.4.2) is

$$\frac{1}{n(t)a^3(t)} = \text{constant} + \int \frac{\langle\sigma v\rangle dt}{a^3(t)} ,$$

---

[4] B. W. Lee and S. Weinberg, *Phys. Rev. Lett.* **39**, 165 (1977); D. D. Dicus, E. W. Kolb, and V. L. Teplitz, *Phys. Rev. Lett.* **39**, 168 (1977); E. W. Kolb and K. A. Olive, *Phys. Rev.* D **33**, 1202 (1986).

or, in other words,

$$n(t)a^3(t) = \frac{n(t_1)a^3(t_1)}{1 + n(t_1)\,a^3(t_1)\int_{t_1}^{t}\langle\sigma v\rangle\,a^{-3}(t')\,dt'}\,,$$

where t_1 is any convenient time chosen late enough so that the creation term in Eq. (3.4.1) may be neglected for $t' > t_1$.

The important point here is that the integral in the denominator converges for $t \to \infty$. The denominator $a^3(t)$ increases like $t^{3/2}$ when the energy density is dominated by relativistic particles, and even faster later, when it is dominated by non-relativistic particles and/or vacuum energy. Also, if annihilation is possible from states of zero orbital angular momentum then σv approaches a constant for low energies, so its thermal average $\langle\sigma v\rangle$ approaches a constant for low temperatures, and hence for late times. The contribution of states of higher orbital angular momentum decreases with decreasing temperature, so if s wave annihilation is forbidden by selection rules the integral converges even faster. Because the integral converges, the particle number in a co-moving volume a^3 approaches a finite limit:

$$n(t)a^3(t) \to \frac{n(t_1)a^3(t_1)}{1 + n(t_1)\,a^3(t_1)\int_{t_1}^{\infty}\langle\sigma v\rangle\,a^{-3}(t')\,dt'}\,. \qquad (3.4.3)$$

Let us assume that the annihilation of L particles and antiparticles took place during a time when the density of the universe was dominated by relativistic particles, so that $a \propto 1/T$, and the time is given by Eq. (3.1.14):

$$dt = -2\sqrt{\frac{3}{16\pi\,G\mathcal{N}\,a_B}}\frac{dT}{T^3}\,,$$

or, using the formula $a_B = \pi^2 k_B^4/15$ with $\hbar = c = 1$,

$$dt = -\sqrt{\frac{45}{4\pi^3\,G\mathcal{N}}}\,m_L^{-2}\frac{dx}{x^3} \qquad (3.4.4)$$

where $x \equiv k_B T/m_L$. Eq. (3.4.1) therefore takes the form

$$\frac{du(x)}{dx} = B\left[u^2(x) - u_{eq}^2(x)\right] \qquad (3.4.5)$$

where $u(x)$ is the dimensionless quantity of interest

$$u \equiv n/(k_B T)^3\,, \qquad (3.4.6)$$

$u_{eq}(k_B T/m_L)$ is its equilibrium value, and B is the dimensionless parameter

$$B = \sqrt{\frac{45}{4\pi^3\,G\mathcal{N}}}\,m_L\langle\sigma v\rangle\,. \qquad (3.4.7)$$

We see that the left-over value of $n/(k_B T)^3$ depends only on B, and on the spin of the L particles, which we need to know to give a formula for $u_{eq}(x)$:

$$u_{eq}(x) = \frac{2s_L + 1}{(2\pi)^3} \int_0^\infty \frac{4\pi y^2 \, dy}{\exp \sqrt{x^{-2} + y^2} \pm 1} \tag{3.4.8}$$

where $y \equiv p/k_B T$, and as usual the sign is $+1$ for fermions and -1 for bosons.

For instance, for $k_B T \ll m_L$ the heavy particles are non-relativistic, and the mean value of the low-energy annihilation cross section times velocity is a constant,

$$\langle \sigma v \rangle = G_{wk}^2 m_L^2 \mathcal{F}/2\pi , \tag{3.4.9}$$

where $G_{wk} = 1.1664 \times 10^{-5} \text{ GeV}^{-2}$ is the weak coupling constant, and $\mathcal{F}$ is a fudge factor to take account of the number of annihilation channels and the details of the interaction responsible for the annihilation. This gives

$$B = 1.59 \times 10^8 \left(m_L \, [\text{GeV}] \right)^3 \mathcal{F} \mathcal{N}^{-1/2} . \tag{3.4.10}$$

The solution of Eq. (3.4.5) for various values of B and $s_L = 1/2$ is shown in Figure 3.2. We see that $u(x)$ drops steeply from a constant B-independent value for $x > 1$ to a constant B-dependent value for $x < 0.01$.

For m_L in the GeV range B is quite large, and for such values of B (and $s_L = 1/2$) the asymptotic value of u is well approximated by[4]

$$u(0) \simeq 6.1 \, B^{-0.95} . \tag{3.4.11}$$

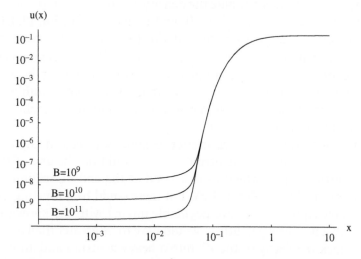

Figure 3.2: The function $u(x) \equiv n/(k_B T)^3$ vs $x \equiv k_B T/m_L$, for a dark matter particle of spin $1/2$ and mass m_L, with various values of the parameter B defined by Eq. (3.4.7).

(A value $u(0) \propto B^{-1}$ is what we would expect if the second term in the denominator in Eq. (3.4.3) were much greater than unity.) The present mass density of these heavy particles and antiparticles is then (recalling that electron–positron annihilation increases the photon temperature by a factor $(11/4)^{1/3}$)

$$\rho_L = 2m_L u(0)(4/11)(k_B T_{\gamma 0})^3$$
$$= \left(2.15 \times 10^{-28} \text{g/cm}^3\right) \left(m_L(\text{GeV})\right)^{-1.85} \left(\mathcal{F}/\sqrt{\mathcal{N}}\right)^{-0.95}, \quad (3.4.12)$$

or in other words

$$\Omega_L \equiv 8\pi G \rho_L/3 H_0^2 = 11.5\, h^{-2} \left(m_L(\text{GeV})\right)^{-1.85} \left(\mathcal{F}/\sqrt{\mathcal{N}}\right)^{-0.95}. \quad (3.4.13)$$

Note that Ω_L is a *decreasing* function of m_L, because heavy L particles annihilate more effectively than light ones. If we assume that the left-over heavy particles make up most of the cosmic mass density, then $\Omega_M \simeq \Omega_L$, and so

$$m_L(\mathcal{F}/\sqrt{\mathcal{N}})^{0.51} \simeq 3.7\,(\Omega_M h^2)^{-.54} \text{GeV} . \quad (3.4.14)$$

which is 10 GeV for $\Omega_M h^2 = 0.15$. Otherwise this provides a *lower* bound on the mass of these particles.

Incidentally, similar arguments apply to the annihilation of nucleons and antinucleons, if the universe has zero net baryon number. The difference here of course is that instead of $\langle \sigma v \rangle$ being of order $G_F^2 m_L^2$, it is roughly of order m_π^{-2}. We can thus estimate the density of left-over baryon–antibaryon pairs by taking $m_L = m_N$, $\mathcal{N} \approx 10$, and replacing $\mathcal{F}$ in Eq. (3.4.13) with $2\pi m_\pi^{-2}/G_F^2 m_N^2 = 2.7 \times 10^{12}$, so that Eq. (3.4.13) would give a baryon–antibaryon density parameter $\Omega_{B+\bar{B}} h^2 \approx 6 \times 10^{-11}$. This is much less than the present observed density parameter of baryons, ruling out the possibility that the baryons around us are just those that missed being annihilated in an initially baryon–antibaryon symmetric universe, and then somehow became segregated from the antibaryons.

Returning to the cold dark matter, originally it was thought that the L particles might be heavy neutrinos. They could not be any of the three known neutrino types, which as discussed at the end of Section 3.1 have masses at most of the order of 1 eV, but there could be a fourth generation of very heavy leptons that have negligible mixing with the known leptons. The mass of a new heavy neutrino would have to be greater than $m_Z/2 = 45$ GeV, to block the decay of the Z^0 into a heavy neutrino and antineutrino, which if it occurred would destroy the present excellent agreement between theory and experiment for the total decay rate of the Z^0 particle. If the

L particles were heavy neutrinos, then the fudge factor $\mathcal{F}$ in Eq. (3.4.9) would be of the order of the number of types of particles into which the L particles might annihilate. If we assume very roughly that $\mathcal{F} \approx \mathcal{N} \approx 100$, then Eq. (3.4.13) gives $\Omega_L h^2 \leq 10^{-3}$. So it does not seem that the cold dark matter could consist of a new heavy neutrino.

The most plausible candidate for the L particle is one of the new particles required by supersymmetry.[5] In many supersymmetric theories there is a multiplicatively conserved quantum number R, which takes the value $+1$ for all the known particles of the Standard Model, and -1 for their super-symmetric partners. (Multiplicative conservation means that the *product* of the Rs for all the particles in the final state of any reaction is the same as for the initial state.) Among other things, this conservation law tells us that the lightest particle with $R = -1$ (which is often called the LSP, for "lightest supersymmetric particle") is *stable*, although two of these particles could annihilate into ordinary particles with $R = +1$. To judge which is the lightest particle with $R = -1$, it is necessary to distinguish between two possible pictures of supersymmetry breaking.[6] In both pictures supersymmetry is spontaneously broken by non-perturbative effects in some hidden sector of particles with a large typical mass M_S, which interact through some strong force that is not felt by the known particles of the Standard Model or their superpartners. The supersymmetry breaking in the hidden sector also gives the gravitino (the superpartner of the graviton with spin $3/2$) a mass $m_g \approx \sqrt{G} M_S^2$.

In one picture of supersymmetry breaking, the breakdown of super-symmetry is communicated to the particles of the Standard Model and their superpartners by the electroweak and ordinary strong forces of the Standard Model. In this case M_S would have to be of the order of 100 GeV to 100 TeV, and the gravitino mass would be at most of order 1 eV. The lightest particle with $R = -1$ would be the gravitino, which would be too light to furnish the cold dark matter. Even if gravitinos were once in equilibrium with other particles, and did not annihilate or decay, by the same arguments as in Section 3.1 their number density now would be less than the number density of photons by a factor of order $2/\mathcal{N}$, where $\mathcal{N}$ is the effective number of relativistic particle states at the time that gravitinos went out of thermal equilibrium. Photons with typical energy $k_B T_{\gamma 0}$ have an energy density parameter given by Eq. (2.1.8) as $\Omega_\gamma h^2 = 2.47 \times 10^{-5}$, so in order for gravitinos to furnish cold dark matter with $\Omega_M h^2 \simeq 0.15$, the mass of the gravitino would have to be of order

---

[5]This and other possibilities are discussed in a comprehensive review by G. Bertone, D. Hooper, and J. Silk, *Phys. Rep.* **405**, 279 (2005) [hep-ph/0404175].

[6]See QTF, Vol. III, Sec. 28.3.

$0.15 \, \mathcal{N} \, k_B T_{\gamma 0} / 2 \Omega_\gamma h^2$, which for $\mathcal{N} \approx 100$ is roughly 70 eV.[7] Thus in this picture of supersymmetry breaking, gravitinos are not likely to be the cold dark matter, though this is not absolutely ruled out. The next-to-lightest particles with $R = -1$ would all presumably decay over billions of years into gravitinos and ordinary particles, so they could not furnish the cold dark matter either. (The rate of decay into relativistic gravitinos with helicity $\pm 3/2$ would be suppressed by a factor G, but particles with $R = -1$ could decay into gravitinos with helicity $\pm 1/2$ with the decay rate suppressed only by a factor M_S^{-2} and by some powers of the gauge couplings of the Standard Model.)

In the other picture, supersymmetry breaking is mediated by the gravitational field and its superpartners, and because these interactions are very weak M_S must be correspondingly large, of order 10^{11} to 10^{13} GeV. The gravitino mass would be of the same order of magnitude as the masses of the superpartners of the known particles of the Standard Model, so it might or might not be the lightest particle with $R = -1$.

If the lightest particle with $R = -1$ is not a gravitino, then it could be either a sneutrino, the spin 0 superpartner of the neutrino, or a neutralino, the spin 1/2 superpartner of some mixture of the neutral gauge and scalar bosons of the Standard Model.[8] Whichever of these is the lightest, the decay process will involve the exchange of a particle with $R = -1$, having a mass $\tilde{M}$ that is expected to be of the order of a TeV or so. Hence its annihilation amplitude (aside from factors of 2 and π) will be of order $g^2 / \tilde{M}^2$, where g is a typical electroweak coupling. This is smaller than the weak coupling constant G_{wk} by a factor of order $m_W^2 / \tilde{M}^2$, so the fudge factor $\mathcal{F}$ in the annihilation rate constant (3.4.9) will be of order $m_W^4 \mathcal{N}_A / \tilde{M}^4 \approx 10^{-4} \mathcal{N}_A$, where $\mathcal{N}_A$ is the number of annihilation channels. Taking $\mathcal{N}_A / \sqrt{\mathcal{N}} \approx 1$, Eq. (3.4.14) tells us that to furnish cold dark matter with $\Omega_M h^2 \simeq 0.15$, the lightest particle with $R = -1$ (if not the gravitino) would have to have a mass of order 1 TeV. This is similar to estimates of the masses of the superpartners of the particles of the Standard Model in typical supersymmetric models, a circumstance that greatly encourages the hope that the particles of dark matter will be found to be created in experiments at high energy accelerators.

Cosmological considerations rule out the possibility that the lightest particle with $R = -1$ in this picture of supersymmetry breaking is the gravitino. Unlike the superpartners of the particles of the Standard Model, the

---

[7]H. Pagels and J. R. Primack, *Phys. Rev. Lett.* **48**, 223 (1982) first remarked that in order for the gravitino not to give $\Omega_M \gg 1$, its mass would have to be less than about 1 keV.

[8]S. Weinberg, *Phys. Rev. Lett.* **50**, 387 (1983).

gravitino has a two-body annihilation *amplitude* proportional to two factors[9] of the gravitational coupling constant $\sqrt{G}$, so the annihilation *rate* is proportional to G^2, and is therefore much too small for annihilation to play any significant role in reducing the number density of gravitinos. If, because of particle annihilations after gravitino decoupling, the number density of gravitinos were now, say, 1% of the number density of photons, then, as we have seen, in order for their mass density now not to exceed cosmological bounds their mass would have to be less than roughly 100 eV. This is quite inconsistent with the gravitino masses expected in theories of gravity mediated supersymmetry breaking. But if the gravitinos are not the lightest particles with $R = -1$ then they can decay, reducing their present mass density to acceptable values even if their masses are quite high.[10] The coupling of the gravitino to other fields is proportional to $\sqrt{G}$, so on dimensional grounds the decay rate Γ_g of a gravitino at rest is roughly of the order of Gm_g^3. This is to be compared with the rate of expansion of the universe, which at temperature T is of order $\sqrt{G(k_BT)^4}$. (We are here ignoring factors of order 10–100, including those involving non-gravitational coupling constants and the number of particle species.) When the cosmic temperature drops to the value $k_BT \approx m_g$ at which gravitinos become non-relativistic, the ratio of their decay rate to the expansion rate is of order $\sqrt{G}m_g = m_g/m_{\text{Planck}} \ll 1$, so gravitino decay becomes significant only after this time, when the gravitinos are highly non-relativistic. As we have seen, their number density will be of order $(k_BT)^3$, so their energy density will then be of order $m_g(k_BT)^3$, which is greater than the energy density of order $(k_BT)^4$ of the photons and other relativistic particles in thermal equilibrium at temperature T, and therefore makes the dominant contribution to the cosmic gravitational field that governs the rate of expansion of the universe. The expansion rate under these conditions is therefore of order $\sqrt{Gm_g(k_BT)^3}$, and gravitino decay becomes significant when this equals the gravitino decay rate of order Gm_g^3, and therefore at a temperature

$$k_BT_g \approx G^{1/3}m_g^{5/3} \ .$$

After they decay, their energy must go into the energy of photons and other relativistic particles, so the temperature T_g' after decay is related to the temperature T_g calculated above by the energy conservation condition

---

[9]This does not include a factor of $\sqrt{G}$ in the gravitino mass, because this factor is multiplied by the scale of supersymmetry breaking, which for gravitationally mediated supersymmetry breaking is very large, giving a gravitino mass that is comparable to the mass of other supersymmetric particles.

[10]S. Weinberg, *Phys. Rev. Lett.* **48**, 223 (1982).

$m_g T_g^3 \approx T_g'^4$, and hence

$$k_B T_g' \approx G^{1/4} m_g^{3/2} \, .$$

In particular, since $k_B T_g << m_g$, we have $T_g' >> T_g$. If T_g were less than the temperature $T_n \simeq 0.1$ MeV at which cosmological nucleosynthesis can occur, then gravitinos would still be abundant before nucleosynthesis, giving a higher energy density and hence a faster expansion, so that there would be less time for free neutrons to decay before being incorporated into complex nuclei, and hence more helium would be produced when nucleosynthesis occurs. Also, the ratio of the photon and baryon densities would have been subsequently increased by gravitino decay, so this ratio at the time of nucleosynthesis would have been considerably less than is usually estimated from the present cosmic microwave background temperature, and so nuclear reactions would have incorporated neutrons more completely into helium, and less deuterium would be left today. The present agreement between theory and observation for the cosmic helium and deuterium abundances would thus be destroyed. This problem is avoided if $T_g > 0.1$ MeV, but it can also be avoided under the much weaker condition that $T_g' > 0.4$ MeV, because then after the gravitinos decay the temperature would have been high enough to break up the excess helium and give cosmological nucleosynthesis a fresh start as the universe recoils. This condition requires that $m_g > 10$ TeV. This limit on m_g corresponds to a supersymmetry breaking scale $M_S > 10^{11}$ GeV for $m_g \approx \sqrt{G} M_S^2$.

It may be possible to detect cosmic WIMPs through observation of the recoil of atomic nuclei from which they scatter elastically,[11] as for instance the scattering of heavy neutrinos through the neutral current weak interaction.[12] This is being pursued by a number of collaborations: DAMA,[13] CRESST,[14] EDELWEISS,[15] UK Dark Matter,[16] CDMS,[17] and WARP.[18] Assuming that WIMPs are more or less at rest in the halo of our galaxy, the motion of the solar system through the halo produces a WIMP "wind" with

---

[11] For reviews of current experiments, see Y. Ramachers, *Nucl. Phys. B. Proc. Suppl.* [astro-ph/0211500]; G. Chardin, in *Cryogenic Particle Detection*, ed. C. Ens (Springer, Heidelberg, 2005) [astro-ph/0411503]; R. J. Gaitskell, *Ann. Rev. Nucl. Part. Sci.* **54**, 315 (2004); J. Ellis, K. Olive, Y. Santoso, and V. C. Spanos, *Phys. Rev. D* **71**, 095007 (2005) [hep-ph/0502001]; K. Freese, *Nucl. Instrum. Meth.* **A559**, 337 (2006) [astro-ph/0508279]; L. Baudis, Int. J. Mod. Phys. **A21**, 1925 (2006) [astro-ph/0511805].

[12] A. Drukier and L. Stodolsky, *Phys. Rev. D* **30**, 2295 (1985); M. Goodman and E. Witten, *Phys. Rev. D* **31**, 3059 (1985).

[13] R. Bernabei et al., *Phys. Lett.* **B480**, 23 (2000).

[14] G. Angloher et al., *Astropart. Phys.* **23**, 325 (2005) [astro-ph/0408006].

[15] V. Sanglard et al., *Phys. Rev. D* **71**, 122002 (2005) [astro-ph/0503265].

[16] G. J. Alner et al., *Astropart. Phys.* **23**, 444 (2005).

[17] D. S. Akerib et al., *Phys. Rev. Lett.* **93**, 211301 (2004) [astro-ph/0405033], and astro-ph/0507190.

[18] P. Benetti et al., astro-ph/0701286.

a speed v_w of about 220 km/s. A nucleus of mass Am_N when struck with a WIMP of mass $m_w \gg Am_N$ traveling at this velocity will recoil with a typical velocity of order v_w, and hence a kinetic energy of order $Am_N v_w^2/2 \approx 100\,A$ eV. These recoils can be detected by observing the ionization of atoms struck by the recoiling nucleus, or by detecting light emitted by these atoms, or by detecting vibrations in the crystal lattice of the detector. The mass density ρ_h in the halo of our galaxy near earth is estimated from observations of stellar motions to be about $0.3\,\text{GeV/cm}^3$, giving a number density for WIMP mass m_w of ρ_h/m_w, from which the elastic scattering rate can be calculated for any assumed values of the WIMP mass and the scattering cross section. Failing to observe nuclear recoil events then excludes some region of the m_w–cross-section plane.

The greatest problem in these experiments is distinguishing true WIMP events from background, caused by natural radioactivity and cosmic rays. The best hope for distinguishing events from background is to exploit the motion of the earth around the sun.[19] This orbital motion adds about 15 km/s to the speed of the WIMP wind in summer, and subtracts an equal amount in winter (though this depends on the halo model), so one may expect a 7% seasonal modulation of true WIMP events. The DAMA collaboration reported just such a modulation, but almost all of the region in the m_w–cross-section plane that would account for this observation was subsequently apparently excluded (using a different detection scheme) by the CDMS collaboration. So far, experiments set an upper bound on the effective cross section of about 10^{-42} to 10^{-43} cm^2 for $m_w > 50$ GeV, and much larger for $m_w < 50\,\text{GeV}$, the precise bound depending on assumptions about the distribution of WIMPs in the galactic halo.[20] (For comparison, the effective cross section for the low energy scattering of the neutralinos of supersymmetric theories on nucleons is expected to be less than about 10^{-41} cm^2.)

There is also a possibility of an indirect detection of WIMPs, through observation of gamma rays or other particles produced when pairs of WIMPs annihilate in regions of high WIMP concentration.[21] Gamma rays do not penetrate the earth's atmosphere, so they have to be detected either through the observation of showers of charged particles produced in the atmosphere, as for instance by the Cerenkov radiation associated with these particles, or by gamma ray telescopes carried by balloons or by satellites in orbit above the earth's atmosphere. If WIMP–WIMP annihilation produces

---

[19] A. K. Drukier, K. Freese, and D. N. Spergel, *Phys. Rev.* D **33**, 3495 (1986); K. Freese, J. A. Frieman, and A. Gould, *Phys. Rev.* D **37**, 3388 (1988).

[20] A. Bottino, F. Donato, N. Fornengo, and S. Scopel, Phys. Rev. D **72**, 083521 (2005) [hep-ph/0508270].

[21] For a review, see P. Gondolo, hep-ph/0501134.

just a pair of gamma rays, then each will carry a unique energy equal to the WIMP mass, giving a very clear WIMP signal. So far, there has been no sign of such a monochromatic gamma ray. Of course the annihilation may produce other particles along with gamma rays, or particles whose decay then produces gamma rays, but in either case this would yield a continuum of gamma ray energies, which would be much harder to identify as coming from WIMP annihilation. The Cangaroo-II[22] and HESS[23] atmospheric Cerenkov detector collaborations have both reported continuum sources of gamma rays coming from near the center of our galaxy, but with very different spectra, which can be interpreted as coming from annihilation of WIMPs with a mass of about 1 TeV or 19 TeV, respectively.

For decades there has been evidence of monochromatic gamma rays coming from the galactic center, but at the energy of 511 keV expected from electron–positron annihilation, rather than WIMP–WIMP annihilation. In 1970 a balloon-borne pair of gamma ray detectors found evidence of a gamma ray line around 500 keV coming more-or-less from the direction of the galactic center.[24] The evidence for this has since become much stronger through observations made by the INTEGRAL (International Gamma-Ray Astrophysics Laboratory) satellite,[25] which found a gamma ray line coming from the galactic center with an energy within about 3 keV of 511 keV. It is possible that this is due to the decay of a relatively light WIMP into electron–positron pairs, with the positrons then losing energy by ionization, after which they annihilate with ambient electrons.[26]

There is also a report of an excess of positrons in cosmic rays found by the HEAT balloon experiment,[27] which might or might not come from WIMP annihilation. Very recently it has been suggested that an excess of microwave emission from the direction of the center of our galaxy observed by the WMAP satellite may come from synchrotron emission by relativistic electrons and positrons produced in WIMP annihilation.[28]

It is too early to reach any definite conclusions from any of these experiments about WIMPs as candidates for the particles of dark matter.

---

[22] K. Tsuchiya *et al.*, *Astrophys. J.* **606**, L115 (2004) [astro-ph/0403592].

[23] D. Horns, *Phys. Lett. B* **607**, 225 (2005). But see F. Aharonian, *Phys. Rev. Lett.* **97**, 221102 (2006)

[24] W. N. Johnson, III, F. R. Harnden, Jr., and R. C. Haymes, *Astrophys. J.* **172**, L1 (1972.)

[25] P. Jean *et al.*, *Astron. Astrophys.* **407**, 55 (2003) [astro-ph/0309484]; J. Knödlseder *et al.*, *Astron. Astrophys.* **411**, 457 (2003) [astro-ph/0309442]; E. Churazov *et al.*, Mon. Not. Roy. Astron. Soc. **357**, 1377 (2005) [astro-ph/0411351]; J. Knödlseder *et al.*, *Astron. Astrophys.* **441**, 513 (2005) [astro-ph/0506026]; G. Weidenspointer *et al.*, *Astron. Astrophys.* **450**, 1013 (2006) [astro-ph/0601673].

[26] C. Boehm, D. Hooper, J. Silk, M. Casse, and J. Paul, *Phys. Rev. Lett.* **92**, 1301 (2004) [astro-ph/0309686].

[27] S. W. Barwick *et al.*, *Phys. Rev. Lett.* **75**, 390 (1995); *Astrophys. J.* **482**, L191 (1997).

[28] D. Hooper, D. P. Finkbeiner, and G. Dobler, 0705.3655.

B. Axions and axinos

Axions[29] are light neutral spinless particles that are made necessary[30] by the spontaneous breakdown of a symmetry that first appeared in a model[31] that was proposed to explain why non-perturbative effects of the strong interactions do not violate **CP** invariance. For our present purposes, all we need to know is that the dominant part of the effective action (with $\hbar = c = 1$) that describes the axion field φ takes the form (B.44) with potential $V(\varphi) = m_a^2 \varphi^2 / 2$:

$$ I[\varphi] = \int d^4x \sqrt{-\mathrm{Det}\, g} \left(-\frac{1}{2} g^{\mu\nu} \partial_\mu \varphi \, \partial_\nu \varphi - \frac{1}{2} m_a^2 \varphi^2 \right) . \tag{3.4.15} $$

The axion mass m_a is a complicated function of temperature, but for temperatures well below 10^{12} K it takes a well-known constant value, related to the energy scale M at which the Peccei–Quinn symmetry is broken by

$$ m_{a0} = \frac{F_\pi m_\pi}{M} \frac{\sqrt{m_d m_u}}{m_d + m_u} \simeq \frac{13 \text{ MeV}}{M[\text{GeV}]} . \tag{3.4.16} $$

(Here $F_\pi = 184$ MeV is the pion decay amplitude, and m_d and m_u are the down and up quark masses appearing in the Lagrangian of quantum chromodynamics, for which $m_d/m_u \simeq 1.85$.) In the original Peccei–Quinn model,[31] M was of the order of the electroweak symmetry breaking scale, $M \approx 100$ GeV, but it was soon realized that this is experimentally ruled out. Axion fields interact with ordinary matter through factors $\partial_\mu \varphi / M$, so with $M \geq 100$ GeV axions interact so weakly that they emerge without attenuation from reactor cores or stellar interiors. The production rate of axions is proportional to $1/M^2$, so the failure to observe effects of axion emission from stars or nuclear reactors sets a *lower* bound on M, and hence an upper bound on m_a. In particular, limits on the rate of cooling of red giant stars by axion emission give[32] $M > 10^7$ GeV, while observations of the supernova SN1987A indicate[33] that $M > 10^{10}$ GeV. A generalization of the Peccei–Quinn model was then proposed,[34] in which M is an arbitrary

[29]See QTF, Vol II, Sec. 23.6. For a review, see P. Sikivie, in *Axions – Lecture Notes on Physics*, ed. M. Kuster (Springer-Verlag, to be published) [astro-ph/0610440].

[30]S. Weinberg, *Phys. Rev. Lett.* **40**, 223 (1978); F. Wilczek, *Phys. Rev. Lett.* **40**, 279 (1978).

[31]R. D. Peccei and H. Quinn, *Phys. Rev. Lett.* **38**, 1440 (1977); *Phys. Rev. D* **16**, 1791 (1977).

[32]D. A. Dicus, E. W. Kolb, V. I. Teplitz, and R. V. Wagoner,*Phys. Rev. D* **18**, 1829 (1978); *Phys. Rev. D* **22**, 839 (1980).

[33]For reviews, see M. S. Turner, *Phys. Rep.*. **197**, 67 (1990); G.G. Raffelt, *Phys. Rep.* **198**. 1 (1990); P. Sikivie, ref. 26.

[34]J. E. Kim, *Phys. Rev. Lett.* **43**, 103 (1979); M. Dine, W. Fischler, and M. Srednicki, *Phys. Lett.* **104B**, 199 (1981); M. B. Wise, H. Georgi, and S. L. Glashow, *Phys. Rev. Lett.* **47**, 402 (1981).

parameter. As we shall see, cosmological considerations provide an *upper* bound[35] on M, which together with the above lower bounds leaves a narrow window of allowed values.

The cosmological axion field is supposed to be spatially homogeneous.[36] In a Robertson–Walker metric, the energy density and pressure are given by Eq. (B.66) and (B.67) as

$$\rho_a = \frac{1}{2}\dot{\varphi}^2 + \frac{1}{2}m_a^2\varphi^2 , \qquad p_a = \frac{1}{2}\dot{\varphi}^2 - \frac{1}{2}m_a^2\varphi^2 , \tag{3.4.17}$$

so the equation (1.1.32) of energy conservation (or the Euler–Lagrange equation derived directly from the action (3.4.15)) gives the field equation

$$\ddot{\varphi} + 3H(t)\dot{\varphi} + m_a^2(t)\varphi = 0 . \tag{3.4.18}$$

At early times $H(t) \gg m_a(t)$, so we can ignore $m_a(t)$, and Eq. (3.4.17) has solutions $\varphi = $ constant, and $\varphi \propto 1/a^3$. Rejecting the singular solution, we see that at early times $\varphi(t)$ is frozen at a value φ_0, which (absent fine tuning) would be expected to be of order M. Later, when $H(t)$ dropped below $m_a(t)$, $\varphi(t)$ began a rapid oscillation, so that in this case Eq. (3.4.18) can be solved using the WKB approximation, which gives

$$\varphi(t) \rightarrow \varphi_1 \left(\frac{a(t_1)}{a(t)}\right)^{3/2} \cos\left(\int_0^t m_a(t)\,dt + \alpha\right) , \tag{3.4.19}$$

where t_1 is the time at which $H(t_1) = m_{a0}$, φ_1 is a constant of order $\varphi_0 \approx M$, and α is a phase that cannot be determined without a more detailed study, but which fortunately we do not need to know. (For instance, if $m_a(t)$ has the constant value m_{a0} and the universe is radiation-dominated in the era of interest, then $\varphi_1 = 2^{3/2}\pi^{-1/2}\Gamma(5/4)\varphi_0 = 1.446\varphi_0$, and $\alpha = -3\pi/8$.) Since $a(t)$ for $m_a \gg H$ was varying much more slowly than the phase of the cosine, the energy density at late times is given by Eq. (3.4.17) as

$$\rho_a(t) \rightarrow \frac{1}{2}m_a^2\varphi_1^2 \left(\frac{a(t_1)}{a(t)}\right)^3 . \tag{3.4.20}$$

In order to project this forward to the present, we note that if the universe was radiation-dominated at time t_1, with $\mathcal{N}_1$ the effective number of types of particles with masses much less than $k_B T(t_1)$ (counting each spin state

---

[35] J. Preskill, M. B. Wise, and F. Wilczek, Phys. Lett. **120B**, 127 (1983); L. F. Abbott and P. Sikivie, *Phys. Lett.* **120B**, 127 (1983); M. Dine and W. Fischler, *Phys. Lett.* **120B**, 137 (1983). For reviews, see J. E. Kim, *Phys. Rep.* **150**, 1 (1987); M. S. Turner, *Phys. Rep.* **197**, 68 (1990).

[36] Components with non-zero wave number are doubtless present, but their energy density decays more rapidly than the energy density of the spatially homogeneous coherent field $\varphi(t)$.

of particles and antiparticles separately, and including an extra factor 7/8 for fermions), then the photon temperature at present is

$$T_{\gamma 0} = \mathcal{N}_1^{1/3} T(t_1) a(t_1)/a(t_0) ,$$

the factor $\mathcal{N}_1^{1/3}$ being inserted to take account of the heating of photons by the annihilation of particles and antiparticles between times t_1 and t_0. The temperature $T(t_1)$ may be determined by noting that the expansion rate at time t_1 is

$$m_a \equiv H(t_1) = \sqrt{\frac{8\pi G \mathcal{N}_1 a_B T^4/2}{3}} = \sqrt{\frac{4\pi^3 G \mathcal{N}_1}{45}} (k_B T_1)^2 .$$

Using these results in Eq. (3.4.20) then gives the present axion energy density

$$\rho_a(t_0) \simeq \frac{1}{2} m_a^{1/2} \mathcal{N}_1^{-1/4} \varphi_0^2 \left(\frac{4\pi^3 G}{45} \right)^{3/4} (k_B T_{\gamma 0})^3 . \tag{3.4.21}$$

We expect φ_0 to be of the same order of magnitude as the symmetry breaking scale M, so using Eq. (3.4.16) and ignoring all factors of order unity,

$$\rho_a \approx \frac{F_\pi^2 m_\pi^2 G^{3/4} (k_B T_{\gamma 0})^3}{m_a^{3/2}} \tag{3.4.22}$$

Equivalently, the axion density provides a fraction Ω_a of the critical density given by

$$\Omega_a h^2 \approx \left(m_a/10^{-5} \mathrm{eV} \right)^{-3/2} . \tag{3.4.23}$$

Because the axion field is spatially homogeneous, for $m_a \gg H$ its energy takes the form of massive particles that are essentially at rest. If axions furnish the whole of the cold dark matter, then $m_a \approx 10^{-5}$ eV, corresponding to $M \approx 10^{12}$ GeV. Otherwise, these numbers provide a lower bound on m_a and an upper bound on M.

Axions are much too weakly interacting to be detected in the sort of nuclear recoils looked for in searches for WIMPs. One possibility is to observe the conversion of cosmic axions into photons in intense magnetic fields.[37] This approach has already been used to put a limit on the parameters of axions that would be produced by the sun.[38] The axion field $\varphi(x)$ would be expected to have an interaction with the electromagnetic field of

---

[37]P. Sikivie, *Phys. Rev. Lett.* **51**, 1415 (1983); *Phys. Rev.* D **32**, 2988 (1985).

[38]K. Zioutas *et al.* (CERN Axion Solar Telescope collaboration), *Phys. Rev. Lett.* **94**, 121301 (2005) [hep-ex/0411033].

the form $g_{a\gamma}\,\varphi\,\mathbf{E}\cdot\mathbf{B}$, with $g_{a\gamma}$ of order $\alpha/2\pi\,M$, so that photons in the sun could convert into axions in the presence of the strong electric fields around atomic nuclei, and these solar axions would then convert back into photons in intense laboratory magnetic fields. No such photons were seen, indicating that for axions with mass less than 0.02 eV, $|g_{a\gamma}| < 1.16 \times 10^{-10}$ GeV^{-1}, a much more restrictive limit than provided by earlier experiments of this sort.[39] In a different sort of search,[40] the Axion Dark Matter Experiment, a microwave cavity was used to search for axions in our galactic halo, and put upper limits on the axion density in the narrow mass range (1.98 to 2.17)$\times 10^{-6}$ eV. None of these experiments are in conflict with axion models of dark matter, but a plausible improvement in the sensitivity of this sort of experiment may rule out these models, or perhaps find axions.

In supersymmetric theories, the axion would be partnered with a spin one-half particle, the *axino*, which would probably be the lightest particle with $R = -1$, and hence stable. Axinos could be produced non-thermally,[41] through the decay of other particles with $R = -1$, or thermally.[42] It appears that the axino provides another plausible candidate for the particle of cold dark matter.

---

[39]This and earlier experiments are reviewed by G. G. Raffelt, contribution to *XI International Workshop on Neutrino Telescopes*, hep-ph/0504152.

[40]L. D. Duffy *et al.*, *Phys. Rev.* D **74**, 012006 (2006) [astro-ph/0603108], and earlier references cited therein.

[41]L. Covi, J. E. Kim, and L. Roszkowski, *Phys. Rev. Lett.* **82**, 4180 (1999).

[42]L. Covi, H. B. Kim, J. E. Kim, and L. Roszkowski, *J. High Energy Phys.* **0105**, 033 (2001) [hep-ph/0101009]; A. Brandenburg and F. D. Steffen, *J. Cosm. & Astropart. Phys.* **0408**. 008 (2004) [hep-ph/0405158].

4

Inflation

We can have some confidence in the story of the evolution of the universe from the time of electron–positron annihilation to the present, as told in the previous three chapters. About earlier times, so far we can only speculate. In the past quarter century these speculations have centered on the idea that before the period of radiation domination, during which the Robertson–Walker scale factor $a(t)$ was growing as $\sqrt{t}$, there was an earlier period of *inflation*, when the energy density of the universe was dominated by a slowly varying vacuum energy, and $a(t)$ grew more-or-less exponentially. The possibility of an early exponential expansion had been noticed by several authors,[1] but at first it attracted little attention. It was Alan Guth[2] who incited interest in the possibility of inflation by noting what it was good for.

Guth noticed that, in a model of grand unification he was considering (with Henry Tye), scalar fields could get caught in a local minimum of the potential, which in his work corresponded to a state with an unbroken grand unified symmetry. The energy of empty space would then have remained constant for a while as the universe expanded, which would produce a constant rate of expansion, meaning that $a(t)$ would have grown exponentially. Eventually this inflation would be stopped by quantum-mechanical barrier penetration, after which the scalar field would start rolling down the potential toward a global minimum, corresponding to the present universe. In itself this would have been a result of no great immediate importance. But then it occurred to Guth that the existence of an era of inflation would solve one of the outstanding problems of cosmology, mentioned here in Section 1.5. It is known as the "flatness problem:" Why was the curvature of space was so small in the early universe? Guth soon also discovered that inflation would solve other cosmological puzzles, some of which he had not even realized were puzzles. These problems along with the flatness problem will be discussed in Section 4.1.

As Guth and others soon realized, his version of inflation had a fatal problem, to be described in Section 4.2. Guth's "old inflation" was soon replaced with a "new inflation" model, due to Andrei Linde[3] and Andreas Albrecht and Paul Steinhardt.[4] The essential element introduced by

---

[1] A. A. Starobinsky, *JETP Lett.* **30**, 682 (1979); *Phys. Lett.* **B 91**, 99 (1980); D. Kazanas, *Astrophys. J.* **241**, L59 (1980); K. Sato, *Mon. Not. Roy. Astron. Soc.* **195**, 467 (1981).

[2] A. Guth, Phys. Rev. **D 23**, 347 (1981). Guth tells the story of this work in *The Inflationary Universe: The Quest for a New Theory of Cosmic Origins* (Helix Books/Addison Wesley, 1997).

[3] A. D. Linde, *Phys. Lett.* **B 108**, 389 (1982); **114**, 431 (1982); *Phys. Rev. Lett.* **48**, 335 (1982).

[4] A. Albrecht and P. Steinhardt, *Phys. Rev. Lett.* **48**, 1220 (1982).

theories of new inflation was a nearly exponential expansion during the slow roll of one or more scalar fields down a potential hill, which is the main subject of Section 4.2. This provided a basis for "chaotic inflation" and "eternal inflation" and other variants, some of which are briefly described in Section 4.3.

So far, the details of inflation are unknown, and the whole idea of inflation remains a speculation, though one that is increasingly plausible. Aside from the classic problems that inflation solved at the beginning, it has had one significant experimental success: a prediction of some of the properties of the fluctuations in the cosmic microwave background and large scale structure. We will come to this in Chapter 10, after we take up the evolution of fluctuations in the early universe in Chapters 5 and 6 and the observation of these fluctuations in Chapters 7 and 8 and Section 9.5.

4.1 Three puzzles

In this section we will outline three classic cosmological problems, and work out the extent of the inflation required to solve each of them. For this purpose we will here simply assume that the universe went through an early period of exponential expansion, without worrying yet about how this came about.

A. Flatness

As we saw in Section 1.6 and 1.8, the observed Type Ia supernova redshift–distance relation and measurements of the ages of the oldest stars are consistent with a vanishing spatial curvature parameter Ω_K, though a non-vanishing curvature can be accommodated by changing Ω_M. Including data from the cosmic microwave background temperature fluctuations, discussed in Section 7.2, favors $\Omega_K = 0$. Although there is still room for a small non-zero Ω_K, it seems quite safe to conclude from these observations that $|\Omega_K| < 1$. But Ω_K is just the present value of the dimensionless time dependent curvature parameter $-K/a^2 H^2 = -K/\dot{a}^2$, with K constant. From the time the temperature dropped to about 10^4 K until near the present, $a(t)$ has been increasing as $t^{2/3}$, so $|K|/\dot{a}^2$ has also been increasing as $t^{2/3} \propto T^{-1}$. Thus, if $|\Omega_K| < 1$, then at 10^4 K the curvature parameter $|K|/\dot{a}^2$ could not have been greater than about 10^{-4}. Earlier, $a(t)$ was increasing as $t^{1/2}$, so $|K|/\dot{a}^2$ was increasing as $t \propto T^{-2}$. In order for $|K|/\dot{a}^2$ at 10^4 K to be no greater than about 10^{-4}, it is necessary that $|K|/\dot{a}^2$ was at most about 10^{-16} at the temperature $T \approx 10^{10}$ K of electron–positron annihilation (roughly, the beginning of the period of neutron–proton conversion that results in the observed helium abundance), and even smaller at earlier times.[5] This is not a paradox—there is no reason why the curvature should not have been very small—but it is the sort of thing physicists would like to explain if we can.

What Guth realized was that during inflation $\dot{a}/a$ would have been roughly constant, so $|K|/a^2H^2$ would have been *decreasing* more or less like a^{-2}. So to understand why space was so flat at the beginning of the present big bang it is not necessary to make any arbitrary assumptions; if the radiation–dominated big bang was preceded by a sufficient period of inflation, it would necessarily have started with negligible curvature.

To put this quantitatively, suppose the universe began with a period of inflation during which $a(t)$ increased by some large factor $e^{\mathcal{N}}$, followed by a period of radiation dominance lasting until the time of radiation–matter equality, followed in turn by a period of matter dominance and then a period dominated by vacuum energy. If $|K|/a^2H^2$ had a value of order unity at the beginning of inflation, then at the time t_I of the end of inflation $|K|/a^2H^2$ would have had a value $|K|/a_I^2H_I^2$ of order $e^{-2\mathcal{N}}$ (where a_I and H_I are the Robertson–Walker scale factor and expansion rate at this time), and today we will have

$$|\Omega_K| = \frac{|K|}{a_0^2H_0^2} = e^{-2\mathcal{N}}\left(\frac{a_IH_I}{a_0H_0}\right)^2 , \tag{4.1.1}$$

Thus the flatness problem is avoided if the expansion during inflation has the lower bound

$$e^{\mathcal{N}} > \frac{a_IH_I}{a_0H_0} . \tag{4.1.2}$$

To evaluate this we will make the somewhat risky assumption that not much happens to the cosmic scale factor and expansion rate from the end of inflation to the beginning of the radiation-dominated era, so that

$$a_IH_I \simeq a_1H_1 , \tag{4.1.3}$$

the subscript 1 denoting the beginning of the radiation-dominated era. We can express the expansion a_0/a_1 of the universe since the start of the radiation-dominated era in terms of ratios of expansion rates by noting that over the whole of the radiation and matter-dominated era, the expansion

---

[5]R. H. Dicke and P. J. E. Peebles, in *General Relativity – An Einstein Centenary Survey*, eds. S. Hawking and W. Israel (Cambridge University Press, 1979).

rate was

$$H = \frac{H_{EQ}}{\sqrt{2}}\sqrt{\left(\frac{a_{EQ}}{a}\right)^3 + \left(\frac{a_{EQ}}{a}\right)^4},$$

where $a_{EQ} = a_0\Omega_R/\Omega_M$ and $H_{EQ} = \sqrt{2\Omega_M}H_0(a_0/a_{EQ})^{3/2}$ are the scale factor and expansion rate at matter–radiation equality. Setting $a = a_1 \ll a_{EQ}$ gives

$$H_1 = \frac{H_{EQ}}{\sqrt{2}}\left(\frac{a_{EQ}}{a_1}\right)^2. \tag{4.1.4}$$

Using this relation to eliminate a_1, we can put the bound (4.1.2) in the more useful form

$$e^{\mathcal{N}} > \left(\frac{\Omega_M a_{EQ}}{a_0}\right)^{1/4}\sqrt{\frac{H_1}{H_0}} = \Omega_R^{1/4}\sqrt{\frac{H_1}{H_0}} = \left(\Omega_R\frac{\rho_1}{\rho_{0,crit}}\right)^{1/4} = \frac{[\rho_1]^{1/4}}{0.037\, h\, \mathrm{eV}},$$
$$\tag{4.1.5}$$

where ρ_1 is the energy density at the beginning of the radiation-dominated era, and $\rho_{0,crit} = [3.00 \times 10^{-3}\ \mathrm{eV}]^4 h^2$ is the critical density (1.5.28).

To go further, we need some idea of the energy density at the end of inflation. The success of the theory of cosmological nucleosynthesis shows that ρ_1 cannot be less than the energy density at the time of the beginning of neutron–proton conversion, roughly $[1\ \mathrm{MeV}]^4$, in which case Eq. (4.1.5) with $h = 0.7$ requires that the universe expanded during inflation by at least a factor 4×10^7, or 17 e-foldings. At the other extreme, we would not expect ρ_1 to be greater than the Planck energy density $G^{-2} = [1.22 \times 10^{19}\ \mathrm{GeV}]^4$, in which case Eq. (4.1.5) with $h = 0.7$ would require that the expansion during inflation was at least by a factor 5×10^{29}, or 68 e-foldings. We will see some evidence in Section 10.3 that ρ_1 is of order $[2 \times 10^{16}\ \mathrm{GeV}]^4$, in which case $e^{\mathcal{N}}$ for $h = 0.7$ would have to be at least 8×10^{26}, so that $\mathcal{N} > 62$.

This is the least convincing of the arguments for inflation, because the small value of $|K|/\dot{a}^2$ in the past could be explained by the assumption (one that was often made before anyone heard of inflation) that space is precisely flat, so that $K = 0$ now and always. On the other hand, as we will discuss in Section 4.3, inflation opens up the interesting possibility that the universe in the large is not at all homogeneous and isotropic, and that its apparent flatness of the cosmic metric is just the result of inflation.

It would be quite a coincidence if inflation lasted for precisely the right number of e-foldings so that $|K|/\dot{a}^2$ would have decreased during inflation from an initial value of order unity just enough so that its subsequent increase during the radiation and matter-dominated eras would have brought it back to a value of order unity now. It seems more likely that the present value $|\Omega_K|$ would be either much larger or much smaller than unity, and since observations tell us that it is not much larger than unity, inflationary theories suggest that $|\Omega_K| \ll 1$. (But this sort of reasoning would also suggest that the vacuum energy is much less than the present matter density, which we now know is not the case.)

B.   Horizons

From the beginning the observed high degree of isotropy of the cosmic microwave radiation background posed a problem. Recall that the horizon size in a matter- or radiation-dominated universe is of order t, which, because $a(t)$ has increased as $t^{2/3}$ since the time of last scattering, was of order $d_H \approx H_0^{-1}(1+z_L)^{-3/2}$ at the time of last scattering. (See Eq. (2.6.32).) Also, according to Eq. (2.6.29), the angular diameter distance d_A to the surface of last scattering is of order $H_0^{-1}(1+z_L)^{-1}$, so the horizon at the time of last scattering now subtends an angle of order $d_H/d_A \approx (1+z_L)^{-1/2}$ radians, which for $z_L \simeq 1100$ is about $1.6°$. Therefore in a matter- or radiation-dominated universe no physical influence could have smoothed out initial inhomogeneities and brought points at a redshift z_L that are separated by more than a few degrees to the same temperature, in contradiction with the nearly perfect isotropy of the microwave background at large angular scales observed ever since the background radiation was discovered. Inflation provides an explanation: during the inflationary era the part of the universe that we can observe would have occupied a tiny space, and there would have been plenty of time for everything in this space to be homogenized.

To work out what this means for the expansion during inflation, first recall that as discussed in Section 1.13, the proper horizon size at the time t_L of last scattering is

$$d_H(t_L) \equiv a(t_L) \int_{t_*}^{t_L} \frac{dt}{a(t)} , \tag{4.1.6}$$

with t_* (possibly equal to $-\infty$) the beginning of the era of inflation. We have seen that the contribution to the integral from the radiation and matter-dominated eras is much too small to account for the isotropy of the microwave radiation background, so we will assume that the integral is dominated by an era of inflation. For definiteness, we assume that during inflation $a(t)$ increased exponentially at a rate H_I, so that

$$a(t) = a(t_*) \exp\left(H_I(t - t_*)\right) = a_I \exp\left(- H_I(t_I - t)\right),$$

where t_I is again the time of the end of inflation, and $a_I = a(t_I)$. With $\mathcal{N} \equiv H_I(t_I - t_*)$ the number of e-foldings of expansion during inflation, Eq. (4.1.6) gives

$$d_H(t_L) = \frac{a(t_L)}{a_I H_I} \left[e^{\mathcal{N}} - 1 \right] . \tag{4.1.7}$$

In order to have any hope of solving the horizon problem, we must have $e^{\mathcal{N}} \gg 1$, so we can drop the term -1 in square brackets in Eq. (4.1.7).

To account for the observed high degree of isotropy of the cosmic microwave background at large angular scales we need $d_H(t_L) > d_A(t_L)$, where $d_A(t_L)$ is the angular–diameter distance of the surface of last scattering. According to Eq. (2.6.29),

$$d_A(t_L) \approx \frac{a(t_L)}{H_0 a_0} . \tag{4.1.8}$$

The condition $d_H(t_L) > d_A(t_L)$ for the isotropy of the cosmic microwave background is then

$$e^{\mathcal{N}} > \frac{a_I H_I}{a_0 H_0} . \tag{4.1.9}$$

This is the same as the condition (4.1.2) for the solution of the flatness problem. If we again make the assumption that not much happens between the end of inflation and the beginning of the radiation-dominated era, and use Eq. (4.1.3), then to solve the horizon problem we again need $\mathcal{N} > 17$ if $\rho_1 \simeq [1 \text{ MeV}]^4$, $\mathcal{N} > 62$ if $\rho_1 \simeq [2 \times 10^{16} \text{ GeV}]^4$, and $\mathcal{N} > 68$ if $\rho_1 \simeq G^{-2}$. We will see in Chapter 10 that whether 17 or 62 or 68 e-foldings are needed to solve the horizon problem, it is only that number of e-foldings before the end of inflation that can be explored through observations of nonuniformities in the present universe.

It should be noted that the time t_L of last scattering does not enter in the bound (4.1.9) on $\mathcal{N}$, so this is also the condition that the whole sky at any redshift $z < z_{EQ}$ was within the horizon at the time that light observed now with that redshift left its source. Indeed, Eq. (4.1.9) is also the condition that the horizon size at the present should be greater than the size of the observable universe, which is roughly $1/H_0$.

C. Monopoles

In grand unified theories local symmetry under some simple symmetry group is spontaneously broken at an energy $M \approx 10^{16}$ GeV to the gauge symmetry of the Standard Model under the group $SU(3) \times SU(2) \times U(1)$.

In all such cases, the scalar fields that break the symmetry can be left in twisted configurations that carry non-zero magnetic charge and that cannot be smoothed out through any continuous processes.[6] This poses a problem for some cosmological models.[7] The scalar fields before this phase transition would have necessarily been uncorrelated at distances larger than the horizon distance, the farthest distance that light could have traveled since the initial singularity. At an early time t in the standard big bang theory the horizon distance was of order $t \approx (G(k_B T)^4)^{-1/2}$ (where $G \simeq (10^{19} \text{ GeV})^{-2}$ is Newton's constant), so the number density of monopoles produced at the time that the temperature drops to M/k_B would have been of order $t^{-3} \approx (GM^4)^{3/2}$, which is smaller than the photon density $\approx M^3$ at $T \approx M/k_B$ by a factor of order $(GM^2)^{3/2}$. For $M \approx 10^{16} \text{ GeV}$ this factor is of order 10^{-9}. If monopoles did not find each other to annihilate, then this ratio would remain roughly constant to the present, but with at least 10^9 microwave background photons per nucleon today, this would give at least one monopole per nucleon, in gross disagreement with what is observed.

This potential paradox was one of the factors leading to interest in inflationary cosmological models. In such models, a period of exponential expansion that occurred before the monopoles were produced would have greatly extended the horizon, and an exponential expansion that occurred after the production of monopoles (but before photons were created in a period of reheating) would have greatly reduced the monopole to photon ratio. To be specific, the search for monopoles in iron ore, seawater, etc. shows that there are fewer than 10^{-6} per gram, or about 10^{-30} monopoles per nucleon, and hence fewer than about 10^{-39} monopoles per photon.[8] (With this abundance, even if the monopole mass were as large as 10^{19} GeV, they would make a negligible contribution to the cosmic mass density.) In order for inflation to have reduced the monopole/photon ratio by a factor 10^{-30}, it must have increased the horizon size (at some time before the reheating that creates photons) by a factor 10^{10}. That is, the horizon size $e^{\mathcal{N}}/H_1$ after inflation must be greater than the previous estimate $(GM^4)^{-1/2}$ by at least a factor 10^{10}. For $H_1 \approx (GM^4)^{1/2}$ this requires the number $\mathcal{N}$ of e-foldings to be greater than $\ln 10^{10} = 23$. Of course, another possible solution of the monopole problem is that inflation ends at a temperature below the grand unification scale M, so that there never was a time when the grand unification group was unbroken. An even simpler possibility, which does not rely on inflation, is that there may be no simple gauge group that is

---

[6]For a discussion, see QTF, Vol. II, Sec. 23.3.

[7]Ya. B. Zel'dovich and M. Yu. Khlopov, *Phys. Lett.* **B 79**, 239 (1978); J. Preskill, *Phys. Rev. Lett.* **43**, 1365 (1979). For a review, see J. Preskill, *Annual Rev. Nucl. Part. Science* **34**, 461 (1984).

[8]For a review, see Particle Data Group, *Phys Lett.* **B 582**, 1001 (2004).

spontaneously broken to the gauge group $SU(3) \times SU(2) \times U(1)$ of the Standard Model.

The most serious of the above three problems is the horizon problem. As we have seen, there are possible solutions of the flatness and monopole problems that do not rely on inflation. Also, any number of e-foldings of inflation that solves the horizon problem automatically solves not only the flatness problem, but also the monopole problem. If the radiation-dominated era begins with an energy density $\rho_1 > [10^{15} \text{ GeV}]^4$ then for inflation to solve the flatness and horizon problems we need at least 59 e-foldings of inflation, which is more than enough to avoid the monopole problem, while if $\rho_1 < [10^{15} \text{ GeV}]^4$ GeV then in the usual picture of grand unification there would be no monopoles at all.

4.2 Slow-roll inflation

In Guth's original work, inflation was conceived to be due to a delayed first-order phase transition, in which a scalar field was initially trapped in a local minimum of some potential, and then leaked through the potential barrier and rolled toward a true minimum of the potential. It was soon realized[1] that this idea does not work, because of what has come to be called the graceful exit problem. The transition from the super-cooled initial "false vacuum" phase to the lower energy "true vacuum" phase could not have occurred everywhere simultaneously, but here and there in small bubbles of true vacuum, which rapidly expanded into the background of false vacuum, in which the scalar field would have been still trapped in its local minimum,[2] like water droplets forming in super-cooled water vapor. The trouble is that the latent heat released in the phase transition would have wound up in the bubble walls, leaving the interiors of the bubbles essentially empty, so that the only places where there would be energy that could grow into the present contents of the universe would be highly inhomogeneous and anisotropic. At first Guth thought the bubbles in inflationary cosmologies would have merged, leading to our present more-or-less homogeneous universe, but this could not have happened; because the background false-vacuum space continued to inflate, the bubble walls would have moved too fast away from each other ever to have coalesced.

---

[1] S. W. Hawking, I. G. Moss, and J. M. Stewart, *Phys. Rev.* D **26**, 2681 (1982); A. H. Guth and E. J. Weinberg, *Nucl. Phys.* B **212**, 321 (1983).

[2] For a description of this process and references to the original literature on bubble formation in quantum field theory, see QTF II, Section 23.8.

Guth's version of inflation was soon supplanted by a version due to Linde and to Albrecht and Steinhardt, known as 'new inflation.'[3] Originally new inflation was formulated in a particular model of the breakdown of a grand unified symmetry, using a symmetry-breaking mechanism introduced by Coleman and E. Weinberg.[4] With this mechanism the zero-temperature potential for a scalar field φ is artificially adjusted to have zero second derivative at $\varphi = 0$. One-loop radiative corrections then give a potential equal to a known positive factor times $\varphi^4 \ln(\varphi/M)$, where M is a free constant; changing the value of M amounts to changing the φ^4 coupling constant. This potential has an unstable stationary point at $\varphi = 0$ and a minumum at $\varphi_0 = Me^{-1/4}$. At finite temperature T there is also a quadratic term in the potential, proportional to $T^2\varphi^2$, which makes the stationary point at $\varphi = 0$ into a local minimum. Again the phase transition occurs by forming bubbles, but for low temperature the potential barrier is very small, and so the scalar field in the interior of the bubble starts with φ nearly zero. The field then rolls slowly down the potential, in the manner discussed in Section 1.12, while the universe (including the bubble) undergoes an exponential expansion. Eventually the field energy is converted into ordinary particles, filling the bubble. Our observable universe is supposed to occupy a small part of one such bubble.

The consequences of the new inflationary theories turned out to depend on the slow roll of the scalar field after bubble formation, rather than the process of bubble formation itself. Indeed, the important aspects of inflation do not really require any assumptions about grand unification or the Coleman–E. Weinberg mechanism. All we need to assume is that there is a scalar field φ, known as the *inflaton*, which at some early time takes a value at which the potential $V(\varphi)$ is large but quite flat. The scalar field "rolls" very slowly at first down this potential, so that the Hubble constant decreases only slowly, and the universe experiences a more-or-less exponential inflation before the field changes very much.

To put this quantitatively, recall that the energy density (B.66) and pressure (B.67) of a spatially homogeneous scalar field $\varphi(t)$ with potential $V(\varphi)$ in a Robertson–Walker spacetime take the form (with $\hbar = c = 1$)

$$\rho = \frac{1}{2}\dot{\varphi}^2 + V(\varphi), \quad p = \frac{1}{2}\dot{\varphi}^2 - V(\varphi),$$

so the energy conservation equation $\dot{\rho} = -3H(\rho + p)$ takes the form:

$$\ddot{\varphi} + 3H\dot{\varphi} + V'(\varphi) = 0, \tag{4.2.1}$$

---

[3] A. D. Linde, *Phys. Lett. B* **108**, 389 (1982); **114**, 431 (1982); *Phys. Rev. Lett.* **48**, 335 (1982); A. Albrecht and P. Steinhardt, *Phys. Rev. Lett.* **48**, 1220 (1982).

[4] S. Coleman and E. Weinberg, *Phys. Rev. D* **7**, 1888 (1973).

where $H \equiv \dot{a}/a$ is the time dependent expansion rate, which during the period of scalar field energy dominance is given by

$$H = \sqrt{\frac{8\pi G \rho}{3}} = \sqrt{\frac{8\pi G}{3}\left(\frac{1}{2}\dot{\varphi}^2 + V(\varphi)\right)}. \qquad (4.2.2)$$

From Eqs. (4.2.1) and (4.2.2) we can derive an extremely useful formula for $\dot{H}$. By taking the time derivative of the square of Eq. (4.2.2) and then using Eq. (4.2.1), we have

$$2H\dot{H} = \frac{8\pi G}{3}\left(\dot{\varphi}\ddot{\varphi} + V'(\varphi)\dot{\varphi}\right) = -8\pi G H \dot{\varphi}^2 ,$$

and therefore

$$\dot{H} = -4\pi G \dot{\varphi}^2 . \qquad (4.2.3)$$

Now, in order to have a nearly exponential expansion, the fractional change $|\dot{H}/H|(1/H)$ in H during an expansion time $1/H$ must be much less than unity. That is, we must have

$$|\dot{H}| \ll H^2 . \qquad (4.2.4)$$

With Eqs. (4.2.3) and (4.2.2), this requires that

$$\dot{\varphi}^2 \ll |V(\varphi)| . \qquad (4.2.5)$$

This has the consequence that $p \simeq -\rho$, and also

$$H \simeq \sqrt{\frac{8\pi G V(\varphi)}{3}} . \qquad (4.2.6)$$

Usually it is also assumed that the fractional change $|\ddot{\varphi}/\dot{\varphi}|(1/H)$ in $\dot{\varphi}$ during an expansion time $1/H$ is much less than unity. That is,

$$|\ddot{\varphi}| \ll H|\dot{\varphi}| . \qquad (4.2.7)$$

This has the consequence that we may drop the inertial term $\ddot{\varphi}$ in Eq. (4.2.1), which then becomes

$$\dot{\varphi} = -\frac{V'(\varphi)}{3H} = -\frac{V'(\varphi)}{\sqrt{24\pi G V(\varphi)}} . \qquad (4.2.8)$$

The fractional change of the expansion rate H in an expansion time $1/H$ will then be

$$\frac{|\dot{H}|}{H^2} = \frac{1}{2}\sqrt{\frac{3}{8\pi G}}\left|\frac{V'(\varphi)\dot{\varphi}}{V^{3/2}(\varphi)}\right| = \frac{1}{16\pi G}\left(\frac{V'(\varphi)}{V(\varphi)}\right)^2 , \qquad (4.2.9)$$

so the exponential expansion of the universe will last for many *e*-foldings if

$$\left|\frac{V'(\varphi)}{V(\varphi)}\right| \ll \sqrt{16\pi G} \,. \tag{4.2.10}$$

According to Eq. (4.2.8), the condition on the potential for the inequality (4.2.5) to be satisfied is that

$$\left|\frac{V'(\varphi)}{V(\varphi)}\right| \ll \sqrt{24\pi G} \,, \tag{4.2.11}$$

which is guaranteed by the inequality (4.2.10). Also, Eq. (4.2.8) gives

$$\ddot{\varphi} = -\frac{V''(\varphi)\dot{\varphi}}{3H} + \frac{V'(\varphi)\dot{H}}{3H^2} = \frac{V''(\varphi)V'(\varphi)}{9H^2} - \frac{V'^3}{48\pi G V^2} \,. \tag{4.2.12}$$

The inequality (4.2.10) ensures that the absolute value of the last term on the right-hand side is much less than $|V'(\varphi)|$, so the condition for $|\ddot{\varphi}|$ to be much less than $|V'(\varphi)|$ is that $|V''(\varphi)| \ll 9H^2$, or, in other words,

$$\left|\frac{V''(\varphi)}{V(\varphi)}\right| \ll 24\pi G \,. \tag{4.2.13}$$

Eqs. (4.2.10) and (4.2.13) are the two "flatness" conditions needed to insure the slow roll of both φ and $\dot{\varphi}$. It is possible in principle that the second flatness condition (4.2.13) may not be satisfied for a potential that does satisfy the first flatness condition (4.2.10), but this is unusual, and in particular is not possible for the simple potentials discussed below.

Under these conditions the expansion is generally not strictly exponential, but it can easily be exponentially large. Suppose that during some time interval the field $\varphi(t)$ shifts from an initial value φ_1 to a final value φ_2, with $0 < V(\varphi_2) < V(\varphi_1)$, with both inequalities (4.2.10) and (4.2.13) assumed valid over this range of φ. The Robertson–Walker scale factor will increase during this period by a factor

$$\frac{a(t_2)}{a(t_1)} = \exp\left[\int_{t_1}^{t_2} H\,dt\right] = \exp\left[\int_{\varphi_1}^{\varphi_2} \frac{H\,d\varphi}{\dot{\varphi}}\right]$$
$$\simeq \exp\left[-\int_{\varphi_1}^{\varphi_2} \left(\frac{8\pi G V(\varphi)}{V'(\varphi)}\right) d\varphi\right]. \tag{4.2.14}$$

In this range the potential is positive and decreases as $\varphi(t)$ runs from φ_1 to φ_2, so the argument of the exponential in Eq. (4.2.14) is positive. Condition (4.2.10) tells us that this argument is much greater than $\sqrt{4\pi G}|\varphi_1 - \varphi_2|$, so this flatness condition guarantees that we get a large number of *e*-foldings in any time interval in which φ changes by an amount at least as large as $1/\sqrt{4\pi G} = 3.4 \times 10^{18}$ GeV.

It is important to note that such large values of the scalar field do not necessarily rule out the classical treatment of gravitation on which we have been relying. The condition that allows us to neglect quantum gravitational effects is that the energy density should be much less than the Planck energy density:

$$|V(\varphi)| \ll (4\pi G)^{-2} . \tag{4.2.15}$$

This condition can be satisfied, even if φ is comparable to the Planck mass $\approx G^{-1/2}$, by supposing that $V(\varphi)$ is proportional to a sufficiently small coupling constant. Neither the flatness conditions (4.2.10), (4.2.13) nor the growth (4.2.14) of $a(t)$ for a given change in $\varphi(t)$ depend on the value of such a coupling constant.

Depending on the potential shape, the flatness conditions (4.2.10) and (4.2.13) may provide either conditions on the initial value of the scalar field or on the parameters of the potential itself. As an example of the first sort, consider the power-law potential

$$V(\varphi) = g\varphi^\alpha , \tag{4.2.16}$$

with g and α arbitrary real parameters, except that we assume that $g > 0$ and take $|\alpha|$ larger but not orders of magnitude larger than unity. The flatness conditions (4.2.10) and (4.2.13) are then both satisfied for $|\varphi| \gg 1/\sqrt{4\pi G}$, irrespective of the value of the coupling constant g. The number of e-foldings of expansion for a scalar field that starts at a value φ_1 and ends at a much smaller value is given by Eq. (4.2.14) as $8\pi G\varphi_1^2/\alpha$, so for instance for $\alpha = 4$ we get the 62 e-foldings needed to avoid the horizon problem for inflation ending at a temperature 2×10^{16} GeV$/k_B$ if $|\varphi_1| > \sqrt{31/\pi G}$. On the other hand, for this potential the condition (4.2.15) for the neglect of quantum gravitational effects does put an upper bound on $|g|$. For instance, in the case $\alpha = 4$, the potential (4.2.16) will satisfy condition (4.2.15) if $g \ll (\sqrt{4\pi G}|\varphi|)^{-4}$ so, with $|\varphi|$ just large enough to get 62 e-foldings of exponential expansion, to avoid quantum gravity corrections we need $g \ll 2 \times 10^{-5}$. The need for a very small coupling can be avoided in theories with more than one scalar field, such as "hybrid inflation" theories,[5] in which the effective self-coupling of one scalar field is very small because the other scalar field has a very small mass and hence a small expectation value.

---

[5]A. D. Linde, *Phys. Lett.* B **259**, 38 (1991); *Phys. Rev.* D **49**, 748 (1994) [astro-ph/9307002].

There is a well-known example of a rather different sort for which Eqs. (4.2.1) and (4.2.2) can be solved exactly, the exponential potential:[6]

$$V(\varphi) = g \exp(-\lambda\varphi) \,, \tag{4.2.17}$$

with g and λ arbitrary constants. It is easy to verify that Eqs. (4.2.1) and (4.2.2) are satisfied by[7]

$$\varphi(t) = \frac{1}{\lambda} \ln\left(\frac{8\pi Gg\epsilon^2 t^2}{3 - \epsilon} \right) \,, \tag{4.2.18}$$

and

$$a \propto t^{1/\epsilon} \,, \qquad H = 1/\epsilon t \,, \tag{4.2.19}$$

where ϵ is the positive dimensionless quantity

$$\epsilon \equiv \frac{\lambda^2}{16\pi G} \,. \tag{4.2.20}$$

The flatness conditions (4.2.10) and (4.2.13) here respectively read $\epsilon \ll 1$ and $\epsilon \ll 3/2$, so both are satisfied if and only if

$$\epsilon \ll 1 \,, \tag{4.2.21}$$

with no constraint on the values of either g or φ. This exact solution is useful as a check of approximate calculations for more general potentials. For instance, Eq. (4.2.18) may be compared with the solution of the approximate equation (4.2.8), which for the exponential potential is

$$\varphi(t) = \frac{1}{\lambda} \ln\left(\frac{8\pi Gg\epsilon^2 t^2}{3} \right) \,, \tag{4.2.22}$$

The difference between this and the exact solution (4.2.18) is evidently negligible for $\epsilon \ll 1$. Likewise, the increase in the Robertson–Walker scale factor during a time interval from t_1 to t_2 in which the field drops from φ_1 to φ_2 is

$$\frac{a(t_2)}{a(t_1)} = \left(\frac{t_2}{t_1} \right)^{1/\epsilon} = e^{(\varphi_2 - \varphi_1)\lambda/2\epsilon} \,, \tag{4.2.23}$$

---

[6]L. F. Abbott and M. B. Wise, *Nucl. Phys.* **B 244**, 541 (1984); F. Lucchin and S. Matarrese, *Phys. Rev. D* **32**, 1316 (1985); *Phys. Lett.* **B 164**, 282 (1985); D. H. Lyth and E. D. Stewart, *Phys. Lett.* **B 274**, 168 (1992).

[7]This is an exact solution for all values of t, but it is not the most general solution. Since Eqs. (4.2.1) is a second-order differential equation, it has a two-parameter set of solutions. The particular solution (4.2.18) is an attractor, which the general solutions approach for large t.

which is exactly the same as the "slow-roll" result (4.2.14) for the exponential potential.

It is assumed in these theories that when $V(\varphi)$ dropped sufficiently far, the inequalities (4.2.10) and (4.2.13) were in general no longer be satisfied, and φ began a damped oscillation around the minimum of $V(\varphi)$, which is at the present value φ_0. Eventually φ would have approached close enough to φ_0 so that we can approximate the potential as a quadratic,

$$V(\varphi) = \frac{1}{2}m^2 (\varphi - \varphi_0)^2 . \tag{4.2.24}$$

(In order to account for the small present value of the vacuum energy, it is necessary to assume that, for reasons that remain entirely mysterious, the minimum value of the potential is very close to zero.) This is just like the field theory of spinless particles with mass m and negligible velocity. In order to have ended inflation, there must also be some coupling of the inflaton scalar field to other fields, including the fields of ordinary matter and radiation, so that the energy density of the inflaton field decreased as

$$\rho_\varphi(t) = \rho_\varphi(t_I) \left(\frac{a(t_I)}{a(t)}\right)^3 e^{-\Gamma(t-t_I)} , \tag{4.2.25}$$

where Γ is the rate of decay of the φ quanta into other particles, and t_I is taken at the beginning of the inflaton oscillation and decay. This is known as the period of *reheating*.[8] It is this period in which the entropy observed in the present universe is supposed to be generated.[9]

The energy density ρ_M of the particles into which φ decayed satisfies a conservation equation like Eq. (1.1.32), but corrected to take account of the flow of energy from the inflaton:

$$\dot\rho_M + 3H(\rho_M + p_M) = \Gamma\rho_\varphi . \tag{4.2.26}$$

For definiteness, we will assume that the decay products of the inflaton are highly relativistic, so that $p_M = \rho_M/3$. Then the solution of Eq. (4.2.26) is

$$\rho_M(t) = \frac{\rho_\varphi(t_I)\,\Gamma\,a^3(t_I)}{a^4(t)} \int_{t_I}^t a(t')\, e^{-\Gamma(t'-t_I)}\, dt' \tag{4.2.27}$$

In contemporary models of inflation this is the source of all the matter and radiation in the present universe. (In using this relation it is important

---

[8]The term *reheating* is a historical relic of early theories of inflation in which it was assumed that the zero-temperature slow roll of the inflaton field followed an earlier period of high temperature.

[9]A. D. Dolgov and A. D. Linde, *Phys. Lett.* **116B**, 329 (1982); L. F. Abbott, E. Farhi, and M. B. Wise, *Phys. Lett.* **117B**, 29 (1982).

to keep in mind the possibility of *parameteric amplification* of the density $\rho_\varphi(t_I)$. That is, scalar field expectation values can increase the effective mass of the inflaton field φ, and thereby increase its energy density.[10])

The matter energy density (4.2.27) starts equal to zero at $t = t_I$, then rises at first, and finally falls as the density is attenuated by the expansion of the universe. It is of some interest to find the value of $\rho_M(t)$ at its maximum, because this tells us the maximum temperature ever reached, which controls the kinds of relics—cold dark matter, baryons, monopoles, axions—left over from the early universe. This maximum density can be easily calculated in two extreme cases, for $\Gamma \gg H(t_I)$ and $\Gamma \ll H(t_I)$.

For $\Gamma \gg H(t_I)$, we can express Eq. (4.2.27) as a power series in $H(t_I)/\Gamma$ by repeated integration by parts

$$\rho_M(t) = \rho_\varphi(t_I) \left(\frac{a(t_I)}{a(t)}\right)^4 \left(1 + \frac{H(t_I)}{\Gamma} + \cdots \right) \qquad (4.2.28)$$

We see that in this case ρ_M jumped up almost immediately to the value $\rho_\varphi(t_I)$, and then decreased with the usual a^{-4} factor, so in this case all the energy of the inflaton field at the end of inflation went into ordinary matter and radiation. This was the assumption made in deriving lower bounds on the number of e-foldings of inflation in the previous section.

For $\Gamma \ll H(t_I)$, the maximum value of $\rho_M(t)$ was reached at a time when the exponential $e^{-\Gamma(t'-t_I)}$ in Eq. (4.2.27) had not yet begun to decay. (This will be checked below.) Setting this factor equal to unity, we have

$$\rho_M(t) \simeq \frac{\rho_\varphi(t_I)\, \Gamma\, a^3(t_I)}{a^4(t)} \int_{t_I}^{t} a(t')\, dt' \qquad (4.2.29)$$

At this time the energy density of the universe was still dominated by the inflaton, so $a(t) = a(t_I)(t/t_I)^{2/3}$, and Eq. (4.2.29) becomes

$$\rho_M(t) \simeq \frac{3}{5}\Gamma\, t_I\, \rho_\varphi(t_I) \left(\frac{t_I}{t}\right)^{8/3} \left(\left(\frac{t}{t_I}\right)^{5/3} - 1\right). \qquad (4.2.30)$$

This reached a maximum at $t = (8/3)^{3/5} t_I$, which incidentally confirms that, under the assumption that $\Gamma \ll H(t_I) = 2/3t_I$, the argument of the exponential in Eq. (4.2.27) at this maximum was still negligible. At this maximum, the matter density is

$$\rho_{M,\text{max}} = (3/8)^{8/5}\Gamma\, t_I\, \rho_\varphi(t_I) = 0.139 \left(\Gamma/H(t_I)\right)\rho_\varphi(t_I). \qquad (4.2.31)$$

---

[10]L. Kofman, A. D. Linde, and A. A. Starobinsky, *Phys. Rev. Lett.* **73**, 3195 (1994); *Phys. Rev. D* **56**, 3258 (1997).

In this case the maximum energy density at the beginning of the radiation-dominated era would have been much less than the energy density in the inflaton field at the end of inflation.

4.3 Chaotic inflation, eternal inflation

It was soon realized that some "new inflation" models actually entail an endless production of inflating bubbles.[1] This has come to be called "eternal inflation."[2] To take one example, if the inflaton scalar field at a given point in space once had a value of unstable equilibrium (like at the top of a potential hill) then the probability that the inflaton field was still at this value after a time t decreased as $\exp(-\gamma t)$. However, the volume in which the scalar field had this value was meanwhile increasing as $\exp(+3Ht)$, so as long as $3H > \gamma$ the volume of space that still undergoes inflation eternally increases exponentially.

We have been assuming that the scalar field is initially independent of position, aside from small perturbations, about which more in Chapter 10. Soon after the introduction of new inflation, the possibilities of inflationary theory were greatly expanded and improved when Linde[3] proposed the theory of "chaotic inflation," in which initially one or more scalar fields varied in a random way with position. Here and there one would have found patches of space in which an inflaton field took a nearly uniform value at which the potential satisfied the slow-roll conditions (4.2.10) and (4.2.13), as for instance a value substantially greater than the Planck mass for a power-law potential. Inflation will then have occured in such a patch, provided the patch was initially sufficiently large.

It is necessary to require that the uniform patch be sufficiently large, because the scalar field Lagrangian density given by Eq. (B.63) contains a term involving spatial derivatives, which for a non-uniform scalar field contributes a term $-a^{-2}\nabla^2\varphi$ on the left-hand side of Eq. (4.2.1). In order for this term not to interfere with the slow-roll analysis of the previous section, we need the scale L of *proper* distances over which φ takes a roughly constant initial value to be greater than $|\varphi/V'(\varphi)|^{1/2}$. For instance, for the

[1] P. J. Steinhardt, in *The Very Early Universe — Proceedings of the Nuffield Workshop, 1982*, eds. G. W. Gibbon S. W. Hawking, and S. T. C. Siklos (Cambridge University Press, 1983): 251; A. Vilenkin, *Phys. Rev. D* **27**, 2848 (1983).

[2] For reviews of this and other variants of inflation, see A. Guth, talk given at the *Pritzker Symposium on the Status of Inflationary Cosmology*, January 1999, astro-ph/0002188; A. Linde, *J. Phys. Conf. Ser.* **24**, 151 (2005) [hep-th/0503195].

[3] A.D. Linde, Phys. Lett. **129B**, 177 (1983).

potential $V(\varphi) = g\varphi^4$, this condition gives

$$L \gg \frac{1}{2\sqrt{g}|\varphi|} = \left(\frac{1}{g(4\pi G)^2 \varphi^4}\right)^{1/2} \left(\sqrt{\pi G}\varphi\right) \sqrt{4\pi G} \qquad (4.3.1)$$

The classical field theory condition (4.2.15) makes the first factor much larger than unity, while the slow-roll condition (4.2.10) makes the second factor also much larger than unity, so the patch size must be very much greater than the third factor, which is essentially the Planck length. Such relatively large uniform patches may be quite rare, but that is no argument against this hypothesis, because life can only arise in big bangs that stem from such patches.

A sufficiently large patch will inflate to an enormous size, which to observers deep inside seems highly homogeneous and isotropic. In this way, chaotic inflation solves a puzzle that had not generally been realized to be a puzzle, even when the first inflation theories were being developed. It explains not just why the Robertson–Walker metric in which we find ourselves was remarkably flat in the past; it also explains why we find ourselves in a Robertson–Walker metric at all. Unfortunately, it is hard to see how we will ever observe any part of the universe beyond our inflated patch. The validity of the idea of chaotic inflation will probably have to come from progress in fundamental physics, which may verify the existence of a suitable inflaton field, rather than from astronomical observation.

Even when the scalar field in a patch of the space was large enough to start a slow roll inflation, quantum fluctuations in smaller regions within that patch would subsequently have driven the inflaton field to even higher values, so that these regions will begin an earlier stage of inflation.[4] In this way, chaotic inflation turns out also to be eternal.

---

[4]A. D. Linde, *Mod. Phys. Lett.* **A1**, 81 (1986); *Phys. Lett.* **B 175**, 395 (1986); A. S. Goncharev, A. D. Linde, and V. F. Mukhanov, *Int. J. Mod. Phys.* **A 2**, 561 (1987); A. D. Linde, D. A. Linde, and A. Mezhlumian, *Phys. Rev. D 49*, 1783 (1994) [gr-qc/9306035].

5

General Theory of Small Fluctuations

In most of the work described in the previous chapters, the universe has been treated as isotropic and homogeneous, with a gravitational field described by the Robertson–Walker metric. This is of course just an approximation, which ignores many of the most interesting things in the universe: galaxy clusters, galaxies, stars, us. We now turn to an analysis of these departures from homogeneity and isotropy.

In this chapter we will lay a foundation for this analysis, by deriving the general relativistic equations that govern small fluctuations, and drawing general conclusions about their implications. Chapter 6 will apply this formalism to the evolution of structure, from the radiation-dominated era to near the present. In Chapters 7 and 8 we apply the results obtained in Chapter 6 to the observed fluctuations in the cosmic microwave background and to the growth of structure. Chapter 9 deals with gravitational lensing, which may in the long run provide the best tool for analyzing the large scale structure of dark matter. These chapters are kept at a general level, independent of detailed assumptions about an inflationary era before the radiation-dominated era. Chapter 10 will then explore the implications of inflationary theories for the calculations of Chapters 6 through 9. Some readers may prefer to skip on immediately to Chapter 6, using the present chapter as a source of useful formulas, while others will do better to read these chapters in order.

5.1 Field equations

As an essential feature of the analysis presented here, we assume that during most of the history of the universe all departures from homogeneity and isotropy have been small, so that they can be treated as first-order perturbations.[1] Because the observable universe is nearly homogeneous, and its spatial curvature either vanishes or is negligible until very near the present, we will take the unperturbed metric to have the Robertson–Walker form (1.1.11), with curvature constant $K = 0$. (Effects of a possible finite curvature at times close to the present will be included in Chapters 7 through 9

---

[1] The study of first-order cosmological fluctuations was initiated by E. Lifshitz, *J. Phys. U.S.S.R.* **10**, 116 (1946). Classical second-order corrections were worked out by K. Tomita, *Prog. Theor. Phys.* **37**, 831 (1967); **45**, 1747 (1970); **47**, 416 (1971). For recent work, see K. Tomita, *Phys. Rev. D* **71**, 3504 (2005); N. Bartolo, S. Matarrese, and A. Riotto, *J. Cosm. & Astropart. Phys.* **0606**, 024 (2006) [astro-ph/0604416].

where they are relevant.) The total perturbed metric is then

$$g_{\mu\nu} = \bar{g}_{\mu\nu} + h_{\mu\nu} , \qquad (5.1.1)$$

where $\bar{g}_{\mu\nu}$ is the unperturbed $K = 0$ Robertson–Walker metric

$$\bar{g}_{00} = -1 , \quad \bar{g}_{i0} = \bar{g}_{0i} = 0 , \quad \bar{g}_{ij} = a^2(t)\delta_{ij} , \qquad (5.1.2)$$

and $h_{\mu\nu} = h_{\nu\mu}$ is a small perturbation. (Here and from now on, a bar over any quantity denotes its unperturbed value.) The perturbation to the inverse of a general matrix M is $\delta M^{-1} = -M^{-1}(\delta M)M^{-1}$, so the inverse metric is perturbed by

$$h^{\mu\nu} \equiv g^{\mu\nu} - \bar{g}^{\mu\nu} = -\bar{g}^{\mu\rho}\bar{g}^{\nu\sigma}h_{\rho\sigma} , \qquad (5.1.3)$$

with components

$$h^{ij} = -a^{-4}h_{ij} , \quad h^{i0} = a^{-2}h_{i0} , \quad h^{00} = -h_{00} . \qquad (5.1.4)$$

Note the $-$ sign in the last expression in Eq. (5.1.3); in our notation, the perturbation $\delta g^{\mu\nu}$ to $g^{\mu\nu}$ is *not* given by simply using the unperturbed metric to raise the indices on $\delta g_{\mu\nu}$.

The metric perturbation produces a perturbation to the affine connection

$$\delta\Gamma^{\mu}_{\nu\lambda} = \frac{1}{2}\bar{g}^{\mu\rho}\left[-2h_{\rho\sigma}\bar{\Gamma}^{\sigma}_{\nu\lambda} + \partial_{\lambda}h_{\rho\nu} + \partial_{\nu}h_{\rho\lambda} - \partial_{\rho}h_{\lambda\nu}\right] . \qquad (5.1.5)$$

For $K = 0$, the only non-vanishing components of the unperturbed affine connection are given by Eqs. (1.1.17)–(1.1.19) as

$$\bar{\Gamma}^{i}_{j0} = \bar{\Gamma}^{i}_{0j} = \frac{\dot{a}}{a}\delta_{ij} , \qquad \bar{\Gamma}^{0}_{ij} = a\dot{a}\delta_{ij} .$$

Thus Eq. (5.1.5) gives the components of the perturbed affine connection as

$$\delta\Gamma^{i}_{jk} = \frac{1}{2a^2}\left(-2a\dot{a}\,h_{i0}\,\delta_{jk} + \partial_{k}h_{ij} + \partial_{j}h_{ik} - \partial_{i}h_{jk}\right) \qquad (5.1.6)$$

$$\delta\Gamma^{i}_{j0} = \frac{1}{2a^2}\left(-\frac{2\dot{a}}{a}h_{ij} + \dot{h}_{ij} + \partial_{j}h_{i0} - \partial_{i}h_{j0}\right) \qquad (5.1.7)$$

$$\delta\Gamma^{0}_{ij} = \frac{1}{2}\left(2a\dot{a}\,\delta_{ij}\,h_{00} - \partial_{j}h_{i0} - \partial_{i}h_{j0} + \dot{h}_{ij}\right) \qquad (5.1.8)$$

$$\delta\Gamma^{i}_{00} = \frac{1}{2a^2}\left(2\dot{h}_{i0} - \partial_{i}h_{00}\right) \qquad (5.1.9)$$

$$\delta\Gamma^{0}_{i0} = \frac{\dot{a}}{a}h_{i0} - \frac{1}{2}\partial_{i}h_{00} \qquad (5.1.10)$$

$$\delta\Gamma^{0}_{00} = -\frac{1}{2}\dot{h}_{00} \qquad (5.1.11)$$

In particular, we will need

$$\delta\Gamma^{\lambda}_{\lambda\mu} = \partial_{\mu}\left[\frac{1}{2a^2}h_{ii} - \frac{1}{2}h_{00}\right] .$$

To write the Einstein equations, we need the perturbation to the Ricci tensor

$$\delta R_{\mu\kappa} = \frac{\partial\,\delta\Gamma^{\lambda}_{\mu\lambda}}{\partial x^{\kappa}} - \frac{\partial\,\delta\Gamma^{\lambda}_{\mu\kappa}}{\partial x^{\lambda}}$$

$$+\delta\Gamma^{\eta}_{\mu\nu}\,\bar{\Gamma}^{\nu}_{\kappa\eta} + \delta\Gamma^{\nu}_{\kappa\eta}\,\bar{\Gamma}^{\eta}_{\mu\nu} - \delta\Gamma^{\eta}_{\mu\kappa}\,\bar{\Gamma}^{\nu}_{\nu\eta} - \delta\Gamma^{\nu}_{\nu\eta}\,\bar{\Gamma}^{\eta}_{\mu\kappa} , \quad (5.1.12)$$

with components

$$\delta R_{jk} = -\frac{1}{2}\partial_j\partial_k h_{00} - \left(2\dot{a}^2 + a\ddot{a}\right)\delta_{jk}h_{00} - \frac{1}{2}a\dot{a}\,\delta_{jk}\,\dot{h}_{00}$$

$$+\frac{1}{2a^2}\left(\nabla^2 h_{jk} - \partial_i\partial_j h_{ik} - \partial_i\partial_k h_{ij} + \partial_j\partial_k h_{ii}\right)$$

$$-\frac{1}{2}\ddot{h}_{jk} + \frac{\dot{a}}{2a}\left(\dot{h}_{jk} - \delta_{jk}\dot{h}_{ii}\right) + \frac{\dot{a}^2}{a^2}\left(-2h_{jk} + \delta_{jk}h_{ii}\right) + \frac{\dot{a}}{a}\delta_{jk}\partial_i h_{i0}$$

$$+\frac{1}{2}\left(\partial_j\dot{h}_{k0} + \partial_k\dot{h}_{j0}\right) + \frac{\dot{a}}{2a}\left(\partial_j h_{k0} + \partial_k h_{j0}\right) , \quad (5.1.13)$$

$$\delta R_{0j} = \delta R_{j0} = \frac{\dot{a}}{a}\partial_j h_{00} + \frac{1}{2a^2}\left(\nabla^2 h_{j0} - \partial_j\partial_i h_{i0}\right) - \left(\frac{\ddot{a}}{a} + \frac{2\dot{a}^2}{a^2}\right)h_{j0}$$

$$+\frac{1}{2}\frac{\partial}{\partial t}\left[\frac{1}{a^2}\left(\partial_j h_{kk} - \partial_k h_{kj}\right)\right] , \quad (5.1.14)$$

$$\delta R_{00} = \frac{1}{2a^2}\nabla^2 h_{00} + \frac{3\dot{a}}{2a}\dot{h}_{00} - \frac{1}{a^2}\partial_i\dot{h}_{i0}$$

$$+\frac{1}{2a^2}\left[\ddot{h}_{ii} - \frac{2\dot{a}}{a}\dot{h}_{ii} + 2\left(\frac{\dot{a}^2}{a^2} - \frac{\ddot{a}}{a}\right)h_{ii}\right] . \quad (5.1.15)$$

In general, we can put the Einstein field equations (B.71) in the form

$$R_{\mu\nu} = -8\pi\,G S_{\mu\nu} , \quad (5.1.16)$$

where

$$S_{\mu\nu} = T_{\mu\nu} - \frac{1}{2}g_{\mu\nu}\,g^{\rho\sigma}\,T_{\rho\sigma} . \quad (5.1.17)$$

(A cosmological constant can be accommodated by including a term in $T_{\mu\nu}$ proportional to $g_{\mu\nu}$, with a constant coefficient.) The perturbation to the energy-momentum tensor and metric produces a perturbation to the source tensor $S_{\mu\nu}$:

$$\delta S_{\mu\nu} = \delta T_{\mu\nu} - \frac{1}{2}\bar{g}_{\mu\nu}\delta T^{\lambda}{}_{\lambda} - \frac{1}{2}h_{\mu\nu}\bar{T}^{\lambda}{}_{\lambda} \;. \tag{5.1.18}$$

We are not assuming that the contents of the universe form a perfect fluid, but the rotational and translational invariance of the *unperturbed* energy-momentum tensor $\bar{T}^{\mu\nu}$ require that it takes the perfect fluid form (B.42):

$$\bar{T}_{\mu\nu} = \bar{p}\,\bar{g}_{\mu\nu} + \left(\bar{p} + \bar{\rho}\right)\bar{u}_{\mu}\bar{u}_{\nu} \;, \tag{5.1.19}$$

where $\bar{\rho}(t)$, $\bar{p}(t)$, and $\bar{u}^{\mu}$ are the unperturbed energy density, pressure, and velocity four-vector, respectively, with $\bar{u}^0 = 1$ and $\bar{u}^i = 0$. Also, we use the unperturbed Einstein equations (1.5.17) and (1.5.18) to write $\bar{\rho}$ and $\bar{p}$ in terms of the Robertson–Walker scale factor and its derivatives

$$\bar{\rho} = \frac{3}{8\pi G}\left(\frac{\dot{a}^2}{a^2}\right) \;, \quad \bar{p} = -\frac{1}{8\pi G}\left(\frac{2\ddot{a}}{a} + \frac{\dot{a}^2}{a^2}\right) \;. \tag{5.1.20}$$

It follows in particular that the unperturbed energy-momentum tensor has the trace

$$\bar{T}^{\lambda}{}_{\lambda} = 3\bar{p} - \bar{\rho} = -\frac{3}{4\pi G}\left(\frac{\ddot{a}}{a} + \frac{\dot{a}^2}{a^2}\right) \;.$$

Thus Eq. (5.1.18) gives

$$\delta S_{jk} = \delta T_{jk} - \frac{a^2}{2}\delta_{jk}\delta T^{\lambda}{}_{\lambda} + \frac{3}{8\pi G}\left(\frac{\ddot{a}}{a} + \frac{\dot{a}^2}{a^2}\right)h_{jk} \;, \tag{5.1.21}$$

$$\delta S_{j0} = \delta T_{j0} + \frac{3}{8\pi G}\left(\frac{\ddot{a}}{a} + \frac{\dot{a}^2}{a^2}\right)h_{j0} \;, \tag{5.1.22}$$

$$\delta S_{00} = \delta T_{00} + \frac{1}{2}\delta T^{\lambda}{}_{\lambda} + \frac{3}{8\pi G}\left(\frac{\ddot{a}}{a} + \frac{\dot{a}^2}{a^2}\right)h_{00} \;. \tag{5.1.23}$$

The Einstein equations (5.1.16) thus take the form

$$-8\pi G\left(\delta T_{jk} - \frac{a^2}{2}\delta_{jk}\delta T^{\lambda}{}_{\lambda}\right) = -\frac{1}{2}\partial_j\partial_k h_{00} - \left(2\dot{a}^2 + a\ddot{a}\right)\delta_{jk}h_{00}$$

$$-\frac{1}{2}a\dot{a}\,\delta_{jk}\,\dot{h}_{00} + \frac{1}{2a^2}\left(\nabla^2 h_{jk} - \partial_i\partial_j h_{ik} - \partial_i\partial_k h_{ij} + \partial_j\partial_k h_{ii}\right)$$

$$-\frac{1}{2}\ddot{h}_{jk} + \frac{\dot{a}}{2a}\left(\dot{h}_{jk} - \delta_{jk}\dot{h}_{ii}\right) + \left(\frac{\dot{a}^2}{a^2} + \frac{3\ddot{a}}{a}\right)h_{jk} + \left(\frac{\dot{a}^2}{a^2}\right)\delta_{jk}h_{ii}$$

$$+ \frac{\dot{a}}{a}\delta_{jk}\partial_i h_{i0} + \frac{1}{2}\left(\partial_j \dot{h}_{k0} + \partial_k \dot{h}_{j0}\right) + \frac{\dot{a}}{2a}\left(\partial_j h_{k0} + \partial_k h_{j0}\right), \quad (5.1.24)$$

$$-8\pi G\delta T_{j0} = \frac{\dot{a}}{a}\partial_j h_{00} + \frac{1}{2a^2}\left(\nabla^2 h_{j0} - \partial_j \partial_i h_{i0}\right) + \left(\frac{\dot{a}^2}{a^2} + \frac{2\ddot{a}}{a}\right)h_{j0}$$

$$+ \frac{1}{2}\frac{\partial}{\partial t}\left[\frac{1}{a^2}\left(\partial_j h_{kk} - \partial_k h_{kj}\right)\right], \quad (5.1.25)$$

$$-8\pi G\left(\delta T_{00} + \frac{1}{2}\delta T^\lambda{}_\lambda\right) = \frac{1}{2a^2}\nabla^2 h_{00} + \frac{3\dot{a}}{2a}\dot{h}_{00} - \frac{1}{a^2}\partial_i \dot{h}_{i0}$$

$$+ \frac{1}{2a^2}\left[\ddot{h}_{ii} - \frac{2\dot{a}}{a}\dot{h}_{ii} + 2\left(\frac{\dot{a}^2}{a^2} - \frac{\ddot{a}}{a}\right)h_{ii}\right] + 3\left(\frac{\dot{a}^2}{a^2} + \frac{\ddot{a}}{a}\right)h_{00}. \quad (5.1.26)$$

The components of the energy momentum tensor are subject to the conservation condition that $T^\mu{}_{\nu;\mu} = 0$, which to first order in perturbations gives

$$\partial_\mu \delta T^\mu{}_\nu + \bar{\Gamma}^\mu{}_{\mu\lambda}\delta T^\lambda{}_\nu - \bar{\Gamma}^\lambda{}_{\mu\nu}\delta T^\mu{}_\lambda + \delta\Gamma^\mu{}_{\mu\lambda}\bar{T}^\lambda{}_\nu - \delta\Gamma^\lambda{}_{\mu\nu}\bar{T}^\mu{}_\lambda = 0, \quad (5.1.27)$$

in which the perturbations to the energy-momentum tensor $\delta T^\mu{}_\nu$ with mixed indices can be calculated from

$$\delta T^\mu{}_\nu = \bar{g}^{\mu\lambda}\left[\delta T_{\lambda\nu} - h_{\lambda\kappa}\bar{T}^\kappa{}_\nu\right]. \quad (5.1.28)$$

Setting ν equal to a spatial coordinate index j gives the equation of momentum conservation

$$\partial_0\delta T^0{}_j + \partial_i\delta T^i{}_j + \frac{2\dot{a}}{a}\delta T^0{}_j - a\dot{a}\delta T^j{}_0 - (\bar{\rho} + \bar{p})\left(\frac{1}{2}\partial_j h_{00} - \frac{\dot{a}}{a}h_{j0}\right) = 0, \quad (5.1.29)$$

while setting ν equal to the time coordinate index 0 gives the equation of energy conservation

$$\partial_0\delta T^0{}_0 + \partial_i\delta T^i{}_0 + \frac{3\dot{a}}{a}\delta T^0{}_0 - \frac{\dot{a}}{a}\delta T^i{}_i - \left(\frac{\bar{\rho} + \bar{p}}{2a^2}\right)\left(-\frac{2\dot{a}}{a}h_{ii} + \dot{h}_{ii}\right) = 0. \quad (5.1.30)$$

As remarked in Appendix B, these conservation equations are not independent conditions, but may be derived from the Einstein field equations. However, it is often convenient to use either or both in place of one or two of the field equations. Also, in the frequently encountered case where the constituents of the universe are non-interacting fluids (such as one fluid consisting of cold dark matter and another consisting of ordinary matter and radiation) these conservation equations are satisfied *separately* by each fluid, information that could not be derived from the field equations.

The results obtained so far are repulsively complicated. Fortunately, the spatial isotropy and homogeneity of the unperturbed metric and energy-momentum tensor allow us to simplify these results by decomposing the perturbations into scalars, divergenceless vectors, and divergenceless traceless symmetric tensors, which are not coupled to each other by the field equations or conservation equations. The perturbation to the metric can always be put in the form[2]

$$h_{00} = -E ,$$ (5.1.31)

$$h_{i0} = a \left[\frac{\partial F}{\partial x^i} + G_i \right] ,$$ (5.1.32)

$$h_{ij} = a^2 \left[A\delta_{ij} + \frac{\partial^2 B}{\partial x^i \partial x^j} + \frac{\partial C_i}{\partial x^j} + \frac{\partial C_j}{\partial x^i} + D_{ij} \right] ,$$ (5.1.33)

where the perturbations A, B, C_i, $D_{ij} = D_{ji}$, E, F, and G_i are functions of **x** and t, satisfying the conditions

$$\frac{\partial C_i}{\partial x^i} = \frac{\partial G_i}{\partial x^i} = 0 , \quad \frac{\partial D_{ij}}{\partial x^i} = 0 , \quad D_{ii} = 0 .$$ (5.1.34)

To carry out a similar decomposition of the energy-momentum tensor, we first note that for a perfect fluid we would have

$$T_{\mu\nu} = pg_{\mu\nu} + (\rho + p)u_\mu u_\nu ,$$ (5.1.35)

with

$$g^{\mu\nu} u_\mu u_\nu = -1 ,$$ (5.1.36)

Recalling that $\bar{u}_i = 0$ and $\bar{u}_0 = -1$, we find that the normalization condition Eq. (5.1.36) gives

$$\delta u^0 = \delta u_0 = h_{00}/2 ,$$ (5.1.37)

while δu_i is an independent dynamical variable. (Note that $\delta u^\mu \equiv \delta(g^{\mu\nu} u_\nu)$ is not given by $\bar{g}^{\mu\nu} \delta u_\nu$.) Then the first-order perturbation to the energy-momentum tensor for a perfect fluid is

$$\delta T_{ij} = \bar{p} h_{ij} + a^2 \delta_{ij} \delta p, \quad \delta T_{i0} = \bar{p} h_{i0} - (\bar{\rho} + \bar{p})\delta u_i, \quad \delta T_{00} = -\bar{\rho} h_{00} + \delta\rho.$$ (5.1.38)

---

[2]To see this, we define F, A, and B, as the solutions of $\nabla^2 F = a^{-1}\partial_i h_{i0}$, $\nabla^2 A + \nabla^4 B = a^{-2}\partial_i\partial_j h_{jk}$, and $3A + \nabla^2 B = a^{-2}h_{ii}$, then define C_i as the solution of $\nabla^2 C_i = a^{-2}\partial_j h_{jk} - \partial_i[A + \nabla^2 B]$, and then use Eqs. (5.1.32) and (5.1.33) to define G_i and D_{jk}.

More generally, we can always put the perturbed energy-momentum tensor in a form like that of the perturbed metric. In general, we define $\delta\rho$ just as for a perfect fluid, as the difference between δT_{00} and $-\bar{\rho}h_{00}$, but $\delta\rho$ is *not* necessarily given by varying the temperature and chemical potentials in the formula for ρ that applies in thermal equilibrium. Also, in general we define the velocity perturbation δu_i times $\bar{\rho} + \bar{p}$ as for a perfect fluid, as the difference between $-\delta T_{i0}$ and $\bar{p}h_{i0}$, and we decompose δu_i into the gradient of a scalar velocity potential δu and a divergenceless vector δu_i^V. Finally, we define $a^2\delta p$ as the coefficient of δ_{ij} in the difference between δT_{ij} and $\bar{p}\delta_{ij}$, again without assuming that δp is given by varying the temperature and chemical potentials in the formula for p that applies in thermal equilibrium. The other terms in δT_{ij}, denoted $\partial_i\partial_j\pi^S$, $\partial_i\pi_j^V + \partial_j\pi_i^V$, and π_{ij}^T, represent dissipative corrections to the inertia tensor. That is, we write

$$\delta T_{ij} = \bar{p}\,h_{ij} + a^2\left[\delta_{ij}\delta p + \partial_i\partial_j\pi^S + \partial_i\pi_j^V + \partial_j\pi_i^V + \pi_{ij}^T\right], \quad (5.1.39)$$

$$\delta T_{i0} = \bar{p}\,h_{i0} - (\bar{\rho} + \bar{p})\left(\partial_i\delta u + \delta u_i^V\right), \quad (5.1.40)$$

$$\delta T_{00} = -\bar{\rho}\,h_{00} + \delta\rho, \quad (5.1.41)$$

where π_i^V, π_{ij}^T, and δu_i^V satisfy conditions analogous to the conditions (5.1.34) satisfied by C_i, D_{ij}, and G_i:

$$\partial_i\pi_i^V = \partial_i\delta u_i^V = 0, \quad \partial_i\pi_{ij}^T = 0, \quad \pi_{ii}^T = 0. \quad (5.1.42)$$

To repeat, these formulas can be taken as a definition of the quantities $\delta\rho$, δp, and $\delta u_i \equiv \partial_i\delta u + \delta u_i^V$, as well as of the *anisotropic inertia* terms π^S, π^V, and π^T, which characterize departures from the perfect fluid form of the energy-momentum tensor. The perturbed mixed components (5.1.28) of the energy-momentum tensor, which are needed in the conservation laws, now take the form

$$\delta T^i{}_j = \delta_{ij}\delta p + \partial_i\partial_j\pi^S + \partial_i\pi_j^V + \partial_j\pi_i^V + \pi_{ij}^T,$$

$$\delta T^i{}_0 = a^{-2}(\bar{\rho} + \bar{p})(a\partial_i F + aG_i - \partial_i\delta u - \delta u_i^V), \quad (5.1.43)$$

$$\delta T^0{}_i = (\bar{\rho} + \bar{p})(\partial_i\delta u + \delta u_i^V), \quad \delta T^0{}_0 = -\delta\rho,$$

$$\delta T^\lambda{}_\lambda = 3\delta p - \delta\rho + \nabla^2\pi^S.$$

With these decompositions, and again using Eqs. (5.1.20), the Einstein field equations (5.1.24)–(5.1.26) and conservation equations (5.1.29) and (5.1.30) fall into three classes of coupled equations:

Scalar (compressional) modes

These are the most complicated, involving the eight scalars E, F, A, B, $\delta\rho$, δp, π^S, and δu. The part of Eq. (5.1.24) proportional to δ_{jk} gives

$$-4\pi G a^2 \left[\delta\rho - \delta p - \nabla^2 \pi^S \right] = \frac{1}{2} a\dot{a}\dot{E} + (2\dot{a}^2 + a\ddot{a})E + \frac{1}{2}\nabla^2 A - \frac{1}{2}a^2\ddot{A}$$
$$-3a\dot{a}\dot{A} - \frac{1}{2}a\dot{a}\nabla^2 B + \dot{a}\nabla^2 F . \qquad (5.1.44)$$

The part of Eq. (5.1.24) of the form $\partial_j\partial_k S$ (where S is any scalar) gives

$$\partial_j\partial_k \left[16\pi G a^2 \pi^S + E + A - a^2\ddot{B} - 3a\dot{a}\dot{B} + 2a\dot{F} + 4\dot{a}F \right] = 0 . \quad (5.1.45)$$

The part of Eq. (5.1.25) of the form $\partial_j S$ (where S is again any scalar) gives

$$8\pi G a (\bar{\rho} + \bar{p})\partial_j\delta u = -\dot{a}\partial_j E + a\partial_j\dot{A} . \qquad (5.1.46)$$

Eq. (5.1.26) gives

$$-4\pi G \left(\delta\rho + 3\delta p + \nabla^2 \pi^S \right) = -\frac{1}{2a^2}\nabla^2 E - \frac{3\dot{a}}{2a}\dot{E} - \frac{1}{a}\nabla^2\dot{F} - \frac{\dot{a}}{a^2}\nabla^2 F$$
$$+ \frac{3}{2}\ddot{A} + \frac{3\dot{a}}{a}\dot{A} - \frac{3\ddot{a}}{a}E + \frac{1}{2}\nabla^2\ddot{B} + \frac{\dot{a}}{a}\nabla^2\dot{B} . $$
$$(5.1.47)$$

The part of the momentum conservation condition (5.1.29) that is a derivative ∂_j is

$$\partial_j \left[\delta p + \nabla^2\pi^S + \partial_0[(\bar{\rho} + \bar{p})\delta u] + \frac{3\dot{a}}{a}(\bar{\rho} + \bar{p})\delta u + \frac{1}{2}(\bar{\rho} + \bar{p})E \right] = 0 ,$$
$$(5.1.48)$$

and the energy-conservation condition (5.1.30) is

$$\delta\dot{\rho} + \frac{3\dot{a}}{a}(\delta\rho + \delta p) + \nabla^2 \left[-a^{-1}(\bar{\rho} + \bar{p})F + a^{-2}(\bar{\rho} + \bar{p})\delta u + \frac{\dot{a}}{a}\pi^S \right]$$
$$+ \frac{1}{2}(\bar{\rho} + \bar{p})\partial_0 \left[3A + \nabla^2 B \right] = 0 . \qquad (5.1.49)$$

In Eqs. (5.1.48) and (5.1.49), $\delta\rho$, δp, and π^S are elements of the perturbation to the total energy-momentum tensor, but the same equations apply to each constituent of the universe that does not exchange energy and momentum with other constituents.

Vector (vortical) modes

These involve the four divergenceless vectors G_i, C_i, δu_i^V, and π_i^V. The part of Eq. (5.1.24) of the form $\partial_k V_j$ (where V_j is any vector satisfying $\partial_j V_j = 0$) gives

$$\partial_k\left[16\pi\, Ga^2\pi_j^V - a^2\ddot{C}_j - 3a\dot{a}\dot{C}_j + a\dot{G}_j + 2\dot{a}G_j\right] = 0 , \qquad (5.1.50)$$

while the part of Eq. (5.1.25) of the form V_j (where V_j is again any vector satisfying $\partial_j V_j = 0$) gives

$$8\pi\, G(\bar{\rho}+\bar{p})a\delta u_j^V = \frac{1}{2}\nabla^2 G_j - \frac{a}{2}\nabla^2\dot{C}_j . \qquad (5.1.51)$$

The part of the momentum conservation equation (5.1.29) that takes the form of a divergenceless vector is

$$\nabla^2\pi_j^V + \partial_0[(\bar{\rho}+\bar{p})\delta u_j^V] + \frac{3\dot{a}}{a}(\bar{\rho}+\bar{p})\delta u_j^V = 0 , \qquad (5.1.52)$$

In particular, for a perfect fluid $\pi_i^V = 0$, and Eq. (5.1.52) tells us that $(\bar{\rho}+\bar{p})\delta u_j^V$ decays as $1/a^3$. In this case, both Eqs. (5.1.50) and (5.1.51) imply that the quantity $G_j - a\dot{C}_j$ (which we will see in Section 5.3 is the only physically relevant combination of metric components for vector perturbations) decays as $1/a^2$. Because they decay, vector modes have not played a large role in cosmology.

Tensor (radiative) modes

These involve only the two traceless divergenceless symmetric tensors D_{ij} and π_{ij}^T. There is only one field equation here: the part of Eq. (5.1.24) of the form of a divergenceless traceless tensor is the wave equation for gravitational radiation

$$-16\pi\, Ga^2\pi_{ij}^T = \nabla^2 D_{ij} - a^2\ddot{D}_{ij} - 3a\dot{a}\dot{D}_{ij} . \qquad (5.1.53)$$

The above equations for scalar, vector, and tensor perturbations do not form a complete set. This is in part because we still have the freedom to make changes in the coordinate system, of the same order as the physical perturbations. In Section 5.3 we will see how to remove this freedom by a choice of "gauge."

But even after the gauge has been fixed, the equations for the scalar modes will still not form a complete set, unless the pressure p and anisotropic inertia π^S can be expressed as functions of the energy density ρ. This is the case, for instance, for a constituent of the universe such as cold dark matter, whose particles have negligible velocities, and do not interact with other constituents. For such constituents, we can simply set p and π^S equal to zero. Things are only a little more complicated for constituents of the universe whose particles' velocities cannot be neglected, but which experience collisions that are sufficiently rapid to maintain local thermal equilibrium. In such cases, π^S can be neglected, and if the particles are highly relativistic p is simply $\rho/3$. Even where the particles are only moderately relativistic, the pressure in thermal equilibrium can usually be expressed as a function of ρ and one or more number densities n that satisfy the condition that the current nu^μ is conserved

$$(nu^\mu)_{;\mu} = 0 . \qquad (5.1.54)$$

This condition tells us that the unperturbed number density satisfies $\bar{n} \propto a^{-3}$, while the perturbation satisfies

$$\frac{\partial}{\partial t}\left(\frac{\delta n}{\bar{n}}\right) + \frac{1}{a^2}\nabla^2 \delta u + \frac{1}{2}\left(3\dot{A} + \nabla^2\dot{B}\right) - \frac{1}{a}\nabla^2 F = 0 . \qquad (5.1.55)$$

With p given as a function of ρ and n, after gauge fixing the field equations and conservation equations (5.1.48), (5.1.49), and (5.1.55) form a complete set of equations for the scalar modes.

Similarly, even after gauge fixing, the equations for vector and tensor modes do not form a complete set unless we have formulas for π_i^V and π_{ij}^T, respectively. This is no problem for perfect fluids, for which $\pi_i^V = \pi_{ij}^T = 0$.

In the general case local thermal equilibrium is not maintained, and we must calculate $\delta\rho$, δp, π^S, π_i^V and π_{ij}^T by following changes in the distribution of individual particle positions and momenta, which are governed by Boltzmann equations. The Boltzmann equations for photons and neutrinos are derived in Appendix H, and used in Chapters 6 and 7.

5.2 Fourier decomposition and stochastic initial conditions

In order to simplify our work we want to make full use of the symmetries of the problem. We have already used the rotational symmetry of the field and conservation equations in sorting out perturbations into scalar, vector, and tensor modes, and we will apply rotational symmetry again later in this section. The equations also have a symmetry under translations in space, which can best be exploited by working with the Fourier components of the perturbations. As long as we treat perturbations as infinitesimal, there is no coupling between the Fourier components of different wave number.

Further, only the initial conditions but not the equations themselves depend on the direction of the co-moving wave number $\mathbf{q}$.

Let us first see how this works out for the scalar modes. Because the unperturbed metric, energy density, etc. are independent of position, the general solution for the corresponding perturbed quantities may be written as a superposition of plane wave solutions, whose spatial variation is given by factors $\exp(i\mathbf{q} \cdot \mathbf{x})$, where $\mathbf{q}$ is a real wave vector, over which we must integrate. The coefficient of $\exp(i\mathbf{q} \cdot \mathbf{x})$ in each of the scalar perturbations $A(\mathbf{x}, t)$, $B(\mathbf{x}, t)$, etc. is a solution of coupled ordinary differential equations with time as the independent variable, obtained by replacing $\partial/\partial x^j$ with iq_j everywhere in the equations presented in Section 5.1. The differential equations obtained in this way depend on $q \equiv |\mathbf{q}|$, but not on the direction of $\mathbf{q}$, so the solutions can be written as superpositions of independent q-dependent normal modes, each characterized by a set of perturbations $A_{nq}(t)$, $B_{nq}(t)$, etc., with an overall normalization factor $\alpha_n(\mathbf{q})$ carrying a discrete index n that labels the various independent solutions. These normalization factors depend on the initial conditions, which of course are not rotationally invariant — if they were then there would be no galaxies or stars — so the $\alpha_n(\mathbf{q})$ depend on the direction of $\mathbf{q}$, but the solutions $A_{nq}(t)$, $B_{nq}(t)$ can be chosen to depend only on $q \equiv |\mathbf{q}|$. That is, we write

$$A(\mathbf{x}, t) = \sum_n \int d^3q \, \alpha_n(\mathbf{q}) A_{nq}(t) e^{i\mathbf{q}\cdot\mathbf{x}},$$

$$B(\mathbf{x}, t) = \sum_n \int d^3q \, \alpha_n(\mathbf{q}) B_{nq}(t) e^{i\mathbf{q}\cdot\mathbf{x}}, \qquad (5.2.1)$$

and likewise (with the *same* $\alpha_n(\mathbf{q})$) for E and F as well as for $\delta\rho$, δp, δu, π^S and any other rotational scalars.

Now, as discussed in Section 2.6, we expect the scalar variables $A(\mathbf{x}, t)$, $B(\mathbf{x}, t)$, etc. to be stochastic variables, characterized by averages of their products. The solutions $\{A_{nq}(t), B_{nq}(t), \ldots\}$ are ordinary fixed functions, not stochastic variables, so the stochastic nature of the scalar variables arises from the stochastic character of the initial conditions, embodied in the factors $\alpha_n(\mathbf{q})$. Under the assumption that the scalar variables are governed by Gaussian distributions, of the sort discussed in Appendix E, all averages of scalar quantities can be expressed in terms of bilinear averages $\langle A(\mathbf{x}, t) A(\mathbf{y}, t)\rangle$, $\langle A(\mathbf{x}, t) B(\mathbf{y}, t)\rangle$, etc. Let us consider the average of the product of any two real scalar quantities $X(\mathbf{x}, t)$ and $Y(\mathbf{y}, t)$. It turns out to be very convenient to use Eq. (5.2.1) for X and its complex conjugate for Y, so that

$$\langle X(\mathbf{x}, t) \, Y(\mathbf{y}, t)\rangle = \sum_{nm} \int d^3q \int d^3q' X_{nq}(t) \, Y^*_{mq}(t) \langle \alpha_n(\mathbf{q})\alpha^*_m(\mathbf{q}')\rangle \, e^{i\mathbf{q}\cdot\mathbf{x}} \, e^{-i\mathbf{q}'\cdot\mathbf{y}}.$$

We assume that although the initial conditions are not translationally invariant, they are governed by a translationally invariant probability distribution function, so that $\langle X(\mathbf{x}, t)\, Y(\mathbf{y}, t)\rangle$ should be a function only of $\mathbf{x} - \mathbf{y}$. This immediately tells us that $\langle \alpha_n(\mathbf{q})\alpha_m^*(\mathbf{q}')\rangle$ must be proportional to a delta function, $\delta^3(\mathbf{q} - \mathbf{q}')$. Furthermore, although the initial conditions are not rotationally invariant, we also assume that they are governed by a rotationally invariant probability distribution, so that the function of $\mathbf{q}$ multiplying this delta function can only depend on the magnitude $q \equiv |\mathbf{q}|$ of the wave vector, not on its direction. That is, we can write

$$\langle \alpha_n(\mathbf{q})\alpha_m^*(\mathbf{q}')\rangle = P_{nm}(q)\delta^3(\mathbf{q} - \mathbf{q}')\,, \tag{5.2.2}$$

and so

$$\langle X(\mathbf{x}, t)\, Y(\mathbf{y}, t)\rangle = \sum_{nm} \int d^3q \; X_{nq}(t)\, Y_{mq}^*(t)\, P_{nm}(q)\, \exp(i\mathbf{q} \cdot (\mathbf{x} - \mathbf{y})\,. $$
$$\tag{5.2.3}$$

The task of the theory of cosmological perturbations is twofold: to find the solutions $A_{nq}(t)$, $B_{nq}(t)$, etc. of the differential equations under suitable assumptions about the constituents of the universe, and to calculate the spectral functions $P_{nm}(q)$ in a theory of the origin of fluctuations in the very early universe.

Inspection of Eq. (5.2.2) shows immediately that $P_{mn}(q)$ is a Hermitian matrix

$$P_{nm}^*(q) = P_{mn}(q)\,, \tag{5.2.4}$$

and it is positive, in the sense that

$$\sum_{nm} P_{nm}(q)\xi_n\xi_m^* > 0 \tag{5.2.5}$$

for any set of complex numbers ξ_n (or functions of q) that are not all zero.

In general, for an arbitrary choice of independent solutions, there is no reason why $P_{nm}(q)$ should also be diagonal, so in general there will be interference between the different modes. However, it is sometimes convenient to *choose* the solutions so that $P_{nm}(q)$ is simply equal to δ_{nm}. To see that this can always be done, we recall a theorem of matrix algebra, which says that, because it is positive and Hermitian, $P_{nm}(q)$ can be put in the form

$$P_{nm}(q) = \sum_r Z_{nr}(q)Z_{mr}^*(q)\,, \tag{5.2.6}$$

for some square matrix $Z_{nr}(q)$, with r running over as many values as n. (That is, in matrix notation, $\mathcal{P} = ZZ^\dagger$.) We can then redefine the independent solutions, defining

$$\tilde{A}_{rq}(t) \equiv \sum_n Z_{nr}(q) A_{nq}(t) , \quad \tilde{B}_{rq}(t) \equiv \sum_n Z_{nr}(q) B_{nq}(t) , \text{ etc.,}$$

with a corresponding redefinition of the normalization factors

$$\alpha_n(q) \equiv \sum_r Z_{nr}(q)\, \tilde{\alpha}_r(q)$$

chosen so that

$$A(\mathbf{x}, t) = \sum_r \int d^3q\, \tilde{\alpha}_r(\mathbf{q}) \tilde{A}_{rq}(t) e^{i\mathbf{q}\cdot\mathbf{x}} ,$$

and likewise (with the same $\tilde{\alpha}_r(\mathbf{q})$) for B, E, and F, as well as for $\delta\rho$, δp, δu, etc. The advantage of this is that now the relevant bilinear averages are

$$\langle \tilde{\alpha}_r(\mathbf{q})\tilde{\alpha}_s^*(\mathbf{q}')\rangle = \delta_{rs}\delta^3(\mathbf{q} - \mathbf{q}') . \tag{5.2.7}$$

This is the result that was to be proved. With the solutions $\tilde{A}_{rq}(t)$, etc. chosen so that the $\tilde{\alpha}_r(\mathbf{q})$ satisfy Eq. (5.2.7), the different modes are uncorrelated. That is, any binary average is a sum over the individual modes. For instance,

$$\langle A(\mathbf{x}, t)\, A(\mathbf{y}, t)\rangle = \sum_r \int d^3q\, e^{i\mathbf{q}\cdot(\mathbf{x}-\mathbf{y})} |\tilde{A}_{rq}(t)|^2 . \tag{5.2.8}$$

In this way, the problem of calculating the spectral function $\mathcal{P}_{nm}(q)$ is traded for the problem of finding the correct linear combination of solutions for which Eq. (5.2.7) applies.

Incidentally, there is no problem in calculating the averages $\langle \alpha_n \alpha_m \rangle$ if we know the averages $\langle \alpha_n \alpha_m^* \rangle$, because the α_n satisfy a reality condition. To derive this condition, note that the differential equations for $A_{nq}(t)$, $B_{nq}(t)$, etc. are real, so the set of complex conjugates $A_{nq}^*(t)$, $B_{nq}^*(t)$, etc. of solution n is also a solution, and can therefore be expressed as a set of linear combinations $\sum_m c_{nm}(q)A_{mq}(t)$, $\sum_m c_{nm}(q)B_{mq}(t)$, etc. (Of course, in the case of a real solution, $c_{nm}(q) = \delta_{nm}$.) The functions $A(\mathbf{x}, t)$, $B(\mathbf{x}, t)$ are real, so by taking the complex conjugate of Eqs. (5.2.1) and replacing the integration variable $\mathbf{q}$ with $-\mathbf{q}$ we see that

$$\sum_n \alpha_n^*(\mathbf{q})c_{nm}(q) = \alpha_m(-\mathbf{q}) . \tag{5.2.9}$$

In particular, it follows that

$$\langle \alpha_n(\mathbf{q})\alpha_m(\mathbf{q}')\rangle = \sum_l \mathcal{P}_{nl}(q)c_{lm}(q)\delta^3(\mathbf{q} + \mathbf{q}') .$$

Now let us consider the tensor modes. These are completely charact-
erized by the traceless divergenceless symmetric tensor $D_{ij}(\mathbf{x}, t)$. (The stress
tensor $\pi_{ij}^T(\mathbf{x}, t)$ is not an independent dynamical quantity but, as we will
see in Section 6.6, it is given by a linear functional of $\dot{D}_{ij}$—specifically, the
solution of a linear integral equation with $\dot{D}_{ij}$ as the inhomogeneous term.)
If we write D_{ij} as a Fourier integral

$$D_{ij}(\mathbf{x}, t) = \int d^3q \, e^{i\mathbf{q}\cdot\mathbf{x}} \, \mathcal{D}_{ij}(\mathbf{q}, t) , \qquad (5.2.10)$$

then the Fourier transform $\mathcal{D}$ must satisfy the conditions

$$\mathcal{D}_{ij} = \mathcal{D}_{ji} , \quad \mathcal{D}_{ii} = 0 , \quad q_i \mathcal{D}_{ij} = 0 . \qquad (5.2.11)$$

For a given wave vector $\mathbf{q}$, there are just two independent matrices satisfying
these conditions. For instance, for $\mathbf{q}$ in the three-direction, Eq. (5.2.11)
requires that

$$\mathcal{D}_{11} = -\mathcal{D}_{22} , \quad \mathcal{D}_{12} = \mathcal{D}_{21} , \quad \mathcal{D}_{i3} = \mathcal{D}_{3i} = 0 , \qquad (5.2.12)$$

so all $\mathcal{D}_{ij}$ can be expressed in terms of the two independent components $\mathcal{D}_{11}$
and $\mathcal{D}_{12}$. (These components are frequently denoted $h^+ = \mathcal{D}_{11}$ and $h^\times = \mathcal{D}_{12}$.) It is convenient to classify the possible $\mathcal{D}_{ij}$ by their transformation
properties under a rotation by an angle θ around the three-axis.[1] It is easily
seen that under such a rotation,

$$\mathcal{D}_{11} \to \cos^2\theta \, \mathcal{D}_{11} + \cos\theta \, \sin\theta \, \mathcal{D}_{12} + \sin\theta \, \cos\theta \, \mathcal{D}_{21} + \sin^2\theta \, \mathcal{D}_{22}$$
$$= \cos 2\theta \, \mathcal{D}_{11} + \sin 2\theta \, \mathcal{D}_{12}$$
$$\mathcal{D}_{12} \to \cos^2\theta \, \mathcal{D}_{12} - \cos\theta \, \sin\theta \mathcal{D}_{11} + \sin\theta \, \cos\theta \, \mathcal{D}_{22} - \sin^2\theta \, \mathcal{D}_{21}$$
$$= -\sin 2\theta \, \mathcal{D}_{11} + \cos 2\theta \, \mathcal{D}_{12} ,$$

or more succinctly

$$\mathcal{D}_{11} \mp i\mathcal{D}_{12} \to e^{\pm 2i\theta}\left[\mathcal{D}_{11} \mp i\mathcal{D}_{12}\right] . \qquad (5.2.13)$$

For this reason, the linear combinations $\mathcal{D}_{11} \mp i\mathcal{D}_{12}$ are said to have *helicity*
± 2. (A wave of helicity λ consists of quanta with angular momentum in
the direction of motion equal to $\hbar\lambda$.) We will write $\mathcal{D}_{ij}(\mathbf{q}, t)$ as a sum over
helicities:

$$\mathcal{D}_{ij}(\mathbf{q}, t) = \sum_{\lambda=\pm 2} e_{ij}(\hat{q}, \lambda) \, \mathcal{D}(\mathbf{q}, \lambda, t) , \qquad (5.2.14)$$

---

[1] See, for instance, G&C, Sec. 10.2.

where for $\mathbf{q}$ in the three-direction

$$e_{11}(\hat{z}, \pm 2) = -e_{22}(\hat{z}, \pm 2) = \mp i e_{12}(\hat{z}, \pm 2) = \mp i e_{21}(\hat{z}, \pm 2) = \frac{1}{\sqrt{2}},$$

$$e_{i3} = e_{3i} = 0, \tag{5.2.15}$$

while for $\hat{q}$ in any other direction $e_{ij}(\hat{q}, \pm 2)$ is defined by applying on each of the indices i and j a standard rotation, that takes the three-direction into the direction of $\hat{q}$.

For $\mathbf{q}$ in the three-direction, the combination $\mathcal{D}_{11}(\mathbf{q}, t) \mp i\mathcal{D}_{12}(\mathbf{q}, t)$ is proportional to $\mathcal{D}(\mathbf{q}, \pm 2, t)$, so according to the transformation rule (5.2.13), a rotation by an angle θ around the direction of $\mathbf{q}$ changes $\mathcal{D}(\mathbf{q}, \pm 2, t)$ by a factor $\exp(\pm 2i\theta)$. There is nothing special about the three-direction, and the same is true for $\mathbf{q}$ in any direction.

The quantities $\mathcal{D}(\mathbf{q}, \lambda, t)$ in Eq. (5.2.14) satisfy the λ-independent second-order field equation (5.1.53), with the Laplacian ∇^2 replaced with $-q^2 \equiv -\mathbf{q}^2$:

$$\ddot{\mathcal{D}}(\mathbf{q}, \lambda, t) + 3\frac{\dot{a}}{a}\dot{\mathcal{D}}(\mathbf{q}, \lambda, t) + \frac{q^2}{a^2}\mathcal{D}(\mathbf{q}, \lambda, t) = 16\pi G\pi^T(\mathbf{q}, \lambda, t), \tag{5.2.16}$$

where $\pi^T(\mathbf{q}, \lambda, t)$ is the Fourier transform of the tensor part of the anisotropic inertia tensor:

$$\pi_{ij}^T(\mathbf{x}, t) = \sum_{\lambda = \pm 2} \int d^3q \, e^{i\mathbf{q} \cdot \mathbf{x}} e_{ij}(\hat{q}, \lambda) \pi^T(\mathbf{q}, \lambda, t). \tag{5.2.17}$$

As already mentioned, $\pi^T(\mathbf{q}, \lambda, t)$ is a helicity-independent linear functional of $\dot{\mathcal{D}}(\mathbf{q}, \lambda, t)$. Therefore, just as in the absence of anisotropic inertia, there are two independent solutions (distinguished by a label N) that, aside from normalization factors β, are independent of λ and of the direction of $\mathbf{q}$. Thus we can write

$$\mathcal{D}(\mathbf{q}, \lambda, t) = \sum_N \beta_N(\mathbf{q}, \lambda)\mathcal{D}_{Nq}(t), \tag{5.2.18}$$

and

$$\pi^T(\mathbf{q}, \lambda, t) = \sum_N \beta_N(\mathbf{q}, \lambda)\pi_{Nq}^T(t), \tag{5.2.19}$$

the sum over N running over the labels of the two independent solutions $\mathcal{D}_{Nq}(t)$ of the field equation

$$\ddot{\mathcal{D}}_{Nq}(t) + 3\frac{\dot{a}}{a}\dot{\mathcal{D}}_{Nq}(t) + \frac{q^2}{a^2}\mathcal{D}_{Nq}(t) = 16\pi G\pi_{Nq}^T(t). \tag{5.2.20}$$

Putting together Eqs. (5.2.10), (5.2.14), and (5.2.18), we have

$$D_{ij}(\mathbf{x}, t) = \sum_{N, \lambda} \int d^3q \, e^{i\mathbf{q} \cdot \mathbf{x}} \, e_{ij}(\hat{q}, \lambda) \, \beta_N(\mathbf{q}, \lambda) \, \mathcal{D}_{Nq}(t) \,. \qquad (5.2.21)$$

The normalization factors $\beta_N(\mathbf{q}, \lambda)$ like the $\alpha_n(\mathbf{q})$ are stochastic variables. The translational invariance of the probability distribution governing these factors tells us that $\langle \beta_N(\mathbf{q}, \lambda) \, \beta^*_{N'}(\mathbf{q}', \lambda') \rangle$ is proportional to $\delta^3(\mathbf{q} - \mathbf{q}')$. Under a rotation by an angle θ around the direction of $\mathbf{q}$, the product $\beta_N(\mathbf{q}, \lambda)$ $\beta^*_{N'}(\mathbf{q}, \lambda')$ is changed by a factor $\exp\left(i\theta(\lambda - \lambda')\right)$, so the rotational invariance of the probability distribution requires further that $\langle \beta_N(\mathbf{q}, \lambda) \, \beta^*_{N'}(\mathbf{q}', \lambda') \rangle$ is also proportional to $\delta_{\lambda\lambda'}$. For the most part, we will also assume that the probability distribution is invariant under the space-inversion operator P. This operator reverses the direction of momentum but not of angular momentum, so it changes the sign of the helicity, and therefore with this assumption, aside from the factor $\delta_{\lambda\lambda'}$, the mean value $\langle \beta_N(\mathbf{q}, \lambda) \, \beta^*_{N'}(\mathbf{q}', \lambda') \rangle$ is independent of helicity.[2] We have then

$$\langle \beta_N(\mathbf{q}, \lambda) \, \beta^*_{N'}(\mathbf{q}', \lambda') \rangle = \mathcal{P}^{\mathrm{grav}}_{NN'}(q) \, \delta_{\lambda\lambda'} \, \delta^3(\mathbf{q} - \mathbf{q}') \qquad (5.2.22)$$

Rotational and translational invariance also tell us that there is no correlation between the normalization factors for scalar and tensor modes:

$$\langle \beta_N(\mathbf{q}, \lambda) \, \alpha^*_n(\mathbf{q}') \rangle = 0 \,. \qquad (5.2.23)$$

Just as in the case of scalar modes, it is always possible to choose the two tensor modes so that $\mathcal{P}^{\mathrm{grav}}_{NN'}(q)$ is just $\delta_{NN'}$, but we need to know the probability distribution governing initial conditions in order to decide which linear combination of modes have this property.

The average of a product of two tensor perturbations is given by Eqs. (5.2.21) and (5.2.22) as

$$\langle D_{ij}(\mathbf{x}, t) D_{kl}(\mathbf{y}, t) \rangle = \sum_{NN'} \int d^3q \, \mathcal{P}^{\mathrm{grav}}_{NN'}(q) \, e^{i\mathbf{q} \cdot (\mathbf{x} - \mathbf{y})}$$

$$\times \Pi_{ij,kl}(\hat{q}) \mathcal{D}_{Nq}(t) \mathcal{D}^*_{N'q}(t), \qquad (5.2.24)$$

---

[2]The same conclusion follows if we assume that the probability distribution is invariant under CP, where C is the charge conjugation operator. Neither P nor CP is an exact symmetry of nature, but CP-invariance is a better approximation than P-invariance. The product CPT (where T is the reversal of the direction of time) is an exact symmetry in any quantum field theory, but T-invariance is broken by the expansion of the universe.

where[3]

$$\Pi_{ij,k\ell}(\hat{q}) \equiv \sum_{\lambda} e_{ij}(\hat{q}, \lambda)\, e^*_{k\ell}(\hat{q}, \lambda)$$

$$= \delta_{ik}\delta_{j\ell} + \delta_{i\ell}\delta_{jk} - \delta_{ij}\delta_{k\ell} + \delta_{ij}\hat{q}_k\hat{q}_\ell + \delta_{k\ell}\hat{q}_i\hat{q}_j$$

$$-\delta_{ik}\hat{q}_j\hat{q}_\ell - \delta_{i\ell}\hat{q}_j\hat{q}_k - \delta_{jk}\hat{q}_i\hat{q}_\ell - \delta_{j\ell}\hat{q}_i\hat{q}_k + \hat{q}_i\hat{q}_j\hat{q}_k\hat{q}_\ell. \quad (5.2.25)$$

Formulas like Eqs. (5.2.24) and (5.2.3) will be applied to the cosmic microwave background in Section 7.4.

We will see in Section 5.4 that the anisotropic inertia π_{Nq} becomes negligible during the era when the physical wave number q/a is much less than the expansion rate $\dot{a}/a$. In this case, one tensor mode becomes dominant. In the absence of anisotropic inertia, the gravitational wave equation (5.2.20) in the limit $q/a \ll H$ becomes

$$\ddot{\mathcal{D}}_{Nq}(t) + 3\frac{\dot{a}}{a}\dot{\mathcal{D}}_{Nq}(t) = 0 .$$

This has two obvious solutions:

$$\mathcal{D}_{1q}(t) = 1 , \quad \mathcal{D}_{2q}(t) = \int_t^{\infty} \frac{dt'}{a^3(t')} .$$

(The integral in the second solution converges, because $a(t) \propto t^{1/2}$ in the radiation-dominated era, and grows even faster in the matter-dominated and vacuum-dominated eras.) Since $\mathcal{D}_{2q}(t) \to 0$ at late times, for generic initial conditions the gravitational waves will eventually be dominated by the first solution. Thus in this case, to evaluate bilinear averages like (5.2.24), we only need to know $\mathcal{P}_{11}^{\text{grav}}(q)$. Alternatively, we could take $\mathcal{D}_{1q}(t)$ to be a constant $\mathcal{D}_q^o$ (the o superscript denoting "outside the horizon"), chosen so that $\mathcal{P}_{11}^{\text{grav}}(q) = 1$. In Section 5.4 we will see that similar remarks usually apply to the scalar modes.

5.3   Choosing a gauge

The equations derived in Section 5.1 have two unsatisfactory features. First, even with the simplifications introduced by decomposing the equations into

---

[3]To obtain the final formula for $\Pi_{ij,k\ell}(\hat{q})$, one can use the conditions that $\Pi_{ij,k\ell}(\hat{q})$ is a tensor function of $\hat{q}$, symmetric in i and j and in k and ℓ, that $\Pi_{ij,k\ell}(\hat{q}) = \Pi^*_{k\ell,ij}(\hat{q})$, and that $\hat{q}_i\Pi_{ij,k\ell}(\hat{q}) = 0$, to show that $\Pi_{ij,k\ell}(\hat{q})$ is proportional to the quantity on the right-hand side of Eq. (5.2.25). The constant of proportionality can then be found by using Eq. (5.2.15), letting $\hat{q}$ be in the three-direction and setting $i, j, k,$ and ℓ all equal to 1.

scalar, vector, and tensor modes, the equations for the scalar modes are still fearsomely complicated. Second, among the solutions of these equations are unphysical scalar and vector modes, corresponding to a mere coordinate transformation of the unperturbed Robertson–Walker metric and energy-momentum tensor. We can eliminate the second problem and ameliorate the first by fixing the coordinate system, adopting suitable conditions on the full perturbed metric and/or energy-momentum tensor. We will deal here with the coordinate-dependent perturbations $A(\mathbf{x}, t)$, $B(\mathbf{x}, t)$, etc., but all of the results of this section could be applied just as well to the Fourier components $A_{nq}(t)$, $B_{nq}(t)$, etc., by simply replacing each Laplacian with $-q^2$.

Consider a spacetime coordinate transformation

$$x^\mu \to x'^\mu = x^\mu + \epsilon^\mu(x) , \tag{5.3.1}$$

with $\epsilon^\mu(x)$ small in the same sense that $h_{\mu\nu}$, $\delta\rho$, and other perturbations are small. Under this transformation, the metric tensor will be transformed to

$$g'_{\mu\nu}(x') = g_{\lambda\kappa}(x)\frac{\partial x^\lambda}{\partial x'^\mu}\frac{\partial x^\kappa}{\partial x'^\nu} . \tag{5.3.2}$$

Instead of working with such transformations, which affect the coordinates and unperturbed fields as well as the perturbations to the fields, it is more convenient to work with so-called *gauge transformations*, which affect only the field perturbations. For this purpose, after making the coordinate transformation (5.3.1), we relabel coordinates by dropping the prime on the coordinate argument, and we attribute the whole change in $g_{\mu\nu}(x)$ to a change in the perturbation $h_{\mu\nu}(x)$. The field equations should thus be invariant under the gauge transformation $h_{\mu\nu}(x) \to h_{\mu\nu}(x) + \Delta h_{\mu\nu}(x)$, where

$$\Delta h_{\mu\nu}(x) \equiv g'_{\mu\nu}(x) - g_{\mu\nu}(x) , \tag{5.3.3}$$

with the unperturbed Robertson–Walker metric $\bar{g}_{\mu\nu}(x)$ left unchanged, and corresponding gauge transformations of other perturbations. To first order in $\epsilon(x)$ and $h_{\mu\nu}(x)$, Eq. (5.3.3) is

$$\Delta h_{\mu\nu}(x) = g'_{\mu\nu}(x') - \frac{\partial g_{\mu\nu}(x)}{\partial x^\lambda}\epsilon^\lambda(x) - g_{\mu\nu}(x)$$

$$= -\bar{g}_{\lambda\mu}(x)\frac{\partial \epsilon^\lambda(x)}{\partial x^\nu} - \bar{g}_{\lambda\nu}(x)\frac{\partial \epsilon^\lambda(x)}{\partial x^\mu} - \frac{\partial \bar{g}_{\mu\nu}(x)}{\partial x^\lambda}\epsilon^\lambda(x),$$
$$\tag{5.3.4}$$

or in more detail

$$\Delta h_{ij} = -\frac{\partial \epsilon_i}{\partial x^j} - \frac{\partial \epsilon_j}{\partial x^i} + 2a\dot{a}\delta_{ij}\epsilon_0, \tag{5.3.5}$$

$$\Delta h_{i0} = -\frac{\partial \epsilon_i}{\partial t} - \frac{\partial \epsilon_0}{\partial x^i} + 2\frac{\dot{a}}{a}\epsilon_i, \tag{5.3.6}$$

$$\Delta h_{00} = -2\frac{\partial \epsilon_0}{\partial t}, \tag{5.3.7}$$

with all quantities evaluated at the same spacetime *coordinate* point, and indices now raised and lowered with the Robertson–Walker metric, so that $\epsilon_0 = -\epsilon^0$ and $\epsilon_i = a^2\epsilon^i$. The field equations will be invariant only if the same gauge transformation is applied to all tensors, and in particular to the energy-momentum tensor, so that we must transform $\delta T_{\mu\nu}(x) \rightarrow \delta T_{\mu\nu}(x) + \Delta \delta T_{\mu\nu}(x)$, where $\Delta \delta T_{\mu\nu}$ is given by a formula[1] analogous to Eq. (5.3.4):

$$\Delta \delta T_{\mu\nu} = -\bar{T}_{\lambda\mu}(x)\frac{\partial \epsilon^\lambda(x)}{\partial x^\nu} - \bar{T}_{\lambda\nu}(x)\frac{\partial \epsilon^\lambda(x)}{\partial x^\mu} - \frac{\partial \bar{T}_{\mu\nu}(x)}{\partial x^\lambda}\epsilon^\lambda(x), \tag{5.3.8}$$

or in more detail

$$\Delta \delta T_{ij} = -\bar{p}\left(\frac{\partial \epsilon_i}{\partial x^j} + \frac{\partial \epsilon_j}{\partial x^i}\right) + \frac{\partial}{\partial t}\left(a^2\bar{p}\right)\delta_{ij}\epsilon_0, \tag{5.3.9}$$

$$\Delta \delta T_{i0} = -\bar{p}\frac{\partial \epsilon_i}{\partial t} + \bar{\rho}\frac{\partial \epsilon_0}{\partial x^i} + 2\bar{p}\frac{\dot{a}}{a}\epsilon_i, \tag{5.3.10}$$

$$\Delta \delta T_{00} = 2\bar{\rho}\frac{\partial \epsilon_0}{\partial t} + \dot{\bar{\rho}}\epsilon_0. \tag{5.3.11}$$

Note that we use δ to signify a perturbation, while Δ here denotes the change in a perturbation associated with a gauge transformation.

In order to write these gauge transformations in terms of the scalar, vector, and tensor components introduced in Section 5.1, it is necessary to decompose the spatial part of ϵ^μ into the gradient of a spatial scalar plus a divergenceless vector:

$$\epsilon_i = \partial_i \epsilon^S + \epsilon_i^V, \qquad \partial_i \epsilon_i^V = 0. \tag{5.3.12}$$

Then the transformations (5.3.5)–(5.3.7) and (5.3.9)–(5.3.11) give the gauge transformations of the metric perturbation components defined by

---

[1] The right-hand sides of Eqs. (5.3.4) and (5.3.8) are known as the *Lie derivatives* of the metric and energy-momentum tensor, respectively. For a discussion of Lie derivatives, see G&C, Secs. 10.9 and 12.3.

Eqs. (5.1.31)–(5.1.33):

$$\Delta A = \frac{2\dot{a}}{a}\epsilon_0 , \quad \Delta B = -\frac{2}{a^2}\epsilon^S ,$$

$$\Delta C_i = -\frac{1}{a^2}\epsilon_i^V , \quad \Delta D_{ij} = 0 , \quad \Delta E = 2\dot{\epsilon}_0 , \qquad (5.3.13)$$

$$\Delta F = \frac{1}{a}\left(-\epsilon_0 - \dot{\epsilon}^S + \frac{2\dot{a}}{a}\epsilon^S\right) , \quad \Delta G_i = \frac{1}{a}\left(-\dot{\epsilon}_i^V + \frac{2\dot{a}}{a}\epsilon_i^V\right) ,$$

and of the perturbations (5.1.39)–(5.1.41) to the pressure, energy density, and velocity potential

$$\Delta \delta p = \dot{\bar{p}}\epsilon_0 , \quad \Delta \delta \rho = \dot{\bar{\rho}}\epsilon_0 , \quad \Delta \delta u = -\epsilon_0 . \qquad (5.3.14)$$

The other ingredients of the energy-momentum tensor are gauge invariant:

$$\Delta \pi^S = \Delta \pi_i^V = \Delta \pi_{ij}^T = \Delta \delta u_i^V = 0 . \qquad (5.3.15)$$

Note in particular that the conditions $\pi^S = \pi_i^V = \pi_{ij}^T = 0$ for a perfect fluid and the condition $\delta u_i^V = 0$ for potential (*i.e.*, irrotational) flow are gauge invariant.

For the field equations to be gauge-invariant, similar transformations must of course be made on any other ingredients in these equations. For instance, any four-scalar $s(x)$ for which $s'(x') = s(x)$ under arbitrary four-dimensional coordinate transformations would undergo the change $\Delta \, \delta s(x) \equiv s'(x) - s(x) = s'(x) - s'(x')$, which to first order in perturbations is

$$\Delta \, \delta s(x) = s(x) - s(x') = -\frac{\partial \bar{s}(t)}{\partial x^\mu}\epsilon^\mu(x) = \dot{\bar{s}}(t)\epsilon_0 . \qquad (5.3.16)$$

This applies for instance to the number density n or a scalar field φ. For a perfect fluid both p and ρ are defined as scalars, and the gauge transformations in Eq. (5.3.14) of δp and $\delta \rho$ are other special cases of Eq. (5.3.16). Likewise, for a perfect fluid the gauge transformation in Eq. (5.3.14) of δu can be derived from the vector transformation law of u_μ. Because the gauge transformation properties of $\delta \rho$, δp, δu, etc. do not depend on the conservation laws, Eqs. (5.3.14) and (5.3.15) apply to each individual constituent of the universe in any case in which the energy-momentum tensor is a sum of terms for different constituents of the universe, even if these individual terms are not separately conserved.

We can eliminate the gauge degrees of freedom either by working only with gauge-invariant quantities,[2] or by choosing a gauge. The tensor

---

[2] J. M. Bardeen, *Phys. Rev. D* **22**, 1882 (1980).

quantities π_{ij}^T and D_{ij} appearing in Eq. (5.1.53) are already gauge invariant, and no gauge-fixing is necessary or possible. For the vector quantities π_i^V, δu_i^V, C_i and G_i, we can write Eqs. (5.1.50)–(5.1.52) in terms of the gauge-invariant quantities π_i^V, δu_i^V and $\tilde{G}_i \equiv G_i - a\dot{C}_i$, or we can fix a gauge for these quantities by choosing ϵ_i^V so that either C_i or G_i vanishes. (Note that $\tilde{G}_i$ is the vector field that we showed in Section 5.1 to decay as $1/a^2$ in the absence of anisotropic inertia.) For the scalar perturbations it is somewhat more convenient to fix a gauge. There are several frequently considered possibilities.

A. Newtonian gauge

Here we choose ϵ^S so that $B = 0$, and then choose ϵ_0 so that $F = 0$. Both choices are unique, so that after choosing Newtonian gauge, there is no remaining freedom to make gauge transformations. It is conventional to write E and A in this gauge as

$$E \equiv 2\Phi , \quad A \equiv -2\Psi , \tag{5.3.17}$$

so that (now considering only scalar perturbations) the perturbed metric has components

$$g_{00} = -1 - 2\Phi , \quad g_{0i} = 0 , \quad g_{ij} = a^2 \delta_{ij}[1 - 2\Psi] . \tag{5.3.18}$$

The gravitational field equations (5.1.44)–(5.1.47) then take the form

$$-4\pi Ga^2 \left[\delta\rho - \delta p - \nabla^2 \pi^S \right] = a\dot{a}\dot{\Phi} + (4\dot{a}^2 + 2a\ddot{a})\Phi - \nabla^2\Psi + a^2\ddot{\Psi}$$
$$+ 6a\dot{a}\dot{\Psi} , \tag{5.3.19}$$

$$-8\pi Ga^2 \partial_i\partial_j \pi^S = \partial_i\partial_j[\Phi - \Psi] , \tag{5.3.20}$$

$$4\pi G a (\bar{\rho} + \bar{p})\partial_i \delta u = -\dot{a}\partial_i\Phi - a\partial_i\dot{\Psi} , \tag{5.3.21}$$

$$4\pi G \left(\delta\rho + 3\delta p + \nabla^2 \pi^S \right) = +\frac{1}{a^2}\nabla^2\Phi + \frac{3\dot{a}}{a}\dot{\Phi}$$
$$+ 3\ddot{\Psi} + \frac{6\dot{a}}{a}\dot{\Psi} + \frac{6\ddot{a}}{a}\Phi , \tag{5.3.22}$$

and the equations (5.1.48) and (5.1.49) of momentum and energy conservation become (aside from modes of zero wave number)

$$\delta p + \nabla^2 \pi^S + \partial_0[(\bar{\rho} + \bar{p})\delta u] + \frac{3\dot{a}}{a}(\bar{\rho} + \bar{p})\delta u + (\bar{\rho} + \bar{p})\Phi = 0 , \tag{5.3.23}$$

$$\delta\dot{\rho} + \frac{3\dot{a}}{a}(\delta\rho + \delta p) + \nabla^2 \left[a^{-2}(\bar{\rho} + \bar{p})\delta u + \frac{\dot{a}}{a}\pi^S \right] - 3(\bar{\rho} + \bar{p})\dot{\Psi} = 0 . \tag{5.3.24}$$

239

In particular, Eq. (5.3.20) shows that Φ and Ψ are not physically independent fields; they differ only by a term arising from the anisotropic part of the stress tensor, and in particular they are equal for a perfect fluid, for which $\pi^S = 0$. The perturbation to the number density of a species of particle whose total number is conserved will satisfy relation (5.1.55), which in Newtonian gauge reads

$$\frac{\partial}{\partial t}\left(\frac{\delta n}{\bar{n}}\right) + \frac{1}{a^2}\nabla^2 \delta u - 3\dot{\Psi} = 0 . \tag{5.3.25}$$

Given an equation of state for p as a function of ρ (or, if p depends also on other quantities like n, then given also field equations for those quantities) and given also a formula for π^S as a linear combination of the other perturbations (such as for instance the formula $\pi^S = 0$ for a perfect fluid) we can regard Eqs. (5.3.21), (5.3.23), and (5.3.24) (and, where needed, Eq. (5.3.25)) as equations of motion for Ψ, δu, and $\delta\rho$, respectively, with Φ given in terms of Ψ by Eq. (5.3.20). The remaining equations provide a constraint on the solution of this coupled system of equations. By subtracting $3/a^2$ times Eq. (5.3.19) from Eq. (5.3.22) and then using Eqs. (5.3.20) and (5.3.21) to eliminate π^S and Φ, we find that[2]

$$a^3 \delta\rho - 3Ha^3(\bar{\rho} + \bar{p})\delta u - \left(\frac{a}{4\pi G}\right)\nabla^2\Psi = 0 . \tag{5.3.26}$$

This is a constraint rather than an equation of motion, because the equations of motion (5.3.21), (5.3.23), and (5.3.24) imply that the left-hand side of Eq. (5.3.26) is time-independent, so that Eq. (5.3.26) only has to be imposed as an initial condition.

B. Synchronous gauge

Here we choose ϵ_0 so that $E = 0$, and then choose ϵ^S so that again $F = 0$. Considering only scalar perturbations, the complete perturbed metric is then

$$g_{00} = -1 , \quad g_{0i} = 0 , \quad g_{ij} = a^2\left[(1+A)\delta_{ij} + \frac{\partial^2 B}{\partial x^i \partial x^j}\right] . \tag{5.3.27}$$

In this gauge, the Einstein field equations (5.1.44)–(5.1.47) take the form

$$-4\pi Ga^2\left[\delta\rho - \delta p - \nabla^2\pi^S\right] = \frac{1}{2}\nabla^2 A - \frac{1}{2}a^2\ddot{A} - 3a\dot{a}\dot{A} - \frac{1}{2}a\dot{a}\nabla^2\dot{B}, \tag{5.3.28}$$

$$-16\pi Ga^2\pi^S = A - a^2\ddot{B} - 3a\dot{a}\dot{B} , \tag{5.3.29}$$

$$8\pi G a (\bar{\rho} + \bar{p})\delta u = a\dot{A} , \tag{5.3.30}$$

$$-4\pi G \left(\delta\rho + 3\delta p + \nabla^2\pi^S\right) = \frac{3}{2}\ddot{A} + \frac{3\dot{a}}{a}\dot{A} + \frac{1}{2}\nabla^2\ddot{B} + \frac{\dot{a}}{a}\nabla^2\dot{B} , \tag{5.3.31}$$

and the equations (5.1.48) and (5.1.49) of momentum and energy conservation read

$$\delta p + \nabla^2\pi^S + \partial_0[(\bar{\rho} + \bar{p})\delta u] + \frac{3\dot{a}}{a}(\bar{\rho} + \bar{p})\delta u = 0 , \tag{5.3.32}$$

$$\delta\dot{\rho} + \frac{3\dot{a}}{a}(\delta\rho + \delta p) + \nabla^2\left[a^{-2}(\bar{\rho} + \bar{p})\delta u + \frac{\dot{a}}{a}\pi^S\right]$$
$$+ \frac{1}{2}(\bar{\rho} + \bar{p})\partial_0\left[3A + \nabla^2 B\right] = 0 . \tag{5.3.33}$$

Note that in this gauge the equation of momentum conservation, which furnishes the equation of motion (the Navier–Stokes equation) for an imperfect fluid, does not depend at all on the perturbed metric, while the equation of energy conservation may be written as

$$\delta\dot{\rho} + \frac{3\dot{a}}{a}(\delta\rho + \delta p) + \nabla^2\left[a^{-2}(\bar{\rho} + \bar{p})\delta u + \frac{\dot{a}}{a}\pi^S\right]$$
$$+ (\bar{\rho} + \bar{p})\psi = 0 \tag{5.3.34}$$

where

$$\psi \equiv \frac{1}{2}[3\dot{A} + \nabla^2\dot{B}] = \frac{\partial}{\partial t}\left(\frac{h_{ii}}{2a^2}\right) . \tag{5.3.35}$$

We need A and B separately to calculate the motion of individual particles, but the effect of gravitation on a perfect or imperfect fluid is entirely governed by the quantity ψ. Inspection of the field equation (5.3.31) shows that it provides a differential equation for just this combination of scalar fields:

$$-4\pi G a^2 \left(\delta\rho + 3\delta p + \nabla^2\pi^S\right) = \frac{\partial}{\partial t}\left(a^2\psi\right) \tag{5.3.36}$$

Also, in synchronous gauge the equation (5.1.55) for particle conservation takes the form

$$\frac{\partial}{\partial t}\left(\frac{\delta n}{\bar{n}}\right) + a^{-2}\nabla^2\delta u + \psi = 0 . \tag{5.3.37}$$

Given an equation of state for p as a function of ρ (and perhaps n) and a formula expressing π^S as a linear combination of the other scalar perturbations we can use Eqs. (5.3.32), (5.3.34), (5.3.36) (and perhaps

Eq. (5.3.37)) to find solutions for the three independent perturbations δu, $\delta \rho$, and ψ, respectively. The left-over equations (5.3.28)–(5.3.30) are not needed, for a reason given in Section 5.1: the full set of equations (5.3.28)–(5.3.33) are not independent, because the equations of energy and momentum conservation can be derived from the Einstein field equations.

If we need to know A and B separately we can find them from ψ and $\delta \rho$. By adding 3 times Eq. (5.3.28), plus 1/2 the Laplacian of Eq. (5.3.29), plus a^2 times Eq. (5.3.31), we obtain the simple relation

$$\nabla^2 A = -8\pi \, Ga^2 \delta \rho + 2Ha^2 \psi \, , \tag{5.3.38}$$

where as usual $H \equiv \dot{a}/a$. After A is found in this way, we can find B from A and ψ by solving Eq. (5.3.35).

Synchronous gauge was widely used in early calculations of the evolution of perturbations in cosmology, starting with the ground-breaking work of Lifshitz in 1946.[3] In the 1980s synchronous gauge became unpopular, because of a feature emphasized by Bardeen:[2] even after we impose the conditions $E = F = 0$, we are left with a residual gauge invariance. We can see from Eq. (5.3.13) that E and F are not affected by a gauge transformation with

$$\epsilon_0(\mathbf{x}, t) = -\tau(\mathbf{x}) \, , \quad \epsilon^S(\mathbf{x}, t) = a^2(t)\tau(\mathbf{x}) \int a^{-2}(t) \, dt \, , \tag{5.3.39}$$

where $\tau(\mathbf{x})$ is an arbitrary function of $\mathbf{x}$, but not of t. But under this transformation A and B do change

$$\Delta A = -\frac{2\dot{a}\tau}{a} \, , \quad \Delta B = -2\tau \int a^{-2}(t) \, dt \, . \tag{5.3.40}$$

In particular, the combination (5.3.35) undergoes the change

$$\Delta \psi = -3\tau \frac{d}{dt}\left(\frac{\dot{a}}{a}\right) - a^{-2}\nabla^2 \tau \, . \tag{5.3.41}$$

Also, the changes in the perturbations to the energy density, pressure, and velocity potential are given by Eqs. (5.3.14) and (5.3.39) as

$$\Delta \delta p = -\dot{p}\tau \, , \quad \Delta \delta \rho = -\dot{\rho}\tau \, , \quad \Delta \delta u = \tau \, , \tag{5.3.42}$$

while π^S is invariant. The same transformation rules apply for any one of the individual constituents of the universe. Any scalar perturbation δs such as the number density perturbation δn or a scalar field perturbation

---

[3]E. Lifshitz, *J. Phys. USSR* **10**, 116 (1946). Also see G&C, Sec. 15.10.

$\delta\varphi$ undergoes a change like that of the pressure and density perturbations:

$$\Delta\delta s = -\dot{\bar{s}}\tau \, . \tag{5.3.43}$$

The reader can check that all of the equations (5.3.28)–(5.3.34) and (5.3.37) are invariant under these residual gauge transformations. This being the case, for any solution ψ, δp, $\delta\rho$, δu, δn, etc. of these equations there will be another solution $\psi + \Delta\psi$, $\delta p + \Delta\delta p$, $\delta\rho + \Delta\delta\rho$, $\delta u + \Delta\delta u$, $n + \Delta\delta n$, etc., and since the field equations are linear, this means that $\Delta\psi$, $\Delta\delta p$, $\Delta\delta\rho$, $\Delta\delta u$, $\Delta\delta n$, etc., is also a solution. (For this solution there is no scalar anisotropic inertia, because π^S is gauge invariant.) This is a nuisance, because in finding solutions of the field equations we keep having to check that our solution represents a physical disturbance, not a mere change of gauge.

However, this is not a problem if we can remove the residual gauge symmetry in any natural way. This is the case if the universe contains a fluid (such as cold dark matter) whose individual particles are moving at speeds much less than that of light. In this case, the space-space components T^{ij}_D of the energy-momentum tensor for that fluid are negligible, so we can take $\bar{p}_D = 0$, $\bar{\rho}_D \propto a^{-3}$, and $\delta p_D = \pi^S_D = 0$ in the equation (5.3.32) of momentum conservation for this fluid, which tells us then that δu_D is time-independent. According to Eq. (5.3.42) (which applies separately to each constituent of the universe) a *time-independent* velocity potential δu_D can always be removed in synchronous gauge by a residual gauge transformation, with $\tau = -\delta u_D$. There is then no longer any ambiguity in the choice of gauge. These features make synchronous gauge particularly convenient in dealing with the later stages of cosmological evolution, when cold dark matter plays a prominent role.

C. Newtonian/synchronous conversion

We will find it convenient to do calculations using Newtonian gauge in some eras, and synchronous gauge in others. To connect results for different eras, we need to be able to convert them from one gauge to another.[4]

Suppose first that we begin in Newtonian gauge, and make an infinitesimal coordinate transformation $x^\mu \to x^\mu + \epsilon^\mu$, with $\epsilon_i = \partial_i \epsilon^S$ so as not to induce vector perturbations. According to Eqs. (5.3.13) and (5.3.17), in order to give E the synchronous gauge value $E = 0$, we need

$$\dot{\epsilon}_0 = -\Phi \tag{5.3.44}$$

---

[4]See, for instance, C.-P. Ma and E. Bertschinger, *Astrophys. J.* **455**, 7 (1995).

Then according to Eq. (5.3.13), to keep $F = 0$ we need

$$\frac{\partial}{\partial t}\left(\frac{\epsilon^S}{a^2}\right) = -\frac{\epsilon_0}{a^2} .$$ (5.3.45)

The A and B components of the spatial metric in synchronous gauge are given by Eqs. (5.3.13) and (5.3.17):

$$A = -2\Psi + 2H\epsilon_0 , \qquad B = -\frac{2}{a^2}\epsilon^S ,$$ (5.3.46)

where $H \equiv \dot{a}/a$. In particular, the field ψ is

$$\psi = -3\dot{\Psi} + 3\frac{\partial}{\partial t}\left(H\,\epsilon_0\right) + \frac{\nabla^2\epsilon_0}{a^2} .$$ (5.3.47)

Also, Eq. (5.3.14) allows us to calculate the synchronous gauge pressure perturbation δp^s, energy density perturbation $\delta\rho^s$, and velocity potential δu^s from the corresponding quantities δp, $\delta\rho$, and δu in Newtonian gauge:

$$\delta p^s = \delta p + \epsilon_0\dot{\bar{p}} , \qquad \delta\rho^s = \delta\rho + \epsilon_0\dot{\bar{\rho}} , \qquad \delta u^s = \delta u - \epsilon_0 .$$ (5.3.48)

Given Φ we can calculate ϵ_0 from Eq. (5.3.44), and then given Ψ we can obtain ψ from Eq. (5.3.47) and the synchronous gauge pressure, energy density, and velocity potential perturbations from Eq. (5.3.48). The quantity ϵ_0 is determined by Eq. (5.3.44) only up to a time-independent function of position, so the values of the synchronous gauge quantities A, B, ψ, $\tilde{p}$, $\tilde{\rho}$, and $\delta\tilde{u}$ are only determined up to a residual gauge transformation (5.3.40)–(5.3.42).

Next suppose that we begin in synchronous gauge, with metric fields A and B, and want to convert to Newtonian gauge. According to Eq. (5.3.13), to make g_{ij} proportional to δ_{ij} we need to take

$$\epsilon^S = a^2 B/2 .$$ (5.3.49)

Then to keep $g_{i0} = 0$, Eq. (5.3.13) tells us that we must take

$$\epsilon_0 = -a^2\dot{B}/2 .$$ (5.3.50)

Using Eq. (5.3.13) again together with the definitions (5.3.18), we have then

$$\Phi = \dot{\epsilon}_0 = -\frac{1}{2}\frac{\partial}{\partial t}\left(a^2\dot{B}\right) ,$$ (5.3.51)

$$\Psi = -\frac{1}{2}\left(A + \frac{2\dot{a}}{a}\epsilon_0\right) = -\frac{1}{2}A + \frac{a\dot{a}}{2}\dot{B} .$$ (5.3.52)

In contrast to the previous case, Eqs. (5.3.51) and (5.3.52) give Φ and Ψ uniquely. Not only that – it is easy to see that the results for Φ and Ψ are unaffected if the A and B with which we start are subjected to the residual gauge transformation (5.3.40).

D. Other gauges

In choosing a gauge, it is not necessary to impose conditions only on the scalar fields appearing in the metric tensor.[5] Instead, some of the gauge conditions can impose constraints on the scalars appearing in the energy-momentum tensor. For instance, in *co-moving gauge* we choose ϵ_0 so that $\delta u = 0$ (which for scalar perturbations makes the velocity perturbation δu^i vanish). Where the only "matter" is a single scalar field, as in popular theories of inflation, this means that the time coordinate is defined so that at any given time the scalar field equals its unperturbed value, with all perturbations relegated to components of the metric.[6] In the *constant density gauge* we choose ϵ_0 so that $\delta\rho = 0$. In either case, after fixing ϵ_0 we can make F vanish with a suitable choice of ϵ^S, so that the scalar perturbations still have $g_{i0} = 0$. Note that although this procedure fixes ϵ_0, it only fixes ϵ^S up to terms of the form $a^2(t)\tau(\mathbf{x})$, so these gauges share the drawback of synchronous gauge, of leaving a residual gauge symmetry.

5.4 Conservation outside the horizon

The perturbations that concern us are believed to have originated in quantum fluctuations during an era of inflation in the very early universe, discussed in Chapter 10. That early time and the much later time when these perturbations are observed are separated by a time interval in which the equations governing perturbations are not well known. For instance, at the end of the era of inflation there was a time of so-called reheating, during which the energy of the vacuum was transferred to ordinary matter and radiation, but we have no idea what particles were first created during reheating or how the energy transfer took place. Later, there was presumably a time when some particles effectively stopped interacting with the rest of matter and radiation, and became what is now observed as cold dark matter, but we can only guess when this was and how it happened.

---

[5]A great variety of different gauges are described by H. Kodama and M. Sasaki, *Prog. Theor. Phys. Suppl.* **78**, 1 (1984).

[6]For instance, this gauge was used in calculations of non-Gaussian corrections to cosmological correlations by J. Maldacena, *J. High Energy Phys.* **0305**, 013 (2003) [astro-ph/0210603].

The only reason that we are able to use inflationary theories to make any predictions at all about observable perturbations despite these uncertainties is that the wavelengths of the perturbations that concern us were outside the horizon during a period that extends from well before the end of inflation until relatively near the present (and hence includes the times of reheating and cold dark matter decoupling), in the sense that the physical wave number q/a during this period was much less than the expansion rate $H \equiv \dot{a}/a$. During inflation q/a decreases with time more-or-less exponentially while H is roughly constant, so all perturbations originally had $q/a \gg H$ but except for very short wavelengths eventually have $q/a \ll H$. Then during the radiation and matter-dominated eras a increases like $t^{1/2}$ or $t^{2/3}$ while H falls like $1/t$, so except for the longest wavelengths the perturbations that had $q/a \ll H$ at the end of inflation eventually again have $q/a \gg H$. During the intervening period when perturbations were outside the horizon, the scalar and tensor fluctuations were subject to certain conservation laws, that allow us to connect the distant past to the relatively recent past.

As we shall see below, for scalar modes it is the quantity defined in Newtonian gauge by[1]

$$\mathcal{R}_q \equiv -\Psi_q + H\delta u_q \qquad (5.4.1)$$

that in certain circumstances is conserved outside the horizon. (Recall that δu is the velocity potential for the total energy-momentum tensor.) Equivalently, there is another quantity[2]

$$\zeta_q \equiv -\Psi_q + \frac{\delta\rho_q}{3(\bar{\rho}+\bar{p})} \qquad (5.4.2)$$

that according to the constraint (5.3.26) is related to $\mathcal{R}_q$ by

$$\zeta_q = \mathcal{R}_q - \frac{q^2\Psi_q}{12\pi G(\bar{\rho}+\bar{p})a^2} . \qquad (5.4.3)$$

The difference is of relative order $(q/aH)^2$, so for $q/a \ll H$ the quantity ζ_q is conserved outside the horizon if $\mathcal{R}_q$ is.

The importance of $\mathcal{R}_q$ in the work of ref. 1 is that (as will be shown below) it is conserved outside the horizon in inflation driven by a single scalar

---

[1]The constancy of $\mathcal{R}$ was noted in various special cases by J. M. Bardeen, *Phys. Rev. D* **22**, 1882 (1980); D. H. Lyth, *Phys. Rev. D* **31**, 1792 (1985). For reviews, see J. Bardeen, in *Cosmology and Particle Physics*, eds. Li-zhi Fang and A. Zee (Gordon & Breach, New York, 1988); A. R. Liddle and D. H. Lyth, *Cosmological Inflation and Large Scale Structure* (Cambridge University Press, Cambridge, UK, 2000).

[2]J. M. Bardeen, P. J. Steinhardt, and M. S. Turner, *Phys. Rev. D* **28**, 679 (1983). This quantity was re-introduced by D. Wands, K. A. Malik, D. H. Lyth, and A. R. Liddle, *Phys. Rev. D* **62**, 043527 (2000).

field. But this is not enough for our purposes. After inflation, the universe becomes filled with a number of types of matter and radiation, and $\mathcal{R}_q$ is not necessarily conserved outside the horizon under these circumstances. What we need is a theorem that says that, whatever the constituents of the universe, there is always a solution of the field equations for which $\mathcal{R}_q$ and ζ_q are conserved outside the horizon. Such a solution is called "adiabatic," for reasons explained later. If cosmological fluctuations are described by such a solution during inflation, then they will continue to be described by such a solution and $\mathcal{R}_q$ and ζ_q will remain constant as long as the perturbation is outside the horizon, because this *is* a solution under all circumstances.

This result (and a corresponding result for tensor modes) is contained in the following theorem:[3] Whatever the contents of the universe, there are two independent adiabatic physical scalar solutions of the Newtonian gauge field equations for which the quantity $\mathcal{R}_q$ is time-independent in the limit $q/a \ll H$, and there is one tensor mode for which the tensor amplitude $\mathcal{D}_q$ is time-independent in the limit $q/a \ll H$. In this limit one of the scalar modes has $\mathcal{R}_q \neq 0$; the scalar metric components are

$$\Phi_q(t) = \Psi_q(t) = \mathcal{R}_q \left[-1 + \frac{H(t)}{a(t)} \int_{T}^{t} a(t')\, dt' \right] ; \qquad (5.4.4)$$

the perturbation to any four-scalar $s(x)$ (such as the energy density, pressure, inflaton field, etc.) is given by

$$\delta s_q(t) = -\frac{\mathcal{R}_q \dot{s}(t)}{a(t)} \int_{T}^{t} a(t')\, dt' ; \qquad (5.4.5)$$

and the perturbation to the velocity potential is

$$\delta u_q(t) = \frac{\mathcal{R}_q}{a(t)} \int_{T}^{t} a(t')\, dt' . \qquad (5.4.6)$$

where T is an arbitrary initial time, the same in all integrals. The other scalar mode has $\mathcal{R}_q = 0$, and

$$\Phi_q(t) = \Psi_q(t) = \frac{C_q H(t)}{a(t)} , \quad \delta s_q(t) = -\frac{C_q \dot{s}(t)}{a(t)} , \quad \delta u_q(t) = \frac{C_q}{a(t)} , \quad (5.4.7)$$

where C_q is time-independent. For $q/a \ll H$, the anisotropic inertia components π_q^S, π_{iq}^V, and π_{ijq}^T vanish in both adiabatic scalar modes and in the conserved tensor mode, even when some mean free times are comparable

---

[3]S. Weinberg, *Phys. Rev.* D **67**, 123504 (2003) [astro-ph/0302326]. Also see the appendix of S. Weinberg, *Phys. Rev.* D **69**, 023503 (2004) [astro-ph/0306304].

with the Hubble time. These are physical solutions for scalar as well as tensor modes, because the choice of Newtonian gauge leaves no remaining gauge freedom.

The proof of the theorem is based on the observation that in the special case of a spatially homogeneous universe, the coordinate space Newtonian gauge field equations and dynamical equations for matter and radiation (as well as the condition of spatial homogeneity) are invariant under coordinate transformations that are *not* symmetries of the unperturbed metric.[4] In Newtonian gauge, general first-order spatially homogeneous scalar and tensor perturbations to the metric take the form

$$h_{00} = -2\Phi(t) , \quad h_{i0} = 0 , \quad h_{ij} = -2\delta_{ij}a^2(t)\Psi(t) + a^2(t)D_{ij}(t) ,$$

with D_{ij} subject to the condition that $D_{ii} = 0$. Spatial homogeneity also requires that all pressures, densities, velocity potentials, etc. are functions only of time. The Newtonian gauge field equations for these spatially homogeneous perturbations are necessarily invariant under those gauge transformations of the form (5.3.4)–(5.3.7), (5.3.14)–(5.3.16) that preserve the conditions for Newtonian gauge and spatial homogeneity. Eq. (5.3.7) shows that in order for h_{00} to remain spatially homogeneous, ϵ_0 must be of the form

$$\epsilon_0(\mathbf{x}, t) = \epsilon(t) + \chi(\mathbf{x}) ,$$

so that

$$\Delta\Phi = \dot{\epsilon}$$

Eq. (5.3.6) then shows that in order for the h_{i0} to remain equal to zero, ϵ_i must have the form

$$\epsilon_i(\mathbf{x}, t) = a^2(t)f_i(\mathbf{x}) - a^2(t)\frac{\partial\chi(\mathbf{x})}{\partial x^i}\int\frac{dt}{a^2(t)} .$$

Eq. (5.3.5) then shows that

$$\Delta h_{ij} = -a^2\left(\frac{\partial f_i}{\partial x^j} + \frac{\partial f_j}{\partial x^i}\right) + 2\delta_{ij}a\dot{a}\Big[\epsilon + \chi\Big] - 2\frac{\partial^2\chi}{\partial x^i\,\partial x^j}\int\frac{dt}{a^2} .$$

In order not to introduce any $\mathbf{x}$-dependence in h_{ij}, we must take χ constant, in which case by shifting it into ϵ it can be taken to be zero, and we must also take f_i to have the form $f_i(\mathbf{x}) = \omega_{ij}x^j$, with ω_{ij} a constant matrix. (An $\mathbf{x}$-independent term in f_i would have no effect on the metric or anything

---

[4]In this respect, the theorem proved here is similar to the Goldstone theorem of quantum field theory; see J. Goldstone, *Nuovo Cimento* **9**, 154 (1961); J. Goldstone, A. Salam, and S. Weinberg, *Phys. Rev.* **127**, 965 (1962). The modes for which $\mathcal{R}_q$ or $\mathcal{D}_q$ are constant outside the horizon take the place here of the Goldstone bosons that become free particles for long wavelength.

else, and so can be ignored here.) Then

$$\Delta h_{ij} = -a^2[\omega_{ij} + \omega_{ji}] + 2\delta_{ij}a\dot{a}\epsilon .$$

We compare this with Eq. (5.1.33) (with the Newtonian gauge conditions $B = C_i = 0$ and $A = -2\Psi$), which gives

$$\Delta h_{ij} = -2a^2\delta_{ij}\Delta\Psi + a^2\Delta D_{ij}$$

Matching the terms in Δh_{ij} that are either proportional to δ_{ij} or traceless gives

$$\Delta\Psi = \frac{1}{3}\omega_{ii} - H\epsilon ,$$

$$\Delta D_{ij} = -\omega_{ij} - \omega_{ji} + \frac{2}{3}\delta_{ij}\omega_{kk} .$$

(Note that an antisymmetric term in ω_{ij} would have no effect, because the unperturbed metric is invariant under three-dimensional rotations.) The corresponding gauge transformations of the quantities appearing in the energy-momentum tensor and of general scalars are given by Eqs. (5.3.14)–(5.3.16). Since $\{h_{\mu\nu}, T_{\mu\nu}\}$ and $\{h_{\mu\nu}+\Delta h_{\mu\nu}, T_{\mu\nu}+\Delta T_{\mu\nu}\}$ are both solutions of the field equations and conservation equations, their difference must also be such a solution. We conclude that there is always a spatially homogeneous solution of the Newtonian gauge field and conservation equations, with scalar perturbations

$$\Psi = H\epsilon - \omega_{ii}/3 , \quad \Phi = -\dot{\epsilon} \tag{5.4.8}$$

$$\delta p = -\dot{\bar{p}}\epsilon , \quad \delta\rho = -\dot{\bar{\rho}}\epsilon , \quad \delta u = \epsilon , \quad \pi^S = 0 \tag{5.4.9}$$

and more generally for any four-scalar s

$$\delta s = -\dot{\bar{s}}\epsilon . \tag{5.4.10}$$

(The reader can check that when we drop all spatial gradients, then the perturbations (5.4.8)–(5.4.10) satisfy the Newtonian gauge field equations (5.3.19)–(5.3.22), the conservation laws (5.3.23)–(5.3.24) and (5.3.25), and the constraint (5.3.26).) There is also a spatially homogeneous solution with a tensor perturbation

$$D_{ij} \propto \omega_{ij} - \frac{1}{3}\delta_{ij}\omega_{kk} , \quad \pi^T_{ij} = 0 . \tag{5.4.11}$$

(This is obviously a solution of Eq. (5.1.53), but it includes the not so obvious information that the equations that determine π^T_{ij} necessarily give $\pi^T_{ij} = 0$ for D_{ij} constant, even if some particle collision rates become comparable with H.) Equivalently, Eqs. (5.4.8)–(5.4.11) are solutions for the Fourier transforms of the perturbations with zero wave number.

So far, ϵ is an arbitrary function of time, and ω_{ij} is an unrelated arbitrary constant 4×4 matrix. But these are just gauge modes for zero wave number. On the other hand, if they can be extended to non-zero wave number they become physical modes, since the choice of Newtonian gauge leaves no residual gauge symmetries except for zero wave number. For the tensor modes there is no problem; in this case there are no field equations that disappear for zero wave number, so the solution with D_{ij} time-independent automatically has an extension to a physical mode for non-zero wave number. But matters are more complicated for the scalar modes.

For the scalar modes the field equation (5.3.20) disappears in the limit of zero wave number, so to get a physical mode we must impose on the perturbations the condition

$$\Phi = \Psi \ . \tag{5.4.12}$$

(The condition $\delta u = \epsilon$ that is required for Eq. (5.3.21) to be satisfied for $q \neq 0$ is already satisfied, according to Eq. (5.4.9).) Inserting Eq. (5.4.8) in (5.4.12) gives a differential equation for ϵ:

$$\dot{\epsilon} = -H\epsilon + \omega_{kk}/3 \tag{5.4.13}$$

Also, Eq. (5.4.8) for Ψ and Eq. (5.4.9) for δu give the quantity $\mathcal{R}$ defined by Eq. (5.4.1) the time-independent value

$$\mathcal{R} = \omega_{kk}/3 \ . \tag{5.4.14}$$

There is a general solution of Eq. (5.4.13) for $\epsilon(t)$ with $\mathcal{R} \neq 0$:

$$\epsilon(t) = \frac{\mathcal{R}}{a(t)} \int_T^t a(t') \, dt' \ . \tag{5.4.15}$$

with T arbitrary. Using Eq. (5.4.15) in Eq. (5.4.8) gives the explicit solution (5.4.4) for large wavelengths

$$\Psi = \Phi = \mathcal{R}\left[-1 + \frac{H(t)}{a(t)} \int_T^t a(t') \, dt'\right] \ . \tag{5.4.16}$$

Eq. (5.4.9) gives

$$\frac{\delta p}{\dot{p}} = \frac{\delta \rho}{\dot{\rho}} = -\delta u = -\frac{\mathcal{R}}{a(t)} \int_T^t a(t') \, dt' \ , \tag{5.4.17}$$

and more generally for any four-scalar Eq. (5.4.10) gives

$$\frac{\delta s}{\dot{s}} = -\frac{\mathcal{R}}{a(t)} \int_T^t a(t') \, dt' \ . \tag{5.4.18}$$

Because for any $\mathcal{T}$, Eq. (5.4.15) is a solution of Eq. (5.4.13) with the same value of $\mathcal{R}$, the difference of two of these solutions with different values of $\mathcal{T}$ is also a solution, but with $\mathcal{R} = 0$. For this solution

$$\epsilon(t) = \frac{C}{a(t)}, \tag{5.4.19}$$

with C another constant, and

$$\Psi = \Phi = \frac{CH(t)}{a(t)}, \tag{5.4.20}$$

$$\frac{\delta\rho}{\dot{\rho}} = \frac{\delta p}{\dot{p}} = -\delta u = \frac{\delta s}{\dot{s}} = -\frac{C}{a(t)}, \tag{5.4.21}$$

and likewise for individual constituents of the universe. (Since $a(t)$ increases and $H(t)$ decreases with time, this is a decaying mode, which is usually assumed to play no significant role at late times.) Note in particular that these scalar modes have equal values for $\delta\rho_\alpha/\dot{\rho}_\alpha$ for all individual constituents α of the universe, whether or not energy is separately conserved for these constituents. For this reason, such perturbations are called *adiabatic*.[5] For the same reason, any other solutions are called *entropic*. (Sometimes such other solutions are called zero-curvature modes, but this is misleading, because setting $\Phi = \Psi = 0$ does not usually give a solution at all.)

We have shown that for sufficiently small wave number (in practice, this means $q/a \ll H$) there are always two adiabatic *physical* solutions for scalar perturbations that take the form (5.4.16)–(5.4.18) and (5.4.20)–(5.4.21), with $\mathcal{R}$ and C arbitrary constants, and a physical solution for tensor perturbations that takes the form (5.4.11) with ω_{ij} an arbitrary constant matrix. Since the equations we have solved are homogeneous, it follows that there are also solutions of the same form for which $\mathcal{R}$, C, and ω_{ij} are arbitrary time-independent functions of q. This concludes the proof.[6]

---

[5] For instance, if several constituents (such as an electron–positron plasma, and photons) each have a density and pressure that depends (even when perturbed) only on the temperature, then $\delta\rho_\alpha(T) = \rho'_\alpha(T)\delta T$ and $d\bar{\rho}_\alpha(T)/dt = \rho'_\alpha(T)\dot{T}$, so $\delta\rho_\alpha/\dot{\bar{\rho}}_\alpha = \delta T/\dot{T}$, and likewise for pressure and any other scalars.

[6] The existence of solutions with $\delta s_q/\dot{s}$ equal for all four-scalars s such as energy densities, pressures, etc. (but not the detailed solutions (5.4.4)–(5.4.7)) seems to have been generally accepted for a long time. An intuitive "separate universe" argument for the existence of solutions for which $\delta\rho_{\alpha q}/\dot{\bar{\rho}}_\alpha$ are equal for all constituents α of the universe has been given by D. H. Lyth and D. Wands, *Phys. Rev.* D **68**, 103516 (2003); also see D. Wands, K. A. Malik, D. H. Lyth, and A. R. Liddle, *Phys. Rev.* D **62**, 043627 (2000); A. R. Liddle and D. H. Lyth, *Cosmological Inflation and Large-Scale Structure* (Cambridge University Press, 2000). This reasoning has been extended beyond perturbation theory by D. H. Lyth, K. A. Malik, and M. Sasaki, *J. Cosm. & Astropart. Phys.* **0505**, 004 (2005) [astro-ph/0411220]. But this

This theorem shows that if the perturbations produced during inflation are actually in the modes found above at the end of inflation, then they stay in these modes as long as the perturbations remain outside the horizon, and in particular $\mathcal{R}_q$ and $\mathcal{D}_q$ remain constant, since these modes are solutions of the equations whatever the constituents of the universe may become. In particular, the wavelengths that will interest us are far outside the horizon during the era of reheating that is supposed to follow inflation, so if the scalar perturbations are adiabatic at the end of inflation, then reheating cannot generate entropic perturbations.[7]

The question that is left unanswered by this theorem is whether the scalar perturbations produced during inflation are actually in these modes at the end of inflation. This is often a matter of counting. We know that there are always two independent adiabatic solutions of the differential equations governing the scalar fluctuations, so *if these equations have no more than two independent solutions, then any perturbations must be adiabatic.* As we will see in Chapter 10, this counting shows that all solutions are adiabatic for inflation with a single inflaton field.

But it must not be thought that if observation of the cosmic microwave background reveals purely adiabatic perturbations at the time of last scattering, then the perturbations must have been adiabatic at the end of inflation. In a state of complete local thermal equilibrium in which all conserved quantum numbers vanish (such as is usually assumed to have existed at some early time in theories of cosmological baryonsynthesis or leptonsynthesis, like those discussed in Section 3.3) there are only two scalar degrees of freedom, the temperature and the gravitational potential $\Phi = \Psi$. They are governed by coupled first-order ordinary differential equations, so there are just two independent solutions, which must be adiabatic since there are always at least two adiabatic solutions. Thus whatever happens during inflation, if the universe subsequently spends sufficient time in a state of local thermal equilibrium with no non-zero conserved quantities, then the perturbations become adiabatic, and they remain adiabatic subsequently, even when the conditions of local thermal equilibrium are no longer satisfied.[8]

The use of conservation laws to connect different cosmological eras is not limited to Newtonian gauge. Indeed, both ζ_q and $\mathcal{R}_q$ can be put in

---

sort of argument only shows that there is a solution satisfying this condition for zero wave number. As we have seen, there are indeed many such solutions for zero wave number, most of which have no physical significance because they cannot be extended to finite wave number. The proof presented here shows that the requirement that the solution can be extended to finite wave number yields just two solutions, described by Eqs. (5.4.4)–(5.4.7). It is this requirement that makes it necessary for the infinitesimal redefinition of the time coordinate, used in the "separate universe" argument to generate the solutions for zero wave number, to be accompanied with an infinitesimal rescaling of the space coordinate.

[7]S. Weinberg, *Phys. Rev. D* **70**, 043541 (2004) (astro-ph/0401313).
[8]S. Weinberg, *Phys. Rev. D* **70**, 083522 (2004) (astro-ph/0405397).

a gauge invariant form. It is only necessary to remark that according to Eqs. (5.3.13) and (5.3.14), the quantities $A/2 - H\delta\rho/\dot{\bar{\rho}}$ and $A/2 + H\delta u$ are gauge invariant, and that they reduce in Newtonian gauge to the quantities (5.4.2) and (5.4.1), respectively, so in any gauge

$$\zeta_q = A_q/2 - H\delta\rho_q/\dot{\bar{\rho}} \,, \quad \mathcal{R}_q = A_q/2 + H\delta u_q \,, \qquad (5.4.22)$$

provided of course that A, $\delta\rho$, and δu are all calculated in the same gauge.

For instance, in synchronous gauge Eq. (5.3.38) gives

$$q^2 A_q = 8\pi G a^2 \delta\rho_q - 2H a^2 \psi_q \,, \qquad (5.4.23)$$

so in this gauge ζ_q and $\mathcal{R}_q$ can be expressed in terms of the convenient gravitational variable ψ_q by

$$q^2 \zeta_q = -a^2 H \psi_q + 4\pi G a^2 \delta\rho_q - q^2 H \delta\rho_q/\dot{\bar{\rho}},$$
$$q^2 \mathcal{R}_q = -a^2 H \psi_q + 4\pi G a^2 \delta\rho_q + q^2 H \delta u_q \,, \qquad (5.4.24)$$

of course with the understanding that $\delta\rho_q$ and δu_q are here calculated in synchronous gauge. This result can also be derived directly from the Newtonian gauge formulas (5.4.1) and (5.4.2) by using the rules for transforming from Newtonian to synchronous gauge given in the previous section.

There is a convenient general formula for the rate of change of ζ_q and $\mathcal{R}_q$ that holds for $q/a \ll H$ whether or not the cosmological fluctuations are in an adiabatic mode. To derive this formula, we use the Newtonian gauge energy-conservation law (5.3.24), which for $q/a \ll H$ gives

$$\delta\dot{\rho}_q + 3H(\delta\rho_q + \delta p_q) = 3(\bar{\rho} + \bar{p})\dot{\Psi}_q \,, \qquad (5.4.25)$$

and the corresponding unperturbed conservation law

$$\dot{\bar{\rho}} = -3H(\bar{\rho} + \bar{p}) \,. \qquad (5.4.26)$$

It is then a matter of simple algebra to calculate that for $q/a \ll H$

$$\dot{\zeta}_q = \frac{\dot{\bar{\rho}}\delta p_q - \dot{\bar{p}}\delta\rho_q}{3(\bar{\rho} + \bar{p})^2} \,. \qquad (5.4.27)$$

For $q/a \ll H$, the same formula then also gives $\dot{\mathcal{R}}_q$. It should be noted that according to Eq. (5.3.14) the quantity on the right-hand side is gauge-invariant, so we can check whether or not ζ_q and $\mathcal{R}_q$ are time-independent by evaluating $\dot{\bar{\rho}}\delta p_q - \dot{\bar{p}}\delta\rho_q$ in any gauge.

In particular, if for arbitrary perturbations $\bar{p} + \delta p$ is a function $F(\bar{\rho} + \delta\rho)$ of the perturbed energy density alone, then $\dot{\bar{p}} = F'(\bar{\rho})\dot{\bar{\rho}}$ and $\delta p_q = F'(\bar{\rho})\delta\rho_q$, so the terms in the numerator of the right-hand side of Eq. (5.4.27) cancel, and so ζ_q and $\mathcal{R}_q$ are conserved outside the horizon, *even if the fluctuations are not in an adiabatic mode.*[9] This is the case for instance if all abundant particles are highly relativistic, in which case the total perturbed pressure is $1/3$ the total perturbed energy density, or if they are highly non-relativistic, in which case the total perturbed pressure is negligible. On the other hand, the theorem proved earlier in this section tells us that ζ_q and $\mathcal{R}_q$ are conserved outside the horizon in the adiabatic modes, even in cases (such as inflation with a single scalar field) for which the perturbed pressure is *not* given as a function only of the perturbed energy density.

The proof of Eq. (5.4.27) depended only on the energy conservation equations (5.4.25) and (5.4.26), not on the gravitational field equations. Therefore, if there is a constituent α of the universe that is energetically isolated, in the sense of not exchanging energy with the rest of the matter and radiation of the universe, then for $q/a \ll H$ the quantity

$$\zeta_{\alpha q} \equiv -\Psi_q + \delta\rho_{\alpha q}/3(\bar{\rho}_\alpha + \bar{p}_\alpha) , \qquad (5.4.28)$$

calculated using the energy density ρ_α and pressure p_α of the isolated constituent, satisfies the gauge-invariant relation[10]

$$\dot{\zeta}_{\alpha q} = \frac{\dot{\bar{\rho}}_\alpha \delta p_{\alpha q} - \dot{\bar{p}}_\alpha \delta\rho_{\alpha q}}{3(\bar{\rho}_\alpha + \bar{p}_\alpha)^2} . \qquad (5.4.29)$$

In particular, $\zeta_{\alpha q}$ (and hence also $\mathcal{R}_{\alpha q}$) is conserved outside the horizon if the perturbed pressure of the isolated constituent is a function only of its perturbed energy density.[11] (As we will see in Section 6.1, this is actually a good approximation for each of the individual constituents of the universe at times after the temperature dropped below about 10^{10} K, when neutrinos and antineutrinos no longer had significant interactions with matter and radiation.) In this case, Eq. (5.4.28) can be regarded as a convenient formula for the fractional density fluctuation outside the horizon in Newtonian gauge

$$\frac{\delta\rho_{\alpha q}}{3(\bar{\rho}_\alpha + \bar{p}_\alpha)} = \zeta_{\alpha q} + \Psi_q , \qquad (5.4.30)$$

with $\zeta_{\alpha q}$ time-independent.

---

[9] If the total pressure is a function only of the total energy density, then the three first-order differential equations (5.3.21), (5.3.23) and (5.3.24) form a closed set of equations for $\Psi = \Phi$, δu, and $\delta\rho$, so with the constraint (5.3.26) there must be just two independent solutions for these quantities, which therefore must take the adiabatic form (5.4.4) and (5.4.6), or (5.4.7). But the complete solution is not necessarily adiabatic, because other four-scalars (such as the energy densities or pressures of individual components of the universe) may not be given by Eq. (5.4.9).

[10] S. Bashinsky and U. Seljak, *Phys. Rev. D* **69**, 083002 (2004) [astro-ph/0310198].

[11] D. Wands *et al.*, ref. 2.

There is a corresponding result in synchronous gauge: For long wave-lengths the energy conservation equation (5.3.34) for an energetically isolated component of the universe takes the form

$$\delta\dot\rho^s_{\alpha q} + 3H\left(\delta\rho^s_{\alpha q} + \delta p^s_{\alpha q}\right) + \left(\bar\rho_\alpha + \bar p_\alpha\right)\psi_q = 0 , \qquad (5.4.31)$$

the superscript s denoting synchronous gauge. If also p_α is either negligible or a function only of ρ_α, then this can be written as

$$\frac{d}{dt}\left(\frac{\delta\rho^s_{\alpha q}}{\bar\rho_\alpha + \bar p_\alpha}\right) = -\psi_q ,$$

so that

$$\frac{\delta\rho^s_{\alpha q}}{\bar\rho_\alpha + \bar p_\alpha} = -\int \psi_q \, dt + c_{\alpha q} . \qquad (5.4.32)$$

The integration constants $c_{\alpha q}$ depend on what we take as the lower limit on the integral of ψ_q, but their *differences* have an absolute significance, and are simply related to the differences of the $\zeta_{\alpha q}$. The gauge transformation equation (5.3.14) and the equation $\dot{\bar\rho}_\alpha = -3H(\bar\rho_\alpha + \bar p_\alpha)$ for separate energy conservation shows that for any two components of the universe whose energy is separately conserved, we have

$$\frac{\delta\rho^s_{\alpha q}}{\bar\rho_\alpha + \bar p_\alpha} - \frac{\delta\rho^s_{\beta q}}{\bar\rho_\beta + \bar p_\beta} = \frac{\delta\rho_{\alpha q}}{\bar\rho_\alpha + \bar p_\alpha} - \frac{\delta\rho_{\beta q}}{\bar\rho_\beta + \bar p_\beta} \qquad (5.4.33)$$

Comparing Eqs. (5.4.30) and (5.4.33), we see that

$$c_{\alpha q} - c_{\beta q} = 3\left(\zeta_{\alpha q} - \zeta_{\beta q}\right) . \qquad (5.4.34)$$

In the special case in which the perturbations are adiabatic, according to Eq. (5.4.9) the quantities $\delta\rho_{\alpha q}/(\bar\rho_\alpha + \bar p_\alpha) = -3H\delta\rho_{\alpha q}/\dot{\bar\rho}_\alpha$ are all equal, and hence the $\zeta_{\alpha q}$ are all equal, and in fact equal to ζ_q. According to Eq. (5.4.34), the $c_{\alpha q}$ will then also be all equal. The cosmic microwave background can be used to measure the differences of the $\zeta_{\alpha q}$ or $c_{\alpha q}$ before perturbations re-enter the horizon, and hence to decide whether the cosmological fluctuations are truly adiabatic. So far, as we will see in Section 7.2, it seems they are.

6

Evolution of Cosmological Fluctuations

We will now apply the formalism developed in the previous chapter to work out the evolution of cosmological fluctuations, from a temperature $\approx 10^9$ K when electron positron annihilation is substantially complete and neutrinos have decoupled from matter and radiation, down to the relatively recent time when the matter fluctuations become too large to be treated as first-order perturbations. Our results will be applied to analyze the observed fluctuations in the cosmic microwave background in the next chapter, and large scale structure in Chapter 8.

For reasons discussed in Section 5.4, the connection between the evolution of fluctuations in this era and what happened at earlier times will appear in a few parameters including $\mathcal{R}_q$ (or ζ_q), which are conserved during the many e-foldings of expansion when the perturbations to the various constituents of the universe were still outside the horizon — that is, when the physical wave number q/a was much less than the expansion rate $\dot{a}/a$. The values of these parameters when outside the horizon thus characterize the strength of the various perturbations. In Chapter 10 we will see what can be understood about these parameters from a study of the much earlier era of inflation, and in this way work out what large scale structure and fluctuations in the cosmic microwave background can tell us about the era of inflation.

Section 6.1 presents the equations governing scalar perturbations. These equations take a simple hydrodynamic form for cold dark matter and the baryonic plasma, but for calculations of high accuracy it is necessary to use the Boltzmann equations of kinetic theory to follow the detailed distribution of photons and neutrinos in phase space.[1] These equations are too complicated for an analytic treatment; that is a task for comprehensive computer programs such as CMBfast[2] and CAMB.[3] Unfortunately such computer programs do not lend themselves to an exposition aimed at an

---

[1] P. J. E. Peebles and J. T. Yu, *Astrophys. J.* **162**, 815 (1970); R. A. Sunyaev and Ya. B. Zel'dovich, *Astrophys. Space Sci.* **7**, 3 (1970).

[2] The original code for the Boltzmann equations for photons and neutrinos and the other dynamical and gravitational equations on which CMBfast is based was written by C.-P. Ma and E. Bertschinger, *Astrophys. J.* **455**, 7 (1995) [astro-ph/9506072]. An important element discussed below was added by U. Seljak and M. A. Zaldarriaga, *Astrophys. J.* **469**, 437 (1996).[astro-ph/9603033]. Also see M. Zaldarriaga, U. Seljak, and E. Bertschinger, *Astrophys. J.* **494**, 491 (1998); M. Zaldarriaga and U. Seljak, *Astrophys. J.* **129**, 431 (2000). The program is available on the website www.cmbfast.org.

[3] This program is based on CMBfast. It was written by A. Lewis and A. Challinor, and is available at camb.info/.

understanding of the physical phenomena involved. Therefore in subsequent sections we will present hydrodynamic calculations that are simple
enough to be done analytically, aside from a few numerical integrations,
and yet realistic enough so that the results are a good approximation to the
more accurate results of computer programs.

The general equations and initial conditions for our analytic treatment
of scalar modes are given in Section 6.2. Our analytic treatment of scalar
modes then divides into the study of two wavelength regimes: wavelengths
long enough to have come within the horizon during the matter-dominated
era, to be considered in Section 6.3, and wavelengths short enough to have
come within the horizon during the radiation-dominated era, considered in
Section 6.4. Section 6.5 will show how to interpolate between the solutions
found for these long and short wavelengths. Section 6.6 treats the evolution
of tensor perturbations.

In Chapter 7 the results of this chapter for both scalar and tensor modes
are applied to the anisotropies and polarization of the cosmic microwave
background. Chapter 8 takes the treatment of matter perturbations beyond
the time of last scattering, with results applied to observations of the cosmic
distribution of matter.

6.1 Scalar perturbations – kinetic theory

It seems highly likely that from the beginning of the period of interest here,
from just after $e^+ - e^-$ annihilation at a temperature $T \approx 10^9$ K, down
to the time of last scattering when $T \simeq 3,000$ K, the universe consisted of
just four components: photons, cold dark matter, neutrinos, and a baryonic plasma consisting of free electrons, ions, and neutral atoms. In this
section we will consider the perturbations in scalar modes to each of these
four constituents in turn, adopting for this purpose the synchronous gauge
described in Section 5.3. Each perturbed quantity $X(\mathbf{x}, t)$ (such as $\delta\rho$, δp,
A, B, etc.) is written as a Fourier integral and a sum over modes, as in
Eq. (5.2.1):

$$X(\mathbf{x}, t) = \sum_n \int d^3q \, \alpha_n(\mathbf{q}) \, X_{nq}(t) \, e^{i\mathbf{q}\cdot\mathbf{x}} , \qquad (6.1.1)$$

where $\alpha_n(\mathbf{q})$ is the stochastic parameter for the n-th mode. In particular, the
metric perturbation in synchronous gauge is given by $\delta g_{00} = 0$, $\delta g_{i0} = 0$,
and

$$\delta g_{ij}(\mathbf{x}, t) = a^2(t) \sum_n \int d^3q \, \alpha_n(\mathbf{q}) \left[A_{nq}(t)\delta_{ij} - q_i q_j B_{nq}(t) \right] e^{i\mathbf{q}\cdot\mathbf{x}} , \quad (6.1.2)$$

(Note that q is the co-moving wave number, related to the physical wave number k by $k = q/a$. It is common to define a so that $a = 1$ at the present time, so that the co-moving wave number equals the present value of the physical wave number, in which case the co-moving wave number is often denoted k. We will instead leave the normalization of a arbitrary, and reserve the symbol k for the physical wave number q/a.) In this section we will consider any one mode, dropping the label n; the equations we find will have a number of solutions, which define the various modes.

Cold dark matter

The individual cold dark matter particles are assumed to move too slowly for them to produce an appreciable pressure or anisotropic inertia, and, as shown in Section 5.3, the absence of pressure or anisotropic inertia allows us to adopt a particular synchronous gauge in which the cold dark matter fluid velocity u_D^i vanishes. That is, the coordinate mesh is tied to the dark matter particles in such a way that they remain at rest despite fluctuations in the gravitational field in which they move. Cold dark matter is therefore characterized solely by a total density $\bar{\rho}_D(t) + \delta\rho_D(\mathbf{x}, t)$, with the unperturbed density $\bar{\rho}_D(t)$ simply decreasing as $a^{-3}(t)$, and the Fourier transform of the perturbation $\delta\rho_D(\mathbf{x}, t)$ governed by the equation (5.3.34) of energy conservation with zero pressure and velocity

$$\delta\dot{\rho}_{Dq} + 3H\delta\rho_{Dq} = -\bar{\rho}_D\psi_q .\tag{6.1.3}$$

(Recall that $\psi_q \equiv (3\dot{A}_q - q^2\dot{B}_q)/2$.)

Baryonic plasma

The Coulomb interactions of electrons and atomic nuclei are sufficiently strong so that they act together as a single perfect fluid. In the era of interest both electrons and nuclei are highly non-relativistic, so the baryonic plasma has negligible pressure and anisotropic inertia, and therefore the unperturbed density $\bar{\rho}_B(t)$ goes as $a^{-3}(t)$, and the Fourier transform of the density perturbation $\delta\rho_B(\mathbf{x}, t)$ is governed by the energy-conservation equation (5.3.34), but now with a non-zero velocity potential $\delta u_B(\mathbf{x}, t)$:

$$\delta\dot{\rho}_{Bq} + 3H\delta\rho_{Bq} - (q^2/a^2)\bar{\rho}_B\delta u_{Bq} = -\bar{\rho}_B\psi_q .\tag{6.1.4}$$

On the other hand, Thomson scattering allows the baryonic plasma to exchange momentum with photons, so it is the combination of photons and plasma that satisfies the equation (5.3.32) of momentum conservation, which here reads

$$\delta p_{\gamma q} - q^2\pi_{\gamma q}^S + [\partial_0 + 3H]\left[\bar{\rho}_B\delta u_{Bq} + \frac{4}{3}\bar{\rho}_\gamma\delta u_{\gamma q}\right] = 0 ,\tag{6.1.5}$$

where δp_γ and δu_γ are the pressure perturbation and velocity potential of the photons. We can regard this as the equation of motion of the baryonic plasma, with the photonic quantities $\pi^S_{\gamma q}, p_{\gamma q}, \delta u_{\gamma q}$ calculated as described in the next subsection.

Photons

Well before the time of recombination the density of free electrons was high enough so that photons could be described hydrodynamically: Thomson scattering gave the photons a total momentum locked to that of the baryonic plasma, so that $\delta u_{\gamma q} = \delta u_{B q}$, and a momentum distribution that was isotropic in the co-moving frame, so that $\delta p_{\gamma q} = \delta \rho_{\gamma q}/3$ and $\pi^S_{\gamma q} = 0$. But for a highly accurate treatment of photons around the time of recombination it is necessary to treat them kinetically, studying the distribution of photons in momentum space, and taking account of photon polarization. As discussed in Appendix H, this distribution is an Hermitian *number density matrix* $n^{ij}(\mathbf{x}, \mathbf{p}, t)$, defined so that the number of photons in a space volume $\prod_i dx^i$ with momenta in a momentum-space volume $\prod_i dp_i$, weighted with the probability of their having polarization e^i, is $e_i e_j^* n^{ij}(\mathbf{x}, \mathbf{p}, t) \prod_k dx^k dp_k$, and $p_i n^{ij} = 0$. (The polarization vectors satisfy the normalization condition $g_{ij} e^i e^{j*} = 1$ and the transversality condition $p_i e^i = 0$, and $e_i \equiv g_{ik} e^k$. For further discussion of polarization, see Appendix G.) For small perturbations, this distribution is conveniently written in the form

$$n^{ij}(\mathbf{x}, \mathbf{p}, t) = \frac{1}{2} \bar{n}_\gamma \left(a(t) p^0(\mathbf{x}, \mathbf{p}, t) \right) \left[g^{ij}(\mathbf{x}, t) - \frac{g^{ik}(\mathbf{x}, t) g^{jl}(\mathbf{x}, t) p_k p_l}{[p^0(\mathbf{x}, \mathbf{p}, t)]^2} \right]$$
$$+ \delta n^{ij}(\mathbf{x}, \mathbf{p}, t), \qquad (6.1.6)$$

where

$$p^0(\mathbf{x}, \mathbf{p}, t) \equiv \sqrt{g^{ij}(\mathbf{x}, t) p_i p_j}. \qquad (6.1.7)$$

Here $\bar{n}_\gamma(p)$ is the equilibrium phase space number density

$$\bar{n}_\gamma(p) \equiv \frac{1}{(2\pi)^3} \left[\exp \left(p/k_B a(t) \bar{T}(t) \right) - 1 \right]^{-1}, \qquad (6.1.8)$$

with $\bar{T}(t)$ the unperturbed temperature of the baryonic plasma, and δn^{ij} is a small intrinsic perturbation. The first term in Eq. (6.1.6) is just the distribution matrix for unpolarized photons in thermal equilibrium at temperature $\bar{T}(t)$, written in a general spatial coordinate system, so δn includes the dynamical rather than the purely geometric effect of metric perturbations on the photon distribution. (Note that the factor $a(t)$ in the argument of $\bar{n}_\gamma$ in

Eq. (6.1.6) is canceled by the factor $a(t)$ multiplying $\bar{T}(t)$ in Eq. (6.1.8); this factor is introduced because in the era of interest $\bar{T}(t) \propto a^{-1}(t)$, so that as we have defined it, $\bar{n}_\gamma$ is a time-independent function of its argument.)

Since the sum over polarizations of $e_i e_j^*$ is g_{ij}, the phase space density of photons is $g_{ij}n^{ij}$. Hence the total energy-momentum tensor of the photons is

$$T^\mu{}_\nu = \frac{1}{\sqrt{\text{Det}\,g}} \int \left[\prod_{k=1}^{3} dp_k\right] g_{ij}n^{ij}\frac{p^\mu p_\nu}{p^0} . \tag{6.1.9}$$

The first term in Eq. (6.1.6) contains first-order perturbations arising from the metric perturbations in $p^i = g^{ij}p_j$ and $p^0 = -p_0 = \sqrt{g^{ij}p_ip_j}$, in the metric determinant Det$g$, and in the factor g_{ij} in $g_{ij}n^{ij}$. It is straight-forward though tedious to show directly that all these contributions to $\delta T^\mu{}_\nu$ cancel, but this can be seen more easily by noting that Eq. (6.1.9) shows that the contribution of the first term in Eq. (6.1.6) to $\delta T^i{}_{\gamma j}(\mathbf{x},t)$, $\delta T^0{}_{\gamma j}(\mathbf{x},t)$, $\delta T^j{}_{\gamma 0}(\mathbf{x},t)$, and $\delta T^0{}_{\gamma 0}(\mathbf{x},t)$ are local functions of the three-metric $g_{ij}(\mathbf{x},t)$ that transform under general spatial coordinate trans-formations as a mixed three-tensor, a covariant three-vector, a contravariant three-vector, and a three-scalar, respectively. But there are no non-trivial local functions of the three-metric that transform in this way under spatial coordinate transformations! This leaves only the contribution from the trace of δn^{ij}_γ. That is, the total first-order perturbations to the mixed components of the energy-momentum tensor are

$$\delta T^i{}_{\gamma j}(\mathbf{x},t) = \frac{1}{a^4(t)} \int \left(\prod_{k=1}^{3} dp_k\right) a^2(t)\,\delta n^{kk}_\gamma(\mathbf{x},\mathbf{p},t) \frac{p_ip_j}{\sqrt{p_kp_k}} , \tag{6.1.10}$$

$$\delta T^0{}_{\gamma j}(\mathbf{x},t) = \frac{1}{a^3(t)} \int \left(\prod_{k=1}^{3} dp_k\right) a^2(t)\,\delta n^{kk}_\gamma(\mathbf{x},\mathbf{p},t)\, p_j , \tag{6.1.11}$$

$$\delta T^0{}_{\gamma 0}(\mathbf{x},t) = -\frac{1}{a^4(t)} \int \left(\prod_{k=1}^{3} dp_k\right) a^2(t)\,\delta n^{kk}_\gamma(\mathbf{x},\mathbf{p},t) \sqrt{p_kp_k} . \tag{6.1.12}$$

Evidently, all we need to calculate these perturbations is the integral of $\delta n^{kk}_\gamma(\mathbf{x},\mathbf{p},t)$ over photon energy, weighted with a single factor of energy. We therefore introduce a *dimensionless intensity matrix* $J_{ij}(\mathbf{x},\hat{p},t)$, defined by

$$a^4(t)\,\bar{\rho}_\gamma(t)\,J_{ij}(\mathbf{x},\hat{p},t) \equiv a^2(t) \int_0^\infty \delta n^{ij}_\gamma(\mathbf{x},p\hat{p},t)\,4\pi p^3\,dp , \tag{6.1.13}$$

where $\bar{\rho}_\gamma(t) \equiv a^{-4}(t) \int 4\pi p^3 \bar{n}_\gamma(p)\, dp$ is the unperturbed photon energy density. The components (6.1.10)–(6.1.12) then become

$$\delta T^i_{\gamma j}(\mathbf{x}, t) = \bar{\rho}_\gamma(t) \int \frac{d^2\hat{p}}{4\pi} J_{kk}(\mathbf{x}, \hat{p}, t)\, \hat{p}_i \hat{p}_j , \qquad (6.1.14)$$

$$\delta T^0_{\gamma j}(\mathbf{x}, t) = a(t)\bar{\rho}_\gamma(t) \int \frac{d^2\hat{p}}{4\pi} J_{kk}(\mathbf{x}, \hat{p}, t)\, p_j , \qquad (6.1.15)$$

$$\delta T^0_{\gamma 0}(\mathbf{x}, t) = -\bar{\rho}_\gamma(t) \int \frac{d^2\hat{p}}{4\pi} J_{kk}(\mathbf{x}, \hat{p}, t) . \qquad (6.1.16)$$

As shown in Eq. (H.37) of Appendix H, the perturbation δn^{ij} is governed by the Boltzmann equation

$$\frac{\partial\, \delta n^{ij}(\mathbf{x}, \mathbf{p}, t)}{\partial t} + \frac{\hat{p}_k}{a(t)} \frac{\partial\, \delta n^{ij}(\mathbf{x}, \mathbf{p}, t)}{\partial x^k} + \frac{2\dot{a}(t)}{a(t)} \delta n^{ij}(\mathbf{x}, \mathbf{p}, t)$$

$$- \frac{1}{4a^2(t)} p\bar{n}'_\gamma(p)\hat{p}_k\hat{p}_l \frac{\partial}{\partial t}\left(a^{-2}(t)\delta g_{kl}(\mathbf{x}, t)\right)\left(\delta_{ij} - \hat{p}_i\hat{p}_j\right)$$

$$= -\omega_c(t)\, \delta n^{ij}(\mathbf{x}, \mathbf{p}, t) + \frac{3\omega_c(t)}{8\pi} \int d^2\hat{p}_1$$

$$\times \left[\delta n^{ij}(\mathbf{x}, p\hat{p}_1, t) - \hat{p}_i\hat{p}_k\, \delta n^{kj}(\mathbf{x}, p\hat{p}_1, t) - \hat{p}_j\hat{p}_k\, \delta n^{ik}(\mathbf{x}, p\hat{p}_1, t) \right.$$

$$\left. + \hat{p}_i\hat{p}_j\hat{p}_k\hat{p}_l\, \delta n^{kl}(\mathbf{x}, p\hat{p}_1, t) \right]$$

$$- \frac{\omega_c}{2a^3} p_k\delta u_{Bk}\, \bar{n}'_\gamma(p)\left(\delta_{ij} - \hat{p}_i\hat{p}_j\right) , \qquad (6.1.17)$$

where now $p \equiv \sqrt{p_i p_i}$, $\hat{p} \equiv \mathbf{p}/p$; $\omega_c(t)$ is the collision rate of a photon with electrons in the baryonic plasma; and δu_{Bk} is the peculiar velocity of the baryonic plasma. We can derive a Boltzmann equation for the dimensionless intensity matrix by multiplying Eq. (6.1.17) with $4\pi p^3$ and integrating over $p \equiv \sqrt{p_i p_i}$, using

$$4\pi \int_0^\infty p^4\, \bar{n}'_\gamma(p)\, dp = -16\pi \int_0^\infty p^3\, \bar{n}_\gamma(p)\, dp = -4a^4(t)\rho_\gamma(t) .$$

Writing the intensity matrix and plasma velocity as Fourier transforms

$$J_{ij}(\mathbf{x}, \hat{p}, t) = \int d^3q\, J_{ij}(\mathbf{q}, \hat{p}, t)\, e^{i\mathbf{q}\cdot\mathbf{x}} , \quad \delta u_{Bk}(\mathbf{x}, t) = \int d^3q\, \delta u_{Bi}(\mathbf{q}, t)\, e^{i\mathbf{q}\cdot\mathbf{x}}$$

$$(6.1.18)$$

we find

$$
\frac{\partial J_{ij}(\mathbf{q}, \hat{p}, t)}{\partial t} + i\frac{\hat{p} \cdot \mathbf{q}}{a(t)} J_{ij}(\mathbf{q}, \hat{p}, t)
$$

$$
+ \alpha(\mathbf{q}) \left[\dot{A}_q(t) - (\mathbf{q} \cdot \hat{p})^2 \dot{B}_q(t) \right] \left(\delta_{ij} - \hat{p}_i \hat{p}_j \right)
$$

$$
= -\omega_c(t) J_{ij}(\mathbf{q}, \hat{p}, t) + \frac{3\omega_c(t)}{8\pi} \int d^2\hat{p}_1
$$

$$
\times \left[J_{ij}(\mathbf{q}, \hat{p}_1, t) - \hat{p}_i \hat{p}_k J_{kj}(\mathbf{q}, \hat{p}_1, t) - \hat{p}_j \hat{p}_k J_{ik}(\mathbf{q}, \hat{p}_1, t) \right.
$$

$$
\left. + \hat{p}_i \hat{p}_j \hat{p}_k \hat{p}_l J_{kl}(\mathbf{q}, \hat{p}_1, t) \right]
$$

$$
+ \frac{2\omega_c(t)}{a(t)} \left[\delta_{ij} - \hat{p}_i \hat{p}_j \right] \hat{p}_k \delta u_{Bk}(\mathbf{q}, t) , \tag{6.1.19}
$$

in which we have used Eqs. (5.3.27) and (6.1.2) for the metric perturbation.

The intensity matrix and plasma velocity are proportional to the stochastic parameter $\alpha(\mathbf{q})$ for whatever mode is under consideration, which contains all information about initial conditions. Apart from this factor, there are no preferred directions in the problem, so the coefficient of $\alpha(\mathbf{q})$ in the intensity matrix can be decomposed into a sum of terms proportional to the two symmetric three-dimensional tensors $\delta_{ij} - \hat{p}_i \hat{p}_j$ and $\left(\hat{q}_i - (\hat{q} \cdot \hat{p})\hat{p}_i \right) \left(\hat{q}_j - (\hat{q} \cdot \hat{p})\hat{p}_j \right)$ that vanish when contracted with $\hat{p}_i$ or $\hat{p}_j$, with coefficients that depend on the directions $\hat{q}$ and $\hat{p}$ only through the scalar product $\hat{q} \cdot \hat{p}$. This decomposition is conventionally written as

$$
J_{ij}(\mathbf{q}, \hat{p}, t) = \alpha(\mathbf{q}) \left\{ \frac{1}{2} \left(\Delta_T^{(S)}(q, \hat{q} \cdot \hat{p}, t) - \Delta_P^{(S)}(q, \hat{q} \cdot \hat{p}, t) \right) \left(\delta_{ij} - \hat{p}_i \hat{p}_j \right) \right.
$$

$$
\left. + \Delta_P^{(S)}(q, \hat{q} \cdot \hat{p}, t) \left[\frac{\left(\hat{q}_i - (\hat{q} \cdot \hat{p})\hat{p}_i \right) \left(\hat{q}_j - (\hat{q} \cdot \hat{p})\hat{p}_j \right)}{1 - (\hat{p} \cdot \hat{q})^2} \right] \right\} . \tag{6.1.20}
$$

(The subscripts T and P stand for "temperature" and "polarization." Note that the trace J_{ii}, which is all that appears in the energy-momentum tensor, is proportional solely to $\Delta_T^{(S)}$, but we need to keep track of $\Delta_P^{(S)}$ because it is linked to $\Delta_T^{(S)}$ through the dynamical equations.) Similarly, the integral over $\hat{p}$ appearing on the right-hand side of Eq. (6.1.19) may be expressed in terms of a pair of "source functions" $\Phi(q, t)$ and $\Pi(q, t)$ as

$$
\int \frac{d^2\hat{p}}{4\pi} J_{ij}(\mathbf{q}, \hat{p}, t) = \alpha(\mathbf{q}) \left[\delta_{ij} \Phi(q, t) + \frac{1}{2} \hat{q}_i \hat{q}_j \Pi(q, t) \right] , \tag{6.1.21}
$$

and as usual we write

$$\delta u_{Bi}(\mathbf{q}, t) = i\alpha(\mathbf{q})q_i \delta u_{Bq}(t) .$$ (6.1.22)

Inserting Eqs. (6.1.20)–(6.1.22) in Eq.(6.1.19) yields the coupled Boltzmann equations for $\Delta_T^{(S)}$ and $\Delta_P^{(S)}$:

$$\dot{\Delta}_P^{(S)}(q,\mu,t) + i\left(\frac{q\mu}{a(t)}\right)\Delta_P^{(S)}(q,\mu,t) = -\omega_c(t)\Delta_P^{(S)}(q,\mu,t)$$

$$+ \frac{3}{4}\omega_c(t)(1-\mu^2)\Pi(q,t) ,$$ (6.1.23)

$$\dot{\Delta}_T^{(S)}(q,\mu,t) + i\left(\frac{q\mu}{a(t)}\right)\Delta_T^{(S)}(q,\mu,t) = -\omega_c(t)\Delta_T^{(S)}(q,\mu,t)$$

$$- 2\dot{A}_q(t) + 2q^2\mu^2 \dot{B}_q(t)$$

$$+ 3\omega_c(t)\,\Phi(q,t) + \frac{3}{4}\omega_c(t)(1-\mu^2)\Pi(q,t) + 4iq\mu\omega_c(t)\delta u_{Bq}(t),$$ (6.1.24)

with Φ and Π defined by Eq. (6.1.21).

The usual approach[4] to the solution of these Boltzmann equations is through an expansion of $\Delta_T^{(S)}$ and $\Delta_P^{(S)}$ in partial wave amplitudes:

$$\Delta_T^{(S)}(q,\mu,t) = \sum_{\ell=0}^{\infty} i^{-\ell}(2\ell+1)P_\ell(\mu)\,\Delta_{T,\ell}^{(S)}(q,t)$$ (6.1.25)

$$\Delta_P^{(S)}(q,\mu,t) = \sum_{\ell=0}^{\infty} i^{-\ell}(2\ell+1)P_\ell(\mu)\,\Delta_{P,\ell}^{(S)}(q,t) .$$ (6.1.26)

To derive the Boltzmann equations for the partial wave amplitudes, we use the recursion and normalization relations for Legendre polynomials

$$(2\ell+1)\mu P_\ell(\mu) = (\ell+1)P_{\ell+1}(\mu)+\ell P_{\ell-1}(\mu), \qquad \int_{-1}^{+1} P_\ell^2(\mu)\,d\mu = \frac{2}{2\ell+1}.$$

[4]M. L. Wilson and J. Silk, *Astrophys. J.* **243**, 14 (1981); J. R. Bond and G. Efstathiou, *Astrophys. J.* **285**, L45 (1984); R. Crittenden, J. R. Bond, R. L. Davis, G. Efstathiou, and P. Steinhardt, *Phys. Rev. Lett.* **71**, 324 (1993); C. -P. Ma and E. Bertschinger, *Astrophys. J.* **455**, 7 (1995).

Multiplying Eqs. (6.1.23) and (6.1.24) by $P_\ell(\mu)$ and integrating over μ then gives

$$
\dot{\Delta}_{P,\ell}^{(S)} + \frac{q}{a(2\ell+1)}\left[(\ell+1)\Delta_{P,\ell+1}^{(S)} - \ell\Delta_{P,\ell-1}^{(S)}\right]
$$

$$
= -\omega_c \Delta_{P,\ell}^{(S)} + \frac{1}{2}\omega_c \Pi \left(\delta_{\ell 0} + \frac{\delta_{\ell 2}}{5}\right) , \tag{6.1.27}
$$

$$
\dot{\Delta}_{T,\ell}^{(S)} + \frac{q}{a(2\ell+1)}\left[(\ell+1)\Delta_{T,\ell+1}^{(S)} - \ell\Delta_{T,\ell-1}^{(S)}\right]
$$

$$
= -2\dot{A}_q \delta_{\ell 0} + 2q^2 \dot{B}_q \left(\frac{\delta_{\ell 0}}{3} - \frac{2\delta_{\ell 2}}{15}\right)
$$

$$
- \omega_c \Delta_{T,\ell}^{(S)} + \omega_c \left(3\Phi + \frac{1}{2}\Pi\right)\delta_{\ell 0} + \frac{1}{10}\omega_c \Pi \,\delta_{\ell 2}
$$

$$
- \frac{4}{3}q\omega_c \,\delta u_{Bq}\, \delta_{\ell 1} \tag{6.1.28}
$$

We can express the source functions in terms of $\Delta_{T,\ell}^{(S)}$ and $\Delta_{P,\ell}^{(S)}$ by inserting Eq.(6.1.20) in Eq. (6.1.21), and using the integral formula, that for any function $f(\hat{q}\cdot\hat{p})$,

$$
\frac{1}{4\pi}\int d^2\hat{p}\, f(\hat{q}\cdot\hat{p})\hat{p}_i\hat{p}_j = \mathcal{A}\,\delta_{ij} + \mathcal{B}\,\hat{q}_i\hat{q}_j ,
$$

where

$$
\mathcal{A} = \frac{1}{4}\int_{-1}^{1} d\mu\, f(\mu)\,(1-\mu^2) = \frac{1}{6}\int_{-1}^{1} d\mu\, f(\mu)\left(P_0(\mu) - P_2(\mu)\right)
$$

$$
\mathcal{B} = \frac{1}{4}\int_{-1}^{1} d\mu\, f(\mu)\,(3\mu^2-1) = \frac{1}{2}\int_{-1}^{1} d\mu\, f(\mu)\, P_2(\mu)
$$

(The general form of the integrals is dictated by rotational invariance, while the formulas for the coefficients are found by contracting the integral with δ_{ij} and with $\hat{q}_i\hat{q}_j$, and then solving the resulting pair of linear equations for $\mathcal{A}$ and $\mathcal{B}$.) Using this, we can easily evaluate the two terms in the integral (6.1.21) (here dropping the arguments q and t):

$$
\frac{1}{8\pi}\int d^2\hat{p}\left(\Delta_T^{(S)}(\hat{q}\cdot\hat{p}) - \Delta_P^{(S)}(\hat{q}\cdot\hat{p})\right)(\delta_{ij} - \hat{p}_i\hat{p}_j)
$$

$$
= \int_{-1}^{+1} d\mu \left(\Delta_T^{(S)}(\mu) - \Delta_P^{(S)}(\mu)\right)\left[\left(\frac{1}{6}P_0(\mu) + \frac{1}{12}P_2(\mu)\right)\delta_{ij} - \frac{1}{4}P_2(\mu)\hat{q}_i\hat{q}_j\right]
$$

$$\frac{1}{4\pi} \int d^2\hat{p}\Delta_P^{(S)}(\hat{q}\cdot\hat{p})\frac{\left(\hat{q}_i - (\hat{q}\cdot\hat{p})\hat{p}_i\right)\left(\hat{q}_j - (\hat{q}\cdot\hat{p})\hat{p}_j\right)}{1 - (\hat{q}\cdot\hat{p})^2}$$

$$= \int_{-1}^{+1} d\mu\,\Delta_P^{(S)}(\mu)\left[\left(\frac{1}{12}P_0(\mu) + \frac{1}{6}P_2(\mu)\right)\delta_{ij} + \left(\frac{1}{4}P_0(\mu) - \frac{1}{2}P_2(\mu)\right)\hat{q}_i\hat{q}_j\right]$$

In this way, and using the partial wave expansions (6.1.25) and (6.1.26), we find

$$\Phi = \frac{1}{6}\left[2\Delta_{T,0}^{(S)} - \Delta_{P,0}^{(S)} - \Delta_{T,2}^{(S)} - \Delta_{P,2}^{(S)}\right], \tag{6.1.29}$$

$$\Pi = \Delta_{P,0}^{(S)} + \Delta_{T,2}^{(S)} + \Delta_{P,2}^{(S)}. \tag{6.1.30}$$

In the same way, using Eq. (6.1.20) in Eqs. (6.1.14)–(6.1.16) gives

$$\delta T_{\gamma j}^i(\mathbf{x}, t) = \bar{\rho}_\gamma(t)\int d^3q\,e^{i\mathbf{q}\cdot\mathbf{x}}\alpha(\mathbf{q})\int_{-1}^1 d\mu\,\Delta_T^{(S)}(q,\mu,t)$$

$$\times\left[\frac{1}{6}\left(P_0(\mu) - P_2(\mu)\right)\delta_{ij} + \frac{1}{2}P_2(\mu)\hat{q}_i\hat{q}_j\right],$$

$$\delta T_{\gamma j}^0(\mathbf{x}, t) = a(t)\,\bar{\rho}_\gamma(t)\int d^3q\,e^{i\mathbf{q}\cdot\mathbf{x}}\alpha(\mathbf{q})\frac{\hat{q}_i}{2}\int_{-1}^1 d\mu\,\Delta_T^{(S)}(q,\mu,t)P_1(\mu),$$

$$\delta T_{\gamma 0}^0(\mathbf{x}, t) = -\frac{1}{2}\bar{\rho}_\gamma(t)\int d^3q\,e^{i\mathbf{q}\cdot\mathbf{x}}\alpha(\mathbf{q})\int_{-1}^1 d\mu\,\Delta_T^{(S)}(q,\mu,t)\,P_0(\mu),$$

in which we have used the formula

$$\frac{1}{4\pi}\int d^2\hat{p}f(\hat{q}\cdot\hat{p})\,\hat{p}_i = \frac{\hat{q}_i}{2}\int_{-1}^1 d\mu f(\mu)\,P_1(\mu).$$

Comparing this with the first three of Eqs. (5.1.43), and again using the partial wave expansions (6.1.25) and (6.1.26), we find

$$\delta p_{\gamma q} = \frac{\bar{\rho}_\gamma}{3}\left(\Delta_{T,0}^{(S)} + \Delta_{T,2}^{(S)}\right), \tag{6.1.31}$$

$$q^2\pi_{\gamma q}^S = \bar{\rho}_\gamma\Delta_{T,2}^{(S)}, \tag{6.1.32}$$

$$\delta\rho_{\gamma q} = \bar{\rho}_\gamma\Delta_{T,0}^{(S)}, \tag{6.1.33}$$

$$q\delta u_{\gamma q} = -\frac{3}{4}\Delta_{T,1}^{(S)}. \tag{6.1.34}$$

As a check of Eqs. (6.1.31)–(6.1.33), note that when used in the last of Eqs. (5.1.43) these results give $\delta T_{\gamma\lambda}^\lambda = 0$, a necessary consequence of the masslessness of the photon for any distribution of photon momenta.

Eqs. (6.1.31)–(6.1.34) show that to work out the cosmological evolution of the gravitational field and its effect on other perturbations in scalar modes, all we need to know about photons is $\Delta_{T,\ell}^{(S)}$ for $\ell \leq 2$. But of course the evolution of these three amplitudes is coupled by the Boltzmann equations (6.1.27)–(6.1.30) to both $\Delta_{T,\ell}^{(S)}$ and $\Delta_{P,\ell}^{(S)}$ for all higher ℓ. In computer programs like CMBfast and CAMB, the partial wave expansion is cut off at a sufficiently high value of ℓ; in the latest version of CMBfast, the maximum value of ℓ is taken as $\ell_{max} = 12$, in which case the computer has to solve $2(\ell_{max} + 1) = 26$ coupled ordinary differential equations for each value of q, not counting the other equations that describe the evolution of the baryonic plasma, cold dark matter, neutrinos, and the gravitational field.

As we will see in the next chapter, the interpretation of observations of the cosmic microwave radiation background requires calculation of $\Delta_{T,\ell}^{(S)}$ and $\Delta_{P,\ell}^{(S)}$ for ℓ ranging to values well over 1,000. Originally this was done by a direct use of the coupled Boltzmann equations (6.1.27) and (6.1.28),[4] but this required hours or even days of computer time for each theoretical model. A great improvement was introduced with the suggestion to use instead a formal solution of the Boltzmann equation (6.1.19), in the form of a "line-of-sight" integral,[5] which in matrix notation takes the form

$$
J_{ij}(\mathbf{q}, \hat{p}, t) = \alpha(\mathbf{q}) \int_{t_1}^{t} dt' \exp\left(-i\mathbf{q} \cdot \hat{p} \int_{t'}^{t} \frac{dt''}{a(t'')} - \int_{t'}^{t} dt'' \, \omega_c(t'') \right)
$$

$$
\times \left[-\left(\delta_{ij} - \hat{p}_i \hat{p}_j \right) \left(\dot{A}_q(t') - (\hat{p} \cdot \mathbf{q})^2 B_q(t') \right) \right.
$$

$$
+ \frac{3\omega_c(t')}{2} (\delta_{ij} - \hat{p}_i \hat{p}_j) \Phi(q, t')
$$

$$
+ \frac{3\omega_c(t')}{4} \Pi(q, t') \left(\hat{q}_i - \hat{p}_i (\hat{q} \cdot \hat{p}) \right) \left(\hat{q}_j - \hat{p}_j (\hat{q} \cdot \hat{p}) \right)
$$

$$
\left. + \frac{2\omega_c(t')}{a(t')} [\delta_{ij} - \hat{p}_i \hat{p}_j] \hat{p}_k \delta u_k(\mathbf{q}, t') \right]
$$

$$
+ J_{ij}(\mathbf{q}, \hat{p}, t_1) \exp\left(-i\mathbf{q} \cdot \hat{p} \int_{t_1}^{t} \frac{dt'}{a(t')} - \int_{t_1}^{t} dt' \, \omega_c(t') \right) , \tag{6.1.35}
$$

where t_1 is any arbitrary initial time. If we choose t_1 to be sufficiently early, before recombination, so that $\omega_c(t_1) \gg H(t_1)$, and take t at any time

---

[5]U. Seljak and M. A. Zaldarriaga, *Astrophys. J.* **469**, 437 (1996). [astro-ph/9603033].

after recombination, the final term in Eq. (6.1.35) may be neglected, and we have

$$J_{ij}(\mathbf{q}, \hat{p}, t) = \alpha(\mathbf{q}) \int_{t_1}^{t} dt' \, \exp\left(-iq \cdot \hat{p} \int_{t'}^{t} \frac{dt''}{a(t'')} - \int_{t'}^{t} dt'' \, \omega_c(t'')\right)$$

$$\times \left[-\left(\delta_{ij} - \hat{p}_i \hat{p}_j\right)\left(\dot{A}_q(t') - (\hat{p} \cdot \mathbf{q})^2 \dot{B}_q(t')\right) \right.$$

$$+ \frac{3\omega_c(t')}{2}(\delta_{ij} - \hat{p}_i \hat{p}_j)\Phi(q, t')$$

$$+ \frac{3\omega_c(t')}{4}\Pi(q, t')\left(\hat{q}_i - \hat{p}_i(\hat{q} \cdot \hat{p})\right)\left(\hat{q}_j - \hat{p}_j(\hat{q} \cdot \hat{p})\right)$$

$$\left. + \frac{2\omega_c(t')}{a(t')}[\delta_{ij} - \hat{p}_i \hat{p}_j]\hat{p}_k \delta u_k(\mathbf{q}, t') \right]. \tag{6.1.36}$$

In terms of the temperature and polarization amplitudes defined by the decomposition (6.1.20), the line-of-sight solution reads

$$\Delta_T^{(S)}(q, \mu, t) = \Delta_P^{(S)}(q, \mu, t) + 2 \int_{t_1}^{t} dt' \, \exp\left[-iq\mu \int_{t'}^{t} \frac{dt''}{a(t'')} - \int_{t'}^{t} \omega_c(t'') \, dt'' \right]$$

$$\times \left[-\dot{A}_q(t') + \mu^2 q^2 \dot{B}_q(t') + \frac{3}{2}\omega_c(t') \, \Phi(q, t') + \frac{2i\mu q \omega_c(t')}{a(t')} \delta u_q(t') \right], \tag{6.1.37}$$

$$\Delta_P^{(S)}(q, \mu, t) = \frac{3}{4}(1 - \mu^2) \int_{t_1}^{t} dt' \, \exp\left[-iq\mu \int_{t'}^{t} \frac{dt''}{a(t'')} - \int_{t'}^{t} \omega_c(t'') \, dt'' \right]$$

$$\times \omega_c(t') \, \Pi(q, t') \,, \tag{6.1.38}$$

where δu_q is the scalar velocity potential, defined by $\delta u_k(\mathbf{q}, t) = iq_k u_q(t)$. Once Φ and Π have been calculated from Eqs. (6.1.29) and (6.1.30), Eqs. (6.1.37) and (6.1.38) can be used to calculate $\Delta_{T,\ell}^{(S)}$ and $\Delta_{P,\ell}^{(S)}$ for arbitrarily high values of ℓ. (It is also possible to use Eq. (6.1.36) as a substitute for the partial wave expansion in calculating the source terms Φ and Π. Integrating Eq. (6.1.36) over $\hat{p}$ yields integral equations:[6] expressions for $\Phi(q, t)$ and $\Pi(q, t)$ as integrals from t_1 to t in which the integrand is a linear function of $\Phi(q.t')$ and $\Pi(q, t')$ for $t' < t$. This approach will be applied to tensor modes in Section 6.6.)

---

[6]S. Weinberg, *Phys. Rev. D* **74**, 063517 (2006) [astro-ph/0607076]; D. Baskaran, L. P. Grishchuk, and A. G. Polnarev, *Phys. Rev. D* **74**, 063517 [gr-qc/0605100].

Neutrinos

The number density $n_\nu(\mathbf{x}, \mathbf{p}, t)$ of each species of massless neutrinos (or antineutrinos) in phase space can be conveniently expressed in terms of an intrinsic perturbation $\delta n_\nu(\mathbf{x}, \mathbf{p}, t)$ by a formula like Eq. (6.1.6):

$$n_\nu(\mathbf{x}, \mathbf{p}, t) = \bar{n}_\nu\left(a(t)p^0(\mathbf{x}, \mathbf{p}, t)\right) + \delta n_\nu(\mathbf{x}, \mathbf{p}, t) , \qquad (6.1.39)$$

where $\bar{n}_\nu$ is the equilibrium phase space density of each neutrino species

$$\bar{n}_\nu(p) \equiv \frac{1}{(2\pi)^3}\left[\exp\left(p/k_B a(t)\bar{T}(t)\right) + 1\right]^{-1} . \qquad (6.1.40)$$

As shown in Appendix H, the perturbation $\delta n_\nu(\mathbf{x}, \mathbf{p}, t)$ satisfies the same Boltzmann equation as the photon phase space density $a^2(t)\, n_\gamma^{kk}(\mathbf{x}, \mathbf{p}, t)$, except that for $T \ll 10^{10}\text{K}$, the terms proportional to the collision frequency are absent:

$$0 = \frac{\partial \,\delta n_\nu(\mathbf{x}, \mathbf{p}, t)}{\partial t} + \frac{\hat{p}_k}{a(t)}\frac{\partial \,\delta n_\nu(\mathbf{x}, \mathbf{p}, t)}{\partial x^k}$$

$$-\frac{1}{2}p\bar{n}'_\nu(p)\hat{p}_k\hat{p}_l\frac{\partial}{\partial t}\left(a^{-2}(t)\delta g_{kl}(\mathbf{x}, t)\right) \qquad (6.1.41)$$

The contribution of each species of neutrino to the perturbations to the energy-momentum tensor is given by formulas (6.1.10)–(6.1.12), except that $\delta n_\nu(\mathbf{x}, \mathbf{p}, t)$ appears instead of $a^2(t)\,\delta n_\gamma^{kk}(\mathbf{x}, \mathbf{p}, t)$. Once again, all we need for this purpose is a dimensionless direction-dependent intensity, defined by a formula like Eq. (6.1.13):

$$a^4(t)\bar{\rho}_\nu(t)J(\mathbf{x}, \hat{p}, t) \equiv N_\nu \int_0^\infty \delta n_\nu(\mathbf{x}, \mathbf{p}, t)\, 4\pi p^3 \, dp , \qquad (6.1.42)$$

where N_ν is the number of species of neutrino, counting antineutrinos separately, and $\bar{\rho}_\nu \equiv N_\nu\, a^{-4} \int 4\pi p^3 \bar{n}_\nu(p)\, dp$. Then the total neutrino and antineutrino contribution to the energy-momentum tensor is

$$\delta T^i_{\nu j}(\mathbf{x}, t) = \bar{\rho}_\nu(t) \int \frac{d^2\hat{p}}{4\pi}J(\mathbf{x}, \hat{p}, t)\, \hat{p}_i\hat{p}_j , \qquad (6.1.43)$$

$$\delta T^0_{\nu j}(\mathbf{x}, t) = a(t)\bar{\rho}_\nu(t) \int \frac{d^2\hat{p}}{4\pi}J(\mathbf{x}, \hat{p}, t)\, p_j , \qquad (6.1.44)$$

$$\delta T^0_{\nu 0}(\mathbf{x}, t) = -\bar{\rho}_\nu(t) \int \frac{d^2\hat{p}}{4\pi}J(\mathbf{x}, \hat{p}, t) , \qquad (6.1.45)$$

just as in Eqs. (6.1.14)–(6.1.16). Rotational and translational invariance allow us to express $J(\mathbf{x}, \hat{p}, t)$ as a Fourier integral of the form

$$J(\mathbf{x}, \hat{p}, t) = \int \alpha(\mathbf{q}) \, \Delta_\nu^{(S)}(q, \hat{q} \cdot \hat{p}, t) \, e^{i\mathbf{q} \cdot \mathbf{x}} \, d^3q \qquad (6.1.46)$$

To derive a Boltzmann equation for $\Delta_\nu^{(S)}$, we multiply Eq. (6.1.41) with $4\pi \, |\mathbf{p}|^3$ and integrate over $|\mathbf{p}|$, and find

$$\frac{\partial \Delta_\nu^{(S)}(q, \mu, t)}{\partial t} + i \frac{q\mu}{a(t)} \Delta_\nu^{(S)}(q, \mu, t) = -2\dot{A}_q(t) + 2q^2 \mu^2 \dot{B}_q(t) . \qquad (6.1.47)$$

In computer programs like CMBfast, the Boltzmann equation for neutrinos as well as for photons is solved by a partial wave expansion. One writes

$$\Delta_\nu^{(S)}(q, \mu, t) = \sum_{\ell=0}^{\infty} i^{-\ell}(2\ell + 1) P_\ell(\mu) \, \Delta_{\nu,\ell}^{(S)}(q, t) . \qquad (6.1.48)$$

Inserting this in Eq. (6.1.46) and then in Eqs. (6.1.43)–(6.1.45) and then comparing the results with the first three of Eqs. (5.1.43) gives the perturbed pressure, scalar anisotropic inertia, perturbed energy density, and velocity potential of the neutrinos

$$\delta p_{\nu\,q}(t) = \frac{\bar{\rho}_\nu(t)}{3} \left(\Delta_{\nu,0}^{(S)}(q, t) + \Delta_{\nu,2}^{(S)}(q, t) \right) , \qquad (6.1.49)$$

$$q^2 \pi_{\nu\,q}^S(t) = \bar{\rho}_\nu(t) \, \Delta_{\nu,2}^{(S)}(q, t) , \qquad (6.1.50)$$

$$\delta \rho_{\nu\,q}(t) = \bar{\rho}_\nu(t) \, \Delta_{\nu,0}^{(S)}(q, t) , \qquad (6.1.51)$$

$$q \, \delta u_{\nu\,q}(t) = -\frac{3}{4} \Delta_{\nu,1}^{(S)}(q, t) . \qquad (6.1.52)$$

To derive the Boltzmann equations for the partial wave amplitudes, we multiply Eq. (6.1.47) with $P_\ell(\mu)$ and integrate over μ:

$$\dot{\Delta}_{\nu,\ell}^{(S)}(q, t) + \frac{q}{a(2\ell + 1)} \left[(\ell + 1)\Delta_{\nu,\ell+1}^{(S)}(q, t) - \ell \, \Delta_{\nu,\ell-1}^{(S)}(q, t) \right]$$

$$= -2\dot{A}_q(t)\delta_{\ell 0} + q^2 \dot{B}_q(t) \left(\frac{2\delta_{\ell 0}}{3} - \frac{4\delta_{\ell 2}}{15} \right) . \qquad (6.1.53)$$

In the current version of CMBfast, this equation is cut off at a maximum value of ℓ equal to 25. Instead of relying on a truncated partial wave expansion, it is possible to write a solution (here not merely a formal solution) of

Eq. (6.1.47) as another line-of-sight integral:

$$\Delta_\nu^{(S)}(q,\mu,t) = -2\int_{t_1}^t \exp\left(-iq\mu\int_{t'}^t \frac{dt''}{a(t'')}\right)\left(\dot{A}_q(t') - \mu^2 q^2 \dot{B}_q(t')\right)$$

$$+\Delta_\nu^{(S)}(q,\mu,t_1)\exp\left(-iq\mu\int_{t_1}^t \frac{dt'}{a(t')}\right)\,, \tag{6.1.54}$$

where t_1 is any arbitrary initial time. If t_1 is taken at some time after the neutrinos went out of thermal equilibrium with the baryonic plasma, but early enough so that gravitational field perturbations have not yet had a chance to distort the neutrino distribution, then $\Delta_\nu^{(S)}(q,\mu,t_1)$ arises only from the temperature perturbation at time t_1:

$$\Delta_\nu^{(S)}(q,\mu,t_1) = 4\left[\frac{\delta T_q(t_1)}{\bar{T}(t_1)} + i\frac{\mu q\delta u_q(t_1)}{a(t_1)}\right]\,, \tag{6.1.55}$$

in which the second term in square brackets arises from the Doppler effect due to the streaming of the electron–positron–photon plasma.

The integrals over direction in Eqs. (6.1.43)–(6.1.45) can be done analytically; for this purpose we need the formulas

$$\int \frac{d^2\hat{p}}{4\pi}\hat{p}_i\hat{p}_j\hat{p}_k\hat{p}_l\, e^{-i\hat{p}\cdot\mathbf{v}} = \left(\delta_{ij}\delta_{kl} + \delta_{ik}\delta_{jl} + \delta_{il}\delta_{kj}\right)j_2(v)/v^2$$

$$- \left(\delta_{ij}\hat{v}_k\hat{v}_l + \delta_{ik}\hat{v}_j\hat{v}_l + \delta_{il}\hat{v}_k\hat{v}_j + \delta_{jk}\hat{v}_i\hat{v}_l + \delta_{jl}\hat{v}_k\hat{v}_i + \delta_{kl}\hat{v}_i\hat{v}_j\right)j_3(v)/v$$

$$+ \hat{v}_i\hat{v}_j\hat{v}_k\hat{v}_l j_4(v)\,, \tag{6.1.56}$$

$$\int \frac{d^2\hat{p}}{4\pi}\hat{p}_i\hat{p}_j\hat{p}_k e^{-i\hat{p}\cdot\mathbf{v}} = -i\left(\delta_{ij}\hat{v}_k + \delta_{jk}\hat{v}_i + \delta_{ki}\hat{v}_j\right)j_2(v)/v$$

$$+ i\hat{v}_i\hat{v}_j\hat{v}_k j_3(v)\,, \tag{6.1.57}$$

$$\int \frac{d^2\hat{p}}{4\pi}\hat{p}_i\hat{p}_j\, e^{-i\hat{p}\cdot\mathbf{v}} = \delta_{ij}j_1(v)/v - \hat{v}_i\hat{v}_j\, j_2(v) \tag{6.1.58}$$

$$\int \frac{d^2\hat{p}}{4\pi}\hat{p}_i\, e^{-i\hat{p}\cdot\mathbf{v}} = -i\hat{v}_i\, j_1(v) \tag{6.1.59}$$

$$\int \frac{d^2\hat{p}}{4\pi}e^{-i\hat{p}\cdot\mathbf{v}} = j_0(v)\,, \tag{6.1.60}$$

where $v \equiv |\mathbf{v}|$, and $j_\ell(v)$ is the usual spherical Bessel function. Again comparing the results with the first three of Eqs. (5.1.43), we obtain explicit formulas for the perturbed pressure, scalar anisotropic inertia, perturbed energy density, and velocity potential of the neutrinos:[7]

$$\delta p_{\nu\,q}(t) = -2\bar{\rho}_\nu(t) \int_{t_1}^{t} dt'$$

$$\times \left[\dot{A}_q(t')\, K_1 \left(q \int_{t'}^{t} \frac{dt''}{a(t'')} \right) - q^2 \dot{B}_q(t')\, K_2 \left(q \int_{t'}^{t} \frac{dt''}{a(t'')} \right) \right], \quad (6.1.61)$$

$$q^2 \pi_{\nu\,q}^{S}(t) = -2\bar{\rho}_\nu(t) \int_{t_1}^{t} dt'$$

$$\times \left[\dot{A}_q(t')\, j_2 \left(q \int_{t'}^{t} \frac{dt''}{a(t'')} \right) - q^2 \dot{B}_q(t')\, K_3 \left(q \int_{t'}^{t} \frac{dt''}{a(t'')} \right) \right], \quad (6.1.62)$$

$$q \delta u_{\nu\,q}(t) = \frac{3a(t)}{2} \int_{t_1}^{t} dt'$$

$$\times \left[\dot{A}_q(t')\, j_1 \left(q \int_{t'}^{t} \frac{dt''}{a(t'')} \right) - q^2 \dot{B}_q(t')\, K_4 \left(q \int_{t'}^{t} \frac{dt''}{a(t'')} \right) \right], \quad (6.1.63)$$

$$\delta \rho_{\nu\,q}(t) = -2\bar{\rho}_\nu(t) \int_{t_1}^{t} dt'$$

$$\times \left[\dot{A}_q(t')\, j_0 \left(q \int_{t'}^{t} \frac{dt''}{a(t'')} \right) - q^2 \dot{B}_q(t')\, K_5 \left(q \int_{t'}^{t} \frac{dt''}{a(t'')} \right) \right], \quad (6.1.64)$$

where

$$K_1(v) \equiv j_1(v)/v \,, \qquad\qquad\qquad (6.1.65)$$

$$K_2(v) \equiv j_2(v)/v^2 - j_3(v)/v \,, \qquad (6.1.66)$$

$$K_3(v) \equiv -2j_2(v)/v^2 + 5j_3(v)/v - j_4(v) \,, \qquad (6.1.67)$$

$$K_4(v) \equiv 3j_2(v)/v - j_3(v) \,, \qquad\qquad (6.1.68)$$

$$K_5(v) \equiv j_1(v)/v - j_2(v) \,. \qquad\qquad (6.1.69)$$

Using these formulas, one no longer needs the truncated partial wave expansion for neutrinos.

Gravitation

It only remains to give the equations of motion for the scalar gravitational field components $\dot{A}$ and $\dot{B}$. (Note that A and B themselves are nowhere

---

[7] These formulas are given also for massive neutrinos by S. Weinberg, ref. 6.

needed.) It is convenient to take one of these as Eq. (5.3.31), written as

$$\frac{\partial}{\partial t}\left[a^2\,\psi_q\right] = -4\pi\,Ga^2$$

$$\times \left(\delta\rho_{Dq} + \delta\rho_{Bq} + \delta\rho_{\gamma\,q} + \delta\rho_{v\,q} + 3\delta p_{\gamma\,q} + 3\delta p_{v\,q} - q^2\pi^S_{\gamma\,q} - q^2\pi^S_{v\,q}\right)$$

(6.1.70)

where

$$\psi_q \equiv \frac{1}{2}\left(3\dot{A}_q - q^2\dot{B}_q\right).$$

(6.1.71)

The other can be taken from Eq. (5.3.30)

$$\dot{A}_q = 8\pi\,G\left[\frac{4}{3}\bar{\rho}_\gamma\delta u_{\gamma\,q} + \frac{4}{3}\bar{\rho}_v\delta u_{v\,q} + \bar{\rho}_B\delta u_{B\,q}\right].$$

(6.1.72)

(Recall that we have adopted the particular synchronous gauge in which $\delta u_{D\,q} = 0$.) After solving the first-order differential equation (6.1.70) for ψ_q and using Eq. (6.1.72) to find $\dot{A}_q$, the other component is trivially given by the definition Eq. (6.1.71) as

$$q^2\dot{B}_q = 3\dot{A}_q - 2\psi_q.$$

(6.1.73)

With this, as long as we truncate the partial wave expansions used for photons and neutrinos, we have a closed system of ordinary differential equations for all the perturbations, which can be straightforwardly solved by computer for any given initial conditions.

To find initial conditions, we note that at a sufficiently early time t_1, well before the era of recombination, (say, for $\bar{T}(t_1) > 10^5$ K), the collision rate of photons with the baryonic plasma is so great that photons are in thermal and kinetic equilibrium with the plasma. Under these conditions, the photon distribution δn^{ij} arises only from a perturbation to the temperature in the first term of Eq. (H.31), including the Doppler shift due to the photon streaming velocity, which in equilibrium is the same as the baryonic plasma velocity $\delta \mathbf{u}_B$:

$$\delta n^{ij}_\gamma(\mathbf{x}, \mathbf{p}, t_1) = -\frac{1}{2}a^{-2}(t_1)\left[\delta_{ij} - \hat{p}_i\hat{p}_j\right]\bar{n}'_\gamma(p)\,p\left[\frac{\delta T(\mathbf{x}, t_1)}{\bar{T}(t_1)} + \frac{\hat{p}_k\delta u_{Bk}(\mathbf{x}, t_1)}{a(t_1)}\right].$$

(6.1.74)

(The factor $1/a(t_1)$ in the Doppler term in Eq. (6.1.74) was explained in connection with Eq. (H.13).) Note that for ω_c very large, the coefficients of ω_c on the right-hand side of equation (6.1.17) must cancel, which gives an initial condition consistent with Eq. (6.1.74). Multiplying with $4\pi p^3$ and integrating over p, we find a corresponding condition on J_{ij}:

$$J_{ij}(\mathbf{x},\hat{p},t_1) = 2\Big[\delta_{ij} - \hat{p}_i\hat{p}_j\Big]\left[\frac{\delta T(\mathbf{x},t_1)}{\bar{T}(t_1)} + \frac{\hat{p}_k\delta u_k(\mathbf{x},t_1)}{a(t_1)}\right]. \tag{6.1.75}$$

Note that $J_{ij}(\mathbf{x},\hat{p},t_1)$ receives contributions only from scalar and vector perturbations, not from tensor perturbations.

Similar remarks apply to neutrinos, except that we must go back to an earlier time, when the temperature was a little below 10^{10} K, so that neutrinos were already traveling freely, but not enough time had elapsed for the gravitational field perturbation to have altered the equilibrium form of the neutrino phase space distribution.

This still leaves us with the necessity of stipulating initial values for A_q, B_q, δ_{Dq}, δ_{Bq}, δu_{Bq}, and δT_q. For this, we must go back to a time early enough so that the wave numbers of interest were outside the horizon, in the sense that $q/a \ll H$. In the following section the needed initial values will be worked out for the dominant adiabatic mode, with a normalization expressed in terms of the quantity $\mathcal{R}_q$, given in synchronous gauge outside the horizon by Eq. (5.4.24).

6.2 Scalar perturbations – the hydrodynamic limit

The system of equations described in the previous section is much too complicated to allow an analytic solution. Fortunately, until near the time of recombination the rate of collisions of photons with free electrons was so great that photons were in local thermal equilibrium with the baryonic plasma, and so photons at these times can be treated hydrodynamically, like the plasma and cold dark matter. This approach loses its validity around the time of recombination, but a fair degree of accuracy will be preserved in Section 6.4 by taking into account the damping caused by the growing mean free times in this era. After the time of recombination photons traveled more or less freely, and their path can be followed by solving their equation of motion. Neutrinos are more of a problem, but at very early times perturbations were outside the horizon, so at these times $q^2\pi_{\nu q}^S$ and $q\delta u_{\nu q}$ were negligible and Eqs. (6.1.49)–(6.1.51) show that $\delta\rho_{\nu q} = 3\delta p_{\nu q}$, just as if the neutrinos were in local thermal equilibrium, while at late times the universe became matter dominated and neutrinos made only a small contribution to the cosmic gravitational field.

274

With these justifications, in order to allow an analytic treatment, in this and the next three sections we will adopt a hydrodynamic approach.[1] To be specific, for the most part we will neglect anisotropic inertia, take $\delta u_{\gamma q}$ equal to δu_{Bq}, and take $p_{\gamma q} = \rho_{\gamma q}/3$ and $p_{vq} = \rho_{vq}/3$. This necessarily entails the loss of some accuracy, but our aim in this chapter (and in the next two chapters) is not to calculate the course of cosmic evolution and its observational consequences with the high level of accuracy that would optimize the extraction of cosmological parameters from the latest data on the cosmic microwave background and large scale structure. Rather, we wish here to elucidate the physics of cosmic evolution, and clarify the dependence of observables on cosmological assumptions. Fortunately, the results we obtain from this analytic treatment will turn out in Chapter 7 to yield predictions for anisotropies in the cosmic microwave background that are quite similar to those obtained by comprehensive computer programs, using the full Boltzmann equations described in the previous section. We are thereby reassured that the hydrodynamic approach captures the essence of what is going on in the early universe.

Under the above assumptions, the Fourier transforms with co-moving wave number q of the synchronous gauge perturbations are governed by the gravitational field equation (5.3.36):

$$\frac{d}{dt}\left(a^2\psi_q\right) = -4\pi G a^2\left(\delta\rho_{Dq} + \delta\rho_{Bq} + 2\delta\rho_{\gamma q} + 2\delta\rho_{vq}\right),\qquad(6.2.1)$$

the equations (5.3.34) of energy conservation for each of the four fluids

$$\delta\dot\rho_{\gamma q} + 4H\delta\rho_{\gamma q} - (4q^2/3a^2)\bar\rho_\gamma\delta u_{\gamma q} = -(4/3)\bar\rho_\gamma\psi_q,\qquad(6.2.2)$$
$$\delta\dot\rho_{Dq} + 3H\delta\rho_{Dq} = -\bar\rho_D\psi_q,\qquad(6.2.3)$$
$$\delta\dot\rho_{Bq} + 3H\delta\rho_{Bq} - (q^2/a^2)\bar\rho_B\delta u_{\gamma q} = -\bar\rho_B\psi_q,\qquad(6.2.4)$$
$$\delta\dot\rho_{vq} + 4H\delta\rho_{vq} - (4q^2/3a^2)\bar\rho_v\delta u_{vq} = -(4/3)\bar\rho_v\psi_q,\qquad(6.2.5)$$

and the equations (5.3.32) of momentum conservation for the photon–baryon plasma and the neutrinos:

$$\frac{d}{dt}\left(\left(\frac{4}{3}\bar\rho_\gamma + \bar\rho_B\right)\delta u_{\gamma q}\right) + 3H\left(\left(\frac{4}{3}\bar\rho_\gamma + \bar\rho_B\right)\delta u_{\gamma q}\right) = -(1/3)\delta\rho_{\gamma q},\quad(6.2.6)$$

$$\frac{d}{dt}\left(\bar\rho_v\delta u_{vq}\right) + 3H\bar\rho_v\delta u_{vq} = -(1/4)\delta\rho_{vq}.\qquad(6.2.7)$$

[1] Other analytic or semi-analytic treatments of the evolution of fluctuations have been given by W. Hu and N. Sugiyama, *Astrophys. J.* **444**, 489 (1995); **471**, 542 (1996); V. Mukhanov, *Int. J. Theor. Phys.* **43**, 623 (2004) [astro-ph/0303072]. The treatment given here is in my opinion more transparent though somewhat less accurate than that of Hu and Sugiyama, and (because we allow ourselves the use of a computer to do numerical integrals) it is more accurate than that of Mukhanov.

It is very convenient to rewrite these equations in terms of the dimensionless fractional perturbations[2]

$$\delta_{\alpha q} \equiv \frac{\delta\rho_{\alpha q}}{\bar{\rho}_\alpha + \bar{p}_\alpha} \tag{6.2.8}$$

where α runs over γ, D, B, and ν. Taking into account that $a^4 \bar{\rho}_\gamma$, $a^3 \bar{\rho}_D$, $a^3 \bar{\rho}_B$, and $a^4 \bar{\rho}_\nu$ are all time-independent, Eqs. (6.2.1)–(6.2.7) now read

$$\frac{d}{dt}\left(a^2 \psi_q\right) = -4\pi G a^2 \left(\bar{\rho}_D \delta_{Dq} + \bar{\rho}_B \delta_{Bq} + \frac{8}{3}\bar{\rho}_\gamma \delta_{\gamma q} + \frac{8}{3}\bar{\rho}_\nu \delta_{\nu q}\right), \tag{6.2.9}$$

$$\dot{\delta}_{\gamma q} - (q^2/a^2)\delta u_{\gamma q} = -\psi_q, \tag{6.2.10}$$

$$\dot{\delta}_{Dq} = -\psi_q, \tag{6.2.11}$$

$$\dot{\delta}_{Bq} - (q^2/a^2)\delta u_{\gamma q} = -\psi_q, \tag{6.2.12}$$

$$\dot{\delta}_{\nu q} - (q^2/a^2)\delta u_{\nu q} = -\psi_q, \tag{6.2.13}$$

$$\frac{d}{dt}\left(\frac{(1+R)\delta u_{\gamma q}}{a}\right) = -\frac{1}{3a}\delta_{\gamma q}, \tag{6.2.14}$$

$$\frac{d}{dt}\left(\frac{\delta u_{\nu q}}{a}\right) = -\frac{1}{3a}\delta_{\nu q}, \tag{6.2.15}$$

where $R \equiv 3\bar{\rho}_B/4\bar{\rho}_\gamma$. Eqs. (6.2.1)–(6.2.7) or (6.2.9)–(6.2.15) form a closed system of seven first-order differential equations for ψ_q, the four density perturbations, and the plasma and neutrino velocity potentials, so there must be seven independent solutions.

Before trying to find solutions valid up to the time of recombination, we must first consider the initial conditions to be imposed. These initial conditions will distinguish the different independent solutions. At sufficiently early times the universe was in a radiation-dominated era, when $\bar{\rho}_M \ll \bar{\rho}_R$, where

$$\bar{\rho}_M \equiv \bar{\rho}_D + \bar{\rho}_B, \qquad \bar{\rho}_R \equiv \bar{\rho}_\gamma + \bar{\rho}_\nu, \tag{6.2.16}$$

so that to a good approximation $a \propto \sqrt{t}$ and $8\pi G\bar{\rho}_R/3 = 1/4t^2$, while $R \ll 1$. (This fixes our definition of the zero of time.) If we take $a \propto \sqrt{t}$ and $R \ll 1$, Eqs. (6.2.9)–(6.2.15) become

$$\frac{d}{dt}\left(t\psi_q\right) = -4\pi G t \left(\bar{\rho}_D \delta_{Dq} + \bar{\rho}_B \delta_{Bq} + \frac{8}{3}\bar{\rho}_\gamma \delta_{\gamma q} + \frac{8}{3}\bar{\rho}_\nu \delta_{\nu q}\right), \tag{6.2.17}$$

$$\dot{\delta}_{\gamma q} = \dot{\delta}_{Bq} = -\psi_q + (q^2/a^2)\delta u_{\gamma q}, \tag{6.2.18}$$

---

[2]Note that this differs from a commonly used convention, according to which $\delta_{\alpha q}$ would be defined as $\delta\rho_{\alpha q}/\bar{\rho}_\alpha$.

$$\dot{\delta}_{Dq} = -\psi_q \, , \tag{6.2.19}$$

$$\dot{\delta}_{vq} = -\psi_q + (q^2/a^2)\delta u_{vq} \, , \tag{6.2.20}$$

$$\frac{d}{dt}\left(\frac{\delta u_{\gamma q}}{\sqrt{t}}\right) = -\frac{1}{3\sqrt{t}}\delta_{\gamma q} \, , \tag{6.2.21}$$

$$\frac{d}{dt}\left(\frac{\delta u_{vq}}{\sqrt{t}}\right) = -\frac{1}{3\sqrt{t}}\delta_{vq} \, . \tag{6.2.22}$$

At very early times the perturbation was outside the horizon, in this sense that $q/aH \ll 1$, but we have not yet dropped the terms in Eqs. (6.2.18) and (6.2.20) proportional to q^2, because in some modes there are cancelations in the calculation of the conserved quantity $\mathcal{R}_q$ outside the horizon that require us to take such terms into account. Also, we have not dropped the terms on the right-hand side of Eq. (6.2.17) proportional to $\bar{\rho}_D$ or $\bar{\rho}_B$, because even though we are now assuming that $\bar{\rho}_D$ and $\bar{\rho}_B$ are much less than $\bar{\rho}_\gamma$ and $\bar{\rho}_v$, we want to leave open the possibility of modes in which the *fractional fluctuations* in the dark matter and/or baryon density are much larger than the fractional fluctuations in the photon and neutrino densities.

Mode 1

This is the dominant adiabatic mode — adiabatic, in the sense that all the $\delta_{\alpha q}$ become equal at very early times. (As discussed in Section 5.4, only these modes are present in inflationary theories with a single scalar field, or if the universe was ever earlier in a state of complete local thermal equilibrium with no non-zero conserved quantities.) Inspection of Eqs. (6.2.17)–(6.2.22) shows that, if we make the ansatz,

$$\delta_{\gamma q} = \delta_{Bq} = \delta_{Dq} = \delta_{vq} \equiv \delta_q \, , \qquad \delta u_{\gamma q} = \delta u_{vq} \equiv \delta u_q \, , \tag{6.2.23}$$

and if we now drop the baryon and dark matter-dominated terms in Eq. (6.2.17), and consider times early enough so that we can drop the q^2/a^2 terms in Eqs. (6.2.18) and (6.2.20), then Eqs. (6.2.18)–(6.2.22) and (6.2.17) become

$$\dot{\delta}_q = -\psi_q \, , \tag{6.2.24}$$

$$\frac{d}{dt}\left(\frac{\delta u_q}{\sqrt{t}}\right) = -\frac{1}{3\sqrt{t}}\delta_q \, , \tag{6.2.25}$$

and

$$\frac{d}{dt}\left(t\psi_q\right) = -\frac{1}{t}\delta_q \, . \tag{6.2.26}$$

Combining Eqs. (6.2.24) and (6.2.26) gives a second-order differential equation for δ_q:

$$\frac{d}{dt}\left(t\frac{d}{dt}\delta_q\right) - \frac{1}{t}\delta_q = 0 \,.$$

This has two solutions, with $\delta_q \propto t$ and $\delta_q \propto 1/t$, and for each solution Eqs. (6.2.24) and (6.2.25) give solutions for ψ_q and δu_q. The growing solution has $\delta_q \propto t$ and $\delta u_q \propto t^2$, and gives our first adiabatic mode:

$$\delta_{\gamma q} = \delta_{Bq} = \delta_{vq} = \delta_{Dq} = \frac{q^2 t^2 \mathcal{R}_q^o}{a^2} \,, \tag{6.2.27}$$

$$\psi_q = -\frac{t q^2 \mathcal{R}_q^o}{a^2} \,, \tag{6.2.28}$$

$$\delta u_{\gamma q} = \delta u_{vq} = -\frac{2t^3 q^2 \mathcal{R}_q^o}{9a^2} \,, \tag{6.2.29}$$

We have normalized this mode so that the quantity given by Eq. (5.4.24) as

$$q^2 \mathcal{R}_q \equiv -a^2 H \psi_q + 4\pi G a^2 \delta\rho_q + q^2 H \delta u_q \tag{6.2.30}$$

takes the time-independent value $q^2 \mathcal{R}_q^o$ for $q/a \ll H$ (the superscript o standing for "outside the horizon").

Mode 2

The solution of Eqs. (6.2.24)–(6.2.26) which goes as $\delta_q \propto 1/t$ for $t \to 0$ gives us our second adiabatic solution:

$$\delta_{\gamma q} = \delta_{Bq} = \delta_{vq} = \delta_{Dq} = \epsilon_q/t \,, \quad \psi_q = \epsilon_q/t^2 \,, \quad \delta u_{\gamma q} = \delta u_{vq} = \frac{2\epsilon_q}{3} \,. \tag{6.2.31}$$

with ϵ_q an arbitrary time-independent function of q. The calculation of $\mathcal{R}_q$ for this solution has the problem that the first two terms in Eq. (6.2.30) for $q^2 \mathcal{R}_q$ cancel to zeroth order in $q^2/a^2 H^2$, leaving us with an unknown residue in $q^2 \mathcal{R}_q$ of order $q^2/a^2 H^2$, and hence an unknown term in $\mathcal{R}_q$ of zeroth order in $q^2/a^2 H^2$. Fortunately, in this mode we can find a solution to Eqs. (6.2.17)–(6.2.22) that is valid to all orders in q/aH, as long as

$\rho_M \ll \rho_R$:

$$\delta_{\gamma q} = \delta_{Bq} = \delta_{vq} = \epsilon_q / t , \quad \delta_{Dq} = \frac{\epsilon_q}{t}\left[1 - \frac{q^2}{3H^2 a^2}\ln\left(\frac{q}{Ha}\right)\right] , \quad (6.2.32)$$

$$\psi_q = \frac{\epsilon_q}{t^2}\left[1 + \frac{2q^2 t^2}{3a^2}\right] , \quad\quad (6.2.33)$$

$$\delta u_{\gamma q} = \delta u_{vq} = \frac{2\epsilon_q}{3} . \quad\quad (6.2.34)$$

Using this in Eq. (6.2.30) shows that this mode has $\mathcal{R}_q = 0$ to all orders in q/aH as long as $\bar\rho_M \ll \bar\rho_R$.

The other five modes are non-adiabatic, in the sense that some of the $\delta_{\alpha q}$ are unequal even for $q/a \ll H$. One particularly simple mode can serve as an illustration:

Mode 3

$$\delta_{Dq} = \frac{\epsilon_q \bar\rho_B}{\bar\rho_B + \bar\rho_D} , \quad \delta_{Bq} = -\frac{\epsilon_q \bar\rho_D}{\bar\rho_B + \bar\rho_D} , \quad\quad (6.2.35)$$

$$\psi_q = 0 , \quad \delta_{\gamma q} = \delta_{vq} = 0 , \quad \delta u_{\gamma q} = \delta u_{vq} = 0 , \quad\quad (6.2.36)$$

again with ϵ_q time-independent but otherwise arbitrary. This just amounts to a perturbation in the time-independent ratio of the densities of baryons and dark matter, and is an exact solution for all times. It is an *isocurvature* mode, in both the sense that $\mathcal{R}_q = 0$, and also that $\psi_q = 0$.

$$* * *$$

As already indicated, these results apply only at times early enough so that $\bar\rho_M \ll \bar\rho_R$ (and, for mode 1, $q/a \ll H$). As an aid to extending these early-time solutions to later times, note that for all times before recombination, the difference of Eqs. (6.2.10) and (6.2.12) gives

$$\frac{d}{dt}\left(\delta_{Bq} - \delta_{\gamma q}\right) = 0 . \quad\quad (6.2.37)$$

We see that any solution that satisfies the adiabatic condition, that $\delta_{Bq} = \delta_{\gamma q}$ at early times, when the perturbation is far outside the horizon, will continue thereafter to have

$$\delta_{Bq} = \delta_{\gamma q} . \quad\quad (6.2.38)$$

279

Eqs. (6.2.9)–(6.2.15) are then reduced to

$$\frac{d}{dt}\left(a^2\psi_q\right) = -4\pi\, G a^2\left[\bar{\rho}_D\delta_{Dq} + \left(\bar{\rho}_B + \frac{8}{3}\bar{\rho}_\gamma\right)\delta_{\gamma q} + \frac{8}{3}\bar{\rho}_\nu\delta_{\nu q}\right], \quad (6.2.39)$$

$$\dot{\delta}_{\gamma q} - (q^2/a^2)\delta u_{\gamma q} = -\psi_q\,, \tag{6.2.40}$$

$$\dot{\delta}_{Dq} = -\psi_q\,, \tag{6.2.41}$$

$$\dot{\delta}_{\nu q} - (q^2/a^2)\delta u_{\nu q} = -\psi_q\,, \tag{6.2.42}$$

$$\frac{d}{dt}\left(\frac{(1+R)\delta u_{\gamma q}}{a}\right) = -\frac{1}{3a}\delta_{\gamma q}\,, \tag{6.2.43}$$

$$\frac{d}{dt}\left(\frac{\delta u_{\nu q}}{a}\right) = -\frac{1}{3a}\delta_{\nu q}\,, \tag{6.2.44}$$

where again $R \equiv 3\bar{\rho}_B/4\bar{\rho}_R$. These equations apply to modes 1 and 2 at all times before recombination. We will be chiefly interested in mode 1, since the perturbations of mode 2 decay by a factor $1/t^2$ relative to those of mode 1 during the part of the radiation dominated era when the perturbation is still outside the horizon.

It is not possible to find an analytic solution of even the reduced set of equations (6.2.39)–(6.2.44) that would be valid for all times and wave numbers. They can be treated analytically, however, in two wavelength regimes: *long wavelengths*, for which $q \ll q_{EQ}$, and *short wavelengths*, for which $q \gg q_{EQ}$, where q_{EQ} is the wave number for which $q/a = H$ at matter–radiation equality. Recall that, once inflation is over, q/a decreases more slowly than H, so that for long wavelengths, the wave number is so small that $\bar{\rho}_M$ becomes equal to $\bar{\rho}_R$ when q/a is still much less than H, while for short wavelengths, the wave number is so large that q/a becomes equal to H when $\bar{\rho}_M$ is still much less than $\bar{\rho}_R$. For long wavelengths we will be able to find analytic solutions of Eqs. (6.2.39)–(6.2.44) in both the early era, when the perturbation is outside the horizon, and in the later era, when the expansion is dominated by non-relativistic matter, and patch them together in the era of overlap, when the perturbation is still outside the horizon and the universe is already matter dominated. Conversely, for short wavelengths we will be able to find analytic solutions of these equations in the early era when the universe is radiation dominated and in the later era when the perturbation is deep inside the horizon, and patch them together in the era of overlap, when the universe is still radiation dominated and the perturbation is already deep inside the horizon. The two cases of long and short wavelength are considered in Sections 6.3 and 6.4, respectively.

To calculate the critical wave number q_{EQ} for which $q/a = H$ at matter–radiation equality, we recall that the redshift of matter–radiation equality is given by Eq. (2.2.5) as $1 + z_{EQ} = \Omega_M/\Omega_R = \Omega_M h^2/4.15 \times 10^{-5}$.

Eq. (2.2.9) gives the Hubble rate during the radiation dominated era as $H = 2.1 \times 10^{-20} (1 + z)^2 s^{-1}$, but at radiation–matter equality the contribution of matter to the total energy density makes H larger by a factor $\sqrt{2}$, so for fluctuations that just enter the horizon at matter–radiation equality the physical wave number and Hubble rate at that time are given by Eq. (2.2.9) as

$$q_{EQ}/a_{EQ}=H_{EQ}=\sqrt{2}\times 2.1\times 10^{-20} (1+z_{EQ})^2 s^{-1}=1.72\times 10^{-11}(\Omega_M h^2)^2 s^{-1} .$$

This corresponds to a critical physical wavelength at present given by

$$\lambda_0 \equiv \frac{2\pi}{q_{EQ}/a_0} = \frac{2\pi(1+z_{EQ})}{q_{EQ}/a_{EQ}} = 85 (\Omega_M h^2)^{-1} \text{ Mpc} . \qquad (6.2.45)$$

For comparison, the size of the local supercluster, estimated from the distance between our galaxy and the Virgo cluster, is about 15 Mpc. Perturbations that are now observed to extend over distances that are larger or smaller than λ_{EQ} came within the horizon before or after matter–radiation equality, respectively. Using the present mass density ρ_{M0} given by Eq. (1.5.28) and (1.5.39), the average mass now contained within a sphere of diameter λ_0 is

$$\frac{\pi}{6}\rho_{M0}\lambda_0^3 = 0.9 \times 10^{17} (\Omega_M h^2)^{-2} M_\odot . \qquad (6.2.46)$$

This may be compared with the mass of a large galaxy, about $10^{12} M_\odot$. Thus any perturbation that is relevant to the formation of galaxies or even clusters of galaxies would have been well within the horizon at the time of radiation-matter equality.

We can also identify a corresponding critical multipole order ℓ_{EQ} of anisotropies in the cosmic microwave background. As remarked in Section 2.6, the integral over wave numbers for the multipole coefficient C_ℓ is dominated by co-moving wave numbers of order ℓ/r_L, where r_L is the radial coordinate of the surface of last scattering, and hence the integral for C_ℓ is dominated by wave numbers of order q_{EQ} that just come into the horizon at matter–radiation equality if ℓ is of order $\ell_{EQ} = q_{EQ}r_L$. This can be written

$$\ell_{EQ} = \left(\frac{q_{EQ}}{a_{EQ}}\right)\left(\frac{a_{EQ}}{a_0}\right)\left(\frac{a_0}{a_L}\right)a_L r_L = \frac{H_{EQ}(1+z_L)}{(1+z_{EQ})}a_L r_L ,$$

where a_L is the Robertson–Walker scale factor at last scattering. We recall that $H_{EQ} = \sqrt{2\Omega_M}(1 + z_{EQ})^{3/2} H_0$, and $a_L r_L = d_A$, the angular diameter distance of the surface of last scattering. Also, $1 + z_{EQ} = \Omega_M/\Omega_R$. Hence

$$\ell_{EQ} = \Omega_M \sqrt{2/\Omega_R} H_0 d_A (1 + z_L) . \qquad (6.2.47)$$

For instance, if we take sample parameters $\Omega_M = 0.26$ and $\Omega_\Lambda = 0.74$, then as noted in Section 2.6 $d_A = 3.38 H_0^{-1}(1+z_L)^{-1}$. Taking $\Omega_R = 8.01 \times 10^{-5}$ (corresponding to $T_0 = 2.725\text{K}$ and $H_0 = 72\,\text{km sec}^{-1}\text{Mpc}^{-1}$), Eq. (6.2.47) gives a critical multipole order $\ell_{EQ} = 140$. Multipole coefficients for larger values of ℓ arise only from perturbations that entered the horizon during the radiation dominated era.

In what follows, for both scalar and tensor modes, we will find it convenient to introduce a dimensionless rescaled wave number,

$$\kappa \equiv \frac{\sqrt{2}q}{q_{Eq}} = \frac{(q/a_0)\sqrt{\Omega_R}}{H_0 \Omega_M} = \frac{q/a_0}{0.052\,\Omega_M h^2\,\text{Mpc}^{-1}} . \tag{6.2.48}$$

In the cases of long and short wavelength, we have $\kappa \ll 1$ and $\kappa \gg 1$, respectively.

The calculations of the next three sections are necessarily complicated and perhaps tedious. As a guide, it may help to say that the results at which we are aiming are Eqs. (6.5.15) and (6.5.16) for the perturbations to the dark matter density and gravitational field in the whole of the matter-dominated era, and Eqs. (6.5.17) and (6.5.18) for the perturbations to the photon and baryonic plasma density and velocity potential in the matter-dominated era, up to the time of the decoupling of matter and radiation.

6.3 Scalar perturbations – long wavelengths

We first consider perturbations with wavelengths that are long enough so that they are still outside the horizon at the time of radiation–matter equality. As discussed at the end of the previous section, such perturbations are responsible for multipole moments of the cosmic microwave background anisotropies with $\ell < 140$. Because $q\bar{\rho}_R/aH\bar{\rho}_M$ is constant during the radiation dominated era, when $a \propto \sqrt{t}$ and $H \propto a^{-2}$, and we are here assuming it is much less than one when $\bar{\rho}_M = \bar{\rho}_R$, it follows that for these wavelengths we have

$$\frac{q}{aH} \ll \frac{\bar{\rho}_M}{\bar{\rho}_R} \tag{6.3.1}$$

throughout the radiation-dominated era. (Recall that $\bar{\rho}_M \equiv \bar{\rho}_D + \bar{\rho}_B$ and $\bar{\rho}_R \equiv \bar{\rho}_\gamma + \bar{\rho}_\nu$.) On the other hand, during the matter-dominated era when $a \propto t^{2/3}$ and $H \propto a^{-3/2}$, it is $q^2\bar{\rho}_R/a^2H^2\bar{\rho}_M$ that remains constant, and since this quantity is assumed to be much less than one when $\bar{\rho}_M = \bar{\rho}_R$, for these wavelengths we have

$$\frac{q^2}{a^2 H^2} \ll \frac{\bar{\rho}_M}{\bar{\rho}_R} \tag{6.3.2}$$

throughout the matter-dominated era.

We cannot give a single analytic formula for these perturbations during the whole era from just after electron–positron annihilation until near the present, but fortunately we can find analytic solutions in two eras: first, the era when the perturbations are outside the horizon, and, second, the era when the energy density of the universe is dominated by non-relativistic matter. (The anisotropic inertia due to neutrinos can be neglected in both eras, because it is negligible outside the horizon, and irrelevant when the energy density of neutrinos is much less than that of matter.) For the long wavelengths considered in this section, that are still outside the horizon at the time that the matter density becomes equal to that of radiation, these eras *overlap*. This allows us to take the initial condition in the second era from the results for the first era in the period in which they overlap.

A. Outside the horizon

The perturbations of greatest interest, corresponding to adiabatic modes, are governed by Eqs. (6.2.39)–(6.2.44), with the fractional perturbations $\delta_{\alpha q}$ defined by Eq. (6.2.8), and with $\delta_{Bq} = \delta_{\gamma q}$. For $q/a \ll H$, this gives

$$\frac{d}{dt}\left(a^2 \psi_q\right) = -4\pi\,Ga^2\left[\bar\rho_D \delta_{Dq} + \left(\bar\rho_B + \frac{8}{3}\bar\rho_\gamma\right)\delta_{\gamma q} + \frac{8}{3}\bar\rho_\nu \delta_{\nu q}\right], \qquad (6.3.3)$$

$$\dot\delta_{\gamma q} = \dot\delta_{\nu q} = \dot\delta_{Dq} = -\psi_q\,, \qquad (6.3.4)$$

$$\frac{d}{dt}\left(\frac{(1+R)\delta u_{\gamma q}}{a}\right) = -\frac{1}{3a}\delta_{\gamma q}\,. \qquad (6.3.5)$$

$$\frac{d}{dt}\left(\frac{\delta u_{\nu q}}{a}\right) = -\frac{1}{3a}\delta_{\nu q}\,. \qquad (6.3.6)$$

These can be solved analytically, at least for some of the modes described in the previous section. Most importantly, for the adiabatic solutions with all $\delta_{\alpha q}$ equal outside the horizon, Eqs. (6.3.3) and (6.3.4) become

$$\frac{d}{dt}\left(a^2 \psi_q\right) = -4\pi\,Ga^2\left[\bar\rho_M + \frac{8}{3}\bar\rho_R\right]\delta_q\,, \qquad (6.3.7)$$

$$\dot\delta_q = -\psi_q\,, \qquad (6.3.8)$$

where

$$\delta_q \equiv \delta_{\gamma q} = \delta_{\nu q} = \delta_{Bq} = \delta_{Dq}\,, \qquad (6.3.9)$$

and we recall that $\bar\rho_M \equiv \bar\rho_B + \bar\rho_D$ and $\bar\rho_R \equiv \bar\rho_\gamma + \bar\rho_\nu$. Inserting Eq. (6.3.8) in (6.3.7) then gives a second-order differential equation for δ_q:

$$\frac{d}{dt}\left(a^2\frac{d}{dt}\delta_q\right) = 4\pi\,Ga^2\left[\bar\rho_M + \frac{8}{3}\bar\rho_R\right]\delta_q\,, \qquad (6.3.10)$$

283

To solve Eq. (6.3.10), it is very convenient to replace the dependent variable t with $y \equiv a/a_{\rm EQ} = \bar{\rho}_M/\bar{\rho}_R$, where $a_{\rm EQ}$ is the Robertson–Walker scale factor at matter–radiation equality. Then $\bar{\rho}_M = \rho_{\rm EQ}/y^3$ and $\bar{\rho}_R = \rho_{\rm EQ}/y^4$, where $\rho_{\rm EQ}$ is the common density of matter and radiation when they are equal. Using the Friedmann formula for the expansion rate, we have then

$$\frac{d}{dt} = \frac{H_{\rm EQ}}{\sqrt{2}} \frac{\sqrt{1+y}}{y} \frac{d}{dy}, \tag{6.3.11}$$

and Eq. (6.3.10) becomes

$$y\sqrt{1+y}\frac{d}{dy}\left(y\sqrt{1+y}\frac{d}{dy}\delta_q\right) - \frac{3}{2}\left(y+\frac{8}{3}\right)\delta_q = 0 \tag{6.3.12}$$

This has two independent solutions,

$$\delta^{(a)} = y^{-2}\left(16 + 8y - 2y^2 + y^3\right), \qquad \delta^{(b)} = y^{-2}\sqrt{1+y}.$$

We are looking for a solution that, as found for Mode 1 in the previous section, vanishes like $t \propto y^2$ for $t \to 0$, so we must take our solution to be proportional to $\delta^{(a)} - 16\delta^{(b)}$, which for $y \to 0$ approaches $5y^2/8$. Adjusting the normalization of this solution to match Eq. (6.2.27), we have then

$$\delta_q = \frac{4q^2\mathcal{R}_q^o}{5H_{\rm EQ}^2 a_{\rm EQ}^2 y^2}\left(16 + 8y - 2y^2 + y^3 - 16\sqrt{1+y}\right), \tag{6.3.13}$$

$$\delta u_{\gamma q} = -\frac{\sqrt{2}y}{3H_{\rm EQ}(1+R)}\int_0^y \frac{dy'}{\sqrt{1+y'}}\delta_q(y'), \tag{6.3.14}$$

$$\delta u_{vq} = -\frac{\sqrt{2}y}{3H_{\rm EQ}}\int_0^y \frac{dy'}{\sqrt{1+y'}}\delta_q(y'), \tag{6.3.15}$$

$$\psi_q = \frac{\sqrt{2}q^2\mathcal{R}_q^o}{5H_{\rm EQ}a_{\rm EQ}^2 y^4}\left(2\sqrt{1+y}\left(32+8y-y^3\right) - 64 - 48y\right). \tag{6.3.16}$$

With this normalization factor, $\mathcal{R}_q$ takes the time-independent value $\mathcal{R}_q^o$ outside the horizon.

B. The matter-dominated era

We can also solve Eqs. (6.2.39)–(6.2.44) analytically in the matter-dominated era, when $\bar{\rho}_M \gg \bar{\rho}_R$, whether or not the perturbation is outside the horizon. For simplicity, we will also assume that $\bar{\rho}_B \ll \bar{\rho}_D$. (Their ratio is actually about 0.2.) However, since $\bar{\rho}_\gamma$ is also much less than $\bar{\rho}_D$ in the matter-dominated era, we will not assume that $\bar{\rho}_B$ is negligible compared with $\bar{\rho}_\gamma$ in this era. Recall that in the matter–dominated era[1] $a \propto t^{2/3}$ and $8\pi G\bar{\rho}_M/3 = H^2 = 4/9t^2$. If we keep only the term in Eq. (6.2.39) proportional to $\bar{\rho}_D$, but now make no assumption about the relative magnitude of q/a and H, then Eqs. (6.2.39)–(6.2.44) become

$$\frac{d}{dt}\left(t^{4/3}\psi_q\right) = -\frac{2}{3}t^{-2/3}\delta_{Dq} . \tag{6.3.17}$$

$$\dot{\delta}_{\gamma q} - (q^2/a^2)\delta u_{\gamma q} = -\psi_q \tag{6.3.18}$$

$$\dot{\delta}_{vq} - (q^2/a^2)\delta u_{vq} = -\psi_q \tag{6.3.19}$$

$$\dot{\delta}_{Dq} = -\psi_q , \tag{6.3.20}$$

$$\frac{d}{dt}\left(t^{-2/3}(1+R)\delta u_{\gamma q}\right) = -\frac{1}{3}t^{-2/3}\delta_{\gamma q} , \tag{6.3.21}$$

$$\frac{d}{dt}\left(t^{-2/3}\delta u_{vq}\right) = -\frac{1}{3}t^{-2/3}\delta_{vq} , \tag{6.3.22}$$

where once again $R \equiv 3\bar{\rho}_B/4\bar{\rho}_R \propto a$.

There are two solutions to Eqs. (6.3.17) and (6.3.20), one with $\psi_q \propto t^{-1/3}$ and $\delta_{Dq} \propto t^{2/3}$, the other with $\psi_q \propto t^{-2}$ and $\delta_{Dq} \propto t^{-1}$. To evaluate the coefficients of these solutions in mode 1, we must compare these results with those given by Eqs. (6.3.13) and (6.3.16) in the era where both sets of results apply, the era (which exists because of our assumption of long wavelength) when both $q/a \ll H$ and $\bar{\rho}_R \ll \bar{\rho}_M$, i. e., $y \gg 1$. Eqs. (6.3.13) and (6.3.16) give in mode 1

$$\delta_{Dq} = \frac{4q^2 y \mathcal{R}_q^o}{5H_{EQ}^2 a_{EQ}^2} = \frac{9q^2 t^2 \mathcal{R}_q^o}{10a^2} , \tag{6.3.23}$$

$$\psi_q = \frac{-2\sqrt{2}q^2 \mathcal{R}_q^o}{5H_{EQ}a_{EQ}^2 y^{1/2}} = -\frac{3q^2 t \mathcal{R}_q^o}{5a^2} . \tag{6.3.24}$$

(The final expression in Eqs. (6.3.23)–(6.3.24) is derived using the result that for $\bar{\rho}_R \ll \bar{\rho}_D$, the Hubble rate is $H = 2/3t = H_{EQ}/\sqrt{2}y^{3/2}$, and eliminating H_{EQ}.) For $a \propto t^{2/3}$, these match the solution with $\delta_{Dq} \propto t^{2/3}$ and $\psi_q \propto t^{-1/3}$, while it is easy to see that Mode 2 gives the other solution in the matter-dominated era, the one with $\psi_q \propto t^{-2}$ and $\delta_{Dq} \propto t^{-1}$. We conclude that if Mode 1 is dominant outside the horizon, then the dark

---

[1] The zero of time is chosen in this subsection so that in the matter-dominated era $a \propto t^{2/3}$. In other words, t could here be defined as $2/3H$.

matter density perturbation and gravitational field perturbation are given in the matter-dominated era by Eqs. (6.3.23) and (6.3.24).

This is a good place to pause, and make contact with the more elementary treatment of Section 2.6. Under the assumption that the fluctuation in the Newtonian gravitational potential $\phi(\mathbf{x}, t)$ at around the time of last scattering is dominated by fluctuations in the dark matter density, its Fourier transform is given by using Poisson's equation, with the Laplacian replaced with $-q^2/a^2$:

$$\delta\phi(\mathbf{q}, t) = -4\pi G\left(a^2(t)/q^2\right)\delta\rho_D(\mathbf{q}, t) = -4\pi G\left(a^2(t)/q^2\right)\bar{\rho}_D(t)\delta_{Dq}(t) \ .$$

Using the Mode 1 solution Eq. (6.3.23) and the Friedmann equation $H^2 = (2/3t)^2 = 8\pi G\bar{\rho}_D/3$, we see that $\delta\phi(\mathbf{q}, t)$ takes the time-independent value

$$\delta\phi(\mathbf{q}) = -\frac{3}{5}\alpha(\mathbf{q})\mathcal{R}_q^o \ .$$

Then $\langle\delta\phi(\mathbf{q})\delta\phi(\mathbf{q}')\rangle = \mathcal{P}_\phi(q)\delta^3(\mathbf{q} + \mathbf{q}')$, with the correlation function $\mathcal{P}_\phi$ introduced in Section 2.6 equal to

$$\mathcal{P}_\phi(q) = \frac{9}{25}|\mathcal{R}_q^o|^2 \ .$$

The behavior $\mathcal{P}_\phi(q) = N_\phi^2 q^{-3}$ that was found in Section 2.6 to yield a temperature multipole coefficient $C_\ell = 8\pi N_\phi^2/9\ell(\ell+1)$ thus corresponds to the assumption that $|\mathcal{R}_q^o|^2 = N^2 q^{-3}$, with $N_\phi^2 = 9N^2/25$.

Even if we assume that the gravitational field is dominated by cold dark matter, we must still consider the perturbations to the photon density and the photon–baryon velocity as preparation for calculating the contribution of intrinsic temperature fluctuations and Doppler effect to the cosmic microwave background anisotropies in the next chapter. Equations (6.3.18) and (6.3.21) are a pair of coupled inhomogeneous first-order differential equations for the two unknowns $\delta_{\gamma q}$ and $\delta u_{\gamma q}$, with a forcing term proportional to ψ_q. Surprisingly, with ψ_q given by Eq. (6.3.24), there is a simple exact solution of these equations:

$$\delta_{\gamma q}^{(1)} = \frac{3q^2t^2(1+3R)\mathcal{R}_q^o}{5a^2(t^2q^2/a^2 + 2R)} \ , \quad \delta u_{\gamma q}^{(1)} = -\frac{3t^3q^2\mathcal{R}_q^o}{5a^2(t^2q^2/a^2 + 2R)} \ . \quad (6.3.25)$$

(We will not be concerned with the remaining equations, (6.3.19) and (6.3.22), which are needed only to calculate the neutrino perturbations.)

To the particular solution (6.3.25), we must add a suitable solution of the homogeneous version of Eqs. (6.3.18) and (6.3.21),

$$\dot{\delta}_{\gamma q}^{(2)} = (q^2/a^2)\delta u_{\gamma q}^{(2)}, \qquad \frac{d}{dt}\left(t^{-2/3}(1+R)\delta u_{\gamma q}^{(2)}\right) = -\frac{1}{3}t^{-2/3}\delta_{\gamma q}^{(2)},$$

$$(6.3.26)$$

or, eliminating the velocity potential,

$$\frac{d}{dt}\left(t^{-2/3}(1+R)a^2\frac{d}{dt}\delta_{\gamma q}^{(2)}\right) + \frac{q^2}{3}t^{-2/3}\delta_{\gamma q}^{(2)} = 0, \qquad (6.3.27)$$

with coefficients chosen so that $\delta_{\gamma q}^{(1)} + \delta_{\gamma q}^{(2)}$ matches the solution found earlier outside the horizon, when $q^2 t^2/a^2 \ll 1$.

Using the fact that in the matter dominated era $R \propto a \propto t^{2/3}$, we can find a general solution of Eq. (6.3.27) as a linear combination of the functions

$$F\left(\frac{1}{4} - \frac{1}{4}\sqrt{1-16\eta}, \frac{1}{4} + \frac{1}{4}\sqrt{1-16\eta}, \frac{1}{2}, -R\right),$$

$$\sqrt{R}\, F\left(\frac{3}{4} - \frac{1}{4}\sqrt{1-16\eta}, \frac{3}{4} + \frac{1}{4}\sqrt{1-16\eta}, \frac{3}{2}, -R\right),$$

where F is the Gauss hypergeometric function (also known as $_2F_1$), and η is the quantity

$$\eta \equiv \frac{3q^2 t^2}{4a^2 R},$$

which is time-independent during the matter-dominated era. Unfortunately, this does not provide much insight into the behavior of the solutions. Instead, at this point we will make the further assumption that the wavelength is only moderately long, in the sense that $\eta \gg 1$. That is, although q is small enough so that $q/aH \ll 1$ at matter–radiation equality, we assume that it is sufficiently large so that throughout the matter-dominated era we have

$$R \ll t^2 q^2/a^2 \ll \bar{\rho}_M/\bar{\rho}_R. \qquad (6.3.28)$$

Each term in this inequality is proportional to $t^{2/3}$ during the matter-dominated era, so if Eq. (6.3.28) holds at any time during this era, then it holds throughout it. This assumption will allow us to find solutions of Eq. (6.3.27) in terms of elementary functions. In the opposite case of extremely long wavelengths, for which $t^2 q^2/a^2 \ll R$ throughout the matter-dominated era, we can solve the general solution of Eqs. (6.3.18) and (6.3.21) as a power series in the quantity $t^2 q^2/a^2 R$. We are more interested here in the case of moderately long wavelengths, because as we will see in Section 6.5, it is this case that can be connected to the case of short wavelengths by a smooth extrapolation.

For perturbations satisfying the inequality (6.3.28), when the perturbation is outside the horizon and for some time after it re-enters the horizon, we will have $R \ll 1$. During the period when $R \ll 1$, the homogeneous equations (6.3.26) have the exact solution:

$$\delta_{\gamma q}^{(2)} = c_q \cos(\sqrt{3}qt/a) + d_q \sin(\sqrt{3}qt/a) , \tag{6.3.29}$$

$$\delta u_{\gamma q}^{(2)} = \frac{a}{\sqrt{3}q} \left[-c_q \sin(\sqrt{3}qt/a) + d_q \cos(\sqrt{3}qt/a) \right] , \tag{6.3.30}$$

with c_q and d_q constant. To this, we must add the inhomogeneous solution (6.3.25), in the limit $R \ll q^2 t^2/a^2$:

$$\delta_{\gamma q}^{(1)} = \frac{3(1 + 3R)\mathcal{R}_q^o}{5} , \qquad \delta u_{\gamma q}^{(1)} = -\frac{3t\mathcal{R}_q^o}{5} . \tag{6.3.31}$$

We can evaluate the constants c_q and d_q by requiring that for $qt/a \ll 1$ (which according to Eq. (6.3.28) also implies that $R \ll 1$), the total photon density perturbation $\delta_{\gamma q}^{(1)} + \delta_{\gamma q}^{(2)}$ must approach $\delta_{Dq} \to 9q^2 t \mathcal{R}_q^o/10a^2$. This gives

$$c_q = -\frac{3\mathcal{R}_q^o}{5} , \qquad d_q = 0 , \tag{6.3.32}$$

so that, as long as $R \ll 1$,

$$\delta_{\gamma q}^{(2)} = -\frac{3\mathcal{R}_q^o}{5} \cos(\sqrt{3}qt/a) , \tag{6.3.33}$$

Eventually R becomes non-negligible, but under the assumption (6.3.28), by then qt/a will be much larger than one, and we can solve the homogeneous equations (6.3.26) using the WKB approximation. Inspection of Eq. (6.3.27) suggests that for $qt/a \gg 1$ the density fluctuation will oscillate rapidly, with phase

$$\varphi \equiv \int_0^t \frac{q \, dt}{a\sqrt{3(1 + R)}} = \frac{\sqrt{3}qt}{a\sqrt{R}} \ln\left(\sqrt{R} + \sqrt{1 + R}\right) . \tag{6.3.34}$$

Using φ as the independent variable instead of t, and recalling that during the matter-dominated era $a \propto t^{2/3}$, Eq. (6.3.27) becomes

$$\frac{d^2 \delta_{\gamma q}}{d\varphi^2} + \frac{1}{2} \left(\frac{d \ln(1 + R)}{d\varphi} \right) \frac{d\delta_{\gamma q}}{d\varphi} + \delta_{\gamma q} = 0 .$$

288

We try for a solution of the form $\mathcal{A}e^{\pm i\varphi}$, with $\mathcal{A}$ varying slowly with φ, so that we can neglect the second derivative of $\mathcal{A}$ in the first term on the left and neglect the first derivative of $\mathcal{A}$ in the second term on the left, which is already small because of the factor $d\ln(1+R)/d\varphi$. This gives $d\mathcal{A}/d\varphi \simeq -(\mathcal{A}/4)d\ln(1+R)/d\varphi$, and hence the general WKB solutions of Eq. (6.3.27) are

$$\delta_{\gamma q}^{(2\pm)} \propto (1+R)^{-1/4}\exp(\pm i\varphi) .$$

Clearly, the linear combination of these solutions that merge smoothly with the results for $R \ll 1$ is obtained by replacing the argument of the cosine in Eq. (6.3.33) with φ, and multiplying with $(1+R)^{-1/4}$. Adding the inhomogeneous term (6.3.25), the total photon and baryon fractional density perturbations for moderately long wavelengths in the matter-dominated era are

$$\delta_{\gamma q} = \delta_{Bq} = \frac{3\mathcal{R}_q^o}{5}\left[1 + 3R - (1+R)^{-1/4}\cos\varphi\right] \tag{6.3.35}$$

We can then use Eqs. (6.3.18) and (6.3.24) to calculate the velocity potential

$$\delta u_{\gamma q} = \frac{3t\mathcal{R}_q^o}{5}\left[-1 + \frac{a}{\sqrt{3}qt(1+R)^{3/4}}\sin\varphi\right] . \tag{6.3.36}$$

(Here we neglect a term in square brackets of order Ra^2/t^2q^2.) As a check, note that early in the matter-dominated era, when $qt/a \ll 1$, we have $R \ll 1$, so $\varphi \to \sqrt{3}qt/a \to 0$, and hence Eq. (6.3.35) gives $\delta_{\gamma q} = \delta_{Bq} \to 9q^2t^2\mathcal{R}_q^o/10a^2$, in agreement with Eq. (6.3.23) and the condition that for adiabatic modes all $\delta_{\alpha q}$ are equal outside the horizon. This condition is satisfied by Eq. (6.3.35) even if the inequality (6.3.28) is not satisfied, as long as both $R \ll 1$ and $qt/a \ll 1$.

The results (6.3.35) and (6.3.36) for $\delta_{\gamma q} = \delta_{Bq}$ and $\delta u_{\gamma q}$ in the case of moderately long wavelengths all apply only up to the time of last scattering. On the other hand, to the extent that the energy density after last scattering is dominated by dark matter, δ_{Dq} and ψ_q are unaffected by the decoupling of radiation from the baryonic plasma, and continue to be given by Eq. (6.3.23) and (6.3.24) until either vacuum energy and possibly spatial curvature become significant or the perturbations become too strong to be treated as first-order perturbations.

6.4 Scalar perturbations – short wavelengths

We next consider adiabatic perturbations with wavelengths that are short enough so that they are already well within the horizon at the time of radiation–matter equality. As discussed at the end of Section 6.2, such perturbations are responsible for multipole moments of the cosmic microwave background anisotropies with $\ell \gg 140$, and also for the onset of gravitational condensations that lead to the formation of structures on the scale of galaxies or clusters of galaxies. Following the same reasoning as at the beginning of the previous section, for these wavelengths in the radiation-dominated era

$$\frac{q}{aH} \gg \frac{\bar{\rho}_M}{\bar{\rho}_R}, \tag{6.4.1}$$

and in the matter-dominated era

$$\frac{q^2}{a^2 H^2} \gg \frac{\bar{\rho}_M}{\bar{\rho}_R}, \tag{6.4.2}$$

where as before $\bar{\rho}_M \equiv \bar{\rho}_D + \bar{\rho}_B$ and $\bar{\rho}_R \equiv \bar{\rho}_\gamma + \bar{\rho}_\nu$.

As shown in Section 6.2, because we are considering adiabatic perturbations, the fractional perturbations $\delta_{\alpha q} \equiv \delta\rho_{\alpha q}/(\bar{\rho}_\alpha + \bar{p}_\alpha)$ are subject to the condition $\delta_{Bq} = \delta_{\gamma q}$. Consequently the perturbations are governed by Eqs. (6.2.39)–(6.2.44). Again, we cannot give a single analytic solution of these equations during the whole era from just after electron–positron annihilation until near the present, but fortunately we can find analytic solutions in two *overlapping* eras: first, the era when the energy density of the universe is dominated by radiation (photons and neutrinos), and second, the era when perturbations are well within the horizon.

A. The radiation-dominated era

We can solve Eqs. (6.2.39)–(6.2.44) analytically in the radiation-dominated era, when $\bar{\rho}_R \gg \bar{\rho}_M$. We will assume tentatively that at this time the photon and neutrino density fluctuations on the right-hand side of the gravitational field equation (6.2.39) dominate over the dark matter density fluctuations, an assumption we will check later in this section. In this era $a \propto t^{1/2}$, so Eqs. (6.2.39)–(6.2.44) take the form

$$\frac{d}{dt}\left(t\psi_q\right) = -\frac{32\pi Gt}{3}\left(\bar{\rho}_\gamma \delta_{\gamma q} + \bar{\rho}_\nu \delta_{vq}\right), \tag{6.4.3}$$

$$\dot{\delta}_{\gamma q} - (q^2/a^2)\delta u_{\gamma q} = -\psi_q, \tag{6.4.4}$$

$$\dot{\delta}_{Dq} = -\psi_q, \tag{6.4.5}$$

$$\frac{d}{dt}\left(t^{-1/2}\delta u_{\gamma q}\right) = -\frac{1}{3}t^{-1/2}\delta_{\gamma q}, \tag{6.4.6}$$

$$\dot{\delta}_{vq} - (q^2/a^2)\delta u_{vq} = -\psi_q , \tag{6.4.7}$$

$$\frac{d}{dt}\left(t^{-1/2}\delta u_{vq}\right) = -\frac{1}{3}t^{-1/2}\delta_{vq} . \tag{6.4.8}$$

We are interested in adiabatic solutions for which all $\delta_{\alpha q}$ and $\delta u_{\alpha q}$ become equal at early times, so since the differential equations here are the same for photons and neutrinos, these adiabatic solutions have

$$\delta_{\gamma q} = \delta_{vq} , \qquad \delta u_{\gamma q} = \delta u_{vq} .$$

For these modes, Eq. (6.4.3) now simplifies to

$$\frac{d}{dt}\left(t\psi_q\right) = -\frac{32\pi G\bar{\rho}Rt}{3}\delta_{\gamma q} = -\frac{1}{t}\delta_{\gamma q} .$$

Also, assuming that the cosmological perturbations are in the growing adiabatic mode, Mode 1, they satisfy the initial conditions (6.2.27)–(6.2.30): For $q/aH \ll 1$

$$\delta_{\gamma q} = \delta_{Bq} = \delta_{vq} \to \frac{q^2 t^2 R_q^o}{a^2} , \qquad \delta_{Dq} \to \frac{q^2 t^2 R_q^o}{a^2} , \tag{6.4.9}$$

$$\psi_q \to -\frac{tq^2 R_q^o}{a^2} , \qquad \delta u_{\gamma q} = \delta u_{vq} \to -\frac{2t^3 q^2 R_q^o}{9a^2} , \tag{6.4.10}$$

where R_q^o is the value of R_q outside the horizon. The reader can check that the solution of Eqs. (6.4.3)–(6.4.8) satisfying these initial conditions is[1]

$$\delta_{\gamma q} = \delta_{Bq} = \delta_{vq} = 3R_q^o\left(\frac{2}{\Theta}\sin\Theta - \left(1 - \frac{2}{\Theta^2}\right)\cos\Theta - \frac{2}{\Theta^2}\right) , \tag{6.4.11}$$

$$\psi_q = \frac{3R_q^o}{t}\left(\frac{2}{\Theta}\sin\Theta + \frac{2}{\Theta^2}\cos\Theta - \frac{2}{\Theta^2} - 1\right) , \tag{6.4.12}$$

$$\delta_{Dq} = -6R_q^o\int_0^\Theta\left(\frac{2}{\vartheta^3}\sin\vartheta + \frac{2}{\vartheta^4}\cos\vartheta - \frac{2}{\vartheta^4} - \frac{1}{\vartheta^2}\right)\vartheta\,d\vartheta , \tag{6.4.13}$$

$$\delta u_{\gamma q} = \delta u_{vq} = 4tR_q^o\left(\frac{\sin\Theta}{2\Theta} - \frac{1 - \cos\Theta}{\Theta^2}\right) , \tag{6.4.14}$$

---

[1] By replacing t with Θ as the dependent variable, Eqs. (6.4.3), (6.4.4), and (6.4.6) are put in the form of a parameter-free third-order system of differential equations for $\delta_{\gamma q} = \delta_{Bq} = \delta_{vq}$, $\delta u_{\gamma q}/t = \delta u_{vq}/t$ and $t\psi_q$. After finding the solution that matches the initial conditions (6.4.9), (6.4.10), we can solve Eq. (6.4.5) for δ_{Dq} by an integration. Aside from normalization, this solution for the various fluctuations is equivalent to that given for the Newtonian potential in a different gauge in Eq. (48) of S. Bashinsky & E. Bertschinger *Phys. Rev. D* **65**, 123008 (2002).

where

$$\Theta \equiv \frac{2qt}{\sqrt{3}a} \,. \tag{6.4.15}$$

Note that the fractional perturbations $\delta_{\alpha q}$ are all of the same order of magnitude for moderate values of Θ. This justifies the neglect of the matter terms in Eq. (6.2.39) when $\bar{\rho}_R$ is much greater than $\bar{\rho}_D$ and q/aH is not very much larger than unity. The condition for continuing to neglect the matter terms in Eq. (6.2.39) for perturbations deep inside the horizon, when Θ becomes large, will be discussed below.

B. Deep inside the horizon

We can also find a solution of Eqs. (6.2.39)–(6.2.44) when the wavelength is well within the horizon, in the sense that $q/a \gg H$, whether or not $\bar{\rho}_D$ is negligible compared with $\bar{\rho}_R$. For $q/a \gg H$ we can distinguish two different kinds of solutions: "slow modes," for which time derivatives yield factors of order H, and "fast modes," for which time derivatives acting on the perturbations yield factors of order of q/a, as well as other terms with factors of H instead. Eqs. (6.2.39)–(6.2.44) are a sixth-order system of differential equations, so they have six independent solutions. We are going to identify four independent fast solutions and two independent slow solutions, so we can be sure that there are no solutions other than what we have called fast and slow modes.

1. Fast modes

Up to now we have ignored the neutrino anisotropic inertia, but for the rapidly oscillating fast modes we must take into account its effect of the long neutrino mean free path in damping the neutrino density and velocity perturbations. In considering the fast modes deep inside the horizon, we shall simply assume that this damping allows us to ignore δ_{vq} in Eq. (6.2.39). Turning to the other perturbations in a fast mode with fractional rates of change of order q/a, Eq. (6.2.41) shows that δ_{Dq} is of order $\psi_q/(q/a)$, so the dark-matter term on the right-hand side of Eq. (6.2.39) is of order

$$4\pi G a^2 \bar{\rho}_D \psi_q/(q/a) \le 3H^2 a^2 \psi_q/2(q/a) \,,$$

while the left-hand side of Eq. (6.2.39) is of order $(q/a)a^2\psi_q$, and the dark matter term is therefore less than the left-hand side of Eq. (6.2.39) by a factor less than of order $\le H^2/(q/a)^2$, which deep in the horizon is much less than unity. Dropping the dark matter term in Eq. (6.2.39), we see that $\delta_{\gamma q}$ must be at least of order $(q/a)\psi_q/H^2$. Eq. (6.2.43) then shows that $\delta u_{\gamma q}$ is of order $\delta_{\gamma q}/(q/a)$, and hence at least of order ψ_q/H^2. Both terms on the left-hand side of Eqs. (6.2.42) are then larger than ψ_q by factors of order $q^2/a^2H^2 \gg 1$, so we can drop ψ_q on the right-hand side. (That is, because the wavelength is short, pressure gradients exert a much larger force on the baryon–photon plasma than gravitation.) Hence if we neglect all terms in Eqs. (6.2.40), (6.2.43), (6.2.39), and (6.2.41) that are suppressed by factors $H^2/(q/a)^2$ (but *not* terms arising from derivatives of a that are only suppressed by factors $H/(q/a)$), these equations become

$$\dot{\delta}_{\gamma q} = (q^2/a^2)\delta u_{\gamma q} \tag{6.4.16}$$

$$\frac{d}{dt}\left(\frac{(1+R)\delta u_{\gamma q}}{a}\right) = -\frac{1}{3a}\delta_{\gamma q}, \tag{6.4.17}$$

$$\frac{d}{dt}\left(a^2\psi_q\right) = -\frac{16\pi\,Ga^2}{3}\bar{\rho}_\gamma(R+2)\delta_{\gamma q}, \tag{6.4.18}$$

$$\dot{\delta}_{Dq} = -\psi_q, \tag{6.4.19}$$

where, as before, $R \equiv 3\bar{\rho}_B/4\bar{\rho}_\gamma$. Equations (6.4.16) and (6.4.17) have two independent solutions for $\delta_{\gamma q}$ and $\delta u_{\gamma q}$. Given these solutions, and looking only for fast modes, Eq. (6.4.18) then has a unique solution for ψ_q, and Eq. (6.4.19) then has a unique solution for δ_{Dq}. (Possible constant terms that might be included in the solutions for $a^2\psi_q$ or δ_{Dq} would contribute to the slow modes, not the fast modes.) Together with the two strongly damped solutions of Eqs. (6.2.42) and (6.2.44) for the neutrino perturbations, there are four independent fast modes, as promised.

By eliminating $\delta u_{\gamma q}$ from Eqs. (6.4.16) and (6.4.17), we obtain a second-order differential equation for $\delta_{\gamma q}$ alone

$$\frac{d}{dt}\left(a(1+R)\frac{d}{dt}\delta_{\gamma q}\right) + \frac{q^2}{3a}\delta_{\gamma q} = 0. \tag{6.4.20}$$

If a and R were constant, this would be just the wave equation for a sound wave, with physical wave number q/a and velocity[2] $v_s = 1/\sqrt{3(1+R)}$. With a and R varying at a relatively slow fractional rate H, Eq. (6.4.20) can be solved for $q/a \gg H$ by the WKB approximation. For this purpose, we introduce a new independent variable, the phase $\varphi \equiv q\int_0^t dt/a\sqrt{3(1+R)}$, and rewrite Eq. (6.4.20) as

$$\frac{d^2\delta_{\gamma q}}{d\varphi^2} + \frac{1}{2}\frac{d\ln(1+R)}{d\varphi}\frac{d\delta_{\gamma q}}{d\varphi} + \delta_{\gamma q} = 0.$$

---

[2] Note that the condition of constant entropy gives $d\rho_B/\bar{\rho}_B = d\rho_\gamma/(\bar{\rho}_\gamma + \bar{p}_\gamma) = (3/4)d\rho_\gamma/\bar{\rho}_\gamma$, so $v_s^2 = dp/d\rho = d\rho_\gamma/3(d\rho_\gamma + d\rho_B) = 1/3(1+R)$.

Writing $\delta_{\gamma q} = A \exp(\pm i\varphi)$, and neglecting $d^2 A/d\varphi^2$ in the first term on the left and neglecting $dA/d\varphi$ in the second term (which is already small because of the factor $d\ln(1+R)/d\varphi$), we find $dA/d\varphi = -(A/4)d\ln(1+R)/d\varphi$, so the WKB solutions are[3]

$$\delta_{\gamma q}^{\pm} = (1+R)^{-1/4} \exp\left[\pm iq \int_0^t \frac{dt}{a\sqrt{3(1+R)}}\right] \qquad (6.4.21)$$

There is a further complication that must be taken into account for fast modes, though it is not important for slow modes. The amplitude of a sound wave whose physical wave number k is larger than the inverse mean path of the particles in a relativistic medium is damped by viscosity and heat conduction, at a rate given in general by[4]

$$\Gamma = \frac{k^2}{2(\rho+p)} \left\{ \zeta + \frac{4}{3}\eta + \chi \left(\frac{\partial \rho}{\partial T}\right)_n^{-1} \right.$$

$$\left. \left[\rho + p - 2T \left(\frac{\partial p}{\partial T}\right)_n + v_s^2 T \left(\frac{\partial \rho}{\partial T}\right)_n - \frac{n}{v_s^2} \left(\frac{\partial p}{\partial n}\right)_T \right] \right\}, \qquad (6.4.22)$$

where η, χ, and ζ are the coefficients of shear viscosity, heat conduction, and bulk viscosity, respectively, defined in Appendix B; v_s is the sound speed; n is any number density on which the fluid properties may depend; and subscripts indicate the quantities held constant in taking partial derivatives. For the baryon–photon fluid, for which n is the baryon number density, we have $\rho = \rho_B + \rho_\gamma$, $p = \rho_\gamma/3$, with $\rho_B \propto n$ and $\rho_\gamma \propto T^4$. Also, as we have seen $v_s^2 = 1/\sqrt{3(1+R)}$. Hence, (setting $k = q/a$) the damping rate becomes

$$\Gamma = \frac{3q^2}{8a^2 \bar{\rho}_\gamma (1+R)} \left\{ \zeta + \frac{4}{3}\eta + \frac{\chi TR^2}{3(1+R)} \right\}. \qquad (6.4.23)$$

The viscosity and heat conduction coefficients for photons interacting with a non-relativistic plasma with mean free time $t_\gamma = 1/\sigma_T n_e$ are

$$\eta = \frac{16}{45} \bar{\rho}_\gamma t_\gamma, \quad \chi T = \frac{4}{3} \bar{\rho}_\gamma t_\gamma, \quad \zeta = 0 \qquad (6.4.24)$$

---

[3] P. J. E. Peebles and J. T. Yu, *Astrophys. J.* **162**, 815 (1970). R. A. Sunyaev and Ya. B. Zel'dovich, *Astrophys. & Space Sci.* **7**, 3 (1970).
[4] S. Weinberg, *Astrophys. J.* **168**, 175 (1971).

so in this case the damping rate is[5]

$$\Gamma = \frac{q^2 t_\gamma}{6a^2(1+R)} \left\{ \frac{16}{15} + \frac{R^2}{1+R} \right\}. \tag{6.4.25}$$

The effect is to replace Eq. (6.4.21) for the fast mode amplitudes with

$$\delta_{\gamma q}^\pm = (1+R)^{-1/4} \exp\left[\pm iq \int_0^t \frac{dt}{a\sqrt{3(1+R)}} - \int_0^t \Gamma dt \right]. \tag{6.4.26}$$

This damping of the fast modes is known as *Silk damping*.[6] With this result, $\delta u_{\gamma q}^\pm$ for these fast modes can be obtained from Eq. (6.4.16), $\psi_q^\pm$ can be obtained from Eq. (6.4.18), and then $\delta_{Dq}^\pm$ can be obtained from Eq. (6.4.19).

2. Slow modes

For solutions whose fractional rate of change is of order $H = O(1/t)$, we can run through the same sort of counting of powers of q/aH as for fast modes, but with very different results. From Eqs. (6.2.41), (6.2.43), and (6.2.44), we see that δ_{Dq} is of order ψ_q/H while $\delta_{\gamma q}$ and δ_{vq} are of order $H\delta u_{\gamma q}$ and $H\delta u_{vq}$, respectively. The terms $\dot{\delta}_{\gamma q}$ and $\dot{\delta}_{vq}$ on the left-hand side of Eqs. (6.2.40) and (6.2.42) are then of order $H^2\delta u_{\gamma q}$ and $H^2\delta u_{vq}$, and hence are less than the terms $(q^2/a^2)\delta u_{\gamma q}$ and $(q^2/a^2)\delta u_{vq}$ by factors of order $H^2 a^2/q^2$, and may be dropped, giving instead

$$(q^2/a^2)\delta u_{\gamma q} = (q^2/a^2)\delta u_{vq} = \psi_q, \tag{6.4.27}$$

so Eqs. (6.2.43) and (6.2.44) show that $\delta_{\gamma q}$ and δ_{vq} are of order $(a^2 H/q^2)\psi_q$.

The ratios of the photon and neutrino terms on the right-hand side of Eq. (6.2.39) to the dark matter term are then of order

$$\frac{\text{photons \& neutrinos}}{\text{dark matter}} = O\left[\left(\frac{\bar{\rho}_B + 8\bar{\rho}_R/3}{\bar{\rho}_D} \right) \left(\frac{a^2 H^2}{q^2} \right) \right],$$

---

[5]This damping rate was first calculated by N. Kaiser, *Mon. Not. Roy. Astron. Soc.* **202**, 1169 (1983), and is derived here in Appendix H. The formulas for the shear viscosity and heat conduction coefficients are obtained by comparing formulas (6.4.23) and (6.4.25) for the acoustic damping rate, taking into account that the bulk viscosity vanishes because energy and momentum are transported by relativistic particles; see ref. 4. The damping rate calculated by ref.4 had given the correct values for χ and ζ, but it gave a value for η that was 3/4 the correct value of Kaiser, quoted here in Eq. (6.4.24). This was because its results were based on calculations of L. H. Thomas, *Quart. J. Math. (Oxford)* **1**, 239 (1930), that had assumed isotropic scattering and ignored photon polarization. (The same value for η had been given earlier by C. Misner, *Astrophys. J.* **151**, 431 (1968).) Kaiser's results are calculated using the correct differential cross section for Thomson scattering and take photon polarization into account, and therefore supersede the value for η quoted in ref. 4 and Chapter 15 of G&C.

[6]J. Silk, *Nature* **215**, 1155 (1972).

where as before $\bar{\rho}_R \equiv \bar{\rho}_\gamma + \bar{\rho}_\nu$. The term in the numerator proportional to $\bar{\rho}_B$ contributes much less than unity to this ratio, because $\bar{\rho}_B/\bar{\rho}_D \approx 1/5$ and we are now assuming that $aH/q \ll 1$. According to Eq. (6.4.2), the contribution of the term proportional to $\bar{\rho}_R$ is much less than unity throughout the matter-dominated era. It actually begins to be much less than unity during the radiation-dominated era, when the ratio of radiation density to dark matter density falls below the critical value $(\bar{\rho}_R/\bar{\rho}_D)_{\mathrm{crit}}$, given by

$$\left(\frac{\bar{\rho}_R}{\bar{\rho}_D}\right)_{\mathrm{crit}} = \left(\frac{3}{8}\right)^{1/3} \left(\frac{q}{aH}\frac{\bar{\rho}_R}{\bar{\rho}_D}\right)^{2/3} . \tag{6.4.28}$$

According to Eqs. (6.4.1), the right-hand side is constant and much greater than unity throughout the radiation-dominated era for the wavelengths considered in this section, so $\bar{\rho}_R/\bar{\rho}_D$ will fall below this critical value well before radiation–matter equality. (That is, strong pressure forces keep perturbations to the baryon–photon plasma density small enough so that their effect on the gravitational field is negligible once $\bar{\rho}_\gamma/\bar{\rho}_B$ falls below the critical value (6.4.28), even though the unperturbed radiation density is at first still larger than the dark matter density.) From then on the photon and neutrino terms may be neglected in Eq. (6.2.39) for the slow modes, yielding

$$\frac{d}{dt}\left(a^2 \psi_q\right) = -4\pi G \bar{\rho}_D a^2 \delta_{Dq} . \tag{6.4.29}$$

The remaining equations, (6.2.41), (6.2.43) and (6.2.44) are unchanged:

$$\dot{\delta}_{Dq} = -\psi_q , \tag{6.4.30}$$

$$\frac{d}{dt}\left(\frac{(1+R)\delta u_{\gamma q}}{a}\right) = -\frac{1}{3a}\delta_{\gamma q} , \tag{6.4.31}$$

$$\frac{d}{dt}\left(\frac{\delta u_{\nu q}}{a}\right) = -\frac{1}{3a}\delta_{\nu q} , \tag{6.4.32}$$

Using Eq. (6.4.30) in Eq. (6.4.29) yields a second-order differential equation for δ_{Dq}:[7]

$$\frac{d}{dt}\left(a^2 \frac{d\delta_{Dq}}{dt}\right) = 4\pi G a^2 \bar{\rho}_D \delta_{Dq} . \tag{6.4.33}$$

---

[7] Eq. (6.4.33) was first derived by P. Mészáros, *Astron. Astrophys.* **37**, 225 (1974), who simply ignored fluctuations in the radiation energy density. The argument given here for the neglect of perturbations in the radiation density in Eq. (6.4.33) was given by S. Weinberg, *Astrophys. J.* **581**, 810 (2002). It applies only to the slow mode part of the solution; in the fast mode it is the perturbations in the *dark matter* density that become negligible for small wavelength. This paper also gives comments on other attempts to justify the neglect of perturbations in the radiation density in Eq. (6.4.33).

It is convenient once again to convert the independent variable from t to $y \equiv a/a_{EQ} = \bar{\rho}_M/\bar{\rho}_R$, using the Friedmann equation

$$\frac{\dot{y}^2}{y^2} = \frac{8\pi G}{3}(\bar{\rho}_M + \bar{\rho}_R) = \frac{8\pi G\rho_{EQ}}{3}\left(y^{-3} + y^{-4}\right), \tag{6.4.34}$$

with ρ_{EQ} the values of $\bar{\rho}_M$ and $\bar{\rho}_R$ when they are equal. Then (6.4.33) becomes what is sometimes known as the *Mészáros equation*:

$$y(1+y)\frac{d^2\delta_D}{dy^2} + \left(1 + \frac{3y}{2}\right)\frac{d\delta_D}{dy} - \frac{3}{2}(1-\beta)\,\delta_D = 0\,, \tag{6.4.35}$$

where $\beta \equiv \bar{\rho}_B/\bar{\rho}_M = \Omega_B/\Omega_M$. The independent solutions of Eq. (6.4.35) for $\beta = 0$ were given by Mészáros,[7] and by Groth and Peebles[8]

$$\delta_{Dq}^{(1)} = 1 + \frac{3y}{2}\,, \qquad \delta_{Dq}^{(2)} = \left(1 + \frac{3y}{2}\right)\ln\left(\frac{\sqrt{1+y}+1}{\sqrt{1+y}-1}\right) - 3\sqrt{1+y}\,.$$
$$\tag{6.4.36}$$

Subsequently Hu and Sugiyama[9] gave two independent solutions for general β:

$$\delta_{Dq} \propto (1+y)^{-\alpha_{\pm}} F\left(\alpha_{\pm}, \alpha_{\pm} + \frac{1}{2}, 2\alpha_{\pm} + \frac{1}{2}; \frac{1}{1+y}\right)$$

where F is the Gauss hypergeometric function and

$$\alpha_{\pm} = \frac{1 \pm \sqrt{1 + 24\beta}}{4}\,,$$

In order to obtain our final results in an analytic form, we will continue to drop corrections proportional to $\beta \equiv \bar{\rho}_B/\bar{\rho}_M \approx 1/6$, while keeping those proportional to $R \equiv 3\bar{\rho}_B/4\bar{\rho}_\gamma$, so we shall use the slow solutions (6.4.36) for $\beta = 0$. (Corrections for the finite value of β are discussed in the following section.) From these two solutions, we can find unique corresponding slow solutions for ψ_q, $\delta u_{\gamma q}$, δu_{vq}, $\delta_{\gamma q}$ and δ_{vq} by successive use of Eqs. (6.4.30), (6.4.27), (6.4.31) and (6.4.32). (The neutrino perturbations are of no known observational interest.) We have thus found two slow modes, giving six in all.

C. Matching

Fortunately, for small wavelength there is an overlap in the two eras in which we have found solutions for δ_D, etc., satisfying *both* conditions $q/a \gg \dot{a}/a$

[8]E. J. Groth and P. J. E. Peebles, *Astron. Astrophys.* **41**, 143 (1975).
[9]W. Hu and N. Sugiyama, *Astrophys. J.* **471**, 542 (1996).

and $\bar{\rho}_M \ll \bar{\rho}_R$. In this period the solution (6.4.11)–(6.4.14) found for the radiation-dominated era can be decomposed into a fast and a slow mode. Since the variable Θ defined by Eq. (6.4.15) is here much larger than unity, we have

$$\delta_{\gamma q}^{\text{fast}} = \delta_{Bq}^{\text{fast}} = \delta_{vq}^{\text{fast}} = 3\mathcal{R}_q^o \left(\frac{2}{\Theta} \sin \Theta - \left(1 - \frac{2}{\Theta^2}\right) \cos \Theta \right),$$

$$\rightarrow -3\mathcal{R}_q^o \cos \Theta \tag{6.4.37}$$

$$\psi_q^{\text{fast}} = \frac{3\mathcal{R}_q^o}{t} \left(\frac{2}{\Theta} \sin \Theta + \frac{2}{\Theta^2} \cos \Theta \right) \rightarrow \frac{6\mathcal{R}_q^o}{t\Theta} \sin \Theta, \tag{6.4.38}$$

$$\delta_{Dq}^{\text{fast}} = -6\mathcal{R}_q^o \int_{\Theta_1}^{\Theta} \left(\frac{2}{\vartheta^3} \sin \vartheta + \frac{2}{\vartheta^4} \cos \vartheta \right) \vartheta \, d\vartheta \rightarrow \frac{12\mathcal{R}_q^o}{\Theta^2} \cos \Theta, \tag{6.4.39}$$

$$\delta u_{\gamma q}^{\text{fast}} = \delta u_{vq}^{\text{fast}} = 4t\mathcal{R}_q^o \left(\frac{\sin \Theta}{2\Theta} + \frac{\cos \Theta}{\Theta^2} \right) \rightarrow \frac{2t\mathcal{R}_q^o}{\Theta} \sin \Theta, \tag{6.4.40}$$

and

$$\delta_{\gamma q}^{\text{slow}} = \delta_{Bq}^{\text{slow}} = \delta_{vq}^{\text{slow}} = -\frac{6\mathcal{R}_q^o}{\Theta^2}, \tag{6.4.41}$$

$$\psi_q^{\text{slow}} = -\frac{3\mathcal{R}_q^o}{t} \left(\frac{2}{\Theta^2} + 1 \right) \rightarrow -\frac{3\mathcal{R}_q^o}{t}, \tag{6.4.42}$$

$$\delta_{Dq}^{\text{slow}} = -6\mathcal{R}_q^o \int_0^{\Theta_1} \left(\frac{2}{\vartheta^3} \sin \vartheta + \frac{2}{\vartheta^4} \cos \vartheta - \frac{2}{\vartheta^4} - \frac{1}{\vartheta^2} \right) \vartheta \, d\vartheta$$

$$+ 6\mathcal{R}_q^o \left[-\frac{1}{\Theta^2} + \frac{1}{\Theta_1^2} + \ln \left(\Theta/\Theta_1 \right) \right]$$

$$\rightarrow 6\mathcal{R}_q^o \left(-\frac{1}{2} + \gamma + \ln \Theta \right), \tag{6.4.43}$$

$$\delta u_{\gamma q}^{\text{slow}} = \delta u_{vq}^{\text{slow}} = -\frac{4t\mathcal{R}_q^o}{\Theta^2}, \tag{6.4.44}$$

where again $\Theta \equiv 2qt/\sqrt{3}a$, Θ_1 is any constant in the range $1 \ll \Theta_1 \ll \Theta$, and $\gamma = 0.5772\ldots$ is the Euler constant.[10]

---

[10]To evaluate the asymptotic limit of the integral in Eq. (6.4.43), we can rewrite this integral as the sum of three terms, each of which converges at $\vartheta = 0$:

$$\int_0^{\Theta_1} \left(\frac{2}{\vartheta^3} \sin \vartheta + \frac{2}{\vartheta^4} \cos \vartheta - \frac{2}{\vartheta^4} - \frac{1}{\vartheta^2} \right) \vartheta \, d\vartheta = \int_0^{\Theta_1} \left(\frac{1}{\vartheta^2} \sin \vartheta + \frac{2}{\vartheta^3} \cos \vartheta - \frac{2}{\vartheta^3} \right) d\vartheta$$

$$+ \int_0^{\Theta_1} \left(\frac{\sin \vartheta}{\vartheta} - \frac{1}{1+\vartheta} \right) \frac{d\vartheta}{\vartheta} + \int_0^{\Theta_1} \left(\frac{1}{1+\vartheta} - 1 \right) \frac{d\vartheta}{\vartheta}.$$

Matching the solutions in the radiation-dominated era and deep within the horizon is straightforward for the fast modes. In the radiation-dominated era, when $R \ll 1$, damping is negligible (because the mean free time is very short), and $a \propto \sqrt{t}$, the argument of the cosine in Eq. (6.4.37) may be expressed as the integral

$$q \int_0^t \frac{dt}{a\sqrt{3(1+R)}} \rightarrow \frac{2qt}{a\sqrt{3}} = \Theta \,.$$

Hence the linear combination of the two fast solutions (6.4.26) that fits smoothly with the result (6.4.37) is:

$$\delta_{\gamma q}^{\text{fast}} = -\frac{3\mathcal{R}_q^o}{(1+R)^{1/4}}\, e^{-\int_0^t \Gamma dt}\, \cos\left(\int_0^t \frac{q\,dt}{a\sqrt{3(1+R)}}\right). \qquad (6.4.45)$$

By a successive use of Eqs. (6.4.16), (6.4.18), and (6.4.19) (and ignoring the time dependence of all factors except the rapidly oscillating sines and cosines), we then also find that for $q/a \gg H$:

$$\delta u_{\gamma q}^{\text{fast}} = \frac{a\sqrt{3}\mathcal{R}_q^o}{q(1+R)^{3/4}}\, e^{-\int_0^t \Gamma dt}\, \sin\left(\int_0^t \frac{q\,dt}{a\sqrt{3(1+R)}}\right). \qquad (6.4.46)$$

$$\psi_q^{\text{fast}} = 16\sqrt{3}\pi\, G\bar{\rho}_\gamma (2+R)(1+R)^{1/4}(a/q)\mathcal{R}_q^o e^{-\int_0^t \Gamma dt}\, \sin\left(\int_0^t \frac{q\,dt}{a\sqrt{3(1+R)}}\right), \qquad (6.4.47)$$

$$\delta_{Dq}^{\text{fast}} = 48\pi\, G\bar{\rho}_\gamma (2+R)(1+R)^{3/4}(a/q)^2\mathcal{R}_q^o\, e^{-\int_0^t \Gamma dt}\, \cos\left(\int_0^t \frac{q\,dt}{a\sqrt{3(1+R)}}\right). \qquad (6.4.48)$$

We will use Eqs. (6.4.45)–(6.4.48) in dealing with baryon acoustic oscillations in Section 8.1.

The reader can easily check that in the overlap era, when the universe is radiation dominated, the perturbation is deep inside the horizon, and damping is negligible, Eqs. (6.4.45) and (6.4.46) give the same results for $\delta_{\gamma q}^{\text{fast}}$ and $\delta u_{\gamma q}^{\text{fast}}$ as Eqs. (6.4.37) and (6.4.40). On the other hand, Eqs. (6.4.47) and (6.4.48) give results for ψ_q^{fast} and $\delta_{Dq}^{\text{fast}}$ that differ from

---

For large Θ_1, these three integrals approach the values $-1/2$, $1-\gamma$, and $-\ln \Theta_1$, respectively, giving a total of $1/2 - \gamma - \ln \Theta_1$, which when combined with the term on the second line of Eq. (6.4.43) yields the quoted result.

Eqs. (6.4.38) and (6.4.39) by a factor $\bar{\rho}_\gamma/\bar{\rho}_R$. This is because in deriving Eqs. (6.4.38) and (6.4.39) we treated the neutrinos as a perfect fluid throughout the radiation-dominated era, although this is valid only outside the horizon when anisotropic inertia is negligible, while in deriving Eqs. (6.4.47) and (6.4.48) we assumed that deep inside the horizon neutrino density fluctuations are so damped by anisotropic inertia that they can be neglected as a source of gravitational field perturbations. This discrepancy is a small price to pay for the simplicity gained by these approximations, especially since it will turn out that the slow contributions to ψ_q and δ_{Dq} are much larger than the fast contributions.

Next let us consider the slow modes. Here there is a complication in matching solutions in the radiation-dominated era and deep inside the horizon. In deriving Eqs. (6.4.41)–(6.4.43) we have assumed that in the radiation dominated era we can neglect perturbations in the baryon and dark matter densities as a source of the gravitational field perturbations, but in exploring the solutions deep inside the horizon we found that this assumption is violated for the slow modes once the radiation/dark matter density ratio drops below the limit (6.4.28), even if this ratio is still much larger than unity. Indeed, we can see from Eqs. (6.4.41) and (6.4.43) that for large Θ, $\delta_{Dq}^{\text{slow}}/\delta_{\gamma q}^{\text{slow}} \to -\Theta^2 \ln \Theta$, so the assumption under which Eqs. (6.4.41)–(6.4.44) were derived breaks down once $\Theta^2 \ln \Theta$ becomes comparable to $\bar{\rho}_R/\bar{\rho}_D$, which is close to when $\bar{\rho}_\gamma/\bar{\rho}_D$ falls below the critical value (6.4.28). Therefore we have to interpolate between the results (6.4.41)–(6.4.44), which are valid early in the radiation-dominated era, when $q^2 t^2/a^2 \ll \bar{\rho}_R/\bar{\rho}_D$, and the results later in the radiation-dominated era and in the matter-dominated era, when the Mészáros equation (6.4.35) applies.

This is easiest for the dark matter density perturbation, because its time-dependence turns out to have the same form deep in the horizon in the radiation-dominated era both before and after $\bar{\rho}_R/\bar{\rho}_D$ falls below the critical value (6.4.28). Before this time in the radiation-dominated era $\delta_{Dq}^{\text{slow}}$ is given by Eq. (6.4.43). After this time, the gravitational field perturbation becomes dominated by dark matter, and the dark matter density perturbation becomes a linear combination of the solutions (6.4.36) of the Mészáros equation, which in the radiation-dominated era when $y \ll 1$ become

$$\delta_{Dq}^{(1)} \to 1 , \quad \delta_{Dq}^{(2)} \to -\ln(y/4) - 3 . \tag{6.4.49}$$

The linear combination of these two solutions that fits smoothly with Eq. (6.4.43) is then

$$\delta_{Dq}^{\text{slow}} = 6\mathcal{R}_q^0 \left\{ \left[-\frac{7}{2} + \gamma + \ln\left(\frac{4q\sqrt{2}}{\sqrt{3}H_{\text{EQ}}a_{\text{EQ}}} \right) \right] \delta_{Dq}^{(1)} - \delta_{Dq}^{(2)} \right\} , \tag{6.4.50}$$

where H_{EQ} and a_{EQ} are the expansion rate and Robertson–Walker scale factor at matter–radiation equality. (Here we use $t = 1/2H = y^2/\sqrt{2}H_{EQ}$.) The slow part of the gravitational field, radiation velocity potential, and radiation density perturbations are given by a successive use of Eqs. (6.4.30), (6.4.27), and (6.4.31) as

$$\psi_q^{\text{slow}} = -\delta_{Dq}^{\text{slow}} \tag{6.4.51}$$

$$\delta u_{\gamma q}^{\text{slow}} = -(a^2/q^2)\dot{\delta}_{Dq}^{\text{slow}} , \tag{6.4.52}$$

$$\delta_{\gamma q}^{\text{slow}} = \frac{3a}{q^2}\frac{d}{dt}\left(a(1+R)\frac{d}{dt}\delta_{Dq}^{\text{slow}}\right) . \tag{6.4.53}$$

In particular, in the matter-dominated era we have $y \gg 1$, $\delta_{Dq}^{(1)} \to 3y/2$, and $\delta_{Dq}^{(2)} \to 4/15y^{3/2}$, so Eqs. (6.4.50)–(6.4.53) become

$$\delta_{Dq}^{\text{slow}} \to \frac{9\mathcal{R}_q^0 a}{a_{EQ}}\left[-\frac{7}{2} + \gamma + \ln\left(\frac{4\kappa}{\sqrt{3}}\right)\right] , \tag{6.4.54}$$

$$\psi_q^{\text{slow}} \to -\frac{6\mathcal{R}_q^0 a}{a_{EQ}t}\left[-\frac{7}{2} + \gamma + \ln\left(\frac{4\kappa}{\sqrt{3}}\right)\right] . \tag{6.4.55}$$

$$\delta u_{\gamma q}^{\text{slow}} \to -\frac{6\mathcal{R}_q^0 a^3}{a_{EQ}tq^2}\left[-\frac{7}{2} + \gamma + \ln\left(\frac{4\kappa}{\sqrt{3}}\right)\right] , \tag{6.4.56}$$

$$\delta_{\gamma q}^{\text{slow}} \to \frac{6\mathcal{R}_q^0 a^3(1+3R)}{a_{EQ}t^2q^2}\left[-\frac{7}{2} + \gamma + \ln\left(\frac{4\kappa}{\sqrt{3}}\right)\right] . \tag{6.4.57}$$

We have here again introduced a dimensionless rescaled wave number

$$\kappa \equiv \frac{q\sqrt{2}}{a_{EQ}H_{EQ}} = \frac{(q/a_0)\sqrt{\Omega_R}}{H_0\Omega_M} = \frac{19.3(q/a_0)[\text{Mpc}^{-1}]}{\Omega_M h^2} , \tag{6.4.58}$$

in which we have used the relations $H_{EQ} = \sqrt{2}(H_0\sqrt{\Omega_M})(a_0/a_{EQ})^{3/2}$ and $a_0/a_{EQ} = \Omega_M/\Omega_R$. In terms of the wave number q_{EQ} introduced in Section 6.2, for which the perturbation just comes into the horizon at matter-radiation equality, we have $\kappa = \sqrt{2}q/q_{EQ}$, so the assumption of short wavelength made in this section essentially amounts to the condition that $\kappa \gg 1$.

The full solution up to the time of recombination is given by adding the contributions of the fast mode and slow mode:

$$
\delta_{Dq} \rightarrow \frac{9\mathcal{R}_q^0 a}{a_{EQ}} \left[-\frac{7}{2} + \gamma + \ln\left(\frac{4\kappa}{\sqrt{3}}\right) \right]
$$
$$
+ 48\pi G \bar{\rho}_\gamma (2+R)(1+R)^{3/4} (a/q)^2 \mathcal{R}_q^o
$$
$$
\times e^{-\int_0^t \Gamma dt} \cos\left(\int_0^t \frac{q\,dt}{a\sqrt{3(1+R)}} \right) , \qquad (6.4.59)
$$

$$
\psi_q \rightarrow -\frac{6\mathcal{R}_q^0 a}{a_{EQ}t} \left[-\frac{7}{2} + \gamma + \ln\left(\frac{4\kappa}{\sqrt{3}}\right) \right]
$$
$$
+ 16\sqrt{3}\pi G \bar{\rho}_\gamma (2+R)(1+R)^{1/4} (a/q) \mathcal{R}_q^o
$$
$$
\times e^{-\int_0^t \Gamma dt} \sin\left(\int_0^t \frac{q\,dt}{a\sqrt{3(1+R)}} \right) , \qquad (6.4.60)
$$

$$
\delta u_{\gamma q} \rightarrow -\frac{6\mathcal{R}_q^0 a^3}{a_{EQ}t q^2} \left[-\frac{7}{2} + \gamma + \ln\left(\frac{4\kappa}{\sqrt{3}}\right) \right]
$$
$$
+ \frac{a\sqrt{3}\mathcal{R}_q^o}{q(1+R)^{3/4}} e^{-\int_0^t \Gamma dt} \sin\left(\int_0^t \frac{q\,dt}{a\sqrt{3(1+R)}} \right) ,
$$
$$
\qquad (6.4.61)
$$

$$
\delta_{Bq} = \delta_{\gamma q} \rightarrow \frac{6\mathcal{R}_q^o a^3 (1+3R)}{a_{EQ}t^2 q^2} \left[-\frac{7}{2} + \gamma + \ln\left(\frac{4\kappa}{\sqrt{3}}\right) \right]
$$
$$
- \frac{3\mathcal{R}_q^o}{(1+R)^{1/4}} e^{-\int_0^t \Gamma dt} \cos\left(\int_0^t \frac{q\,dt}{a\sqrt{3(1+R)}} \right) , (6.4.62)
$$

with Γ and κ given by Eqs. (6.4.25) and (6.4.58).

Each of the different perturbations (6.4.59)–(6.4.62) is dominated by either its fast or slow term. First, let us consider δ_{Dq} and ψ_q. Comparison of Eqs. (6.4.48) and (6.4.47) with Eqs. (6.4.54) and (6.4.55) shows that the ratios of the fast and slow contributions to δ_{Dq} and ψ_q are of order

$$
\frac{\delta_{Dq}^{fast}}{\delta_{Dq}^{slow}} \sim \frac{a^2 H^2}{q^2} \left(\frac{\bar{\rho}_R}{\bar{\rho}_M}\right)^2 , \qquad \frac{\psi_q^{fast}}{\psi_q^{slow}} \sim \frac{aH}{q} \left(\frac{\bar{\rho}_R}{\bar{\rho}_M}\right)^2 . \qquad (6.4.63)
$$

We are assuming in this section that the wavelengths are short enough so that perturbations enter the horizon during the radiation-dominated era, so these ratios are much less than one at matter–radiation equality. Subsequently the ratio of fast to slow contributions to δ_{Dq} decreases as $1/a^3$, while for ψ_q the ratio of fast to slow contributions decreases as $1/a^{5/2}$. Hence it is a good approximation to take $\delta_{Dq} = \delta_{Dq}^{slow}$ and $\psi_q = \psi_q^{slow}$ throughout the matter-dominated era. Nevertheless, because the fast terms in Eqs. (6.4.59) and (6.4.60) have an oscillatory dependence on q, we will need to take them into account when we consider baryon acoustic oscillations in Section 8.1.

We also need $\delta_{\gamma q}$ and $\delta u_{\gamma q}$ in calculations of the cosmic microwave background anisotropies. Comparison of Eqs. (6.4.45) and (6.4.46) with Eqs. (6.4.57) and (6.4.56) shows that the ratios of the fast and slow contributions to $\delta_{\gamma q}$ and $u_{\gamma q}$ are (apart from the damping of the fast terms) of order

$$\frac{\delta_{\gamma q}^{\text{fast}}}{\delta_{\gamma q}^{\text{slow}}} \sim \frac{q^2}{a^2 H^2} \frac{\bar{\rho}_R}{\bar{\rho}_M} , \qquad \frac{\delta u_{\gamma q}^{\text{fast}}}{\delta u_{\gamma q}^{\text{slow}}} \sim \frac{q}{aH} \frac{\bar{\rho}_R}{\bar{\rho}_M} . \qquad (6.4.64)$$

These ratios are much larger than unity at horizon entry, and remain so until matter–radiation equality. After that the ratio of fast to slow contributions to $\delta_{\gamma q}$ remains of the same order of magnitude, while for $\delta u_{\gamma q}$ the ratio of fast to slow terms decreases like $1/\sqrt{a}$. Hence, once the wavelength enters the horizon, the slow contribution to $\delta_{\gamma q}$ is nominally smaller than the fast term, while the slow contribution to $\delta u_{\gamma q}$ remains nominally smaller than the fast term until late in the matter-dominated era. Nevertheless, we will keep the slow as well as fast terms here, because they are not affected by the damping that suppresses the fast terms. Also, even though it is relatively small, the slow term in δ_{Rq} will be found in Section 7.2 to produce a characteristic effect in the plot of the cosmic microwave background multipole coefficients C_ℓ vs. ℓ that would not be present with the fast term alone.

6.5 Scalar perturbations – interpolation & transfer functions

In the previous two sections we found analytic results for wavelengths that are long enough to enter the horizon well after matter–radiation equality, or short enough to enter the horizon well before matter–radiation equality. Unfortunately, this leaves out wavelengths that enter the horizon around the time of matter–radiation equality. It is wavelengths of this magnitude that make the dominant contributions to the first acoustic peak at around $\ell = 200$ in the multipole coefficients of the cosmic microwave background anisotropies. In this section we will consider how to construct formulas for the fluctuations that interpolate between the results of Sections 6.3 and 6.4, concentrating on results in the matter-dominated era, which are of the greatest observational interest.

We can get a good clue to this interpolation by first considering the limit of negligible baryon mass density and negligible damping, for which the form of the solution in the matter-dominated era can be found exactly, with no limitations on wavelength. In this limit, and leaving aside the neutrinos, the equations for the perturbations in the matter-dominated era are given for all wavelengths by Eqs. (6.3.17), (6.3.18), (6.3.20) and (6.3.21), with $R = 0$:

$$\frac{d}{dt}\left(t^{4/3}\psi_q\right) = -\frac{2}{3}t^{-2/3}\delta_{Dq} . \tag{6.5.1}$$

$$\dot{\delta}_{\gamma q} - (q^2/a^2)\delta u_{\gamma q} = -\psi_q \tag{6.5.2}$$

$$\dot{\delta}_{Dq} = -\psi_q , \tag{6.5.3}$$

$$\frac{d}{dt}\left(t^{-2/3}\delta u_{\gamma q}\right) = -\frac{1}{3}t^{-2/3}\delta_{\gamma q} , \tag{6.5.4}$$

with $a \propto t^{2/3}$. These have four independent exact solutions, three of which are quite simple:

Solution 1:

$$\delta_{Dq} = \frac{3q^2 t^2}{2a^2} , \qquad \psi_q = -\frac{q^2 t}{a^2} ,$$

$$\delta_{\gamma q} = 1 , \qquad \delta u_{\gamma q} = -t .$$

Solution 2:

$$\delta_{Dq} = \psi_q = 0 ,$$

$$\delta_{\gamma q} = -\cos\left(q\int_0^t \frac{dt}{\sqrt{3}a}\right) , \qquad \delta u_{\gamma q} = \frac{a}{\sqrt{3}q}\sin\left(q\int_0^t \frac{dt}{\sqrt{3}a}\right) .$$

Solution 3:

$$\delta_{Dq} = \psi_q = 0 ,$$

$$\delta_{\gamma q} = \sin\left(q\int_0^t \frac{dt}{\sqrt{3}a}\right) , \qquad \delta u_{\gamma q} = \frac{a}{\sqrt{3}q}\cos\left(q\int_0^t \frac{dt}{\sqrt{3}a}\right) .$$

The fourth solution is more complicated; it decays, with $\delta_{Dq} \propto 1/t$ and $\psi_q \propto 1/t^2$. As we will see, the initial conditions are satisfied by a linear combination of the first three solutions, so the fourth solution will not concern us here.

Without any loss of generality, we can write the linear combination of solutions 1, 2, and 3 that fits the solution at earlier time in a form that simplifies this fit, as[1]

$$\delta_{Dq} = \frac{9q^2 t^2 \mathcal{R}_q^o \mathcal{T}(\kappa)}{10a^2} \tag{6.5.5}$$

$$\psi_q = -\frac{3q^2 t \mathcal{R}_q^o \mathcal{T}(\kappa)}{5a^2} \tag{6.5.6}$$

---

[1]From now on, we define the zero of time so that $a \propto t^{1/2}$ in the radiation-dominated era. This is different from the definition of the zero of time in Solutions 1 and 2, where time is defined so that $a \propto t^{2/3}$ in the matter-dominated era. However, the difference this makes in the integral $\int_0^t dt/a$ is just a constant, which can be absorbed into $\Delta(\kappa)$.

$$\delta_{\gamma q} = \delta_{\nu q} = \frac{3\mathcal{R}_q^o}{5}\left[\mathcal{T}(\kappa) - \mathcal{S}(\kappa)\cos\left(q\int_0^t \frac{dt}{\sqrt{3}a} + \Delta(\kappa)\right)\right], \qquad (6.5.7)$$

$$\delta u_{\gamma q} = \delta u_{\nu q} = \frac{3t\mathcal{R}_q^o}{5}\left[-\mathcal{T}(\kappa) + \mathcal{S}(\kappa)\frac{a}{\sqrt{3}qt}\sin\left(q\int_0^t \frac{dt}{\sqrt{3}a} + \Delta(\kappa)\right)\right],$$

$$(6.5.8)$$

where $\mathcal{S}(\kappa)$, $\mathcal{T}(\kappa)$, and $\Delta(\kappa)$ are time-independent dimensionless functions of the dimensionless rescaled wave number introduced in Sections 6.2 and 6.4:

$$\kappa \equiv \frac{q\sqrt{2}}{a_{EQ}H_{EQ}} = \frac{(q/a_0)\sqrt{\Omega_R}}{H_0\Omega_M} = \frac{19.3(q/a_0)[\text{Mpc}^{-1}]}{\Omega_M h^2}, \qquad (6.5.9)$$

in which a_{EQ} and H_{EQ} are respectively the Robertson–Walker scale factor and expansion rate at matter–radiation equality. These are known as *transfer functions*. (These functions can only depend on κ because they must be independent of the normalization of the spatial coordinates and are dimensionless.) The division we made in the previous two sections between long and short wavelengths can be expressed in terms of the parameter κ. In the matter-dominated era we have $t = 2/3H = (t/3H_{EQ})(a/a_{EQ})^{3/2}$ and $\bar\rho_R/\bar\rho_M = a_{EQ}/a$, so

$$\frac{t^2q^2}{a^2}\frac{\bar\rho_R}{\bar\rho_M} = \frac{4\kappa^2}{9}.$$

Hence t^2q^2/a^2 is much less or much greater than $\bar\rho_M/\bar\rho_R$, according as the parameter κ is much less or much greater than one.

We choose to write the linear combination of solutions 1, 2, and 3 in the form (6.5.5)–(6.5.8) because it leads to simple values for the coefficients $\mathcal{S}(\kappa)$, $\mathcal{T}(\kappa)$, and $\Delta(\kappa)$ for $\kappa \ll 1$. For $\bar\rho_B = 0$, the results (6.3.23), (6.3.24), (6.3.35), (6.3.36) obtained earlier for moderately long wavelengths satisfying $t^2q^2/a^2 \ll \bar\rho_M/\bar\rho_R$, or in other words in the limit $\kappa \ll 1$, are consistent with Eqs. (6.5.5)–(6.5.8), and tell us that in this limit[2]

$$\mathcal{T}(\kappa) \to 1, \quad \mathcal{S}(\kappa) \to 1, \quad \Delta(\kappa) \to 2\kappa/\sqrt{3}. \qquad (6.5.10)$$

(The case of extremely long wavelengths does not arise here, because it requires that $t^2q^2/a^2 \ll R$, and for the present we are taking $R = 0$.)

Similarly, for $\bar\rho_B = 0$ and $\Gamma = 0$, the results (6.4.59)–(6.4.62) obtained earlier for short wavelengths satisfying $t^2q^2/a^2 \gg \bar\rho_M/\bar\rho_R$, or in other words

---

[2] The limit $2\kappa/\sqrt{3}$ for $\Delta(\kappa)$ is the difference for $y \equiv \bar\rho_M/\bar\rho_R \gg 1$ between the phase $\sqrt{3}qt/a = 2\kappa\sqrt{y}/\sqrt{3}$ in Eq. (6.3.34) (for $R \ll 1$), and the phase $q\int_0^t dt/\sqrt{3}a = 2\kappa[\sqrt{1+y}-1]/\sqrt{3}$ in Eqs. (6.5.7) and (6.5.8).

in the limit $\kappa \gg 1$, are consistent with Eqs. (6.5.5)–(6.5.8), and tell us that in this limit[3]

$$\mathcal{T}(\kappa) \to \frac{45}{2\kappa^2}\left[-\frac{7}{2} + \gamma + \ln\left(\frac{4\kappa}{\sqrt{3}}\right)\right], \quad \mathcal{S}(\kappa) \to 5, \quad \Delta(\kappa) \to 0.$$
(6.5.11)

As we shall see in Chapter 8, the decrease of $\mathcal{T}(\kappa)$ for large κ is very important in determining the intensity of fluctuations in the dark matter density at various wavelengths. (We can understand this decrease qualitatively by following the history of the dark matter density fluctuations. According to Eq. (6.2.27), for all wavelengths δ_{Dq} grows like t at early times, when the universe is radiation dominated and the wavelength is outside the horizon. For $\kappa \ll 1$ this growth continues until the universe becomes matter dominated, after which Eqs. (6.3.13) and (6.3.23) tell us that δ_{Dq} grows like $t^{2/3}$ both before and after the wavelength enters the horizon. In contrast, for $\kappa \gg 1$ the growth of δ_{Dq} like t continues only until the wavelength enters the horizon, which in this case is during the radiation-dominated era, after which according to Eq. (6.4.50) δ_{Dq} grows only logarithmically with time until the universe becomes matter dominated, after which according to Eq. (6.4.54) it grows as $t^{2/3}$, just as in the case $\kappa \ll 1$. The end of the period of growth proportional to t is at the time when $q/a \approx 1/t$, which since $a \propto \sqrt{t}$ is at $t \propto 1/q^2$, so the growth until matter–radiation equality is proportional to $\ln q/q^2$, accounting for the asymptotic behavior of $\mathcal{T}(\kappa)$. That is, the smallness of $\mathcal{T}(\kappa)$ for $\kappa \gg 1$ reflects not a decay of δ_{Dq}, but rather the failure of δ_{Dq} to *grow* appreciably during the interval from horizon-entry to matter–radiation equality.)

For values of κ of order unity, we have to find some way of interpolating between these two limiting cases. This can be done almost "by hand"; it turns out that almost any smooth interpolation between the limits $\kappa \ll 1$ and $\kappa \gg 1$ gives reasonable results for the cosmic microwave background anisotropies. For better precision, we need to solve the full coupled equations (6.2.9)–(6.2.15) (still with $\bar\rho_B = 0$) numerically for general wavelengths and general values of $y \equiv \bar\rho_M/\bar\rho_R$, imposing the initial conditions found in Section 6.1 for $y \ll 1$, and then comparing this solution for $y \gg 1$ with Eqs. (6.5.5)–(6.5.8). These equations can be put in a dimensionless form by using y as the independent variable, and writing

$$\delta_{Dq} = \kappa^2 \mathcal{R}_q^0 d(y)/4, \quad \delta_{\gamma q} = \delta_{vq} = \kappa^2 \mathcal{R}_q^0 r(y)/4,$$

$$\psi_q = (\kappa^2 H_{EQ}/4\sqrt{2})\mathcal{R}_q^0 f(y), \quad \delta u_{\gamma q} = \delta u_{vq} = (\kappa^2\sqrt{2}/4 H_{EQ})\mathcal{R}_q^0 g(y).$$

---

[3]This result for $\mathcal{T}(\kappa)$ was given by S. Weinberg, *Astrophys. J.* **581**, 810 (2002) [astro-ph/0207375]. The fact that the Mészáros equation (6.4.35) implies a fall-off of the transfer function for large wave number k like $\ln k/k^2$ had been pointed out by W. Hu and N. Sugiyama, *Astrophys. J.* **444**, 489 (1995).

Eqs. (6.2.9)–(6.2.15) then become

$$\sqrt{1+y}\frac{d}{dy}\left(y^2 f(y)\right) = -\frac{3d(y)}{2} - \frac{4r(y)}{y}$$

$$\sqrt{1+y}\frac{d}{dy}d(y) = -yf(y) ,$$

$$\sqrt{1+y}\frac{d}{dy}r(y) - \frac{\kappa^2}{y}g(y) = -yf(y) ,$$

$$\sqrt{1+y}\frac{d}{dy}\left(\frac{g(y)}{y}\right) = -\frac{1}{3}r(y) .$$

In this notation, the initial conditions (6.2.27)–(6.2.29) read

$$d(y) \rightarrow r(y) \rightarrow y^2 ,$$

$$f(y) \rightarrow -2 , \qquad g(y) \rightarrow -\frac{y^4}{9} .$$

When evaluated for $y \gg 1$, the numerical solutions of these equations[4] match Eqs. (6.5.5)–(6.5.8), with transfer functions given in Table 6.1. Inspection of this table shows that these numerical results agree with the analytic results (6.5.10) and (6.5.11) in the limits $\kappa \ll 1$ and $\kappa \gg 1$, respectively, although κ must be quite large before the asymptotic results (6.5.11) are reached.

As we will see in the following two chapters, the microwave background anisotropies and the correlation function of large scale structure are given by integrals involving one or more of these transfer functions, so it will be useful to give approximate analytic formulas for these functions. Dicus[5] has found "fitting formulas," which to a good approximation agree with the asymptotic formulas (6.5.10) and (6.5.11) for $\kappa \ll 1$ and $\kappa \gg 1$, and (except for $\Delta(\kappa)$ at large κ) generally match the numerical results of Table 6.1 at intermediate values of κ to better than 2%:

$$T(\kappa) \simeq \frac{\ln[1+(0.124\,\kappa)^2]}{(0.124\,\kappa)^2}\left[\frac{1+(1.257\,\kappa)^2+(0.4452\,\kappa)^4+(0.2197\,\kappa)^6}{1+(1.606\,\kappa)^2+(0.8568\,\kappa)^4+(0.3927\,\kappa)^6}\right]^{1/2} ,$$

$$\text{(6.5.12)}$$

$$S(\kappa) \simeq \left[\frac{1+(1.209\,\kappa)^2+(0.5116\,\kappa)^4+5^{1/2}(0.1657\,\kappa)^6}{1+(0.9459\,\kappa)^2+(0.4249\,\kappa)^4+(0.1657\,\kappa)^6}\right]^2 , \qquad \text{(6.5.13)}$$

---

[4] I thank D. Dicus for this numerical calculation.
[5] D. Dicus, private communication.

Table 6.1: The Scalar Transfer Functions

κ	S	T	Δ	κ	S	T	Δ
0.1	1.0167	0.9948	0.1207	10	3.9895	0.1608	0.3270
0.2	1.0551	0.9780	0.2240	11	4.0546	0.1440	0.3147
0.3	1.1147	0.9569	0.3156	12	4.1172	0.1298	0.2962
0.4	1.1891	0.9339	0.3852	13	4.1841	0.1178	0.2850
0.5	1.2680	0.9101	0.4423	14	4.2175	0.1075	0.2747
0.6	1.3529	0.8860	0.4800	15	4.2676	0.0985	0.2604
0.7	1.4388	0.8620	0.5148	16	4.3135	0.0907	0.2541
0.8	1.5195	0.8384	0.5336	17	4.3336	0.0838	0.2438
0.9	1.6081	0.8154	0.5531	18	4.3796	0.0777	0.2339
1.	1.6801	0.7930	0.5637	19	4.4043	0.0723	0.2296
1.2	1.8330	0.7502	0.5784	20	4.4233	0.0675	0.2195
1.4	1.9777	0.7104	0.5854	25	4.5271	0.0496	0.1920
1.6	2.1126	0.6734	0.5842	30	4.6051	0.0383	0.1713
1.8	2.2354	0.6391	0.5782	35	4.6650	0.0305	0.1542
2.	2.3451	0.6074	0.5700	40	4.7087	0.0249	0.1396
2.5	2.5895	0.5378	0.5537	45	4.7389	0.0209	0.1276
3.	2.7839	0.4798	0.5334	50	4.7605	0.0177	0.1182
3.5	2.9473	0.4311	0.5094	55	4.7794	0.0153	0.1111
4.	3.0970	0.3898	0.4854	60	4.7992	0.0134	0.1053
4.5	3.2346	0.3545	0.4659	65	4.8192	0.0118	0.0997
5.	3.3506	0.3241	0.4509	70	4.8365	0.0105	0.0940
5.5	3.4114	0.2976	0.4367	75	4.8487	0.0094	0.0885
6.	3.5181	0.2726	0.4203	80	4.8563	0.0084	0.0838
6.5	3.5953	0.2531	0.4029	85	4.8622	0.0077	0.0803
7.	3.6754	0.2361	0.3884	90	4.8695	0.0070	0.0776
7.5	3.7473	0.2200	0.3782	95	4.8792	0.0064	0.0751
8.	3.8015	0.2056	0.3695	100	4.8895	0.0059	0.0722
8.5	3.8432	0.1927	0.3590				
9.	3.8865	0.1810	0.3465				
9.5	3.9380	0.1704	0.3350				

$$\Delta(\kappa) \simeq \left[\frac{(0.1585\,\kappa)^2 + (0.9702\,\kappa)^4 + (0.2460\,\kappa)^6}{1 + (1.180\,\kappa)^2 + (1.540\,\kappa)^4 + (0.9230\,\kappa)^6 + (0.4197\,\kappa)^8} \right]^{1/4} .$$

$$(6.5.14)$$

These transfer functions are shown in Figure 6.1. A fitting formula for $T(\kappa)$ that includes the effects of neutrino anisotropic inertia and fits the

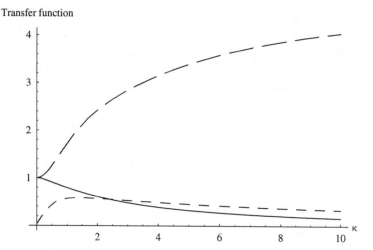

Figure 6.1: The transfer functions $T(\kappa)$ (solid curve), $S(\kappa)$ (long dashes), and $\Delta(\kappa)$ (short dashes), as functions of the rescaled wave number κ.

CAMB numerical results is given by Eisenstein and Hu.[6] Because of the neglect of neutrino anisotropic inertia, the results for $T(\kappa)$ of Eq. (6.5.12) and Table 6.1 are about 4% too low.

Now we have to consider how to take damping and the non-zero ratio of baryon to photon density into account. Eqs. (6.3.23), (6.3.24) and (6.4.59), (6.4.60) show that in the limits of either short or long wavelength, the leading terms in δ_{Dq} and ψ_q are unaffected by either damping or baryons, so in leading order we can simply use Eqs. (6.5.5) and (6.5.6):

$$\delta_{Dq} = \frac{9q^2t^2\mathcal{R}_q^o T(\kappa)}{10a^2} \qquad (6.5.15)$$

$$\psi_q = -\frac{3q^2t\mathcal{R}_q^o T(\kappa)}{5a^2} \qquad (6.5.16)$$

Damping affects the short-wavelength results (6.4.61) and (6.4.62) for $\delta u_{\gamma q}$ and $\delta_{\gamma q}$ by multiplying the sines and cosines with a factor $\exp(-\int_0^t \Gamma dt)$. This factor is absent in the long-wavelength results (6.3.36) and (6.3.35) for these perturbations, but this factor is essentially unity anyway for long wavelengths, because $\Gamma \propto q^2$. Hence for all wavelengths we can take damping into account by simply multiplying the sines and cosines in Eqs. (6.5.7) and (6.5.8) by $\exp(-\int_0^t \Gamma dt)$. Finally, in both the short-wavelength results (6.4.62) and (6.4.61) and the long-wavelength results (6.3.35) and (6.3.36),

[6]D. J. Eisenstein and W. Hu, *Astrophys. J.* **496**, 605 (1998).

the effect of the non-zero baryon density is to multiply the non-oscillatory term in $\delta_{\gamma q}$ with $1 + 3R$, multiply the cosine in $\delta_{\gamma q}$ with $(1 + R)^{-1/4}$, and multiply the sine in $\delta u_{\gamma q}$ with $(1 + R)^{-3/4}$, so it is highly plausible that a finite ratio of baryon density to photon density can be taken into account at all wavelengths by making the same alterations in Eqs. (6.5.7) and (6.5.8):

$$\delta_{\gamma q} = \delta_{Bq} = \frac{3\mathcal{R}_q^o}{5}[\mathcal{T}(\kappa)(1 + 3R)$$

$$-(1 + R)^{-1/4}e^{-\int_0^t \Gamma dt}\mathcal{S}(\kappa)\cos\left(\int_0^t \frac{q\,dt}{a\sqrt{3(1 + R)}} + \Delta(\kappa)\right)],$$
(6.5.17)

$$\delta u_{\gamma q} = \delta_{Bq} = \frac{3\mathcal{R}_q^o}{5}[-t\mathcal{T}(\kappa)$$

$$+\frac{a}{\sqrt{3}q(1 + R)^{3/4}}e^{-\int_0^t \Gamma dt}\mathcal{S}(\kappa)\sin\left(\int_0^t \frac{q\,dt}{a\sqrt{3(1 + R)}} + \Delta(\kappa)\right)],$$
(6.5.18)

To repeat, these results agree with the results of Section 6.4 for short wavelength, they agree with the results of Section 6.3 for long wavelength (for which Γ is negligible), and they agree with the results found in this section for all wavelengths when damping and the baryon density are neglected.

We must say a little more about the function $\mathcal{T}(\kappa)$, which is commonly known as *the* transfer function. It is conventional to write this transfer function as a function of a variable Q:

$$Q \equiv \frac{q}{a_0} \times \frac{1\,\text{Mpc}}{\Omega_M h^2}\,,$$
(6.5.19)

which according to Eq. (6.5.9) is the same as $\kappa/19.3$. In these terms, Eqs. (6.5.10) and (6.5.11) for $\mathcal{T}(\kappa)$ now read

$$\mathcal{T}(\kappa) \to \begin{cases} 1 & Q \to 0 \\ \frac{\ln(2.40\,Q)}{(4.07\,Q)^2} & Q \to \infty \end{cases}$$
(6.5.20)

A numerical solution of the equations for growth of dark matter density fluctuations has been fit for large Q with the formula[7]

$$\mathcal{T}_{\text{BBKS}}(\kappa) \simeq \frac{\ln(1 + 2.34Q)}{2.34Q}$$

$$\times \left[1 + 3.89Q + (16.1Q)^2 + (5.46Q)^3 + (6.71Q)^4\right]^{-1/4}.$$
(6.5.21)

[7]J. M. Bardeen, J. R. Bond, N. Kaiser, & A. S. Szalay, *Astrophys. J.* **304**, 15 (1986).

This goes to $\ln(2.34Q)/(3.96\,Q)^2$ for large Q, in excellent agreement with the analytic result (6.5.20). However, it should be noted that although the BBKS transfer function (6.5.21) is correctly normalized at $Q = 0$, it does not give a good account of the behavior of the transfer function for $Q \ll 1$. The power series expansion of Eq. (6.5.21) contains odd as well as even terms in Q, which is not consistent with the requirement that it must be an analytic function of the three-vector $\mathbf{q}$. The Dicus fitting formula (6.5.12) was constructed to be analytic in the three-vector $\mathbf{q}$ at $\mathbf{q} = 0$, and in fact fits the numerical results of Table 6.1 better than the BBKS transfer function (6.5.21). In our analysis of cosmic microwave background anisotropies, we will use the fitting formula (6.5.12) instead of Eq. (6.5.21).

All this has been for $\beta \equiv \bar{\rho}_B/\bar{\rho}_M \ll 1$, though we have now taken into account a non-negligible ratio $R = 3\bar{\rho}_B/4\bar{\rho}_\gamma$. The corrections to the transfer function for finite values of β are roughly 10%; they have been calculated analytically[8] for $\kappa \gg 1$ to lowest order in β, and numerically[9] for general κ and for selected values of β to all orders in β. In using the transfer function in calculations of large scale structure, it is common to use a simple modification of the fitting formula. Peacock and Dodds[10] proposed in effect that the same fitting formula could be used for finite baryon density, but with $\kappa = (k\sqrt{\Omega_R}/H_0\Omega_M)\exp(\Omega_B)$. This worked well for the limited range of cosmological parameters studied, but it is physically impossible for the transfer function to have this sort of dependence on Ω_B. There is no way that the physical processes during the radiation-dominated era that are responsible for the transfer function to know anything about the time at which we happen to measure cosmological parameters like $\bar{\rho}_M$. Aside from κ, the transfer function can only depend on quantities such as $\beta \equiv \bar{\rho}_B/\bar{\rho}_M = \Omega_B/\Omega_M$, which is constant, or $\Omega_M h^2$, which (for a known present radiation temperature) tells us the matter density at any given temperature. Indeed, Sugiyama[11] pointed out that the correction factor $\exp(\Omega_B)$ actually works well only for values of Ω_M close to unity. As an alternative that would apply for smaller values of Ω_M, he proposed a correction factor $\exp(\Omega_B + \Omega_B/\Omega_M)$, which is physically impossible for the same reason as $\exp(\Omega_B)$. Another difficulty with all these suggestions is that the baryon correction must disappear for small wave number, because in this case pressure forces are negligible, and baryons behave just like cold dark matter. A baryon correction that satisfies all these physical criteria has been proposed by Eisenstein and Hu:[6] The transfer function is evaluated

---

[8]S. Weinberg, ref. 2.

[9]J. A. Holtzman, *Astrophys. J. Suppl.* **71**, 1 (1989).

[10]J. A. Peacock and S.J. Dodds, *Mon. Not. Roy. Astron. Soc.* **267**, 1020 (1994).

[11]N. Sugiyama, *Astrophys. J.* **100**, 281 (1995).

with κ taken as

$$\kappa = \left(\frac{k\sqrt{\Omega_R}}{H_0\Omega_M}\right)\left[\alpha + \frac{1-\alpha}{1+(0.43\,k\,s)^4}\right]^{-1} , \qquad (6.5.22)$$

where

$$\alpha = 1 - 0.328\,\ln(431\,\Omega_M h^2)\,\beta + 0.38\,\ln(22.3\,\Omega_M h^2)\,\beta^2 ,$$

and s is the acoustic horizon at the time $t = t_L$ of last scattering, projected (as is k) to the present:

$$s \equiv a(t_0) \int_0^{t_L} \frac{v_s\,dt}{a(t)} = (1+z_L)d_H ,$$

with d_H given by Eq. (2.6.32).

6.6 Tensor perturbations

We next turn to the tensor modes. These are considerably simpler to study than the scalar modes, so what took five sections to analyze for the scalar modes will be treated here in just one section. As in the case of scalar modes, we begin by setting down the full set of equations used in computer programs like CMBfast, and then move on to approximations.

A. Cold dark matter and baryonic plasma

As already mentioned, the particles of both the cold dark matter and baryonic plasma move too slowly to contribute any anisotropic inertia. In tensor modes there are no perturbations to densities or streaming velocities, so there are no perturbations to either the cold dark matter or baryonic plasma that need to be followed here.

B. Gravitation

According to Section 5.1, in tensor modes the gravitational perturbation takes the form:

$$\delta g_{ij}(\mathbf{x}, t) = a^2(t)\,D_{ij}(\mathbf{x}, t) , \qquad (6.6.1)$$

with $D_{ij}(\mathbf{x}, t)$ satisfying the wave equation (5.1.53):

$$\ddot{D}_{ij} + 3H\dot{D}_{ij} - a^{-2}\nabla^2 D_{ij} = 16\pi\,G\pi_{ij}^T , \qquad (6.6.2)$$

and the trace and transversality conditions

$$D_{ii} = 0 , \qquad \partial_i D_{ij} = 0 . \qquad (6.6.3)$$

We will see that the anisotropic inertia tensor π_{ij}^T is a linear functional of $\dot{D}_{ij}$, so Eq. (6.6.2) has two independent solutions. During the period when the perturbation is outside the horizon, the anisotropic inertia tensor π_{ij}^T and the $a^{-2}\nabla^2 D_{ij}$ term are negligible, and Eq. (6.6.2) becomes

$$\ddot{D}_{ij} + 3H\dot{D}_{ij} = 0 .$$

Hence outside the horizon one of the solutions is constant, while the other decays as $\int a^{-3}dt$, which in the radiation dominated era goes as $t^{-1/2}$. For all interesting wave numbers the perturbation remains outside the horizon during many e-foldings of cosmic expansion, so the decaying mode becomes negligible, and we can consider only the other mode. The metric perturbation and anisotropic inertia can therefore be put in the form

$$D_{ij}(\mathbf{x}, t) = \sum_{\lambda=\pm 2} \int d^3q \, e^{i\mathbf{q}\cdot\mathbf{x}} \beta(\mathbf{q}, \lambda) \, e_{ij}(\hat{q}, \lambda) \, \mathcal{D}_q(t) , \qquad (6.6.4)$$

$$\pi_{ij}^T(\mathbf{x}, t) = \sum_{\lambda=\pm 2} \int d^3q \, e^{i\mathbf{q}\cdot\mathbf{x}} \beta(\mathbf{q}, \lambda) \, e_{ij}(\hat{q}, \lambda) \, \pi_q^T(t) . \qquad (6.6.5)$$

where $\beta(\mathbf{q}, \lambda)$ is a stochastic parameter for the single non-decaying mode with wave number $\mathbf{q}$ and helicity λ; $e_{ij}(\hat{q}, \lambda)$ is the corresponding polarization tensor, defined in Section 5.2, with $e_{ii} = q_i e_{ij} = 0$; and $\mathcal{D}_q(t)$ is the solution of the wave equation

$$\ddot{\mathcal{D}}_q + 3H\dot{\mathcal{D}}_q + a^{-2}q^2\mathcal{D}_q = 16\pi G\pi_q^T . \qquad (6.6.6)$$

We will return later to the solution of this equation.

C. Photons

The Boltzmann equation for the photon density matrix perturbation $\delta n^{ij}(\mathbf{x}, t)$ in the tensor mode is given by Eq. (H.35) as

$$\frac{\partial \, \delta n_\gamma^{ij}(\mathbf{x}, \mathbf{p}, t)}{\partial t} + \frac{\hat{p}_k}{a(t)} \frac{\partial \, \delta n_\gamma^{ij}(\mathbf{x}, \mathbf{p}, t)}{\partial x^k} + \frac{2\dot{a}(t)}{a(t)} \delta n_\gamma^{ij}(\mathbf{x}, \mathbf{p}, t)$$

$$- \frac{1}{4a^2(t)}p\bar{n}_\gamma'(p)\hat{p}_k\hat{p}_l \, \dot{D}_{kl}(\mathbf{x}, t)\left(\delta_{ij} - \hat{p}_i\hat{p}_j\right)$$

$$= -\omega_c(t)\, \delta n_\gamma^{ij}(\mathbf{x}, \mathbf{p}, t) + \frac{3\omega_c(t)}{8\pi}\int d^2\hat{p}_1$$

$$\times \left[\delta n_\gamma^{ij}(\mathbf{x}, p\hat{p}_1, t) - \hat{p}_i\hat{p}_k \, \delta n_\gamma^{kj}(\mathbf{x}, p\hat{p}_1, t) - \hat{p}_j\hat{p}_k \, \delta n_\gamma^{ik}(\mathbf{x}, p\hat{p}_1, t)\right.$$

$$\left. + \hat{p}_i\hat{p}_j\hat{p}_k\hat{p}_l \, \delta n_\gamma^{kl}(\mathbf{x}, p\hat{p}_1, t)\right], \qquad (6.6.7)$$

in which we have used Eq. (6.6.1) for the metric perturbation. As in the case of scalar modes, instead of $\delta n_\gamma^{ij}(\mathbf{x}, \mathbf{p}, t)$, we will concentrate on the fractional intensity matrix defined by the analog of Eq. (6.1.13):

$$a^4(t)\,\bar{\rho}_\gamma(t)\,J_{ij}(\mathbf{x}, \hat{p}, t) \equiv a^2(t) \int_0^\infty \delta n_\gamma^{ij}(\mathbf{x}, p\hat{p}, t)\,4\pi p^3\,dp \ . \tag{6.6.8}$$

We seek a solution in the form

$$J_{ij}(\mathbf{x}, \hat{p}, t) = \sum_{\lambda=\pm 2} \int d^3q\, e^{i\mathbf{q}\cdot\mathbf{x}}\, \beta(\mathbf{q}, \lambda)\, J_{ij}(\mathbf{q}, \hat{p}, t, \lambda) \ . \tag{6.6.9}$$

The Boltzmann equation (6.6.7) then takes the form

$$\frac{\partial\, J_{ij}(\mathbf{q}, \hat{p}, t, \lambda)}{\partial t} + i\frac{\mathbf{q}\cdot\hat{p}}{a(t)} J_{ij}(\mathbf{q}, \hat{p}, t, \lambda)$$

$$+\, \hat{p}_k\hat{p}_l\, e_{kl}(\hat{q}, \lambda)\, \dot{D}_q(t)\left(\delta_{ij} - \hat{p}_i\hat{p}_j\right)$$

$$= -\omega_c(t)\, J_{ij}(\mathbf{q}, \hat{p}, t, \lambda) + \frac{3\omega_c(t)}{2}$$

$$\times \Big[\, \mathcal{J}_{ij}(\mathbf{q}, t, \lambda) - \hat{p}_i\hat{p}_k\, \mathcal{J}_{kj}(\mathbf{q}, t, \lambda) - \hat{p}_j\hat{p}_k\, \mathcal{J}_{ik}(\mathbf{q}, t, \lambda)$$

$$+\, \hat{p}_i\hat{p}_j\hat{p}_k\hat{p}_l\, \mathcal{J}_{kl}(\mathbf{q}, t, \lambda)\,\Big] \ , \tag{6.6.10}$$

where

$$\mathcal{J}_{ij}(\mathbf{q}, t, \lambda) \equiv \int \frac{d^2\hat{p}}{4\pi}\, J_{ij}(\mathbf{q}, \hat{p}, t, \lambda) \ . \tag{6.6.11}$$

Furthermore, because $J_{ij}(\mathbf{q}, \hat{p}, t, \lambda)$ for a given helicity λ must be a linear combination of the polarization tensor components $e_{kl}(\hat{q}, \lambda)$ with the same λ, while $\hat{q}_k\, e_{kl}(\hat{q}, \lambda)$ and $e_{kk}(\hat{q}, \lambda)$ both vanish, the only possible form of $\mathcal{J}_{ij}(\mathbf{q}, t, \lambda)$ allowed by rotational invariance is just $e_{ij}(\hat{q}, \lambda)$ times some function of $q \equiv |\mathbf{q}|$ and t. This relation is conventionally written

$$\mathcal{J}_{ij}(\mathbf{q}, t, \lambda) = -\frac{2}{3}e_{ij}(\hat{q}, \lambda)\, \Psi(q, t) \ . \tag{6.6.12}$$

Rotational invariance allows the intensity matrix perturbation to be written in the form

$$J_{ij}(\mathbf{q}, \hat{p}, t, \lambda) = \frac{1}{2}\left(\delta_{ij} - \hat{p}_i\hat{p}_j\right)\hat{p}_k\hat{p}_l\, e_{kl}(\hat{q}, \lambda)$$

$$\times \left(\Delta_T^{(T)}(q, \hat{p}\cdot\hat{q}, t) + \Delta_P^{(T)}(q, \hat{p}\cdot\hat{q}, t)\right)$$

$$+\left(e_{ij}(\hat{q}, \lambda) - \hat{p}_i\hat{p}_k e_{kj}(\hat{q}, \lambda) - \hat{p}_j\hat{p}_k e_{ik}(\hat{q}, \lambda) + \hat{p}_i\hat{p}_j\hat{p}_k\hat{p}_l\, e_{kl}(\hat{q}, \lambda)\right)$$

$$\times \Delta_P^{(T)}(q, \hat{p}\cdot\hat{q}, t) \tag{6.6.13}$$

(Here the superscript T stands for "tensor," while the subscript T stands for "temperature." The coefficients are chosen so that J_{ii} is proportional to Δ_T, and the polarization is proportional to Δ_P.) A third term proportional to $(\hat{q}_i - \hat{p}_i(\hat{p} \cdot \hat{q}))(\hat{q}_j - \hat{p}_j(\hat{p} \cdot \hat{q}))\hat{p}_k\hat{p}_l e_{kl}$ would be allowed by symmetry principles, but is not generated in the Boltzmann equation by Thomson scattering. Using Eq. (6.6.13) in Eq. (6.6.10) yields separate Boltzmann equations for $\Delta_T^{(T)}$ and $\Delta_P^{(T)}$:

$$\frac{\partial}{\partial t}\Delta_T^{(T)}(q,\mu,t) + i\, a^{-1}(t)\, q\mu\, \Delta_T^{(T)}(q,\mu,t)$$
$$= -2\dot{D}_q(t) - \omega_c(t)\, \Delta_T^{(T)}(q,\mu,t) + \omega_c(t)\, \Psi(q,t)\,, \quad (6.6.14)$$

$$\frac{\partial}{\partial t}\Delta_P^{(T)}(q,\mu,t) + i\, a^{-1}(t)q\mu\, \Delta_P^{(T)}(q,\mu,t)$$
$$= -\omega_c(t)\, \Delta_P^{(T)}(q,\mu,t) - \omega_c(t)\, \Psi(q,t)\,. \quad (6.6.15)$$

The functions $\Delta_T^{(T)}$ and $\Delta_P^{(T)}$ may be expanded in Legendre polynomials

$$\Delta_T^{(T)}(q,\hat{p}\cdot\hat{q},t) = \sum_{\ell=0}^{\infty} i^{-\ell}(2\ell+1)\, P_\ell(\hat{q}\cdot\hat{p})\, \Delta_{T,\ell}^{(T)}(q,t) \quad (6.6.16)$$

$$\Delta_P^{(T)}(q,\hat{p}\cdot\hat{q},t) = \sum_{\ell=0}^{\infty} i^{-\ell}(2\ell+1)\, P_\ell(\hat{q}\cdot\hat{p})\, \Delta_{P,\ell}^{(T)}(q,t)\,. \quad (6.6.17)$$

Using the familiar recursion relation

$$z\, P_\ell(z) = \frac{\ell+1}{2\ell+1}P_{\ell+1}(z) + \frac{\ell}{2\ell+1}P_{\ell-1}(z)\,,$$

we find that the Boltzmann equations (6.6.14) and (6.6.15) now read

$$\dot{\Delta}_{T,\ell}^{(T)} + \frac{q}{a(2\ell+1)}\left((\ell+1)\Delta_{T,\ell+1}^{(T)} - \ell\Delta_{T,\ell-1}^{(T)}\right)$$
$$= \left(-2\dot{D}_q + \omega_c\Psi\right)\delta_{\ell,0} - \omega_c\Delta_{T,\ell}^{(T)} \quad (6.6.18)$$

$$\dot{\Delta}_{P,\ell}^{(T)} + \frac{q}{a(2\ell+1)}\left((\ell+1)\Delta_{P,\ell+1}^{(T)} - \ell\Delta_{P,\ell-1}^{(T)}\right)$$
$$= -\omega_c\Psi\,\delta_{\ell,0} - \omega_c\Delta_{P,\ell}^{(T)}\,. \quad (6.6.19)$$

To calculate the source term $\Psi(q,t)$ in terms of partial waves, one first integrates Eq. (6.6.13) over $\hat{p}$, using the formulas

$$\int d^2\hat{p}\, f(\hat{p}\cdot\hat{q})\,\hat{p}_i\hat{p}_k e_{jk}(\hat{q}) = \frac{1}{2}e_{ij}(\hat{q})\int d^2\hat{p}\, f(\hat{p}\cdot\hat{q})\left(1-(\hat{p}\cdot\hat{q})^2\right)$$
$$\int d^2\hat{p}\, f(\hat{p}\cdot\hat{q})\,\hat{p}_i\hat{p}_j\hat{p}_k\hat{p}_l e_{kl}(\hat{q}) = \frac{1}{4}e_{ij}(\hat{q})\int d^2\hat{p}\, f(\hat{p}\cdot\hat{q})\left(1-(\hat{p}\cdot\hat{q})^2\right)^2\,,$$

where f is any function of $\hat{p} \cdot \hat{q}$ and e_{ij} is any symmetric traceless matrix function of $\hat{q}$ with $\hat{q}_i e_{ij} = 0$ and $e_{ii} = 0$. The integral gives

$$\Psi(q,t) = -\frac{3}{2} \int \frac{d^2\hat{p}}{4\pi} \left[-\frac{1}{8} \left(1 - (\hat{p} \cdot \hat{q})^2\right)^2 \Delta_T^{(T)}(q, \hat{p} \cdot \hat{q}, t) \right.$$

$$\left. + \left((\hat{p} \cdot \hat{q})^2 + \frac{1}{8}\left(1 - (\hat{p} \cdot \hat{q})^2\right)^2 \right) \Delta_P^{(T)}(q, \hat{p} \cdot \hat{q}, t) \right].$$

(6.6.20)

Inserting the partial wave expansions (6.6.16) and (6.6.17) then gives:[1]

$$\Psi(q,t) = \frac{1}{10} \Delta_{T,0}^{(T)}(q,t) + \frac{1}{7} \Delta_{T,2}^{(T)}(q,t) + \frac{3}{70} \Delta_{T,4}^{(T)}(q,t) - \frac{3}{5} \Delta_{P,0}^{(T)}(q,t)$$

$$+ \frac{6}{7} \Delta_{P,2}^{(T)}(q,t) - \frac{3}{70} \Delta_{P,4}^{(T)}(q,t).$$

(6.6.21)

Eqs. (6.6.18) and (6.6.19) thus form a closed system of coupled differential equations for the partial wave amplitudes produced by a given gravitational field perturbation $\mathcal{D}_q(t)$. Of course, their solution requires a truncation of the partial wave expansion at some maximum ℓ.

The solution is of interest in itself, because we can measure anisotropies in the cosmic microwave background. It will be applied to these anisotropies in Chapter 7. It is also needed in calculations of the tensor anisotropic inertia. In tensor modes, the only non-vanishing contribution of photons to the energy-momentum tensor is to the space-space component (6.1.10)

$$\delta T^i_{\gamma j}(\mathbf{x}, t) = a^{-4}(t) \int d^3p \, a^2(t) \delta n^{kk}(\mathbf{x}, \mathbf{p}, t) \, p\hat{p}_i\hat{p}_j$$

$$= \bar{\rho}_\gamma(t) \sum_\lambda \int d^3q \, \beta(\mathbf{q}, \lambda) \, e^{i\mathbf{q} \cdot \mathbf{x}} \int \frac{d^2\hat{p}}{4\pi} J_{kk}(\mathbf{q}, \hat{p}, t, \lambda) \hat{p}_i\hat{p}_j$$

$$= \bar{\rho}_\gamma(t) \sum_\lambda \int d^3q \, \beta(\mathbf{q}, \lambda) \, e^{i\mathbf{q} \cdot \mathbf{x}} e_{kl}(\hat{q}, \lambda)$$

$$\times \int \frac{d^2\hat{p}}{4\pi} \Delta_T^{(T)}(q, \hat{p} \cdot \hat{q}, t) \hat{p}_i\hat{p}_j\hat{p}_k\hat{p}_l$$

$$= \bar{\rho}_\gamma(t) \sum_\lambda \int d^3q \, \beta(\mathbf{q}, \lambda) \, e^{i\mathbf{q} \cdot \mathbf{x}} e_{ij}(\hat{q}, \lambda)$$

$$\times \frac{1}{4} \int \frac{d^2\hat{p}}{4\pi} \Delta_T^{(T)}(q, \hat{p} \cdot \hat{q}, t) \left(1 - (\hat{p} \cdot \hat{q})^2\right)^2.$$

(6.6.22)

---

[1] R. Crittenden, J. R. Bond, R. L. Davis, G. Efstathiou, and P. J. Steinhardt, *Phys. Rev. Lett.* **71**, 324 (1993) [astro-ph/9303014].

For tensor modes there is no pressure perturbation, so according to Eq. (5.1.43), this is the same as the anisotropic inertia tensor $\pi^T_{\gamma\,ij}(\mathbf{x}, t)$. Comparing Eq. (6.6.22) with Eq. (6.6.5) thus gives

$$\pi^T_{\gamma\,q}(t) = \frac{\bar{\rho}_\gamma(t)}{4} \int \frac{d^2\hat{p}}{4\pi} \Delta^{(T)}_T(q, \hat{p} \cdot \hat{q}, t) \left(1 - (\hat{p} \cdot \hat{q})^2\right)^2$$

$$= 2\bar{\rho}_\gamma(t) \left[\frac{1}{15} \Delta^{(T)}_{T,0}(q, t) + \frac{2}{21} \Delta^{(T)}_{T,2}(q, t) + \frac{1}{35} \Delta^{(T)}_{T,4}(q, t)\right].$$

(6.6.23)

Experience shows that to accurately calculate the partial wave amplitudes up to $\ell = 4$, which appear in Eqs. (6.6.21) and (6.6.23), one needs to solve the Boltzmann equations for the partial wave amplitudes up to larger values of ℓ, up to $\ell = 10$. Once Ψ is calculated in this way, we can calculate $J_{ij}(\mathbf{q}, \hat{p}, t, \lambda)$ for very much higher values of ℓ by using the "line of sight" solution of Eq. (6.6.10):

$$J_{ij}(\mathbf{q}, \hat{p}, t, \lambda) = \int_{t_1}^t dt' \, \exp\left(-i\mathbf{q} \cdot \hat{p} \int_{t'}^t \frac{dt''}{a(t'')} - \int_{t'}^t dt'' \, \omega_c(t'')\right)$$

$$\times \left[- \hat{p}_k \hat{p}_l \left(\delta_{ij} - \hat{p}_i \hat{p}_j\right) e_{kl}(\hat{q}, \lambda) \dot{\mathcal{D}}_q(t') - \omega_c(t') \Psi(q, t')\right.$$

$$\left. \left(e_{ij}(\hat{q}, \lambda) - \hat{p}_i \hat{p}_k e_{kj}(\hat{q}, \lambda) - \hat{p}_j \hat{p}_k e_{ik}(\hat{q}, \lambda) + \hat{p}_i \hat{p}_j \hat{p}_k \hat{p}_l e_{kl}(\hat{q}, \lambda)\right)\right]$$

(6.6.24)

where t_1 is any time that is early enough before recombination so that $\omega_c(t_1) \gg H(t_1)$, which allows us to drop a term proportional to $J_{ij}(\mathbf{q}, \hat{p}, t_1, \lambda)$. In terms of the temperature and polarization amplitudes defined by the decomposition (6.6.13), the line-of-sight integrals read

$$\Delta^{(T)}_T(q, \mu, t) = -\Delta^{(T)}_P(q, \mu, t) - 2 \int_{t_1}^t dt' \, \exp\left[-iq\mu \int_{t'}^t \frac{dt''}{a(t'')}\right.$$

$$\left. - \int_{t'}^t \omega_c(t'') \, dt''\right] \times \mathcal{D}_q(t') \,,$$

(6.6.25)

$$\Delta^{(T)}_P(q, \mu, t) = - \int_{t_1}^t dt' \, \exp\left[-iq\mu \int_{t'}^t \frac{dt''}{a(t'')} - \int_{t'}^t \omega_c(t'') \, dt''\right]$$

$$\times \omega_c(t') \Psi(q, t') \,.$$

(6.6.26)

This line of sight integral also provides an alternative to the use of the truncated partial wave expansion.[2] We can derive an integral equation for

---

[2] S. Weinberg, *Phys. Rev. D* **74**, 063517 (2006) [astro-ph/0607076]; D. Baskaran, L. P. Grishchuk, and A. G. Polnarev, *Phys. Rev. D* **74**, 083008 (2006) [gr-qc/0605100].

$\Psi(q, t)$ by simply analytically integrating Eq. (6.6.24) over $\hat{p}$. Equating the coefficients of e_{ij} on both sides gives the integral equation

$$
\Psi(q, t) = \frac{3}{2} \int_{t_1}^{t} dt' \exp\left[- \int_{t'}^{t} \omega_c(t'') \, dt'' \right]
$$
$$
\times \left[-2\dot{D}_q(t') K \left(q \int_{t'}^{t} \frac{dt''}{a(t'')} \right) + \omega_c(t') F \left(q \int_{t'}^{t} \frac{dt''}{a(t'')} \right) \Psi(q, t') \right],
$$

(6.6.27)

where $K(v)$ and $F(v)$ are the functions

$$
K(v) \equiv j_2(v)/v^2 , \qquad F(v) \equiv j_0(v) - 2j_1(v)/v + 2j_2(v)/v^2 . \qquad (6.6.28)
$$

Eq. (6.6.27) can be solved efficiently either by iteration, or by numerical recipes appropriate for integral equations of the Volterra type. Once $\Psi(q, t)$ is calculated in this way, the complete photon intensity matrix can be obtained by a numerical integration in Eq. (6.6.24).

Neutrinos

The Boltzmann equation for the perturbation $\delta n_v(\mathbf{x}, \mathbf{p}, t)$ to the neutrino phase space density is given by Eq. (H.14) as

$$
\frac{\partial \delta n_v(\mathbf{x}, \mathbf{p}, t)}{\partial t} + \frac{\hat{p}_i}{a(t)} \frac{\partial \delta n_v(\mathbf{x}, \mathbf{p}, t)}{\partial x^i} = \frac{p\bar{n}'(p)}{2} \hat{p}_i \hat{p}_j \dot{D}_{ij}(\mathbf{x}, t) , \qquad (6.6.29)
$$

in which we have used Eq. (6.6.1) for the metric perturbation. As in the case of scalar modes, instead of δn_v, we find it more convenient to deal with a dimensionless intensity perturbation J, defined by the analog of Eq. (6.1.42):

$$
a^4(t)\bar{\rho}_v(t) J(\mathbf{x}, \hat{p}, t) \equiv N_v \int_0^{\infty} \delta n_v(\mathbf{x}, \mathbf{p}, t) \, 4\pi p^3 \, dp , \qquad (6.6.30)
$$

where N_v is the number of species of neutrino, counting antineutrinos separately, and $\bar{\rho}_v \equiv N_v \, a^{-4} \int 4\pi p^3 \bar{n}_v(p)$. This satisfies a Boltzmann equation

$$
\frac{\partial J(\mathbf{x}, \hat{p}, t)}{\partial t} + \frac{\hat{p}_i}{a(t)} \frac{\partial J(\mathbf{x}, \hat{p}, t)}{\partial x^i} = -2\hat{p}_i \hat{p}_j \dot{D}_{ij}(\mathbf{x}, t) , \qquad (6.6.31)
$$

with D_{ij} given by Eq. (6.6.4). We will be able to find a solution in the form

$$
J(\mathbf{x}, \hat{p}, t) = \sum_{\lambda=\pm 2} \int d^3q \, e^{i\mathbf{q}\cdot\mathbf{x}} \beta(\mathbf{q}, \lambda) \, e_{ij}(\hat{q}, \lambda) \hat{p}_i \hat{p}_j \Delta_v^{(T)}(q, \hat{p} \cdot \hat{q}, t) . \qquad (6.6.32)
$$

Then Eq. (6.6.29) becomes an equation for $\Delta_\nu^{(T)}$:

$$\dot{\Delta}_\nu^{(T)}(q, \mu, t) + \frac{iq\mu}{a(t)} \Delta_\nu^{(T)}(q, \mu, t) = -2\dot{D}_q(t) \qquad (6.6.33)$$

This can be solved by a partial wave expansion

$$\Delta_\nu^{(T)}(q, \mu, t) = \sum_\ell i^{-\ell} (2\ell + 1) \Delta_{\nu,\ell}^{(T)}(q, t) , \qquad (6.6.34)$$

with $\Delta_{\nu,\ell}^{(T)}(q, t)$ satisfying

$$\dot{\Delta}_{\nu,\ell}^{(T)} + \frac{q}{a(2\ell + 1)} \left((\ell + 1) \Delta_{\nu,\ell+1}^{(T)} - \ell \Delta_{T,\ell-1}^{(T)} \right) = -2\dot{D}_q(t) \delta_{\ell 0} . \quad (6.6.35)$$

This of course needs to be truncated at some more-or-less arbitrary maximum value of ℓ. But instead we can find a direct solution of Eq. (6.6.33), as a line of sight integral

$$\Delta_\nu^{(T)}(q, \mu, t) = -2 \int_{t_1}^t dt' \, \exp\left(-iq\mu \int_{t'}^t \frac{dt''}{a(t'')} \right) \dot{D}_q(t') . \qquad (6.6.36)$$

We take t_1 soon enough after the decoupling of neutrinos at $T \approx 10^{10}$ K so that at this time the distribution of neutrinos is still essentially that of local thermal equilibrium. In this case the perturbation to this distribution arises only from the perturbations δT_ν and $\delta \mathbf{u}_\nu$ to the neutrino temperature and streaming velocity, which do not have tensor components, so we do not need to include an initial value term $\Delta_\nu^{(T)}(q, \mu, t_1)$ on the right-hand side of Eq. (6.6.36).

It will be a long while before anyone measures the angular distribution of cosmic neutrinos, so the only use to be made of calculations of δn_ν is in calculating components of the energy momentum tensor. In the tensor mode, the only non-vanishing component is $\delta T_{\nu j}^i$, given by the same formula (6.6.22) as for photons, except that $\Delta_\nu^{(T)}$ appears instead of $\Delta_T^{(T)}$:

$$\delta T_{\nu j}^i(\mathbf{x}, t) = a^{-4}(t) \int d^3p \, \delta n_\nu(\mathbf{x}, \mathbf{p}, t) \, p \hat{p}_i \hat{p}_j$$

$$= \bar{\rho}_\nu(t) \sum_\lambda \int d^3q \, \beta(\mathbf{q}, \lambda) \, e^{i\mathbf{q}\cdot\mathbf{x}} \int \frac{d^2\hat{p}}{4\pi} J(\mathbf{q}, \hat{p}, t, \lambda) \hat{p}_i \hat{p}_j$$

$$= \bar{\rho}_\nu(t) \sum_\lambda \int d^3q \, \beta(\mathbf{q}, \lambda) \, e^{i\mathbf{q}\cdot\mathbf{x}} e_{kl}(\hat{q}, \lambda)$$

$$\times \int \frac{d^2\hat{p}}{4\pi} \Delta_\nu^{(T)}(q, \hat{p}\cdot\hat{q}, t) \hat{p}_i \hat{p}_i \hat{p}_j \hat{p}_k$$

$$= \bar{\rho}_\nu(t) \sum_\lambda \int d^3q \, \beta(\mathbf{q}, \lambda) \, e^{i\mathbf{q}\cdot\mathbf{x}} e_{ij}(\hat{q}, \lambda)$$

$$\times \frac{1}{4} \int \frac{d^2\hat{p}}{4\pi} \, \Delta_\nu^{(T)}(q, \hat{p}\cdot\hat{q}, t) \left(1 - (\hat{p}\cdot\hat{q})^2\right)^2 . \qquad (6.6.37)$$

This is the neutrino contribution to the anisotropic inertia tensor π_{ij}^T. Comparing this with Eq. (6.6.5) then gives

$$\pi_{\nu q}^T(t) = \frac{\bar{\rho}_\nu(t)}{4} \int \frac{d^2\hat{p}}{4\pi} \, \Delta_\nu^{(T)}(q, \hat{p}\cdot\hat{q}, t) \left(1 - (\hat{p}\cdot\hat{q})^2\right)^2 . \qquad (6.6.38)$$

As for photons, this can be evaluated using the partial wave expansion:

$$\pi_{\nu q}^T(t) = 2\bar{\rho}_\nu(t) \left[\frac{1}{15} \Delta_{\nu,0}^{(T)}(q,t) + \frac{2}{21} \Delta_{\nu,2}^{(T)}(q,t) + \frac{1}{35} \Delta_{\nu,4}^{(T)}(q,t) \right]. \qquad (6.6.39)$$

But for neutrinos it is easier to use the explicit solution (6.6.36). Using Eqs. (6.6.36) and (6.1.55) in the third expression of Eq. (6.6.35) and comparing with Eq. (6.6.5) gives[3]

$$\pi_{\nu q}^T(t) = -4\bar{\rho}_\gamma(t) \int_{t_1}^t dt' \, K \left(q \int_{t'}^t \frac{dt''}{a(t'')} \right) \dot{D}_q(t') , \qquad (6.6.40)$$

where

$$K(v) \equiv j_2(v)/v^2 = -\frac{\sin v}{v^3} - \frac{3 \cos v}{v^4} + \frac{3 \sin v}{v^5} . \qquad (6.6.41)$$

We will use this below in calculating the decay of gravitational waves that exit from the horizon in the radiation-dominated era.

We have found a complete set of differential equations for the tensor perturbations. Now we shall turn to the calculation of $\mathcal{D}_q(t)$, which provides the essential input in the calculation of tensor anisotropies in the cosmic microwave background. We begin by neglecting the tensor anisotropic inertia π_{ij}^T, returning to a consideration of its effects at the end of this section. The anisotropic inertia is negligible during most of the history of the universe, when the cosmic energy density is dominated by one or more perfect fluids, and it is never very large. With this approximation, the field equation (6.6.6) governing the Fourier components of the tensor component D_{ij} of the metric perturbation is simply

$$\ddot{D}_q(t) + 3\frac{\dot{a}}{a}\dot{D}_q(t) + \frac{q^2}{a^2}D_q(t) = 0 . \qquad (6.6.42)$$

---

[3]S. Weinberg, *Phys. Rev. D* **69**, 023503 (2004) [astro-ph/0306304].

As already mentioned earlier in this section, at very early times when $q/a \ll \dot{a}/a$ one solution is constant while the other decays, so as long as this era lasts sufficiently long we can neglect the decaying solution, and take our initial condition that $\mathcal{D}_q(t)$ goes to a constant $\mathcal{D}_q^o$ at early times.

To treat the evolution of $\mathcal{D}_q(t)$ at later times, it is convenient once again to change the independent variable from t to $y \equiv a/a_{\mathrm{EQ}} = \bar{\rho}_M/\bar{\rho}_R$, where a_{EQ} is the value of the Robertson–Walker scale factor at matter–radiation equality. Assuming that the energy density of the universe is governed by radiation and non-relativistic matter,[4] we can put Equation (6.3.11) in the form

$$\frac{H_{\mathrm{EQ}}\, dt}{\sqrt{2}} = \frac{y\, dy}{\sqrt{1+y}},\qquad (6.6.43)$$

where H_{EQ} is the expansion rate at matter–radiation equality. Then Eq. (6.6.42) becomes

$$(1+y)\frac{d^2\mathcal{D}_q}{dy^2} + \left(\frac{2(1+y)}{y} + \frac{1}{2}\right)\frac{d\mathcal{D}_q}{dy} + \kappa^2 \mathcal{D}_q = 0 \qquad (6.6.44)$$

where κ is the dimensionless rescaled wave number (6.5.9):

$$\kappa \equiv \frac{\sqrt{2}q}{a_{\mathrm{EQ}} H_{\mathrm{EQ}}} = \frac{(q/a_0)\sqrt{\Omega_R}}{H_0 \Omega_M} = \frac{19.3(q/a_0)[\mathrm{Mpc}^{-1}]}{\Omega_M h^2}. \qquad (6.6.45)$$

We also have the initial condition, that $\mathcal{D}_q \to \mathcal{D}_q^o$ for $y \ll 1$. As in the case of scalar modes, we can find analytic solutions in two extreme cases, for $\kappa \ll 1$ and $\kappa \gg 1$.

Consider first the case $\kappa \gg 1$. In this limit we already have $q/a \gg \dot{a}/a$ at matter–radiation equality, so horizon entry occurs early in the radiation-dominated era. Hence for $\kappa \gg 1$ there are two overlapping eras; an era when the universe is radiation dominated extending to early times when the perturbation is outside the horizon, and an era when the perturbation is well inside the horizon extending to the present, in both of which we will be able to find analytic solutions. Because these eras overlap, we will be able to match the analytic solutions in the era of overlap, and in this way relate the gravitational wave amplitude at late times to the initial condition when the perturbation is outside the horizon.

First, in the era when the universe is radiation dominated $y \ll 1$, and Eq. (6.6.44) becomes

$$\frac{d^2\mathcal{D}_q}{dy^2} + \left(\frac{2}{y}\right)\frac{d\mathcal{D}_q}{dy} + \kappa^2 \mathcal{D}_q = 0. \qquad (6.6.46)$$

---

[4]The evolution of the tensor amplitude under more general assumptions is considered by L. A. Boyle and P. J. Steinhardt, astro-ph/0512014.

The solution which approaches a constant $\mathcal{D}_q^o$ for $y \to 0$ is

$$D_q = \frac{\sin(\kappa y)}{\kappa y} \, \mathcal{D}_q^o \, . \tag{6.6.47}$$

Next, consider the case of perturbations deep inside the horizon, that is, for which $q/a \gg H$. In this case we can find solutions using the WKB approximation for a completely arbitrary dependence of $a(t)$ on time, so that our results will apply even in the presence of vacuum energy, whether or not it is constant. For this purpose, we must put Eq. (6.6.42) into a standard form with no first derivative term[5] by introducing a new independent variable,

$$x \equiv \int a^{-3}(t) \, dt \, ,$$

so that Eq. (6.6.42) becomes

$$\frac{d^2 \mathcal{D}_q}{dx^2} + q^2 a^4 \mathcal{D}_q = 0 \, . \tag{6.6.48}$$

We write $\mathcal{D}_q = A \exp(\pm iq \int a^2 \, dx)$, and keep only terms in Eq. (6.6.48) of order q^2 and q. This gives $dA/dx = -(A/2a^2)da^2/dx$, so $A \propto 1/a$, and the WKB solutions for $\mathcal{D}_q$ are $a^{-1} \exp(\pm iq \int a^2 \, dx) = a^{-1} \exp(\pm iq \int a^{-1} \, dt)$. The factor $1/a$ gives a gravitational wave energy density that decreases as a^{-4}, a factor of a^{-1} representing the redshift of individual gravitons, and a factor a^{-3} arising from the dilution of gravitons as the universe expands.[6]

To find the correct linear combination of the two WKB solutions, we have to match them to the solution (6.6.47) in the radiation-dominated era, by considering these solutions in the intermediate range where both $q/a \gg H$ *and* $y \ll 1$. In the radiation-dominated era vacuum energy is presumably negligible, so here $q/aH = \kappa \, a_{EQ} H_{EQ}/\sqrt{2}aH = \kappa y$, and for $\kappa \gg 1$ there does exist a range of y for which $\kappa y \gg 1$ even though $y \ll 1$. In this range of y, we have

$$q \int \frac{dt}{a} = q \int \frac{da}{Ha^2} = \kappa \int \frac{y \, da}{a} = \kappa y \, ,$$

---

[5]The standard way of eliminating the first derivative term in Eq. (6.6.42) is to change the dependent instead of the independent variable. For instance, see V. S. Mukhanov, H. A. Feldman, and R. H. Brandenberger, *Phys. Rep.* **215**, 203 (1992); K. Ng and A. Speliotopoulos, *Phys. Rev. D* **52**, 2112 (1995); and more recently, J. R. Pritchard and M. Kamionkowski, *Ann. Phys.* **318**, 2 (2005). The results obtained in this section are much simpler than with this standard method, because by changing the independent rather than the dependent variable we encounter no turning point in the WKB solution.

[6]See G&C, Sec. 15.10.

so (recalling that $y \propto a$) the linear combination of $a^{-1} \exp(+iq \int a^{-1} \, dt)$
and $a^{-1} \exp(-iq \int a^{-1} \, dt)$ that matches Eq. (6.6.47) where both are valid is

$$\mathcal{D}_q = \frac{1}{\kappa y} \sin \left(q \int_0^t \frac{dt'}{a(t')} \right) \mathcal{D}_q^o . \tag{6.6.49}$$

To repeat, this is the solution deep in the horizon, whatever the contents of
the universe. For times near the present, this is

$$\mathcal{D}_q = \frac{\sqrt{\Omega_R H_0^2}}{k} \sin \left(\eta + k(t - t_0) \right) \mathcal{D}_q^o , \tag{6.6.50}$$

where $k \equiv q/a_0$, and $\eta \equiv k a_0 \int_0^{t_0} dt/a(t)$.

It is possible that cosmological gravitational waves might be detected
directly.[7] Tensor modes detectable in this way would certainly have $\kappa \gg 1$,
the case we have been considering, so that they would enter the horizon
much earlier than the time of recombination, and hence could provide an
opportunity for a direct observation of the universe at very early times, and
even for the exploration of physics at higher energies than can be reached by
conventional particle accelerators. The existing ground-based Laser Interferometer Gravitational-Wave Observatory (LIGO) operates at about 100
Hz; perturbations with this frequency re-entered the horizon when the cosmic temperature was about 10^8 GeV, but LIGO does not have the sensitivity required to detect cosmological gravitational waves. But cosmological
gravitational waves might be detected by space-borne laser interferometers.
For instance, two detectors of cosmological gravitational waves have been
under consideration: the Big Bang Observer in the U.S., and the Deci-
hertz Laser Interferometer Gravitational Wave Observatory in Japan. Both
operate at frequencies around 0.01 to 0.1 Hz, and according to Eq. (6.6.43),
a gravitational wave with a frequency of $qc/2\pi a_0 = 10^{-2}$ Hz would have
$\kappa = 1.3 \times 10^{15}/\Omega_M h^2 \gg 1$. The wavelengths to which such detectors would
be sensitive would short enough to have come into the horizon when the
cosmic temperature was 10^4 to 10^5 GeV. In this case, the changes in the
time dependence of $a(t)$ associated with changes in the equation of state of
matter at various annihilation thresholds produce distinctive features in the
spectrum of the tensor modes.[8] For reasonable assumptions about the pri-

---

[7]For recent studies, see N. Seto, S. Kawamura, and T. Nakamura, *Phys. Rev. Lett.* **87**, 221103
(2001) [astro-ph/0108011]; A. Buonanno, gr-qc/0303085; A. Cooray, astro-ph/0503118; T. L. Smith,
M. Kamionkowski, and A. Cooray, *Phys. Rev. D* **73**, 023504 (2006) [astro-ph/0506422]; G. Efstathiou
and S. Chongchitnan, *Prog. Theor. Phys. Suppl.* **163**, 204 (2006) [astro-ph/0603118]; B. Friedman, A.
Cooray, and A. Melchiorri, *Phys. Rev. D* **74**, 123509 (2006) [astro-ph/0610220]. Also see NASA web
page universe.nasa.gov/program/bbo.html.

[8]Y. Watanabe and E. Komatsu, *Phys. Rev. D* **73**, 123515 (2006) [astro-ph/0604176].

mordial tensor spectrum, it seems likely[9] that a tensor perturbation strong enough to be detected directly at short wavelengths would be detected also indirectly at longer wavelengths through its effect on the polarization of the cosmic microwave background, to be discussed in Section 7.4.

In calculating the effect of cosmological gravitational waves on the cosmic microwave background, we need also to consider values of κ of order unity or less. Let us therefore turn to the other case that can be treated analytically, the case $\kappa \ll 1$. Here we still have $q/a \ll \dot{a}/a$ at matter–radiation equality, so the perturbation remains outside the horizon and $\mathcal{D}_q(t)$ remains equal to the constant value $\mathcal{D}_q^0$ until well into the matter-dominated era, when $y \gg 1$. In the limit $y \gg 1$, and whatever the value of κ, Eq. (6.6.42) becomes

$$y\frac{d^2 \mathcal{D}_q}{dy^2} + \frac{5}{2}\frac{d\mathcal{D}_q}{dy} + \kappa^2 \mathcal{D}_q = 0 \,. \tag{6.6.51}$$

This has two independent solutions, which can be written as functions of $\kappa\sqrt{y}$:

$$\frac{2j_1(2\kappa\sqrt{y})}{\kappa\sqrt{y}} = \frac{-\cos\left(2\kappa\sqrt{y}\right)}{\kappa^2 y} + \frac{\sin\left(2\kappa\sqrt{y}\right)}{2\kappa^3 y^{3/2}},$$

$$-\frac{2n_1(2\kappa\sqrt{y})}{\kappa\sqrt{y}} = \frac{\sin\left(2\kappa\sqrt{y}\right)}{\kappa^2 y} + \frac{\cos\left(2\kappa\sqrt{y}\right)}{2\kappa^3 y^{3/2}} \,. \tag{6.6.52}$$

In the matter-dominated era, $q/aH = \kappa a_{EQ} H_{EQ}/\sqrt{2}aH = \kappa\sqrt{y}$, so we must impose the condition that the solution should approach the constant value $\mathcal{D}_q^0$ for $\kappa\sqrt{y} \ll 1$. In order to satisfy this condition, we must exclude the second solution, which becomes singular for $\kappa\sqrt{y} \to 0$. In this limit the first solution approaches the constant $4/3$, so the correct solution when $\kappa \ll 1$ is

$$\mathcal{D}_q \to \frac{3\,\mathcal{D}_q^0}{4}\left(\frac{-\cos\left(2\kappa\sqrt{y}\right)}{\kappa^2 y} + \frac{\sin\left(2\kappa\sqrt{y}\right)}{2\kappa^3 y^{3/2}}\right) \,. \tag{6.6.53}$$

(Note that although Eq. (6.6.51) applies only for $y \gg 1$, Eq. (6.6.53) is valid for $\kappa \ll 1$ and any y, because it correctly gives $\mathcal{D}_q(t) = \mathcal{D}_q^0$ as long as $\kappa\sqrt{y} \ll 1$, which for $\kappa \ll 1$ applies until $y \gg 1$.)

The interpolation between the cases $\kappa \gg 1$ and $\kappa \ll 1$ is simplest in the matter-dominated era, when $y \gg 1$. This is only a fair approximation at the time of the decoupling of matter and radiation, when $y \approx 3$, but it

---

[9]Smith, Kamionkowski, and Cooray, ref. 7.

becomes a good approximation thereafter. In this case the tensor modes are governed by Eq. (6.6.51), and the general solution is a linear combination of the two solutions (6.6.52), which for any κ can conveniently be written

$$
\mathcal{D}_q = \frac{3 \mathcal{D}_q^o \, \mathcal{U}(\kappa)}{4\kappa^2} \left(\frac{-\cos\left(2\kappa\sqrt{y} + \Xi(\kappa)\right)}{y} + \frac{\sin\left(2\kappa\sqrt{y} + \Xi(\kappa)\right)}{2\kappa y^{3/2}} \right) ,
$$

(6.6.54)

where $\mathcal{U}$ and Ξ are real dimensionless functions of κ. From Eq. (6.6.53) we see that for $\kappa \ll 1$,

$$
\mathcal{U}(\kappa) \to 1 , \quad \Xi(\kappa)/\kappa^3 \to 0 ,
$$

(6.6.55)

which is why we chose to write the physical solution in the form (6.6.54). Also, for $y \gg 1$ Eq. (6.6.49) becomes

$$
\mathcal{D}_q = \frac{\sin\left(2\kappa(\sqrt{1+y} - 1)\right)}{\kappa y} \mathcal{D}_q^o ,
$$

(6.6.56)

and therefore for $\kappa \gg 1$,

$$
\mathcal{U}(\kappa) \to \frac{4\kappa}{3} , \quad \Xi(\kappa) \to \frac{\pi}{2} - 2\kappa .
$$

(6.6.57)

The values of $\mathcal{U}(\kappa)$ and $\Xi(\kappa)$ for general κ must be found by a computer calculation of the solution of Eq. (6.6.44). The results[10] are presented in Table 6.2. It can be seen that the analytic asymptotic limits (6.6.55) and (6.6.57) agree quite well with the computer calculation for $\kappa \ll 1$ and $\kappa \gg 1$, respectively.

Now let us consider the effect of anisotropic inertia.[11] The anisotropic inertia tensor is the sum of the contributions from photons and neutrinos, but photons are have a short mean free time before the era of recombination, and make only a small contribution to the total energy density afterwards, so their contribution to the anisotropic inertia is small. This leaves neutrinos (including antineutrinos), which have been travelling essentially without collisions[12] since the temperature dropped below

---

[10] D. Dicus, private communication.

[11] S. Weinberg, ref. 3. For results of an earlier computer calculation, see J. R. Bond, in *Cosmology and Large Scale Structure, Les Houches Session LX*, eds. R. Schaeffer, J. Silk, and J. Zinn-Justin (Elsevier, Amsterdam, 1996).

[12] For very short wavelengths that entered the horizon before the time of electron–positron annihilation, it is necessary to take into account collisions of neutrinos with each other and with electrons and positrons. This is considered by M. Lattanzi and G. Montani, *Mod. Phys. Lett. A* **20**, 2607 (2005) [astro-ph/0508364].

Table 6.2: The Tensor Transfer Functions

κ	$\mathcal{U}$	Ξ	κ	$\mathcal{U}$	Ξ
0.001	1.0000	-2.991×10^{-11}	1.8	2.990	-2.418
0.002	1.0000	-9.557×10^{-10}	2.0	3.250	-2.781
0.003	1.0000	-7.239×10^{-9}	2.5	3.906	-3.706
0.004	1.0000	-9.236×10^{-8}	3.0	4.565	-4.646
0.005	1.0001	-9.236×10^{-8}	3.5	5.222	-5.597
0.006	1.0001	-2.286×10^{-7}	4.0	5.886	-6.555
0.007	1.0002	-4.908×10^{-7}	4.5	6.551	-7.518
0.008	1.0002	-9.498×10^{-7}	5.0	7.216	-8.484
0.009	1.0003	-1.697×10^{-6}	5.5	7.878	-9.453
0.01	1.0004	-2.847×10^{-6}	6.0	8.541	-10.425
0.02	1.0016	-7.848×10^{-5}	6.5	9.208	-11.398
0.03	1.0037	-4.655×10^{-4}	7.0	9.875	-12.372
0.04	1.0065	-1.397×10^{-3}	7.5	10.538	-13.347
0.05	1.0095	-2.809×10^{-3}	8.0	11.201	-14.324
0.06	1.0128	-4.397×10^{-3}	8.5	11.869	-15.302
0.07	1.0165	-6.023×10^{-3}	9.0	12.538	-16.279
0.08	1.0210	-7.929×10^{-3}	10.0	13.863	-18.237
0.09	1.0260	-1.044×10^{-2}	11.0	15.202	-20.196
0.1	1.0310	-1.352×10^{-2}	12.0	16.526	-22.157
0.2	1.0960	-5.802×10^{-2}	13.0	17.867	-24.119
0.3	1.1800	-0.1293	14.0	19.190	-26.081
0.4	1.2766	-0.2223	15.0	20.532	-28.044
0.5	1.3816	-0.3327	16.0	21.854	-30.001
0.6	1.4926	-0.4568	17.0	23.197	-31.973
0.7	1.6079	-0.5919	18.0	24.52	-33.94
0.8	1.7265	-0.7357	19.0	25.86	-35.90
0.9	1.848	-0.8866	20.0	27.18	-37.87
1.0	1.970	-1.043	21.0	28.53	-39.83
1.1	2.095	-1.205	22.0	29.85	-41.80
1.2	2.220	-1.370	23.0	31.19	-43.77
1.4	2.474	-1.171	24.0	32.51	-45.73
1.6	2.731	-2.061	25.0	33.86	-47.70

about 10^{10} K, and which make up a good fraction of the energy density of the universe until cold dark matter becomes important, at a temperature about 10^4 K. The tensor part of the anisotropic inertia tensor is given by Eq. (6.6.40), so the gravitational wave equation (6.6.6) now becomes an integro-differential equation[13]

$$\ddot{D}_q(t) + 3\frac{\dot{a}}{a}\dot{D}_q(t) + \frac{q^2}{a^2}D_q(t) = -64\pi \, G\bar{\rho}_\nu(t) \int_0^t K\left(q\int_{t'}^t \frac{dt''}{a(t'')}\right)\dot{D}_q(t')\,dt',$$

(6.6.58)

with $K(v)$ given by Eq. (6.6.41). Note that despite the presence of anisotropic inertia, for $q/a \ll H$ this has a solution with $D_q(t)$ time-independent, in accordance with the general theorem of Section 5.4.

This wave equation becomes particularly simple for wavelengths short enough to enter the horizon during the radiation-dominated era (though well after e^+–e^- annihilation), that is, for $\kappa \gg 1$. We will define the zero of time so that in this era $a \propto \sqrt{t}$. It is convenient now to write D_q as a function of the variable

$$u \equiv q \int_0^t \frac{dt'}{a(t')} = \frac{2qt}{a(t)},$$

(6.6.59)

instead of t. Using the Friedmann equation $8\pi \, G\bar{\rho}/3 = H^2 = 1/4t^2$, the gravitational wave equation (6.6.58) in the radiation-dominated era becomes

$$\frac{d^2}{du^2}D_q(u) + \frac{2}{u}\frac{d}{du}D_q(u) + D_q(u) = -\frac{24 f_\nu}{u^2}\int_0^u K(u-u')\frac{d\,D_q(u')}{du'}\,du',$$

(6.6.60)

where

$$f_\nu \equiv \frac{\bar{\rho}_\nu}{\bar{\rho}_\nu + \bar{\rho}_\gamma} = \frac{3\cdot(7/8)\cdot(4/11)^{4/3}}{1+3\cdot(7/8)\cdot(4/11)^{4/3}} = 0.4052\,.$$

(6.6.61)

Late in the radiation-dominated era, the factor $1/u^2$ makes the right-hand side negligible, so $D_q(u)$ approaches a solution of the homogeneous equation. In general, one might expect a linear combination of $\sin u/u$ and $\cos u/u$, but in fact no $\cos u/u$ term appears in the solution.[14] A numerical solution of Eq. (6.6.60) shows that if $D_q(u)$ takes the value D_q^o for $u \ll 1$ then for $u \gg 1$ (but still in the radiation-dominated era)

$$D_q(u) \to D_q^o\,\alpha\,\frac{\sin(u)}{u},$$

(6.6.62)

---

[13]The lower bound on this integral should in principle be taken as the time of neutrino decoupling, at a temperature of about 10^{10} K, but it is a good approximation to take it at a time $t = 0$, defined by writing the scale factor during the radiation-dominated era as $a \propto t^{1/2}$.

[14]D. Dicus and W. Repko, *Phys. Rev.* **72**, 088302(2005) [astro-ph/0509096] have shown analytically that no $\cos u/u$ term appears in this solution, and have given an analytic solution as a rapidly convergent sum of spherical Bessel functions. The absence of a $\cos u/u$ term was shown very generally on causality grounds by S. Bashinsky, astro-ph/0505502.

where $\alpha = 0.8026$. This then serves instead of Eq. (6.6.47) as an initial condition for the subsequent evolution of the gravitational wave amplitude, so that later, during the matter-dominated era, the effect of damping on the amplitude (neglecting δ) is simply to multiply the result given in Eq. (6.6.54) by a factor α.[15]

The damping of gravitational waves of longer wavelength is considerably more complicated. Because of a shift in the phase of the oscillation, the effect of anisotropic inertia on the amplitude of gravitational waves at the time of decoupling is a sensitive function of wave number, and for some wave numbers can even be an enhancement instead of a damping, but typically the amplitude is damped for $\kappa = O(1)$ by roughly 5%.

Because the damping effect is small anyway for $\kappa \ll 1$, it will be an adequate approximation for all wavelengths to take the gravitational wave amplitude in the matter dominated era to be given by multiplying the result (6.6.54) with a factor $\alpha(\kappa)$:

$$
\mathcal{D}_q = \frac{3\mathcal{D}_q^o \mathcal{U}(\kappa)\alpha(\kappa)}{4\kappa^2}\left(\frac{-\cos\left(2\kappa\sqrt{y}+\Xi(\kappa)\right)}{y} + \frac{\sin\left(2\kappa\sqrt{y}+\Xi(\kappa)\right)}{2\kappa y^{3/2}}\right),
$$

(6.6.63)

with $\alpha(\kappa)$ some function of κ that rises smoothly from $\alpha(\kappa) = 0.8026$ for $\kappa \gg 1$ to $\alpha(\kappa) \simeq 1$ for $\kappa \ll 1$. For instance, we can take $\alpha(\kappa) \simeq (1+.8026\kappa)/(1+\kappa)$. All observable effects of cosmological gravitational waves will be reduced by this factor $\alpha(\kappa)$.

[15]The effect of possible neutrino masses and/or chemical potentials is considered by K. Ichiki, M. Yamaguchi, and J. Yokayama, *Pub. Astron. Soc. Pacific* **119**, 30 (2007) [hep-ph/0611121].

Anisotropies in the Microwave Sky

We will now return to the theory of anisotropies in the cosmic microwave background, introduced in Section 2.6. In Section 7.1 we derive general formulas for the observed temperature fluctuation. Then in Sections 7.2 and 7.3 we combine these results with the analysis of cosmic evolution in Chapter 6 and introduce a series of approximations that simplify the evaluation of the multipole coefficients for scalar and tensor modes, respectively. Section 7.4 deals with the polarization of the microwave background.

7.1 General formulas for the temperature fluctuation

In this section we will derive general formulas for the contribution of scalar and tensor modes to the observed temperature fluctuation. We begin by carrying the solution of the Boltzmann equations in synchronous gauge given in Section 6.1 forward to the present. When implemented with computer programs such as CMBfast or CAMB, the approach provides numerical results of great accuracy, but neither the derivation of the formulas for temperature fluctuations nor the computer programs are physically very transparent. We will then show how these results can be simplified by making the approximation of a sharp transition from thermal equilibrium to complete transparency at a moment t_L of last scattering. This ignores the scattering of photons by matter that becomes reionized at a redshift of order 10. For temperature correlations, the corrections due to reionization are very simple for multipole orders ℓ greater than about 20, and will be included in Section 7.2. In Section 7.2 we will also partly make up for the approximation of a sharp drop from opacity to transparency by including effects of viscous damping during this transition, and by including effects of averaging over t_L. (At the end of this section we will show how the same simplified results can also be obtained in a more general gauge by following photon trajectories from t_L to the present, with no need to use the Boltzmann equation.) The results of the sudden-decoupling approximation obtained here will be used together with other approximations in Section 7.2 to derive analytic expressions for the temperature multipole coefficients, that require computer calculations only to carry out a single numerical integration.

Because the proper energy density of black body radiation is proportional to the fourth power of the temperature, the fractional perturbation in temperature of radiation coming from a direction $\hat{n}$ is one-fourth the fractional perturbation in the proper energy density of photons travelling in direction $\hat{p} = -\hat{n}$ at our position $\mathbf{x} = 0$ and time $t = t_0$. Eqs. (6.1.13) or (6.6.8) thus give

$$\frac{\Delta T(\hat{n})}{T_0} = \frac{1}{4} J_{ii}(\mathbf{x} = 0, -\hat{n}, t_0) ,\qquad (7.1.1)$$

where $J_{ij}(\mathbf{x}, \hat{p}, t)$ is the fractional density matrix, defined by Eqs. (6.1.13) for scalar modes and (6.6.8) for tensor modes. Using the decompositions (6.1.18), (6.1.20) for scalar modes or (6.6.9), (6.6.13) for tensor modes, the scalar and tensor contributions to the temperature fluctuation are

$$\left(\frac{\Delta T(\hat{n})}{T_0}\right)^{(S)} = \frac{1}{4} \int d^3q \, \alpha(\mathbf{q}) \, \Delta_T^{(S)}(q, -\hat{q} \cdot \hat{n}, t_0) ,\qquad (7.1.2)$$

and

$$\left(\frac{\Delta T(\hat{n})}{T_0}\right)^{(T)} = \frac{1}{4} \sum_{\lambda=\pm 2} \int d^3q \, \beta(\mathbf{q}, \lambda) \, \hat{n}_k \hat{n}_l \, e_{kl}(\hat{q}, \lambda) \Delta_T^{(T)}(q, -\hat{q} \cdot \hat{n}, t_0) ,$$

$$(7.1.3)$$

where $\alpha(\mathbf{q})$ and $\beta(\mathbf{q}, \lambda)$ are the stochastic parameters for whatever modes are assumed to dominate the scalar and tensor perturbations, respectively; $e_{kl}(\hat{q}, \lambda)$ is the polarization tensor defined in Section 6.6 for a gravitational wave with wave number $\mathbf{q}$ and helicity λ; and $\Delta_T^{(S)}(q, \mu, t)$ and $\Delta_T^{(T)}(q, \mu, t)$ are amplitudes appearing in the decompositions (6.1.20) and (6.6.13) of the scalar and tensor contributions to the fractional density matrix J_{ij}.

In order to display the angular dependence of the temperature shift, we use the line-of-sight integrals (6.1.37) and (6.1.38) for scalar modes and (6.6.25) and (6.6.26) for tensor modes, which give, for the scalar modes

$$\left(\frac{\Delta T(\hat{n})}{T_0}\right)^{(S)} = \frac{1}{4} \int d^3q \, \alpha(\mathbf{q})$$

$$\times \int_{t_1}^{t_0} dt \, \exp\left[i\mathbf{q} \cdot \hat{n} \int_t^{t_0} \frac{dt'}{a(t')} - \int_t^{t_0} dt' \, \omega_c(t')\right]$$

$$\times \left[-2\dot{A}_q(t) + 2(\mathbf{q} \cdot \hat{n})^2 \dot{B}_q(t) + 3\omega_c(t)\Phi(q, t) \right.$$

$$-4i(\mathbf{q} \cdot \hat{n}) \, \omega_c(t) \, \delta u_{Bq}(t)/a(t)$$

$$\left. +\frac{3}{4}\left(1 - (\hat{q} \cdot \hat{n})^2\right) \omega_c(t) \, \Pi(q, t) \right] ,\qquad (7.1.4)$$

and for the tensor modes

$$
\left(\frac{\Delta T(\hat{n})}{T_0}\right)^{(T)} = \frac{1}{4} \sum_{\lambda=\pm 2} \int d^3q \, \beta(\mathbf{q}, \lambda) \, \hat{n}_k \hat{n}_l \, e_{kl}(\hat{q}, \lambda)
$$

$$
\times \int_{t_1}^{t_0} dt \, \exp\left[i\mathbf{q} \cdot \hat{n} \int_t^{t_0} \frac{dt'}{a(t')} - \int_t^{t_0} dt' \, \omega_c(t') \right]
$$

$$
\times \left[-2\dot{\mathcal{D}}_q(t) + \omega_c(t) \, \Psi(q, t) \right].
\tag{7.1.5}
$$

As a reminder: t_1 is any time sufficiently early before recombination so that any photon present then would have been scattered many times before the present; $\omega_c(t)$ is the photon collision rate at time t; $A_q(t)$ and $B_q(t)$ are the scalar fields in the perturbation to the metric in synchronous gauge, defined in Section 5.2; $\Phi(q, t)$ and $\Pi(q, t)$ are the scalar source functions, defined by Eq. (6.1.21); $\delta u_{Bq}(t)$ is the scalar velocity potential of the baryonic plasma; $\mathcal{D}_q(t)$ is the gravitational wave amplitude, defined in Section 5.2; and $\Psi(q, t)$ is the tensor source function, defined by Eqs. (6.6.11) and (6.6.12).

Eq. (7.1.4) does not give the expression for the scalar temperature fluctuation in its most convenient form. When we pass to the limit of a sharp moment of last scattering, its terms will not correspond to the decomposition of the fluctuation into Sachs–Wolfe, Doppler, intrinsic, and integrated Sachs–Wolfe terms, discussed in Section 2.6. The individual terms in Eq. (7.1.4) are not even invariant under the limited class of gauge transformations (5.3.39)–(5.3.42) that preserve the conditions for synchronous gauge. We will therefore rewrite Eq. (7.1.4) by using the identity

$$
\exp\left(i\mathbf{q} \cdot \hat{n} \int_t^{t_0} \frac{dt'}{a(t')} \right) (\mathbf{q} \cdot \hat{n})^2 \dot{B}_q(t) =
$$

$$
- \exp\left(i\mathbf{q} \cdot \hat{n} \int_t^{t_0} \frac{dt'}{a(t')} \right) \frac{d}{dt}\left(a^2(t) \ddot{B}_q(t) + a(t)\dot{a}(t) \, \dot{B}(t) \right)
$$

$$
+ \frac{d}{dt}\left\{ \exp\left(i\mathbf{q} \cdot \hat{n} \int_t^{t_0} \frac{dt'}{a(t')} \right) \right.
$$

$$
\left. \times \left[a^2(t) \ddot{B}_q(t) + a(t)\dot{a}(t) \dot{B}_q(t) + ia(t)\mathbf{q} \cdot \hat{n} \dot{B}_q(t) \right] \right\}
\tag{7.1.6}
$$

Using this in Eq. (7.1.4) and integrating by parts, yields our final formula for the scalar temperature fluctuation

$$
\left(\frac{\Delta T(\hat{n})}{T_0}\right)^{(S)} = \left(\frac{\Delta T(\hat{n})}{T_0}\right)^{(S)}_{\text{early}} + \left(\frac{\Delta T(\hat{n})}{T_0}\right)^{(S)}_{\text{ISW}},
\tag{7.1.7}
$$

where the first, "early," term receives contributions only from times with an appreciable free electron density, before recombination and after reionization,

$$
\left(\frac{\Delta T(\hat{n})}{T_0}\right)^{(S)}_{\text{early}} = \int d^3q\, \alpha(\mathbf{q})
$$
$$
\times \int_{t_1}^{t_0} dt\, \exp\left[i\mathbf{q}\cdot\hat{n}\int_t^{t_0}\frac{dt'}{a(t')} - \int_t^{t_0} dt'\,\omega_c(t')\right]
$$
$$
\times \omega_c(t)\left[\frac{3}{4}\Phi(q,t) + \frac{3}{16}\left(1-(\hat{q}\cdot\hat{n})^2\right)\Pi(q,t)\right.
$$
$$
-\frac{1}{2}a^2(t)\ddot{B}_q(t) - \frac{1}{2}a(t)\dot{a}(t)\dot{B}_q(t)
$$
$$
\left. - i(\mathbf{q}\cdot\hat{n})\left(\delta u_{Bq}(t)/a(t) + a(t)\dot{B}_q(t)/2\right)\right],
\tag{7.1.8}
$$

and the second, "integrated Sachs–Wolfe term," receives contributions from the whole period from t_1 to the present,

$$
\left(\frac{\Delta T(\hat{n})}{T_0}\right)^{(S)}_{\text{ISW}} = -\frac{1}{2}\int d^3q\, \alpha(\mathbf{q})\int_{t_1}^{t_0} dt
$$
$$
\times \exp\left[i\mathbf{q}\cdot\hat{n}\int_t^{t_0}\frac{dt'}{a(t')} - \int_t^{t_0} dt'\,\omega_c(t')\right]
$$
$$
\times \frac{d}{dt}\left[A_q(t) + a^2(t)\ddot{B}_q(t) + a(t)\dot{a}(t)\dot{B}_q(t)\right].
\tag{7.1.9}
$$

(Here we are ignoring a surface term in the integration by parts at $t = t_0$, because it is linear in $\hat{n}$, and therefore just contributes to the $\ell = 0$ and $\ell = 1$ partial waves. This "late" term is calculated by a different method later in this section, and given in Eq.(7.1.38).)

The integrated Sachs–Wolfe term (7.1.9) represents the effect of changing gravitational fields during the passage of the microwave photons from last scattering to the present. As already noted in Section 2.6, the ISW term would vanish if the gravitational field from last scattering to the present (or more precisely, at times t when the transparency $\exp\left(-\int_t^{t_0}\omega_c(t')dt'\right)$ was non-negligible) arose solely from the density of cold matter. (This feature provides another reason, apart from gauge invariance, for the rearrangement of terms that led to Eq. (7.1.7).) Under this approximation, Eqs. (6.5.15) and (6.5.16) show that $\delta_{Dq}\propto t^{2/3}$ and $\psi_q\propto t^{-1/3}$. Eq. (5.3.38) gives $q^2 A_q = 8\pi a^2\bar{\rho}_D\delta_{Dq} - 2Ha^2\psi_q$, both terms of which are constant in the matter-dominated era, so $\dot{A}_q$ does not contribute to Eq. (7.1.9). With $\dot{A}_q$

negligible, $\psi_q = -q^2 B_q$, so $\dot{B}_q \propto t^{-1/3}$, and therefore both terms $a^2 \ddot{B}_q$ and $a\dot{a}\dot{B}_q$ in Eq. (7.1.9) are time-independent, and therefore do not contribute to the ISW effect either. For this reason the ISW term (7.1.9) is relatively small. The early-time ISW effect depends sensitively on the ratio of matter to radiation at last scattering, which is an aid in using observations of temperature fluctuations to measure $\Omega_M h^2$. The late-time ISW effect receives its main contribution from times near the present, when the dark matter density falls below the vacuum energy density. Anisotropies subtend larger angles when viewed nearby than from great distances, so the late-time ISW term in the temperature anisotropy contributes to the temperature multipole coefficients C_ℓ only for relatively small ℓ, say $\ell < 20$. It is the integrated Sachs–Wolfe effect that causes the predicted values of $\ell(\ell + 1)C_\ell$ to rise as ℓ drops from around 20 to smaller values. This effect has been difficult to see in present data on the cosmic microwave background temperature fluctuations.[1]

It is the "early" term (7.1.8) that makes the largest contribution to the scalar temperature fluctuation for $\ell > 20$. The terms in Eq. (7.1.8) proportional to $\mathbf{q} \cdot \hat{n}$ represent the Doppler effect,[2] while the other terms give the combined effect of gravitational time dilation and intrinsic temperature fluctuations.

Let's pause to check invariance under the limited set of gauge transformations that preserve the conditions for synchronous gauge. These transformations induce the changes $\Delta \dot{B}_q = -2\tau/a^2$ and $\Delta \delta u_{Bq} = \tau$, where τ is an arbitrary function of $\mathbf{x}$, so the Doppler term proportional to $i\mathbf{q} \cdot \hat{n}$ in

---

[1] The ISW effect can be detected through its correlation with inhomogeneities in the distribution of matter (which are also linear in $\alpha(\mathbf{q})$), as suggested by R. G. Crittenden and N. Turok, *Phys. Rev. Lett.* **76**, 575 (1996) [astro-ph/9510072]. This effect has been seen in the correlation of data from the WMAP satellite (discussed in the next section) with various surveys, by P. Fosalba and E. Gaztañaga, *Mon. Not. Roy. Astron. Soc.* **350**, L37 (2004) [astro-ph/0305468]; P. Fosalba, E. Gaztañaga, and F. Castander, *Astrophys. J.* **597**, L89 (2003) [astro-ph/0307249]; N. Ashfordi, Y-S. Loh, and M. A. Strauss, astro-ph/0308260; S. P. Boughn and R. G. Crittenden, *New Astron. Rev.* **49**, 75 (2005) [astro-ph/0404470]; N. Padmanabhan, C. M. Hirata, U. Seljak, D. Schlegel, J. Brinkmann and D. P. Schneider, *Phys. Rev. D* **72**, 043525 (2005) [astro-ph/0410360]. More recent cross-correlations of anisotropies seen by the three-year Wilkinson Microwave Anisotropy Probe (discussed in the next section) with galaxies in the Sloan Digital Sky Survey and with radio galaxy data from the NRAO VLA Sky Survey give $0.7 \leq \Omega_V \leq 0.82$ and $0.3 \leq \Omega_V \leq 0.8$ at a 95% confidence level, respectively; see A. Cabré *et al.*, astro-ph/0603690; D. Pietrobon, A. Balbi, and D. Marinucci, *Phys. Rev. D* **74**, 043524 (2006) [astro-ph/0606475]. Unfortunately, cosmic variance limits the accuracy with which this approach can be used to study the time dependence of the vacuum energy.

[2] In a gauge in which cold dark matter remains at rest, the effect of gravitational perturbations on the cold dark matter particles is canceled by the definition of surfaces of equal time, so in this gauge the velocity perturbations of the baryonic plasma arise solely from pressure forces, not from gravitational forces. This is why the gravitational term proportional to $\dot{B}_q$ appears accompanying the plasma velocity potential in Eq. (7.1.8); it represents the velocity that in a different gauge would be given to the photon–baryon plasma by gravitational forces.

Eq. (7.1.8) is separately gauge invariant. Also,[3] $\Delta\Phi = -(4/3)\dot{\bar{T}}\tau/\bar{T} = (4/3)\dot{a}\tau/a$ and $\Delta\Pi = 0$, so the other terms in Eq. (7.1.8) are also gauge invariant; and $\Delta A = -2\dot{a}\tau/a$, so also Eq. (7.1.9) is invariant under this limited set of gauge transformations.

Our results so far are exact, aside from the use of first-order perturbation theory and the assumption that photons interact only through purely elastic Thomson scattering. (In Eqs. (7.1.2) and (7.1.3) we are assuming that the scalar and tensor fluctuations are each dominated by a single mode, but it would be trivial to introduce a sum over modes in these expressions.) These results are equivalent to those given by Seljak and Zaldarriaga,[4] which are used in computer programs like CMBfast and CAMB. At the cost of only a small loss of numerical accuracy, they can be greatly simplified if we now make the approximation of a sharp transition from thermal equilibrium to perfect transparency at a definite time t_L.

The integrand of the "early" contribution (7.1.8) to the scalar temperature perturbation contains a factor $P(t)$ equal to the probability distribution of the last photon scattering

$$P(t) = \omega_c(t) \exp\left(-\int_t^{t_0} dt'\, \omega_c(t')\right). \qquad (7.1.10)$$

Assuming that $\omega_c(t)$ drops sharply at time t_L from a value much greater than the expansion rate to zero, the function $P(t)$ is non-zero only in a narrow interval around t_L. But $P(t)$ is a normalized probability distribution:

$$\int_{t_1}^{t_0} P(t)\, dt = 1 \qquad (7.1.11)$$

so the integral over t in Eq. (7.1.8) can be evaluated by dropping the factor $P(t)$ and setting $t = t_L$:

$$\left(\frac{\Delta T(\hat{n})}{T_0}\right)_{\text{early}}^{(S)} \simeq \int d^3q\, \alpha(\mathbf{q})\, \exp\left[i\mathbf{q}\cdot\hat{n}\int_{t_L}^{t_0}\frac{dt'}{a(t')}\right]$$
$$\times\left[\frac{3}{4}\Phi(q, t_L) + \frac{3}{16}\left(1 - (\hat{q}\cdot\hat{n})\right)^2\Pi(q, t_L)\right]$$

---

[3]This follows from the rule (5.3.42), that the change $\Delta\delta s$ in the perturbation δs to a scalar s with unperturbed value $\bar{s}$ is $\Delta s = -\dot{\bar{s}}\tau$. In using this rule, we note that the unperturbed photon distribution is isotropic and unpolarized, so that, of the terms in Eqs. (6.1.29) and (6.1.30) for Φ and Π, the only one that has an unperturbed value is $\Delta_{T,0}^{(S)}/3$. For the purposes of assessing the gauge transformation property of this term, we must define the unperturbed value of $\Delta_{T,0}^{(S)}$ so that its perturbation is the fractional photon density fluctuation $4\delta T/\bar{T}$, and so its unperturbed value is $4\ln\bar{T}$.

[4]U. Seljak and M. Zaldarriaga, *Astrophys. J.* **459**, 437 (1996) [astro-ph/9603033].

$$-\frac{1}{2}a^2(t)\ddot{B}_q(t_L) - \frac{1}{2}a(t_L)\dot{a}(t_L)\dot{B}_q(t_L)$$

$$-i(\mathbf{q} \cdot \hat{n})\Big(\delta u_q(t_L)/a(t_L) + a(t_L)\dot{B}_q(t_L)/2\Big)\Big]. \tag{7.1.12}$$

Under the same assumption of a rapid drop in $\omega_c(t)$ from a large value for $t < t_L$ to a negligible value for $t > t_L$, the factor $\exp\Big(-\int_t^{t_0} dt'\, \omega_c(t')\Big)$ in Eq. (7.1.9) rises sharply from zero for $t < t_L$ to unity for $t > t_L$, so Eq (7.1.9) becomes

$$\left(\frac{\Delta T(\hat{n})}{T_0}\right)^{(S)}_{\text{ISW}} \simeq -\frac{1}{2} \int d^3q\, \alpha(\mathbf{q}) \int_{t_L}^{t_0} dt\, \exp\Big[i\mathbf{q} \cdot \hat{n} \int_t^{t_0} \frac{dt'}{a(t')}\Big]$$

$$\times \frac{d}{dt}\Big[A_q(t) + a^2(t)\ddot{B}_q(t) + a(t)\dot{a}(t)\dot{B}_q(t)\Big]. \tag{7.1.13}$$

The same approximation applied to the tensor contribution (7.1.5) gives

$$\left(\frac{\Delta T(\hat{n})}{T_0}\right)^{(T)} = \frac{1}{4} \sum_{\lambda \simeq \pm 2} \int d^3q\, \beta(\mathbf{q}, \lambda)\, \hat{n}_k \hat{n}_l\, e_{kl}(\hat{q}, \lambda)$$

$$\times\Big[-2\int_{t_L}^{t_0} dt\, \dot{\mathcal{D}}_q(t) \exp\Big(i\mathbf{q} \cdot \hat{n} \int_t^{t_0} \frac{dt'}{a(t')}\Big)$$

$$+\Psi(q, t_L) \exp\Big(i\mathbf{q} \cdot \hat{n} \int_{t_L}^{t_0} \frac{dt'}{a(t')}\Big)\Big]. \tag{7.1.14}$$

Further, in local thermal equilibrium photons are unpolarized and have an isotropic momentum distribution, so $\Delta^{(S)}_{P,\ell}$, $\Delta^{(T)}_{P,\ell}$, and $\Delta^{(T)}_{T,\ell}$ vanish for all ℓ, and $\Delta^{(S)}_{T,\ell}$ vanishes for all ℓ except $\ell = 0$, so that the formulas (6.1.29), (6.1.30), and (6.6.21) for the source functions give $\Pi = \Psi = 0$ and $\Phi = \Delta^{(S)}_{T,0}/3 = 4\delta T/3\bar{T}$ in local thermal equilibrium.[5] The assumption of a sharp drop from a very high to a very low photon collision frequency then

---

[5]This can be seen formally by taking the limit $\omega_c \to \infty$ in the Boltzmann equations (6.1.27), (6.1.28), (6.6.18), and (6.6.19). This gives

$$\Delta^{(S)}_{P,\ell} = \frac{1}{2}\Pi(\delta_{\ell 0} + \delta_{\ell 2}), \quad \Delta^{(S)}_{T,\ell} = \Big(3\Phi + \frac{1}{2}\Pi\Big)\delta_{\ell 0} + \frac{1}{2}\Pi\delta_{\ell 2},$$

$$\Delta^{(T)}_{T,\ell} = \delta_{\ell 0}\Psi, \quad \Delta^{(T)}_{P,\ell} = -\delta_{\ell 0}\Psi.$$

The formulas (6.1.29), (6.1.30), and (6.6.21) for the source functions then read $\Phi = \Phi - \Pi/12$, $\Pi = 3\Pi/5$, and $\Psi = 7\Psi/10$, which require that $\Pi = \Psi = 0$ and $\Phi = \Delta^{(S)}_{T,0}/3$.

allows Eqs. (7.1.12) and (7.1.14) to be further simplified to

$$
\left(\frac{\Delta T(\hat{n})}{T_0}\right)^{(S)}_{\text{early}} \simeq \int d^3q\, \alpha(\mathbf{q})\, \exp\left[i\mathbf{q}\cdot\hat{n}\int_{t_L}^{t_0}\frac{dt'}{a(t')}\right]
$$
$$
\times\left[F(q) + i(\hat{q}\cdot\hat{n})G(q)\right], \tag{7.1.15}
$$

where $F(q)$ and $G(q)$ are the form factors

$$
F(q) = \frac{\delta T_q(t_L)}{\bar{T}(t_L)} - \frac{1}{2}a^2(t)\ddot{B}_q(t_L) - \frac{1}{2}a(t_L)\dot{a}(t_L)\dot{B}_q(t_L) \tag{7.1.16}
$$

$$
G(q) = -q\left(\delta u_{\gamma q}(t_L)/a(t_L) + a(t_L)\dot{B}_q(t_L)/2\right), \tag{7.1.17}
$$

and

$$
\left(\frac{\Delta T(\hat{n})}{T_0}\right)^{(T)} = -\frac{1}{2}\sum_{\lambda\simeq\pm2}\int d^3q\, \beta(\mathbf{q},\lambda)\,\hat{n}_k\hat{n}_l\, e_{kl}(\hat{q},\lambda)
$$
$$
\times\int_{t_L}^{t_0} dt\, \dot{\mathcal{D}}_q(t)\, \exp\left(i\mathbf{q}\cdot\hat{n}\int_t^{t_0}\frac{dt'}{a(t')}\right). \tag{7.1.18}
$$

* * *

We will now make the same approximation, of a sudden drop in opacity at t_L, and use it to derive formulas for the scalar and tensor temperature fluctuations by following photon trajectories, without needing to use the Boltzmann equation formalism of Section 6.1. Because it is easy, we will carry out this derivation in a more general class of gauges. After this derivation, we will check that it yields the results (7.1.15) and (7.1.13) for scalar fluctuations in synchronous gauge, and (7.1.18) for tensor fluctuations. The reader who is comfortable with the derivation of these formulas using the Boltzmann equation in synchronous gauge may want to skip the rest of this section.

We start with some general remarks, that apply equally to scalar and tensor perturbations, and that for scalar perturbations apply in any gauge in which $g_{i0} = 0$, including both Newtonian and synchronous gauge. Continuing to neglect a possible unperturbed spatial curvature, we write the perturbed metric in any gauge with $g_{i0} = 0$ in the form

$$
g_{00} = -1 - E(\mathbf{x}, t), \quad g_{i0} = 0, \quad g_{ij} = a^2(t)\delta_{ij} + h_{ij}(\mathbf{x}, t). \tag{7.1.19}
$$

A light ray travelling toward the center of the Robertson–Walker coordinate system from the direction $\hat{n}$ will have a co-moving radial coordinate r related to t by

$$
0 = g_{\mu\nu}dx^\mu dx^\nu = -\left(1 + E(r\hat{n}, t)\right)dt^2 + \left(a^2(t) + h_{rr}(r\hat{n}, t)\right)dr^2, \tag{7.1.20}
$$

or in other words

$$\frac{dr}{dt} = -\left(\frac{a^2 + h_{rr}}{1 + E}\right)^{-1/2} \simeq -\frac{1}{a} + \frac{h_{rr}}{2a^3} - \frac{E}{2a} . \tag{7.1.21}$$

(To first order in perturbations we don't have to worry about a deflection of the ray from the radial direction, because $\bar{g}_{r\theta} = \bar{g}_{r\phi} = 0$, so that any deflection would have to be of first order, and its effect in the term $h_{ij}dx^i dx^j$ would therefore be of second order.)

We now make the approximation that the transition of cosmic matter from opacity to transparency occurred suddenly at a time t_L of last scattering, at a red shift $1 + z_L \simeq 1090$. With this approximation, the relevant first-order solution of Eq. (7.1.21) is

$$r(t) = s(t) + \int_{t_L}^{t} \frac{dt'}{a(t')} N\left(s(t')\hat{n}, t'\right) , \tag{7.1.22}$$

where

$$N(\mathbf{x}, t) \equiv \frac{1}{2}\left[\frac{h_{rr}(\mathbf{x}, t)}{a^2(t)} - E(\mathbf{x}, t)\right] , \tag{7.1.23}$$

and $s(t)$ is the zeroth order solution for the radial coordinate which has the value r_L at $t = t_L$:

$$s(t) = r_L - \int_{t_L}^{t} \frac{dt'}{a(t')} = \int_{t}^{t_0} \frac{dt'}{a(t')} . \tag{7.1.24}$$

In particular, if the ray reaches $r = 0$ at a time t_0, then Eq. (7.1.22) gives

$$0 = s(t_0) + \int_{t_L}^{t_0} \frac{dt}{a(t)} N\left(s(t)\hat{n}, t\right) = r_L + \int_{t_L}^{t_0} \frac{dt}{a(t)}\left(N\left(s(t)\hat{n}, t\right) - 1\right) . \tag{7.1.25}$$

A time interval δt_L between the departure of successive light wave crests at the time t_L of last scattering produces a time interval δt_0 between arrival of successive crests at t_0 given by the variation of Eq. (7.1.25):

$$0 = \frac{\delta t_L}{a(t_L)}\left[1 - N\left(r_L\hat{n}, t_L\right) + \int_{t_L}^{t_0} \frac{dt}{a(t)}\left(\frac{\partial N(r\hat{n}, t)}{\partial r}\right)_{r=s(t)}\right]$$
$$+ \delta t_L \, \delta u_\gamma^r(r_L\hat{n}, t_L) + \frac{\delta t_0}{a(t_0)}\left[-1 + N(0, t_0)\right] . \tag{7.1.26}$$

(The term on the right-hand side involving the radial velocity δu_γ^r of the photon gas or photon–electron–nucleon fluid arises from the change with time of the radial coordinate r_L of the light source in Eq. (7.1.25). We don't consider the variation of the argument $s(t)\hat{n}$ in N, because to zeroth order r_L and t_L are related in such a way that $s(t) = 0$ for all r_L, so its variation with r_L is of first order, and the effect of this variation on N would be of second order.) The total rate of change of the quantity $N(s(t)\hat{n}, t)$ in Eq. (7.1.25) is

$$\frac{d}{dt}N\big(s(t)\hat{n}, t\big) = \left(\frac{\partial}{\partial t}N(r\hat{n}, t)\right)_{r=s(t)} - \frac{1}{a(t)}\left(\frac{\partial N(r\hat{n}, t)}{\partial r}\right)_{r=s(t)} ,$$

so Eq. (7.1.26) may be written

$$0 = \frac{\delta t_L}{a(t_L)}\left[1 - N(0, t_0) + \int_{t_L}^{t_0} dt \left\{\frac{\partial}{\partial t}N(r\hat{n}, t)\right\}_{r=s(t)}\right]$$

$$+\delta t_L \, \delta u_\gamma^r (r_L\hat{n}, t_L) + \frac{\delta t_0}{a(t_0)}\Big[-1 + N(0, t_0)\Big]. \qquad (7.1.27)$$

This gives the ratio of the coordinate time interval between crests when emitted and received, but what we want is the ratio of the proper time intervals

$$\delta \tau_L = \sqrt{1 + E(r_L, t_L)}\, \delta t_L , \quad \delta \tau_0 = \sqrt{1 + E(0, t_0)}\, \delta t_0 , \qquad (7.1.28)$$

which to first order gives the ratio of the received and emitted frequencies as

$$\frac{\nu_0}{\nu_L} = \frac{\delta \tau_L}{\delta \tau_0} = \frac{a(t_L)}{a(t_0)}\left[1 + \frac{1}{2}\Big(E(r_L\hat{n}, t_L) - E(0, t_0)\Big)\right.$$

$$- \int_{t_L}^{t_0}\left\{\frac{\partial}{\partial t}N(r\hat{n}, t)\right\}_{r=s(t)} dt$$

$$\left. - a(t_L)\, \delta u_\gamma^r (r_L\hat{n}, t)\right]. \qquad (7.1.29)$$

The temperature observed at the present time t_0 coming from direction $\hat{n}$ is then

$$T(\hat{n}) = (\nu_0/\nu_L)\Big(\bar{T}(t_L) + \delta T(r_L\hat{n}, t_L)\Big) ,$$

where now we have added a term δT to take account of the intrinsic temperature fluctuation at time t_L. Likewise, in the absence of perturbations the temperature observed in all directions would be

$$T_0 = \Big(a(t_L)/a(t_0)\Big)\bar{T}(t_L) ,$$

so the fractional shift in the radiation temperature observed coming from direction $\hat{n}$ from its unperturbed value is[6]

$$
\frac{\Delta T(\hat{n})}{T_0} \equiv \frac{T(\hat{n}) - T_0}{T_0} = \frac{v_0}{a(t_L)v_L/a(t_0)} - 1 + \frac{\delta T(r_L\hat{n}, t_L)}{\bar{T}(t_L)}
$$

$$
= \frac{1}{2}\left(E(r_L\hat{n}, t_L) - E(0, t_0)\right) - \int_{t_L}^{t_0} dt\, \left\{\frac{\partial}{\partial t}N\left(r\hat{n}, t\right)\right\}_{r=s(t)}
$$

$$
- a(t_L)\,\delta u^r_\gamma(r_L\hat{n}, t_L) + \frac{\delta T(r_L\hat{n}, t_L)}{\bar{T}(t_L)}\ . \tag{7.1.30}
$$

Because tensor and scalar perturbations are uncorrelated, we will treat their contribution to C_ℓ independently.

For scalar perturbations in any gauge with $h_{i0} = 0$, the metric perturbation is given by Eqs. (5.1.31)–(5.1.33) as

$$
h_{00} = -E\ , \qquad h_{ij} = a^2\left[A\delta_{ij} + \frac{\partial^2 B}{\partial x^i \partial x^j}\right] \tag{7.1.31}
$$

Also for scalar perturbations the radial photon fluid velocity is given in terms of a velocity potential δu_γ as

$$
\delta u^r_\gamma = \bar{g}^{r\mu}\frac{\partial \delta u_\gamma}{\partial x^\mu} = \frac{1}{a^2}\frac{\partial \delta u_\gamma}{\partial r} \tag{7.1.32}
$$

Thus Eq. (7.1.30) gives the scalar contribution to the temperature fluctuation

$$
\left(\frac{\Delta T(\hat{n})}{T_0}\right)^S = \frac{1}{2}\left(E(r_L\hat{n}, t_L) - E(0, t_0)\right) - \int_{t_L}^{t_0} dt\, \left\{\frac{\partial}{\partial t}N\left(r\hat{n}, t\right)\right\}_{r=s(t)}
$$

$$
- \frac{1}{a(t_L)}\left(\frac{\partial \delta u_\gamma(r\hat{n}, t_L)}{\partial r}\right)_{r=r_L} + \frac{\delta T(r_L\hat{n}, t_L)}{\bar{T}(t_L)}\ , \tag{7.1.33}
$$

where now

$$
N = \frac{1}{2}\left[A + \frac{\partial^2 B}{\partial r^2} - E\right]\ . \tag{7.1.34}
$$

Eq. (7.1.33) is not in the form that is most useful for our purposes. The total temperature fluctuation given by Eq. (7.1.33) is invariant with respect to the limited class of gauge transformations that leave $g_{i0} = 0$, but this is not true of its individual terms, including even the integral of $(\partial N/\partial t)_{r=s(t)}$.

---

[6]This result is essentially that first found by R. K. Sachs and A. M. Wolfe, *Astrophys. J.* **147**, 73 (1967).

It will be much more convenient to rewrite Eq. (7.1.33) in a way that leaves the integral term separately gauge invariant.

For this purpose, we make use of an identity corresponding to Eq. (7.1.6):

$$
\left\{\frac{\partial^2 \dot{B}(r\hat{n}, t)}{\partial r^2}\right\}_{r=s(t)}
$$

$$
= -\frac{d}{dt}\left[\left\{a^2(t)\dddot{B}(r\hat{n}, t) + a(t)\dot{a}(t)\ddot{B}(r\hat{n}, t) + a(t)\frac{\partial \dot{B}(r\hat{n}, t)}{\partial r}\right\}_{r=s(t)}\right]
$$

$$
+ \left\{\frac{\partial}{\partial t}\left(a^2(t)\dddot{B}(r\hat{n}, t) + a(t)\dot{a}(t)\ddot{B}(r\hat{n}, t)\right)\right\}_{r=s(t)}.
$$

Together with Eq. (7.1.34), this gives the integrand in Eq. (7.1.33) as

$$
\left\{\frac{\partial}{\partial t}N\left(r\hat{n}, t\right)\right\}_{r=s(t)}
$$

$$
= -\frac{1}{2}\frac{d}{dt}\left[\left\{a^2(t)\dddot{B}(r\hat{n}, t) + a(t)\dot{a}(t)\ddot{B}(r\hat{n}, t) + a(t)\frac{\partial \dot{B}(r\hat{n}, t)}{\partial r}\right\}_{r=s(t)}\right]
$$

$$
+ \frac{1}{2}\left\{\frac{\partial}{\partial t}\left(a^2(t)\dddot{B}(r\hat{n}, t) + a(t)\dot{a}(t)\ddot{B}(r\hat{n}, t)\right.\right.
$$

$$
\left.\left.+ A(r\hat{n}, t) - E(r\hat{n}, t)\right)\right\}_{r=s(t)}. \tag{7.1.35}
$$

The scalar fractional temperature fluctuation (7.1.33) may therefore be written

$$
\left(\frac{\Delta T(\hat{n})}{T_0}\right)^S = \left(\frac{\Delta T(\hat{n})}{T_0}\right)^S_{\text{early}} + \left(\frac{\Delta T(\hat{n})}{T_0}\right)^S_{\text{late}} + \left(\frac{\Delta T(\hat{n})}{T_0}\right)^S_{\text{ISW}}, \tag{7.1.36}
$$

where

$$
\left(\frac{\Delta T(\hat{n})}{T_0}\right)^S_{\text{early}} = -\frac{1}{2}a^2(t_L)\ddot{B}(r_L\hat{n}, t_L) - \frac{1}{2}a(t_L)\dot{a}(t_L)\dot{B}(r_L\hat{n}, t_L)
$$

$$
+ \frac{1}{2}E(r_L\hat{n}, t_L) + \frac{\delta T(r_L\hat{n}, t_L)}{\bar{T}(t_L)}
$$

$$
- a(t_L)\left[\frac{\partial}{\partial r}\left(\frac{1}{2}\dot{B}(r\hat{n}, t_L) + \frac{1}{a^2(t_L)}\delta u_\gamma(r\hat{n}, t_L)\right)\right]_{r=r_L}, \tag{7.1.37}
$$

340

$$\left(\frac{\Delta T(\hat{n})}{T}\right)^S_{\text{late}} = \frac{1}{2}a^2(t_0)\ddot{B}(0, t_0) + \frac{1}{2}a(t_0)\dot{a}(t_0)\dot{B}(0, t_0)$$

$$-\frac{1}{2}E(0, t_0) + a(t_0)\left[\frac{\partial}{\partial r}\left(\frac{1}{2}\dot{B}(r\hat{n}, t_0) + \frac{\delta u_\gamma(r\hat{n}, t_0)}{a^2(t_0)}\right)\right]_{r=0},$$

$$(7.1.38)$$

$$\left(\frac{\Delta T(\hat{n})}{T}\right)^S_{\text{ISW}} = -\frac{1}{2}\int_{t_L}^{t_0} dt\left[\frac{\partial}{\partial t}\left(a^2(t)\ddot{B}(r\hat{n}, t) + a(t)\dot{a}(t)\dot{B}(r\hat{n}, t)\right.\right.$$

$$\left.\left. + A(r\hat{n}, t) - E(r\hat{n}, t)\right)\right]_{r=s(t)}. \qquad (7.1.39)$$

The "late" term (7.1.38) (which was ignored earlier in this section) is the sum of a direction-independent term and a term proportional to $\hat{n}$, which has been added to represent the anisotropy due to the local cosmic gravitational field and velocity. It only affects terms in the multipole expansion of the temperature correlation function with $\ell = 0$ and $\ell = 1$, so it can be ignored if from now on we consider only multipole orders $\ell \geq 2$.

Now let us check the separate gauge invariance of the ISW, Doppler, and the remaining combined terms, at least for the limited class of gauge transformations that preserve the condition $g_{i0} = 0$, or in the notation of Eq. (5.1.32), $F = 0$. Eq. (5.3.13) tells us that these gauge transformations have

$$\epsilon_0 = -a^2\frac{\partial}{\partial t}\left(\frac{\epsilon^S}{a^2}\right),$$

and so $\Delta\dot{B} = 2\epsilon_0/a^2$. According to Eqs. (5.3.13) and (5.3.16), a gauge transformation therefore shifts the terms $-a^2\ddot{B}/2$, $-a\dot{a}\dot{B}/2$, $E/2$, and $\delta T/\bar{T}$ in the combined gravitational and intrinsic temperature fluctuations by the amounts $-a^2\partial(\epsilon_0/a^2)/\partial t$, $-\dot{a}\epsilon_0/a$, ϵ_0, and $\epsilon_0\dot{\bar{T}}/\bar{T} = -\epsilon_0\dot{a}/a$, giving no net change; it shifts the terms $a\dot{B}/2$ and $\delta u_\gamma/a$ in the Doppler contribution by ϵ_0/a and $-\epsilon_0/a$, giving no net change; and it shifts the four terms $a^2\ddot{B}$, $2a\dot{a}\dot{B}$, A, and $-E$ in the integrand of the integrated Sachs–Wolfe term by the amounts $2\dot{\epsilon}_0 - 4\epsilon_0\dot{a}/a$, $2\epsilon_0\dot{a}/a$, $2\epsilon_0\dot{a}/a$, and $-2\dot{\epsilon}_0$, respectively, giving no net change. Thus the integrated Sachs–Wolfe term, the Doppler term, and the combined gravitational and intrinsic temperature terms are separately gauge invariant. In particular, both Newtonian and synchronous gauge will give the same results for each of these three contributions.

Let us now assume that from last scattering to the present the scalar contributions to the fluctuations are dominated by a *single* mode, so that any perturbation $X(\mathbf{x}, t)$ (such as B, E, δu_γ, or δT) can be written as

$$X(\mathbf{x}, t) = \int d^3q\, \alpha(\mathbf{q})\, e^{i\mathbf{q}\cdot\mathbf{x}} X_q(t) \qquad (7.1.40)$$

with $\alpha(\mathbf{q})$ a stochastic variable (the same for all X), normalized so that

$$\langle \alpha(\mathbf{q})\, \alpha^*(\mathbf{q}') \rangle = \delta^3(\mathbf{q} - \mathbf{q}') \,. \qquad (7.1.41)$$

Then Eqs. (7.1.37) and (7.1.39) give

$$\left(\frac{\Delta T(\hat{n})}{T_0} \right)^S_{\text{early}} = \int d^3q\, \alpha(\mathbf{q})\, e^{i\mathbf{q}\cdot\hat{n}r(t_L)} \left(F(q) + i\hat{q}\cdot\hat{n}\, G(q) \right), \qquad (7.1.42)$$

$$\left(\frac{\Delta T(\hat{n})}{T_0} \right)^S_{\text{ISW}} = -\frac{1}{2} \int_{t_0}^{t_L} dt \int d^3q\, \alpha(\mathbf{q})\, e^{i\mathbf{q}\cdot\hat{n}s(t)} \frac{d}{dt}\left(a^2(t)\ddot{B}_q(t) \right.$$

$$\left. + a(t)\dot{a}(t)\dot{B}_q(t) + A_q(t) - E_q(t) \right), \qquad (7.1.43)$$

where

$$F(q) = -\frac{1}{2}a^2(t_L)\ddot{B}_q(t_L) - \frac{1}{2}a(t_L)\dot{a}(t_L)\dot{B}_q(t_L) + \frac{1}{2}E_q(t_L) + \frac{\delta T_q(t_L)}{\bar{T}(t_L)} \qquad (7.1.44)$$

$$G(q) = -q\left(\frac{1}{2}a(t_L)\dot{B}_q(t_L) + \frac{1}{a(t_L)}\delta u_{\gamma q}(t_L) \right). \qquad (7.1.45)$$

As we have seen, the form factors $F(q)$ and $G(q)$ and the integrand of the ISW term are separately invariant under gauge transformations that leave g_{i0} equal to zero.

In synchronous gauge $E_q = 0$, so Eqs. (7.1.42)–(7.1.45) are the same as the previously derived results (7.1.15), (7.1.13), (7.1.17), and (7.1.18). In Newtonian gauge $B = 0$ and $E = 2\Phi$, so the form factors are

$$F(q) = \Phi_q(t_L) + \frac{\delta T_q(t_L)}{\bar{T}(t_L)} \,, \qquad (7.1.46)$$

$$G(q) = -\frac{1}{a(t_L)}\delta u_{\gamma q}(t_L) \,. \qquad (7.1.47)$$

(Of course, δT_q and $\delta u_{\gamma q}$ are different in Newtonian gauge from what they are in synchronous gauge.) Also, the integrated Sachs–Wolfe term is

$$\left(\frac{\Delta T(\hat{n})}{T_0} \right)_{\text{ISW}} = -2 \int_{t_L}^{t_0} dt \int d^3q\, \alpha(\mathbf{q})\, e^{i\mathbf{q}\cdot\hat{n}s(t)} \dot{\Phi}_q(t) \,. \qquad (7.1.48)$$

It is left as an exercise for the reader to show that Φ_q is constant during the matter dominated era. Therefore, as we saw earlier in synchronous gauge, the integrated Sachs–Wolfe effect receives contributions only from departures from strict matter dominance.

Finally, we consider the tensor contribution to temperature fluctuations. In the approximation of a sudden transition from thermal equilibrium to transparency at time t_L, the only contribution of tensor perturbations to the observed temperature anisotropy comes from the term $h_{rr}/2a^2$ in the definition (7.1.23) of N, which according to Eq. (5.1.33) contains a term $D_{rr}/2$. Using this in Eq. (7.1.29) gives the tensor contribution to the temperature fluctuation

$$\left(\frac{\Delta T(\hat{n})}{T_0}\right)^T = -\frac{\hat{n}_i \hat{n}_j}{2} \int_{t_L}^{t_0} \left\{\frac{\partial}{\partial t} D_{ij}(r\hat{n}, t)\right\}_{r=s(t)} dt \ . \tag{7.1.49}$$

The gravitational wave amplitude $D_{ij}(\mathbf{x}, t)$ can be expressed as a Fourier integral (5.2.21)

$$D_{ij}(\mathbf{x}, t) = \sum_{\lambda=\pm 2} \int d^3q \ e^{i\mathbf{q}\cdot\mathbf{x}} \ e_{ij}(\hat{q}, \lambda)\beta(\mathbf{q}, \lambda)\,\mathcal{D}_q(t) \ , \tag{7.1.50}$$

where $\mathcal{D}_q(t)$ is the dominant solution of the wave equation (5.2.16):

$$\ddot{\mathcal{D}}_q(t) + 3\frac{\dot{a}}{a}\dot{\mathcal{D}}_q(t) + \frac{q^2}{a^2}\mathcal{D}_q(t) = 16\pi G\,\pi_q^T(t) \ , \tag{7.1.51}$$

$e_{ij}(\lambda, \hat{q})$ are polarization tensors defined in Section 5.2, and $\beta(\mathbf{q}, \lambda)$ is a stochastic variable chosen to satisfy Eq. (5.2.22). The tensor mode contribution to the temperature fluctuation is then

$$\left(\frac{\Delta T(\hat{n})}{T_0}\right)^T = -\frac{1}{2}\sum_{\lambda=\pm 2} \int d^3q \ \hat{n}_i \hat{n}_j e_{ij}(\lambda, q)\beta(\mathbf{q}, \lambda) \int_{t_L}^{t_0} dt \ e^{is(t)\mathbf{q}\cdot\hat{n}}\,\dot{\mathcal{D}}_q(t) \ . \tag{7.1.52}$$

This is the same as our previously derived result (7.1.18). We will return to the tensor term in C_ℓ in Section 7.3, after we have studied the scalar term in the next section.

7.2 Temperature multipole coefficients: Scalar modes

We will now apply the results of the previous section to the calculation of the contribution $C^S_{TT,\ell}$ of scalar modes to the multipole coefficients of temperature–temperature angular correlations:

$$C_{TT,\ell} = \frac{1}{4\pi} \int d^2\hat{n} \int d^2\hat{n}' \ P_\ell(\hat{n} \cdot \hat{n}') \langle \Delta T(\hat{n}) \, \Delta T(\hat{n}') \rangle \ , \tag{7.2.1}$$

where $\Delta T(\hat{n})$ is the stochastic variable giving the departure from the mean of the temperature observed in the direction $\hat{n}$, and $\langle\ldots\rangle$ denotes an average over the position of the observer, or equivalently, over the sequence of accidents that led to the particular pattern of temperature fluctuations we observe. (We are including a label TT to distinguish this multipole coefficient from those in temperature–polarization or polarization–polarization correlations, which are the subject of Section 7.4.) Of course this is not what is observed; the observed quantity is

$$C^{\text{obs}}_{TT,\ell} = \frac{1}{4\pi}\int d^2\hat{n}\int d^2\hat{n}' P_\ell(\hat{n}\cdot\hat{n}')\,\Delta T(\hat{n})\,\Delta T(\hat{n}')\,,$$

but as shown in Section 2.6, the cosmic variance, the mean square fractional difference between this and Eq. (7.2.1), is $2/2\ell+1$, and therefore may be neglected for $\ell\gg1$. In this section we will consider only the contribution $C^S_{TT,\ell}$ of scalar modes to $C_{TT,\ell}$; as we saw in Section 5.2, tensor and scalar modes do not interfere, so we can take up the contribution of tensor modes separately in the following section.

First let's apply the results that were obtained in the previous section by using the kinetic theory approach described in Section 6.1. The use of the Boltzmann equation yields formulas (7.1.7)–(7.1.9) for the temperature fluctuation. To calculate the coefficients in a partial wave expansion of the temperature fluctuation, we use the familiar expansion (2.6.16) of a plane wave in Legendre polynomials, together with the addition theorem for spherical harmonics:

$$e^{i\hat{q}\cdot\hat{n}\rho} = 4\pi\sum_{\ell=0}^{\infty}\sum_{m=-\ell}^{\ell} i^\ell j_\ell(\rho)\,Y_\ell^m(\hat{n})\,Y_\ell^{m*}(\hat{q})\,. \tag{7.2.2}$$

Using this in Eqs. (7.1.7)–(7.1.9), and replacing factors of $i\hat{q}\cdot\hat{n}$ with derivatives of the spherical Bessel function j_ℓ, the scalar contribution to the temperature fluctuation observed in a direction $\hat{n}$ is given by

$$\left(\Delta T(\hat{n})\right)^{(S)} = \sum_{\ell m} a^S_{T,\ell m}\,Y_\ell^m(\hat{n})\,, \tag{7.2.3}$$

where

$$a^S_{T,\ell m} = 4\pi i^\ell T_0\int d^3q\,\alpha(\mathbf{q})\,Y_\ell^{m*}(\hat{q})\int_{t_1}^{t_0}dt$$
$$\times\left[j_\ell\!\left(q\,r(t)\right)F(q,t)+j_\ell'\!\left(q\,r(t)\right)G(q,t)+j_\ell''\!\left(q\,r(t)\right)H(q,t)\right]\,. \tag{7.2.4}$$

Here $\alpha(\mathbf{q})$ is the stochastic parameter for the dominant scalar mode, normalized to satisfy Eq. (5.2.7):

$$\langle \alpha(\mathbf{q})\, \alpha^*(\mathbf{q}') \rangle = \delta^3(\mathbf{q} - \mathbf{q}') ; \tag{7.2.5}$$

t_1 is any time sufficiently early before recombination so that a photon present then would have scattered many times before the present; $r(t)$ is the radial coordinate of a point from which light emitted at time t would reach us at the present time t_0;

$$r(t) \equiv \int_t^{t_0} \frac{dt'}{a(t')} ; \tag{7.2.6}$$

and $F(q, t)$, $G(q, t)$, and $H(q, t)$ are time-dependent form-factors, given by

$$F(q, t) = \exp\left[-\int_t^{t_0} \omega_c(t')\, dt' \right]$$

$$\times \left\{ \omega_c(t) \left[\frac{3}{4}\Phi(q, t) + \frac{3}{16}\Pi(q, t) \right.\right.$$

$$\left. -\frac{1}{2}a^2(t)\ddot{B}_q(t) - \frac{1}{2}a(t)\dot{a}(t)\dot{B}_q(t) \right]$$

$$\left. -\frac{1}{2}\frac{d}{dt}\left[A_q(t) + a^2(t)\ddot{B}_q(t) + a(t)\dot{a}(t)\dot{B}_q(t) \right] \right\}, \tag{7.2.7}$$

$$G(q, t) = -q\,\omega_c(t) \exp\left[-\int_t^{t_0} \omega_c(t')\, dt' \right]$$

$$\times \left[\delta u_{Bq}(t)/a(t) + a(t)\dot{B}_q(t)/2 \right] , \tag{7.2.8}$$

$$H(q, t) = \frac{3}{16}\omega_c(t) \exp\left[-\int_t^{t_0} \omega_c(t')\, dt' \right] \Pi(q, t) . \tag{7.2.9}$$

As a reminder: $\omega_c(t)$ is the photon collision frequency at time t; $A_q(t)$ and $B_q(t)$ are the scalar fields in the perturbation to the metric in synchronous gauge, defined in Section 5.2; $\Phi(q, t)$ and $\Pi(q, t)$ are the scalar source functions, defined by Eq. (6.1.21); and $\delta u_{Bq}(t)$ is the scalar velocity potential of the baryonic plasma. A subscript T has been appended to the $a_{\ell m}$ introduced in Section 2.6, to indicate that these are partial wave coefficients in the temperature rather than the polarization, and a superscript S has been included to distinguish scalar from tensor contributions.

Together with the orthonormality property (2.6.19) of Legendre polynomials, Eqs. (7.2.1) and (7.2.3)–(7.2.5) give the scalar multipole coefficients[1]

$$C_{TT,\ell}^S = 16\pi^2 T_0^2 \int_0^\infty q^2 \, dq$$

$$\times \left| \int_{t_1}^{t_0} dt \left[j_\ell\left(q\,r(t)\right) F(q,t) \right. \right.$$

$$\left. \left. + j_\ell'\left(q\,r(t)\right) G(q,t) + j_\ell''\left(q\,r(t)\right) H(q,t) \right] \right|^2 . \qquad (7.2.10)$$

This formula gives results of high accuracy, but the computer calculations used to calculate the multipole coefficients $C_{TT,\ell}^S$ in this way are not particularly revealing. Instead, we will apply a series of approximations that lead to a simple analytic formula for the $C_{TT,\ell}^S$.

First, we will neglect the integrated Sachs–Wolfe effect, given by the last term in Eq. (7.2.7). This effect is important only for relatively small values of ℓ, where cosmic variance intrudes on measurements of $C_{TT,\ell}^S$.

Next, we assume a sudden transition from perfect opacity to perfect transparency at a definite time t_L,[2] and a single dominant mode of perturbation. The fractional temperature fluctuation then takes the form (7.1.15):

$$\left(\frac{\Delta T(\hat{n})}{T_0} \right) = \int d^3q \, \alpha(\mathbf{q}) \, e^{i\mathbf{q}\cdot\hat{n}r_L} \left(F(q) + i\hat{q}\cdot\hat{n}\,G(q) \right), \qquad (7.2.11)$$

which was also derived at the end of Section 7.1 by following photon trajectories after the time of last scattering. (Here $r_L = r(t_L)$ is the co-moving radius of the surface of last scattering.) For the present we will not use formulas (7.1.16) and (7.1.17) for the form factors $F(q)$ and $G(q)$, but will proceed for general form factors, returning later to the specific form factors (7.1.16) and (7.1.17).

Using Eq. (7.2.2) in Eq. (7.2.11), and replacing $i\hat{q}\cdot\hat{n}$ in the Doppler term with $\partial/\partial(qr_L)$, we again have the partial-wave expansion (7.2.3), but this

---

[1] The derivatives of the spherical Bessel functions in Eq. (7.2.10) can be expressed as time derivatives, and then integrating by parts this can be written at the integral of a single form factor times $j_\ell\left(qr(t)\right)$. The result is equivalent to Eq. (16) of M. Zaldarriaga and U. Seljak, *Phys. Rev. D* **55**, 1830 (1997) [astro-ph/9609170], except for a difference of normalization: the source functions Π and Φ used here are 4 times those of Zaldarriaga and Seljak.

[2] To correct for whatever inaccuracy is introduced by this approximation, we will later include the effect of damping of acoustic oscillations before t_L, and average the temperature fluctuation over the time of last scattering.

time with

$$a_{T,\ell m}^{S} = 4\pi i^{\ell} T_0 \int d^3 q \, \alpha(\mathbf{q}) \, Y_{\ell}^{m*}(\hat{q}) \left[j_{\ell}(qr_L)F(q) + j_{\ell}'(qr_L)G(q) \right]. \quad (7.2.12)$$

Inserting this in Eq. (7.2.1) and using Eq. (7.2.5) gives the multipole coefficients

$$C_{TT,\ell}^{S} = \frac{1}{2\ell+1} \sum_{m=-\ell}^{\ell} |a_{\ell m}|^2 = 16\pi^2 T_0^2 \int_0^{\infty} q^2 \, dq$$

$$\times \left[j_{\ell}(qr_L)F(q) + j_{\ell}'(qr_L)G(q) \right]^2 \quad (7.2.13)$$

This is a standard result, but it does not provide transparent information about the dependence of $C_{TT,\ell}^{S}$ on ℓ. For this purpose, we now make a further approximation: we specialize to the most interesting case of large ℓ, where cosmic variance can be neglected. In this case, we can use an approximate formula for the spherical Bessel functions:[3]

$$j_{\ell}(\rho) \to \begin{cases} \cos b \, \cos\left[\nu(\tan b - b) - \pi/4 \right] / \nu\sqrt{\sin b} & \rho > \nu \\ \\ 0 & \rho < \nu, \end{cases}$$
$$(7.2.14)$$

where $\nu \equiv \ell + 1/2$, and $\cos b \equiv \nu/\rho$, with $0 \le b \le \pi/2$. This approximation is valid for $|\nu^2 - \rho^2| \gg \nu^{4/3}$. Hence for $\ell \gg 1$, this formula can be used over most of the ranges of integration in Eq. (7.2.13). Furthermore, for $\rho > \nu \gg 1$ the phase $\nu(\tan b - b)$ in Eq. (7.2.14) is a very rapidly increasing function of ρ, so the derivative acting on the spherical Bessel function in Eq. (7.2.13) can be taken to act chiefly on this phase:

$$j_{\ell}'(\rho) \to \begin{cases} -\cos b \sqrt{\sin b} \, \sin\left[\nu(\tan b - b) - \pi/4 \right] / \nu & \rho > \nu \\ \\ 0 & \rho < \nu, \end{cases}$$
$$(7.2.15)$$

(Letting the derivative act on the factor $1/\sqrt{\sin b}$ in Eq. (7.2.14) introduces an apparent divergence in the integral at $\rho = \nu$, but this divergence is spurious; for ρ very close to ν the approximation (7.2.14) breaks down, and

---

[3] See, *e.g.*, I. S. Gradshteyn & I. M. Ryzhik, *Table of Integrals, Series, and Products*, translated, corrected and enlarged by A. Jeffrey (Academic Press, New York, 1980): formula 8.453.1.

there is no singularity.) Using these limits in Eq. (7.2.13) and changing the variable of integration from q to $b = \cos^{-1}(v/qr_L)$ gives

$$
\begin{aligned}
C^S_{TT,\ell} = \frac{16\pi^2 v}{r_L^3} \int_0^{\pi/2} \frac{db}{\cos^2 b} \\
\times \Bigg[F\left(\frac{v}{r_L \cos b}\right) \cos[v(\tan b - b) - \pi/4] \\
- \sin b \, G\left(\frac{v}{r_L \cos b}\right) \sin[v(\tan b - b) - \pi/4] \Bigg]^2 . \quad (7.2.16)
\end{aligned}
$$

For $v \gg 1$ the functions $\cos[v(\tan b - b) - \pi/4]$ and $\sin[v(\tan b - b) - \pi/4]$ oscillate very rapidly, so $\cos^2[v(\tan b - b) - \pi/4]$ and $\sin^2[v(\tan b - b) - \pi/4]$ average to $1/2$, while $\cos[v(\tan b - b) - \pi/4] \sin[v(\tan b - b) - \pi/4]$ averages to zero. Dropping the distinction between ℓ and $v = \ell + 1/2$, and changing the variable of integration again, from b to $\beta = 1/\cos b$, Eq. (7.2.16) then becomes[4]

$$
\begin{aligned}
\ell(\ell+1)C^S_{TT,\ell} = \frac{8\pi^2 \ell^3}{r_L^3} \int_1^\infty \frac{\beta \, d\beta}{\sqrt{\beta^2 - 1}} \\
\times \left[F^2\left(\frac{\ell\beta}{r_L}\right) + \frac{\beta^2 - 1}{\beta^2} G^2\left(\frac{\ell\beta}{r_L}\right) \right]. \quad (7.2.17)
\end{aligned}
$$

We will see that the form factors $F(q)$ and $G(q)$ fall off rapidly for large q, in part because $|\mathcal{R}_q^o|^2$ decreases more or less like q^{-3}, so the integral over β converges at $\beta = \infty$, and in fact is dominated by small values of β. The integral of the F^2 term thus receives its greatest contribution from $\beta \approx 1$, or in other words, for $q \approx \ell/r_L$, or $q/a_L \approx \ell/d_A$, where $d_A = r_L a_L$ is the angular diameter distance of the surface of last scattering. On the other hand, the factor $\beta^2 - 1$ multiplying G^2 in Eq. (7.2.17) kills the contribution of β values very close to unity, so the Doppler term proportional to G^2 makes a relatively small contribution to the multipole coefficients.

Even without a detailed calculation of the form factors $F(q)$ and $G(q)$, we know that they depend on the baryon and total matter densities at last scattering, which for a given present microwave background temperature can be expressed in terms of $\Omega_B h^2$ and $\Omega_M h^2$, but since spatial curvature and dark energy are (presumably) negligible at last scattering, the form factors cannot depend on Ω_K or Ω_Λ or H_0. Thus there is a high

---

[4]A more rigorous but rather more complicated derivation was given by S. Weinberg, *Phys. Rev. D* **64**, 123512 (2001) [astro-ph/0103281]. The contribution of $F(q)$ had earlier been calculated by J. R. Bond, "Theory and Observations of the Cosmic Background Radiation," in *Cosmology and Large Scale Structure*, eds. R. Schaeffer, J. Silk, M. Spiro and J. Zinn-Justin (Elsevier, 1996), Section 5.1.3.

degree of degeneracy here; all dependence on Ω_K or Ω_Λ or H_0 can only enter in the single parameter r_L, the Robertson–Walker radius of the surface of last scattering. If we assume as commonly done that $\Omega_K = 0$, so that $\Omega_\Lambda \simeq 1 - \Omega_M$, then temperature anisotropies can tell us the values of H_0 as well as Ω_M and Ω_B, but they cannot tell us that $\Omega_K = 0$ unless we have information about H_0 from other source. Likewise, they cannot distinguish quintessence from a constant vacuum energy. Furthermore, r_L and ℓ appear in $\ell(\ell + 1)C^S_{TT,\ell}$ only in the ratio ℓ/r_L, so the values of Ω_K, Ω_Λ, and H_0 can only effect the *scale* of the ℓ-dependence of $\ell(\ell + 1)C^S_{TT,\ell}$. For instance, the values of these parameters can affect the positions of the peaks in $\ell(\ell+1)C^S_{TT,\ell}$ by a common factor, but cannot affect their heights.

We next make the approximation that the gravitational field perturbations at last scattering are dominated by perturbations in the dark matter density. We already remarked in connection with the ISW effect in the previous section that in this case $\dot{A}_q$ vanishes (because each term on the right-hand side of Eq. (5.3.38) is time-independent). The field ψ_q whose evolution we followed in Chapter 6 is defined in general as $\psi_q \equiv (3\dot{A}_q - q^2\dot{B}_q)/2$, so here $\dot{B}_q = -2\psi_q/q^2$, and since Eq. (6.5.16) gives $\psi_q \propto t^{-1/3}$, we have also $\ddot{B}_q = 2\psi_q/3tq^2$. Also, $a \propto t^{2/3}$, so Eqs. (7.1.16) and (7.1.17) give the form factors as

$$F(q) = \frac{1}{3}\delta_{\gamma q}(t_L) + \frac{a^2(t_L)\psi_q(t_L)}{3q^2 t_L} \tag{7.2.18}$$

$$G(q) = -q\delta u_{\gamma q}(t_L)/a(t_L) + a(t_L)\psi_q(t_L)/q, \tag{7.2.19}$$

in which we have used $\delta T_q/\bar{T} = \delta\rho_{\gamma q}/4\bar{\rho}_\gamma = \delta_{\gamma q}/3$.

As our next approximation, we shall use the results (6.5.16), (6.5.17), and (6.5.18) of our analysis of cosmic evolution in Chapter 6, in which we neglected the baryon/dark matter density ratio:

$$\psi_q(t_L) = -\frac{3q^2 t_L \mathcal{R}_q^o \mathcal{T}(\kappa)}{5\, a_L^2}, \tag{7.2.20}$$

$$\delta_{\gamma q}(t_L) = \frac{3\mathcal{R}_q^o}{5}\left[\mathcal{T}(\kappa)(1 + 3R_L) \right.$$

$$-(1 + R_L)^{-1/4}\, e^{-\int_0^{t_L}\Gamma dt}\, \mathcal{S}(\kappa)$$

$$\left. \times \cos\left(\int_0^{t_L}\frac{q\,dt}{a(t)\sqrt{3(1 + R(t))}} + \Delta(\kappa)\right)\right], \tag{7.2.21}$$

$$\delta u_{\gamma q}(t_L) = \frac{3\mathcal{R}_q^o}{5}\left[-t_L \mathcal{T}(\kappa) \right.$$

$$+\frac{a_L}{\sqrt{3}q(1+R_L)^{3/4}} e^{-\int_0^{t_L} \Gamma dt}\, \mathcal{S}(\kappa)$$

$$\left. \times \sin\left(\int_0^{t_L} \frac{q\,dt}{a(t)\sqrt{3(1+R(t))}} + \Delta(\kappa) \right) \right],$$

$$(7.2.22)$$

Here $R(t) \equiv 3\bar{\rho}_B(t)/4\bar{\rho}_\gamma(t)$, $R_L \equiv R(t_L)$, and $a_L \equiv a(t_L)$; $\mathcal{T}(\kappa)$, $\mathcal{S}(\kappa)$, and $\Delta(\kappa)$ are the transfer functions defined and calculated in Section 6.5; $\kappa \equiv \sqrt{2}q/q_{EQ}$, where q_{EQ} is the wave number that comes into the horizon at matter–radiation equality; and $\Gamma(t)$ is the acoustic damping rate (6.4.25). Using Eqs. (7.2.20)–(7.2.22) in Eqs. (7.2.18) and (7.2.19) gives the form factors as

$$F(q) = \frac{\mathcal{R}_q^o}{5}\left[3\mathcal{T}(\kappa)R_L \right.$$

$$-(1+R_L)^{-1/4} e^{-\int_0^{t_L} \Gamma dt}\, \mathcal{S}(\kappa)$$

$$\left. \times \cos\left(\int_0^{t_L} \frac{q\,dt}{a(t)\sqrt{3(1+R(t))}} + \Delta(\kappa) \right) \right], \qquad (7.2.23)$$

$$G(q) = -\frac{\sqrt{3}\mathcal{R}_q^o}{5(1+R_L)^{3/4}} e^{-\int_0^{t_L} \Gamma dt}\, \mathcal{S}(\kappa)$$

$$\times \sin\left(\int_0^{t_L} \frac{q\,dt}{a(t)\sqrt{3(1+R(t))}} + \Delta(\kappa) \right). \qquad (7.2.24)$$

Note that the "slow" terms have canceled in $G(q)$, and would have canceled in $F(q)$ if it were not for a finite baryon/photon density ratio at last scattering.

Now we must take up a complication that arises only for the "fast" part of the form factors in the case of short wavelengths. We have been assuming that the opacity of the universe drops to zero instantaneously at a time t_L of last scattering, but of course the drop takes place during some finite interval of time, over which the form factors must be averaged. This makes little difference for the contribution of the slow modes, but for large wave numbers the cosines and sines in Eqs. (7.2.23) and (7.2.24) are rapidly oscillating functions of t_L, so the fast terms can be significantly reduced by this averaging. This is similar to what in other contexts is called *Landau damping*, except that usually Landau damping arises from a spread in the frequency of an oscillation, while here it is produced by a spread in the moment at which the oscillating amplitude is observed.

To continue with our analytic treatment, since the probability of last scattering is a sharply peaked function of time, we can approximate it as a Gaussian: the probability that last scattering occurs between t and $t + dt$ will be taken in the form

$$P(t)dt = \frac{\exp[-(t - t_L)^2/2\sigma_t^2] \, dt}{\sigma_t \sqrt{2\pi}} \ . \tag{7.2.25}$$

In Eqs. (7.2.23) and (7.2.24), we must make the replacements

$$\begin{Bmatrix} \cos\left(\int_0^{t_L} \omega \, dt + \Delta\right) \\ \sin\left(\int_0^{t_L} \omega \, dt + \Delta\right) \end{Bmatrix} \rightarrow \int_{-\infty}^{+\infty} P(t) \, dt \begin{Bmatrix} \cos\left(\int_0^t \omega \, dt + \Delta\right) \\ \sin\left(\int_0^t \omega \, dt + \Delta\right) \end{Bmatrix}, \tag{7.2.26}$$

where $\omega = q/a\sqrt{3(1 + R)}$. For a sharply peaked distribution function $P(t)$, we can do these integrals by expanding the arguments of sines and cosines to first order in $t - t_L$:

$$\int_0^t \omega \, dt \simeq \int_0^{t_L} \omega \, dt + \omega_L (t - t_L) \ .$$

The integrals (7.2.26) are now easily done

$$\int_{-\infty}^{+\infty} P(t) \, dt \begin{Bmatrix} \cos\left(\int_0^t \omega \, dt + \Delta\right) \\ \sin\left(\int_0^t \omega \, dt + \Delta\right) \end{Bmatrix}$$

$$\simeq \exp(-\omega_L^2 \sigma_t^2/2) \begin{Bmatrix} \cos\left(\int_0^{t_L} \omega \, dt + \Delta\right) \\ \sin\left(\int_0^{t_L} \omega \, dt + \Delta\right) \end{Bmatrix} . \tag{7.2.27}$$

Thus, the whole effect of this averaging is to introduce an additional damping factor $\exp(-\omega_L^2 \sigma_t^2/2)$ in the fast part of the form factors. Both Γ and ω_L^2 are proportional to q^2, so we may write

$$\int_0^{t_L} \Gamma \, dt + \omega_L^2 \sigma_t^2/2 = q^2 d_D^2/a_L^2 \ , \tag{7.2.28}$$

where d_D is a *damping length*, given by Eqs. (6.4.25) and (7.2.27) as

$$d_D^2 = d_{\text{Silk}}^2 + d_{\text{Landau}}^2 , \tag{7.2.29}$$

$$d_{\text{Silk}}^2 = a_L^2 \int_0^{t_L} \frac{t_\gamma}{6a^2(1 + R)} \left\{ \frac{16}{15} + \frac{R^2}{(1 + R)} \right\} dt , \tag{7.2.30}$$

$$d_{\text{Landau}}^2 = \frac{\sigma_t^2}{6(1 + R_L)} , \tag{7.2.31}$$

with $t_\gamma = 1/\omega_c$ the photon mean free time, and $R \equiv 3\bar{\rho}_B/4\bar{\rho}_\gamma$.

To evaluate the Silk damping term, we recall that $R \propto a$, so

$$t_\gamma = \frac{1}{n_e \sigma_T c} = \frac{R^3}{n_{B0} R_0^3 (1 - Y) X \sigma_T c}, \tag{7.2.32}$$

where $R_0 = 3\Omega_B/4\Omega_\gamma$ is the present value of R, $Y \simeq 0.24$ is the fraction of nucleons in the form of un-ionized helium around the time of last scattering, $n_{B0} = 3H_0^2 \Omega_B/8\pi G m_N$ is the present number density of baryons, and $X(R)$ is the fractional ionization, calculated in Section 2.3. Also,

$$dt = \frac{dR}{R H_0 \sqrt{\Omega_M (R_0/R)^3 + \Omega_R (R_0/R)^4}} = \frac{R \, dR}{H_0 \sqrt{\Omega_M} R_0^{3/2} \sqrt{R_{\text{EQ}} + R}} \tag{7.2.33}$$

where $R_{\text{EQ}} \equiv \Omega_R R_0/\Omega_M = 3\Omega_R \Omega_B/4\Omega_M \Omega_\gamma$ is the value of R at matter–radiation equality. Putting this all together, the Silk damping length is given by

$$d_{\text{Silk}}^2 = \frac{R_L^2}{6(1 - Y) n_{B0} \sigma_T c H_0 \sqrt{\Omega_M} R_0^{9/2}}$$

$$\times \int_0^{R_L} \frac{R^2 \, dR}{X(R)\,(1 + R)\sqrt{R_{\text{EQ}} + R}} \left\{ \frac{16}{15} + \frac{R^2}{(1 + R)} \right\} \tag{7.2.34}$$

Also, the standard deviation σ_t in the time of last scattering is related to the standard deviation σ in the temperature of last scattering, calculated in Section 2.3, by $\sigma_t = 3 t_L \sigma/2 T_L$, so

$$d_{\text{Landau}}^2 = \frac{3\sigma^2 t_L^2}{8 T_L^2 (1 + R_L)}. \tag{7.2.35}$$

The form factors (7.2.23) and (7.2.24) may now be written as explicit functions of wave number

$$F(q) = \frac{\mathcal{R}_q^o}{5} \left[3T(q d_T/a_L) R_L \right.$$

$$\left. - (1 + R_L)^{-1/4} S(q d_T/a_L)\, e^{-q^2 d_D^2/a_L^2} \cos\left(q d_H/a_L + \Delta(q d_T/a_L) \right) \right], \tag{7.2.36}$$

$$G(q) = -\frac{\sqrt{3} \mathcal{R}_q^o}{5(1 + R_L)^{3/4}}\, e^{-q^2 d_D^2/a_L^2} S(q d_T/a_L) \sin\left(q d_H/a_L + \Delta(q d_T/a_L) \right), \tag{7.2.37}$$

where d_T is a length defined by writing the argument of the transfer functions as $\kappa = q d_T/a_L$, so that Eq. (6.5.9) gives

$$d_T = \frac{\sqrt{\Omega_R}}{(1+z_L)H_0\Omega_M} = \frac{0.0177}{\Omega_M h^2}\ \text{Mpc}\ . \tag{7.2.38}$$

Also, d_H is the acoustic horizon distance at last scattering, given by

$$d_H \equiv a_L \int_0^{t_L} \frac{dt}{a\sqrt{3(1+R)}}$$

$$= \frac{2}{H_0\sqrt{3R_L\Omega_M}(1+z_L)^{3/2}}\ \ln\left(\frac{\sqrt{1+R_L}+\sqrt{R_{EQ}+R_L}}{1+\sqrt{R_{EQ}}}\right)\ , \tag{7.2.39}$$

where again $R_{EQ} = 3\Omega_R\Omega_B/4\Omega_M\Omega_\gamma$ and $R_L = 3\Omega_B/4\Omega_\gamma(1+z_L)$.

Before using these form factors in Eq. (7.2.17), there is one more complication that needs to be mentioned. At a redshift z_{reion} of order 10, the neutral hydrogen left over from the time of recombination becomes reionized by ultraviolet light from the first generation of massive stars. The photons of the cosmic microwave background have a small but non-negligible probability $1 - \exp(-\tau_{\text{reion}})$ (where τ_{reion} is the optical depth of the reionized plasma) of being scattered by the electrons set free by this reionization. The temperature anisotropy ΔT is then the sum of two terms. One arises from photons that are not scattered by the reionized hydrogen, and is just equal to the anisotropy we have calculated times the probability $\exp(-\tau_{\text{reion}})$ of no scattering. The other term arises from scattered photons, and since this scattering occurs at redshifts much less than $z_L \simeq 1,090$, we see the anisotropies at a smaller distance, and hence at much lower values of ℓ. Thus the effect on $C_{TT,\ell}^S$ of scattering by the reionized plasma is simply to multiply $C_{TT,\ell}^S$ by a factor $\exp(-2\tau_{\text{reion}})$, except for very small values of ℓ, where in any case cosmic variance interferes with the interpretation of observations. This means that observations of temperature anisotropies alone cannot effectively disentangle the reionization probability from the over-all scale of the function $\mathcal{R}_q^o$ that characterizes primordial fluctuations; they can only tell us the value of $|\mathcal{R}_q^o|^2\exp(-2\tau_{\text{reion}})$. The measurement of the polarization of microwave photons (discussed in Section 7.4) produced by scattering after reionization suggests that $\exp(-2\tau_{\text{reion}}) \approx 0.8$.

It is conventional to parameterize the quantity $\mathcal{R}_q^o$ in a form equivalent to

$$|\mathcal{R}_q^o|^2 = N^2 q^{-3}\left(\frac{q/a_0}{k_{\mathcal{R}}}\right)^{n_S-1} \tag{7.2.40}$$

with n_S perhaps varying with wave number. This parameterization is convenient, because then

$$\left(\frac{\ell}{r_L}\right)^3 |\mathcal{R}_{\beta\ell/r_L}^o|^2 = \frac{N^2}{\beta^3}\left(\frac{\ell\beta}{k_\mathcal{R} d_A(1+z_L)}\right)^{n_S-1},$$

and it will turn out that n_S is not very different from 1. It is only $N^2 k_\mathcal{R}^{1-n_S}$ that enters in the normalization of $|\mathcal{R}_q^o|^2$, so the choice of $k_\mathcal{R}$ is arbitrary; it is conventional to take it as $k_\mathcal{R} = 0.05$ Mpc^{-1}.

We conclude from all this that for reasonably large values of ℓ (say, $\ell > 20$), where we can ignore cosmic variance and the integrated Sachs–Wolfe effect, use the large ℓ approximations that led to Eq. (7.2.17), and treat the effect of reionization as a simple factor $\exp(-2\tau_{\text{reion}})$, the quantity usually quoted as giving the scalar contribution to the multipole coefficients is

$$\frac{\ell(\ell+1)C_{TT,\ell}^S}{2\pi} = \frac{4\pi T_0^2 N^2 e^{-2\tau_{\text{reion}}}}{25}\int_1^\infty d\beta \left(\frac{\beta\ell}{\ell_\mathcal{R}}\right)^{n_S-1}$$

$$\times\left\{\frac{1}{\beta^2\sqrt{\beta^2-1}}\left[3\mathcal{T}(\beta\ell/\ell_T)R_L\right.\right.$$

$$\left.-(1+R_L)^{-1/4}\mathcal{S}(\beta\ell/\ell_T)\,e^{-\beta^2\ell^2/\ell_D^2}\cos\left(\beta\ell/\ell_H + \Delta(\beta\ell/\ell_T)\right)\right]^2$$

$$\left.+\frac{3\sqrt{\beta^2-1}}{\beta^4(1+R_L)^{3/2}}e^{-2\beta^2\ell^2/\ell_D^2}\mathcal{S}^2(\beta\ell/\ell_T)\sin^2\left(\beta\ell/\ell_H + \Delta(\beta\ell/\ell_T)\right)\right\},$$

$$(7.2.41)$$

where

$$\ell_D \equiv d_A/d_D\,,\quad \ell_T \equiv d_A/d_T\,,\quad \ell_H \equiv d_A/d_H\,,\quad \ell_\mathcal{R} = (1+z_L)k_\mathcal{R} d_A\,,$$

$$(7.2.42)$$

and again d_A is the angular diameter distance of the surface of last scattering:

$$d_A \equiv r_L a_L = \frac{1}{\Omega_K^{1/2} H_0(1+z_L)}\sinh\left[\Omega_K^{1/2}\int_{1/(1+z_L)}^1 \frac{dx}{\sqrt{\Omega_\Lambda x^4 + \Omega_K x^2 + \Omega_M x}}\right],$$

$$(7.2.43)$$

with $\Omega_\mathcal{R}$ neglected, and $\Omega_K = 1 - \Omega_\Lambda - \Omega_M$. If we assume as discussed earlier that the integral over β is dominated by values $\beta \approx 1$, neglect the Doppler term, and for the moment neglect the term proportional to the transfer function $\mathcal{T}$, then Eq. (7.2.41) shows that $C_{TT,\ell}^S$ has peaks at $\chi(\ell) =$

$\pi, 2\pi, 3\pi$, etc., where $\chi(\ell) = \ell/\ell_H + \Delta(\ell/\ell_T)$ is the phase of the cosine in Eq. (7.2.41) for $\beta = 1$. The presence of the positive term $3TR_L$ arising from the inertia of the baryonic plasma enhances and slightly shifts the peaks for $\chi = \pi, 3\pi$, etc., where the cosine is negative, and reduces the peaks for $\chi = 2\pi, 4\pi$, etc., where the cosine is positive. For large ℓ the Silk and Landau damping factor $\exp(-\beta^2\ell^2/d_D^2)$ damps out all these peaks. As we will soon see, the results of numerical calculations exhibit this expected pattern, of a sequence of decreasing peaks, with odd peaks somewhat enhanced over the even peaks.

We can now read off the dependence of the quantities appearing in $C_{TT,\ell}^S$ on various cosmological parameters. Taking as fixed the well established value of the present microwave temperature, (which yields values for $\Omega_\gamma h^2$ and $\Omega_R h^2 = \Omega_\gamma h^2 + \Omega_\nu h^2$), and also fixing the values of t_L, z_L, and σ, which are only weakly dependent on other cosmological parameters, we see that

- $R_L \propto \Omega_B h^2$.

- The integral in Eq. (7.2.34) for d_{Silk}^2 is a complicated but not very sensitive function of $\Omega_B h^2$ and $\Omega_M h^2$. Aside from the integral, d_{Silk}^2 is proportional to $(\Omega_B h^2)^{-7/2}(\Omega_M h^2)^{-1/2}$.

- d_{Landau}^2 depends on $\Omega_B h^2$ through a factor $(1 + R_L)^{-1}$.

- $d_T \propto (\Omega_M h^2)^{-1}$

- Aside from a slowly varying logarithm, $d_H \propto (\Omega_B h^2)^{-1/2}(\Omega_M h^2)^{-1/2}$.

- Only d_A depends on H_0, Ω_Λ, Ω_M, Ω_B, or Ω_K apart from a dependence on $\Omega_B h^2$ and $\Omega_M h^2$. (For any observationally allowed values of Ω_Λ or Ω_K, the effects of a constant vacuum energy or of spatial curvature would be quite negligible at times before recombination, so their effect on $C_{TT,\ell}^S$ is limited to their influence on the propagation of light after recombination, that is, on d_A.)

To see how well the various approximations we have made work in practice, we shall calculate $C_{TT,\ell}^S$ for a realistic set of values for cosmological parameters taken from a fit[5] to data on the microwave background from the CBI, ACBAR, and first-year WMAP observations (about which more below). These are the same parameters that have been used in a full-scale computer calculation whose results are readily available,[6] so that we will

[5]D.N. Spergel *et al.*, *Astrophys. J. Suppl.* **148**, 175 (2003) [astro-ph/0302209].

[6]http://lambda.gsfc.nasa.gov/data/map/powspec/wmap_lcdm_pl_model_yr1_v1.txt

conveniently be able to compare Eq. (7.2.41) with an accurate numerical calculation. The cosmological parameters of this set are

$$\Omega_M h^2 = 0.13299 , \quad \Omega_B h^2 = 0.02238 , \quad h = 0.71992 , \quad \Omega_\Lambda = 1 - \Omega_M .$$
$$(7.2.44)$$

and[7]

$$n_S = 0.95820 , \quad k_{\mathcal{R}} = 0.05 \text{ Mpc}^{-1} , \quad N^2 = 1.736 \times 10^{-10} ,$$
$$e^{-2\tau_{\text{reion}}} = 0.80209 .$$
$$(7.2.45)$$

We take $T_0 = 2.725\text{K}$, which yields $\Omega_\gamma h^2 = 2.47 \times 10^{-5}$, and we assume three flavors of massless neutrinos, which gives $\Omega_R h^2 = 1.6813\Omega_\gamma h^2 = 4.15 \times 10^{-5}$. We will also adopt the parameters describing recombination calculated in Section 2.3 (which are not very sensitive to other cosmological parameters) for $\Omega_B h^2 = 0.02$ and $\Omega_M h^2 = 0.15$; in particular,

$$1 + z_L = 1,090 , \quad \sigma = 262 \text{ K} , \quad t_L = 370,000 \text{ yrs.} \quad (7.2.46)$$

From the values (7.2.44) of $\Omega_M h^2$ and $\Omega_B h^2$, we find

$$R_0 = 679.6 , \quad R_L = 0.6234 , \quad R_{\text{EQ}} = 0.2121. \quad (7.2.47)$$

Then Eqs. (7.2.38), (7.2.39), (7.2.43) give

$$d_T = 0.1331 \text{ Mpc} , \quad d_H = 0.1351 \text{ Mpc} , \quad d_A = 12.99 \text{ Mpc} , \quad (7.2.48)$$

while the damping lengths are given by Eqs. (7.2.34), (7.2.35) and (7.2.29):

$$d_{\text{Silk}} = 0.006555 \text{ Mpc} , \quad d_{\text{Landau}} = 0.004809 \text{ Mpc} , \quad d_D = 0.008130 \text{ Mpc} .$$
$$(7.2.49)$$

Finally, the parameters (7.2.42) appearing in Eq. (7.2.41) are

$$\ell_T = 97.60 , \quad \ell_H = 96.15 , \quad \ell_D = 1598 , \quad \ell_{\mathcal{R}} = 708 \quad (7.2.50)$$

while the factor multiplying the integral is

$$\frac{4\pi T_0^2 N^2 e^{-2\tau_{\text{reion}}}}{25} = 519.7 \ \mu\text{K}^2 . \quad (7.2.51)$$

---

[7]The parameter N^2 is related to what is often given as an "amplitude" A by $4\pi N^2 \equiv 20,000\pi A/9T_{\gamma 0}(\mu K)^2 = 2.95 \times 10^{-9}A$. For instance, see L. Verde *et al.*, *Astrophys. J. Suppl.* **148**, 195 (2003) [astro-ph/0302218]. It should be noted that, because of different conventions used in writing Fourier integrals, the quantity $\mathcal{R}_k$ in this paper is equal to what we have defined as $\mathcal{R}^o_{a_0 k}$ times a factor $(2\pi)^{3/2}$. Reference 6 uses $A = 0.73935$, corresponding to the value of N^2 given in Eq. (7.2.45).

Using these values, the result of using Eq. (7.2.41) with the parameter set of reference 6 is

$$\frac{\ell(\ell+1)C_{TT,\ell}^{S}}{2\pi} = 519.7\ \mu K^2 \int_1^\infty d\beta\ \left(\frac{\beta\ell}{708}\right)^{-0.0418}$$

$$\times \left\{\frac{1}{\beta^2\sqrt{\beta^2-1}}\left[3(0.6234)\mathcal{T}(\beta\ell/97.6)\right.\right.$$

$$\left.-(1.6234)^{-1/4}\mathcal{S}(\beta\ell/97.6)\,e^{-(\beta\ell/1598)^2}\cos\left(\beta\ell/96.15+\Delta(\beta\ell/97.6)\right)\right]^2$$

$$+\frac{3\sqrt{\beta^2-1}}{\beta^4(1.6234)^{3/2}}e^{-2(\beta\ell/1598)^2}\mathcal{S}^2(\beta\ell/97.6)$$

$$\times \sin^2\left(\beta\ell/96.15+\Delta(\beta\ell/97.6)\right)\right\}.$$

$$(7.2.52)$$

The integral over β converges very rapidly, and can be done with a cut-off at $\beta = 5$; raising the cut-off to $\beta = 50$ has a negligible effect.

The results are shown in Figure 7.1, in comparison with the more accurate Boltzmann hierarchy calculation of reference 6, based on Eq. (7.2.10). Evidently the hydrodynamic calculation does quite well; like the computer calculation of reference 6, it shows a high first peak, followed by two nearly equal lower peaks, followed by a decaying tail punctuated by successively lower peaks. To give a quantitative comparison, Tables 7.1 and 7.2 show the

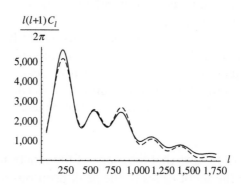

Figure 7.1: The scalar multipole coefficient $\ell(\ell+1)C_\ell^S/2\pi$ in square microKelvin, *vs.* ℓ, for the cosmological parameters given in Eqs. (7.2.44) and (7.2.45). The hydrodynamic approximation (7.2.41) is indicated by the dashed curve, while for comparison the solid curve gives the more accurate large scale computer calculation of reference 6, based on Eq. (7.2.10).

Table 7.1: Comparison of the results for the values of ℓ at the first five peaks in $\ell(\ell + 1)C^S_{TT,\ell}/2\pi$, as given by Eq. (7.2.52) and as given by reference 6.

Eq. (7.2.52):	220	541	821	1134	1425
reference 6:	219	536	813	1127	1425

Table 7.2: Comparison of the results for the values of $\ell(\ell + 1)C^S_{TT,\ell}/2\pi$ (in μK^2) at the peaks listed in Table 7.1, as given by Eq. (7.2.52) and as given by reference 6.

Eq. (7.2.52):	5155	2694	2783	1126	746
Reference 6:	5591	2525	2451	1221	806

peak positions and heights calculated using Eq. (7.2.52) and those given by the results of reference 6. In all cases, the peak heights given by Eq. (7.2.52) agree with the more accurate computer results to within 10%, while the results of Eq. (7.2.52) for the peak positions are almost embarrassingly good, in no case being off by more than 1%.

Among other things, the general success of the calculations of this section shows that the evolution of cosmological perturbations is primarily hydrodynamic, in the sense that it can be well described by the equations of hydrodynamics without the full apparatus of coupled Boltzmann equations used in computer calculations like those of reference 6. The Boltzmann equation is implicit in our calculations, because in calculating the Silk damping rate we have used standard values for the shear viscosity and heat conductivity that were obtained by using the Boltzmann equation for photons in an ionized gas, but evidently not much is lost by not solving the Boltzmann equation over again in a cosmological context.

To the extent that the general formula (7.2.41) has been validated by this comparison, we can use it to see what can be learned from measurements of $C_{TT,\ell}$:

- Eq. (7.2.41) shows that the *shape* of the function $\ell(\ell + 1)C^S_{TT,\ell}$, as for instance the set of *ratios* of the peak positions or of the peak heights, depends on only four quantities: d_D, d_H, d_T, and n_S. Also, with the present radiation temperature and the number of massless neutrinos fixed, the three lengths d_D, d_H, and d_T depend only on the two cosmological parameters $\Omega_B h^2$ and $\Omega_M h^2$, which can therefore be found from the shape of the function $\ell(\ell + 1)C^S_{TT,\ell}$.

358

- Measurement of the scale of this function's ℓ-dependence (as for instance, the ℓ value of any one peak) depends on these two parameters, but also on d_A, which is a function not only of $\Omega_B h^2$ and $\Omega_M h^2$ but also of h and Ω_Λ. It is therefore impossible, without taking the integrated Sachs–Wolfe effect into account, to use measurements of $C_{TT,\ell}^S$ alone to make separate determinations of both h and Ω_Λ. It is often simply assumed that the universe is spatially flat, in which case to a good approximation $\Omega_\Lambda = 1 - \Omega_M$, and then measurements of $C_{TT,\ell}^S$ can be used to determine h as well as $\Omega_B h^2$ and $\Omega_M h^2$.

- Measurement of the magnitude of the function $\ell(\ell+1)C_{TT,\ell}$ for $\ell \gg 1$ (as for instance measurement of the height of any one peak) only tells us the quantity $N^2 \exp(-2\tau_{\text{reion}})$, so as already remarked we cannot use it to make a separate determination of N^2 or τ_{reion}. This ambiguity is resolved by measurements of the polarization of the cosmic microwave background, discussed in Section 7.4.

The original discovery of microwave background anisotropy was made by the COBE collaboration, and is discussed in Section 2.6. This only gave information about the anisotropy for relatively small ℓ, well below the position of the first acoustic peak at $\ell \approx 200$. This discovery was followed by a series of balloon-borne and ground-based observations,[8] which gave definite evidence for the first acoustic peak, and some data on higher peaks, extending in the case of the CBI collaboration up to values of ℓ beyond the position of the fifth acoustic peak, at $\ell \approx 1400$. The accuracy of these measurements up to about $\ell \approx 600$ was then greatly improved by observations made by a remarkable satellite mission, known as the Wilkinson Microwave Anisotropy Probe, or WMAP.[9] The WMAP satellite was launched on June 30, 2001, made loops around the moon to pick up kinetic energy from the Moon's motion, and finally reached an orbit about the equilibrium point known as L2. This point orbits the Sun at the speed needed to keep it about 1.5×10^6 km from Earth, on the other side of the earth from the sun, a location chosen to isolate the instrument from microwave radiation from the sun, earth, or moon. The satellite carries two back-to-back 1.4×1.6 meter

---

[8]The collaborations are ARCHEOPS: A. Benoit *et al., Astron. Astrophys.* **399**, L19, L25 (2003) [astro-ph/0210305, 0210306]; CDMP & MAT/TOCO: A. Miller *et al., Astrophys. J. Suppl.* **140**, 115 (2002) [astro-ph/0108030]; BOOMERANG: J. E. Ruhl *et al., Astrophys. J.* **599**, 786 (2003) [astro-ph/0212229]; MAXIMA: A. T. Lee *et al. Astrophys. J.* **561**, L1 (2001) ; DASI: N. W. Halverson *et al., Astrophys. J.* **568**, 38 (2002) [astro-ph/0104489]; CBI: T. J. Pearson *et al., Astrophys. J.* **591**, 556 (2003) [astro-ph/0205388]; ACBAR: C. L. Kuo *et al., Astrophys. J.* **600**, 32 (2004) [astro-ph/0212289]; and VSA: C. Dickinson *et al., Mon. Not. Roy. Astron. Soc.* **353**, 732 (2004) [astro-ph/0402498].

[9]C. L. Bennett *et al., Astrophys. J. Suppl.* **148**, 1 (2003) [astro-ph/0302207]. Aspects of this mission are treated in detail in other articles in the same volume of Astrophys. J. Suppl.

microwave receivers, cooled by thermal radiators to about 90 K, and measures the polarization of the microwave background as well as differences in its temperature over the whole sky.

After a year of observation, $C_{TT,\ell}$ had been measured out to $\ell \simeq 600$ with small errors arising mostly from cosmic variance and foreground emission.[10] These measurements were fit to the results of the "ΛCDM" model we have been studying, with zero curvature; the contents of the universe supposed to consist of photons, three flavors of massless neutrinos, baryonic matter, cold dark matter, and a constant vacuum energy; and a primordial spectrum of purely adiabatic fluctuations given by Eq. (7.2.40), with n_S constant. The values of cosmological parameters derived from this fit were:[11]

- $\Omega_B h^2 = 0.024 \pm 0.001$

- $\Omega_M h^2 = 0.14 \pm 0.02$

- $h = 0.72 \pm 0.05$

- $|N|^2 = (2.1 \pm 0.2) \times 10^{-10}$

- $n_S = 0.99 \pm 0.04$

- $\tau_{\text{reion}} = 0.166^{+0.076}_{-0.071}$

(Measurements of polarization were used here chiefly to measure the optical depth τ_{reion} of the reionized plasma,[12] which is needed to obtain $|N|^2$ from the value of $|N|^2 \exp(-2\tau_{\text{reion}})$ given by the measured values of $C^S_{TT,\ell}$.)

In March 2006 the WMAP group announced the results of the second and third years of observation.[13] The results are shown together with a best fit of $\ell(\ell+1)C^S_{TT,\ell}/2\pi$ to the ΛCDM model (with zero tensor anisotropies) in Figure 7.2. This fit gave the parameters:[14]

- $\Omega_B h^2 = 0.02229 \pm 0.00073$

- $\Omega_M h^2 = 0.1277^{+0.0080}_{-0.0079}$

- $h = 0.732^{+0.031}_{-0.032}$

- $|N|^2 = (1.93 \pm 0.12) \times 10^{-10}$

---

[10]G. Hinshaw *et al.*, *Astrophys. J.* **148**, 135 (2003) [astro-ph/0302217].

[11]D.N. Spergel *et al.*, ref. 5. Errors given here represent a 68% confidence range.

[12]A. Kogut *et al.*, *Astrophys. J.* **148**, 161 (2003) [astro-ph/0302213].

[13]Temperature and polarization results are given by G. Hinshaw *et al.*, *Astrophys. J. Suppl. Ser.* **170**, 288 (2007) [astro-ph/0603451] and L. Page *et al.*, *Astrophys. J. Suppl. Ser.* **170**, 335 (2007) [astro-ph/0603450], respectively.

[14]D. N. Spergel *et al.*, *Astrophys. J. Suppl. Ser.* **170**, 377 (2007) [astro-ph/0603449].

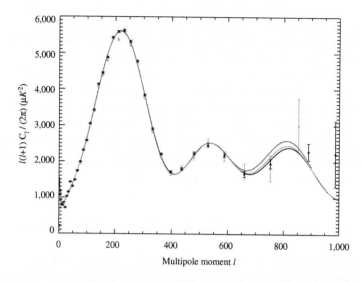

Figure 7.2: Comparison of theory and WMAP observations for the multipole coefficient $\ell(\ell+1)C_\ell/2\pi$ in square microKelvin *vs.* ℓ, from D. N. Spergel *et al.*, astro-ph/0603449. The solid data points are the three-year WMAP data, and the light gray data points are the first year WMAP data. The bottom curve is the best fit to the three-year WMAP data. The top curve is the best fit to the first-year WMAP data, and the middle curve is the best fit to the first-year WMAP data combined with data from CBI and ACBAR.

- $n_S = 0.958 \pm 0.016$

- $\tau_{\text{reion}} = 0.089 \pm 0.030$

The new results are consistent with those found before, but evidently there has been a significant improvement in the precision with which these parameters are known. This increased precision has now revealed the important information that n_s is very likely somewhat less than unity, as expected on the basis of typical inflationary theories, to be discussed in Chapter 10. It is also important that the optical depth of the reionized plasma found by polarization measurements is considerably less than previously found, suggesting a more plausible time of reionization.[15] Because the magnitude of $C_{TT,\ell}^S$ is proportional to $|N|^2 \exp(-2\tau_{\text{reion}})$, the reduction in τ_{reion} has led to a corresponding reduction in the estimated value of $|N|^2$; there has been very little change in the reported value of $|N|^2 \exp(-2\tau_{\text{reion}})$.

---

[15]For the implications of the three-year WMAP measurement of the plasma optical depth for theories of early star formation, see M. A. Alvarez, P. R. Shapiro, K. Ahn, and I. T. Iliev, *Astrophys. J.* **644**, L101 (2006) [astro-ph/0604447]; Z. Haiman and G. L. Bryan, *Astrophys. J.* **650**, 7 (2006) [astro-ph/0603541]; T. R. Choudhury and A. Ferrara, *Mon. Not. Roy. Astron. Soc.* **371**, L55 (2006) [astro-ph/0603617]; M. Shull and A. Venkatesan, astro-ph/0702323.

It is truly satisfying to see that the values of h, $\Omega_M h^2$, and $\Omega_B h^2$ obtained from the cosmic microwave background anisotropies confirm the values obtained by the very different methods discussed in Chapters 1 and 3. The values of $|N|^2$ and n_s are also in good agreement with values inferred from observations of large scale structure, discussed in the next chapter. The general agreement of theory and observation, both for the microwave background anisotropies alone and for the microwave background anisotropies in conjunction with other observations, goes far to confirm the general assumptions of the cosmological model, including cold dark matter and dark energy, that we have been using.

7.3 Temperature multipole coefficients: Tensor modes

We next consider the contribution of cosmological gravitational radiation to the temperature multipole coefficients.[1] According to Eq. (7.1.5), in the case of a single dominant tensor mode the tensor contribution to the fractional temperature fluctuation is

$$
\left(\frac{\Delta T(\hat{n})}{T_0} \right)^{(T)} = -\frac{1}{2} \sum_{\lambda=\pm 2} \int d^3 q \, \beta(\mathbf{q}, \lambda) \, \hat{n}_k \hat{n}_l \, e_{kl}(\hat{q}, \lambda)
$$

$$
\times \int_{t_1}^{t_0} dt \, \exp\left[i\mathbf{q} \cdot \hat{n} \int_t^{t_0} \frac{dt'}{a(t')} \right] d(q, t) , \quad (7.3.1)
$$

where for brevity we have now introduced the quantity

$$
d(q, t) \equiv \exp\left[-\int_t^{t_0} dt' \, \omega_c(t') \right] \left(\dot{\mathcal{D}}_q(t) - \frac{1}{2} \omega_c(t) \Psi(q, t) \right) . \quad (7.3.2)
$$

As a reminder: $\beta(\mathbf{q}, \lambda)$ is the stochastic parameter for the mode that does not decay outside the horizon, which is assumed to dominate the tensor perturbation; $e_{kl}(\hat{q}, \lambda)$ is the polarization tensor defined in Section 5.2 for a gravitational wave with wave number $\mathbf{q}$ and helicity λ; t_1 is any time sufficiently early before recombination so that any photon present then would have been scattered many times before the present; $\omega_c(t)$ is the photon collision rate at time t; $\mathcal{D}_q(t)$ is the gravitational wave amplitude, defined in Section 5.2; and $\Psi(q, t)$ is the tensor source function, defined by Eqs. (6.6.11) and (6.6.12).

---

[1] For an early calculation of the first few multipole moments, see V. A. Rubakov, M. V. Sazhin, and A. V. Veryaskin, *Phys. Lett.* **115B**, 189 (1982). A general formula was given by R. Fabbri and M. D. Pollock, *Phys. Lett.* **125B**, 445 (1983).

Again using Eq. (7.2.2) to expand the exponential in spherical harmonics, the tensor temperature fluctuation has the expansion

$$\left(\Delta T(\hat{n})\right)^{(T)} = \sum_{\ell m} a_{T,\ell m}^T \, Y_\ell^m(\hat{n}) \,, \tag{7.3.3}$$

where

$$a_{T,\ell m}^T = T_0 \sum_{\lambda = \pm 2} \int d^3q \, \beta(\mathbf{q}, \lambda) f_{\ell m}(\mathbf{q}, \lambda) \,, \tag{7.3.4}$$

$$f_{\ell m}(\mathbf{q}, \lambda) \equiv -2\pi \int d^2\hat{n} \, Y_\ell^{m*}(\hat{n}) \, \hat{n}_k \hat{n}_l \, e_{kl}(\hat{q}, \lambda) \sum_{LM} i^L \, Y_L^M(\hat{n}) \, Y_L^{M*}(\hat{q})$$

$$\times \int_{t_1}^{t_0} dt \, j_L\!\left(q\,r(t)\right) d(q, t) \,, \tag{7.3.5}$$

and $r(t)$ is again the radial coordinate of a point from which light emitted at time t would reach us at the present:

$$r(t) \equiv \int_t^{t_0} \frac{dt'}{a(t')} \,. \tag{7.3.6}$$

It will be very convenient first to calculate $f_{\ell m}(\mathbf{q}, \lambda)$ for $\mathbf{q}$ in the three-direction $\hat{z}$. In this case we have $Y_L^M(\hat{q}) = \delta_{M0}\sqrt{2L + 1/4\pi}$. Also, using Eq. (5.2.15), for $\hat{n} = (\sin\theta \, \cos\phi, \, \sin\theta \, \sin\phi, \, \cos\theta)$ we have

$$\hat{n}_i \hat{n}_j \, e_{ij}(\hat{z}, \pm 2) = \frac{1}{\sqrt{2}} \sin^2\theta \, e^{\pm 2i\phi} = 4\sqrt{\frac{\pi}{15}} \, Y_2^{\pm 2}(\hat{n}) \,. \tag{7.3.7}$$

The integral over $\hat{n}$ is given by a special case of the general formula[2]

$$\int d^2\hat{n} \, Y_L^M(\hat{n}) \, Y_\Lambda^\mu(\hat{n}) \, Y_\ell^{m*}(\hat{n}) = \sqrt{\frac{(2\Lambda + 1)(2\ell + 1)}{4\pi(2L + 1)}}$$

$$\times C_{\ell\Lambda}(L, M; m, -\mu) \, C_{\ell\Lambda}(L, 0; 0, 0) \,, \tag{7.3.8}$$

where $C_{\ell\Lambda}(L, M; m, \mu)$ is the usual Clebsch–Gordan coefficient for combining angular momentum quantum numbers ℓ, m and Λ, μ to form

---

[2]See, e.g., J. M. Blatt and V. F. Weisskopf, *Theoretical Nuclear Physics* (John Wiley & Sons, New York, 1952): Appendix A, Eq. (5.11).

angular momentum quantum numbers L, M. Hence

$$f_{\ell m}(q\hat{z}, \pm 2) = -2\sqrt{\frac{\pi(2\ell+1)}{3}} \sum_L i^L C_{\ell 2}(L, 0; \pm 2, \mp 2) C_{\ell 2}(L, 0; 0, 0)$$

$$\times \delta_{m,\pm 2} \int_{t_1}^{t_0} dt \, j_L\left(q\, r(t)\right) d(q, t) . \tag{7.3.9}$$

In our case we have $\Lambda = 2$, $\mu = \pm 2$, and $M = 0$, so the relevant Clebsch–Gordan coefficients are

$$C_{\ell 2}(\ell+2, 0; 0, 0) = \sqrt{\frac{3(\ell+2)(\ell+1)}{2(2\ell+1)(2\ell+3)}}$$

$$C_{\ell 2}(\ell+2, 0; \pm 2, \mp 2) = \frac{1}{2}\sqrt{\frac{(\ell-1)\ell}{(2\ell+1)(2\ell+3)}}$$

$$C_{\ell 2}(\ell, 0; 0, 0) = -\sqrt{\frac{\ell(\ell+1)}{(2\ell-1)(2\ell+3)}}$$

$$C_{\ell 2}(\ell, 0; \pm 2, \mp 2) = \sqrt{\frac{3(\ell-1)(\ell+2)}{2(2\ell-1)(2\ell+3)}}$$

$$C_{\ell 2}(\ell-2, 0; 0, 0) = \sqrt{\frac{3\ell(\ell-1)}{2(2\ell-1)(2\ell+1)}}$$

$$C_{\ell 2}(\ell-2, 0; \pm 2, \mp 2) = \frac{1}{2}\sqrt{\frac{(\ell+1)(\ell+2)}{(2\ell-1)(2\ell+1)}} ,$$

while $C_{\ell 2}(\ell \pm 1, 0; 0, 0) = 0$. Putting this together shows that for **q** in the three-direction, the non-vanishing values of the quantity (7.3.5) are[3]

$$f_{\ell m}(q\hat{z}, \pm 2) = i^\ell \sqrt{\frac{\pi(2\ell+1)(\ell+2)!}{2(\ell-2)!}} \delta_{m,\pm 2} \int dt \, d(q, t)$$

$$\times \left[\frac{j_{\ell+2}\left(q\, r(t)\right)}{(2\ell+1)(2\ell+3)} + \frac{2j_\ell\left(q\, r(t)\right)}{(2\ell-1)(2\ell+3)} + \frac{j_{\ell-2}\left(q\, r(t)\right)}{(2\ell-1)(2\ell+1)} \right].$$

$$\tag{7.3.10}$$

---

[3] This is essentially the result originally obtained by L.F. Abbott and M. B. Wise, *Nucl. Phys.* **B 244**, 541 (1984), and A. A. Starobinsky, *Sov. Astron. Lett.* **11**(3), 133 (1985), generalized to an arbitrary gravitational wave amplitude $D_q(t)$ and including the correction proportional to Ψ in Eq. (7.3.2).

This can be greatly simplified by iterating the familiar recursion relation

$$j_\ell(\rho)/\rho = [j_{\ell-1}(\rho) + j_{\ell+1}(\rho)]/(2\ell+1) ,$$

which gives

$$\frac{j_\ell(\rho)}{\rho^2} = \frac{j_{\ell-2}(\rho)}{(2\ell+1)(2\ell-1)} + \frac{2j_\ell(\rho)}{(2\ell+3)(2\ell-1)} + \frac{j_{\ell+2}(\rho)}{(2\ell+1)(2\ell+3)} . \quad (7.3.11)$$

Using this in Eq. (7.3.10), we have

$$f_{\ell m}(q\hat{z}, \pm 2) = i^\ell \sqrt{\frac{\pi(2\ell+1)(\ell+2)!}{2(\ell-2)!}} \delta_{m,\pm 2} \int_{t_1}^{t_0} dt\, d(q,t) \frac{j_\ell\big(qr(t)\big)}{q^2 r^2(t)} . \quad (7.3.12)$$

The amplitude (7.3.5) can now be found for a general direction of $\mathbf{q}$ by performing a standard rotation $S(\hat{q})$ that takes the three-axis into the direction $\hat{q}$. (An explicit formula for $S(\hat{q})$ will be given in the next section; it is not needed here.) This gives

$$f_{\ell m}(\mathbf{q}, \lambda) = \sum_{\pm} D^{(\ell)}_{m,m'}\big(S(\hat{q})\big) f_{\ell m'}(q\hat{z}, \lambda) . \quad (7.3.13)$$

The coefficients (7.3.4) are then

$$a^T_{T,\ell m} = T_0 i^\ell \sqrt{\frac{\pi(2\ell+1)(\ell+2)!}{2(\ell-2)!}} \sum_{\pm} \int d^3q\, \beta(\mathbf{q}, \pm 2)\, D^{(\ell)}_{m,\pm 2}\big(S(\hat{q})\big)$$

$$\times \int_{t_1}^{t_0} dt\, d(q,t) \frac{j_\ell\big(qr(t)\big)}{q^2 r^2(t)} . \quad (7.3.14)$$

We can now easily calculate the multipole coefficients of the temperature correlation function, defined by

$$\langle a^T_{T,\ell m} a^{T*}_{T,\ell'm'}\rangle = \delta_{\ell\ell'}\delta_{mm'} C^T_{TT,\ell} . \quad (7.3.15)$$

For this purpose, we recall that the stochastic parameter $\beta(\mathbf{q}, \lambda)$ is normalized so that

$$\langle \beta(\mathbf{q}, \lambda)\, \beta^*(\mathbf{q}, \lambda)\rangle = \delta_{\lambda\lambda'}\delta^3(\mathbf{q} - \mathbf{q}') . \quad (7.3.16)$$

In the calculation of $C^T_{TT,\ell}$ we encounter an angular integral

$$\int d^2\hat{q}\, D^{(\ell)}_{m,\pm 2}\big(S(\hat{q})\big) D^{(\ell')}_{m',\pm 2}\big(S(\hat{q})\big)^* .$$

This matrix can readily be seen to commute with all rotation matrices, so that it must be proportional to $\delta_{mm'}\delta_{\ell\ell'}$ with an m-independent coefficient. To calculate this coefficient, we can set $m = m'$ and $\ell = \ell'$ and sum m from $-\ell$ to $+\ell$; using the unitarity of the matrix $D_{m,n}^{(\ell)}\big(S(\hat{q})\big)$, this must equal $\int d^2\hat{q} = 4\pi$, so

$$\int d^2\hat{q}\, D_{m,\pm 2}^{(\ell)}\big(S(\hat{q})\big) D_{m',\pm 2}^{(\ell')}\big(S(\hat{q})\big)^* = \frac{4\pi}{2\ell + 1}\delta_{mm'}\delta_{\ell\ell'} . \qquad (7.3.17)$$

Using this and Eqs. (7.3.14) and (7.3.16) in Eq. (7.3.15), we see that the sum over helicities in Eq. (7.3.14) now just yields a factor 2, so[4]

$$C_{TT,\ell}^T = \frac{4\pi^2(\ell + 2)!T_0^2}{(\ell - 2)!}\int_0^\infty \frac{dq}{q^2}\left|\int_{t_1}^{t_0} dt\, d(q,t)\,\frac{j_\ell\big(qr(t)\big)}{r^2(t)}\right|^2 . \qquad (7.3.18)$$

(This formula can be obtained more easily by a direct calculation of the temperature correlation function, without calculating the $a_{T,\ell m}^T$, but we will need the $a_{T,\ell m}^T$ in the next section to find the correlation between polarization and temperature anisotropies.)

It remains to calculate $\mathcal{D}_q(t)$ and the source function $\Psi(q,t)$, As we saw in Section 6.6, if the gravitational wave amplitude $\mathcal{D}_q(t)$ is written as a function of $\kappa \equiv q\sqrt{2}/H_{EQ}a_{EQ}$ and $y \equiv a(t)/a_{EQ}$ instead of q and t (the subscript EQ referring to the time of radiation–matter equality), then aside from an over-all factor $\mathcal{D}_q^o$ (equal to the value of $\mathcal{D}_q(t)$ outside the horizon), the amplitude $\mathcal{D}_q(t)$ is independent of any other cosmological parameters. We write this as

$$\mathcal{D}_q(t) = \mathcal{D}_q^o\, D(\kappa, y) . \qquad (7.3.19)$$

For a spectrum of gravitational waves that is scale invariant outside the horizon, the quantity $q^3|\mathcal{D}_q^o|^2$ is a constant. To take account of more general possibilities, it is conventional to write it as proportional to a power n_T of q, or equivalently,

$$q^3|\mathcal{D}_q^o|^2 = N_T^2\left(\frac{q/a_0}{k_D}\right)^{n_T} , \qquad (7.3.20)$$

where k_D is an arbitrary wave number, often taken as $k_D = 0.002$ Mpc^{-1}, and N_T is a constant analogous to the constant N that describes the strength

---

[4]This result is derived in a different way by M. Zaldarriaga and U. Seljak, *Phys. Rev. D* **55**, 1830 (1997) [astro-ph/9609170]. Their result is the same as Eq. (7.3.18), if we take their undefined gravitational wave amplitude h to be $\mathcal{D}_q(t)/2$, with their stochastic parameter normalized so that $P_h = 1$.

of the adiabatic scalar mode. The conventional ratio r of tensor and scalar modes is $r \equiv 4|N_T|^2/|N|^2$. Also, until vacuum energy becomes important, we have $dt = \sqrt{2}y\, dy/H_{EQ}\sqrt{1+y}$, so we can write

$$q \int \frac{dt}{a(t)} = \kappa \int \frac{dy}{\sqrt{1+y}} = 2\kappa\sqrt{1+y} \,.$$

The wave equation (6.6.58) then gives

$$(1+y)\frac{\partial^2 D(\kappa,y)}{\partial y^2} + \left(\frac{1}{2} + \frac{2(1+y)}{y}\right)\frac{\partial D(\kappa,y)}{\partial y} + \kappa^2 D(\kappa,y)$$
$$= -\frac{24f_\nu}{y^2}\int_0^y K\left(2\kappa\,[\sqrt{1+y} - \sqrt{1+y'}]\right)\frac{\partial D(\kappa,y')}{\partial y'}\,dy' \,, \quad (7.3.21)$$

where $f_\nu \equiv \bar{\rho}_\nu/\bar{\rho}_R = 0.4052$ and $K(v) \equiv j_2(v)/v^2$, with the initial condition $D(\kappa,0) = 1$. Having found $D(\kappa,y)$ in this way, the source function $\Psi(q,t)$ is calculated in computer programs such as CMBfast and CAMB by using Eq. (6.6.21), with the amplitudes $\Delta_{T,\ell}^{(T)}$ and $\Delta_{P,\ell}^{(T)}$ found by a numerical solution of the coupled Boltzmann equations truncated at some maximum value of ℓ. There is an easier alternative procedure, based on the integral equation (6.6.27):[5]

$$\Psi(q,t) = \frac{3}{2}\int_{t_1}^t dt' \exp\left[-\int_{t'}^t \omega_c(t'')\,dt''\right]$$
$$\times \left[-2\dot{D}_q(t')K\left(q\int_{t'}^t \frac{dt''}{a(t'')}\right) + \omega_c(t')F\left(q\int_{t'}^t \frac{dt''}{a(t'')}\right)\Psi(q,t')\right],$$

$$(7.3.22)$$

where $F(v) \equiv j_0(v) - 2j_1(v)/v + 2j_2(v)/v^2$. We can also put the differential optical depth in the form

$$d\tau(y) \equiv \omega_c\,dt = n_e\sigma_T c\,dt = \frac{A}{y^2\sqrt{1+y}}X(y)\,dy \,,$$

where $X(y)$ is the fractional hydrogen ionization calculated in Section 2.3, which depends on $\Omega_B h^2$ and $\Omega_M h^2$ as well as y, and A is the dimensionless constant

$$A \equiv 0.76\frac{3\,\Omega_B\,\Omega_M\,H_0\,\sigma_T\,c}{8\pi\,Gm_p\Omega_R^{3/2}} = 1.9646 \times 10^5\,(\Omega_M h^2)\,(\Omega_B h^2)\,.$$

---

[5] S. Weinberg, *Phys. Rev.* D **74**, 063517 (2006) [astro-ph/0607076]; D. Baskaran, L. P. Grishchuk, and A. G. Polnarev, *Phys. Rev.* D **74**, 083005 (2006) [gr-qc/0605100].

(It is assumed that the baryons are 76% hydrogen, with the rest un-ionized in the era of interest.) It follows that the source function takes the form

$$\Psi(q, t) = \mathcal{D}_q^o S(\kappa, y) \tag{7.3.23}$$

where S depends only on $\Omega_M h^2$ and $\Omega_B h^2$ as well as κ and y, and satisfies the integral equation

$$S(\kappa, y) = \frac{3}{2} \int_0^y dy' \exp\left[-\int_{y'}^y \frac{d\tau(y'')}{dy''} dy'' \right]$$

$$\times \left[-2 \frac{dD(\kappa, y')}{dy'} K\left(2\kappa[\sqrt{1+y} - \sqrt{1+y'}] \right) \right.$$

$$\left. + \frac{d\tau(y')}{dy'} F\left(2\kappa[\sqrt{1+y} - \sqrt{1+y'}] \right) S(\kappa, y') \right]. \tag{7.3.24}$$

The multipole coefficient calculated from Eqs. (7.3.18)–(7.3.24) is shown by the solid curve in Figure 7.3. The dashed line in Figure 7.3 shows the result of simplifying the expression for the multipole coefficients in the case $\ell \gg 1$, using the approximate formula (7.2.14) for the spherical Bessel

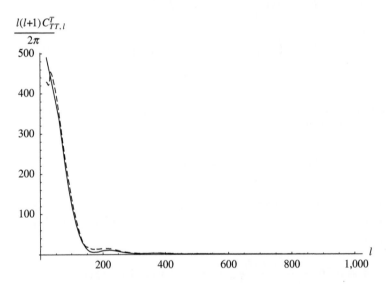

Figure 7.3: The tensor temperature–temperature multipole coefficient $\ell(\ell+1)C_{TT,\ell}^T/2\pi$ in square microKelvin, vs. ℓ, from ref. 6. The solid line is the result of using the essentially exact formula (7.3.18); the dashed line gives the result of the large-ℓ approximation (7.3.25). The cosmological parameters used in these calculations are described in the text.

function in Eq. (7.3.18), which gives[6]

$$\ell(\ell+1)C_{TT,\ell}^T \to 4\pi^2 T_0^2 \int_0^\infty q^2\, dq$$

$$\times \left| \int_{r(t)>\ell/q} dt\, d(q,t) \left\{ \frac{\cos^3 b}{\sqrt{\sin b}} \cos\left[\ell(\tan b - b) - \pi/4\right] \right\}_{\cos b=\ell/q\,r(t)} \right|^2 .$$

(7.3.25)

The source function $\Psi(q,t)$ is calculated in both cases using the integral equation (7.3.22), instead of using the equivalent formula (6.6.21) together with the truncated Boltzmann hierarchy (6.6.18), (6.6.19). Both of these curves are calculated with a cosmological model that in most respects is consistent with current observations: We assume $\Omega_K = 0$, and constant vacuum energy, with $\Omega_B = 0.0432$, $\Omega_M = 0.257$, $\Omega_\Lambda = 0.743$, $h = 0.72$, and $T_0 = 2.725K$. In calculating the photon collision frequency, we use the Recfast recombination code,[7] with helium abundance $Y = 0.24$. The parameters in Eq. (7.3.20) are taken as $N_T^2 = 4.68 \times 10^{-11}$ (corresponding approximately to the WMAP3 value of the scalar amplitude N quoted in the previous section, and $r = 1$) and $n_T = 0$. Reionization is ignored in this calculation. To take into account a different value of r or a finite optical depth τ of the reionized plasma, for $\ell > 10$ it is only necessary to multiply the multipole coefficients shown in Figure 7.3 by $r\exp(-2\tau)$.

(In the approximation of a sharp transition at time t_L from thermal equilibrium, in which $\omega_c(t) \gg H(t)$ and $\Psi(q,t) = 0$, to perfect transparency, in which $\omega_c(t) = 0$, Eq. (7.3.25) becomes

$$\ell(\ell+1)C_{TT,\ell}^T \to 4\pi^2 T_0^2 \int_0^\infty q^2\, dq$$

$$\times \left| \int_{r_L>r(t)>\ell/q} dt\, \dot{D}_q(t) \left\{ \frac{\cos^3 b}{\sqrt{\sin b}} \cos\left[\ell(\tan b - b) - \pi/4\right] \right\}_{\cos b=\ell/q\,r(t)} \right|^2 .$$

(7.3.26)

Because of the spread of values of $\tan b - b$, we cannot here make the sort of further simplification that we made for scalar temperature fluctuations in going from Eq. (7.2.16) to Eq. (7.2.17).)

---

[6]R. Flauger and S. Weinberg, *Phys. Rev.* D **75**, 123505 (2007) [astro-ph/0703179].

[7]S. Seager, D. D. Sasselov, and D. Scott, *Astrophys. J.* **523**, L1 (1999) [astro-ph/9909275]; *Astrophys. J. Suppl.* **128**, 407 (2000) [astro-ph/9912182].

Figure 7.3 shows that $C^T_{TT,\ell}$ has an ℓ-dependence very different from that of the scalar contribution calculated in the previous section, so the agreement between existing observations and theory including only scalar perturbations sets an upper bound on the strength of the tensor perturbations. Under the assumption (as suggested by theories of inflation discussed in Chapter 10) that $\mathcal{D}^0_q$ and $\mathcal{R}^0_q$ have about the same q-dependence, the WMAP collaboration has concluded[8] after three-years of operation that $r < 0.55$, at a 95% confidence level, where $r \equiv 4|\mathcal{D}^0_q|^2/|\mathcal{R}^0_q|^2$. This limit arises chiefly from temperature rather than polarization measurements, because of the much greater signal to noise ratio of the temperature measurements; the limit on r from polarization measurements alone is $r < 2.2$.

Strictly speaking, the limit $r < 0.55$ does rely on polarization measurements, which are used to determine the optical thickness of the reionized plasma. This is needed to determine the value of the slope parameter n_S for scalar modes, which in turn is needed in subtracting the scalar contribution to $C_{TT,\ell}$ from the observed values in order to set a limit on the tensor contribution. But cosmic variance sets a limit to the accuracy with which this subtraction can be made, and in the long run the best upper limits or the actual detection of tensor modes will come directly from polarization measurements, to which we now turn.

7.4 Polarization

Observations of the cosmic microwave background reveal not only its intensity in various directions, but also its polarization. The microwave background is expected to be polarized because of its scattering by free electrons,[1] such as were present around the time of recombination, or during the later period of reionization due to ultraviolet light from the first generation of stars. Polarization measurements have become of importance in learning when reionization began, and in disentangling the effects of reionization from the primordial intensity of fluctuations, and they may become even more important in future, in revealing the effects of gravitational waves

---

[8]L. Page *et al.*, *Astrophys. J. Suppl. Ser.* **170**, 335 (2007) [astro-ph/0603450].

[1]M. J. Rees, *Astrophys. J.* **153**, L1 (1968). Polarization correlations at small angular separation were considered in early papers on the polarization of the cosmic microwave background; see e.g. A. G. Polnarev, *Sov. Astron.* **29**, 607 (1985); R. Crittenden, R. L. Davis, and P. J. Steinhardt, *Astrophys. J. Lett.* **417**, L13 (1993); D. Coulson, R. Crittenden, and N. Turok, *Phys. Rev. Lett.* **73**, 2390 (1994); A. Kosowsky, *Ann. Phys. (N.Y.)* **246**, 49 (1996). Then all-sky analyses were given by U. Seljak and M. Zaldarriaga, *Phys. Rev. Lett.* **78**, 2054 (1997) [astro-ph/9609169]; M. Zaldarriaga and U. Seljak, *Phys. Rev. D* **55**, 1830 (1997) [astro-ph/9609170], and by M. Kamionkowski, A. Kosowsky, and A. Stebbins, *Phys. Rev. Lett.* **78**, 2058 (1997) [astro-ph/9609132]; *Phys. Rev. D.* **55**, 7368 (1997) [astro-ph/9611125].

produced during inflation.

In Section 6.1 we introduced a dimensionless photon intensity perturbation matrix $J_{ij}(\mathbf{x}, \hat{p}, t)$, defined by Eq. (6.1.13). It is Hermitian, transverse in the sense that

$$\hat{p}^i\, J_{ij}(\mathbf{x}, \hat{p}, t) = \hat{p}^j\, J_{ij}(\mathbf{x}, \hat{p}, t) = 0 \, , \qquad (7.4.1)$$

and, as already noted in Section 7.1, its trace at $\mathbf{x} = 0$ and $t = t_0$ is related to the fractional photon temperature fluctuation seen from earth at present in any direction $\hat{n}$ by

$$\frac{\Delta T(\hat{n})}{T_0} = \frac{1}{4} J_{ii}(0, -\hat{n}, t_0) \, . \qquad (7.4.2)$$

For $\hat{n} = -\hat{p}$ in the three-direction $\hat{z}$, any matrix with these properties can be put in the form

$$J_{ij}(0, -\hat{z}, t_0) = \frac{2}{T_0} \begin{pmatrix} \Delta T(\hat{z}) + Q(\hat{z}) & U(\hat{z}) - iV(\hat{z}) & 0 \\ U(\hat{z}) + iV(\hat{z}) & \Delta T(\hat{z}) - Q(\hat{z}) & 0 \\ 0 & 0 & 0 \end{pmatrix}, \qquad (7.4.3)$$

where Q, U, and V are three real functions of direction with the dimensions of temperature, known as *Stokes parameters*.[2] In this case the Stokes parameters can be expressed in terms of the matrix J_{ij}, as

$$Q(\hat{z}) \pm iU(\hat{z}) = \frac{T_0}{2} e_{\pm i}(\hat{z})\, e_{\pm j}(\hat{z})\, J_{ij}(0, -\hat{z}, t_0) \, , \qquad (7.4.4)$$

$$V(\hat{z}) = \frac{T_0}{4} e_{- i}(\hat{z})\, e_{+ j}(\hat{z}) \left(J_{ij}(0, -\hat{z}, t_0) - J_{ji}(0, -\hat{z}, t_0) \right) , \qquad (7.4.5)$$

where $e_{\pm}(\hat{z}) = (1, \pm i, 0)/\sqrt{2}$ are the polarization vectors for a photon coming from the three-direction. Accordingly, for a photon coming from an arbitrary direction $\hat{n}$, we define the Stokes parameters as

$$Q(\hat{n}) \pm iU(\hat{n}) \equiv \frac{T_0}{2} e_{\pm i}(\hat{n})\, e_{\pm j}(\hat{n})\, J_{ij}(0, -\hat{n}, t_0) \, , \qquad (7.4.6)$$

$$V(\hat{n}) \equiv \frac{T_0}{4} e_{- i}(\hat{n})\, e_{+ j}(\hat{n}) \left(J_{ij}(0, -\hat{n}, t_0) - J_{ji}(0, -\hat{n}, t_0) \right) , \qquad (7.4.7)$$

where $e_{\pm}(\hat{n})$ are the polarization vectors for a photon coming from the $\hat{n}$-direction. Writing $\hat{n}$ in terms of polar and azimuthal angles θ and ϕ as

$$\hat{n} = (\sin\theta\, \cos\phi, \ \sin\theta\, \sin\phi, \ \cos\theta) \, , \qquad (7.4.8)$$

---

[2]The Stokes parameters are sometimes defined with extra constant factors to give them the dimensions of intensity rather than temperature.

we can take

$$\mathbf{e}_\pm(\hat{n}) = (\hat{\theta} \pm i\hat{\phi})/\sqrt{2} \,, \qquad (7.4.9)$$

where $\hat{\theta}$ and $\hat{\phi}$ are orthogonal unit vectors in the plane perpendicular to $\hat{n}$, in the directions of increasing θ and ϕ, respectively:

$$\hat{\theta} \equiv (\cos\theta\,\cos\phi,\ \cos\theta\,\sin\phi,\ -\sin\theta)\,, \quad \hat{\phi} \equiv (-\sin\phi,\ \cos\phi\,,0)\,.$$
$$(7.4.10)$$

It is the Stokes parameters that are measured in observations of the microwave background. The scattering of light by non-relativistic electrons does not produce circular polarization, and therefore we expect that all microwave background photons are linearly polarized, so that J_{ij} is real, and therefore $V = 0$. In this case the fractional intensity perturbation $e^i e^j J_{ij}$ reaches a maximum value $2(\Delta T + \sqrt{U^2 + Q^2})/T_0$ for a real polarization vector $\mathbf{e}$ in a direction $\hat{\theta}\cos\xi + \hat{\phi}\sin\xi$, where $\tan 2\xi = U/Q$. For a polarization vector in the orthogonal direction, $e^i e^j J_{ij}$ takes its minimum value $2(\Delta T - \sqrt{U^2 + Q^2})/T_0$, giving a total fractional intensity perturbation $4\Delta T/T_0$, as is necessary since the photon energy density goes as T^4.

Now we have to face a complication: As we have defined them, the Stokes parameters are not rotational scalars. That is, under an arbitrary rotation $x_i \rightarrow x'_i = R_{ij}x_j$ of the three-dimensional coordinate system, we do *not* have $Q'(\hat{n}') = Q(\hat{n})$ or $U'(\hat{n}') = U(\hat{n})$. This is because the polarization vectors are not really three-vectors. We note that

$$\mathbf{e}_{\pm\,i}(\hat{n}) = S_{ij}(\hat{n})\mathbf{e}_{\pm\,j}(\hat{z})\,, \qquad (7.4.11)$$

where $S_{ij}(\hat{n})$ is a standard rotation that takes the three-axis into the direction $\hat{n}$. For $\hat{n} = (\sin\theta\,\cos\phi, \sin\theta\,\sin\phi, \cos\theta)$:

$$S_{ij}(\hat{n}) \equiv \begin{pmatrix} \cos\theta\,\cos\phi & -\sin\phi & \sin\theta\,\cos\phi \\ \cos\theta\,\sin\phi & \cos\phi & \sin\theta\,\sin\phi \\ -\sin\theta & 0 & \cos\theta \end{pmatrix}\,, \qquad (7.4.12)$$

and again $\mathbf{e}_\pm(\hat{z}) = (1, \pm i, 0)/\sqrt{2}$. For an arbitrary rotation R, we can write

$$\mathbf{e}_{\pm\,i}(R\hat{n}) \equiv S_{ij}(R\hat{n})\mathbf{e}_{\pm\,j}(\hat{z}) = \left[R\,S(\hat{n})\right]_{ik}\left[S^{-1}(\hat{n})R^{-1}S(R\hat{n})\right]_{kj}\mathbf{e}_{\pm\,j}(\hat{z})\,.$$

Now, $S(R\hat{n})$ takes the three-axis into the direction $R\hat{n}$, R^{-1} takes this into the direction $\hat{n}$, and then $S^{-1}(\hat{n})$ takes this back to the direction of the three-axis, so $S^{-1}(\hat{n})R^{-1}S(R\hat{n})$ leaves the three-direction invariant, and must therefore be a rotation by some angle $\psi(R, \hat{n})$ around the three-direction. Acting on $\mathbf{e}_\pm(\hat{z})$, this yields a factor $e^{\pm i\psi}$, and so

$$\mathbf{e}_{\pm i}(R\hat{n}) = e^{\pm i\psi(R,\hat{n})} R_{ij}\mathbf{e}_{\pm j}(\hat{n}) . \tag{7.4.13}$$

The matrix J_{ij} is an ordinary three-tensor, in the sense that an arbitrary rotation R takes J_{ij} into J'_{ij}, with

$$J'_{ij}(R\hat{n}) = R_{ik}R_{il}J_{kl}(\hat{n}) ,$$

Because $R^T R = 1$, a rotation R subjects the Stokes parameters (7.4.6) to the transformation $Q \rightarrow Q'$, $U \rightarrow U'$, with

$$Q'(\hat{n}') \pm iU'(\hat{n}') = e^{\pm 2i\psi(R,\hat{n})} [Q(\hat{n}) \pm iU(\hat{n})] . \tag{7.4.14}$$

For this reason, if we were to expand the Stokes parameters in a series of ordinary spherical harmonics, as we do for scalars like the temperature fluctuation, then the expansion coefficients would not transform under rotations according to the usual representations of the rotation group. Instead, we expand the Stokes parameters $Q(\hat{n})$ and $U(\hat{n})$ seen in a direction $\hat{n}$ in a series of functions[3] $\mathcal{Y}_\ell^m(\hat{n})$ with the same dependence on the polarization vectors as the Stokes parameters themselves:

$$Q(\hat{n}) + iU(\hat{n}) = \sum_{\ell=2}^{\infty} \sum_{m=-\ell}^{\ell} a_{P,\ell m}\, \mathcal{Y}_\ell^m(\hat{n}) , \tag{7.4.15}$$

$$\mathcal{Y}_\ell^m(\hat{n}) \equiv 2\sqrt{\frac{(\ell-2)!}{(\ell+2)!}}\, e_{+i}(\hat{n})\, e_{+j}(\hat{n})\, \tilde{\nabla}_i\, \tilde{\nabla}_j\, Y_\ell^m(\hat{n}) , \tag{7.4.16}$$

where $\tilde{\nabla}$ is the angular part of the gradient operator:

$$\tilde{\nabla} \equiv \hat{\theta}\frac{\partial}{\partial\theta} + \frac{\hat{\phi}}{\sin\theta}\frac{\partial}{\partial\phi} . \tag{7.4.17}$$

The subscript "P" (for "polarization") on the coefficients $a_{P,\ell m}$ is introduced here to distinguish them from the coefficients entering in the expansion (2.6.1) of the temperature fluctuation, which in this chapter are denoted $a_{T,\ell m}$. Under a rotation R, the ordinary spherical harmonics transform as

$$Y_\ell^m(R\hat{n}) = \sum_{m'} D_{m'm}^{(\ell)}(R^{-1})\, Y_\ell^{m'}(\hat{n}) , \tag{7.4.18}$$

---

[3] These are a special case of functions introduced in a study of gravitational radiation by E. T. Newman and R. Penrose, *J. Math. Phys.* **7**, 863 (1966). In their notation, which is used in some recent papers on microwave background polarization, the function $\mathcal{Y}_\ell^m$ is denoted $_2Y_{\ell,m}$.

where $D^{(\ell)}(R)$ are the unitary irreducible matrices[4] of rank $2\ell+1$ that form a representation of the rotation group, in the sense that $D^{(\ell)}(R_1)D^{(\ell)}(R_2) = D^{(\ell)}(R_1 R_2)$. It follows then from Eqs. (7.4.13), (7.4.16) and (7.4.18) that

$$\mathcal{Y}_\ell^m(R\hat{n}) = e^{2i\psi(R,\hat{n})} \sum_{m'} D^{(\ell)}_{m'm}(R^{-1})\,\mathcal{Y}_\ell^{m'}(\hat{n}) , \tag{7.4.19}$$

and hence also

$$\mathcal{Y}_\ell^{m*}(R\hat{n}) = e^{-2i\psi(R,\hat{n})} \sum_{m'} D^{(\ell)}_{mm'}(R)\,\mathcal{Y}_\ell^{m'}(\hat{n})^* . \tag{7.4.20}$$

For this reason, $\mathcal{Y}_\ell^m$ and its complex conjugate are known as spherical harmonics of spin 2 and -2, respectively. Just as for ordinary spherical harmonics, it follows from these transformation properties that $\int d\Omega\, \mathcal{Y}_\ell^{m*}\mathcal{Y}_{\ell'}^{m'}$ vanishes except for $\ell = \ell'$ and $m = m'$, and in that case is independent of m. The factor $2\sqrt{(\ell-2)!/(\ell+2)!}$ is inserted in Eq. (7.4.16) to make these functions satisfy the orthonormality condition

$$\int d\Omega\, \mathcal{Y}_\ell^{m*}\mathcal{Y}_{\ell'}^{m'} = \delta_{\ell\ell'}\delta_{mm'} , \tag{7.4.21}$$

just like the ordinary spherical harmonics Y_ℓ^m. Direct evaluation of Eq. (7.4.16) gives[5]

$$\mathcal{Y}_\ell^m(\hat{n}) = \sqrt{\frac{(\ell-2)!}{(\ell+2)!}}\left[\left(\frac{\partial}{\partial\theta}+i\csc\theta\frac{\partial}{\partial\phi}\right)^2 - \cot\theta\left(\frac{\partial}{\partial\theta}+i\csc\theta\frac{\partial}{\partial\phi}\right)\right]Y_\ell^m(\hat{n})$$

$$= e^{im\phi}\sqrt{\frac{(\ell-2)!}{(\ell+2)!}\frac{(\ell-|m|)!}{(\ell+|m|)!}\frac{2\ell+1}{4\pi}}\left[-\ell(\ell+1)P_\ell^{|m|}(\mu)\right.$$

$$\left.+\frac{2(m+\mu)(m-\ell\mu)}{1-\mu^2}P_\ell^{|m|}(\mu)\right.$$

---

[4]See, e.g., A. R. Edmonds, *Angular Momentum in Quantum Mechanics*, (Princeton University Press, Princeton, 1957): Chapter 4; M. E. Rose, *Elementary Theory of Angular Momentum* (John Wiley & Sons, New York, 1957): Chapter IV; L. D. Landau and E. M. Lifshitz, *Quantum Mechanics – Non Relativistic Theory*, 3rd edn. (Pergamon Press, Oxford, 1977): Section 58; Wu-Ki Tung, *Group Theory in Physics* (World Scientific, Singapore, 1985): Sections 7.3 and 8.1. Note that the rotation matrices we use are appropriate for the representation furnished by the spherical harmonics we use, with phases such that $Y_\ell^m(\hat{n})^* = Y_\ell^{-m}(\hat{n})$. The corresponding angular momentum matrices $\mathbf{J}^{(\ell)}$ are not the usual ones, for which the elements of $J_1^{(\ell)}\pm iJ_2^{(\ell)}$ are real and positive. If we changed the phase of the spherical harmonics so that $Y_\ell^m(\hat{n})^* = (-1)^m Y_\ell^{-m}(\hat{n})$, then the rotation matrices would be generated by conventional angular momentum matrices, with the elements of $J_1^{(\ell)}\pm iJ_2^{(\ell)}$ real and positive, but then we would have to introduce phases into the definition (7.4.25) of the E and B-type partial wave amplitudes, different from those used by Zaldarriaga and Seljak in ref. 1.

[5]The final expression is taken from Zaldarriaga and Seljak, ref. 1, while the first expression, which is derived here from our definition (7.4.16), is their definition of the spin +2 weighted spherical harmonic.

$$+\frac{2(\ell + |m|)(m + \mu)}{1 - \mu^2} P_{\ell-1}^{|m|}(\mu)\Bigg)\Bigg]$$

$$= e^{im\phi}\left[\frac{(\ell+m)!(\ell-m)!}{(\ell+2)!(\ell-2)!}\frac{2\ell+1}{4\pi}\right]^{1/2}\sin^{2\ell}(\theta/2)(-1)^{(m-|m|)/2}$$

$$\times \sum_r \binom{\ell-2}{r}\binom{\ell+2}{r+2-m}$$

$$\times(-1)^{\ell-r+m}\cot^{2r+2-m}(\theta/2)\,, \tag{7.4.22}$$

where $\mu \equiv \cos\theta$ and $P_\ell^m(\mu)$ are the usual associated Legendre functions. (Here we will only need the first expression in Eq. (7.4.22).) Unlike the ordinary spherical harmonics, the $\mathcal{Y}_\ell^m$ do not satisfy any simple reality condition. Rather, the complex conjugate of Eq. (7.4.15) gives

$$Q(\hat{n}) - iU(\hat{n}) = \sum_{\ell=2}^{\infty}\sum_{m=-\ell}^{\ell} a_{P,\ell m}^* \, \mathcal{Y}_\ell^m(\hat{n})^* \,, \tag{7.4.23}$$

where

$$\mathcal{Y}_\ell^m(\hat{n})^* = 2\sqrt{\frac{(\ell-2)!}{(\ell+2)!}}\, e_{-i}(\hat{n})\, e_{-j}(\hat{n})\, \tilde{\nabla}_i\, \tilde{\nabla}_j\, Y_\ell^{-m}(\hat{n})\,, \tag{7.4.24}$$

in which we have used the reality property of the spherical harmonics employed here, $Y_\ell^{m*} = Y_\ell^{-m}$.

Because $\mathcal{Y}_\ell^m$ does not satisfy any simple reality condition, neither does $a_{P,\ell m}$. Instead, we can define coefficients

$$a_{E,\ell m} \equiv -\left(a_{P,\ell m} + a_{P,\ell-m}^*\right)\Big/2\,, \quad a_{B,\ell m} \equiv i\left(a_{P,\ell m} - a_{P,\ell-m}^*\right)\Big/2\,. \tag{7.4.25}$$

This is a useful decomposition, because of the properties of the coefficients under space inversion. If we reverse all three coordinate axes, then $\theta \rightarrow \pi - \theta$, while $\phi \rightarrow \phi \pm \pi$, so $\hat{\theta} \rightarrow \hat{\theta}$ and $\hat{\phi} \rightarrow -\hat{\phi}$. It follows then that the polarization vectors (7.4.9) of definite helicity are interchanged under space inversion; that is, $\mathbf{e}_\pm(-\hat{n}) = \mathbf{e}_\mp(\hat{n})$, and therefore according to Eq. (7.4.6), under space inversion $Q(\hat{n}) \rightarrow Q(-\hat{n})$ while $U(\hat{n}) \rightarrow -U(-\hat{n})$ and $V(\hat{n}) \rightarrow -V(-\hat{n})$. (The reader will later be able to check these space-inversion properties of Q and U, by noting that these are the changes that are produced if replace the stochastic parameters $\alpha(\mathbf{q})$ and $\beta(\mathbf{q},\lambda)$ in Eqs. (7.4.31) and (7.4.40) with their space-inversion transforms $\alpha(-\mathbf{q})$ and $\beta(-\mathbf{q},-\lambda)$, respectively.) Thus by applying a space inversion to Eq. (7.4.15), we find that space inversion takes the partial wave amplitudes $a_{P,\ell m}$ into

$a'_{P,\ell m}$, where

$$Q(-\hat{n}) - iU(-\hat{n})$$

$$= 2\sum_{\ell=2}^{\infty}\sum_{\ell=0}^{\ell}\sum_{m=-\ell}^{\ell} a'_{P,\ell m} \sqrt{\frac{(\ell-2)!}{(\ell+2)!}}\, e_{+i}(\hat{n})\, e_{+j}(\hat{n})\, \tilde{\nabla}_i \tilde{\nabla}_j\, Y_\ell^m(\hat{n})$$

$$= 2(-1)^\ell \sum_{\ell=0}^{\infty}\sum_{m=-\ell}^{\ell} a'_{P,\ell m} \sqrt{\frac{(\ell-2)!}{(\ell+2)!}}\, e_{-i}(-\hat{n})\, e_{-j}(-\hat{n})\, \tilde{\nabla}_i \tilde{\nabla}_j\, Y_\ell^m(-\hat{n})$$

$$= (-1)^\ell \sum_{\ell=2}^{\infty}\sum_{m=-\ell}^{\ell} a'_{P,\ell m}\, \mathcal{Y}_\ell^{-m}(-\hat{n})^* \,,$$

the factor $(-1)^\ell$ coming from the reflection property $Y_\ell^m(-\hat{n}) \to (-1)^\ell Y_\ell^m(\hat{n})$. Comparing this with Eq. (7.4.23), we see that space inversion changes $a_{P,\ell m}$ into $a'_{P,\ell m} = (-1)^\ell a^*_{P,\ell-m}$, so it changes $a_{E,\ell m}$ (and also the corresponding temperature multipole coefficient $a_{T,\ell m}$) by a sign $(-1)^\ell$, while it changes $a_{B,\ell m}$ by a sign $-(-1)^\ell$.

These coefficients are stochastic variables, governed by a probability distribution that is presumably invariant under space inversion, so there can be no bilinear correlation between polarization fluctuations of B and E-type, or between temperature fluctuations and polarization fluctuations of B type, though there can be correlations between temperature fluctuations and polarization fluctuations of E type. Taking into account also the rotational invariance of the probability distribution, the only non-vanishing bilinear averages are of the form[6]

$$\langle a^*_{T,\ell m} a_{T,\ell' m'} \rangle = C_{TT,\ell}\, \delta_{\ell,\ell'}\, \delta_{m,m'} \,, \qquad (7.4.26)$$

$$\langle a^*_{T,\ell m} a_{E,\ell' m'} \rangle = C_{TE,\ell}\, \delta_{\ell,\ell'}\, \delta_{m,m'} \,, \qquad (7.4.27)$$

$$\langle a^*_{E,\ell m} a_{E,\ell' m'} \rangle = C_{EE,\ell}\, \delta_{\ell,\ell'}\, \delta_{m,m'} \,, \qquad (7.4.28)$$

$$\langle a^*_{B,\ell m} a_{B,\ell' m'} \rangle = C_{BB,\ell}\, \delta_{\ell,\ell'}\, \delta_{m,m'} \,. \qquad (7.4.29)$$

All $C_{XY,\ell}$ coefficients have the same dimensions, of square temperature. By their definition, the a_T, a_E, and a_B all satisfy the reality conditions

$$a^*_{T,\ell m} = a_{T,\ell-m}\,, \quad a^*_{E,\ell m} = a_{E,\ell-m}\,, \quad a^*_{B,\ell m} = a_{B,\ell-m}\,, \qquad (7.4.30)$$

---

[6]Here $a_{T,\ell m}$ is the coefficient $a_{\ell m}$ in Eq. (2.6.1). The multipole coefficients $C_{XY,\ell}$ defined here are the same as those defined by Zaldarriaga and Seljak, ref. 1. They are related to coefficients defined by Kamionkowski, Kosowsky, and Stebbins, ref. 1, by $C_\ell^G = C_{\ell,EE}I^2/2T_0^2$, $C_\ell^C = C_{\ell,BB}I^2/2T_0^2$, $C_\ell^{TG} = -C_{TE,\ell}I/\sqrt{2}T_0^3$, where I is the radiation intensity. (The superscripts G and C stand for "gradient" and "curl", because the pattern of polarization vectors of type E or B resemble the pattern of velocity in potential or solenoidal flow, respectively.)

and hence the coefficients $C_{XY,\ell}$ are all real, while the $C_{XX,\ell}$ are all also positive. The results of measurements of correlations between temperature and/or polarization fluctuations in different directions are generally reported in terms of the $C_{XY,\ell}$.

Another simplification provided by the decomposition of the Stokes parameters into E and B terms is that, as we will see, scalar perturbations can only contribute to the E terms. This result is important, because it means that any sign of a primordial B-type polarization[7] will be clear evidence for cosmological gravitational waves, of the sort that we shall see in Chapter 10 are expected to be produced during inflation.

We now turn to the calculation of the multipole coefficients (7.4.27)–(7.4.29). The hydrodynamic treatment that worked reasonably well for temperature correlation functions is not well suited to the treatment of polarization, so we will rely on the more accurate kinetic theory outlined in Section 6.1. Here we must distinguish between scalar and tensor modes. As already mentioned, these do not interfere, so the multipole coefficients can all be written as a sum of a scalar and a tensor term, denoted by superscripts S and T:

$$C_{XY,\ell} = C^S_{XY,\ell} + C^T_{XY,\ell} \,. \tag{7.4.31}$$

A. Scalar modes

For scalar modes, the matrix J_{ij} in Eq. (7.4.6) is given by the Fourier integral (6.1.18) and the line-of-sight integral (6.1.36):

$$J_{ij}(0, -\hat{n}, t_0) = \int d^3q \, \alpha(\mathbf{q}) \int_{t_1}^{t_0} dt \, \exp\left(-i\mathbf{q}\cdot\hat{p}\int_t^{t_0} \frac{dt'}{a(t')} - \int_t^{t_0} dt' \, \omega_c(t')\right)$$

$$\times \left[-\left(\delta_{ij} - \hat{n}_i\hat{n}_j\right)\left(\dot{A}_q(t') - (\hat{n}\cdot\mathbf{q})^2\dot{B}_q(')\right)\right.$$

$$+ \frac{3\omega_c(t)}{2}(\delta_{ij} - \hat{n}_i\hat{n}_j)\Phi(q,t)$$

$$+ \frac{3\omega_c(t)}{4}\Pi(q,t)\left(\hat{q}_i - \hat{n}_i(\hat{q}\cdot\hat{n})\right)\left(\hat{q}_j - \hat{n}_j(\hat{q}\cdot\hat{n})\right)$$

$$\left.+ \frac{2\omega_c(t)}{a(t)}[\delta_{ij} - \hat{n}_i\hat{n}_j]\,\hat{n}_k\delta u_k(\mathbf{q},t)\right] \,.$$

---

[7]Weak lensing by foreground objects converts the E-type polarization produced by scalar perturbations into a B-type polarization that may be large enough to interfere with measurements of the primordial B-type polarization; M. Zaldarriaga and U. Seljak, *Phys. Rev. D* **58**, 023003 (1998)[astro-ph/9803150]. For a review, see A. Lewis and A. Challinor, *Phys. Rep.* **429**, 1 (2006) [astro-ph/0601594].

Since $\hat{n}_i e_{+i}(\hat{n}) = 0$ and $e_{+i}(\hat{n}) e_{+i}(\hat{n}) = 0$, the only term in the integrand that contributes to the Stokes parameters is the one proportional to $\hat{q}_i \hat{q}_j$:

$$Q^S(\hat{n}) + iU^S(\hat{n}) = \frac{3T_0}{8} e_{+i}(\hat{n}) e_{+j}(\hat{n}) \int d^3q \, \hat{q}_i \hat{q}_j \, \alpha(\mathbf{q})$$

$$\times \int_{t_1}^{t_0} dt \, P(t) \, \Pi(q, t) \exp\left(i\mathbf{q} \cdot \hat{n} \int_t^{t_0} \frac{dt'}{a(t')} \right) , \qquad (7.4.32)$$

where $P(t) \, dt$ is the probability that the last scattering occurs between t and $t + dt$:

$$P(t) = \omega_c(t) \exp\left(- \int_t^{t_0} \omega_c(t') \, dt' \right) . \qquad (7.4.33)$$

As a reminder: $\alpha(\mathbf{q})$ is the stochastic parameter for whatever mode (presumably the non-decaying adiabatic mode) is assumed to dominate the scalar perturbations; $\omega_c(t)$ is the photon collision rate; t_1 is any time taken early enough so that a photon present at that time would suffer many collisions before the present; and $\Pi(q, t)$ is a source function, given by Eq. (6.1.30) in terms of the partial wave amplitudes of the temperature and polarization perturbation amplitudes:

$$\Pi(q, t) = \Delta_{T,2}^{(S)}(q, t) + \Delta_{P,0}^{(S)}(q, t) + \Delta_{P,2}^{(S)}(q, t) . \qquad (7.4.34)$$

To use Eq. (7.4.31) to calculate the *EE* and *TE* multipole coefficients, we recall the familiar formula

$$\exp\left(i\hat{q} \cdot \mathbf{v} \right) = 4\pi \sum_{\ell, m} i^\ell j_\ell(v) \, Y_\ell^m(\hat{v}) \, Y_\ell^{m*}(\hat{q}) ,$$

where here $\mathbf{v} = \hat{n} q \int_t^{t_0} dt'/a(t')$. Acting on this, we can replace $\hat{q}_i$ with $-i\partial/\partial v_i$. Since $\hat{v} = \hat{n}$, we can write $\partial/\partial v_i = \hat{n}_i \partial/\partial v + \tilde{\nabla}_i/v$, where $\tilde{\nabla}$ is the angular gradient operator (7.4.17) acting on $\hat{n}$. Since $\hat{n}_i e_{+i} = 0$ and $e_{+i} e_{+i} = 0$, Eq. (7.4.31) then reads

$$Q^S(\hat{n}) + iU^S(\hat{n}) = -\frac{3\pi T_0}{2} e_{+i}(\hat{n}) e_{+j}(\hat{n}) \tilde{\nabla}_i \tilde{\nabla}_j \sum_{\ell m} i^\ell Y_\ell^m(\hat{n}) \int d^3q \, \alpha(\mathbf{q})$$

$$\times Y_\ell^{m*}(\hat{q}) \int_{t_1}^{t_0} dt \, P(t) \, \Pi(q, t) \frac{j_\ell\left(q \, r(t) \right)}{q^2 r(t)^2} , \qquad (7.4.35)$$

where here $r(t)$ is the radial coordinate of a point from which light emitted at time t would just reach us at present

$$r(t) = \int_t^{t_0} \frac{dt'}{a(t')} . \qquad (7.4.36)$$

(Note that $\tilde{\nabla}_i \hat{n}_j$ doesn't contribute here, because it equals $\delta_{ij} - \hat{n}_i \hat{n}_j$, which vanishes when contracted with $e_{+i}e_{+j}$.) Comparing Eq. (7.4.34) with Eqs. (7.4.15) and (7.4.16), we see that the scalar contribution to $a_{P,\ell m}$ is

$$a_{P,\ell m}^S = -\frac{3\pi \, T_0 i^\ell}{4} \sqrt{\frac{(\ell+2)!}{(\ell-2)!}} \int d^3q \, \alpha(\mathbf{q}) \, Y_\ell^{m*}(\hat{q}) \int_{t_1}^{t_0} dt \qquad (7.4.37)$$

$$\times \, P(t) \, \Pi(q,t) \frac{j_\ell\big(q\,r(t)\big)}{q^2 \, r^2(t)} \qquad (7.4.38)$$

To check the reality properties of $a_{P,\ell m}$, we note that Eqs. (6.1.18) and (6.1.21) require that $\alpha^*(\mathbf{q})\Pi^*(q,t) = \alpha(-\mathbf{q})\Pi(q,t)$, while $Y_\ell^m(\hat{q}) = (-1)^\ell Y_\ell^{-m*}(-\hat{q})$, and of course $i^{\ell*} = (-1)^\ell i^\ell$, so $a_{P,\ell m}^{S*} = a_{P,\ell\,-m}^S$. Inspection of Eq. (7.4.25) then shows that, as promised, the scalar modes contribute only an E-type polarization

$$a_{E,\ell m}^S = -a_{P,\ell m}^S, \qquad a_{B,\ell m}^S = 0. \qquad (7.4.39)$$

We can now give formulas for the scalar contribution to the EE and TE multipole coefficients defined by Eqs. (7.4.28) and (7.4.27). Recalling the normalization condition (7.2.5) of the scalar stochastic parameter $\alpha(\mathbf{q})$, we see from Eqs. (7.4.36) and (7.4.37) that[8]

$$C_{EE,\ell}^S = \frac{9\pi^2 T_0^2}{16} \frac{(\ell+2)!}{(\ell-2)!} \int_0^\infty q^2 \, dq \left| \int_{t_1}^{t_0} dt \, P(t) \, \Pi(q,t) \frac{j_\ell\big(q\,r(t)\big)}{q^2 \, r^2(t)} \right|^2.$$

$$(7.4.40)$$

Combining Eq. (7.4.36) with the general formula (7.2.4) for the scalar contribution to the temperature partial wave amplitudes, we find in the same way that

$$C_{TE,\ell}^S = -3\pi^2 T_0^2 \sqrt{\frac{(\ell+2)!}{(\ell-2)!}} \int_0^\infty q^2 \, dq$$

$$\times \int_{t_1}^{t_0} dt \, P(t) \, \Pi(q,t) \frac{j_\ell\big(q\,r(t)\big)}{q^2 \, r^2(t)}$$

$$\times \int_{t_1}^{t_0} \Big[j_\ell\big(q\,r(t)\big) F(q,t) + j_\ell'\big(q\,r(t)\big) G(q,t) + j_\ell''\big(q\,r(t)\big) H(q,t) \Big],$$

$$(7.4.41)$$

---

[8]This is the same as Eq. (17) of Zaldarriaga and Seljak, ref. 1, except that Π here is 4 times their Π.

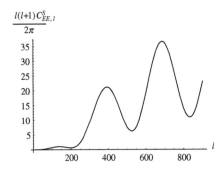

Figure 7.4: The multipole coefficient $\ell(\ell+1)C_{EE,\ell}^{S}/2\pi$ in square microKelvin, *vs.* ℓ, for the cosmological parameters given in Eqs. (7.2.44) and (7.2.45).

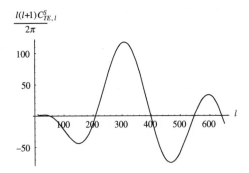

Figure 7.5: The multipole coefficient $\ell(\ell+1)C_{TE,\ell}^{S}/2\pi$ in square microKelvin, *vs.* ℓ, for the cosmological parameters given in Eqs. (7.2.44) and (7.2.45).

where $F(q,t)$, $G(q,t)$, and $H(q,t)$ are the quantities (7.2.7)–(7.2.9). The values of $C_{EE,\ell}^{S}$ and $C_{TE,\ell}^{S}$, calculated[9] for the cosmological parameters (7.2.44)–(7.2.45), are shown in Figures 7.4 and 7.5. Comparing these figures with each other and with Figure (7.3), we see that $C_{EE,\ell}^{S} \ll C_{TE,\ell}^{S} \ll C_{TT,\ell}^{S}$. The microwave background polarization is small, because the universe goes swiftly at the time of recombination from a state of nearly perfect thermal equilibrium to one of nearly perfect transparency, and photons are not polarized in thermal equilibrium.

Eqs. (7.4.38) and (7.4.39) must be used for maximum numerical accuracy, but we can get a good idea of the ℓ dependence of the multipole coefficients for $\ell \gg 1$ by using much simpler approximate versions. The function $P(t)$ in Eq. (7.4.38) is sharply peaked at a time t_L of last scattering, and it has

---

[9]http://lambda.gsfc.nasa.gov/data/map/powspec/wmap_lcdm_pl_model_yr1_v1.txt

$\int dt \, P(t) = 1$, so to a fair approximation we can do the integral over time by simply setting $t = t_L$ and dropping the factor $P(t)$. Also, by using the same approximations that led us from Eq. (7.2.13) to Eq. (7.2.17), we find that for $\ell \gg 1$

$$C_{EE,\ell}^{S} \approx \frac{9\pi^2\ell^3}{32r_L^3} \int_1^\infty \frac{d\beta}{\beta\sqrt{\beta^2-1}} \Pi^2\left(\frac{\ell\beta}{r_L}, t_L\right) .$$

This does not give a reliable result for the magnitude of $C_{EE,\ell}^{S}$, because the function $\Pi(q, t)$ varies rapidly around the time of recombination, but since this function falls off rapidly with q, so that the integral is dominated by $\beta \approx 1$, this approximation does suggest that the peaks and valleys in $C_{EE,\ell}^{S}$ arise chiefly from a factor $\Pi^2(\ell/r_L, t_L)$. On the other hand, as remarked in Section 7.2, the peaks and valleys in $C_{TT,\ell}^{S}$ arise chiefly from a factor $F^2(\ell/r_L)$, and by the same reasoning, the peaks and valleys in $C_{TE,\ell}^{S}$ arise chiefly from a factor $F(\ell/r_L)\Pi(\ell/r_L, t_L)$.

To get an idea of the q-dependence of $\Pi(q, t_L)$, we note first that, to the extent that the photon polarization arises solely from the last scattering, $\Pi(q, t)$ would be given by just the first term $\Delta_{T,2}^{(S)}(q, t)$ in Eq. (7.4.33). This is because the differential cross section for the scattering of an unpolarized photon of initial momentum $\mathbf{p}$ with $|\mathbf{p}| \ll m_e$ by a non-relativistic electron to yield a photon with real polarization vector $\mathbf{e}$ is[10]

$$\frac{d\sigma}{d\Omega} = \frac{e^4}{32\pi^2 m_e^2}\left(1 - (\hat{p} \cdot \mathbf{e})^2\right) .$$

If the distribution of initial momenta $\mathbf{p}$ were spherically symmetric then the average of this cross section over initial directions would be independent of $\mathbf{e}$, so the final photon would be unpolarized. But for a direction-dependent phase space density $n_{\gamma,L}(\mathbf{p})$ of photons with momentum $\mathbf{p}$ at the position and time of last scattering, the intensity of photons after last scattering coming from direction $\hat{n}$ and having polarization vector $\mathbf{e}$ (with $\mathbf{e}^2 = 1$ and $\mathbf{e} \cdot \hat{n} = 0$) is proportional to

$$\int d^3p \, p \, n_{\gamma,L}(\mathbf{p})\left(1 - (\hat{p} \cdot \mathbf{e})^2\right) .$$

The part of J_{ij} that contributes to the Stokes parameters is thus proportional to the traceless part of $\int d^3p \, p \, n_{\gamma,L}(\mathbf{p})\hat{p}_i\hat{p}_j$, which in turn is proportional to the $\ell = 2$ part of $\Delta_T^{(S)}$ evaluated at the time and position of last scattering.

---

[10]See, e.g., QTF, Eq. (8.7.40).   This was first calculated by O. Klein and Y. Nishina, Z. Phys. **52**, 853 (1929).

The additional terms $\Delta^{(S)}_{P,0}(q,t) + \Delta^{(S)}_{P,2}(q,t)$ in $\Pi(q,t)$ take account of whatever polarization the photon may already have acquired by the time of last scattering, and can be expected to be relatively small since photons must be unpolarized under conditions of rapid photon scattering. But then, according to Eq. (6.1.32),

$$\Pi(q,t_L) \simeq \Delta^{(S)}_{T,2}(q,t) \simeq q^2 \pi^S_q(t_L)/\bar{\rho}_\gamma(t_L) .$$

Recall that, according to the first line of Eq. (5.1.43), $\partial_i \partial_j \pi^S(\mathbf{x},t)$ is the term in the scalar part of $\delta T^i{}_j$ that is not proportional to δ_{ij}, while Eq. (B.50) tells us that for short mean free times, this term equals[11]

$$\partial_i \partial_j \pi^S = a^{-2} \eta_\gamma \partial_i \partial_j \delta\tilde{u} ,$$

where η_γ is the shear viscosity due to photon momentum transport, given in terms of the photon mean free time t_γ by[12] $\eta_\gamma = \frac{16}{45}\bar{\rho}_\gamma t_\gamma$, and $\delta\tilde{u}(\mathbf{x},t)$ is the gauge-invariant velocity potential[13]

$$\delta\tilde{u} \equiv \delta u - aF + \frac{a^2 \dot{B}}{2} ,$$

with F and B metric perturbations defined by Eqs. (5.1.32) and (5.1.33). Also, Eq. (7.1.45) (which was derived for a gauge in which the metric component F vanishes) gives the coefficient of the stochastic parameter $\alpha(\mathbf{q})$ in the Fourier transform of the gauge-invariant velocity potential in terms of the form factor $G(q)$ as $\delta\tilde{u}_q(t_L) = -a(t_L)G(q)/q$. We conclude then that

$$\Pi(q,t_L) \simeq \frac{16\,q^2}{45\,a^2(t_L)}\bar{t}_\gamma\, \delta\tilde{u}_q(t_L) \simeq -\frac{16\,q}{45\,a(t_L)}\bar{t}_\gamma\, G(q) .$$

where $\bar{t}_\gamma$ is some appropriate average of $t_\gamma(t)$ during the era of recombination. Thus the peaks and valleys in $C^S_{EE,\ell}$ and in $|C^S_{TE,\ell}|$ are more-or-less the same as those in $|G(q/r_L)|^2$ and $|F(q/r_L)G(q/r_L)|$, respectively, and we recall that the peaks and valleys in $C^S_{TT,\ell}$ are essentially those in $|F(q/r_L)|^2$. Eqs. (7.2.23) and (7.2.24) give

$$F(q) \simeq \frac{\mathcal{R}^o_q}{5}\left[3R_L T(\kappa) - (1+R_L)^{-1/4}\, e^{-\int_0^{t_L} \Gamma dt}\, S(\kappa)\, \cos\chi(q)\right] ,$$

---

[11] Here we use the relation $\delta u^i{}_{;j} = \delta u_j{}^i = a^{-2}\delta u_{i;j} - a^{-4}h_{ij}a\dot{a}$.

[12] N. Kaiser, *Mon. Not. Roy. Astron. Soc.* **202**, 1169 (1983). For a discussion, see footnote 5 of Section 6.4.

[13] We recognize $\delta\tilde{u}$ as the same velocity potential that appears in Eq. (7.1.37) as the Doppler contribution to the temperature shift in the class of gauges with vanishing metric component F. In the special case of Newtonian gauge we have $F = B = 0$, and $\delta\tilde{u} = \delta u$.

$$G(q) = -\frac{\sqrt{3}\mathcal{R}_q^o}{5(1+R_L)^{3/4}} e^{-\int_0^{t_L} \Gamma dt} S(\kappa) \sin \chi(q) ,$$

where $\chi(q)$ is an approximately linear function of q:

$$\chi(q) \equiv \int_0^{t_L} \frac{q\,dt}{a\sqrt{3(1+R)}} + \Delta(\kappa) ,$$

and the argument κ of the transfer functions S, T, and Δ is proportional to q. As remarked in Section 7.2, the peaks in $C_{TT,\ell}^S$ are thus roughly at $\chi(\ell/r_L) = \pi, 2\pi, 3\pi, \ldots$ On the other hand, we expect the peaks in $C_{EE,\ell}^S$ to be roughly at the peaks in $\left|\sin\left(\chi(\ell/r_L)\right)\right|$, or in other words, at $\chi(\ell/r_L) = \pi/2, 3\pi/2, 5\pi/2, \ldots$, and the peaks in $|C_{TE,\ell}^S|$ to be roughly at the peaks in $\left|\cos\left(\chi(\ell/r_L)\right)\sin\left(\chi(\ell/r_L)\right)\right|$, or in other words, at $\chi(\ell/r_L) = \pi/4, 3\pi/4, 5\pi/4, \ldots$ In particular, we expect a peak in $C_{EE,\ell}^S$ before the first peak in $C_{TT,\ell}^S$, followed by one peak in $C_{EE,\ell}^S$ between successive peaks in $C_{TT,\ell}^S$, while there should be two peaks in $|C_{TE,\ell}^S|$ before the first peak in $C_{TT,\ell}^S$, followed by two peaks in $|C_{TE,\ell}^S|$ between successive peaks in $C_{TT,\ell}^S$. This is precisely the pattern seen in computer calculations.[14] For instance, for a plausible set of cosmological parameters there are indeed two peaks in $|C_{TE,\ell}^S|$ (at $\ell = 36$ and $\ell = 144$) below the first peak in $C_{TT,\ell}^S$ at $\ell = 230$.

The E-type polarization and temperature–polarization correlation were first detected by the Degree Angular Scale Interferometer (DASI) collaboration.[15] Then the Wilkinson Microwave Anisotropy Probe (WMAP) collaboration measured the coefficients $C_{TE,\ell}$ with good accuracy over a range of multipole orders from $\ell = 2$ to $\ell \simeq 570$.[16] The results for $\ell > 10$ are in good agreement with the concordance model, assuming primordial fluctuations with a spectrum close to the Harrison–Zel'dovich form $\mathcal{R}_q^o \propto q^{-3/2}$, and a constant vacuum energy along with cold dark matter, baryonic matter, photons, and massless neutrinos, and using the same values of the cosmological parameters $\Omega_B h^2$, $\Omega_M h^2$, h, and N as used in fitting the model to measurements of temperature anisotropies. In particular, there is clear evidence of the second expected peak in $|C_{TE,\ell}|$, at around $\ell \simeq 140$, and the third expected peak, at $\ell \simeq 300$. More recently, the QUaD collaboration[17] has reported preliminary results from the first season of operation of a

---

[14]For instance, see http://lambda.gsfc.nasa.gov/data/map/powspec/wmap_1cdm_pl_model_vr1_v1.txt.

[15]J. Kovac et al., Nature **420**, 772 (2002).

[16]A. Kogut et al., Astrophys. J. Suppl. **148**, 161 (2003).

[17]P. Ade et al., 0705.2359.

microwave polarimeter at the South Pole, which show clear evidence for the first peak in $C_{EE,\ell}$ at $\ell \simeq 350$, about where it is expected from scattering around the time of recombination. But if reionization were not included in the cosmological model, there would be a clear discrepancy between its predictions and the data for $C_{TE,\ell}$ for $\ell < 10$. This discrepancy is attributed to polarization caused by scattering in a plasma that has been reionized by ultraviolet light from a first generation of stars.

The most important application of polarization measurements so far has been in working out the history of the reionization of intergalactic matter at redshifts much less than $z_L \simeq 1090$, which contributes to the last-scattering probability distribution $P(t)$. Because nearby events subtend large angles, the additional terms in $C_{EE,\ell}$ and $C_{TE,\ell}$ due to reionization are negligible except for relatively small ℓ, in fact $\ell < 10$. After three years of operation, the WMAP collaboration[18] found on the basis of the *EE* correlation alone that if reionization is sudden and complete at a redshift z_r, then $z_r = 10.9^{+2.7}_{-2.3}$, corresponding to an optical depth $\tau = 0.09 \pm 0.03$. This result is in line with expectations of the onset of star formation, and it has played an essential role in the use of temperature correlations to determine the magnitude of $\mathcal{R}_q^o$ and to set an upper limit on $\mathcal{D}_q^o$, discussed in Sections 7.2 and 7.3, respectively. Scattering by the reionized plasma produces a peak in $C_{EE,\ell}$ much like the peak at $\ell \simeq 140$, but shifted to much smaller ℓ, around $\ell \simeq 4$, because it occurs at an angular diameter distance much smaller than the angular diameter distance of the era of recombination. The WMAP measurements and other measurements of $C_{EE,\ell}$ for larger values of ℓ are in good agreement with the theoretical formula (7.4.38) for $C_{EE,\ell}^S$, as shown in Figure 7.6, and future measurements are expected to further reduce the uncertainties in this comparison.

B. Tensor modes

For tensor modes, the matrix J_{ij} in Eq. (7.4.4) is given by the Fourier integral (6.6.9) and line-of-sight integral (6.6.24):

$$J_{ij}(\mathbf{x}, \hat{p}, t) = \sum_{\lambda = \pm 2} \int d^3q \, e^{i\mathbf{q}\cdot\mathbf{x}} \beta(\mathbf{q}, \lambda)$$

$$\times \int_{t_1}^{t} dt' \exp\left(-i\mathbf{q} \cdot \hat{p} \int_{t'}^{t} \frac{dt''}{a(t'')} - \int_{t'}^{t} dt'' \, \omega_c(t'')\right)$$

---

[18] L. Page *et al.*, *Astrophys. J. Suppl. Ser.* **170**, 335 (2007) [astro-ph/0603450].

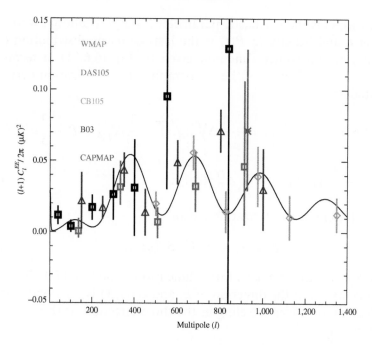

Figure 7.6:   Observed values of the multipole coefficient $(\ell + 1)C_{EE,\ell}/2\pi$ in square microKelvin, *vs.* ℓ, from L. Page *et al.*, astro-ph/0603450. (Note that $C_{EE,\ell}/2\pi$ is multiplied here with $\ell + 1$, rather than the usual $\ell(\ell + 1)$.) Dark squares are the WMAP data; the triangles are BOOMERanG data; the lighter squares are the DASI data; the diamonds are the CBI data; and the asterisk is the CAPMAP datum. The solid curve is the theoretical curve for $(\ell + 1)C^S_{EE,\ell}/2\pi$, with cosmological parameters taken from a fit to WMAP temperature and low-ℓ polarization data.

$$\times \left[-\hat{p}_k\hat{p}_l \left(\delta_{ij} - \hat{p}_i\hat{p}_j\right) e_{kl}(\hat{q}, \lambda) \, \dot{\mathcal{D}}_q(t')\right.$$

$$- \omega_c(t') \, \Psi(q, t')\left(e_{ij}(\hat{q}, \lambda) - \hat{p}_i\hat{p}_k e_{kj}(\hat{q}, \lambda)\right.$$

$$\left.\left. - \hat{p}_j\hat{p}_k e_{ik}(\hat{q}, \lambda) + \hat{p}_i\hat{p}_j\hat{p}_k\hat{p}_l e_{kl}(\hat{q}, \lambda)\right)\right] ,$$

Again using the conditions $\hat{n}_i e_{+i}(\hat{n}) = 0$ and $e_{+i}(\hat{n})e_{+i}(\hat{n}) = 0$, we see that the only term in the integrand here that contributes to the Stokes parameters in Eq. (7.4.4) is the one proportional to $e_{ij}\Psi$. This term gives

$$Q^T(\hat{n}) + iU^T(\hat{n}) = -\frac{1}{2}T_0 \, e_{+i}(\hat{n})e_{+j}(\hat{n}) \sum_{\lambda=\pm2} \int d^3q \, \beta(\mathbf{q}, \lambda)$$

$$\times e_{ij}(\hat{q}, \lambda) \int_{t_1}^{t_0} dt \exp\left(i\mathbf{q} \cdot \hat{n} \int_t^{t_0} \frac{dt'}{a(t')}\right) P(t) \, \Psi(q, t) . \qquad (7.4.42)$$

where $\beta(\mathbf{q}, \lambda)$ is the stochastic parameter for a gravitational wave of wave number $\mathbf{q}$ and helicity λ; $P(t)$ is the last-scattering distribution (7.4.32); and $\Psi(q, t)$ is a source function, given by Eq. (6.6.21) in terms of the partial wave amplitudes of the temperature and polarization perturbation amplitudes:

$$\Psi(q, t) = \frac{1}{10}\Delta_{T,0}^{(T)}(q, t) + \frac{1}{7}\Delta_{T,2}^{(T)}(q, t) + \frac{3}{70}\Delta_{T,4}^{(T)}(q, t) - \frac{3}{5}\Delta_{P,0}^{(T)}(q, t)$$

$$+ \frac{6}{7}\Delta_{P,2}^{(T)}(q, t) - \frac{3}{70}\Delta_{P,4}^{(T)}(q, t) . \tag{7.4.43}$$

Also, $e_{ij}(\hat{q}, \lambda)$ is the polarization tensor

$$e_{ij}(\hat{q}, \lambda) = S_{ik}(\hat{q}) \, S_{jl}(\hat{q}) \, e_{kl}(\hat{z}, \lambda) , \tag{7.4.44}$$

where $S_{ij}(\hat{q})$ is the standard three-dimensional rotation (7.4.12) that takes the three-axis into the direction of $\hat{q}$, and $e_{kl}(\hat{z}, \lambda)$ is the polarization tensor (5.2.15) for waves traveling in the three-direction, with non-vanishing components

$$e_{11}(\hat{z}, \pm 2) = -e_{22}(\hat{z}, \pm 2) = \mp i \, e_{12}(\hat{z}, \pm 2) = \mp i \, e_{21}(\hat{z}, \pm 2) = \frac{1}{\sqrt{2}} .$$

$$\tag{7.4.45}$$

We will first show how to find the tensor multipole coefficients $C_{EE,\ell}^T$, $C_{BB,\ell}^T$, and $C_{TE,\ell}^T$ for a given source function $\Psi(q, t)$, and then report the results of a numerical evaluation using $\Psi(q, t)$ calculated as described in Section 7.3.

We begin by deriving a formula for the coefficients $a_{P,\ell m}^T$ in the expansion of the Stokes parameters in spin-weighted spherical harmonics:

$$Q^T(\hat{n}) + iU^T(\hat{n}) = \sum_{\ell m} a_{P,\ell m}^T \, \mathcal{Y}_\ell^m(\hat{n}) . \tag{7.4.46}$$

Using Eq. (7.4.40) and the orthonormality property (7.4.21) of the $\mathcal{Y}_\ell^m$, we have

$$a_{P,\ell m}^T = \int d^2\hat{n} \, \mathcal{Y}_\ell^{m*}(\hat{n}) \Big(Q(\hat{n}) + iU(\hat{n}) \Big)^T$$

$$= -\frac{T_0}{2} \sum_\lambda \int d^3q \, \beta(\mathbf{q}, \lambda) \int_{t_1}^{t_0} dt \, P(t) \, \Psi(q, t) \, g_{\ell m}\Big(q \, r(t), \hat{q}, \lambda \Big) ,$$

$$\tag{7.4.47}$$

7.4   Polarization

where $r(t)$ is the unperturbed co-moving radial coordinate of a photon at time t that reaches us at time t_0:

$$r(t) \equiv \int_t^{t_0} \frac{dt'}{a(t')} ,$$

(7.4.48)

and

$$g_{\ell m}(\rho, \hat{q}, \lambda) \equiv \int d^2\hat{n}\, \mathcal{Y}_\ell^{m*}(\hat{n})\, e^{i\rho\hat{q}\cdot\hat{n}} \sum_{ij} e_{+i}(\hat{n})e_{+j}(\hat{n})e_{ij}(\hat{q}, \lambda) .$$

(7.4.49)

It is convenient first to calculate the amplitude $g_{\ell m}(\rho, \hat{q}, \lambda)$ for $\hat{q}$ along the direction $\hat{z}$ of the three-axis. Using the graviton polarization tensor (7.4.43) and the photon polarization vector (7.4.9), a straightforward calculation gives

$$e_{+i}(\hat{n})e_{+j}(\hat{n})e_{ij}(\hat{z}, \pm2) = \frac{1}{\sqrt{2}}\left(e_{+1}(\hat{n}) \pm ie_{+2}(\hat{n})\right)^2 = \frac{1}{2\sqrt{2}}e^{\pm2i\phi}(1 \mp \cos\theta)^2,$$

(7.4.50)

where as usual ϕ and θ are the azimuthal and polar angles of the direction $\hat{n}$. Eq. (7.4.22) shows that $\mathcal{Y}_\ell^{m*}(\hat{n})$ is $e^{-im\phi}$ times a function of θ, so the ϕ integral in Eq. (7.4.47) vanishes unless $m = \lambda = \pm2$, in which case it just gives a factor 2π. For $\mathcal{Y}_\ell^{\pm2}$ we will use the top line of Eq. (7.4.22), which gives

$$\mathcal{Y}_\ell^{\pm2}(\theta, \phi) = \sqrt{\frac{(\ell-2)!}{(\ell+2)!}}\partial_\pm Y_\ell^{\pm2}(\theta, \phi) = e^{\pm2i\phi}\sqrt{\frac{2\ell+1}{4\pi}\frac{(\ell-2)!}{(\ell+2)!}}\partial_\pm P_\ell^2(\cos\theta) ,$$

(7.4.51)

where $\partial_\pm$ are the differential operators

$$\partial_\pm \equiv \left(\frac{d}{d\theta} \mp 2\csc\theta\right)^2 - \cot\theta\left(\frac{d}{d\theta} \mp 2\csc\theta\right)$$

$$= (1-\mu^2)\frac{d}{d\mu^2} \pm 4\frac{d}{d\mu} + \frac{4(1\pm\mu)}{1-\mu^2} ,$$

(7.4.52)

where $\mu \equiv \cos\theta$, and $P_\ell^2(\cos\theta)$ is the usual associated Legendre polynomial

$$P_\ell^2(\mu) = (1-\mu^2)\frac{d^2}{d\mu^2}P_\ell(\mu) .$$

(7.4.53)

Putting this together, for $\mathbf{q}$ in the three-direction $\hat{z}$, Eq. (7.4.47) becomes

$$g_{\ell m}(\rho, \hat{z}, \pm2) = \delta_{m,\pm2}\sqrt{\frac{\pi(2\ell+1)}{8}\frac{(\ell-2)!}{(\ell+2)!}}\int_{-1}^{+1}d\mu\left[\partial_\pm P_\ell^2(\mu)\right]e^{i\rho\mu}(1\mp\mu)^2.$$

(7.4.54)

387

We use (7.4.51) for the associated Legendre polynomial, use integration by parts to let the derivatives in $\partial_\pm$ act on $(1 \mp \mu)^2 e^{i\rho\mu}$, and apply the formula

$$\frac{\partial^2}{\partial\mu^2}\left((1-\mu^2)(1\mp\mu)^2 e^{i\rho\mu}\right) + \left(\mp 4\frac{\partial}{\partial\mu} + \frac{4(1\pm\mu)}{1-\mu^2}\right)(1\mp\mu)^2 e^{i\rho\mu}$$

$$= (1-\mu^2)\left(12 \mp 8i\rho(1\mp\mu) - \rho^2(1\mp\mu)^2\right)e^{i\rho\mu} .$$

Replacing μ with $-i\partial/\partial\rho$, we find

$$g_{\ell m}(\rho, \hat{z}, \pm 2) = \delta_{m,\pm 2}\sqrt{\frac{\pi(2\ell+1)}{8}\frac{(\ell-2)!}{(\ell+2)!}}\int_{-1}^{+1} d\mu \, P_\ell(\mu)$$

$$\times \frac{\partial^2}{\partial\mu^2}\left[12 \mp 8i\rho(1\mp\mu) - \rho^2(1\mp\mu)^2\right](1-\mu^2)^2 \, e^{i\rho\mu}$$

$$= -\delta_{m,\pm 2}\sqrt{\frac{\pi(2\ell+1)}{8}\frac{(\ell-2)!}{(\ell+2)!}}\int_{-1}^{+1} d\mu \, P_\ell(\mu)$$

$$\times \left[12 + 8\rho\frac{\partial}{\partial\rho} - \rho^2 + \rho^2\frac{\partial^2}{\partial\rho^2} \mp 8i\rho \mp 2i\rho^2\frac{\partial}{\partial\rho}\right]$$

$$\times \left(1 + \frac{\partial^2}{\partial\rho^2}\right)^2 \rho^2 e^{i\rho\mu}$$

The integral over μ is now simply

$$\int_{-1}^{+1} d\mu \, e^{i\rho\mu}P_\ell(\mu) = 2i^\ell j_\ell(\rho) .$$

A straightforward though tedious calculation using the defining differential equation $j_\ell''(\rho) + (2/\rho)j_\ell'(\rho) + (1 - \ell(\ell+1)/\rho^2))j_\ell(\rho) = 0$ gives

$$\left(1 + \frac{\partial^2}{\partial\rho^2}\right)^2 \rho^2 j_\ell(\rho) = \frac{(\ell+2)!}{(\ell-2)!}\frac{j_\ell(\rho)}{\rho^2} ,$$

so finally

$$g_{\ell m}(\rho, \hat{z}, \pm 2) = -2i^\ell \delta_{m,\pm 2}\sqrt{\frac{\pi(2\ell+1)}{8}}$$

$$\times \left[12 + 8\rho\frac{\partial}{\partial\rho} - \rho^2 + \rho^2\frac{\partial^2}{\partial\rho^2} \mp 8i\rho \mp 2i\frac{\partial}{\partial\rho}\right]\frac{j_\ell(\rho)}{\rho^2} .$$

$$(7.4.55)$$

Just as in Eq. (7.3.13) for temperature fluctuations, the amplitude for $\mathbf{q}$ in a general direction is given by applying the standard rotation $S(\hat{q})$ that takes the three-axis into the direction of $\mathbf{q}$:

$$g_{\ell m}(\rho, \hat{q}, \pm 2) = \sum_{m'} D^{(\ell)}_{m,m'}\left(S(\hat{q})\right) g_{\ell m'}(\rho, \hat{z}, \pm 2) , \tag{7.4.56}$$

where $D^{(\ell)}$ is the spin-ℓ unitary representation of the rotation group. Using Eqs. (7.4.45), (7.4.54), and (7.4.53), we have then

$$a^T_{P,\ell m} = T_0\, i^\ell\, \sqrt{\frac{\pi(2\ell+1)}{8}} \sum_{\pm} \int d^3q\, \beta(\mathbf{q}, \pm 2)\, D^{(\ell)}_{m,\pm 2}\left(S(\hat{q})\right)$$

$$\times \int_{t_1}^{t_0} dt\, P(t)\, \Psi(q, t)$$

$$\times \left\{ \left[12 + 8\rho\frac{\partial}{\partial\rho} - \rho^2 + \rho^2 \frac{\partial^2}{\partial\rho^2} \mp 8i\rho \mp 2i\rho^2\frac{\partial}{\partial\rho} \right] \frac{j_\ell(\rho)}{\rho^2} \right\}_{\rho = q\, r(t)} .$$
$$\tag{7.4.57}$$

In order to separate the E and B terms in $a^T_{P,\ell m}$, we need the reality properties of β and $D^{(\ell)}$. First, note that

$$e_{ij}(\hat{q}, \pm 2) = \frac{1}{\sqrt{2}}\left(S_{i1}(\hat{q}) \pm iS_{i2}(\hat{q}) \right)\left(S_{j1}(\hat{q}) \pm iS_{j2}(\hat{q}) \right)$$

and

$$S_{i1}(\hat{q}) \pm iS_{i2}(\hat{q}) = \left(\cos\theta \cos\phi \mp i \sin\phi, \cos\theta \sin\phi \pm i \cos\phi, -\sin\theta \right),$$

so

$$e^*_{ij}(\hat{q}, \pm 2) = e_{ij}(\hat{q}, \mp 2) = e_{ij}(-\hat{q}, \pm 2) .$$

With the gravitational field dominated by a single mode, $\mathcal{D}^*_q(t)$ must be proportional to $\mathcal{D}_q(t)$, so that by absorbing any phase in $\beta(\mathbf{q}, \pm 2)$ we can choose $\mathcal{D}_q(t)$ to be real. The reality of $D_{ij}(\mathbf{x}, t)$ then requires that

$$\beta^*(\mathbf{q}, \pm 2) = \beta(-\mathbf{q}, \pm 2) . \tag{7.4.58}$$

Also, by writing $D^{(\ell)}\left(S(\hat{q})\right) = \exp\left(-i\phi J^{(\ell)}_3\right) \exp\left(-i\theta J^{(\ell)}_2\right)$ where $J^{(\ell)}_i$ are the angular momentum matrices[19] for angular momentum ℓ, we can

---

[19] As noted in footnote 4, the phases in these angular momentum matrices depend on the phase convention chosen for the spherical harmonics. They are related by $\int d^2\hat{n}\, Y^{m'}_\ell(\hat{n})^* \mathbf{L} Y^m_\ell(\hat{n}) = \mathbf{J}^\ell_{m'm}$, where $\mathbf{L} \equiv -i\mathbf{x} \times \nabla$ is the orbital angular momentum operator.

show that

$$D^{(\ell)}_{m,\pm2}\left(S(\hat{q})\right)^* = (-1)^\ell D^{(\ell)}_{-m,\pm2}\left(S(-\hat{q})\right) . \tag{7.4.59}$$

The sign $(-1)^\ell$ here cancels the same sign in $i^{-\ell} = (-1)^\ell i^\ell$. It follows that the whole effect of taking the complex conjugate of $a^T_{P,\ell m}$, and changing m to $-m$, is to replace the differential operator acting on $j_\ell(\rho)/\rho^2$ with its complex conjugate:

$$a^{T*}_{P,\ell\,-m} = T_0\, i^\ell \sqrt{\frac{\pi(2\ell+1)}{8}} \sum_\pm \int d^3q\, \beta(\mathbf{q},\pm2)\, D^{(\ell)}_{m,\pm2}\left(S(\hat{q})\right)$$

$$\times \int_{t_1}^{t_0} dt\, P(t)\, \Psi(q,t) \tag{7.4.60}$$

$$\times \left\{ \left[12 + 8\rho\frac{\partial}{\partial\rho} - \rho^2 + \rho^2\frac{\partial^2}{\partial\rho^2} \pm 8i\rho \pm 2i\rho^2\frac{\partial}{\partial\rho} \right] \frac{j_\ell(\rho)}{\rho^2} \right\}_{\rho=q\,r(t)} .$$

The definitions (7.4.25) then give

$$a^T_{E,lm} = -i^\ell\, T_0 \sqrt{\frac{\pi(2\ell+1)}{8}} \sum_\pm \int d^3q\, \beta(\mathbf{q},\pm2)\, D^{(\ell)}_{m,\pm2}\left(S(\hat{q})\right)$$

$$\times \int_{t_1}^{t_0} dt\, P(t)\, \Psi(q,t)$$

$$\times \left\{ \left[12 + 8\rho\frac{\partial}{\partial\rho} - \rho^2 + \rho^2\frac{\partial^2}{\partial\rho^2} \right] \frac{j_\ell(\rho)}{\rho^2} \right\}_{\rho=q\,r(t)} . \tag{7.4.61}$$

$$a^T_{B,lm} = i^\ell\, T_0 \sqrt{\frac{\pi(2\ell+1)}{8}} \sum_\pm \pm \int d^3q\, \beta(\mathbf{q},\pm2)\, D^{(\ell)}_{m,\pm2}\left(S(\hat{q})\right)$$

$$\times \int_{t_1}^{t_0} dt\, P(t)\, \Psi(q,t) \left\{ \left[8\rho + 2\rho^2\frac{\partial}{\partial\rho} \right] \frac{j_\ell(\rho)}{\rho^2} \right\}_{\rho=q\,r(t)} \tag{7.4.62}$$

We can now calculate the multipole coefficients, using the stochastic averages

$$\langle \beta(\mathbf{q},\lambda)\, \beta^*(\mathbf{q}',\lambda') \rangle = \delta^3(\mathbf{q}-\mathbf{q}')\delta_{\lambda\lambda'} , \tag{7.4.63}$$

and the unitarity relation (7.3.17):

$$\int d^2\hat{q}\, D^{(\ell)}_{m,\pm2}\left(S(\hat{q})\right) D^{(\ell')}_{m',\pm2}\left(S(\hat{q})\right)^* = \frac{4\pi}{2\ell+1}\delta_{mm'}\delta_{\ell\ell'} .$$

The sum over helicities ±2 just gives a factor 2 in the EE and BB correlations, while as expected in the EB correlation there is a complete cancelation, due to the $\pm$ sign in Eq. (7.4.60). The tensor mode contributions to the polarization multipole coefficients are then[20]

$$C_{EE,\ell}^T = \pi^2 T_0^2 \int_0^\infty q^2 \, dq$$

$$\times \left| \int_{t_1}^{t_0} dt \, P(t) \, \Psi(q,t) \left\{ \left[12 + 8\rho \frac{\partial}{\partial \rho} - \rho^2 + \rho^2 \frac{\partial^2}{\partial \rho^2} \right] \frac{j_\ell(\rho)}{\rho^2} \right\}_{\rho=q\,r(t)} \right|^2$$

$$\tag{7.4.64}$$

$$C_{BB,\ell}^T = \pi^2 T_0^2 \int_0^\infty q^2 \, dq$$

$$\times \left| \int_{t_1}^{t_0} dt \, P(t) \, \Psi(q,t) \left\{ \left[8\rho + 2\rho^2 \frac{\partial}{\partial \rho} \right] \frac{j_\ell(\rho)}{\rho^2} \right\}_{\rho=q\,r(t)} \right|^2 . \tag{7.4.65}$$

The results (7.4.59) and (7.4.60), together with Eq. (7.3.14), allow us also to calculate the multipole coefficients in the correlation between temperature and polarization. In C_{TB}^T the $\pm$ sign in front of the integral over wave numbers in Eq. (7.4.60) produces a cancelation between the two terms in the sum over helicities ± 2, so that as expected $C_{TB}^T = 0$. On the other hand, in C_{TE}^T the sum over helicities ± 2 just gives a factor 2, so that Eqs. (7.3.14), (7.4.61) and (7.4.59) give

$$C_{TE,\ell}^T = -2\pi^2 T_0^2 \sqrt{\frac{(\ell+2)!}{(\ell-2)!}} \int_0^\infty q^2 \, dq$$

$$\times \int_{t_1}^{t_0} dt \, P(t)\Psi(q,t) \left\{ \left[12 + 8\rho \frac{\partial}{\partial \rho} - \rho^2 + \rho^2 \frac{\partial^2}{\partial \rho^2} \right] \frac{j_\ell(\rho)}{\rho^2} \right\}_{\rho=q\,r(t)}$$

$$\times \int_{t_1}^{t_0} dt' \, d(q,t') \frac{j_\ell\big(qr(t')\big)}{q^2 r^2(t')} , \tag{7.4.66}$$

where

$$d(q,t) \equiv \left[\dot{\mathcal{D}}_q(t) - \frac{1}{2}\omega_c(t) \right] \exp\left(-\int_t^{t_0} \omega_c(t') \, dt' \right) . \tag{7.4.67}$$

The calculation of the gravitational wave amplitude $\mathcal{D}_q(t)$ and source function $\Psi(q,t)$ is described in the previous section. For the same cosmological parameters as assumed there, Eqs. (7.4.62) and (7.4.63) give the EE

---

[20]These formulas, and also Eq. (7.4.64), are equivalent to those of Zaldarriaga and Seljak, ref. 1.

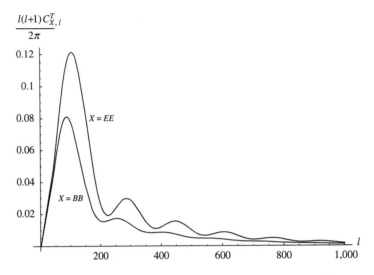

Figure 7.7: The multipole coefficients $\ell(\ell+1)C^T_{EE,\ell}/2\pi$ and $\ell(\ell+1)C^T_{BB,\ell}/2\pi$ in square microKelvin, *vs.* ℓ, for cosmological parameters given in Section 7.3, from ref. 21. The results agree to about 1% with those obtained (adopting the same cosmological parameters) by employing the computer programs CMBfast or CAMB, in which, instead of using the integral equation (7.3.22) as done here, the Boltzmann hierarchy is truncated at a maximum value of ℓ. (The results of these two computer programs also differ from each other by about the same amount.)

and **BB** multipole coefficients shown in Figure 7.7. (We will come to the TE coefficients below.)

The results (7.4.62)–(7.4.64) provide a basis for highly accurate computer calculations of the tensor multipole coefficients, but a casual inspection of these formulas does not provide much insight to the qualitative behavior of these coefficients as functions of ℓ. In particular, looking at Eqs. (7.4.62) and (7.4.63), we could hardly guess that $C^T_{BB,\ell} < C^T_{EE,\ell}$ for all $\ell > 15$, or that $C^T_{BB,\ell}$ and $C^T_{EE,\ell}$ approach each other for $\ell < 100$, as shown for one set of cosmological parameters in Figure 7.7.

These results become much simpler and more transparent for large ℓ. Recall that for $\rho^2 - \nu^2 \gg \nu^{4/3}$ (where $\nu \equiv \ell + 1/2$), the spherical Bessel functions have the well-known asymptotic behavior[21]

$$j_\ell(\rho) \to \frac{\cos b}{\nu\sqrt{\sin b}} \cos\left[\nu(\tan b - b) - \pi/4\right], \qquad (7.4.68)$$

---

[21] I. S. Gradshteyn & I. M. Ryzhik, *Table of Integrals, Series, and Products*, translated, corrected and enlarged by A. Jeffrey (Academic Press, New York, 1980): formula 8.453.1. The same approximation is used by J. R. Pritchard and M. Kamionkowski, *Ann. Phys.* **318**, 2 (2005) [astro-ph/0412581], but their subsequent approximations are very different from those made here and in ref. 21.

where $\cos b \equiv v/\rho$, with $0 \leq b \leq \pi/2$. On the other hand, for $v^2 - \rho^2 \gg v^{4/3}$, $j_\ell(\rho)$ is exponentially small. In the range in which $|\rho^2 - v^2| < v^{4/3}$ neither approximation is valid, but $j_\ell(\rho)$ is a smooth function of ρ in this range, without the singularity at $\rho = v$ that might be suggested by the factor $1/\sqrt{\sin b}$ in Eq. (7.4.66). For $v \gg 1$ this range contributes only a small part of the range of integration, and we would expect to be able to use the approximation (7.4.66). For $\rho^2 - v^2 \gg v^{4/3} \gg 1$, the dominant contributions to derivatives of $j_\ell(\rho)/\rho^2$ come from terms in which the derivative acts only on the cosine in Eq. (7.4.66), so that

$$\left[12 + 8\rho \frac{\partial}{\partial \rho} - \rho^2 + \rho^2 \frac{\partial^2}{\partial \rho^2} \right] \frac{j_\ell(\rho)}{\rho^2} \rightarrow -j_\ell(\rho) + j_\ell''(\rho)$$

$$\rightarrow -\frac{(1 + \sin^2 b) \cos b}{v\sqrt{\sin b}} \cos\left[v(\tan b - b) - \pi/4 \right] \tag{7.4.69}$$

$$\left[8\rho + 2\rho^2 \frac{\partial}{\partial \rho} \right] \frac{j_\ell(\rho)}{\rho^2} \rightarrow 2j_\ell'(\rho) \rightarrow -\frac{2\sqrt{\sin b} \cos b}{v} \sin\left[v(\tan b - b) - \pi/4 \right]$$

$$\tag{7.4.70}$$

(Letting the derivatives act on $1/\sqrt{\sin b}$ would produce a non-integrable singularity at $b = 0$, but this is spurious, because the asymptotic formula (7.4.66) breaks down for ρ very near v, where in fact there is no singularity.) Then Eqs. (7.4.62)–(7.4.64) become, for $v \equiv \ell + 1/2 \gg 1$,[22]

$$C_{EE,\ell}^T = \frac{\pi^2 T_0^2}{v^2} \int_0^\infty q^2 \, dq$$

$$\times \left| \int_{r(t) > v/q} dt \, P(t) \, \Psi(q, t) \right.$$

$$\times \left. \left\{ \frac{(1 + \sin^2 b) \cos b}{\sqrt{\sin b}} \cos\left[v(\tan b - b) - \pi/4 \right] \right\}_{\cos b = v/q \, r(t)} \right|^2$$

$$C_{BB,\ell}^T = \frac{\pi^2 T_0^2}{v^2} \int_0^\infty q^2 \, dq \left| \int_{r(t) > v/q} dt \, P(t) \, \Psi(q, t) \right.$$

$$\times \left. \left\{ 2\sqrt{\sin b} \cos b \sin\left[v(\tan b - b) - \pi/4 \right] \right\}_{\cos b = v/q \, r(t)} \right|^2,$$

$$\tag{7.4.71}$$

---

[22]R. Flauger and S. Weinberg, *Phys. Rev. D* **75**, 123505 (2007) [astro-ph/0703179].

$$C^T_{TE,\ell} = -\frac{2\pi^2 T_0^2}{v^2} \int_0^\infty q^2 \, dq$$

$$\times \int_{r(t)>v/q} dt \, P(t)\Psi(q,t)$$

$$\times \left\{ \frac{(1+\sin^2 b)\cos b}{\sqrt{\sin b}} \cos\left[v(\tan b - b) - \pi/4\right] \right\}_{\cos b = v/q \, r(t)}$$

$$\times \int_{r(t')>v/q} dt' \, d(q,t')$$

$$\times \left\{ \frac{\cos^3 b'}{\sqrt{\sin b'}} \cos\left[v(\tan b' - b') - \pi/4\right] \right\}_{\cos b' = v/q \, r(t')} \qquad . \quad (7.4.72)$$

This approximate result for $C^T_{TE,\ell}$ is compared with the exact result (7.4.64) in Figure 7.8.

For $C^T_{EE,\ell}$ and $C^T_{BB,\ell}$, we can usefully make a further approximation. The quantity $b \equiv \cos^{-1}\left(v/qr(t)\right)$ does not vary appreciably within the relatively narrow range of times t in which the last-scattering probability $P(t)$ is appreciable, so we can set $r(t)$ equal to $r_L \equiv r(t_L)$ everywhere except in the phase $v(\tan b - b)$, which for $v \gg 1$ *does* vary over a wide range in this time interval. Furthermore, because this phase varies over a wide range, the

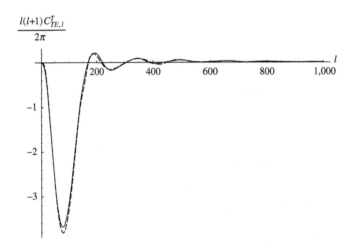

Figure 7.8: The multipole coefficient $\ell(\ell + 1)C^T_{TE,\ell}/2\pi$ in square microKelvin, *vs.* ℓ, from ref. 21. The solid curve is taken from the exact formula (7.4.64); the dashed curve represents the approximation (7.4.71). In both cases, the method of calculating the gravitational wave amplitude and the source function, and the assumed cosmological parameters, are those described in Section 7.3.

difference between $\cos[\nu(\tan b - b) - \pi/4]$ and $\sin[\nu(\tan b - b) - \pi/4]$ is inconsequential, and we may as well replace both with $\cos[\nu(\tan b - b)]$. Then Eqs. (7.4.69) and (7.4.70) become[22]

$$\ell(\ell + 1)C_{EE,\ell}^T \to \pi^2 T_0^2 \int_{\nu/r_L}^\infty q^2 \, dq \, \{(1 + \sin^2 b_L)^2 \cos^2 b_L\}_{\cos b_L = \nu/qr_L}$$

$$\times \left| \int_{r(t)>\nu/q} dt \, P(t) \, \Psi(q, t) \left\{ \frac{\cos\left[\nu(\tan b - b)\right]}{\sqrt{\sin b}} \right\}_{\cos b = \nu/q\, r(t)} \right|^2 ,$$

$$(7.4.73)$$

$$\ell(\ell + 1)C_{BB,\ell}^T \to \pi^2 T_0^2 \int_{\nu/r_L}^\infty q^2 \, dq \, \{4 \sin^2 b_L \cos^2 b_L\}_{\cos b_L = \nu/q\, r_L}$$

$$\times \left| \int_{r(t)>\nu/q} dt \, P(t) \, \Psi(q, t) \left\{ \frac{\cos\left[\nu(\tan b - b)\right]}{\sqrt{\sin b}} \right\}_{\cos b = \nu/q\, r(t)} \right|^2 .$$

$$(7.4.74)$$

(We have not set $b = b_L$ in the factors $1/\sqrt{\sin b}$ in both integrals over t, in order to avoid a divergence in the integration over q at $q = \nu/r_L$.) The results of using these approximate formulas are compared with the results of using the exact formulas (7.4.62) and (7.4.63) in Figures 7.9 and 7.10.

Our approximate result for $C_{BB,\ell}^T$ agrees with the exact result to about 1%, which is good enough for any practical purpose. The approximate result for $C_{EE,\ell}^T$ is not quite as good, agreeing with the exact result only to about 14%, but these approximations are evidently good enough to use them to draw qualitative conclusions. One immediate conclusion is that, since $(1 + \sin^2 b_L) \geq 4 \sin^2 b_L$ for all real b_L, we have $C_{EE,\ell}^T > C_{BB,\ell}^T$. Also, for ℓ small enough so that the wave number ℓ/r_L comes into the horizon before matter–radiation equality, say $\ell < 100$, for which $\Psi(\ell/r_L, t_L)$ is small, the integrals over q are dominated by values for which $\cos b$ is small, in which case $(1 + \sin^2 b)^2 \simeq 4 \sin^2 b$, and hence $C_{EE,\ell}^T \simeq C_{BB,\ell}^T$. As already noted, both properties are evident (for at least one set of cosmological parameters) in Figure 7.7.

The smallness of $C_{BB,\ell}^T$ makes it a difficult target for future observations, but the detection of the primordial BB mode would be of great importance to cosmology, as it would provide clear evidence of cosmological gravitational waves. The expectations for such tensor modes in theories of inflation are discussed in Chapter 10.

$$* \, * \, *$$

$$\frac{l(l+1)C^T_{BB,l}}{2\pi}$$

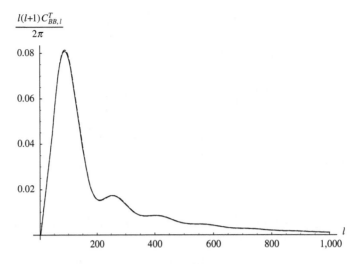

Figure 7.9: The multipole coefficient $\ell(\ell+1)C^T_{BB,\ell}/2\pi$ in square microKelvin, *vs.* ℓ, from ref. 22. The solid curve is taken from the exact formula (7.4.63); the dashed curve represents the approximation (7.4.70). In both cases, the method of calculating the gravitational wave amplitude and the source function, and the assumed cosmological parameters, are those described in Section 7.3.

$$\frac{l(l+1)C^T_{EE,l}}{2\pi}$$

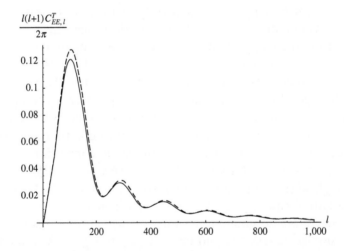

Figure 7.10: The multipole coefficient $\ell(\ell+1)C^T_{EE,\ell}/2\pi$ in square microKelvin, *vs.* ℓ, from ref. 22. The solid curve is taken from the exact formula (7.4.62); the dashed curve represents the approximation (7.4.69). In both cases, the method of calculating the gravitational wave amplitude and the source function, and the assumed cosmological parameters, are those described in Section 7.3.

The multipole coefficients $C_{EE,\ell}$, $C_{BB,\ell}$, and $C_{TE,\ell}$ may be measured by using data on correlations among the Stokes parameters and temperature fluctuations in comparison with general theoretical formulas for these correlation functions,[23] analogous to the general formula (2.6.4) for the temperature correlation function. To derive these formulas, we start by noting from Eqs. (7.4.15) and (7.4.25) that

$$Q(\hat{n}) = -\frac{1}{2}\sum_{\ell m} a_{E,\ell m}\Big[\mathcal{Y}_\ell^m(\hat{n}) + \mathcal{Y}_\ell^{-m}(\hat{n})^*\Big]$$
$$-\frac{i}{2}\sum_{\ell m} a_{B,\ell m}\Big[\mathcal{Y}_\ell^m(\hat{n}) - \mathcal{Y}_\ell^{-m}(\hat{n})^*\Big], \qquad (7.4.75)$$

$$U(\hat{n}) = \frac{i}{2}\sum_{\ell m} a_{E,\ell m}\Big[\mathcal{Y}_\ell^m(\hat{n}) - \mathcal{Y}_\ell^{-m}(\hat{n})^*\Big]$$
$$-\frac{1}{2}\sum_{\ell m} a_{B,\ell m}\Big[\mathcal{Y}_\ell^m(\hat{n}) + \mathcal{Y}_\ell^{-m}(\hat{n})^*\Big]. \qquad (7.4.76)$$

From Eqs. (7.4.28) and (7.4.29), we then find the correlation functions

$$\langle Q(\hat{n})\, Q(\hat{n}')\rangle = \frac{1}{2}\mathrm{Re}\sum_\ell C_{EE,\ell}\Big(F_\ell(\hat{n},\hat{n}') + G_\ell(\hat{n},\hat{n}')\Big)$$
$$+\frac{1}{2}\mathrm{Re}\sum_\ell C_{BB,\ell}\Big(F_\ell(\hat{n},\hat{n}') - G_\ell(\hat{n},\hat{n}')\Big), \qquad (7.4.77)$$

$$\langle U(\hat{n})\, U(\hat{n}')\rangle = \frac{1}{2}\mathrm{Re}\sum_\ell C_{EE,\ell}\Big(F_\ell(\hat{n},\hat{n}') - G_\ell(\hat{n},\hat{n}')\Big)$$
$$+\frac{1}{2}\mathrm{Re}\sum_\ell C_{BB,\ell}\Big(F_\ell(\hat{n},\hat{n}') + G_\ell(\hat{n},\hat{n}')\Big), \qquad (7.4.78)$$

$$\langle Q(\hat{n})\, U(\hat{n}')\rangle = \frac{1}{2}\mathrm{Im}\sum_\ell C_{EE,\ell}\Big(-F_\ell(\hat{n},\hat{n}') + G_\ell(\hat{n},\hat{n}')\Big)$$
$$+\frac{1}{2}\mathrm{Im}\sum_\ell C_{BB,\ell}\Big(-F_\ell(\hat{n},\hat{n}') - G_\ell(\hat{n},\hat{n}')\Big), \qquad (7.4.79)$$

where

$$F_\ell(\hat{n},\hat{n}') \equiv \sum_m \mathcal{Y}_\ell^m(\hat{n})\,\mathcal{Y}_\ell^m(\hat{n}')^*, \qquad (7.4.80)$$

$$G_\ell(\hat{n},\hat{n}') \equiv \sum_m \mathcal{Y}_\ell^m(\hat{n})\,\mathcal{Y}_\ell^{-m}(\hat{n}'). \qquad (7.4.81)$$

---

[23]General formulas for the correlation functions are presented by M. Zaldarriaga, *Astrophys. J.* **503**, 1 (1998).

Also, using Eq. (2.6.2) (with $a_{\ell m}$ now written as $a_{T,\ell m}$) and Eqs. (7.4.27), (7.4.74) and (7.4.75), the correlation functions between temperature fluctuations and Stokes parameters are

$$\langle \Delta T(\hat{n})\, Q(\hat{n}') \rangle = -\sum_{\ell} C_{TE,\ell}\, \mathrm{Re}H_\ell(\hat{n},\hat{n}') \,, \qquad (7.4.82)$$

$$\langle \Delta T(\hat{n})\, U(\hat{n}') \rangle = \sum_{\ell} C_{TE,\ell}\, \mathrm{Im}H_\ell(\hat{n},\hat{n}') \,, \qquad (7.4.83)$$

where

$$H_\ell(\hat{n},\hat{n}') \equiv \sum_{m} Y_\ell^m(\hat{n})\, \mathcal{Y}_\ell^m(\hat{n}')^* \,, \qquad (7.4.84)$$

with Y_ℓ^m the ordinary spherical harmonic.

To calculate the functions $F_\ell(\hat{n},\hat{n}')$ and $G_\ell(\hat{n},\hat{n}')$, we note first (for instance, by inspection of the last line of Eq. (7.4.22)) that, for $\hat{n}$ in the three-direction $\hat{z}$ with $\theta = \phi = 0$,

$$\mathcal{Y}_\ell^m(\hat{z}) = \delta_{m,-2}\sqrt{\frac{2\ell+1}{4\pi}} \,. \qquad (7.4.85)$$

Then, using the definitions (7.4.16) and (7.4.11) and the transformation rule (7.4.18), we can express $\mathcal{Y}_\ell^m$ as an element of a unitary rotation matrix[24]

$$\mathcal{Y}_\ell^m(\hat{n}) = \sum_{m'} D_{m'm}^{(\ell)}\left(S^{-1}(\hat{n})\right)\mathcal{Y}_\ell^{m'}(\hat{z}) = \sqrt{\frac{2\ell+1}{4\pi}}D_{-2,m}^{(\ell)}\left(S^{-1}(\hat{n})\right), \quad (7.4.86)$$

where $S(\hat{n})$ is the rotation (7.4.12) that takes $\hat{z}$ into $\hat{n}$. Using Eq. (7.4.57), it follows then also that

$$\mathcal{Y}^{-m}(\hat{n}) = (-1)^\ell \mathcal{Y}^m(-\hat{n})^* \,. \qquad (7.4.87)$$

Hence, by using the group multiplication property $D^{(\ell)}(S)D^{(\ell)}(S') = D^{(\ell)}(SS')$ and the unitarity of $D^{(\ell)}(S)$, we obtain *addition theorems*[25] that give us the functions needed in our formulas (7.4.76)–(7.4.78) for the correlation functions

$$F_\ell(\hat{n},\hat{n}') = \frac{2\ell+1}{4\pi}D_{-2,-2}^{(\ell)}\left(S^{-1}(\hat{n})S(\hat{n}')\right), \qquad (7.4.88)$$

$$G_\ell(\hat{n},\hat{n}') = (-1)^\ell \frac{2\ell+1}{4\pi}D_{-2,-2}^{(\ell)}\left(S^{-1}(\hat{n})S(-\hat{n}')\right). \qquad (7.4.89)$$

[24] E. Newman and J. Penrose, ref. 3; J. N. Goldberg, A. J. MacFarlane, E. T. Newman, F. Rorlich, and E. C. G. Sudarshan, *J. Math. Phys.* **8**, 2155 (1967); K. S. Thorne, *Rev. Mod. Phys.* **52**, 299 (1980).

[25] General addition theorems for spin-weighted spherical harmonics are given by W. Hu and M. White, *Phys. Rev. D* **56**, 596 (1997) [astro-ph/9702170].

To calculate the function $H_\ell(\hat{n}, \hat{n}')$, we recall that the ordinary spherical harmonics can also be written as elements of a rotation matrix. From the formula $Y_\ell^m(\hat{z}) = \sqrt{(2\ell+1)/4\pi}\, \delta_{m0}$ for the case $\theta = \phi = 0$ and the transformation rule (7.4.18), we have

$$Y_\ell^m(\hat{n}) = \sqrt{\frac{2\ell+1}{4\pi}}\, D_{0,m}^{(\ell)}\!\left(S^{-1}(\hat{n})\right).$$

Hence, again using the group multiplication property of the $D^{(\ell)}$ matrices, we find

$$H_\ell(\hat{n}, \hat{n}') = \frac{2\ell+1}{4\pi} D_{0,-2}^{(\ell)}\!\left(S^{-1}(\hat{n})S(\hat{n}')\right). \tag{7.4.90}$$

Note incidentally that

$$F_\ell(-\hat{n}, -\hat{n}') = \sum_m \mathcal{Y}_\ell^{-m}(\hat{n})^* \, \mathcal{Y}_\ell^{-m}(\hat{n}') = F_\ell^*(\hat{n}, \hat{n}'), \tag{7.4.91}$$

and

$$G_\ell(-\hat{n}, -\hat{n}') = \sum_m \mathcal{Y}_\ell^{-m}(\hat{n})^* \, \mathcal{Y}_\ell^m(\hat{n}')^* = G_\ell^*(\hat{n}, \hat{n}'). \tag{7.4.92}$$

Also, using the property $Y_\ell^m(-\hat{n}) = (-1)^\ell\, Y_\ell^{-m}(\hat{n})^*$,

$$H_\ell(-\hat{n}, -\hat{n}') = \sum_m Y_\ell^{-m}(\hat{n})^* \, \mathcal{Y}_\ell^{-m}(\hat{n}') = H_\ell^*(\hat{n}, \hat{n}'). \tag{7.4.93}$$

Hence the correlation functions have the reflection properties

$$\langle Q(\hat{n})\, Q(\hat{n}')\rangle = \langle Q(-\hat{n})\, Q(-\hat{n}')\rangle, \tag{7.4.94}$$

$$\langle U(\hat{n})\, U(\hat{n}')\rangle = \langle U(-\hat{n})\, U(-\hat{n}')\rangle, \tag{7.4.95}$$

$$\langle Q(\hat{n})\, U(\hat{n}')\rangle = -\langle Q(-\hat{n})\, U(-\hat{n}')\rangle, \tag{7.4.96}$$

$$\langle \Delta T(\hat{n})\, Q(\hat{n}')\rangle = \langle \Delta T(-\hat{n})\, Q(-\hat{n}')\rangle, \tag{7.4.97}$$

$$\langle \Delta T(\hat{n})\, U(\hat{n}')\rangle = -\langle \Delta T(-\hat{n})\, U(-\hat{n}')\rangle, \tag{7.4.98}$$

as expected from the space inversion properties of the Stokes parameters and temperature fluctuation and the assumed space-inversion invariance of the probability distribution over which we average.

It may be noted that, in measuring the correlation functions of the Stokes parameters observed in directions $\hat{n}$ and $\hat{n}'$, observers commonly choose the system of polar coordinates so that $\hat{n}$ and $\hat{n}'$ are on the same meridian; that is, so that $\phi = \phi'$. (This is usually expressed as the condition that the first of the two polarization vectors that are used to define the Stokes parameter should be aligned with the great circle between $\hat{n}$ and $\hat{n}'$,[26] but we have defined these

---

[26] Kamionkowski, Kosowsky, and Stebbins, ref. 1.

two polarization vectors always to lie in the directions of increasing θ and ϕ, so it amounts to the same thing.) This has the advantage that the functions $F_\ell(\hat{n}, \hat{n}')$, $G_\ell(\hat{n}, \hat{n}')$ and $H_\ell(\hat{n}, \hat{n}')$ are then all real, and depend only on $\theta - \theta'$. To see this, note that by using the representation

$$D^{(\ell)}\Big(S(\hat{n})\Big) = \exp\Big(-i\phi J_3^{(\ell)}\Big)\exp\Big(-i\theta J_2^{(\ell)}\Big)$$

of the standard rotation from $\hat{z}$ to $\hat{n}$ in terms of the angular momentum matrices $J_i^{(\ell)}$, we can rewrite Eqs. (7.4.87)–(7.4.89) as

$$F_\ell(\hat{n}, \hat{n}') = \frac{2\ell+1}{4\pi}\Bigg[\exp\Big(i\theta J_2^{(\ell)}\Big)\exp\Big(i\phi J_3^{(\ell)}\Big)$$
$$\times \exp\Big(-i\phi' J_3^{(\ell)}\Big)\exp\Big(-i\theta' J_2^{(\ell)}\Big)\Bigg]_{-2,-2} \, ,$$

$$G_\ell(\hat{n}, \hat{n}') = (-1)^\ell\frac{2\ell+1}{4\pi}\Bigg[\exp\Big(i\theta J_2^{(\ell)}\Big)\exp\Big(i\phi J_3^{(\ell)}\Big)$$
$$\times \exp\Big(-i[\phi' + \pi]J_3^{(\ell)}\Big)\exp\Big(-i[\pi - \theta']J_2^{(\ell)}\Big)\Bigg]_{-2,-2} \, ,$$

$$H_\ell(\hat{n}, \hat{n}') = \frac{2\ell+1}{4\pi}\Bigg[\exp\Big(i\theta J_2^{(\ell)}\Big)\exp\Big(i\phi J_3^{(\ell)}\Big)$$
$$\times \exp\Big(-i\phi' J_3^{(\ell)}\Big)\exp\Big(-i\theta' J_2^{(\ell)}\Big)\Bigg]_{0,-2} \, .$$

If $\phi = \phi'$, then

$$F_\ell(\hat{n}, \hat{n}') = \frac{2\ell+1}{4\pi}\Bigg[\exp\Big(i(\theta - \theta')J_2^{(\ell)}\Big)\Bigg]_{-2,-2} \, ,$$

so $F_\ell(\hat{n}, \hat{n}')$ for $\phi = \phi'$ depends only on $\theta - \theta'$. Also, $iJ_2^{(\ell)}$ is a real matrix, so $F_\ell(\hat{n}, \hat{n}')$ for $\phi = \phi'$ is also real. As to G_ℓ, for $\phi = \phi'$ we have

$$G_\ell(\hat{n}, \hat{n}') = (-1)^\ell\frac{2\ell+1}{4\pi}$$
$$\times \Bigg[\exp\Big(i\theta J_2^{(\ell)}\Big)\exp\Big(-i\pi J_3^{(\ell)}\Big)\exp\Big(-i[\pi - \theta']J_2^{(\ell)}\Big)\Bigg]_{-2,-2}$$
$$= (-1)^\ell\frac{2\ell+1}{4\pi}$$

$$\times \left[\exp\left(i\theta J_2^{(\ell)} \right) \exp\left(+i[\pi - \theta'] J_2^{(\ell)} \right) \exp\left(-i\pi J_3^{(\ell)} \right) \right]_{-2,-2}$$

$$= (-1)^\ell \frac{2\ell + 1}{4\pi} \left[\exp\left(i(\theta - \theta' + \pi) J_2^{(\ell)} \right) \right]_{-2,-2},$$

so $G_\ell(\hat{n}, \hat{n}')$ for $\phi = \phi'$ is real because $iJ_2^{(\ell)}$ is a real matrix, and depends only on $\theta - \theta'$. Finally, $H_\ell(\hat{n}, \hat{n}')$ is real and depends only on $\theta - \theta'$, for the same reasons as for $F_\ell(\hat{n}, \hat{n}')$. We conclude then that if the coordinate system is chosen so that $\phi = \phi'$, the correlation functions $\langle Q(\hat{n}) \, Q(\hat{n}') \rangle$ and $\langle U(\hat{n}) \, U(\hat{n}') \rangle$ and $\langle \Delta T(\hat{n}) \, Q(\hat{n}') \rangle$ depend only on the angle $\theta - \theta'$ between $\hat{n}$ and $\hat{n}'$, while $\langle Q(\hat{n}) \, U(\hat{n}') \rangle$ and $\langle \Delta T(\hat{n}) \, U(\hat{n}') \rangle$ vanish. Also, $\langle Q(\hat{n}) \, Q(\hat{n}') \rangle$ and $\langle U(\hat{n}) \, U(\hat{n}') \rangle$ are obviously symmetric between $\hat{n}$ and $\hat{n}'$, so for $\phi = \phi'$ they actually depend only on $|\theta - \theta'|$; that is, on the angle between $\hat{n}$ and $\hat{n}'$.

8

The Growth of Structure

In Chapter 6 we followed the evolution of small perturbations through the radiation and matter dominated eras, up to the time of decoupling, when radiation no longer interacted effectively with matter. Now we will continue the story past the time of decoupling. In Section 8.1 we will follow the perturbations while they remained small, continuing our linear analysis. This era is increasingly becoming accessible to observation, as studies of the cosmic matter distribution are pushed to larger redshifts. As we shall see, data on the distribution of matter fluctuations already provides an important extension of results from cosmic microwave background anisotropies to smaller wavelengths, and it is hoped that eventually it may provide information of the effect of dark energy on cosmic expansion.

Of course, eventually the perturbations in the matter density became strong enough for the linear approximation to break down, as shown vividly by the existence of stars and galaxies and galaxy clusters. It is believed that these structures were formed in a two-step process.[1] First, in regions where the density was a little larger than average, the cold dark matter and baryonic matter together expanded more slowly than the universe as a whole, eventually reaching a minimum density and then recontracting. This scenario is discussed in Section 8.2. If an overdense region was sufficiently large then as shown in Section 8.3 its baryonic matter collapsed along with its cold dark matter. Then in a second stage, after this collapse, the baryonic matter lost its energy through radiative cooling, and it condensed into protogalaxies consisting of clouds of gas that eventually form stars. The cold dark matter particles could not lose their energy through radiative cooling, so they remained in large more-or-less spherical halos around these galaxies. We will not attempt a proper treatment of this second stage, which involves complications of astrophysics and mathematics that deserve a treatise to themselves.

8.1 Linear perturbations after recombination

After the disappearance of almost all free electrons the baryonic plasma decoupled from photons, and behaved like just another form of cold dark matter. (Effects of pressure at small wavelengths are discussed in

---

[1] S. D. M. White and M. J. Rees, *Mon. Not. Roy. Astron. Soc.* **183**, 341 (1978).

Section 8.3.) With the gravitational field perturbations dominated by fluctuations in the total density of cold matter, the Newtonian treatment of
Appendix F is applicable, and tells us that the fractional density perturbation eventually grew as $a \propto t^{2/3}$. However, we need a relativistic analysis
to connect the constant wave-number dependent factor in this growth with
the strength of primordial fluctuations, characterized by the quantity $\mathcal{R}_q^0$, to
describe certain small but interesting oscillations in the density fluctuations
arising from the interaction of baryons and radiation before decoupling,
and to carry this analysis forward to near the present, when vacuum energy
became important.

In our relativistic analysis baryons and cold dark matter are treated
somewhat differently, because we continue to use a synchronous gauge in
which the cold dark matter velocity potential but not the baryon velocity
potential vanishes. After decoupling the baryon velocity potential was no
longer locked to the photon velocity potential by Thomson scattering, so to
obtain the equation for baryon conservation, we may use Eq. (5.3.34) with
$\bar{p}_B$, δp_{Bq}, and π_{Bq}^S all vanishing (because baryons move slowly), and find

$$\dot{\delta\rho}_{Bq} + 3H\delta\rho_{Bq} - \frac{q^2}{a^2}\bar{\rho}_B\delta u_{Bq} = -\bar{\rho}_B\psi_q \, ,$$

or, dividing by $\bar{\rho}_B \propto a^{-3}$,

$$\dot{\delta}_{Bq} - (q^2/a^2)\delta u_{Bq} = -\psi_q \tag{8.1.1}$$

where as before, $\delta_{Bq} \equiv \delta\rho_{Bq}/\bar{\rho}_B$. For cold dark matter there is no velocity
potential, so we again have Eq. (6.2.11) for cold dark matter
conservation:

$$\dot{\delta}_{Dq} = -\psi_q \tag{8.1.2}$$

It is only our choice of gauge that makes the velocity potential of the dark
matter rather than the baryons vanish, so it is best to think of δu_{Bq} as the
relative velocity potential of the baryons and dark matter.

The equation for baryon momentum conservation is given by
Eq. (5.3.32) with zero pressure and anisotropic inertia,

$$\partial_0(\bar{\rho}_B\delta u_{Bq}) + 3H\bar{\rho}_B\delta u_{Bq} = 0 \, ,$$

so, dividing by $\bar{\rho}_B \propto a^{-3}$, we have simply:

$$\delta\dot{u}_{Bq} = 0 \, . \tag{8.1.3}$$

(As we will see in Section 8.3, our neglect of the baryon pressure after
recombination is justified except for the smallest wavelengths.) Finally, if
we neglect the contribution of photon and neutrino density fluctuations

to the gravitational field perturbation, then the gravitational field equation (5.3.36) becomes

$$\frac{d}{dt}\left(a^2\psi_q\right) = -4\pi \, Ga^2\bar{\rho}_M\Big((1-\beta)\delta_{Dq} + \beta\delta_{Bq}\Big).$$

(8.1.4)

where β is again the constant

$$\beta \equiv \frac{\bar{\rho}_B}{\bar{\rho}_M} = \frac{\Omega_B}{\Omega_M} \approx 1/6.$$

(8.1.5)

We are keeping terms of order β for the present, because as we shall see they lead to effects that although small are quite distinctive.

To derive the initial conditions for these equations, we note that at times sufficiently early (say, $z > 1$), at which vacuum energy as well as curvature made a negligible contribution to the expansion rate, we have $a \propto t^{2/3}$, and $4\pi \, G\bar{\rho}_M = 2/3t^2$, so Eq. (8.1.4) gives

$$\frac{d}{dt}\left(t^{4/3}\psi_q\right) = -\frac{2}{3}t^{-2/3}\Big((1-\beta)\delta_{Dq} + \beta\delta_{Bq}\Big).$$

(8.1.6)

The reader can easily check that the general solution of Eqs. (8.1.1)–(8.1.3) and (8.1.6) is

$$\psi_q = a_q t^{-1/3} + b_q t^{-2} + \beta\delta u_{Bq}q^2/a^2 \,,$$

(8.1.7)

$$\delta_{Bq} = -\frac{3}{2}a_q t^{2/3} + b_q t^{-1} + (1-\beta)c_q - 3(1-\beta)t(q^2/a^2)\delta u_{Bq} \,,$$

(8.1.8)

$$\delta_{Dq} = -\frac{3}{2}a_q t^{2/3} + b_q t^{-1} - \beta c_q + 3\beta t(q^2/a^2)\delta u_{Bq} \,,$$

(8.1.9)

where a_q, b_q, c_q, and δu_{Bq} are constants that must be found by matching these solutions to the values of ψ_q, δ_{Bq}, δ_{Dq}, and δu_{Bq} at the time of decoupling of matter and radiation. For late times, we need only keep the leading terms in this solution, with coefficient a_q; the other terms are suppressed relative to these by factors $t^{-5/3}$, $t^{-2/3}$, and t^{-1}. To calculate the coefficient a_q, we note that $\beta\delta_{Bq} + (1-\beta)\delta_{Dq} - t\psi_q + \beta t(q^2/a^2)\delta u_{Bq} = -5a_q t^{2/3}/2$. Setting t here equal to the time t_L of decoupling[2] (when to a good approximation

---

[2] The subscript L stands for "last scattering," and indicates the time of the decoupling of radiation from matter, associated with the recombination of hydrogen. (To use a subscript R or D for this time might produce confusion with the subscripts R and D that we continue to use to denote radiation and dark matter.) Of course, this is not the moment of last scattering; some photons of the microwave background were scattered again when hydrogen became reionized at much later times. Strictly speaking, we should take t_L here as the time during recombination when a typical electron stops exchanging appreciable momentum with the photons, rather than the slightly earlier time when a typical photon stops exchanging appreciable momentum with the electrons. Because R_L is not very different from unity, there is little difference between these times.

$\delta_{Bq} = \delta_{\gamma q}$ and $\delta u_{Bq} = \delta u_{\gamma q}$) then gives for $t \gg t_L$

$$\delta_{Dq}(t) \to \delta_{Bq}(t) \to -\frac{3}{2}a_q t^{2/3}$$

$$\to \frac{3}{5}\left(\frac{t}{t_L}\right)^{2/3}\left[\beta\delta_{\gamma q}(t_L) + (1-\beta)\delta_{Dq}(t_L) - t_L\psi_q(t_L)\right.$$

$$\left. + \beta t_L(q^2/a_L^2)\delta u_{\gamma q}(t_L)\right]. \tag{8.1.10}$$

Note that, even though the fractional density perturbations of the baryons and dark matter were quite different at decoupling, they approached each other at late times thereafter, an assumption we made in analyzing the observations of X-rays from clusters of galaxies in Section 1.9. It follows that the fractional fluctuation δ_{Mq} in the total mass density approaches

$$\delta_{Mq} \equiv \frac{\bar{\rho}_D\delta_{Dq} + \bar{\rho}_B\delta_{Bq}}{\bar{\rho}_D + \bar{\rho}_B} \to \delta_{Dq} \to \delta_{Bq} \tag{8.1.11}$$

Since these fractional density perturbations all eventually became equal, we will concentrate from now on $\delta_{Mq}(t)$. We now want to carry our calculation forward to the present, when dark energy may no longer be neglected. According to Eqs. (8.1.1)–(8.1.4), under the approximation of neglecting fluctuations in the photon and neutrino energy densities, $\delta_{Mq}(t)$ satisfies the second-order differential equation

$$\frac{d}{dt}\left[a^2\frac{d}{dt}\delta_{Mq}\right] = 4\pi Ga^2\bar{\rho}_M\delta_{Mq} . \tag{8.1.12}$$

The most important consequence of this equation, together with Eq. (8.1.10), is that whatever we assume about dark energy, well after recombination when the terms in $\delta_{Mq}(t)$ that decay as $1/t$ have died away, the dependence of $\delta_{Mq}(t)$ on q and t factorizes:

$$\delta_{Mq}(t) = \Delta(q)F(t) \tag{8.1.13}$$

where

$$\Delta(q) = \beta\delta_{\gamma q}(t_L) + (1-\beta)\delta_{Dq}(t_L)$$
$$-t_L\psi_q(t_L) + \beta t_L(q^2/a_L^2)\delta u_{\gamma q}(t_L) , \tag{8.1.14}$$

and $F(t)$ satisfies the differential equation

$$\frac{d}{dt}\left[a^2\frac{d}{dt}F\right] = 4\pi Ga^2\bar{\rho}_M F , \tag{8.1.15}$$

with the initial condition, that well after recombination, until dark energy becomes important,

$$F(t) \to \frac{3}{5}\left(\frac{t}{t_L}\right)^{2/3} . \qquad (8.1.16)$$

For instance, if we assume that the vacuum energy is constant, then the Friedmann equation for $a(t)$ reads

$$\frac{\dot{a}}{a} = \sqrt{\frac{8\pi G}{3}\left(\rho_\Lambda + \bar{\rho}_M\right)} = H_0\sqrt{\Omega_\Lambda}\sqrt{1 + 1/x} , \qquad (8.1.17)$$

where

$$x \equiv \frac{\rho_\Lambda}{\bar{\rho}_M} = \frac{\Omega_\Lambda}{\Omega_M}\left(\frac{a}{a_0}\right)^3 . \qquad (8.1.18)$$

Using x instead of t as the independent variable allows us to put Eq. (8.1.15) in a parameter-free form

$$\sqrt{x(1 + x)}\frac{d}{dx}\left(x^{7/6}\sqrt{1 + x}\frac{dF}{dx}\right) = \frac{1}{6x^{1/3}}F . \qquad (8.1.19)$$

The growing solution[3] that becomes proportional to $t^{2/3}$ for $x \ll 1$ is

$$F \propto \sqrt{\frac{1 + x}{x}}\int_0^x \frac{du}{u^{1/6}(1 + u)^{3/2}} .$$

Eq. (8.1.16) requires that for $x \ll 1$

$$F \to \frac{3}{5}\left(\frac{a}{a_L}\right) = \frac{3}{5}(1 + z_L)\left(\frac{x\Omega_M}{\Omega_\Lambda}\right)^{1/3} ,$$

so we can write

$$F(t) = \frac{3}{5}\left(\frac{a(t)}{a_L}\right)C\left(\frac{\Omega_\Lambda}{\Omega_M}\left(\frac{a(t)}{a_0}\right)^3\right) , \qquad (8.1.20)$$

where $C(x)$ is a correction factor, normalized so that $C(0) = 1$:

$$C(x) \equiv \frac{5}{6}x^{-5/6}\sqrt{1 + x}\int_0^x \frac{du}{u^{1/6}(1 + u)^{3/2}} . \qquad (8.1.21)$$

Numerical values of $C(x)$ are given in Table 8.1.

---

[3]H. Martel, *Astrophys. J.* **377**, 7 (1991).

Table 8.1: Values of the function $C(x)$, giving the suppression of the growth of matter fluctuations by dark energy as a function of $x \equiv (\Omega_\Lambda / \Omega_M)(a/a_0)^3$.

x	$C(x)$	x	$C(x)$
0	1	1.0	0.8725
0.1	0.9826	1.5	0.8314
0.2	0.9667	2.0	0.7981
0.3	0.9520	2.5	0.7702
0.5	0.9256	3.0	0.7462
0.7	0.9025	3.5	0.7254

We see that at all scales and times, dark energy suppresses the growth of density fluctuations.

It is $\Delta(q)$ that contains information about conditions at decoupling. In place of $\Delta(q)$, it is conventional to introduce a *power spectral function* $P(k)$, defined as a function of the present value $k \equiv q/a_0$ of the physical wave number, by[4]

$$P(k) \equiv (2\pi)^3 a_0^3 F^2(t_0) |\Delta(a_0 k)|^2 . \qquad (8.1.22)$$

Most surveys of large scale structure report their results in terms of $P(k)$.

Now we must consider how these surveys are used to measure $\Delta(q)$ or $P(k)$. We recall that for a single dominant scalar mode (and assuming spatial flatness), the fractional density perturbation in coordinate space takes the form of a Fourier transform like Eq. (5.2.1):

$$\delta_M(\mathbf{x}, t) = \int d^3q \, \alpha(\mathbf{q}) \, \delta_{Mq}(t) \, e^{i\mathbf{q}\cdot\mathbf{x}} = F(t) \int d^3q \, \alpha(\mathbf{q}) \, \Delta(q) \, e^{i\mathbf{q}\cdot\mathbf{x}} \quad (8.1.23)$$

where $\alpha(\mathbf{q})$ is a stochastic variable, normalized so that

$$\langle \alpha(\mathbf{q}) \alpha^*(\mathbf{q}') \rangle = \delta^3(\mathbf{q} - \mathbf{q}') . \qquad (8.1.24)$$

The quantity $\Delta(q)$ can be found either from measurements of the correlation of matter density perturbations at different points, or more directly, from an angular average of the square of a Fourier integral of the matter density perturbation over the survey volume.

---

[4]The factor $(2\pi)^3$ is inserted here because the position-space perturbations are usually written as Fourier transforms with an extra factor of $(2\pi)^{-3/2}$, and the factor a_0^3 is included because these three-dimensional Fourier transforms are usually written as integrals over the present value of the physical wave number, $k \equiv q/a_0$, rather than the co-moving wave number q. If the reader wishes, the co-moving coordinates could be normalized so that $a_0 = 1$, in which case all qs in this section could be replaced with ks.

According to the ergodic theorem discussed in Appendix D, as long as the survey volume V is large compared with the volume over which density fluctuations are correlated, the ensemble averages of products of density fluctuations can be found from an average of these products over the survey volume, In particular, the two-point correlation function can be found from

$$\langle \delta_M(\mathbf{x}, t)\, \delta_M(\mathbf{y}, t) \rangle = \frac{1}{V} \int_V d^3z\, \delta_M(\mathbf{x} + \mathbf{z}, t)\, \delta_M(\mathbf{y} + \mathbf{z}, t) \ . \quad (8.1.25)$$

Of course, we measure density fluctuations as functions of redshifts and angular positions rather than of three-dimensional positions and times, so the correlation function actually measured is the correlation of the fractional matter density perturbation observed at a redshift z and direction $\hat{n}$ with the perturbation observed at a redshift z' and direction $\hat{n}'$:

$$\xi(z, \hat{n}; z', \hat{n}') \equiv \left\langle \delta_M\Big(r(z)\hat{n}, t(z)\Big)\, \delta_M\Big(r(z')\hat{n}', t(z')\Big) \right\rangle , \quad (8.1.26)$$

where $r(z)$ and $t(z)$ are the Robertson–Walker co-moving radial coordinate and emission time associated with a redshift z. Using Eqs. (8.1.23), (8.1.24), and (8.1.13), this is

$$\xi(z, \hat{n}; z', \hat{n}') = \int d^3q\, \delta_{Mq}(t(z))\, \delta^*_{Mq}(t(z')) \exp\Big(i\mathbf{q} \cdot (r(z)\hat{n} - r(z')\hat{n}')\Big)$$

$$= F\Big(t(z)\Big) F\Big(t(z')\Big) \int d^3q\, |\Delta(q)|^2 \, \exp\Big(i\mathbf{q} \cdot (r(z)\hat{n} - r(z')\hat{n}')\Big) .$$

$$(8.1.27)$$

In terms of $P(k)$, this reads

$$\xi(z, \hat{n}; z', \hat{n}') = \frac{1}{2\pi^2} \frac{F\Big(t(z)\Big) F\Big(t(z')\Big)}{F^2(t_0)} \int_0^\infty k\, P(k)\, dk$$

$$\times \frac{\sin\Big(k \big| d_S(z)\hat{n} - d_S(z')\hat{n}' \big| \Big)}{\big| d_S(z)\hat{n} - d_S(z')\hat{n}' \big|} , \quad (8.1.28)$$

where $d_S(z)$ is a convenient *structure distance*, related to $r(z)$ and to the angular diameter and luminosity distances at a redshift z by:

$$d_S(z) \equiv a_0 r(z) = (1 + z)d_A(z) = (1 + z)^{-1} d_L(z) .$$

Eq. (8.1.28) can be used to measure the power spectral function if we know $d_S(z)$, or to measure $d_S(z)$ if we know something about features in the power spectral function. In particular, for observations at relatively low redshift, we can use $d_S(z) \simeq z/H_0$ and $t(z) \simeq t_0$, in which case Eqs. (8.1.28) gives the correlation function as

$$\xi(z, \hat{n}; z', \hat{n}') = \frac{H_0}{2\pi^2} \int_0^\infty k\, P(k)\, dk\, \frac{\sin\left((k/H_0)\left|z\hat{n} - z'\hat{n}'\right|\right)}{\left|z\hat{n} - z'\hat{n}'\right|} , \qquad (8.1.29)$$

Measurements of the shape of this correlation function can evidently tell us about the dependence of the power spectral function on k/H_0, rather than k itself.

It is more common in large surveys to measure $P(k)$ from the angular average of the square of a Fourier integral of the matter density perturbation over the survey volume. Define a Fourier transform as an integral over the co-moving survey volume V:

$$\delta_{M\mathbf{Q}}^V(t) \equiv \frac{1}{\sqrt{V}(2\pi)^{3/2}} \int_V d^3x\, e^{-i\mathbf{Q}\cdot\mathbf{x}} \delta_M(\mathbf{x}, t) . \qquad (8.1.30)$$

(The reason for this normalization will soon be made clear.) Using Eq. (8.1.23), this is

$$\delta_{M\mathbf{Q}}^V(t) = \int d^3q\, \alpha(\mathbf{q}) F_V(\mathbf{q} - \mathbf{Q}) \delta_{Mq}(t) , \qquad (8.1.31)$$

where

$$F_V(\mathbf{q}) \equiv \frac{1}{\sqrt{V}(2\pi)^{3/2}} \int_V d^3x\, e^{i\mathbf{q}\cdot\mathbf{x}} . \qquad (8.1.32)$$

It is plausible, and will be shown formally at the end of this section, that as long as the co-moving survey volume V contains many co-moving wavelengths $2\pi/q$, the angular average of $|\delta_{M\mathbf{Q}}^V|^2$ is the same as its ensemble average

$$\frac{1}{4\pi} \int d^2\hat{Q} \left|\delta_{M\mathbf{Q}}^V(t)\right|^2 = \left\langle \frac{1}{4\pi} \int d^2\hat{Q} \left|\delta_{M\mathbf{Q}}^V(t)\right|^2 \right\rangle . \qquad (8.1.33)$$

Using Eqs. (8.1.31) and (8.1.24), this is

$$\frac{1}{4\pi} \int d^2\hat{Q} \left|\delta_{M\mathbf{Q}}^V(t)\right|^2 = \frac{1}{4\pi} \int d^2\hat{Q} \int d^3q\, |F_V(\mathbf{q} - \mathbf{Q})|^2 \left|\delta_{Mq}(t)\right|^2 . \qquad (8.1.34)$$

Now, for large V, $F_V(\mathbf{q} - \mathbf{Q})$ approaches $((2\pi)^{3/2}/\sqrt{V})\delta^3(\mathbf{q} - \mathbf{Q})$, so

$$|F_V(\mathbf{q} - \mathbf{Q})|^2 \rightarrow \frac{(2\pi)^{3/2}}{\sqrt{V}} \delta^3(\mathbf{q} - \mathbf{Q}) F_V(0) = \delta^3(\mathbf{q} - \mathbf{Q}) , \qquad (8.1.35)$$

and therefore

$$\frac{1}{4\pi} \int d^2\hat{Q} \, \left|\delta^V_{M\mathbf{Q}}(t)\right|^2 = \left|\delta_{M\mathbf{Q}}(t)\right|^2 , \tag{8.1.36}$$

where $Q \equiv |\mathbf{Q}|$.

Again, we measure density fluctuations as functions of direction and redshift rather than position and time. Using Eq. (8.1.30) in Eq. (8.1.36) and expressing the result in terms of the power spectral function (8.1.22), we have

$$P(k) = \frac{1}{4\pi V} \int d^2\hat{Q} \left| \int_V d^2\hat{n} \, dz \, d^2_S(z) \right.$$

$$\left. \times \, d'_S(z) \frac{F(t_0)}{F(t(z))} \delta_M(r(z)\hat{n}, t(z)) \, e^{-i\hat{n}\cdot\hat{Q}kd_S(z)} \right|^2 , \tag{8.1.37}$$

where $V \equiv a_0^3 \mathcal{V}$ is the physical survey volume. As in the case of Eq. (8.1.28) for the correlation function, we can use this formula either to calculate $P(k)$ from data for surveys with relatively low redshifts, or to find $d_S(z)$ from larger redshift surveys combined with information about $P(k)$ from other sources.

Let's now consider the calculation of $\Delta(q)$ and the power spectral function. We begin by neglecting all terms of order β in Eq. (8.1.14), so that

$$\Delta(q) = \delta_{Dq}(t_L) - t_L \psi_q(t_L) . \tag{8.1.38}$$

We will return at the end of this section to the very interesting effects associated with the small baryon density. Using Eqs. (6.5.5) and (6.5.6), this is

$$\Delta(q) = \frac{3q^2 t_L^2 \mathcal{R}^o_q \mathcal{T}(\kappa)}{2a_L^2} = \frac{2q^2 \mathcal{R}^o_q \mathcal{T}(\kappa)}{3H_L^2 a_L^2} , \tag{8.1.39}$$

where $\mathcal{T}$ is the dimensionless transfer function given in Table 6.1, and $\kappa/\sqrt{2}$ is the ratio of the wave number q to the wave number q_{EQ} that comes into the horizon just at matter–radiation equality. (The second expression is derived using $t_L = 2/3H_L$, where $H_L = \sqrt{\Omega_M} H_0 (1 + z_L)^{3/2}$ is the Hubble rate at decoupling if radiation is neglected. This is more accurate than using the actual age of the universe at decoupling, because Eqs. (8.1.38) and (8.1.39) were derived using a definition of the zero of time for which $a \propto t^{2/3}$ during the matter-dominated era.) Using Eq. (8.1.20), the power spectral function (8.1.22) is then

$$P(k) = \frac{4(2\pi)^3 a_0^3 C^2 (\Omega_\Lambda / \Omega_M)}{25\Omega_M^2 H_0^4} \mathcal{R}^{o\,2}_{ka_0} k^4 \mathcal{T}^2 (\sqrt{2}k/k_{EQ}) , \tag{8.1.40}$$

where, according to Eq. (6.4.58),

$$k_{EQ} \equiv q_{EQ}/a_0 = \sqrt{2}H_0\Omega_M/\sqrt{\Omega_R} = [13.6 \text{ Mpc}]^{-1}\Omega_M h^2 . \quad (8.1.41)$$

In particular, if we suppose that $\mathcal{R}_q \simeq Nq^{-3/2}(q/q_*)^{(n_S-1)/2}$ with some spectral index n_S and constant N (which for $n_S \neq 1$ depends on the arbitrary choice of the reference wave number q_*), we would have

$$P(k) = \frac{4(2\pi)^3 N^2 C^2(\Omega_\Lambda/\Omega_M)}{25 \, \Omega_M^2 H_0^4 k_*^{n_S-1}} k^{n_S} T^2(\sqrt{2}k/k_{EQ}) , \quad (8.1.42)$$

where $k_* \equiv q_*/a_0$. The shape of this function for $n_S = 1$ is shown in Figure 8.1.

Knowledge of the power spectral function allows us to calculate the mean square value σ^2 of the fractional density fluctuation:

$$\sigma^2(z) = \xi(z,\hat{n};z,\hat{n}) = \frac{1}{2\pi^2}\left(\frac{F(t(z))}{F(t_0)}\right)^2 \int_0^\infty P(k) \, k^2 \, dk , \quad (8.1.43)$$

With $P(k)$ given by Eq. (8.1.42), for any plausible n_S this integral is convergent at $k = 0$, where $T(\kappa) \rightarrow 1$, but for $n_S \geq 1$ it diverges at $k \rightarrow \infty$ like $\int k^{n_S-2} \ln^2 k \, dk$. (Recall that, as we saw in Section 6.5, $T(\kappa) \propto \ln k/k^2$ for $k \rightarrow \infty$. On the other hand, the sine factor in the integrand in Eq. (8.1.27) makes $\xi(z,\hat{n};z',\hat{n}')$ finite for $z' \neq z$ or $\hat{n}' \neq \hat{n}$.)

In order to avoid the ultraviolet divergence in σ^2, it is common instead to express the intensity of the primordial fluctuations in terms of the mean square value σ_R^2 of the average of the fractional density perturbation over a

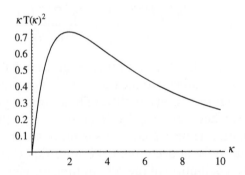

Figure 8.1: Shape of the power spectral function. This plot gives $\kappa T^2(\kappa)$ as a function of κ, where $T(\kappa)$ is the transfer function discussed in Section 6.5, $\kappa = \sqrt{2}k/k_{EQ}$, and $k_{EQ} = [13.6 \text{ Mpc}]^{-1}\Omega_M h^2$ is the wave number that just comes into the horizon at radiation–matter equality. Eq. (8.1.42) shows that for $n_S = 1$, the power spectral function $P(k)$ is proportional to $\kappa T^2(\kappa)$.

sphere of co-moving radius R/a_0:

$$\sigma_R^2(z) \equiv \left\langle \left(\frac{3a_0^3}{4\pi R^3} \int_{a_0|\mathbf{x}|<R} d^3x \; \delta_M\left(\mathbf{x}, t(z)\right) \right)^2 \right\rangle . \tag{8.1.44}$$

(Because of the translation invariance of the average, it would make no difference if we wrote the argument of δ_M as $\mathbf{x}+\mathbf{y}$, with $\mathbf{y}$ any fixed coordinate vector.) Again using Eqs. (8.1.22) and (8.1.23), and now also Eq. (8.1.20), we find

$$\sigma_R^2(z) = \frac{1}{2\pi^2(1+z)^2} \left(\frac{C\left(\Omega_\Lambda/\Omega_M(1+z)^3\right)}{C\left(\Omega_\Lambda/\Omega_M\right)} \right)^2 \int_0^\infty P(k)|f(kR)|^2 k^2 \, dk,$$

$$\tag{8.1.45}$$

where $f(kR)$ is the *top hat distribution function*

$$f(kR) \equiv \frac{3a_0^3}{4\pi R^3} \int_{a_0|\mathbf{x}|<R} d^3x \; e^{i\mathbf{k}\cdot\mathbf{x}a_0} = \frac{3}{(kR)^3}\left(\sin kR - kR \cos(kR) \right) .$$

$$\tag{8.1.46}$$

In particular, if we take $P(k)$ to be given by Eq. (8.1.42), then

$$\sigma_R^2(z) = \frac{16\pi N^2}{25 k_*^{n_S-1} \Omega_M^2 H_0^4} \frac{C^2\left(\Omega_\Lambda/\Omega_M(1+z)^3\right)}{(1+z)^2}$$

$$\times \int_0^\infty |\mathcal{T}(\sqrt{2}k/k_{\mathrm{EQ}})|^2 \, |f(kR)|^2 \, k^{2+n_S} \, dk . \tag{8.1.47}$$

The top hat function has $f(0) = 1$, so there is no change in the infrared convergence of the integral for the mean square fluctuation, but $|f(kR)|^2$ decays as $9\cos^2(kR)/(kR)^4$ for $k \to \infty$, which is fast enough to remove the ultraviolet divergence for $n_S < 5$.

Many observations of the distribution of matter in the universe are commonly expressed in terms of a quantity called σ_8, which is the value of $\sigma_R(z)$ for $z = 0$ and $R = 8\,h^{-1}$ Mpc. In calculating σ_8, we can evaluate the transfer function using the Dicus fitting formula (6.5.12), with the Eisenstein–Hu baryonic correction (6.5.22). Using the parameters $h = 0.72$, $\Omega_M h^2 = 0.14$, $\Omega_B h^2 = 0.024$ and $|N|^2 = 2.1 \times 10^{-10}$ which as discussed in Section 7.2 were found at the end of the first year of **WMAP** observations, and taking the present radiation temperature as $T_{\gamma,0} = 2.725$ K which yields $\Omega_R h^2 = 4.15 \times 10^{-5}$, Eq. (8.1.47) gives $\sigma_8 = 0.92$, in agreement with the result $\sigma_8 = 0.919$ found by the **WMAP** collaboration, which was calculated using the CMBfast computer program.[5] (The three-year WMAP data[6] gives a smaller value, $\sigma_8 = 0.761^{+0.049}_{-0.048}$, because the reduced optical depth inferred from three-year polarization data requires a smaller value of $|N|^2$ to give temperature correlations about the same magnitude. The reduction in $|N|^2$ also yields a corresponding decrease in the value of σ_8 calculated from Eq. (8.1.47).)

Even though Eq. (8.1.43) gives a divergent integral for $n_S \geq 1$, it suggests that $P(k)k^3$ can be used as a measure of the strength of the fluctuations of co-moving wave number $q = a_0 k$. With $P(k)$ given by Eq. (8.1.42) with $n_S \simeq 1$, we have $F^2(t)P(k)k^3 \propto t^{4/3}k^4|T(k/k_{EQ})|^2$. This is a monotonically increasing function of k, so we can conclude that *it is the perturbations of large co-moving wave number and hence small mass that become strong first*, with small condensations merging into pre-galactic dark matter haloes and then ultimately into clusters of galaxies. This "bottom-up" picture of structure formation is the reverse of the "top-down" picture long advocated by Zel'dovich,[7] according to which very large condensations form first and then fragment into condensations on the scale of clusters of galaxies and finally into individual galaxies. The bottom-up picture is supported by the observation that the commonest galaxies are dwarf spheroidals, and that our galaxy and the Andromeda Nebula M31 each have about 20 smaller satellite galaxies.

Returning now to the power spectral function, inspection of Table 6.1 shows that the function $\kappa|T(\kappa)|^2$ has a maximum value of 0.74, reached at $\kappa = 2.0$. Since Eq. (6.4.58) gives k proportional to κ, with $k = \Omega_M h^2 \kappa / 19.3$ Mpc, $P(k)$ for $n_S = 1$ has a maximum at this value of κ, corresponding to

$$k_{\max} = 0.10 \, \Omega_M h^2 \, \text{Mpc}^{-1}. \tag{8.1.48}$$

Taking $C(\Omega_\Lambda/\Omega_M) = 0.767$ and $n_S = 1$, the value of the spectral function at its maximum is

$$P_{\max} = 7.2 \times 10^{13} \left(\Omega_M h^2 \right)^{-1} |N|^2 \, \text{Mpc}^3 . \tag{8.1.49}$$

---

[5]D. N. Spergel *et al.*, *Astrophys. J. Suppl.* **148**, 175 (2003). The result $\sigma_8 = 0.9$ quoted in this reference was rounded off from $\sigma_8 = 0.919$, in order to reflect uncertainties in the input parameters.

[6]D. N. Spergel *et al.*, *Astrophys. J. Suppl. Ser.* **170**, 377 (2007) [astro-ph/0603449].

[7]Ya. B. Zel'dovich, *Soviet Scientific Reviews, Section E: Astrophys. and Space Physics Reviews* **3**, 1 (1984); S. F. Shandarin and Ya. B. Zel'dovich, *Rev. Mod. Phys.* **61**, 185 (1989); and earlier references cited therein.

There are a number of measurements of the matter distribution at low redshift that have been used to calculate $P(k)$:

- There have been several surveys of galaxy positions and redshifts, of which the two most recent and detailed are the Sloan Digital Sky Survey,[8] which will include about 800,000 galaxies, and the 2dF Survey,[9] which at its completion in 2003 had included 220,000 galaxies. Results of the Sloan Digital Sky Survey are shown in Figure 8.2.

  These two surveys give values of $P(k)$ in fair agreement with each other, for values of k ranging from about $0.015\,h\,\mathrm{Mpc}^{-1}$ to $0.2\,h\,\mathrm{Mpc}^{-1}$. (Wave numbers are given in these surveys in units of $h\,\mathrm{Mpc}^{-1}$ because distances are inferred from redshifts, and are therefore proportional to h^{-1}.) The measured $P(k)$ has a shape consistent with the result $\propto k|\mathcal{T}(\kappa)|^2$ expected from Eq. (8.1.42) for $n_S \simeq 1$, and in particular seems to reach a maximum at $k \approx 0.02\,h\,\mathrm{Mpc}^{-1}$, in good agreement with Eq. (8.1.48) for $\Omega_M \approx 0.3$ and $h \approx 0.7$. The value at this maximum is measured to be $P_{\max} \approx 5 \times 10^4\,h^{-3}\,\mathrm{Mpc}^3$. Comparison of this result with Eq. (8.1.49) indicates a value $|N| \approx 2 \times 10^{-5}$ for the strength of primordial fluctuations, if we take $\Omega_M \approx 0.3$ and $h \approx 0.7$.

- The counts of numbers of "virialized" clusters of galaxies (like the Coma cluster) as a function of redshift gives information about the distribution of mass with distance.[10] (See Section 1.11.) Their results yield values for σ_8 ranging from 0.66 to about 1, and can be interpreted as giving a value $P(k) \approx 6 \times 10^3\,h^{-3}\mathrm{Mpc}^3$ for $k \approx 0.1\,h\,\mathrm{Mpc}^{-1}$, which falls on the curve provided by the above galaxy surveys.

- Other information about the distribution of mass with distance comes from correlations between the positions of intergalactic regions of higher than average density, revealed through Lyman α absorption of light from distant quasars that passes through these regions.[11] (See Section 1.10.) Their results yield values for $P(k)$ for k between about $0.1\,h\,\mathrm{Mpc}^{-1}$ and $6\,h\,\mathrm{Mpc}^{-1}$, which lie on the curve provided by the

---

[8]D. G. York *et al.*, *Astron. J.* **120**, 1579 (2000); M. Tegmark *et al.*, *Astrophys. J.* **606**, 702 (2004) [astro-ph/0310725]; *Phys. Rev. D* **69**, 103501 (2004) [astro-ph/0310723]. The latest data release at the time of writing is analyzed by W. J. Percival *et al.*, *Astrophys. J.* **657**, 645 (2007) [astro-ph/0608636] and M. Tegmark *et al.*, *Phys. Rev. D* **74**, 123507 (2006) [astro-ph/0608632].

[9]W. J. Percival et al., *Mon. Not. Roy. Astron. Soc.* **327**, 1297 (2001); M. Colless *et al.*, *Mon. Not. Roy. Astron. Soc.* **328**, 1039 (2001); M. Colless *et al.* (the 2dFGRS team), astro-ph/0306581. The final data set is analyzed in S. Cole *et al.* (the 2dF-GRS Team), *Mon. Not. Roy. Astron. Soc.* **362**, 505 (2005) [astro-ph/0501174].

[10]For a summary with references to the original literature, see Table V of M. Tegmark *et al.*, *Phys. Rev. D* **69**, 103501 (2004).

[11]U. Seljak *et al.*, *Phys. Rev. D* **71**, 103515 (2005).

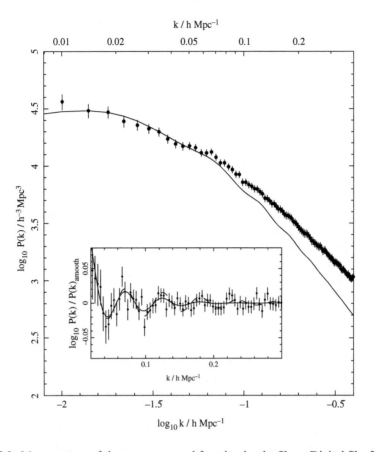

Figure 8.2: Measurement of the power spectral function by the Sloan Digital Sky Survey, from W. J. Percival *et al.*, astro-ph/0608636. Dark circles show values of $h^3 P(k)$, inferred from the survey of galaxy positions and redshifts, with distances calculated from redshifts using assumed cosmological parameters $\Omega_M = 0.24$, $\Omega_\Lambda = 0.76$. Vertical bars indicate $1 - \sigma$ errors. The solid curve is calculated using linear perturbation theory, with cosmological parameters taken from the **WMAP** third-year temperature and polarization data: $h = 0.73$, $\Omega_M = 0.24$, $\Omega_\Lambda = 0.76$, $\Omega_B/\Omega_M = 0.174$. The normalization of this curve is taken from a fit to the data for k between $0.01\,h\,\mathrm{Mpc}^{-1}$ and $0.06\,h\,\mathrm{Mpc}^{-1}$. The departure of the data from this theoretical curve for large k is attributed to non-linear effects on the growth of perturbations. The inset shows the effect of baryon acoustic oscillations, discussed at the end of this section. Data points give the ratio of the measured power spectral function to its value when smoothed to eliminate the oscillations. The solid curve shows the expected ratio, calculated using parameters from the third year WMAP data.

galaxy surveys for $k < 0.2\,h\,\mathrm{Mpc}^{-1}$, and extend this curve to larger values of k, again in agreement with the expected shape $\propto k|\mathcal{T}(\kappa)|^2$.

- Yet more information about the distribution of mass comes from weak lensing of galactic images.[12] (See Section 9.5). Their results yield

---

[12]See, e.g., H. Hoekstra *et al.*, *Astrophys. J.* **647**, 116 (2006).

values of σ_8 ranging from 0.67 to 0.97, and can be interpreted as giving values of $P(k)$ for k in the neighborhood of $0.3\,h\,\text{Mpc}^{-1}$, that fall on the curve provided by the galaxy surveys.

These measurements are subject to various uncertainties, which raise highly technical issues of astrophysics. In relying on redshifts to give distances, these measurements are vulnerable to complications arising from the peculiar velocities of galaxies or galaxy clusters or Lyman α clouds. The theory used to interpret these measurements assumes that the concentrations of dark matter can be treated as small perturbations, so it is necessary to avoid using data for values of k where fluctuations have become non-linear. Finally, in interpreting the distribution of baryonic matter in galaxies or clusters of galaxies or Lyman α clouds in terms of the total mass density, it is necessary either to assume that this introduces no bias, or else to know what the bias is. Only the weak lensing technique directly measures fluctuations in the *total* mass density. For this reason, although the other techniques can give good information about the *shape* of the function $P(k)$, they leave its overall normalization rather uncertain. The results from the galaxy surveys quoted above are for "no bias," that is, with the distribution of baryonic matter before perturbations become nonlinear assumed to trace the distribution of cold dark matter.

One of the striking things to emerge from the study of the cosmic microwave background described in the previous chapter is that these results for $P(k)$ obtained from studies of large scale structure agree with the strength of primordial fluctuations found from the cosmic microwave background. That is, the primordial fluctuation strength $q^3|\mathcal{R}_q|^2$ seems to be roughly constant over a wide range of physical wave numbers $k \equiv q/a_0$, from the values $\approx 10^{-3}\,\text{Mpc}^{-1}$ probed by observations of the cosmic microwave background down to all but the smallest multipole orders, to values more than $1\,\text{Mpc}^{-1}$ probed by studies of large scale structure. Indeed, even before the advent of the COBE measurements discussed in Section 2.6, the study of the large scale structure of matter had led to the expectation that the cosmic microwave background would show fractional temperature fluctuations of order 10^{-5}, corresponding to $|N| \approx 10^{-5}$.

The measurements discussed so far were all at small or moderate redshifts, where the structure distance in Eq. (8.1.28) is not very different from the linear approximation $d_S(z) \simeq z/H_0$. It is widely hoped that measurements of the matter correlation function ξ at larger redshifts will provide a determination of the functional form of $d_S(z)$ (or equivalently of $d_A(z)$) beyond this approximation, which could illuminate the time dependence of the vacuum energy. But Eq. (8.1.28) shows that in the integral for $\xi(z,\hat{n};z',\hat{n}')$ it is only wave numbers $k \overset{>}{\sim} 1/r$ for which the integrand is sensitive to the value of $r \equiv |d_S(z)\hat{n} - d_S(z')\hat{n}'|$. If $r \ll 1/k_{EQ}$ (where k_{EQ} is the wave number that comes into the horizon just at radiation–matter equality) then for $k \overset{>}{\sim} 1/r$ and $n_S \simeq 1$, the approximate formula (8.1.42) for $P(k)$ together with the asymptotic formula (6.5.11) for the transfer function give $P(k) \sim k^{-3}[\ln k + O(1)]^2$, so the part of the integral (8.1.28) for the correlation function that depends on r has the r-dependence

$$\int_{1/r}^{\infty} \frac{dk}{rk^2} \sin(ka_0 r)[\ln k + O(1)]^2 ,$$

which varies only logarithmically with r. Thus in order to use measurements of the matter correlation function to learn something about the z-dependence of $d_S(z)$, we need either to carry the measurements of the correlation function to redshifts and angles that are sufficiently different so that $|d_S(z)\hat{n} - d_S(z')\hat{n}'|$ is at least of order $1/k_{EQ} = 19.3\,\text{Mpc}/\Omega_M h^2$, or take advantage of the small departures of the transfer function from the asymptotic formula (6.5.11) for $k > k_{EQ}$, or else take advantage of small departures from our formula (8.1.42) for $P(k)$ at $k > k_{EQ}$.

Such departures from Eq. (8.1.42) for $P(k)$ are provided by terms of order $\beta \equiv \Omega_B/\Omega_M$, produced by baryon acoustic oscillations before the time of decoupling.[13] To see the effect of these oscillations, we must return to Eq. (8.1.14), and now look at the "fast" terms in the various perturbations, which oscillate with wave number for large wave number. In estimating the order of magnitude of these terms, we take $\bar{\rho}_B(t_L)$ and $\bar{\rho}_\gamma(t_L)$ to be of the same order of magnitude, so that $R(t_L)$ is of order unity, and $8\pi G\bar{\rho}_\gamma(t_L)$ is of order $\beta H^2(t_L)$. In estimating orders of magnitude we also do not distinguish between a_L and a_{EQ}, which only differ by a factor of order 3. Then, keeping track only of the small factors β and $a_L H_L/q$, Eqs. (6.4.45)–(6.4.48) show that, aside from a common factor $\mathcal{R}_q^o \exp(-\int_0^{t_L} \Gamma dt)$, the

---

[13]The influence of baryon acoustic oscillations before decoupling on the matter distribution after decoupling was recognized by P. J. E. Peebles and J. T. Yu, *Astrophys. J.* **162**, 815 (1970); R. A. Sunyaev and Ya. B. Zel'dovich, *Astrophys. Space Sci.* **7**, 3 (1970); J. R. Bond and G. Efstathiou, *Astrophys. J.* **285**, L45 (1984); J. A. Holtzmann, *Astrophys. J. Suppl.* **71**, 1 (1989); W. Hu and N. Sugiyama, *Astrophys. J.* **471**, 30 (1996); D. J. Eisenstein and W. Hu, *Astrophys. J.* **496**, 605 (1998). I believe that the first to suggest using redshift surveys to measure $d_A(z)$ by observation of baryon acoustic oscillations were D. J. Eisenstein, W. Hu, and M. Tegmark, *Astrophys. J.* **504**, L57 (1998), and that the suggestion to use this method to measure the evolution of dark energy is due to D. J. Eisenstein, in *Next Generation Wide-Field Multi-Object Spectroscopy*, eds. M. Brown and A. Dey (ASP Conference Series, vol. 280, 2002): p. 35.

fast terms in $\beta\delta_{\gamma q}(t_L)$, $(1-\beta)\delta_{Dq}(t_L)$, and $t_L\psi_q(t_L)$ are of the order of β, $\beta H_L^2 a_L^2/q^2$, and $\beta H_L a_L/q$, respectively, while $\beta t_L (q^2/a_L^2)\delta u_{\gamma q}(t_L)$ is of order $\beta(q/a_L H_L)$, and is therefore dominant (aside from Silk damping) for $q/a_L H_L \gg 1$. Keeping only this term, we see that the oscillating part of $\Delta(q)$ is approximately

$$\Delta^{\text{fast}}(q) \simeq \beta t_L (q^2/a_L^2)\delta u_{\gamma q}^{\text{fast}}(t_L)$$

$$\simeq \frac{2\beta q \mathcal{R}_q^o}{\sqrt{3}a_L H_L (1+R_L)^{3/4}} \exp\left(-\int_0^{t_L}\Gamma dt\right) \sin\left(qd_H/a_L\right),$$
$$(8.1.50)$$

where as usual $R \equiv 3\bar{\rho}_B/\bar{\rho}_\gamma$; d_H is the acoustic horizon distance (7.2.39) at decoupling

$$d_H \equiv a_L \int_0^{t_L} \frac{dt}{a\sqrt{3(1+R)}}$$

$$= \frac{2}{H_0\sqrt{3R_L\Omega_M}(1+z_L)^{3/2}} \ln\left(\frac{\sqrt{1+R_L}+\sqrt{R_{\text{EQ}}+R_L}}{1+\sqrt{R_{\text{EQ}}}}\right);$$
$$(8.1.51)$$

and, for reasons explained earlier, we have replaced t_L in Eq. (8.1.14) with $2/3H_L$. Thus the ratio of $P(k)$ to the smooth curve given by Eq. (8.1.42) will have bumps at $k \simeq \pi k_H/2$, $3\pi k_H/2$, $5\pi k_H/2\ldots$, where $k_H = 1/d_H(1+z_L)$. These bumps have been seen in the spectral function inferred from observations by both the Sloan Digital Sky Survey[14] and the 2dF Galaxy Redshift Survey.[15] (See the inset in Figure 8.2.) The proportionality of d_H to $\Omega_M^{-1/2}$ has allowed a determination of Ω_M by matching the observed and predicted positions of these bumps[14,15]. The result[16] is $\Omega_M = 0.256^{+0.029}_{-0.024}$, or with a larger sample,[17] $\Omega_M = 0.24 \pm 0.02$. Unfortunately, these measurements are at moderate redshifts ($0.16 < z < 0.47$ for the Sloan survey, and $z < 0.3$ for the 2dF Survey), and although they give evidence for dark energy, they do not yet provide information about its time dependence.

---

[14]D. J. Eisenstein *et al.*, *Astrophys. J.* **633**, 560 (2005) [astro-ph/0501171].

[15]S. Cole *et al.*, *Mon. Not. Roy. Astron. Soc.* **362**, 505 (2005).

[16]W. J. Percival *et al.*, *Astrophys. J.* **657**, 51 (2007) [astro-ph/0608635].

[17]M. Tegmark *et al.*, *Phys. Rev.* **74**, 123507 (2006) [astro-ph/0608632].

$$* * *$$

As promised, we close by estimating the cosmic variance incurred in using Eq. (8.1.33). The mean square fractional error in this formula is

$$\Delta_V(Q,t) \equiv \left\langle \left[\int \frac{d^2\hat{Q}}{4\pi} \left(\left| \delta^V_{M\,Q\hat{Q}}(t) \right|^2 - \left\langle \left| \delta^V_{M\,Q\hat{Q}}(t) \right|^2 \right\rangle \right) \right]^2 \right\rangle$$

$$\times \left(\left\langle \left| \delta^V_{M\,Q\hat{Q}}(t) \right|^2 \right\rangle \right)^{-2} = \delta^{-4}_{MQ}(t) \int \frac{d^2\hat{Q}}{4\pi} \int \frac{d^2\hat{Q}'}{4\pi}$$

$$\times \left[\left\langle \left| \delta^V_{M\,Q\hat{Q}}(t) \right|^2 \left| \delta^V_{M\,Q\hat{Q}'}(t) \right|^2 \right\rangle - \left\langle \left| \delta^V_{M\,Q\hat{Q}}(t) \right|^2 \right\rangle \left\langle \left| \delta^V_{M\,Q\hat{Q}'}(t) \right|^2 \right\rangle \right].$$

$$(8.1.52)$$

Assuming Gaussian statistics, this is

$$\Delta_V(Q,t) = \delta^{-4}_{MQ}(t) \int \frac{d^2\hat{Q}}{4\pi} \int \frac{d^2\hat{Q}'}{4\pi}$$

$$\times \left[\left| \left\langle \delta^V_{M\,Q\hat{Q}}(t)\, \delta^{V*}_{M\,Q\hat{Q}'}(t) \right\rangle \right|^2 + \left| \left\langle \delta^V_{M\,Q\hat{Q}}(t)\, \delta^V_{M\,Q\hat{Q}'}(t) \right\rangle \right|^2 \right].$$

Using Eqs. (8.1.31) and (8.1.24) (and recalling that $\alpha(\mathbf{q}) = \alpha^*(-\mathbf{q})$ and $\delta_{Mq}(t)$ is real) then gives

$$\Delta_V(Q,t) = 2\delta^{-4}_{MQ}(t) \int \frac{d^2\hat{Q}}{4\pi} \int \frac{d^2\hat{Q}'}{4\pi}$$

$$\times \left| \int d^3q\, F_V(\mathbf{q} - Q\hat{Q})\, F^*_V(\mathbf{q} - Q\hat{Q}') \delta^2_{Mq}(t) \right|^2.$$

The large volume limit $F_V(\mathbf{q} - \mathbf{Q}) \to ((2\pi)^{3/2}/\sqrt{V})\delta^3(\mathbf{q} - \mathbf{Q})$ then gives the cosmic variance as

$$\Delta_V(Q,t) = \frac{2(2\pi)^3}{V} \int \frac{d^2\hat{Q}}{4\pi} \int \frac{d^2\hat{Q}'}{4\pi} F^2_V(Q\hat{Q} - Q\hat{Q}').$$

Now, $F^2_V(Q\hat{Q} - Q\hat{Q}')$ takes the value $V/(2\pi)^3$ for $Q|\hat{Q} - \hat{Q}'| \ll V^{-1/3}$, and vanishes exponentially for $Q|\hat{Q} - \hat{Q}'| \gg V^{-1/3}$, so aside from factors of order unity,

$$\Delta_V(Q,t) \approx \frac{1}{Q^2 V^{2/3}}. \tag{8.1.53}$$

Thus the mean square fractional error in using Eq. (8.1.33) vanishes as $Q^{-2}V^{-2/3}$, and becomes negligible when the survey volume contains many wavelengths. The numerical factor depends on the shape of the survey volume. For instance, if the survey volume is a sphere of co-moving radius R, then for $QR \gg 1$,

$$\Delta_V(Q, t) \to \frac{9}{4Q^2R^2} . \qquad (8.1.54)$$

8.2 Nonlinear growth

The study of the growth of perturbations beyond the linear approximation presents formidable mathematical difficulties. It is usually pursued by the use of computer simulations, which are beyond the scope of this book. To allow an analytic treatment, we can adopt a generalization of an idealization of nonlinear growth originally due to Peebles.[1] With Peebles, we will consider a fluctuation to have an overdensity $\Delta\rho_M$ (that is, a total density $\bar{\rho}_M + \Delta\rho_M$ greater than the cosmic average $\bar{\rho}_M$) that is uniform within a finite sphere.[2] According to the Birkhoff theorem,[3] the metric and the equations of motion of a freely falling test particle inside the sphere are independent of what is happening outside the sphere, and are therefore the same as in a homogeneous isotropic universe, described by a Robertson–Walker metric, with a density $\bar{\rho}_M(t) + \Delta\rho_M(t)$, and a curvature constant that is not in general equal to the cosmological curvature constant K. In a Robertson–Walker metric with curvature constant K, the scale factor $a(t)$ satisfies the Friedmann equation (1.5.19):

$$\dot{a}^2(t) + K = \frac{8\pi\, Ga^2(t)}{3}\left(\bar{\rho}_M(t) + \rho_V\right). \qquad (8.2.1)$$

Likewise, the scale factor $\mathcal{A}(t)$ of the Robertson–Walker metric inside the fluctuation will satisfy a Friedmann equation:

$$\dot{\mathcal{A}}^2(t) + K + \Delta K = \frac{8\pi\, G\mathcal{A}^2(t)}{3}\left(\bar{\rho}_M(t) + \Delta\rho_M(t) + \rho_V\right), \qquad (8.2.2)$$

where $K + \Delta K$ is the curvature constant of the interior metric. We are including in the total density both a non-relativistic mass density, and a

---

[1] P. J. E. Peebles, *Astrophys. J.* **147**, 859 (1967).

[2] J. E. Gunn and J. R. Gott, *Astrophys. J.* **176**, 1 (1972) took into account the infall of matter from outside a sphere of uniform overdensity, which amounts to treating the overdensity of the sphere and its surroundings as a step function of radius. A non-vanishing vacuum energy was incorporated in the Gunn–Gott model by H. Martel, P. R. Shapiro, and S. Weinberg, *Astrophys. J.* **492**, 29 (1998), who considered fluctuations consisting of a ball with a uniform overdensity surrounded by a spherical shell with a uniform underdensity.

[3] G&C, Sec. 11.7.

vacuum density ρ_V (not included by Peebles) which for want of any contrary evidence is taken to be time-independent.

We will only consider the case of vanishing cosmological curvature,[4] $K = 0$. We are primarily interested in fluctuations that have a chance to stop expanding and recollapse to high density, so we shall take $\Delta K > 0$. The total matter density obeys the conservation law

$$\bar{\rho}_M(t) + \Delta\rho_M(t) \propto \mathcal{A}^{-3}(t) , \qquad (8.2.3)$$

while the unperturbed density satisfies

$$\bar{\rho}_M(t) \propto a^{-3}(t) . \qquad (8.2.4)$$

With $K = 0$, the normalization of a is arbitrary; we will find it convenient to use Eqs. (8.2.3) and (8.2.4) to normalize a so that

$$\mathcal{A}^3(t)\left(\bar{\rho}_M(t) + \Delta\rho_M(t)\right) = a^3(t)\bar{\rho}_M(t) . \qquad (8.2.5)$$

In order to provide initial conditions for this problem, we must first consider times that are sufficiently early in the matter-dominated era so that $\Delta\rho_M(t)$ and $\Delta\mathcal{A}(t) \equiv \mathcal{A}(t) - a(t)$ *can* be treated as small perturbations, and the vacuum energy may be neglected. As we saw in Section 1.5, in this case Eqs. (8.2.1) and (8.2.4) have the solution

$$a \propto t^{2/3} , \qquad \bar{\rho}_M = \frac{1}{6\pi\, Gt^2} . \qquad (8.2.6)$$

To first order in the perturbations around this solution, Eq. (8.2.2) becomes

$$2\dot{a}\Delta\dot{\mathcal{A}} + \Delta K = \frac{8\pi\, Ga^2\bar{\rho}_M}{3}\left(\frac{2\Delta\mathcal{A}}{a} + \frac{\Delta\rho_M}{\bar{\rho}_M}\right) .$$

or, using Eq. (8.2.6)

$$\frac{\Delta K}{a^2} = \frac{4}{9t^2}\left(\frac{2\Delta\mathcal{A}}{a} + \frac{\Delta\rho_M}{\bar{\rho}_M}\right) - \frac{4}{3t}\frac{\Delta\dot{\mathcal{A}}}{a} .$$

Also, to first order in perturbations, Eq. (8.2.5) gives at early times

$$\frac{\Delta\mathcal{A}}{a} = -\frac{\Delta\rho_M}{3\bar{\rho}_M} ,$$

---

[4]The case of negative cosmological curvature is considered by B. Freivogel, M. Kleban, M. N. Martinez, and L. Susskind, *J. High Energy Phys.* **0603**, 039 (2006) [hep-th/0505232].

so $\Delta\mathcal{A}$ is governed at early times by the first-order differential equation

$$\frac{\Delta K}{a^2} = -\frac{4}{9t^2}\frac{\Delta\mathcal{A}}{a} - \frac{4}{3t}\frac{\Delta\dot{\mathcal{A}}}{a} \, . \qquad (8.2.7)$$

This has solutions for $t \to 0$ of the form $\Delta\mathcal{A} \propto t^{4/3}$ and $\Delta\mathcal{A} \propto t^{-1/3}$. Assuming that enough time has passed for the solution $\Delta\mathcal{A} \propto t^{-1/3}$ to die away, we have $\Delta\mathcal{A} \propto t^{4/3}$, and so

$$\Delta K = -\frac{20}{9}\lim_{t\to 0} t^{-2}a(t)\Delta\mathcal{A}(t) = \frac{40\pi G}{9}\lim_{t\to 0} a^2(t)\Delta\rho_M(t) \, . \quad (8.2.8)$$

In characterizing the initial strength of a fluctuation within a co-moving radius r, we note from Eq. (8.2.8) that at early times $\Delta\rho_M(t) \propto a^{-2}(t) \propto \bar{\rho}_M^{2/3}(t)$, so we can define a time-independent quantity, which we shall call the *initial fluctuation strength*:

$$\rho_1 \equiv \lim_{t\to 0} \frac{\Delta\rho_M^3(t)}{\bar{\rho}_M^2(t)} \, . \qquad (8.2.9)$$

Then Eq. (8.2.8) can be written

$$\begin{aligned}
\Delta K &= \frac{40\pi G}{9} a^2(t)\bar{\rho}_M^{2/3}(t)\rho_1^{1/3} \\
&= \frac{40\pi G}{9} \mathcal{A}^2(t)\Big(\bar{\rho}_M(t) + \Delta\rho_M(t)\Big)^{2/3}\rho_1^{1/3} \qquad (8.2.10)
\end{aligned}$$

Note that we do not have to take the limit $t \to 0$ here, because $a^2\bar{\rho}_M^{2/3}$ is time-independent.

Now let's consider the development of the fluctuation at later times, when it can no longer be treated as a small perturbation. Using Eq. (8.2.10) allows us to write the Friedmann equation (8.2.2) (with $K = 0$) as

$$\dot{\mathcal{A}}^2 = \frac{8\pi G\mathcal{A}^2}{9}\left[3(\bar{\rho}_M + \Delta\rho_M + \rho_V) - 5(\bar{\rho}_M + \Delta\rho_M)^{2/3}\rho_1^{1/3}\right],$$

or, using Eq. (8.2.3),

$$(\bar{\rho}_M + \Delta\rho_M)^{-2}\left(\frac{d}{dt}(\bar{\rho}_M + \Delta\rho_M)\right)^2 = 8\pi G$$

$$\times\left(3(\bar{\rho}_M + \Delta\rho_M + \rho_V) - 5(\bar{\rho}_M + \Delta\rho_M)^{2/3}\rho_1^{1/3}\right). \quad (8.2.11)$$

The right-hand side vanishes at a total mass density $\bar{\rho}_M + \Delta\rho_M = \rho_c$ satisfying

$$3(\rho_c + \rho_V) = 5\rho_c^{2/3}\rho_1^{1/3} \, . \qquad (8.2.12)$$

For $\rho_c > 0$, the quantity $5\rho_c^{2/3}\rho_1^{1/3} - 3\rho_c$ takes values from $-\infty$ to $500\rho_1/243$, so Eq. (8.2.12) has a solution if and only if the fluctuation is strong enough so that[5]

$$\rho_1 \geq \frac{729}{500}\rho_V \,. \tag{8.2.13}$$

As long as Eq. (8.2.12) has a solution, the density $\bar{\rho}_M(t) + \Delta\rho_M(t)$ will drop until it reaches the value ρ_c, and then increase again to infinity. The total time elapsed for this expansion and collapse is given by Eq. (8.2.11) as

$$t_c = \frac{2}{\sqrt{8\pi G}} \int_{\rho_c}^{\infty} \frac{d\rho}{\rho\sqrt{3(\rho+\rho_V) - 5\rho^{2/3}\rho_1^{1/3}}} \,. \tag{8.2.14}$$

(The factor 2 appears here because it takes the same time to contract from the minimum density ρ_c to infinite density as it does to expand from infinite to minimum density.) For example, in the limit $\rho_1 \gg \rho_V$, this gives

$$t_c \rightarrow \frac{9\pi}{5^{3/2}\sqrt{8\pi G\rho_1}} \,. \tag{8.2.15}$$

If linear perturbation theory held up to this time then instead of becoming infinite, in this case the fractional density perturbation at time t_c would be

$$\frac{\Delta\rho_M}{\bar{\rho}_M} = \left(\frac{\rho_1}{\bar{\rho}_M}\right)^{1/3} = \rho_1^{1/3}(6\pi Gt_c^2)^{1/3} = \left(\frac{243\pi^2}{500}\right)^{1/3} = 1.686 \,.$$

We see that gravitational collapse occurs at a time when linear perturbation theory would predict a fractional density perturbation large enough to make obvious its own invalidity.

Of course, different fluctuations will have different initial strengths ρ_1 and co-moving radii R. In order to make contact between this analysis and observation, it is convenient to make use of an approach due to Press and Schechter.[6] By inverting Eq. (8.2.14), we can calculate the minimum initial fluctuation strength $\rho_1(t_c)$ required for collapse at or before a time t_c. For instance, for fluctuations that are sufficiently strong to collapse before vacuum energy becomes important, Eq. (8.2.15) gives $\rho_1(t_c) = 81\pi/1000Gt_c^2$. To calculate the probability that a random point in space will be in a fluctuation this strong, we assume that at an early time t, before non-linearities become significant, the probability $P_{t,R}(\Delta\rho_M)\,d\Delta\rho_M$ that the average density within a co-moving sphere of radius R is increased by a density excess

[5]S. Weinberg, *Phys. Rev. Lett.* **59**, 2607 (1987).
[6]W. H. Press and P. Schechter, *Astrophys. J.* **239**, 1 (1974).

between $\Delta\rho_M$ and $\Delta\rho_M + d\Delta\rho_M$ is given by the Gaussian distribution

$$P_{t,R}(\Delta\rho_M)\, d\Delta\rho_M = \frac{d\Delta\rho_M}{\sqrt{2\pi}\,\sigma_R(z_t)\bar{\rho}_M(t)}\,\exp\left(-\frac{\Delta\rho_M^2}{2\sigma_R^2(z_t)\bar{\rho}_M^2(t)}\right). \quad (8.2.16)$$

For a Harrison–Zel'dovich spectrum of fluctuations with $\mathcal{R}_q = Nq^{-3/2}$, the standard deviation σ_R for the fractional fluctuation $\Delta\rho_M/\bar{\rho}_M$ averaged over a radius R is given by Eq. (8.1.47), which for $n_S = 1$ reads

$$\sigma_R^2(z) = \frac{16\pi N^2}{25\Omega_M^2 H_0^4}\,\frac{C^2\left(\Omega_\Lambda/\Omega_M(1+z)^3\right)}{(1+z)^2}$$
$$\int_0^\infty |\mathcal{T}(\sqrt{2}k/k_{\mathrm{EQ}})|^2\,|f(kR)|^2\,k^3\,dk\,. \quad (8.2.17)$$

where f is the top hat distribution function (8.1.46); $\mathcal{T}$ is the scalar transfer function, defined and calculated in Section 6.5; and k_{EQ} is the wave number (8.1.41) that comes into the horizon at matter–radiation equality. Using Eq. (8.2.9), this can be expressed as a time-independent probability $\tilde{P}_R(\rho_1)\, d\rho_1$ for fluctuations averaged over a co-moving sphere of radius R to have a initial strength between ρ_1 and $\rho_1 + d\rho_1$:

$$\tilde{P}_R(\rho_1)\, d\rho_1 = \frac{d\rho_1}{3\sqrt{2\pi}\,\rho_1^{2/3}\tilde{\sigma}_R}\,\exp\left(-\frac{\rho_1^{2/3}}{2\tilde{\sigma}_R^2}\right) \quad (8.2.18)$$

where $\tilde{\sigma}_R$ is a time-independent quantity, given by

$$\tilde{\sigma}_R^2 \equiv \lim_{z\to\infty} \bar{\rho}_M^{2/3}(t(z))\sigma_R^2(z)$$
$$= \frac{16\pi N^2}{25}\left(\frac{3}{8\pi G}\right)^2 \bar{\rho}_{M0}^{-4/3}\int_0^\infty |\mathcal{T}(\sqrt{2}k/k_{\mathrm{EQ}})|^2\,|f(kR)|^2\,k^3\,dk\,.$$
$$(8.2.19)$$

Because we are averaging over a sphere of co-moving radius R, it is only fluctuations whose co-moving radii are *greater* than R that contribute to this probability. Integrating, we see that at early times, before the fluctuations become strong, the probability that a random point in space is in a fluctuation with initial strength greater than ρ_1 and co-moving radius greater than R is

$$P(> \rho_1, > R) = \int_{\rho_1}^\infty \frac{d\rho}{3\sqrt{2\pi}\,\rho^{2/3}\tilde{\sigma}_R}\,\exp\left(-\frac{\rho^{2/3}}{2\tilde{\sigma}_R^2}\right)$$
$$= \frac{1}{2}\left[1 - \mathrm{erf}\left(\rho_1^{1/3}/\sqrt{2}\tilde{\sigma}_R\right)\right]\,, \quad (8.2.20)$$

where erf (y) is the usual error function

$$\text{erf}(y) \equiv \frac{1}{\sqrt{2\pi}} \int_{-y\sqrt{2}}^{y\sqrt{2}} dx \, \exp(-x^2/2) . \tag{8.2.21}$$

At early times the universe has uniform matter density, so Eq. (8.2.20) also gives the fraction of all matter that is in fluctuations with initial strength greater than ρ_1 and co-moving radius greater than R. If we now set ρ_1 in Eq. (8.2.20) equal to the critical initial strength $\rho_1(t)$ for collapse by a time t, then we find that at time t, when nonlinearities have become important, the fraction of all matter that is in collapsed structures with co-moving radii greater than R will be

$$F(> R, t) = \frac{1}{2} \left[1 - \text{erf}\left(\rho_1^{1/3}(t) / \sqrt{2} \tilde{\sigma}_R \right) \right] . \tag{8.2.22}$$

The mass in a sphere of co-moving radius R is the time-independent quantity $4\pi \bar\rho_M(t) a^3(t) R^3/3$, so at time t the number density $n(M, t) \, dM$ of collapsed structures with mass between M and $M + dM$ is given by

$$\begin{aligned}
n(M, t) &= -\frac{\bar\rho_M(t)}{2M} \frac{d}{dM} \text{erf}\left(\rho_1^{1/3}(t) / \sqrt{2} \tilde{\sigma}_{R(M)} \right) \\
&= \frac{\rho_1^{1/3}(t) \, \bar\rho_M(t)}{M \sqrt{2\pi}} \left| \frac{d \, \tilde{\sigma}_{R(M)}^{-1}}{dM} \right| \exp\left(-\frac{\rho_1^{2/3}(t)}{2 \tilde{\sigma}_{R(M)}^2} \right) , \tag{8.2.23}
\end{aligned}$$

where $R(M) \equiv (3M/4\pi a^3 \bar\rho_M)^{1/3}$. (Press and Schechter somewhat arbitrarily multiplied this by a factor 2, to take account of the matter in regions with a negative density fluctuation.[7]) These collapsed structures eventually furnish the halos of cold dark matter that surround galaxies in the present universe.[8]

In using Eq. (8.2.23), we need to know the mass dependence of $\tilde{\sigma}_{R(M)}$. For large masses and radii, the integral (8.2.19) is dominated by low values of the wave number k, for which the transfer function $T(\kappa)$ is close to unity, so Eq. (8.2.19) gives $\tilde{\sigma}_R^2 \propto R^{-4}$, and so $\tilde{\sigma}_{R(M)}^2 \propto M^{-4/3}$. (For a primordial fluctuation spectrum $\mathcal{R}_q \propto q^{(-4+n_S)/2}$, we would have a factor $k^{2+n_S} dk$ in place of $k^3 dk$ in Eq. (8.2.19), which would give $\tilde{\sigma}_R^2 \propto R^{-3-n_S}$, and so

---

[7] For a derivation of this factor of 2, see J. R. Bond, S. Cole, G. Efstathiou, and N. Kaiser, *Astrophys. J.* **379**, 440 (1991).

[8] The effects of non-spherical collapse on the mass function $n(M, t)$ is considered by P. Monaco, *Astrophys. J.* **447**, 23 (1995); J. Lee and S. F. Shandarin, *Astrophys. J.* **500**, 14 (1998); R. K. Sheth and G. Tormen, *Mon. Not. Roy. Astron. Soc.* **308**, 119 (1999); R. K. Sheth, H. J. Mo, and G. Tormen, *Mon. Not. Roy. Astron. Soc.* **323**, 1 (2001).

$\tilde{\sigma}^2_{R(M)} \propto M^{-(3+n_S)/3}$.) The argument of the exponential in Eq. (8.2.23) in this case is thus proportional for $n_S = 1$ to $M^{4/3}$, giving a rapid fall-off of the number density for large mass. On the other hand, for small masses and radii, the integral (8.2.19) is dominated by large values of the wave number k, for which according to Eq. (6.5.11) the transfer function $T(\kappa)$ falls off like $\ln k/k^2$. The integral (8.2.19) then varies only logarithmically with R, and so the number density $n(M)$ goes for small M more-or-less like M^{-2}. The detailed M-dependence predicted by Eq. (8.2.23) is in reasonable agreement with the results of large computer simulations of the evolution of cold dark matter.[9]

8.3 Collapse of baryonic matter

Until now we have supposed that the baryonic matter of the universe, which after recombination consisted chiefly of neutral hydrogen and helium, had negligible pressure. In this case baryonic matter just followed along with cold dark matter in its expansion and possible recontraction. Actually, as we saw in Section 2.3, the baryonic matter retained a small residual ionization even after the nominal era of recombination, providing enough electrons for Compton scattering of photons of the cosmic microwave background to keep the temperature of baryonic matter equal to the temperature of the microwave background until the redshift dropped below about 150. Small overdense regions did not have enough of a gravitational field to overcome the baryonic pressure, so their baryonic matter did not collapse along with the cold dark matter. This led to relatively small clumps of cold dark matter that now do not contain galaxies, and are therefore undetectable except for their gravitational effects. The question is, how small did a clump of matter have to be for its baryonic component to have resisted gravitational collapse?

Before the existence of cold dark matter was generally accepted, this question was addressed in a simple theory due originally to James Jeans.[1] According to this theory, small perturbations either oscillate or grow according to whether their wave number is greater or less than a critical wave number $k_J = \sqrt{4\pi G\bar{\rho}_B}/v_s$ (where v_s is the speed of sound) so a clump is too small to collapse if its mass is less than the *Jeans mass*, given by $\bar{\rho}_B(2\pi/k_J)^3$. This theory naturally (given its date) was not originally set in the context of an expanding universe, but not much changes

---

[9]V. Springel *et al.*, *Nature* **435**, 629 (2005) [astro-ph/0504097].

[1]J. Jeans, *Phil. Trans. Roy. Soc.* **199A**, 49 (1902), and *Astronomy and Cosmogony* (2nd ed., first published by Cambridge University Press in 1928; reprinted by Dover Publications, New York, 1961), pp. 345–350.

when the expansion is taken into account.[2] But as we shall see, it turns out that results are quite different when we include the effects of cold dark matter.

We are considering only non-relativistic matter during an era in which radiation contributed little to the gravitational field, so we can apply the Newtonian cosmological theory described in Appendix F. By following the same reasoning that led there to Eqs. (F.15) and (F.16), but now including both baryonic matter with a squared sound speed $\partial p_B/\partial \rho_B = v_s^2$ and cold dark matter with zero pressure, we find that the velocity potential perturbations δu_B and δu_D and the density perturbations $\delta \rho_B$ and $\delta \rho_D$ for co-moving wave number $\mathbf{q}$ are governed by the equations of continuity

$$\frac{d\delta\rho_D}{dt} + 3H\delta\rho_D - a^{-1}\bar{\rho}_D\,\mathbf{q}^2\delta u_D = 0 \;. \tag{8.3.1}$$

$$\frac{d\delta\rho_B}{dt} + 3H\delta\rho_B - a^{-1}\bar{\rho}_B\,\mathbf{q}^2\delta u_B = 0 \;. \tag{8.3.2}$$

and the Euler equations

$$\frac{d\delta u_D}{dt} + H\delta u_D = \frac{4\pi Ga}{\mathbf{q}^2}\Big[\delta\rho_D + \delta\rho_B\Big] \;, \tag{8.3.3}$$

$$\frac{d\delta u_B}{dt} + H\delta u_B = \frac{4\pi Ga}{\mathbf{q}^2}\Big[\delta\rho_D + \delta\rho_B\Big] - \frac{v_s^2}{a\bar{\rho}_B}\delta\rho_B \;, \tag{8.3.4}$$

As in the relativistic case, it is convenient to introduce the fractional density perturbations $\delta_n \equiv \delta\rho_n/\bar{\rho}_n$. Using the Friedmann result that both unperturbed densities $\bar{\rho}_n$ go as $a^{-3} \propto t^{-2}$, and eliminating the velocity potentials, these equations then become

$$\ddot{\delta}_D + \frac{4}{3t}\dot{\delta}_D = \frac{2}{3t^2}\Big[\beta\delta_B + (1-\beta)\delta_D\Big] \;, \tag{8.3.5}$$

$$\ddot{\delta}_B + \frac{4}{3t}\dot{\delta}_B = -\frac{2\alpha}{3t^2}\delta_B + \frac{2}{3t^2}\Big[\beta\delta_B + (1-\beta)\delta_D\Big] \;, \tag{8.3.6}$$

where α and β are defined by

$$\alpha \equiv \frac{3\mathbf{q}^2v_s^2t^2}{2a^2} = \frac{\mathbf{q}^2v_s^2}{4\pi G\bar{\rho}_Ma^2} \;, \quad \beta \equiv \frac{\bar{\rho}_B}{\bar{\rho}_M} = \frac{\Omega_B}{\Omega_M} \simeq 0.17 \;, \tag{8.3.7}$$

and $\bar{\rho}_M \equiv \bar{\rho}_D + \bar{\rho}_B$.

Note that α was constant during the era (roughly for $z > 150$) when baryonic matter had the same temperature as radiation, because $v_s^2 \propto T \propto a^{-1}$. Thus we can find power-law solutions[3] to Eqs. (8.3.5) and (8.3.6) that

---

[2]G&C, Sec. 15.9.
[3]Approximate power law solutions were given by P. J. E. Peebles, *Astrophys. J.* **277**, 470 (1984).

apply during this era. We set

$$\delta_D \propto t^\nu , \qquad \delta_B = \xi \, \delta_D ,\qquad (8.3.8)$$

where ν and ξ are time-independent (but **q**-dependent) quantities to be determined. Eqs. (8.3.5) and (8.3.6) then become

$$\nu(\nu - 1) + \frac{4\nu}{3} = \frac{2}{3}\Big[\beta\xi + (1 - \beta)\Big] ,\qquad (8.3.9)$$

$$\nu(\nu - 1) + \frac{4\nu}{3} + \frac{2\alpha}{3} = \frac{2}{3}\Big[\beta + (1 - \beta)/\xi\Big] .\qquad (8.3.10)$$

Eliminating ξ yields a quartic equation for ν. This generically has four different solutions, so there are four independent power-law solutions of the fourth-order system of differential equations (8.3.5) and (8.3.6), which therefore form a complete set of solutions.

The general solutions of Eqs. (8.3.9) and (8.3.10) are too complicated to be illuminating, but we can find useful approximate solutions if we take into account the small value of β. In the limit of very small β, there are two baryon-poor solutions for which $\xi < 1$:

$$\nu = 2/3 , \qquad \xi = \frac{1}{1 + \alpha}\qquad (8.3.11)$$

$$\nu = -1 , \qquad \xi = \frac{1}{1 + \alpha}\qquad (8.3.12)$$

and two baryon-rich solutions for which $\xi \gg 1$:

$$\nu = -\frac{1}{6} \pm \sqrt{\frac{1}{36} - \frac{2\alpha}{3}} , \quad \xi = -\frac{1 + \alpha}{\beta} .\qquad (8.3.13)$$

All of these solutions decay with time, except for the first solution (8.3.11), in which the baryon and cold dark matter fractional density perturbations both grow as $t^{2/3}$. This solution therefore dominates at late times. To first order in β, the power-law exponent and baryon fraction in this mode are

$$\nu = 2/3 - \frac{2\beta\alpha}{5(1 + \alpha)} , \qquad \xi = \frac{1}{1 + \alpha} - \frac{\beta\alpha^2}{(1 + \alpha)^3} .$$

The fractional baryonic corrections to ν and ξ are a maximum for very short wavelengths, for which they take values $3\beta/5 \simeq 10\%$ and $\beta \simeq 17\%$, respectively.

There is nothing here like a critical wave number marking a transition from growth to oscillation, as in the classic one-component Jeans theory. The growing mode (8.3.11) grows for all wave numbers, with a growth rate that for small β depends very little on wave number. This of course is because the pressure of baryonic matter could not prevent clumps of cold dark matter from becoming increasingly denser than average. What does depend on the wave number is the fraction ξ of the baryons that follow the growing condensation of the cold dark matter. This can most conveniently be expressed in terms of the total mass M (dark and baryonic matter) in a cubic physical wavelength $2\pi a/q$:

$$M \equiv \bar{\rho}_M \left(\frac{2\pi a}{q} \right)^3 \tag{8.3.14}$$

According to the formulas for ξ and α given by Eqs. (8.3.11) and (8.3.7), the fraction of baryons that collapsed along with the cold dark matter can be written

$$\frac{\delta\rho_B/\bar{\rho}_B}{\delta\rho_D/\bar{\rho}_D} \equiv \xi = \frac{1}{1 + (M_J/M)^{2/3}} , \tag{8.3.15}$$

where M_J is a sort of Jeans mass

$$M_J = \left(\frac{\pi}{G} \right)^{3/2} \frac{v_s^3}{\bar{\rho}_M^{1/2}} . \tag{8.3.16}$$

Baryons collapsed along with cold dark matter for clumps of mass much greater than M_J, while clumps of mass much less than M_J are largely free of baryonic matter.

The speed of sound for a gas of hydrogen and helium atoms at temperature T is $v_s = (5k_B T/3\mu m_N)^{1/2}$, where μ is the mean molecular weight, which for a helium abundance by weight of 24% is $\mu = 1.22$. Since we are considering an era in which $T = T_\gamma$, and $T_\gamma^3/\bar{\rho}_M$ equals its present value $T_{\gamma 0}^3/\bar{\rho}_{M0}$, the Jeans mass can be written

$$M_J = \left(\frac{5\pi k_B T_{\gamma 0}}{3\mu m_N G} \right)^{3/2} \bar{\rho}_{M0}^{-1/2} = 2.02 \times 10^5 \, (\Omega_M h^2)^{-1/2} \, M_\odot \tag{8.3.17}$$

For $\Omega_M h^2 = 0.13$, this is $6 \times 10^5 \, M_\odot$, corresponding to a baryonic mass $\beta M_J \simeq 10^5 \, M_\odot$.

After the redshift dropped to about 150, the Compton scattering of the cosmic microwave background by the residual ionized hydrogen no longer kept the temperature of baryonic matter equal to the radiation temperature. According to Eq. (1.1.23), the kinetic energy and hence the temperature of the hydrogen and helium atoms then dropped like a^{-2}. Additional baryons then began to fall into the already growing clumps of cold dark matter, until the baryons were heated again by the energy released in gravitational collapse, and eventually by the first generation of stars. The increased baryonic temperature inhibited the further accretion of baryonic matter, and may have resulted in the expulsion of some baryonic matter from the clump. This is all quite complicated,[4] but it does not change the conclusion, that a clump of cold dark matter whose mass is less than M_J will not contain a full complement of baryonic matter.

---

[4]P. R. Shapiro, M.L. Giroux, and A. Babul, *Astrophys. J.* **427**, 25 (1994).

9

Gravitational Lenses

In 1979 Walsh, Carswell, and Weymann[1] noticed a pair of quasars at the same redshift, about 1.4, separated by just $6''$. The similarity of the spectrum of the two quasars suggested to them that there was really just one quasar, now known as Q0957+561, split into two images by the deflection of light in the gravitational field of an intervening massive body. This suggestion was verified with the discovery of a galaxy with a redshift 0.36 between the lines of sight to the quasar. Such gravitational lenses had already been studied theoretically by many authors, and now a serious search for them was put in train.[2] Many more lensing galaxies were discovered, generally elliptical field galaxies, and also some lensing clusters of galaxies, which generally produce arc-like images.[3]

Strong gravitational lensing has been used to search for dark objects, to explore the structure of galaxy clusters, and to measure the Hubble constant. Weak lensing offers great promise in measuring the correlation function of density fluctuations. Lensing is a large subject; in this chapter we will give only an overview of its cosmological applications.

9.1 Lens equation for point masses

We will first consider the gravitational lens provided by a point mass, and later take up more detailed models. To analyze the splitting of images by a point mass, suppose that the lines from the earth to a point source and the earth to the lensing mass in a Robertson–Walker coordinate system centered on the earth are separated by a small angle α. (See Figure 9.1.)

This is the angle that there would be between the images of the source and lens, if their were no gravitational deflection of light. Because of the deflection of light by the lens, there is a different angle β between the actual images of the source and lens. We need to derive a *lens equation*, which gives the relation between β and α.

In the coordinate system centered on the earth, the light ray from the source follows a path from the source to the neighborhood of the lens that

---

[1] D. Walsh, R. F. Carswell, and R. J. Weymann, *Nature* **279**, 381 (1979).

[2] For a comprehensive review, see *Gravitational Lenses*, by P. Schneider, J. Ehlers, and E. E. Falco (Springer-Verlag, Berlin, 1992). A more recent review is given by G. Soucail, *Proceedings of the XX Texas Symposium on Relativistic Astrophysics*, Austin, December 2000.

[3] R. Lynds and V. Petrosian, *Bull. Amer. Astron. Soc.* **18**, 1014 (1986); G. Soucail, B. Fort, Y. Mellier, and J. P. Picat, *Astron. Astrophys.* **172**, L14 (1987); G. Soucail *et al.*, *Astron. Astrophys.* **191**, L19 (1988).

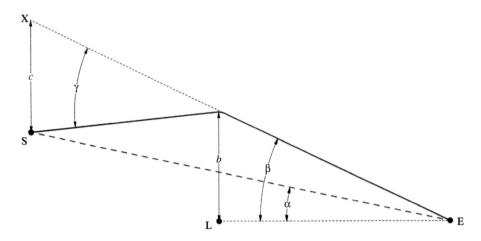

Figure 9.1: Quantities referred to in the derivation of the lens equation, (9.1.5). The bent solid line is the path of a photon from the source S past the lens L to the Earth E. The point X is the apparent position of the source. The transverse distances b and c from the lens to the light path and from the source to its apparent position are greatly exaggerated, as are the angles α, β, and γ. This is drawn for the case $K = 0$; in a Robertson–Walker coordinate system centered on the Earth, the path from the source to the bend would be curved for $|K| \neq 0$.

is curved for $K \neq 0$; is bent by the gravitational field of the lens; and then follows a straight line to the earth. The proper distance b of this path from the lens at its closest approach (which we assume to be much less than the cosmological scale $1/H_0$) is

$$b = \beta \, d_A(EL) \,, \tag{9.1.1}$$

where $d_A(EL)$ is the "angular diameter distance" of the lens as seen from the earth. As discussed in Section 1.4, in general the angular diameter distance $d_A(PQ)$ of a point Q as seen from a point P is the ratio h/θ of a proper length h at Q (normal to the line PQ) to the angle θ subtended at P by this length. It is given by

$$d_A(PQ) = a(t_Q) r_P(Q) \,, \tag{9.1.2}$$

where $r_P(Q)$ is the radial coordinate of Q in the Robertson–Walker coordinate system centered at P (in which rays of light received at P all travel on straight lines), and t_Q is the time the light leaves or arrives at Q. From Figure 9.1, we can see that the line segment SX from the true to the apparent position of the source has a proper length c given by

$$\gamma \, d_A(LS) = c = \left(\beta - \alpha \right) d_A(ES) \,, \tag{9.1.3}$$

where γ is the angle of deflection of the light ray near the lens as seen from the lens, given by general relativity as[4]

$$\gamma = \frac{4MG}{b}, \qquad (9.1.4)$$

and $d_A(LS)$ and $d_A(ES)$ are the angular diameter distances of the source from the lens and from the earth, respectively. From Eqs. (9.1.1), (9.1.3) and (9.1.4) we have

$$\left(\beta - \alpha\right)\beta = \frac{\gamma \, d_A(LS)}{d_A(ES)} \frac{b}{d_A(EL)} = \frac{4MG \, d_A(LS)}{d_A(ES) \, d_A(EL)} \equiv \beta_E^2. \qquad (9.1.5)$$

This is our lens equation. It is a quadratic equation for β, the two solutions giving the directions of the two images into which the point source is split. All the effects of the large scale spacetime geometry and the expansion of the universe are contained in the angular diameter distances $d_A(EL)$, $d_A(ES)$, and $d_A(LS)$. (In general these are independent distances; for $K \neq 0$ we do *not* have $d_A(ES) = d_A(EL) + d_A(LS)$.)

The lens equation (9.1.5) has two roots,

$$\beta_\pm = \frac{\alpha}{2} \pm \sqrt{\frac{\alpha^2}{4} + \beta_E^2}. \qquad (9.1.6)$$

The angle α is not observed, because we cannot remove the lens. If all we measure is the angular separation between the two images of the source, then all we can learn is an upper bound on the mass of the lensing galaxy:

$$|\beta_+ - \beta_-|^2 \geq 4\beta_E^2. \qquad (9.1.7)$$

For instance, if $d_A(EL) = d_A(LS) = 100$ Mpc, $d_A(ES) = 200$ Mpc, and the sources are separated by $1''$, then (remembering that $M_\odot G = 1.475$ km and 1 Mpc $= 3.09 \times 10^{19}$ km), we find that $M \leq 6 \times 10^9$ solar masses.

On the other hand, if the lensing galaxy and the two images are all observed, we can measure the angles $\beta_\pm$ between each image and the lensing galaxy, but with α unknown the best use we can make of Eq. (9.1.6) is to eliminate α by multiplying the roots

$$\beta_+ \beta_- = -\beta_E^2. \qquad (9.1.8)$$

(The minus sign just means that the two images are on opposite sides of the lensing galaxy.) This allows us to calculate the mass M, but if the distances

---

[4]G & C, Section 8.5.

are calculated from measurements of redshifts, then they scale with $1/H_0$, so what we really calculate in this way is MGH_0. (As we saw in Section 1.9, velocity dispersions and angular diameters also only tell us the value of MGH_0.)

In the special case where the lensing galaxy lies directly on the line between the source and the earth, the problem has cylindrical symmetry around this line of sight, and we get an *Einstein ring* rather than a pair of images. The angular radius of the Einstein ring is the value of β given by setting $\alpha = 0$ in Eq. (9.1.5), so it is just the angle β_E, which is why we label it with a subscript E. Einstein rings have been observed for a number of radio sources, starting with the source MG1131+0456.[5]

9.2 Magnification: Strong lensing and microlensing

The various images that are produced by a gravitational lens will not all have the same apparent luminosity. The apparent luminosity is the power received per receiving area, so now we need to consider light paths that end at various points on the telescope receiving area. For this purpose it is helpful to refer positions in the receiving area relative to some fixed point Y, which we can conveniently take on the axis of symmetry of the problem — the line extending from a point source (or a luminous point on an extended source) through the lens and past the earth. (See Figure 9.2.) We can think of the distance h of a point on the telescope mirror from this line as a function of θ, the angle at the source between the light ray to the point on the mirror and the fixed line (in the Robertson–Walker coordinate system centered on the source) from the source through the lens. The fraction of all light that is emitted between polar angles θ and $\theta + d\theta$ (with $\theta \ll 1$) and azimuthal angles ϕ and $\phi + d\phi$ (measured at the source, around the fixed line to the lens) is $\theta\, d\theta\, d\phi/4\pi$, while the receiving area between in the rectangle with height dh and width $h\, d\phi$ is $h\, dh\, d\phi$, so the apparent luminosity is

$$\ell = \left| \frac{L\theta\, d\theta\, d\phi/4\pi}{h\, dh\, d\phi\, (1+z_S)^2} \right| = \frac{L}{4\pi\,(1+z_S)^2} \left| \frac{\theta\, d\theta}{h\, dh} \right| . \qquad (9.2.1)$$

(The factor $(1 + z_S)^{-2}$ accounts for the reduction of energy of individual photons and the reduction in the rate at which photons are emitted from the source.) From Figure 9.2, we see that

$$h = d_A(SE)\chi = d_A(SE)d_A(EL)\alpha/d_A(SL) , \qquad (9.2.2)$$

---

[5]G. H. Chen, C. S. Kochanek, and J. N. Hewitt, *Astrophys. J.* **447**, 62 (1995).

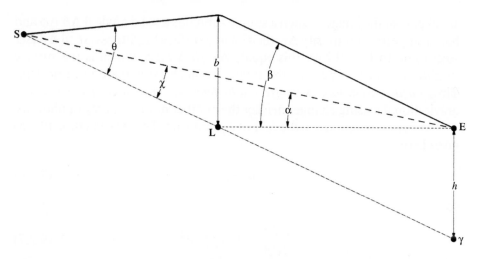

Figure 9.2: Quantities referred to in the derivation of the magnification formula, (9.2.4). The bent solid line is the path of a photon from the source S past the lens L to a point E on a telescope mirror. The point Y is a fixed point near the earth on the line from the source through the lens. The transverse distances b and h from the lens to the light path and from the point Y to the point E where the photon arrives at the telescope mirror are greatly exaggerated, as are the angles α, β, χ, and θ. As in Figure 9.1, this is drawn for the case $K = 0$; in a Robertson–Walker coordinate system centered on the earth, the path from the source to the bend would be curved for $|K| \neq 0$.

where χ is the angle between the line from the source to the point on the mirror and the line from the source to the lens, and

$$\theta = b/d_A(SL) = \beta\, d_A(EL)/d_A(SL) . \qquad (9.2.3)$$

(As in Section 9.1, for any points P and Q, $d_A(PQ)$ is the angular diameter distance of Q as seen from P.) Hence Eq. (9.2.1) gives

$$\ell = \ell_0 \left| \frac{\beta\, d\beta}{\alpha\, d\alpha} \right| , \qquad (9.2.4)$$

where ℓ_0 is the luminosity that would be observed in the absence of the lens:

$$\ell_0 = \frac{L}{4\pi\,(1 + z_S)^2 d_A^2(SE)} . \qquad (9.2.5)$$

(Note that $r_S(E) = r_E(S)$, so Eq. (9.1.2) gives $d_A(SE) = (1 + z_S)d_A(ES)$. According to Eq. (1.4.12), the luminosity distance of the source as seen from the earth is $d_L(S) = (1 + z_S)^2 d_A(ES) = (1 + z_S)d_A(SE)$, which is the distance whose square appears in Eq. (9.2.5).)

Eq. (9.2.4) is a very general result, which applies to lenses and sources of all types. It has a particularly useful consequence for extended sources. If a small part of the image of such a source subtends a solid angle $\beta \, \Delta\beta \, \Delta\phi$ and has an apparent luminosity $\Delta\ell$, then it has a surface brightness $\Delta\ell/\beta \, \Delta\beta \, \Delta\phi$. According to Eq. (9.2.4), this equals $\Delta\ell_0/\alpha \, \Delta\alpha \, \Delta\phi$, which is the surface brightness that this part of the image would have if there were no lens. Thus *surface brightness is unaffected by lensing.* This of course is because gravitational lensing changes neither the number nor the energy of photons.

Let us now specialize to the case of a point lens. The lens equation (9.1.5) gives here

$$\alpha = \beta - \beta_E^2/\beta \, , \tag{9.2.6}$$

so

$$\frac{\alpha \, d\alpha}{\beta \, d\beta} = 1 - \beta_E^4/\beta^4 \tag{9.2.7}$$

The luminosity (9.2.4) is then

$$\ell = \frac{\ell_0}{\left|1 - \beta_E^4/\beta^4\right|} \, . \tag{9.2.8}$$

When the distance[1] a of the lens from the line joining the source and observer is small, we have $\alpha \ll \beta_E$, so the two solutions (9.1.6) for β become

$$\beta_{\pm} \to \pm\beta_E + \tfrac{1}{2}\alpha \, . \tag{9.2.9}$$

In this case both images are amplified by a factor $|\beta_E/2\alpha|$. On the other hand, when the distance a of the lens from the line joining the source and observer is large, we have $\alpha \gg \beta_E$, so the two solutions (9.1.6) for β become

$$\beta_+ \to \alpha \gg \beta_E \, , \qquad -\beta_- \to \beta_E^2/\alpha \ll \beta_E \, . \tag{9.2.10}$$

Under these conditions, the "$-$" ray becomes invisible, while the "$+$" ray has the normal brightness expected without gravitational deflection.

Roughly speaking, therefore, a point mass can only produce noticeable *strong lensing*, with more than one image, if it has $\alpha \leq \beta_E$, i.e., if it lies within a proper distance a_{max} of the line between the source and the observer, given by

$$a_{max} = \beta_E d_A(EL) = \sqrt{\frac{4MG d_A(LS) d_A(EL)}{d_A(ES)}} \, . \tag{9.2.11}$$

---

[1] Here a is a proper distance transverse to the line of sight, and is not related to the Robertson–Walker scale factor, which is distinguished in this section by always writing it with a time argument, as $a(t)$.

Using Eq. (9.1.2), this is

$$
a_{\max} = \sqrt{\frac{4MGr_L(S)r_E(L)a(t_L)}{r_E(S)}} = \sqrt{\frac{4MGr_L(S)r_E(L)a(t_E)}{r_E(S)}} \frac{1}{\sqrt{1+z_L}} ,
$$

(9.2.12)

where t_E is the time the light signal reaches the earth (elsewhere in this book called t_0), while z_L is the redshift of the lens

$$
1 + z_L \equiv a(t_E)/a(t_L) .
$$

(9.2.13)

The radial coordinate $r_L(S)$ of the source in the Robertson–Walker coordinate system centered on the lens can be calculated in terms of the radial coordinates of both the source and lens in the Robertson–Walker coordinate system centered on the earth by using the condition that the three-dimensional *proper* distance between the source and the lens is equal in these two coordinate systems

$$
\int_0^{r_L(S)} \frac{dr}{\sqrt{1 - Kr^2}} = \int_{r_E(L)}^{r_E(S)} \frac{dr}{\sqrt{1 - Kr^2}} .
$$

(9.2.14)

This gives

$$
r_L(S) = r_E(S)\sqrt{1 - Kr_E^2(L)} - r_E(L)\sqrt{1 - Kr_E^2(S)} .
$$

(9.2.15)

If at time t there are $n(t, M)$ objects per proper volume with mass between M and $M + dM$, then (using Eq. (1.1.12)) the total number of objects that can produce a detectable splitting of the image of the source S is

$$
N_S = \int_0^{r_E(S)} \frac{dr_E(L)a(t_L)}{\sqrt{1 - Kr_E^2(L)}} \int_0^\infty \pi a_{\max}^2 \, n(t_L, M) \, dM
$$

$$
= 4\pi Ga^2(t_E) \int_0^{r_E(S)} \frac{dr_E(L)\,\rho(t_L)}{(1 + z_L)^2 \sqrt{1 - Kr_E^2(L)}} \frac{r_L(S)r_E(L)}{r_E(S)} ,
$$

(9.2.16)

where t_L is the time that the light from the source reaches the lens, and $\rho(t)$ is the mass density of possible lenses at time t:

$$
\rho(t) \equiv \int_0^\infty n(t, M)M \, dM .
$$

(9.2.17)

Also, in the integrand $r_E(L)$ and t_L are related as usual by

$$
\int_0^{r_E(L)} \frac{dr}{\sqrt{1 - Kr^2}} = \int_{t_L}^{t_E} \frac{dt}{a(t)} .
$$

(9.2.18)

and $1 + z_L \equiv a(t_E)/a(t_L)$. If there is no evolution of the density of lensing galaxies, then $\rho(t_L) = \rho(t_E)(1 + z_L)^3$, and Eq. (9.2.16) reads

$$N_S = 4\pi \, Ga^2(t_E)\rho(t_E) \int_0^{r_E(S)} \frac{dr_E(L)\,(1 + z_L)}{\sqrt{1 - Kr_E^2(L)}} \left(\frac{r_L(S)r_E(L)}{r_E(S)} \right) \qquad (9.2.19)$$

Where $N_S \ll 1$ we can ignore the possibility of multiple lensing, and interpret N_S as the probability that the image of the source S is appreciably modified by a lens near the line of sight.

For sources with $z_S \ll 1$, Eq. (9.2.19) simplifies to

$$N_S = 4\pi \, Ga^2(t_E)\rho(t_E) \int_0^{r_E(S)} dr_E(L) \left(\frac{\big(r_E(S) - r_E(L)\big)r_E(L)}{r_E(S)} \right)$$

$$= \frac{2\pi \, Gr_E^2(S)a^2(t_E)\rho(t_E)}{3} = \frac{2\pi \, Gz_S^2 \, \rho(t_E)}{3H_0^2} = \frac{z_S^2 \Omega_L}{4} , \qquad (9.2.20)$$

where $\Omega_L \equiv 8\pi\rho(t_E)/3H_0^2$ is the fraction of the critical mass that at present is provided by the lensing objects. We see that even for Ω_L as large as $\Omega_M \approx 0.3$, the probability of strong lensing is small for nearby sources with $z_S \ll 1$, but the probability should become appreciable for sources at cosmological distances.[2]

For z_S of order unity or larger, the strong lensing probability turns out to depend sensitively on the cosmological model. Since the integral (9.2.19) is complicated in general, we will consider here just two extreme cases, of a flat universe dominated either by vacuum energy or by non-relativistic matter.

De Sitter model ($\Omega_\Lambda = 1, \Omega_K = \Omega_M = \Omega_R = 0$)

Here $a(t)/a(t_E) = \exp\left(H_0(t - t(E))\right)$, so

$$r_E(L) = \int_{t_L}^{t_E} \frac{dt}{a(t)} = \frac{1}{a(t_E)H_0} \left[\exp\left(H_0(t_E - t_L)\right) - 1 \right] .$$

We can invert this, and find

$$a(t_L) = \frac{a(t_E)}{1 + a(t_E)H_0r_E(L)}$$

---

[2]W. H. Press and J. E. Gunn, *Astrophys. J.* **185**, 397 (1973).

Then Eq. (9.2.19) becomes

$$N_S = 4\pi\, G\rho(t_E)a^2(t_E)$$
$$\times \int_0^{r_E(S)} \left(\frac{r_E(L)(r_E(S)-r_E(L))}{r_E(S)}\right)\left(1+a(t_E)H_0 r_E(L)\right)dr_E(L)\,.$$

It is convenient to change the variable of integration from $r_E(L)$ to the redshift z_L of the lens, which is given by

$$z_L = \left(\frac{a(t_E)}{a(t_L)}\right) - 1 = a(t_E)H_0 r_E(L)$$

so

$$N_S = \frac{4\pi\, G\rho(t_E)}{H_0^2}\int_0^{z_S}\left(\frac{z_L(z_S-z_L)}{z_S}\right)\left(1+z_L\right)dz_L$$
$$= \frac{\pi\, G\rho(t_E)}{3H_0^2}\left(z_S^3 + 2z_S^2\right) = \frac{\Omega_L}{8}\left(z_S^3 + 2z_S^2\right). \tag{9.2.21}$$

Einstein–de Sitter model ($\Omega_M = 1$, $\Omega_K = \Omega_\Lambda = \Omega_R = 0$)

Here $a(t) = a(t_E)(t/t_E)^{2/3}$ and $1 + z_L = (t_E/t_L)^{2/3}$, so

$$r_E(L) = \int_{t_L}^{t_E}\frac{dt}{a(t)} = 3a^{-1}(t_E)\left(t_E - t_E^{2/3}t_L^{1/3}\right)$$
$$= 3a^{-1}(t_E)t_E\left(1-(1+z_L)^{-1/2}\right)\,.$$

Hence Eq. (9.2.19) here becomes

$$N_S = \frac{18\pi\, G\rho(t_E)t_E^2}{1-(1+z_S)^{-1/2}}\int_0^{z_S} dz_L\,\left(1-(1+z_L)^{-1/2}\right)(1+z_L)^{-1/2}$$
$$\times\left((1+z_L)^{-1/2}-(1+z_S)^{-1/2}\right)$$
$$= \frac{8\pi\, G\rho(t_E)}{H_0^2}\left(-4+\frac{\sqrt{1+z_S}+1}{\sqrt{1+z_S}-1}\ln\left(1+z_S\right)\right)$$
$$= 3\Omega_L\left(-4+\frac{\sqrt{1+z_S}+1}{\sqrt{1+z_S}-1}\ln\left(1+z_S\right)\right)\,. \tag{9.2.22}$$

This grows like z_S^2 for $z_S \ll 1$, in agreement with Eq. (9.2.20), but it flattens out for $z_S \gg 1$, growing only logarithmically with z_S.

We see that the probability of a given source being lensed increases much more rapidly with the redshift of the source for a vacuum-energy dominated model than for a matter-dominated model. For instance, for $z_S = 2$, Eq. (9.2.21) gives $N_S = 4\,\Omega_L$ for a vacuum-dominated model, while Eq. (9.2.22) gives $N_S = 0.30\,\Omega_L$ for a matter-dominated model.

Comparison of lensing theory (under the assumptions $K = 0$ and constant vacuum energy) with the number density of strong gravitational lenses measured as a function of redshift by the Cosmic Lens All Sky Survey[3] has given the result that

$$\Omega_M = 0.31^{+0.27}_{-0.14} \ (68\% \ \text{stat.}) \ ^{+0.12}_{-0.1} \ (\text{syst.}) \ .$$

This is not yet competitive with other measurements of Ω_M, but it shows definite evidence of the effect of dark energy on the number of observed strong lenses.

The assumption that the lens can be approximated as a point mass is appropriate for so-called *microlensing* by stars in our galaxy. Note that for $d_A(EL) \approx 10^3$ pc, $d_A(LS) \approx d_A(ES)$, and $MG \approx 1\,\text{km} \approx 3 \times 10^{-14}$ pc, the effective radius $\sqrt{4MG d_A(EL) d_A(LS)/d_A(ES)}$ of the point mass lens is of order 10^{-5} pc, which is much larger than the size of even a large star. Typically in microlensing observations a change is detected in the luminosity of a distant point source (such as a star outside our galaxy) as the star moves past the line of sight to the source. For a star that moves at 100 km/sec transverse to the line of sight, the time within which the source will be within the effective radius 10^{-5} pc of the star is about a month, which is convenient for monitoring changes in apparent luminosity. Between 13 and 17 microlensing events toward the Large Magellanic Cloud (LMC) have been seen by the MACHO collaboration,[4] and another three toward less crowded fields of the LMC by the EROS collaboration.[5] Of these, only four of the lenses have been identified, and all of them are in the LMC itself, rather than the halo of our galaxy.[6] This suggests though it does not prove that the mass of the halo does not consist of dark objects with the masses of typical stars.[7] Microlensing has also been extensively used in searches for dark stars and extra-solar planets by the Microlensing

---

[3] K.-H. Chae *et al.*, *Phys. Rev. Lett.* **89**, 151301 (2002) [astro-ph/0209602].

[4] C. Alcock *et al.*, *Astrophys. J.* **542**, 281 (2000).

[5] T. Laserre *et al.*, *Astron. and Astrophys.* **355**, L39 (2000).

[6] N. W. Evans, in *IDM 2002: The 4th International Workshop on the Identification of Dark Matter*, eds. N. Spooner and V. Kudryavtsev (World Scientific) [astro-ph/0211302].

[7] For discussion of this issue, see A. F. Zakharov, *Publ. Astron. Obs. Belgrade* **74**, 1 (2002) [astro-ph/0212009]; K. C. Sahu, proceedings of the STSci symposium *Dark Universe: Matter, Energy, and Gravity* [astro-ph/0302325].

Observations in Astrophysics collaboration[8] and the Optical Gravitational Lensing Experiment collaboration.[9]

9.3 Extended lenses

The point source model is not at all valid for clusters of galaxies, and it is only marginally valid for individual galaxies. (Note that for $d_A(LS) \approx d_A(EL) \approx 10^{10}$ pc, $d_A(ES) \approx 2 \times 10^{10}$ pc, and $MG \approx 10^{12}$ km ≈ 0.03 pc, the effective radius $\sqrt{4MGd_A(EL)d_A(LS)/d_A(ES)}$ of the point mass lens is of order 5×10^4 pc, which is close to the size of the spherical halo of our own galaxy.) To deal with light rays that pass through the galaxy, it has become common to approximate the massive halos of these galaxies as spheres of matter in "isothermal" equilibrium, with a ratio of pressure p to mass density ρ given by the mean square value $\langle v^2 \rangle$ of any one component of star velocity, assumed to be equal throughout the lens:

$$p(r) = \rho(r)\langle v^2 \rangle . \tag{9.3.1}$$

We have already seen in Section 1.9 that the solution of the equation of hydrostatic equilibrium with this equation of state behaves at a large proper distance r from the center of the lens as

$$\rho(r) = \frac{2\rho(0)}{9(r/r_0)^2} = \frac{\langle v^2 \rangle}{2\pi G r^2} , \tag{9.3.2}$$

which gives the mass contained within a sphere of radius r as

$$\mathcal{M}(r) = \frac{2 r \langle v^2 \rangle}{G} . \tag{9.3.3}$$

The solution departs from this result for small r, but most of the mass of the lens is at distances from the center where it is a good approximation.

The rate of change of the unit vector $\hat{\mathbf{u}}$ giving the direction of a ray of light in a gravitational field $\mathbf{g} = -\nabla\phi$ is given in the post-Newtonian approximation by[1]

$$\frac{d\hat{\mathbf{u}}}{dt} = -2\,\hat{\mathbf{u}} \times (\hat{\mathbf{u}} \times \mathbf{g}) . \tag{9.3.4}$$

---

[8]P. Yock *et al.*, Proceedings of the ninth Marcel Grossman meeting, Rome, July 2000 [astro-ph/0007317].

[9]A. Udalski *et al.*, *Acta Astron.* **54**, 313 (2004) [astro-ph/0411543].

[1]See, e.g., G&C, Eq. (9.2.7). A factor -2 that was missing on the right-hand side of this equation has been supplied.

For any spherically symmetric non-relativistic distribution of matter, we have

$$\mathbf{g(r)} = -\frac{G\mathcal{M}(r)}{r^3}\mathbf{r} ,\tag{9.3.5}$$

which for the outer parts of an isothermal sphere becomes

$$\mathbf{g(r)} = -\frac{2\langle v^2 \rangle}{r^2}\mathbf{r} ,\tag{9.3.6}$$

For a light ray that passes the center of the lens at a distance b of closest approach, the light ray direction $\hat{\mathbf{u}}$ remains very close to a fixed direction. The small change in direction then has a magnitude

$$\gamma = 4\langle v^2 \rangle \int_{-\infty}^{\infty} \frac{b \, dx}{x^2 + b^2} = 4\pi \langle v^2 \rangle ,\tag{9.3.7}$$

and is independent of b. (We are taking the speed of light to be unity, so that $\langle v^2 \rangle^{1/2}$ is dimensionless, equal to the rms velocity in units of the speed of light.) The direction of this change is toward the center of the lens, so light can arrive at the earth from the source along two different rays, which pass on opposite sides of the lens center, and are separated in angle by $8\pi \langle v^2 \rangle$. The "+" ray, which passes the lens on the side toward the source, makes an angle β_+ at the earth with the ray from the earth to the lens given again by Eq. (9.1.3), while the '−' ray makes an angle with the earth–lens ray β_- (now taken positive), given by replacing α with $-\alpha$ in Eq. (9.1.3), so

$$\gamma d_A(LS) = (\beta_\pm \mp \alpha) \, d_A(ES)\tag{9.3.8}$$

which with Eq. (9.3.7) gives us our lens equation

$$\beta_\pm = \pm \alpha + \beta_E ,\tag{9.3.9}$$

where now

$$\beta_E = \frac{4\pi \langle v^2 \rangle d_A(LS)}{d_A(ES)} .\tag{9.3.10}$$

Here again for a lens on the line of sight from the earth to the source we have $\alpha = 0$, and the two images become an Einstein ring with angular radius β_E. Because we have now defined $\beta_\pm$ to be positive the lens equation gives two images only if $\alpha < \beta_E$, which requires that the proper distance $a = d_A(EL)\alpha$ of the lens from the line between the earth and the source be less than a maximum value

$$a_{\max} = \beta_E d_A(EL) = \frac{4\pi \langle v^2 \rangle d_A(LS)d_A(EL)}{d_A(ES)}\tag{9.3.11}$$

The magnification of the two images is again given by Eq. (9.2.4)

$$\ell = \ell_0 \left| \frac{\beta \, d\beta}{\alpha \, d\alpha} \right| ,$$

but now we must use this with the lens equation (9.3.9), which gives

$$\ell_\pm = \ell_0 \left| \frac{\beta_\pm}{\alpha} \right| = \ell_0 \left| \frac{\beta_\pm}{\beta_\pm - \beta_E} \right| . \tag{9.3.12}$$

The magnification becomes infinite as the lens approaches the line of sight between the earth and the source, in which case $\alpha \to 0$, $\beta_\pm \to \beta_E$, and the image pair becomes an Einstein ring. On the other hand, for a lens approaching the maximum distance (9.3.11) from the earth–source line, we have $\alpha \to \beta_E$, so $\beta_+ \to 2\alpha$ and $\beta_- \to 0$. According to Eq. (9.3.12), the '−' ray disappears in this limit, while the apparent luminosity of the '+' ray is doubled.

The number of possible isothermal spherical lenses that are actually able to split the image of a point source is now given by

$$N_S = \int_0^{r_{E(S)}} \frac{dr_E(L) a(t_L)}{\sqrt{1 - K r_E^2(L)}} \pi a_{\max}^2 \, n(t_L)$$

$$= 16\pi^3 \langle v^2 \rangle^2 \int_0^{r_{E(S)}} dr_E(L) \, \frac{n(t_L) a^3(t_L)}{\sqrt{1 - K r_E^2(L)}} \left(\frac{r_L(S) r_E(L)}{r_E(S)} \right)^2 , \tag{9.3.13}$$

where $n(t)$ is the proper number density of these lenses at time t. For simplicity, we have here taken $\langle v^2 \rangle$ the same for all lenses. If we also now take $K = 0$, so that Eq. (9.2.15) gives $r_S(L) = r_E(S) - r_E(L)$, and also assume no evolution of the population of lensing galaxies, so that $n(t) a^3(t)$ is constant, then

$$N_S = 16\pi^3 \langle v^2 \rangle^2 n(t_E) a(t_E)^3 \int_0^{r_{E(S)}} \left(\frac{\big(r_E(S) - r_E(L)\big) r_E(L)}{r_E(S)} \right)^2 dr_E(L)$$

$$= \frac{8\pi^2 \langle v^2 \rangle^2 n(t_E) a(t_E)^3 r_E^3(S)}{15}$$

$$= \frac{8\pi^2 \langle v^2 \rangle^2 n(t_E)}{15 H_0^3} \mathcal{F}(z_S) , \tag{9.3.14}$$

where

$$\mathcal{F}(z) \equiv \big(a(t_E) H_0 r_E(S) \big)^3 = \left(\int_1^{1+z} \frac{dy}{\sqrt{\Omega_\Lambda + \Omega_M y^3 + \Omega_R y^4}} \right)^3 . \tag{9.3.15}$$

This shows that the chance of a source being lensed depends strongly on Ω_Λ. For instance, for $\Omega_\Lambda = \Omega_R = 0$ and $\Omega_M = 1$ we have $\mathcal{F}(z) = 8(1 - 1/\sqrt{1+z})^3$, which approaches the constant 8 for large z, while for $\Omega_M = \Omega_R = 0$ and $\Omega_\Lambda = 1$ we have $\mathcal{F}(z) = z^3$. The lensing probability for a source at redshift $z = 3$ is 27 times greater if the cosmic energy density is vacuum dominated than if it is matter dominated. From direct observations of galaxies, the constant in Eq. (9.3.15) has been estimated as[2]

$$\frac{8\pi^2 \langle v^2 \rangle^2 n(t_E)}{15 H_0^3} \simeq 0.02 \,. \tag{9.3.16}$$

so for $\Omega_M = \Omega_R = 0$ and $\Omega_\Lambda = 1$ the probability of lensing should become large when $z \approx 3.7$.

Lensing probabilities for quasars of large redshift can provide sensitive limits on Ω_Λ.[3] At one time it was thought on the basis of early surveys that strong lensing statistics ruled out the possibility that a constant vacuum energy made a dominant contribution to producing a spatially flat universe.[4] Since then, many groups have carried out such studies, with corrections for finite core radii, selection effects, etc., to set limits on Ω_Λ in flat cosmologies with $\Omega_\Lambda + \Omega_M = 1$. They find $\Omega_\Lambda < 0.9$,[5] $\Omega_\Lambda < 0.7$[6] $\Omega_\Lambda = 0.64^{+0.15}_{-0.26}$,[7] $\Omega_\Lambda < 0.74$,[8] $\Omega_\Lambda < 0.79$,[9] $\Omega_\Lambda = 0.7^{+0.1}_{-0.2}$,[10] and $\Omega_\Lambda \approx 0.45$–0.75.[11]

The deflection $\Delta\theta$ calculated in the isothermal sphere model of gravitational lenses is relevant only if the light path actually passes through the sphere. For galaxies of relatively small mass, we may have one or both of the deflected rays passing outside the galactic radius R. The deflection of light depends only on the Newtonian gravitational potential, and, as Newton showed, the gravitational potential outside a spherically symmetric distribution of matter is just the same as if all the mass were concentrated at the center of symmetry, so the motion of a light ray outside the galaxy is described by the point mass lens model considered earlier. If the light ray passes far outside the effective radius of the galaxy we have so-called *weak lensing*, which is typically discovered statistically, through a correlation in

[2]E. L. Turner, J. P. Ostriker, and J. R. Gott, *Astrophys. J.* **284**, 1 (1984).

[3]E. L. Turner, *Astrophys. J.* **242**, L135 (1980).

[4]E. L. Turner, *Astrophys. J.* **365**, L43 (1990).

[5]M. Fukugita and E. L. Turner, *Mon. Not. Royal Astron. Soc.* **253**, 99 (1991).

[6]D. Maoz and H-W. Rix, *Astrophys. J.* **416**, 425 (1993)

[7]M. Im, R. E. Griffiths, and K. U. Ratnatunga, *Astrophys. J.* **475**, 457 (1997).

[8]E. E. Falco, C. S. Kochanek, and J. A. Muñoz, *Astrophys. J.* **494**, 47 (1998).

[9]A. R. Cooray, J. M. Quashnock, and M. C. Miller, *Astrophys. J.* **511**, 562 (1999) [astro-ph/9806080]; A. R. *Cooray, Astron. Astrophys.* **342**, 353 (1999) [astro-ph/9811448].

[10]M. Chiba and Y. Yoshii, *Astrophys. J.* **510**, 42 (1999) [astro-ph/9808321].

[11]Y-C. N. Cheng and L. M. Krauss, *Astrophys. J.* **511**, 612 (1999) [astro-ph/9810392].

the orientation of the images of several lensed galaxies, rather than by the study of individual sources. This is the subject of Section 9.5.

9.4 Time delay

In addition to light rays being bent by the gravitational fields of intervening objects, they are also delayed,[1] so that a fluctuation in the distant light source appears at different times on earth in the several lensed images of the source.[2] There are two effects here.

First, there is a *geometrical time delay*, caused by the increased length of the total light path from the source to the earth. Since this arises over very long distances, it can be calculated by idealizing the light path as a geodesic of the Robertson–Walker metric from the source to the point P of closest approach to the lens, where the light path is bent, followed by a similar geodesic from that point to the earth. The time t_E that a light signal that leaves the source at time t_S arrives at the earth is given by

$$\int_{t_S}^{t_E} \frac{dt}{a(t)} = \sigma_{SP} + \sigma_{PE} \tag{9.4.1}$$

where σ_{SP} and σ_{PE} are the proper lengths (the integrals of $(\tilde{g}_{ij}dx^i dx^j)^{1/2}$) along the paths from the source to P and from P to the earth, respectively. The time t_{E0} that the light would arrive at the earth if there were no gravitational deflection is given simply by

$$\int_{t_S}^{t_{E0}} \frac{dt}{a(t)} = \sigma_{SE} , \tag{9.4.2}$$

where σ_{SE} is the proper length along the geodesic from the source to the earth. Hence the geometric time delay (which is always very short compared to a Hubble time) is

$$\Delta t_{\text{geom}} \equiv t_E - t_{E0} = a(t_E)\left(\sigma_{SP} + \sigma_{PE} - \sigma_{SE}\right). \tag{9.4.3}$$

Now, it is easy to calculate proper lengths along geodesics that end at the earth:

$$\sigma_{SE} = \int_0^{r_E(S)} \frac{dr}{\sqrt{1 - Kr^2}} , \qquad \sigma_{PE} = \int_0^{r_E(P)} \frac{dr}{\sqrt{1 - Kr^2}} , \tag{9.4.4}$$

---

[1] The time delay of radar reflections from planets and of radio signals from artificial satellites caused by the gravitational potential of the sun has provided a fourth test of general relativity, as first proposed and measured by I. Shapiro, *Phys. Rev. Lett.* **13**, 789 (1964). For a discussion, see G&C, Section 8.7.

[2] S. Refsdal, *Mon. Not. Roy. Astron. Soc.* **128**, 307 (1964).

where $r_E(S)$ and $r_E(P)$ are the usual radial coordinates (in a Robertson–Walker coordinate system centered on the earth) of the source S and the point P of closest approach of the light to the lens. To calculate σ_{SP} we have to do a little more work. By using Eq. (1.1.17) for the spatial affine connection Γ^i_{jl}, we see that the equation (1.1.26) of a spatial geodesic is

$$\frac{d^2 x^i}{d\sigma^2} + K x^i = 0 \,. \tag{9.4.5}$$

The general solution is

$$x^i(\sigma) = \begin{cases} A^i \cos\sigma + B^i \sin\sigma \,, & K = +1 \\ A^i + B^i \sigma \,, & K = 0 \\ A^i \cosh\sigma + B^i \sinh\sigma \,, & K = -1 \,, \end{cases} \tag{9.4.6}$$

where A^i and B^i are constants characterizing different paths. These constants are subject to a normalization condition

$$1 = \tilde{g}_{ij} \frac{dx^i}{d\sigma} \frac{dx^j}{d\sigma} = \left(\frac{d\mathbf{x}}{d\sigma}\right)^2 + K \frac{(\mathbf{x} \cdot d\mathbf{x}/d\sigma)^2}{1 - K\mathbf{x}^2} \,,$$

which gives

$$(1 - K\mathbf{A}^2)(1 - \mathbf{B}^2) = K(\mathbf{A} \cdot \mathbf{B})^2 \,. \tag{9.4.7}$$

We can determine the constants $\mathbf{A}$ and $\mathbf{B}$ for the geodesic from the source to the point P by requiring that $x^i(0) = r_E(S)\hat{n}_{ES}$ and $x^i(\sigma_{SP}) = r_E(P)\hat{n}_{EP}$, where $\hat{n}_{ES}$ and $\hat{n}_{EP}$ are the unit vectors from the earth to the source or point P, respectively. Imposing the normalization condition (9.4.7) then gives

$$\sigma_{SP}$$
$$= \begin{cases} \cos^{-1}\left[r_E(P)r_E(S)\cos\theta + \sqrt{1 - r_E(P)^2}\sqrt{1 - r_E(S)^2} \right] & K = +1 \\ \sqrt{r_E^2(P) + r_E^2(S) - 2r_E(P)r_E(S)\cos\theta} & K = 0 \\ \cosh^{-1}\left[-r_E(P)r_E(S)\cos\theta + \sqrt{1 + r_E(P)^2}\sqrt{1 + r_E(S)^2} \right] & K = -1 \,, \end{cases} \tag{9.4.8}$$

where θ is the angle between the directions from the earth to the source and to the point P. This result applies for a geodesic triangle with arbitrary angles, but of course we are interested here in the case where θ is very small. In this case, Eq. (9.4.8) becomes

$$\sigma_{SP} = \begin{cases} \sigma_{ES} - \sigma_{EP} + \frac{r_E(P)r_E(S)\theta^2}{2\sin(\sigma_{ES} - \sigma_{EP})} & K = +1 \\ \sigma_{ES} - \sigma_{EP} + \frac{r_E(P)r_E(S)\theta^2}{2(r_E(S) - r_E(P))} & K = 0 \\ \sigma_{ES} - \sigma_{EP} + \frac{r_E(P)r_E(S)\theta^2}{2\sinh(\sigma_{ES} - \sigma_{EP})} & K = -1 \,, \end{cases}$$

In the terms proportional to θ^2 we can ignore the separation of the lens from the point P of closest approach, taking $r_E(P) = r_E(L)$; $\theta = \beta - \alpha$; and (using Eq. (9.2.14)) $\sigma_{ES} - \sigma_{EP} = \sigma_{LS}$, with errors that would introduce terms of higher order in θ. Thus, in all three cases

$$\sigma_{SP} = \sigma_{ES} - \sigma_{EP} + \frac{r_E(L)r_E(S)\theta^2}{2r_L(S)} . \tag{9.4.9}$$

The geometric time delay (9.4.3) is then

$$\Delta t_{\text{geom}} = \frac{a(t_E)r_E(L)r_E(S)\,(\beta - \alpha)^2}{2r_L(S)} . \tag{9.4.10}$$

Using Eq. (9.1.2), we can write this in terms of angular diameter distances and the redshift of the lens:

$$\Delta t_{\text{geom}} = \frac{(1 + z_L)\,d_A(EL)d_A(ES)\,(\beta - \alpha)^2}{2d_A(LS)} . \tag{9.4.11}$$

The details of the lens enter here only in the lens equation, which for each image gives the unobservable angle α in terms of the observable angle β.

There is also a *potential time delay*, caused directly by the motion of the light through the gravitational potential of the lens. The calculation of these time differences is generally done on a case-by-case basis, using detailed models of the lensing galaxy rather than either the point source or isothermal sphere models. Here we will consider only the case of a general spherically symmetric lens. In a coordinate system centered on the lens with line element in the 'standard' form

$$d\tau^2 = B(r)dt^2 - A(r)dr^2 - r^2 d\theta^2 - r^2 \sin^2\theta d\varphi^2 , \tag{9.4.12}$$

the time required for light to travel from a large coordinate distance r to a distance b of closest approach to the lens and then out again to r is[3]

$$2t(r, b) = 2 \int_b^r \left(\frac{A(r)/B(r)}{1 - (b/r)^2 \left(B(r)/B(b) \right)} \right)^{1/2} dr . \tag{9.4.13}$$

For the weak gravitational fields that concern us here, we can use the post-Newtonian approximation,[4] which gives

$$A(r) = 1 + 2r\phi'(r) , \qquad B(r) = 1 + 2\phi(r) , \tag{9.4.14}$$

---

[3]G&C, Eq. (8.7.2)

[4]G&C, Eqs. (9.1.57) and (9.1.60). It is necessary to redefine the radial coordinate in order to put the line element given by the post-Newtonian approximation in the form (9.4.12).

where $\phi(r)$ is the Newtonian gravitational potential. Then to first order in ϕ

$$2t(r,b) = 2\sqrt{r^2 - b^2} + 2\int_b^r \left(1 - \left(\frac{b}{r}\right)^2\right)^{-1/2} \left(r\phi'(r) - \phi(r)\right) dr$$

$$+2\int_b^r \left(1 - \left(\frac{b}{r}\right)^2\right)^{-3/2} \left(\frac{b}{r}\right)^2 \left(\phi(r) - \phi(b)\right) dr . \quad (9.4.15)$$

We will be interested in this in the case $r \to \infty$, because the effects of the finite distance between the lens and the source and earth are already included in the geometric time delay. The second integral in Eq. (9.4.15) converges for $r \to \infty$, but the first integral diverges. Noting that for any kind of lens of mass M, the Newtonian potential at large distances goes as $-MG/r$, we can put the limit of Eq. (9.4.15) for $r \to \infty$ in the form

$$2t(t,b) \to 2r + 2MG\ln(2r) + f(b) , \quad (9.4.16)$$

where

$$f(b) = -2MG\ln b + 2\int_b^\infty \left(1 - \left(\frac{b}{r}\right)^2\right)^{-1/2} \left(r\phi'(r) - \phi(r) - \frac{2MG}{r}\right) dr$$

$$+2\int_b^\infty \left(1 - \left(\frac{b}{r}\right)^2\right)^{-3/2} \left(\frac{b}{r}\right)^2 \left(\phi(r) - \phi(b)\right) dr . \quad (9.4.17)$$

The first two terms in Eq. (9.4.16) diverge for $r \to \infty$, but they are independent of b, so when we calculate the time difference between the arrival of fluctuations in different images of the source we need only to take account of the differences in $f(b)$ for various values of b. But this would give the delay in the time used in the metric (9.4.12), which is the time told by clocks that are far enough from the lens to ignore the lens's gravitational potential, but close enough to ignore cosmological effects. Because of the cosmological redshift, the time delay observed on earth is lengthened by a factor $1 + z_L$, so the potential time delay is

$$\Delta t_{\text{pot}} = (1 + z_L)f(b) . \quad (9.4.18)$$

For instance, if the lens is a point mass then $\phi = -MG/r$, so the first integral in Eq. (9.4.17) vanishes and the second integral is independent of b, so the gravitational potential of the lens produces a difference in time between the arrival of fluctuations in images of the source at angles β_1 and β_2 to the image of the lens, given by

$$\Delta t_{\text{pot}}(\beta_1) - \Delta t_{\text{pot}}(\beta_2) = 2MG(1 + z_L)\ln(b_2/b_1)$$
$$= 2MG(1 + z_L)\ln(\beta_2/\beta_1).$$

Evidently, the potential time delay will generally be of order MG. As shown for instance in Eq. (9.1.5), if $d_A(EL)$, $d_A(ES)$, and $d_A(LS)$ are all of the same order d, then for strong lensing the squared angle $(\beta - \alpha)^2$ is generally of order MG/d, so the geometric time delay (9.4.11) is also of order MG. For a galaxy of 10^{11} solar masses this is 5.7 days. Measurement of this delay can be used to measure the mass of the lens, without needing to know the Hubble constant. We have seen that the measurement of the angles between the source image and the lens can tell us $H_0 MG$, while the measurement of these angles and time delays can tell us MG, so the combination of angular and time-delay measurements can yield a value of H_0. Here is a list of several lenses that have been used in this way:

- QSO 0957+561: Two images separated by 6.1", $z_L = 1.41$, $z_S = 0.36$, with time difference 417 ± 3 days (95%), gave $H_0 = 64 \pm 13$ km sec^{-1} Mpc^{-1},[5] subsequently recalculated[6] as 77^{+29}_{-24} km sec^{-1} Mpc^{-1}.

- B 0218+357: Two images separated by 0.335" and an Einstein ring, $z_S = 0.96$, $z_L = 0.68$, time difference 10.5 ± 0.4 days, gave $H_0 = 69^{+13}_{-19}$ km sec^{-1} Mpc^{-1}.[7]

- PKS 1830-211: Two images separated by 1.0", plus an Einstein ring, $z_L = 0.89$, time difference 26^{+4}_{-5} days (8.6 GHz).[8] Source redshift measured as $2.507 \pm .002$, gave $H_0 = 65^{+15}_{-9}$ km sec^{-1} Mpc^{-1} for $\Omega_M = 1$, $\Omega_\Lambda = 0$, $H_0 = 76^{+19}_{-10}$ km sec^{-1} Mpc^{-1} for $\Omega_M = 0.3$, $\Omega_\Lambda = 0.7$.[9]

- B 1608+656: four lensed images, with three time differences from four to ten weeks, gave $H_0 = 75$ km sec^{-1} Mpc^{-1} $\pm 10\%$.[10]

A table of time delay measurements up to 2003 is given in the review article of Kochanek and Schecter.[11] In 2006, a survey[12] of time delays in ten gravitational lenses (under the assumptions that $\Omega_M = 0.3$ and $\Omega_\Lambda = 0.7$, and with a number of parameters for each lens found by requiring that all give the same Hubble constant) gave $H_0 = 75^{+7}_{-12}$ km sec^{-1} Mpc^{-1}.

[5]T. Kundic *et al.*, *Astrophys. J.* **482**, 75 (1997) [astro-ph/9610162].

[6]G. Bernstein and P. Fischer, *Astron. J.* **118**, 14 (1999).

[7]A. D. Biggs *et al.*, *Mon. Not. Roy. Astron. Soc.* **304**, 349 (1999) [astro-ph/9811282.

[8]J. E. J. Lovell *et al.*, Astrophys. J. **508**, L51 (1998) [astro-ph/9809301]

[9]C. Lidman *et al.*, *Astrophys. J.* **514**, L57 (1999) [astro-ph/9902317].

[10]L. V. E. Koopmans, T. Treu, C. D. Fassnacht, R. D. Blandford, and G. Surpi, *Astrophys. J.* **599**, 70 (2003).

[11]C. S. Kochanek and P. L. Schecter, *Carnegie Observatories Astrophysics Series*, Vol. 2, ed. W.L. Freedman (Cambridge University Press, 2003) [astro-ph/0306040]: Table 1.1.

[12]P. Saha, J. Coles, A. V. Maccio', and L. L. R. Williams, *Astrophys. J.* **660**, L17 (2006) [astro-ph/0607240].

9.5 Weak lensing

For cosmology, the most promising application of gravitational lenses probably lies in surveys of *weak lensing*, the study of the distortion of the images of distant galaxies by numerous small deflections of light as it passes from the galaxies to us through a slightly inhomogeneous distribution of matter. Consider a ray of light from a point source on a distant galaxy at co-moving coordinates $r_S \hat{n}$, where $\hat{n}$ is a unit vector. Whether the intervening lenses are galaxies or clusters of galaxies or concentrations of intergalactic matter, we can think of the total deflection of the light ray as the sum of deflections caused by encounters with point lenses L of mass m_L, at positions $\mathbf{x}_L$. (Since we will be considering only terms of first order in the total deflection, we can think of the lenses as individual particles, even if they are aggregated into extended objects.) For weak lensing, the angle α_L between the directions to the source and the lens is much larger than the "Einstein ring" parameter β_{EL} for lens L, so as we have seen there is only a single image, given by taking the $+$ sign in Eq. (9.1.6), which with $\alpha \ll \beta_E$ gives the amount of the deflection caused by the lens as $\beta_{+L} - \alpha_L = \beta_{EL}^2/\alpha_L$. We decompose the co-moving lens coordinate vector $\mathbf{x}_L$ into components parallel and perpendicular to the light ray:

$$\mathbf{x}_L = r_L \hat{n} + \mathbf{y}_L , \quad r_L \equiv \hat{n} \cdot \mathbf{x}_L , \quad \mathbf{y}_L \equiv \mathbf{x}_L - \hat{n}(\hat{n} \cdot \mathbf{x}_L) . \tag{9.5.1}$$

The lens L deflects the light ray in a direction $-\hat{y}_L$, so the total deflection caused by all the lenses is

$$\Delta \hat{n} = -\sum_L \hat{y}_L \beta_{EL}^2/\alpha_L = -\sum_L \hat{y}_L \frac{4 M_L G \, d_A(LS)}{d_A(ES) d_A(EL) \, \alpha_L} , \tag{9.5.2}$$

where for any points P and Q, $d_A(PQ)$ is the angular diameter distance of Q as seen from P. According to the definition of the angular diameter distance $d_A(EL)$ of the lens as seen from the earth, the angle between the directions to the source and the lens is $\alpha_L = a_L |\mathbf{y}_L|/d_A(EL)$, where a_L is the Robertson–Walker scale factor at the time that the light passes the lens L. Also, it is convenient to use Eq. (9.1.2) to write $d_A(LS) = a(t_S)r(r_L, r_S)$ and $d_A(ES) = a(t_S)r_S$, where t_S is the time the light leaves the source, and $r(r_L, r_S)$ and r_S are the radial coordinates of the source in Robertson–Walker coordinate systems centered on the lens and the earth, respectively. According to Eq. (9.2.15),

$$r(r_L, r_S) = r_S\sqrt{1 - Kr_L^2} - r_L\sqrt{1 - Kr_S^2} . \tag{9.5.3}$$

The deflection is then

$$\Delta \hat{n} = -\sum_L \mathbf{y}_L \frac{4 M_L G\, r(r_L, r_S)}{r_S a_L |\mathbf{y}_L|^2} \,. \tag{9.5.4}$$

This is not in itself a useful result, because we do not generally know where the image of a point on the source would be if there were no lenses. It is more interesting to consider the variation of the deflection with position of the ray origin. Suppose we consider a small change $\boldsymbol{\theta}$ in the undeflected direction $\hat{n}$ to the ray origin, with $\boldsymbol{\theta}$ perpendicular to $\hat{n}$ and $|\boldsymbol{\theta}| \ll 1$. The change in the vector $\mathbf{y}_L$ from the light ray to lens L at the point of closest approach due to the displacement of the source is then

$$\delta \mathbf{y}_L = -\boldsymbol{\theta}(\hat{n} \cdot \mathbf{x}_L) - \hat{n}(\boldsymbol{\theta} \cdot \mathbf{x}_L)$$

Dropping the term proportional to $\hat{n}$, we see that to first order the change in the deflection of the image normal to the line of sight is

$$\Delta \theta_a = \sum_a M_{ab}(r_S, \hat{n})\, \theta_b \,, \tag{9.5.5}$$

with a and b running over two orthogonal directions normal to $\hat{n}$, and $M_{ab}(r_S, \hat{n})$ the 2×2 *shear matrix* for images of a source at a distance r_S:

$$M_{ab}(r_S, \hat{n}) = \sum_L \frac{4 M_L G\, r(r_L, r_S)\, r_L}{r_S a_L} \left(\frac{\delta_{ab}}{|\mathbf{y}_L|^2} - \frac{2 y_{La} y_{Lb}}{|\mathbf{y}_L|^4} \right). \tag{9.5.6}$$

It is conventional to write this matrix as

$$M = \begin{pmatrix} \kappa + \gamma_1 & \gamma_2 \\ \gamma_2 & \kappa - \gamma_1 \end{pmatrix}, \tag{9.5.7}$$

with κ called the *convergence* and γ_i known as the *shear field*.

We may conveniently rewrite the shear matrix in terms of the perturbation to the total Newtonian potential due to the lenses:[1]

$$\delta\phi(\mathbf{x}, t) = -\sum_L \frac{M_L G}{a_L |\mathbf{x} - \mathbf{x}_L(t)|} = -\sum_L \frac{M_L G}{a_L \left((r - r_L(t))^2 + (\mathbf{y} - \mathbf{y}_L(t))^2 \right)^{1/2}},$$

$$\tag{9.5.8}$$

---

[1] We can neglect the effect of lenses at a cosmological distance from the light ray, so the denominator, including the factor a_L, is the proper distance to the lens L. We are here representing the complete perturbation to the mass density as a set of lensing point masses, but since these may be particles as well as galaxies, this introduces no loss of generality.

where $\mathbf{x} = r\hat{n} + \mathbf{y}$ with $\hat{n} \cdot \mathbf{y} = 0$. (We are here explicitly taking into account the fact that the lenses may have a time dependence, as for instance from peculiar motion or cosmological evolution. In Eq. (9.5.6) we are using the abbreviations

$$r_L \equiv r_L(t_{r_L}) , \qquad \mathbf{y}_L \equiv \mathbf{y}_L(t_{r_L}) ,$$

where t_r is the time that a light wave that reaches us at the present moment was at a radial coordinate r, given by $r = \int_0^{t_r} dt/a(t)$.) In particular, the second transverse derivatives of the potential on the light ray are

$$\left[\frac{\partial^2}{\partial y_a \partial y_b} \delta\phi(r\hat{n} + \mathbf{y}, t) \right]_{\mathbf{y}=0} = \sum_L \frac{M_L G}{a_L} \left(\frac{\delta_{ab}}{\left((r - r_L(t))^2 + |\mathbf{y}_L(t)|^2 \right)^{3/2}} \right.$$

$$\left. - \frac{3 y_{La}(t) y_{Lb}(t)}{\left((r - r_L(t))^2 + |\mathbf{y}_L(t)|^2 \right)^{5/2}} \right) ,$$

The functions $\left((r-r_L)^2 + |\mathbf{y}_L|^2 \right)^{-3/2}$ and $\left((r-r_L)^2 + |\mathbf{y}_L|^2 \right)^{-5/2}$ are sharply peaked at $r = r_L$, so the integral of this second derivative times any smooth function $f(r)$ (smooth, in the sense that it varies little when r varies by an amount of order $|\mathbf{y}_L|$) is

$$\int f(r) \, dr \left[\frac{\partial^2}{\partial y_a \partial y_b} \delta\phi(r\hat{n} + \mathbf{y}, t) \right]_{\mathbf{y}=0, \, t=t_r}$$

$$= \sum_L \frac{f(r_L) M_L G}{a_L} \left(\frac{2\delta_{ab}}{|\mathbf{y}_L|^2} - \frac{4 y_{La} y_{Lb}}{|\mathbf{y}_L|^4} \right) .$$

Comparing this with Eq. (9.5.6), we see that the shear matrix M_{ab} for a source at radial coordinate r_S is

$$M_{ab}(r_S, \hat{n}) = 2 \int_0^{r_S} \frac{r(r, r_S) r}{r_S} \left[\frac{\partial^2}{\partial y_a \partial y_b} \delta\phi(r\hat{n} + \mathbf{y}, t) \right]_{\mathbf{y}=0, \, t=t_r} dr . \quad (9.5.9)$$

So measurement of the shear matrix can yield information about perturbations to the gravitational potential by masses al a the line of sight. In particular, by contracting the indices a and b, we see that

$$\kappa = \frac{1}{2} \mathrm{Tr} M = \int_0^{r_S} \frac{r(r, r_S) r}{r_S} \left[\left(\nabla^2 - \frac{\partial^2}{\partial r^2} \right) \delta\phi(r\hat{n} + \mathbf{y}, t) \right]_{\mathbf{y}=0, \, t=t_r} dr .$$

If the lensing is due to a collection of bodies (such as a cluster of galaxies), all at about the same radial coordinate r_L, then $\delta\phi$ falls off rapidly for large values of $|r - r_L|$, so that the factor $r(r, r_S)r$ can be replaced with $r(r_L, r_S)r_L$, and the integral of $\partial^2\delta\phi/\partial^2 r$ can then be set equal to zero. Using the Poisson equation $a^{-2}\nabla^2\delta\phi = 4\pi G \delta\rho_M$, we can write the resulting expression in terms of the density fluctuation $\delta\rho_M$:

$$\kappa = \frac{4\pi G a^2 (t_{r_L}) d_A(LS) d_A(EL)}{d_A(ES)} \int \delta\rho_M(r\hat{n}, t_{r_L}) a(t_{r_L}) dr , \qquad (9.5.10)$$

the integral being taken over a range of r passing through the lens. Hence, assuming that we know the angular diameter distances appearing in Eq. (9.5.10), a measurement of the value of κ for sources seen in some direction can tell us the total mass density of a cluster of lensing masses that lie along that line of sight, projected onto the plane normal to the line of sight.

Now let's consider how the lensing of the images of galaxies can be used to measure the shear matrix. In order to deal with galaxies that are irregularly shaped, it is convenient to describe the shape of a galaxy by a quadrupole matrix:

$$Q_{ab} \equiv \frac{\int d^2\theta \, \mathcal{L}(\boldsymbol{\theta})\theta_a\theta_b}{\int d^2\theta \, \mathcal{L}(\boldsymbol{\theta})} . \qquad (9.5.11)$$

Here the integral is over the transverse displacement $\boldsymbol{\theta}$ of the direction of points on the image of the galaxy from some central point; the indices a, b, etc. run over two orthogonal directions in the plane of this image; and $\mathcal{L}$ is the surface brightness—the apparent luminosity per solid angle—of the image. (In order for the integral in the numerator to converge, it may be necessary to replace $\mathcal{L}$ by some function of $\mathcal{L}$, such as one that equals $\mathcal{L}$ when $\mathcal{L}$ is above some threshold brightness, and otherwise vanishes. This has no effect on the following analysis.) As remarked in Section 9.2, the surface brightness of any point is the same as would be seen in the absence of lensing from the same point on the source, so

$$\mathcal{L}\Big((1 + M)\boldsymbol{\theta}\Big) = \mathcal{L}^s\Big(\boldsymbol{\theta}\Big) , \qquad (9.5.12)$$

where $\mathcal{L}^s$ is the surface brightness of the source in the absence of lensing. Introducing a new variable of integration $\boldsymbol{\theta}'$ in Eq. (9.5.11) by writing $\theta_a = (1 + M)_{ab}\theta'_b$ and using Eq. (9.5.12), we have

$$\int d^2\theta \, \mathcal{L}(\boldsymbol{\theta})\theta_a\theta_b = (1 + M)_{ac}(1 + M)_{bd} \, \mathrm{Det}(1 + M) \int d^2\theta' \, \mathcal{L}^s(\boldsymbol{\theta}')\theta'_c\theta'_d ,$$

and likewise

$$\int d^2\theta \, \mathcal{L}(\boldsymbol{\theta}) = \mathrm{Det}(1 + M) \int d^2\theta' \, \mathcal{L}^s(\boldsymbol{\theta}') ,$$

The determinant cancels in the ratio, so (dropping primes) we have

$$Q_{ab} = (1+M)_{ac}(1+M)_{bd}Q^s_{cd} , \qquad (9.5.13)$$

where Q^s is the quadrupole matrix in the absence of lensing

$$Q^s_{ab} \equiv \frac{\int d^2\theta \; \mathcal{L}^s(\boldsymbol{\theta})\theta_a\theta_b}{\int d^2\theta \; \mathcal{L}^s(\boldsymbol{\theta})} . \qquad (9.5.14)$$

For simplicity of presentation, we will limit ourselves here to the case of lensing that is sufficiently weak so that all elements of M have $|M_{ab}| \ll 1$, the case of greatest interest in cosmology. In this case (9.5.13) becomes

$$Q_{ab} = Q^s_{ab} + M_{ac}Q^s_{cb} + M_{db}Q^s_{ad} . \qquad (9.5.15)$$

Of course, we would not know the unlensed quadrupole matrix for any particular galaxy. But if we have a sample of galaxies in about the same direction and with about the same redshift, so that the shear matrix is the same for all the galaxies in this sample, then we can learn something about this shear matrix by making the reasonable assumption that the orientations of the galaxies are uncorrelated. There are at least two ways that this can be done.

Standard method[2]
In the standard method, one considers the *ellipticity matrices*

$$X_{ab} \equiv Q_{ab}/\mathrm{Tr}Q , \qquad X^s_{ab} \equiv Q^s_{ab}/\mathrm{Tr}Q^s , \qquad (9.5.16)$$

which are normalized to have unit trace. To first order in M, Eq. (9.5.15) gives

$$X_{ab} = X^s_{ab} + M_{ac}X^s_{cd} + M_{bd}X^s_{ad} - 2X^s_{ab}\mathrm{Tr}\Big(MX^s\Big) . \qquad (9.5.17)$$

If the orientation of galaxies is random, then the average of X^s_{ab} over a sufficiently large sample will be proportional to δ_{ab}, and since these matrices are defined to have unit trace, it follows that the coefficient of proportionality must be just $1/2$:

$$\langle X^s_{ab}\rangle = \frac{1}{2}\delta_{ab} . \qquad (9.5.18)$$

To deal with the term in Eq. (9.5.17) that is quadratic in X^s, we also need the average of $X^s_{ab}X^s_{cd}$. The random orientation of the galaxies and the symmetry of the product requires this average to be a linear combination of $\delta_{ab}\delta_{cd}$

---

[2] See, e.g., M. Bartelmann and P. Schneider, *Phys. Rep.* **340**, 291 (2001).

and $\delta_{ac}\delta_{bd} + \delta_{ad}\delta_{bc}$, and since X^s has unit trace, this linear combination must take the form

$$\langle X^s_{ab} X^s_{cd} \rangle = \frac{1}{4} \left[\delta_{ab}\delta_{cd} + \xi \left(\delta_{ac}\delta_{bd} + \delta_{ad}\delta_{bc} - \delta_{ab}\delta_{cd} \right) \right] , \qquad (9.5.19)$$

where, contracting a, b with c, d in Eq. (9.5.19),

$$\xi = \left\langle \mathrm{Tr}\left(X^{s2} \right) \right\rangle - 1/2 . \qquad (9.5.20)$$

(For instance, if the galaxies in our sample are all spheres, then X^s has both eigenvalues equal to $1/2$, and $\xi = 0$, while if they are all extremely prolate ellipsoids then one eigenvalue is unity and the other is zero, and $\xi = 1/2$. More generally, $0 \le \xi \le 1/2$.) Taking the average of Eq. (9.5.17) and using Eqs. (9.5.18) and (9.5.19), we have

$$\langle X_{ab} \rangle = \frac{1}{2}\delta_{ab} + (1 - \xi) \left(M_{ab} - \frac{1}{2}\delta_{ab}\mathrm{Tr}M \right) . \qquad (9.5.21)$$

We do not observe the average of $\mathrm{Tr}(X^{s2})$ that appears in our formula (9.5.20) for ξ, but since ξ appears only in (9.5.21) multiplying a term that is already of first order in M, in calculating $\langle X_{ab} \rangle$ to first order in M we only need ξ to zeroth order, and to this order we can replace X^s in Eq. (9.5.20) with the observed ellipticity matrix X:

$$\xi = \left\langle \mathrm{Tr}\left(X^2 \right) \right\rangle - 1/2 . \qquad (9.5.22)$$

This can be measured from the observed shapes of the galaxies in our sample, and Eq. (9.5.21) can then be used to find the traceless part of the shear matrix from a measurement of the average of the observed ellipticity matrix X.

A word on formalism: In the literature, in place of the real ellipticity matrix, one often encounters a complex ellipticity parameter

$$\chi \equiv X_{11} - X_{22} + 2iX_{12} , \qquad (9.5.23)$$

and in place of Eq. (9.5.21) one finds a formula relating the complex shear parameter $\gamma \equiv \gamma_1 + i\gamma_2$ to averages of functions of the ellipticity parameter[3]

$$\langle \chi \rangle = 2\gamma - 2\langle \chi \, \mathrm{Re}(\chi^*\gamma) \rangle . \qquad (9.5.24)$$

Using Eq. (9.5.19), it is easy to see that the average on the right is

$$\langle \chi \mathrm{Re}(\chi^*\gamma) \rangle = \xi\gamma , \qquad (9.5.25)$$

---

[3] P. Schneider and C. Seitz, *Astron. Astrophys.* **294**, 411 (1995). This is Eq. (4.16) of ref. 2, in the limit $|\gamma_i| \ll 1$ and $|\kappa| \ll 1$.

and that therefore the relation (9.5.24) between χ and γ reads

$$\left\langle \frac{Q_{11} - Q_{22} + 2iQ_{12}}{Q_{11} + Q_{22}} \right\rangle = 2(1 - \xi)(\gamma_1 + i\gamma_2) , \qquad (9.5.26)$$

which is the same as Eq. (9.5.21).

Alternative method
It is possible to avoid the need to measure the parameter ξ for a sample of source galaxies, by considering averages of the quadrupole matrix Q_{ab} itself, rather than the matrix X_{ab} normalized to have unit trace. The random orientation of galaxies in our sample of sources tells us that the average of unlensed quadrupole matrices takes the form

$$\langle Q_{ab}^s \rangle = \frac{1}{2} \bar{Q} \delta_{ab} , \qquad (9.5.27)$$

with $\bar{Q}$ an unknown positive constant. Eq. (9.5.15) then gives

$$\langle Q_{ab} \rangle = \bar{Q} \left[\frac{1}{2} \delta_{ab} + M_{ab} \right] . \qquad (9.5.28)$$

We can eliminate the unknown $\bar{Q}$ by dividing by the average of the trace of the quadrupole moment. Eq. (9.5.28) and its trace give, to first order in M,

$$\frac{\langle Q_{ab} \rangle}{\langle \mathrm{Tr} Q \rangle} = \frac{1}{2} \delta_{ab} + M_{ab} - \frac{1}{2} \delta_{ab} \mathrm{Tr} M \qquad (9.5.29)$$

That is,

$$\gamma_1 + i\gamma_2 = \frac{\langle Q_{11} - Q_{22} + 2iQ_{12} \rangle}{2 \langle Q_{11} + Q_{22} \rangle} . \qquad (9.5.30)$$

Thus we can calculate the traceless part of M_{ab} — that is, the shear field — from a measurement of the average quadrupole tensor, with no need to make a separate measurement of a parameter like ξ.

The difference between the standard and alternative methods is not just one of formalism, for they involve different kinds of averages. In the alternative method galaxies are weighted proportionally to the area of their images, while in the standard method all galaxies are weighted equally. Thus in the alternative method one does not have to worry so much about missing galaxies of small apparent area.

Frequently it is not practical to measure the redshifts of the individual galaxies whose images are distorted by weak lensing. In such cases we must assume that there is some probability distribution $\mathcal{N}(r_S)$ for these galaxies to be at a radial coordinate r_S, with $\int_0^\infty \mathcal{N}(r_S)\, dr_S = 1$. Using the alternative method described above, we can still use Eq. (9.5.29) to calculate the traceless part of the shear matrix from the average observed quadrupole matrix, but now the average of the quadrupole matrix is calculated using all galaxies along a given line of sight, and the shear matrix is replaced with an effective shear matrix

$$M_{ab}(\hat{n}) = \int_0^\infty \mathcal{N}(r_S) \, M_{ab}(r_S, \hat{n}) \, dr_S$$

$$= 2 \int_0^\infty \mathcal{N}(r_S) \, dr_S \int_0^{r_S} \frac{r(r, r_S) \, r}{r_S} \left[\frac{\partial^2}{\partial y_a \partial y_b} \delta\phi(r\hat{n} + \mathbf{y}, t_r) \right]_{\mathbf{y}=0} dr$$

$$= \int_0^\infty \left[\frac{\partial^2}{\partial y_a \partial y_b} \delta\phi(r\hat{n} + \mathbf{y}, t_r) \right]_{\mathbf{y}=0} g(r) \, dr , \qquad (9.5.31)$$

where

$$g(r) \equiv 2 \int_r^\infty \frac{r(r, r_S) \, r}{r_S} \mathcal{N}(r_S) \, dr_S . \qquad (9.5.32)$$

Similarly, using the standard method we can still calculate the traceless part of the effective shear matrix from the mean ellipticity matrix, provided there is no evolution in the shape of galaxies. In this case, Eq. (9.5.21) still applies, but with the shear matrix given by Eq. (9.5.31), and with the average squared ellipticity matrix in our formula (9.5.22) for the ξ parameter calculated by averaging over all galaxies along the line of sight. But if galaxy shapes evolve, then Eq. (9.5.31) must be modified accordingly: Instead of a factor $1 - \xi$ multiplying the shear matrix in Eq. (9.5.21), it is necessary to use Eq. (9.5.22) to calculate a function $\xi(r_S)$ from the average of the squared ellipticity matrix of galaxies at a radial coordinate r_S, and include a factor $1 - \xi(r_S)$ in the integrand of Eq. (9.5.31). This complication does not occur with the alternative method.

Whichever method we use to extract the shear matrix from measurements of weak lensing, we now must face a problem. We would like to use weak lensing measurements to learn about the density perturbation $\delta\rho$, but as shown in Eq. (9.5.10), it is only the convergence κ that is related in any simple way to the density perturbation, while Eqs. (9.5.21) and (9.5.29) show that measurements of weak lensing tell us about the shear field γ_i, not κ. Fortunately, although there is no simple relation between γ_i and κ, there is a simple relation among their gradients.[4] From Eq. (9.5.9), we have

$$\frac{\partial}{\partial \hat{n}_c} M_{ab}(r_S, \hat{n}) = 2 \int_0^{r_S} \frac{r(r, r_S)}{r_S} \left[\frac{\partial^3}{\partial y_a \partial y_b \partial y_c} \delta\phi(r\hat{n} + \mathbf{y}, t) \right]_{\mathbf{y}=0, \, t=t_r} dr .$$

$$(9.5.33)$$

---

[4]N. Kaiser, *Astrophys. J.* **439**, L1 (1995).

Similarly, from Eq. (9.5.31), we have

$$\frac{\partial}{\partial \hat{n}_c} M_{ab}(\hat{n}) = \int_0^\infty \left[\frac{\partial^3}{\partial y_a \partial y_b \partial y_c} \delta\phi(r\hat{n} + \mathbf{y}, t_r) \right]_{\mathbf{y}=0} \frac{g(r)}{r} \, dr \qquad (9.5.34)$$

The important point is that both expressions are completely symmetric among a, b, and c. Thus the divergence of the traceless part of the shear matrix is proportional to the gradient of the trace:

$$\frac{\partial}{\partial \hat{n}_a} \left(M_{ab} - \frac{1}{2} \delta_{ab} \operatorname{Tr} M \right) = \frac{1}{2} \frac{\partial}{\partial \hat{n}_b} \operatorname{Tr} M \,, \qquad (9.5.35)$$

where M can be either $M(r_S, \hat{n})$ or $M(\hat{n})$. In other words,

$$\frac{\partial \kappa}{\partial \hat{n}_1} = \frac{\partial \gamma_1}{\partial \hat{n}_1} + \frac{\partial \gamma_2}{\partial \hat{n}_2} \,, \qquad \frac{\partial \kappa}{\partial \hat{n}_2} = \frac{\partial \gamma_2}{\partial \hat{n}_1} - \frac{\partial \gamma_1}{\partial \hat{n}_2} \,. \qquad (9.5.36)$$

Thus κ can be calculated from the shear field γ_i, up to an $\hat{n}$-independent constant. As mentioned earlier, the values of κ obtained in this way can be used to measure the projected mass density of lensing bodies along the line of sight. Starting in the late 1990s, numerous groups using telescopes of moderate size have detected shear due to weak lensing.[5] These measurements, together with the above analysis, have been used to map out the distribution of all matter, dark as well as baryonic, in various clusters of galaxies, such as the bullet cluster 1E0657-558 described in Section 3.4.

Now let us turn to the application of weak lensing to find the distribution of inhomogeneities in the cosmological mass density. In using weak lensing to study the large scale distribution of matter in the universe, we are not so much interested in the shear matrix in any one direction, as in its distribution around the sky. Fluctuations in the gravitational potential have random sign, so the average over the sky of the shear matrix is zero. It is most useful to consider instead the average of the product of shear matrix elements along two different lines of sight, with "average" understood in the same sense as used in the analysis of fluctuations in the cosmic microwave background: either an average over positions from which the sky might be observed, or an average over the particular series of accidents that lead to a particular distribution of cosmic matter. Neither average corresponds to what we observe, but the difference vanishes for large multipole orders, for

[5]For references to these observations and a brief general review, see D. Munshi and P. Valageas, Roy. Soc. London Trans. Ser. A **363**, 2675 (2001) [astro-ph/0509216]. Comprehensive reviews of weak lensing are given by P. Schneider and M. Bartelmann, ref. 2; A. Refrigier, Ann. Rev. Astron. Astrophys. **41**, 645 (2003); P. Schneider, in *Gravitational Lensing: Strong, Weak, and Micro*, eds. G. Meylan *et al.* (Springer-Verlag, Berlin, 2006): 269 [astro-ph/0509252].

the same reason as explained in Section 2.6 for the multipole coefficients of anisotropies in the cosmic microwave background.

Since we are now considering different directions, it is convenient to write the definition (9.5.7) of the convergence and shear fields more explicitly. If we write $\hat{n}$ in terms of polar and azimuthal angles as

$$\hat{n} = (\sin \vartheta \, \cos \varphi, \sin \vartheta \, \sin \varphi, \cos \vartheta) \,, \tag{9.5.37}$$

and take the two orthogonal directions normal to $\hat{n}$ as

$$\hat{\vartheta} = (\cos \vartheta \, \cos \varphi, \cos \vartheta \, \sin \varphi, - \sin \vartheta), \quad \hat{\varphi} = (- \sin \varphi, \cos \varphi) \tag{9.5.38}$$

then

$$\kappa (\hat{n}) + \gamma_1 (\hat{n}) = \hat{\vartheta}_i \hat{\vartheta}_j \, M_{ij} (\hat{n}) \,, \tag{9.5.39}$$
$$\kappa (\hat{n}) - \gamma_1 (\hat{n}) = \hat{\varphi}_i \hat{\varphi}_j \, M_{ij} (\hat{n}) \,, \tag{9.5.40}$$
$$\gamma_2 (\hat{n}) = \hat{\vartheta}_i \hat{\varphi}_j \, M_{ij} (\hat{n}) \,. \tag{9.5.41}$$

We are here using ϑ and φ instead of the usual θ and ϕ for the polar and azimuthal angles, to avoid confusion with other uses of the symbols θ and ϕ in this section. Also, since we will now be considering different directions $\hat{n}$, we have promoted the two-valued indices a, b, etc. to three-valued indices i, j, etc. Eq. (9.5.31) for the 2×2 shear matrix M_{ab} has accordingly been extended to give a 3×3 matrix

$$M_{ij} (\hat{n}) \equiv \int_0^\infty \left[\frac{\partial^2}{\partial x_i \partial x_j} \delta\phi (\mathbf{x}, t) \right]_{\mathbf{x}=r\hat{n}, \, t=t_r} g(r) \, dr \,. \tag{9.5.42}$$

Equivalently, we can write

$$\gamma (\hat{n}) \equiv \gamma_1 (\hat{n}) + i\gamma_2 (\hat{n}) = e_{+i} (\hat{n}) \, e_{+j} (\hat{n}) \, M_{ij} (\hat{n}) \,, \tag{9.5.43}$$

where, as in Section 7.4, $\mathbf{e}_+ \equiv (\hat{\vartheta} + i\hat{\varphi})/\sqrt{2}$. Also, since $\hat{\vartheta}$ and $\hat{\varphi}$ span the space normal to $\hat{n}$,

$$\kappa (\hat{n}) = \frac{1}{2} \left(\delta_{ij} - \hat{n}_i \hat{n}_j \right) M_{ij} (\hat{n}) \,. \tag{9.5.44}$$

The expansion of the convergence in spherical harmonics follows along almost the same lines as the expansion of microwave background temperature fluctuations described in Sections 2.6 and 7.2. We write the perturbation in the gravitational potential as

$$\delta\phi (\mathbf{x}, t) = \int d^3 q \, \alpha(\mathbf{q}) \delta\phi_q (t) e^{i\mathbf{q}\cdot\mathbf{x}} \,, \tag{9.5.45}$$

where $\alpha(\mathbf{q})$ is the stochastic parameter for scalar perturbations, normalized so that

$$\langle \alpha(\mathbf{q})\alpha^*(\mathbf{q}')\rangle = \delta^3(\mathbf{q} - \mathbf{q}') . \tag{9.5.46}$$

Then Eqs.(9.5.42), (9.5.44), and (9.5.45) give

$$\kappa(\hat{n}) = -\frac{1}{2}\int d^3q\,\alpha(\mathbf{q})\int_0^\infty g(r)\,\delta\phi_q(t_r)\left(q^2 - (\mathbf{q}\cdot\hat{n})^2\right)e^{i\mathbf{q}\cdot\hat{n}r}\,dr .$$

The quantity $-(\mathbf{q}\cdot\hat{n})^2$ can be replaced with $\partial^2/\partial r^2$, and for the exponential we can use the familiar formula

$$e^{i\mathbf{q}\cdot\hat{n}r} = 4\pi\sum_{\ell m} i^\ell\,j_\ell(qr)\,Y_\ell^m(\hat{n})\,Y_\ell^{m*}(\hat{q}) . \tag{9.5.47}$$

This gives the partial wave expansion for the convergence

$$\kappa(\hat{n}) = \sum_{\ell m} a_{\kappa,\ell m}\,Y_\ell^m(\hat{n}) , \tag{9.5.48}$$

with coefficients

$$a_{\kappa,\ell m} = -2\pi i^\ell \int d^3q\,q^2\,\alpha(\mathbf{q})\,Y_\ell^{m*}(\hat{q})$$
$$\times \int_0^\infty g(r)\,\delta\phi_q(t_r)\left(j_\ell(qr) + j_\ell''(qr)\right)dr . \tag{9.5.49}$$

This can be used together with Eq. (9.5.46) to calculate the correlation of the convergence with itself, or with microwave background temperature fluctuations or any other scalar perturbations. In particular, for the correlation of the convergence with itself, we have

$$\langle a_{\kappa,\ell m}\,a_{\kappa,\ell'm'}^*\rangle = \delta_{\ell\ell'}\delta_{mm'}C_{\kappa\kappa,\ell} , \tag{9.5.50}$$

where $C_{\kappa\kappa,\ell}$ is the multipole coefficient

$$C_{\kappa\kappa,\ell} = 4\pi^2\int_0^\infty q^6\,dq\left|\int_0^\infty g(r)\,\delta\phi_q(t_r)\left(j_\ell(qr) + j_\ell''(qr)\right)dr\right|^2 . \tag{9.5.51}$$

On the other hand, comparing Eqs. (9.5.43) and (7.4.6), we see that the shear components $\gamma_1(\hat{n})$ and $\gamma_2(\hat{n})$ involve the polarization $e_+(\hat{n})$ in much the same way as the Stokes parameters $Q(\hat{n})$ and $U(\hat{n})$, so for the same reason as in Section 7.4, it is necessary to expand the complex shear parameter $\gamma_1(\hat{n}) + i\gamma_2(\hat{n})$ in the spin 2 spherical harmonics $\mathcal{Y}_\ell^m(\hat{n})$, defined by Eq. (7.4.16):

$$\mathcal{Y}_\ell^m(\hat{n}) \equiv 2\sqrt{\frac{(\ell-2)!}{(\ell+2)!}}\,e_{+i}(\hat{n})\,e_{+j}(\hat{n})\,\tilde{\nabla}_i\,\tilde{\nabla}_j\,Y_\ell^m(\hat{n}) , \tag{9.5.52}$$

where $\tilde{\nabla}$ is the angular part of the gradient operator:

$$\tilde{\nabla} \equiv \hat{\vartheta} \frac{\partial}{\partial \vartheta} + \frac{\hat{\varphi}}{\sin \vartheta} \frac{\partial}{\partial \varphi} \ . \tag{9.5.53}$$

Using Eqs. (9.5.42), (9.5.43), (9.5.45), and (9.5.47), we have

$$\gamma_1(\hat{n}) + i\gamma_2(\hat{n}) = \sum_{\ell m} a_{\gamma,\ell m} \mathcal{Y}_\ell^m(\hat{n}) \tag{9.5.54}$$

with coefficients

$$a_{\gamma,\ell m} = 2\pi i^\ell \sqrt{\frac{(\ell+2)!}{(\ell-2)!}} \int d^3 q \ \alpha(\mathbf{q}) \ Y_\ell^m(\hat{q})^* \int_0^\infty \delta\phi_q(t_r) j_\ell(qr) g(r) r^{-2} dr \ . \tag{9.5.55}$$

The reality of $\delta\phi(\mathbf{x}, t)$ requires that $\alpha^*(\mathbf{q})\delta\phi_q^*(t) = \alpha(-\mathbf{q})\delta\phi_q(t)$, while $Y_\ell^{m*}(\hat{q}) = (-1)^\ell Y_\ell^{-m}(-\hat{q})$, so $a_{\gamma,\ell m} = a_{\gamma,\ell -m}^*$. That is, the shear produced by perturbations in the mass density is of "E" rather than "B" type, in the sense of Eq. (7.4.25). Thus any observation of B-type shear, for which $a_{\gamma,\ell m} = -a_{\gamma,\ell -m}^*$, would be a sign of lensing caused by something other than density perturbations, such as gravitational waves. Unfortunately, the lensing due to gravitational waves produced in inflation is much too small to be observed.[6]

Using (9.5.46) and the orthonormality of the spin 2 spherical harmonics,

$$\langle a_{\gamma,\ell m} a_{\gamma,\ell'm'}^* \rangle = \delta_{\ell\ell'} \delta_{mm'} C_{\gamma\gamma,\ell} \ , \tag{9.5.56}$$

where the multipole coefficients in shear–shear correlations are

$$C_{\gamma\gamma,\ell} = \frac{4\pi^2(\ell+2)!}{(\ell-2)!} \int_0^\infty q^2 \, dq \left| \int_0^\infty \delta\phi_q(t_r) j_\ell(qr) g(r) r^{-2} \, dr \right|^2 \ . \tag{9.5.57}$$

The Poisson equation $\nabla^2 \delta\phi = 4\pi G a^2 \delta\rho_M$ relates the quantity $\delta\phi_q(t)$ appearing in Eqs. (9.5.51) and (9.5.57) the power spectral function $P(q/a_0)$ of fractional density fluctuations introduced in Section 8.1:

$$q^4 |\delta\phi_q(t)|^2 = (4\pi G)^2 a^4(t) \bar{\rho}_M^2(t) |\delta_{Mq}(t)|^2$$

$$= (4\pi G)^2 a^4(t) \bar{\rho}^2(t) P(q/a_0) \left(\frac{F(t)}{F(t_0)} \right)^2$$

$$= \frac{9\Omega_M^2 H_0^4 a_0^3}{4a^2(t)(2\pi)^3} P(q/a_0) \left(\frac{F(t)}{F(t_0)} \right)^2 \ , \tag{9.5.58}$$

with $F(t)$ the function (8.1.20), and $a_0 \equiv a(t_0)$.

---

[6] S. Dodelson, E. Rozo, and A. Stebbins, *Phys. Rev. Lett.* **91**, 021301 (2003).

It is possible to give less complicated formulas for the multipole coefficients in the most interesting case, of $\ell \gg 1$. Let us first consider the $\kappa\kappa$ correlation. We can evaluate the integral over r in Eq. (9.5.51) by recalling that for $\ell \gg 1$, if qr differs from $\ell + 1/2$ by more than an amount of order $\ell^{-1/3}$, the spherical Bessel function $j_\ell(qr)$ is exponentially small for $qr < \ell + 1/2$ and rapidly oscillating for $qr > \ell + 1/2$, while (unlike the source functions in microwave background anisotropies) all other ingredients in the integral vary slowly with r. For $\ell \gg 1$, we can therefore replace r with ℓ/q in $g(r)$ and t_r:

$$
C_{\kappa\kappa,\ell} \to 4\pi^2 \int_0^\infty dq\, q^6 |\delta\phi_q(t_{\ell/q})|^2\, g^2(\ell/q) \left| \int_0^\infty \left(j_\ell(qr) + j_\ell''(qr) \right) dr \right|^2 ,
$$
(9.5.59)

The integral of $j_\ell(qr)$ gives[7]

$$
\int_0^\infty j_\ell(qr)\, dr = \frac{\sqrt{\pi}}{2q} \frac{\Gamma\left(\frac{\ell+1}{2}\right)}{\Gamma\left(\frac{\ell+2}{2}\right)} \to \frac{1}{q}\sqrt{\frac{\pi}{2\ell}} .
$$
(9.5.60)

On the other hand, since for large ℓ the contributions from $r \ll \ell/q$ and $r \gg \ell/q$ are strongly suppressed, we can drop the integral over r of $j_\ell''(qr)$. Thus Eq. (9.5.58) gives, for $\ell \gg 1$,

$$
\begin{aligned}
C_{\kappa\kappa,\ell} &\to \frac{2\pi^3}{\ell} \int_0^\infty dq\, q^4 |\delta\phi_q(t_{\ell/q})|^2\, g^2(\ell/q) \\
&= 2\pi^3 \ell^4 \int_0^\infty dr\, r^{-5} \left|\delta\phi_{\ell/r}(t_r)\right|^2\, g^2(r) \\
&= \frac{9\Omega_M^2 H_0^4 a_0^3}{16} \int_0^\infty dr\, \frac{g^2(r)}{a^2(t_r)r^2} P(\ell/a_0 r) \left(\frac{F(t_r)}{F(t_0)}\right)^2 .
\end{aligned}
$$
(9.5.61)

The large ℓ limit of the $\gamma\gamma$ multipole coefficient is precisely the same. In this limit, the factor $(\ell+2)!/(\ell-2)!$ in Eq. (9.5.57) becomes ℓ^4, and the factor r^{-2} in the integral over r in Eq. (9.5.57) can be replaced with $(q/\ell)^2$, so for large ℓ the ratio of the integrand of the q-integral in Eq. (9.5.57) to that in Eq. (9.5.51) (dropping $j_\ell''(qr)$) is

$$
\frac{\ell^4 q^2 (q/\ell)^4}{q^6} = 1 .
$$

---

[7] This is a special case of formula 6.561.14 of I. S. Gradshteyn and I. M. Ryzhik, *Table of Integrals, Series, and Products*, ed. A. Jeffrey (Academic Press, New York, 1980).

Hence for large ℓ,

$$C_{\gamma\gamma,\ell} \to C_{\kappa\kappa,\ell} \to \frac{9\Omega_M^2 H_0^4 a_0^3}{16} \int_0^\infty dr \, \frac{g^2(r)}{a^2(t_r)r^2} P(\ell/a_0 r) \left(\frac{F(t_r)}{F(t_0)}\right)^2 .$$

$$(9.5.62)$$

Thus we get just the same information by measuring the shear multipole coefficients at large ℓ as we would get if we could measure the convergence multipole coefficients at the same ℓ.

For a given weighting function $g(r)$ defined by Eq. (9.5.32), observation of the ℓ dependence of $C_{\gamma\gamma,\ell}$ (or $C_{\kappa\kappa,\ell}$) provides a measurement of the k dependence of the power spectral function $P(k)$. It should be noted that if the integral receives its main contribution from radial coordinates r such that vacuum energy has not yet become important by the times t_r, then according to Eq. (8.1.20), $F(t) \propto a(t)$, so Eq. (9.5.62) simplifies further, to

$$C_{\gamma\gamma,\ell} \to C_{\kappa\kappa,\ell} \to \frac{9\Omega_M^2 H_0^4 a_0}{16} \int_0^\infty dr \, \frac{g^2(r)}{r^2} P(\ell/a_0 r) .$$

$$(9.5.63)$$

This is the formula usually quoted.[8]

We can express the correlation functions for the shear components and microwave background temperature fluctuations by taking over results we have already found for the correlation functions of microwave background polarization and temperature fluctuations. Comparing Eq. (9.5.54) with Eqs. (7.4.15) and (7.4.25), and comparing Eq. (9.5.56) with Eq. (7.4.28), we see that we can obtain the correlation functions for the shear components from Eqs. (7.4.76)–(7.4.78) by replacing Q and U with γ_1 and γ_2 and replacing $C_{EE,\ell}$ with $C_{\gamma\gamma,\ell}$ and (since shear is purely of E-type) dropping $C_{BB,\ell}$:

$$\langle \gamma_1(\hat{n}) \, \gamma_1(\hat{n}') \rangle = \frac{1}{2} \sum_\ell C_{\gamma\gamma,\ell} \, \text{Re}\Big(F_\ell(\hat{n}, \hat{n}') + G_\ell(\hat{n}, \hat{n}') \Big) , \qquad (9.5.64)$$

$$\langle \gamma_2(\hat{n}) \, \gamma_2(\hat{n}') \rangle = \frac{1}{2} \sum_\ell C_{\gamma\gamma,\ell} \, \text{Re}\Big(F_\ell(\hat{n}, \hat{n}') - G_\ell(\hat{n}, \hat{n}') \Big) , \qquad (9.5.65)$$

$$\langle \gamma_1(\hat{n}) \, \gamma_2(\hat{n}') \rangle = \frac{1}{2} \sum_\ell C_{\gamma\gamma,\ell} \, \text{Im}\Big(- F_\ell(\hat{n}, \hat{n}') + G_\ell(\hat{n}, \hat{n}') \Big) , \qquad (9.5.66)$$

---

[8]D. J. Bacon, A. R. Refregier and R. S. Ellis, *Mon. Not. Roy. Astron. Soc.* **318**, 625 (2000). Correlation functions for elements of the shear matrix were studied by N. Kaiser, *Astrophys. J.* **498**, 26 (1998).

where F_ℓ and G_ℓ are functions given by Eqs. (7.4.87) and (7.4.88):

$$F_\ell(\hat{n}, \hat{n}') = \frac{2\ell+1}{4\pi} D^{(\ell)}_{-2,-2}\left(S^{-1}(\hat{n})S(\hat{n}')\right), \qquad (9.5.67)$$

$$G_\ell(\hat{n}, \hat{n}') = (-1)^\ell \frac{2\ell+1}{4\pi} D^{(\ell)}_{-2,-2}\left(S^{-1}(\hat{n})S(-\hat{n}')\right), \qquad (9.5.68)$$

with $D^{(\ell)}$ the irreducible unitary representation of the rotation group for angular momentum ℓ, and $S(\hat{n})$ the standard rotation (7.4.12) that takes the three-direction with $\vartheta = \varphi = 0$ into the direction $\hat{n}$. Also, if we write the microwave background temperature fluctuation as in Eq. (2.6.1)

$$\Delta T(\hat{n}) = \sum_{\ell m} a_{T,\ell m} Y_\ell^m(\hat{n}),$$

and use rotational invariance to define

$$\langle a_{T,\ell m} a^*_{\gamma,\ell' m'} \rangle = \delta_{\ell \ell'} \delta_{m m'} C_{T\gamma,\ell}, \qquad (9.5.69)$$

then we can obtain the correlation functions for the shear components with microwave background temperature fluctuations from Eqs. (7.4.81) and (7.4.82) by again replacing Q and U with γ_1 and γ_2 and replacing $C_{TE,\ell}$ with $-C_{T\gamma,\ell}$:

$$\langle \Delta T(\hat{n}) \gamma_1(\hat{n}') \rangle = \sum_\ell C_{T\gamma,\ell} \, \mathrm{Re} H_\ell(\hat{n}, \hat{n}'), \qquad (9.5.70)$$

$$\langle \Delta T(\hat{n}) \gamma_2(\hat{n}') \rangle = -\sum_\ell C_{T\gamma,\ell} \, \mathrm{Im} H_\ell(\hat{n}, \hat{n}'), \qquad (9.5.71)$$

where H_ℓ is the function (7.4.89):

$$H_\ell(\hat{n}, \hat{n}') = \frac{2\ell+1}{4\pi} D^{(\ell)}_{0,-2}\left(S^{-1}(\hat{n})S(\hat{n}')\right). \qquad (9.5.72)$$

For comparison, we mention also that the convergence correlation function is given by the analog of the temperature correlation function (2.6.4):

$$\langle \kappa(\hat{n})\kappa(\hat{n}') \rangle = \sum_\ell \frac{2\ell+1}{4\pi} C_{\kappa\kappa,\ell} P_\ell(\hat{n} \cdot \hat{n}'). \qquad (9.5.73)$$

The measurement of $C_{\gamma\gamma,\ell}$ or (with the aid of Eq. (9.5.36) or (9.5.62)) of $C_{\kappa\kappa,\ell}$ can be used not only to learn about the power spectral function $P(k)$, but also to put constraints on the cosmological parameters that enter in

Eq. (9.5.32) for $g(r)$. So far, shear measurements from the Canada–France–Hawaii Telescope Wide Synoptic Legacy Survey have been used[9] to set a value for the parameter σ_8 discussed in Section 8.1: under the assumption that $\Omega_M = 0.3$, it is found that $\sigma_8 = 0.85 \pm 0.06$. Although this parameter is also measured in studies of large scale structure, the weak lensing measurement has the advantage of not depending on the use of luminous sources as a tracer of dark matter. The same group also finds from cosmic shear data alone that the ratio $w \equiv \bar{p}/\bar{\rho}$, if assumed constant, is less than -0.8 at 68% confidence. Another group,[10] combining shear measurements from the CTIO lensing survey with cosmic microwave background data and measurements of Type Ia supernova redshift and luminosities, has found that $\sigma_8 = 0.81^{+0.15}_{0.21}$, and that (if constant) $w = -0.89^{+0.16}_{-0.21}$, with 95% confidence. The application of weak lensing surveys to cosmology has really just begun.

9.6 Cosmic strings

The spontaneous breakdown of symmetries in the early universe can produce linear discontinuities in fields, known as cosmic strings.[1] Cosmic strings are also common in modern string theories.[2] Unless we are unlucky enough to have a cosmic string slice through the solar system, the only way that a cosmic string can be discovered seems to be through its action as a gravitational lens.[3]

At a sufficient distance r from a string, the gravitational field becomes Newtonian. For a long straight string the solution of Laplace's equation with cylindrical symmetry has a gravitational potential of the form

$$\phi = -2G\sigma \ln r + C , \tag{9.6.1}$$

where σ and C are constants of integration. If the string is non-relativistic, then σ is the string's mass per length. At a sufficient distance from the string the post-Newtonian approximation applies, and the direction $\hat{u}$ of a light

---

[9]H. Hoekstra *et al.*, *Astrophys. J.* **647**, 116 (2006) [astro-ph/0511089].

[10]M. Jarvis, B. Jain, G. Bernstein and D. Dolney, *Astrophys. J.* **644**, 71 (2006) [astro-ph/0502243].

[1]T. W. B. Kibble, *J. Phys. A* **9**, 1387 (1976); A. Vilenkin and E. P. S. Shellard, *Cosmic Strings and Other Topological Defects* (Cambridge University press, Cambridge, 1994).

[2]E. Witten, *Phys. Lett. B* **153**, 243 (1985); E. J. Copeland, R. C. Myers, and J. Polchinski, *J. High Energy Phys.* **0406**, 013 (2005) [hep-th/0312067].

[3]The relativistic calculation of lensing by a cosmic string is due to J. R. Gott, *Astrophys. J.* **288**, 422 (1985). The anisotropy in the cosmic microwave background due to cosmic strings was calculated by N. Kaiser and A. Stebbins, Nature **310**, 391 (1984).

ray is governed by the equation[4]

$$\frac{d\hat{u}}{dt} = -2\hat{u} \times (\hat{u} \times \nabla\phi) \,. \tag{9.6.2}$$

As long as the deflection of the light ray is small, it can be calculated as

$$\delta\hat{u} = -2\hat{u}_0 \times \left(\hat{u}_0 \times \int dt\nabla\phi\right) \,, \tag{9.6.3}$$

where $\hat{u}_0$ is the photon's initial direction. An elementary integration shows that the light ray is deflected toward the string in a direction perpendicular to both the string and the light ray, by an angle

$$|\delta\hat{u}| = \frac{4\pi\sigma G}{\sin\theta} \,, \tag{9.6.4}$$

where θ is the angle between the directions of the string and the light ray. Because the deflection is toward the string, it is in the opposite direction for rays passing the string on opposite sides, so that the image of a source behind the string is split into two parts. It is noteworthy that neither the direction nor the magnitude of this deflection depends on the distance of the light ray from the string, as long as the distance is large enough to allow the use of the non-relativistic formulas (9.6.1) and (9.6.2).

In 2003, general interest in cosmic strings was heightened by the discovery of what seemed at first to be a plausible candidate for lensing by a cosmic string. A pair of images of elliptical galaxies separated by about 1.8 arcseconds was found to have substantially the same redshift, $z = 0.46$, and the same spectra. The images were not distorted in the way that would be expected for lensing of a single galaxy by a more-or-less spherical source, but are consistent with lensing by a cosmic string.[5] Subsequently an excess of gravitationally lensed objects was found in the neighborhood of this string candidate,[6] lending further support to the view that this was the image of a single galaxy lensed by a cosmic string. But in 2006 this interpretation had to be abandoned, when observations at the Hubble Space Telescope revealed that this was in fact a pair of interacting elliptical galaxies, not the result of any sort of lensing.[7] This episode illustrates how difficult it will be to detect cosmic strings through their lensing action.

---

[4]See footnote 1 of Section 9.3.
[5]M. V. Sazhin *et al.*, *Mon. Not. Roy. Astron. Soc.* **343**, 353 (2003) [astro-ph/0302547].
[6]M. V. Sazhin *et al.*, astro-ph/0406516.
[7]M. V. Sazhin *et al.*, *Mon. Not. Roy. Astron. Soc.* **376**, 1731 (2007) [astro-ph/0611744].

10

Inflation as the Origin of Cosmological Fluctuations

The most exciting aspect of the inflationary cosmological theories described in Chapter 4 is that they provide a natural quantum mechanical mechanism for the origin of the cosmological fluctuations observed in the cosmic microwave background and in the large scale structure of matter,[1] and that may in the future be observed in gravitational waves.[2] We have seen in Chapter 6 that the magnitude and wavelength dependence of adiabatic scalar and tensor fluctuations depend on initial conditions only through the quantities $\mathcal{R}_q^o$ and $\mathcal{D}_q^o$, respectively. As given by Eq. (5.4.24), $\mathcal{R}_q^o$ is the value of the gauge-invariant quantity $\mathcal{R}_q(t) \equiv A_q(t)/2 + H(t)\delta u_q(t)$ outside the horizon, that is for $q/a(t) \ll H(t)$, where for adiabatic fluctuations $\mathcal{R}_q$ is time-independent. Likewise, $\mathcal{D}_q^o$ is the time-independent value of the gravitational wave amplitude $\mathcal{D}_q(t)$ for $q/a(t) \ll H(t)$. During the matter- or radiation-dominated eras $a(t)$ increased like $t^{2/3}$ or $t^{1/2}$, respectively, while $H(t) \equiv \dot{a}(t)/a(t)$ decreased like $1/t$, so any wavelength will be found outside the horizon if we go back early enough in one or the other of these eras. But during the period of inflation that is supposed to precede the radiation-dominated era, $H(t)$ was roughly (perhaps very roughly) constant, while $a(t)$ increased more-or-less exponentially, so even if a perturbation was outside the horizon at the end of inflation it would inevitably have been found deep inside the horizon sufficiently early in the era of inflation. At these very early times fields oscillated much more quickly than the universe expanded, and their quantum fluctuations therefore were essentially just what they would be in ordinary Minkowski spacetime. In this chapter we will follow adiabatic scalar fluctuations and tensor fluctuations from this very early era, through the epochs when fluctuations of various wavelengths exited the horizon, to the time when $\mathcal{R}_q(t)$ and $\mathcal{D}_q(t)$ reached the constant values $\mathcal{R}_q^o$ and $\mathcal{D}_q^o$ that are measured in the cosmic microwave background and in the large scale structure of matter, and that may some day be measured through the direct detection of cosmological gravitational radiation.

---

[1] S. V. Mukhanov and G. V. Chibisov, *Sov. Phys. JETP Lett.* **33**, 532 (1981); S. Hawking, *Phys. Lett.* **115B**, 295 (1982); A. A. Starobinsky, *Phys. Lett.* **117B**, 175 (1982); A. Guth and S.-Y. Pi, *Phys. Rev. Lett.* **49**, 1110 (1982); J. M. Bardeen, P. J. Steinhardt, and M. S. Turner, *Phys. Rev. D* **28**, 679 (1983); W. Fischler, B. Ratra, and L. Susskind, *Nucl. Phys. B* **259**, 730 (1985).

[2] The cosmological generation of gravitational waves was considered by L. P. Grishchuk, *Sov. Phys. JETP* **40**, 409 (1974); A. A. Starobinsky, *Sov. Phys. JETP Lett.* **30**, 682 (1979), and calculated in the context of inflation by V. A. Rubakov, M. V. Sazhin, and A. V. Veryaskin, Phys. Lett. **115B**, 189 (1982).

10.1 Scalar fluctuations during inflation

We will first consider the simplest model of inflation, with a single real "inflaton" field $\varphi(x)$, and an action (B.63):

$$I_\varphi = \int d^4x \; \sqrt{-\text{Det}g} \left[-\frac{1}{2} g^{\mu\nu} \frac{\partial \varphi}{\partial x^\mu} \frac{\partial \varphi}{\partial x^\nu} - V(\varphi) \right] , \qquad (10.1.1)$$

involving an arbitrary real potential $V(\varphi)$. This is not the only possibility; we will take up the question of its plausibility at the end of this section.

In line with the observed isotropy and homogeneity of the universe on the average, we take the scalar field as an unperturbed term $\bar{\varphi}(t)$ that depends only on time, plus a small perturbation $\delta\varphi(\mathbf{x}, t)$:

$$\varphi(\mathbf{x}, t) = \bar{\varphi}(t) + \delta\varphi(\mathbf{x}, t) . \qquad (10.1.2)$$

Similarly, as in Chapters 5–7, the metric is given by the unperturbed Robertson–Walker metric $\bar{g}_{\mu\nu}(t)$ (with $K = 0$) plus a small perturbation $h_{\mu\nu}(\mathbf{x}, t)$

$$g_{\mu\nu}(\mathbf{x}, t) = \bar{g}_{\mu\nu}(t) + h_{\mu\nu}(\mathbf{x}, t) . \qquad (10.1.3)$$

The energy momentum tensor of the scalar field is shown in Appendix B to take the perfect fluid form, with an energy density, pressure, and velocity four-vector given by Eqs. (B.66)–(B.68) as

$$\rho = -\frac{1}{2} g^{\mu\nu} \frac{\partial \varphi}{\partial x^\mu} \frac{\partial \varphi}{\partial x^\nu} + V(\varphi)$$

$$p = -\frac{1}{2} g^{\mu\nu} \frac{\partial \varphi}{\partial x^\mu} \frac{\partial \varphi}{\partial x^\nu} - V(\varphi)$$

$$u^\mu = -\left[-g^{\rho\sigma} \frac{\partial \varphi}{\partial x^\rho} \frac{\partial \varphi}{\partial x^\sigma} \right]^{-1/2} g^{\mu\tau} \frac{\partial \varphi}{\partial x^\tau} .$$

The energy-density and pressure of the unperturbed scalar field are then

$$\bar{\rho} = \frac{1}{2} \dot{\bar{\varphi}}^2 + V(\bar{\varphi}) , \qquad \bar{p} = \frac{1}{2} \dot{\bar{\varphi}}^2 - V(\bar{\varphi}) , \qquad (10.1.4)$$

while the unperturbed velocity four-vector has components

$$\bar{u}^i = 0 , \qquad \bar{u}^0 = 1 . \qquad (10.1.5)$$

The Friedmann equation (with zero spatial curvature) is here

$$H^2 = \frac{8\pi G}{3} \left(\frac{1}{2} \dot{\bar{\varphi}}^2 + V(\bar{\varphi}) \right) . \qquad (10.1.6)$$

and the energy conservation condition $\dot{\bar{\rho}} = -3H(\bar{\rho} + \bar{p})$ yields the field equation of the unperturbed scalar field

$$\ddot{\bar{\varphi}} + 3H\dot{\bar{\varphi}} + V'(\bar{\varphi}) = 0 , \qquad (10.1.7)$$

where as usual $H(t) \equiv \dot{a}(t)/a(t)$.

The perturbation $\delta g_{\mu\nu} \equiv h_{\mu\nu}$ to the metric will be taken in the Newtonian gauge form (5.3.18):

$$h_{00} = -2\Psi , \qquad h_{0i} = 0 , \qquad h_{ij} = -2a^2\delta_{ij}\Psi . \qquad (10.1.8)$$

(The scalar anisotropic inertia term π^S, as well as π^V and π^T, vanishes in scalar field theories, so according to Eq. (5.3.20) the other scalar gravitational perturbation Φ here equals Ψ.) The perturbations to the pressure, energy density, and velocity three-vector are given by the terms in Eqs. (B.66)–(B.68) that are of first order in perturbations:

$$\delta\rho = \dot{\bar{\varphi}}\delta\dot{\varphi} + V'(\bar{\varphi})\delta\varphi - \Psi\dot{\bar{\varphi}}^2 , \qquad (10.1.9)$$

$$\delta p = \dot{\bar{\varphi}}\delta\dot{\varphi} - V'(\bar{\varphi})\delta\varphi - \Psi\dot{\bar{\varphi}}^2 , \qquad (10.1.10)$$

$$\delta u_i = \frac{\partial\delta u}{\partial x^i} \quad \text{where} \quad \delta u = -\frac{\delta\varphi}{\dot{\bar{\varphi}}} . \qquad (10.1.11)$$

The field equations may be taken as the Einstein equation Eq. (5.3.21) and the energy conservation condition (5.3.24), which here take the form

$$\dot{\Psi} + H\Psi = 4\pi G \dot{\bar{\varphi}} \, \delta\varphi , \qquad (10.1.12)$$

$$\delta\ddot{\varphi} + 3H\delta\dot{\varphi} + \frac{\partial^2 V(\bar{\varphi})}{\partial\bar{\varphi}^2}\delta\varphi - \left(\frac{\nabla^2}{a^2}\right)\delta\varphi = -2\Psi\frac{\partial V(\bar{\varphi})}{\partial\bar{\varphi}} + 4\dot{\Psi}\dot{\bar{\varphi}}, \qquad (10.1.13)$$

while the constraint (5.3.26) is here

$$\left(\dot{H} - \frac{\nabla^2}{a^2}\right)\Psi = 4\pi G\left(-\dot{\bar{\varphi}}\delta\dot{\varphi} + \ddot{\bar{\varphi}}\delta\varphi\right) . \qquad (10.1.14)$$

(In deriving Eq. (10.1.14), we make use of the convenient relation $\dot{H} = -4\pi G\dot{\bar{\varphi}}^2$, which follows from Eqs. (10.1.6) and (10.1.7).) The field equation (5.3.20) has already been accounted for by setting $\Phi = \Psi$, while the remaining field equations (5.3.19) and (5.3.22) and the momentum conservation equation (5.3.23) just repeat the information contained in Eqs. (10.1.12)–(10.1.14).

Let us first consider the plane wave solutions of Eqs. (10.1.12)–(10.1.14), in which $\delta\varphi$ and Ψ are of the form $\exp(i\mathbf{q} \cdot \mathbf{x})\delta\varphi_q(t)$ and $\exp(i\mathbf{q} \cdot \mathbf{x})\Psi_q(t)$, respectively. We will return later to the issue of how to put these solutions together in forming the perturbations $\delta\varphi(\mathbf{x}, t)$ and $\Psi(\mathbf{x}, t)$. The time dependence of the plane wave solutions is given by Eqs. (10.1.12) and (10.1.13), with $-\nabla^2$ replaced with $q^2 \equiv \mathbf{q}^2$:

$$\dot{\Psi}_q + H\Psi_q = 4\pi G\,\dot{\bar{\varphi}}\,\delta\varphi_q \,, \tag{10.1.15}$$

$$\delta\ddot{\varphi}_q + 3H\delta\dot{\varphi}_q + \frac{\partial^2 V(\bar{\varphi})}{\partial\bar{\varphi}^2}\delta\varphi_q + \left(\frac{q^2}{a^2}\right)\delta\varphi_q$$

$$= -2\,\Psi_q\,\frac{\partial V(\bar{\varphi})}{\partial\bar{\varphi}} + 4\,\dot{\Psi}_q\,\dot{\bar{\varphi}} \,, \tag{10.1.16}$$

while the constraint (10.1.14) now reads

$$\left(\dot{H} + \frac{q^2}{a^2}\right)\Psi_q = 4\pi G\left(-\dot{\bar{\varphi}}\,\delta\dot{\varphi}_q + \ddot{\bar{\varphi}}\,\delta\varphi_q\right) . \tag{10.1.17}$$

At sufficiently early times q/a will be much larger than H or $\partial^2 V(\bar{\varphi})/\partial\bar{\varphi}^2$, so we can look for WKB solutions, for which time derivatives of fields yield factors of order q/a, of the form

$$\delta\varphi_q(t) \rightarrow f(t)\,\exp\left(-iq\int_{t_*}^{t}\frac{dt'}{a(t')}\right),$$

$$\Psi_q(t) \rightarrow g(t)\,\exp\left(-iq\int_{t_*}^{t}\frac{dt'}{a(t')}\right), \tag{10.1.18}$$

where $f(t)$ and $g(t)$ vary much more slowly than the argument of the exponential, and t_* is arbitrary. The exponential factor is chosen so that the terms in Eq. (10.1.16) of second order in q/a should cancel. Eqs. (10.1.15) and (10.1.17) are both satisfied to leading order in q/a if we take

$$g/f = 4i\pi G\dot{\bar{\varphi}}a/q \,. \tag{10.1.19}$$

The terms in Eq. (10.1.16) of first order in q/a then give $2\dot{f} + 2Hf = 0$, so $f \propto 1/a$. For reasons that will soon become clear, we will choose the common constant factor in both f and g so that $f = 1/(2\pi)^{3/2}\sqrt{2q}\,a$. With this normalization, $\delta\varphi_q(t)$ and $\Psi_q(t)$ are defined as the solution of Eqs. (10.1.15)–(10.1.17) that satisfies the initial condition, that for $a(t) \rightarrow 0$,

$$\delta\varphi_q(t) \rightarrow \frac{1}{(2\pi)^{3/2}a(t)\sqrt{2q}}\,\exp\left(-iq\int_{t_*}^{t}\frac{dt'}{a(t')}\right), \tag{10.1.20}$$

$$\Psi_q(t) \rightarrow \frac{4i\pi G\dot{\bar{\varphi}}(t)}{(2\pi)^{3/2}\sqrt{2q^3}}\,\exp\left(-iq\int_{t_*}^{t}\frac{dt'}{a(t')}\right) . \tag{10.1.21}$$

The complex conjugate $\delta\varphi_q^*(t)$, $\Psi_q^*(t)$ is another independent solution, and since the system of equations (10.1.15)–(10.1.17) is second order, these are the only solutions.

In general, the fields $\delta\varphi(\mathbf{x}, t)$ and $\Psi(\mathbf{x}, t)$ satisfying Eqs. (10.1.12)–(10.1.14) can be written as superpositions of these two solutions, which the reality of the fields requires to take the form

$$\delta\varphi(\mathbf{x}, t) = \int d^3q \left[\delta\varphi_q(t) e^{i\mathbf{q}\cdot\mathbf{x}} \alpha(\mathbf{q}) + \delta\varphi_q^*(t) e^{-i\mathbf{q}\cdot\mathbf{x}} \alpha^*(\mathbf{q}) \right] \quad (10.1.22)$$

$$\Psi(\mathbf{x}, t) = \int d^3q \left[\Psi_q(t) e^{i\mathbf{q}\cdot\mathbf{x}} \alpha(\mathbf{q}) + \Psi_q^*(t) e^{-i\mathbf{q}\cdot\mathbf{x}} \alpha^*(\mathbf{q}) \right] . \quad (10.1.23)$$

Now we must say something about the coefficients $\alpha(\mathbf{q})$ and $\alpha^*(\mathbf{q})$. For this purpose, we use the canonical commutation relations of the fields. The interaction of the scalar field with gravitation makes these commutation relations rather complicated, but they become simple at very early times. For any given q, we can find a time sufficiently early so that the expansion rate H is negligible compared with q/a. For such early times, both $\dot{\bar{\varphi}}$ and $\ddot{\bar{\varphi}}$ on the right-hand side of Eq. (10.1.17) become negligible, so Ψ_q becomes negligible, as can also be seen by noting from Eqs. (10.1.18) and (10.1.19) that the ratio of Ψ_q to $\delta\varphi_q$ vanishes like $a(t)$ for $a(t) \to 0$. At such early times, as far as $\alpha(\mathbf{q})$ and $\alpha^*(\mathbf{q})$ are concerned, we can find the canonical commutation relations by using Eq. (10.1.1) with $g_{\mu\nu}$ taken as the unperturbed Robertson–Walker metric:

$$I_\varphi = \int d^4x \, \mathcal{L}_\varphi , \quad \mathcal{L}_\varphi = a^3(t) \left[\frac{1}{2} \left(\frac{\partial\varphi}{\partial t} \right)^2 - \frac{1}{2a^2(t)} \frac{\partial\varphi}{\partial x^i} \frac{\partial\varphi}{\partial x^i} - V(\varphi) \right] ,$$

The canonical conjugate to the field φ is then $\pi = \partial\mathcal{L}_\varphi / \partial\dot{\varphi} = a^3\dot{\varphi}$. Since the unperturbed fields are c-numbers, the commutators of perturbations are the same as the commutators of the fields themselves. This gives, for very early times,

$$\left[\delta\varphi(\mathbf{x}, t), \, \delta\varphi(\mathbf{y}, t) \right] = 0 , \quad \left[\delta\varphi(\mathbf{x}, t), \, \delta\dot{\varphi}(\mathbf{y}, t) \right] = i a^{-3}(t) \, \delta^3(\mathbf{x} - \mathbf{y}) .$$

$$(10.1.24)$$

With $\delta\varphi_q(t)$ normalized to satisfy Eq. (10.1.20) for $a(t) \to 0$, these commutation relations imply that $\alpha(\mathbf{q})$ and $\alpha^*(\mathbf{q})$ behave as conventionally normalized annihilation and creation operators

$$\left[\alpha(\mathbf{q}), \, \alpha(\mathbf{q}') \right] = 0 , \quad \left[\alpha(\mathbf{q}), \, \alpha^*(\mathbf{q}') \right] = \delta^3(\mathbf{q} - \mathbf{q}') . \quad (10.1.25)$$

It may come as a surprise that the same annihilation and creation operators appear in the scalar field and gravitational perturbations, but it should be kept in mind that Ψ does not represent gravitational radiation, whose quanta are created and annihilated by independent operators. Rather, Ψ is an auxiliary field, given by Eq. (10.1.14) as a functional of the inflaton field $\delta\varphi$, in much the same way that in the Coulomb gauge quantization of quantum electrodynamics, the time-component of the vector potential is a functional of the charged matter fields.[3]

It may be noted in passing that the scalar field we have constructed is in accord with the Principle of Equivalence of general relativity. As long as we do not concern ourselves with co-moving wave numbers below some infrared limit Q (as for example by confining measurements to a cube of co-moving volume less than $(2\pi/Q)^3$), then at times for which $Q/a \gg H$ the form of the scalar field $\delta\varphi(\mathbf{x}, t)$ ought to be essentially the same as a free massless real scalar field $\varphi(\mathbf{x}, t)$ in ordinary Minkowskian space-time:

$$\int \frac{d^3k}{(2\pi)^{3/2}\sqrt{2k}} \left[A(\mathbf{k})e^{i\mathbf{k}\cdot a\mathbf{x}} \exp\left(-i\int k\,dt\right) + A^*(\mathbf{k})e^{-i\mathbf{k}\cdot a\mathbf{x}} \exp\left(i\int k\,dt\right) \right],$$

where $A(\mathbf{k})$ and $A^*(\mathbf{k})$ are annihilation and creation operators, satisfying the familiar commutation relations

$$[A(\mathbf{k}), A(\mathbf{k}')] = 0 , \qquad [A(\mathbf{k}), A^*(\mathbf{k}')] = \delta^3(\mathbf{k} - \mathbf{k}') .$$

Note that the space coordinate appearing in the exponentials is $a\mathbf{x}$, because in the $K = 0$ Robertson–Walker metric this is the vector that measures proper distances. It follows that $\mathbf{k}$ is related to the time-independent co-moving wave number vector $\mathbf{q}$ by $\mathbf{k} = \mathbf{q}/a$, so it is time-dependent, which is why we had to write the time-dependence factors as $\exp\left(\mp i\int k\,dt\right)$ rather than $\exp(\mp ikt)$. Also, $\delta^3(\mathbf{k} - \mathbf{k}') = a^3\delta^3(\mathbf{q} - \mathbf{q}')$, so we can define operators $\alpha(\mathbf{q})$ that satisfy (10.1.25) as $\alpha(\mathbf{q}) = a^{-3/2}A(\mathbf{k})$. Changing the variable of integration from $\mathbf{k}$ to $\mathbf{q}$ then gives the field for $a(t) \to 0$ as

$$\int \frac{d^3q/a^3}{(2\pi)^{3/2}\sqrt{2q/a}} \left[a^{3/2}\alpha(\mathbf{q})e^{i\mathbf{q}\cdot\mathbf{x}} \exp\left(-i\int q\,dt/a\right) \right.$$

$$\left. + a^{3/2}\alpha^*(\mathbf{q})e^{-i\mathbf{q}\cdot\mathbf{x}} \exp\left(i\int q\,dt/a\right) \right],$$

in agreement with Eqs. (10.1.20) and (10.1.22).

Finally, we have to choose the quantum state of the inflaton field during inflation. Though there are other possibilities, the simplest and most natural

---

[3] See QTF, Vol. I, Eq. (8.2.9).

assumption is that the state of the universe during inflation is the vacuum $|0\rangle$, defined so that[4]

$$\alpha(\mathbf{q})|0\rangle = 0 , \qquad \langle 0|0\rangle = 1 . \tag{10.1.26}$$

One other possibility is that it is a linear combination of α and α^* rather than α that annihilates the state $|0\rangle$.[5] Another is that inflation takes place in the presence of a thermal distribution of inflatons.[6]

But although not certain, the assumption (10.1.26) is at least plausible. The initial condition (10.1.20), which picks out a particular solution of the second-order system (10.1.15)–(10.1.17), is imposed at an early time, at which $q/a \gg H$. At such times, we can treat the action as if it were nearly time-independent, so there exists a Hamiltonian operator H which to a good approximation generates the time-dependence of the fields:

$$[H, \delta\varphi(\mathbf{x}, t)] \simeq -i\dot{\varphi}(\mathbf{x}, t) .$$

According to Eq. (10.1.18), for $q/a \gg H$ the time dependence of the coefficient function $\varphi_q(t)$ is given approximately by

$$\dot{\varphi}_q \simeq -i(q/a)\varphi_q ,$$

so

$$[H, \alpha(\mathbf{q})] \simeq -(q/a)\, \alpha(\mathbf{q}) .$$

Hence if a state $|\psi\rangle$ is an eigenstate of H with energy E, then $\alpha(\mathbf{q})|\psi\rangle$ is an eigenstate of H with a lower energy $\simeq E - q/a$, unless $|\psi\rangle$ is the state $|0\rangle$ for which $\alpha(\mathbf{q})|0\rangle = 0$, which is therefore the state of lowest energy. Just as in ordinary laboratory physics, we expect any other state to decay into the state $|0\rangle$ of lowest energy, although there remains a question whether the decay occurs rapidly enough to be effective in the period before horizon exit.[7]

In the state satisfying Eq. (10.1.26), quantum averages $\langle 0|\delta\varphi(\mathbf{x}_1, t)\delta\varphi(\mathbf{x}_2, t) \cdots |0\rangle$ (as well as those also involving $\Psi(\mathbf{x}, t)$) may be calculated by moving all annihilation operators $\alpha(\mathbf{q})$ to the right and all creation operators $\alpha^*(\mathbf{q})$ to the left, picking up commutators when a $\alpha(\mathbf{q})$ is moved to the right past a $\alpha^*(\mathbf{q}')$ or a $\alpha^*(\mathbf{q})$ is moved to the left past a $\alpha(\mathbf{q}')$. The result is then given by *Wick's theorem*:[8] The quantum averages of products of $\delta\varphi$s and/or

---

[4]This state is often called the *Bunch-Davies vacuum*: see T. S. Bunch and P. C. W. Davies, *Proc. Roy. Soc. Ser. A* **360**, 117 (1978).

[5]E. Mottola, *Phys. Rev. D* **31**, 754 (1985); B. Allen, *Phys. Rev. D* **32**, 3136 (1985).

[6]K. Bhattacharya, S. Mohanty, and R. Rangarajan, *Phys. Rev. Lett.* **96**, 121302 (2006).

[7]This is studied by C. Armendariz-Picon, *J. Cosm. & Astropart. Phys.* **0702**, 031 (2007) [astro-ph/0612288].

[8]See QTF, Vol. I, Sec. 9.1.

Ψs are Gaussian in the sense of Appendix E (except that we must keep track of the order of operators), with the pairings given by vacuum expectation values of products of the paired fields:

$$\langle 0|\delta\varphi(\mathbf{x}, t)\delta\varphi(\mathbf{y}, t)|0\rangle = \int d^3q \; |\delta\varphi_q(t)|^2 \, e^{i\mathbf{q}\cdot(\mathbf{x}-\mathbf{y})} \,, \qquad (10.1.27)$$

$$\langle 0|\Psi(\mathbf{x}, t)\Psi(\mathbf{y}, t)|0\rangle = \int d^3q \; |\Psi_q(t)|^2 \, e^{i\mathbf{q}\cdot(\mathbf{x}-\mathbf{y})} \,, \qquad (10.1.28)$$

$$\langle 0|\delta\varphi(\mathbf{x}, t)\Psi(\mathbf{y}, t)|0\rangle = \int d^3q \; \delta\varphi_q(t)\Psi_q^*(t) \, e^{i\mathbf{q}\cdot(\mathbf{x}-\mathbf{y})} \,, \qquad (10.1.29)$$

$$\langle 0|\Psi(\mathbf{x}, t)\delta\varphi(\mathbf{y}, t)|0\rangle = \int d^3q \; \Psi_q(t) \, \delta\varphi_q^*(t) \, e^{i\mathbf{q}\cdot(\mathbf{x}-\mathbf{y})} \,. \qquad (10.1.30)$$

For instance,

$$\langle 0|\delta\varphi(\mathbf{w}, t)\Psi(\mathbf{x}, t)\Psi(\mathbf{y}, t)\delta\varphi(\mathbf{z}, t)|0\rangle$$
$$= \langle 0|\delta\varphi(\mathbf{w}, t)\Psi(\mathbf{x}, t)|0\rangle \langle 0|\Psi(\mathbf{y}, t)\delta\varphi(\mathbf{z}, t)|0\rangle$$
$$+ \langle 0|\delta\varphi(\mathbf{w}, t)\Psi(\mathbf{y}, t)|0\rangle \langle 0|\Psi(\mathbf{x}, t)\delta\varphi(\mathbf{z}, t)|0\rangle$$
$$+ \langle 0|\delta\varphi(\mathbf{w}, t)\delta\varphi(\mathbf{z}, t)|0\rangle \langle 0|\Psi(\mathbf{x}, t)\Psi(\mathbf{y}, t)|0\rangle$$

These are quantum averages, not averages over an ensemble of classical field configurations. We see this most clearly in Eqs. (10.1.29) and (10.1.30), which give complex results for the averages of products of real fields, and consequently also depend on the order of the fields.[9] Just as in the measurement of a spin in the laboratory, some sort of decoherence must set in; the field configurations must become locked into one of an ensemble of classical configurations, with ensemble averages given by the quantum expectation values calculated as in Eqs. (10.1.27)–(10.1.30) It is not apparent just how this happens, but it is clear that decoherence cannot occur until expectation values of products of real fields become real, which for free fields will also imply that the expectation values do not depend on the order of the fields. As we shall see, this happens after perturbations leave the horizon, when the various functions $\varphi_q(t)$ and $\Psi_q(t)$ become dominated by a single solution of the field equations, which (since the field equations are real) is necessarily real up to a possible complex factor. Once the universe becomes classical in this sense, we can invoke the Ergodic Theorem of Appendix D to interpret averages over ensembles of possible classical universes as averages over the position of the observer in *our* universe.

For any given potential $V(\bar{\varphi})$, it is always possible to find $\varphi_q(t)$ and $\Psi_q(t)$ by numerically solving Eqs. (10.1.15)–(10.1.17), subject to the initial

---

[9] For a discussion of this point, see D. H. Lyth and D. Seery, astro-ph/0607647.

conditions (10.1.20) and (10.1.21). But this is complicated, and gives more information than we can use. Between the time of inflation and the present there intervenes a so-called reheating period when the energy of the inflaton field φ is converted into ordinary matter and radiation. We know essentially nothing about this process, so the solutions for $\delta\varphi$ and Ψ during inflation do not have an immediate interpretation in terms of observations of the present universe. Fortunately, the reheating era (and other ill-understood eras) occur when all cosmological fluctuations of observational interest are outside the horizon. The one use that we can make of the solutions for the fields during inflation is to calculate some quantity that becomes conserved outside the horizon, and that thus provides an initial condition for the evolution of perturbations after they re-enter the horizon.

Here we will concentrate on the quantity $\mathcal{R}$ discussed in Section 5.4, defined in Newtonian gauge by

$$\mathcal{R} \equiv -\Psi + H\delta u . \tag{10.1.31}$$

With a single scalar field this is conserved outside the horizon during inflation, because as shown in Section 5.4 there are always two solutions for which $\mathcal{R}$ is constant for $q/a \ll H$, and as already noted the equations for inflation with a single scalar field only have two independent solutions. Using Eqs. (10.1.11) and (10.1.22)–(10.1.23) we find that during inflation

$$\mathcal{R}(\mathbf{x}, t) = \int d^3q \left[\mathcal{R}_q(t) \, e^{i\mathbf{q}\cdot\mathbf{x}}\alpha(\mathbf{q}) + \mathcal{R}_q^*(t) \, e^{-i\mathbf{q}\cdot\mathbf{x}}\alpha^*(\mathbf{q}) \right] , \tag{10.1.32}$$

where

$$\mathcal{R}_q = -\Psi_q - H\delta\varphi_q/\dot{\bar{\varphi}} . \tag{10.1.33}$$

Of course, the quantum averages of products of $\mathcal{R}(\mathbf{x}, t)$s are Gaussian in the same sense as those of $\Psi_q(t)$ and $\delta\varphi_q(t)$, with pairings given by the expectation value

$$\langle 0|\mathcal{R}(\mathbf{x}, t)\mathcal{R}(\mathbf{y}, t)|0\rangle = \int d^3q \, e^{i\mathbf{q}\cdot(\mathbf{x}-\mathbf{y})} \left|\mathcal{R}_q(t)\right|^2 \tag{10.1.34}$$

Instead of calculating $\Psi_q(t)$ and $\delta\varphi_q(t)$ and then using the results to calculate $\mathcal{R}_q(t)$, it is much more convenient to solve a differential equation for $\mathcal{R}_q(t)$ itself. This equation can be derived in Newtonian gauge with some trouble from Eqs. (10.1.15)–(10.1.17), but it is more easily derived in a different gauge, defined by the conditions

$$\delta\varphi_q = 0 , \quad B_q = 0$$

Inspection of Eqs. (B.66)–(B.68) shows that in this gauge

$$\delta\rho = \delta p = \frac{1}{2}h_{00}\dot{\bar{\varphi}}^2 = -\frac{1}{2}E\dot{\bar{\varphi}}^2 , \quad \delta u = 0 .$$

477

Also, as mentioned in Appendix B, the energy-momentum tensor for a single real scalar field has the perfect fluid form, so the anisotropic inertia π_{ij} vanishes. The gravitational field equations (5.1.44) and (5.1.46) and the energy conservation equation (5.1.49) then give

$$0 = H\dot{E} + 2(3H^2 + \dot{H})E + a^{-2}\nabla^2 A - \ddot{A} - 6H\dot{A} + 2a^{-1}H\nabla^2 F \,,$$

$$0 = -HE + \dot{A}$$

$$0 = -\frac{1}{2}\frac{\partial}{\partial t}\left(E\dot{H}\right) - 3H\dot{H}E - a^{-1}\dot{H}\nabla^2 F + \frac{3}{2}\dot{H}\dot{A} \,,$$

in which we have again used the relation $\dot{H} = -4\pi G|\dot{\varphi}|^2$. Eliminating E and F yields a differential equation for A:

$$\ddot{A} + \left(3H - \frac{2\dot{H}}{H} + \frac{\ddot{H}}{\dot{H}}\right)\dot{A} - \frac{1}{a^2}\nabla^2 A = 0 \,.$$

The gauge invariant formula (5.4.22) tells us that in this gauge $\mathcal{R} = A/2$, so the same equation applies to $\mathcal{R}$. Going over to its Fourier transform, this gives what is sometimes known as the *Mukhanov–Sasaki equation*:[10]

$$\frac{d^2\mathcal{R}_q}{d\tau^2} + \frac{2}{z}\frac{dz}{d\tau}\frac{d\mathcal{R}_q}{d\tau} + q^2\mathcal{R}_q = 0 \,, \tag{10.1.35}$$

where τ is the conformal time

$$\tau \equiv \int_{t_*}^{t}\frac{dt'}{a(t')} \,, \tag{10.1.36}$$

with t_* an arbitrary time, to be chosen later, and

$$z \equiv \frac{a\dot{\varphi}}{H} \,. \tag{10.1.37}$$

The initial condition is provided by returning to Newtonian gauge, and using Eqs. (10.1.20) and (10.1.21) in Eq. (10.1.33). Assuming that $a(t)\dot{\varphi}^2(t)/H(t)$ vanishes in the limit $a(t) \to 0$, only the term in Eq. (10.1.33) proportional to $\delta\varphi_q$ contributes in this limit, and we find that for $a(t) \to 0$:

$$\mathcal{R}_q(t) \to -\frac{H(t)}{(2\pi)^{3/2}\sqrt{2q}\,a(t)\,\dot{\varphi}(t)}\exp(-iq\tau) \tag{10.1.38}$$

---

[10]V. S. Mukhanov, *JETP Lett.* **41**, 493 (1986); S. Sasaki, *Prog. Theor. Phys.* **76**, 1036 (1986); V. S. Mukhanov, H. A. Feldman, and R.H. Brandenberger, *Phys. Rep.* **215**, 203 (1992); E. D. Stewart and D. H. Lyth, *Phys. Lett. B* **302**, 171 (1993).

For a given potential we must integrate Eq. (10.1.35) out from $a = 0$ to beyond the horizon, where $q/a \ll H$. In this limit we can drop the $q^2 \mathcal{R}_q$ term in Eq. (10.1.35), which then has two solutions, a dominant solution with $\mathcal{R}_q$ a non-zero constant, and a solution for which $\mathcal{R}_q$ approaches zero with $d\mathcal{R}_q/d\tau$ decaying as $1/z^2$. It is the constant limit $\mathcal{R}_q^o$ of $\mathcal{R}_q(t)$ outside the horizon that we need. (As we saw in Section 5.3, the value of the quantity ζ_q that is sometimes used in analyses of cosmological fluctuations is given far outside the horizon by $\zeta_q^o = \mathcal{R}_q^o$.)

It is striking that within the scope of the general assumptions made here, it is not necessary to make any arbitrary assumptions about the strength of cosmological fluctuations. For any given potential, we need only solve equation (10.1.35) with the initial condition (10.1.38), and carry the solution forward in time to when $\mathcal{R}_q(t)$ reaches its constant value $\mathcal{R}_q^o$ outside the horizon.

On what features of the potential does $\mathcal{R}_q^o$ depend? The initial behavior (10.1.38) of $\mathcal{R}_q(t)$ deep inside the horizon is independent of the nature of the potential, while outside the horizon $\mathcal{R}_q(t)$ simply becomes constant. Thus $\mathcal{R}_q^o$ *can depend only on the behavior of the potential* $V(\varphi)$ *for values of* φ *near the value taken by* $\bar{\varphi}(t)$ *at the time the perturbation leaves the horizon.*

This has an important implication for the part of the era of inflation that can be revealed through observations of scalar fluctuations. As we have just seen, in observing a fluctuation with co-moving wave number q, we learn about the time t_q of horizon exit during inflation, when $q/a(t_q) = H(t_q)$. To put this another way, the number of e-foldings $\mathcal{N}(q)$ between the time t_q that we learn about in observing a perturbation of wave number q and the beginning of the radiation-dominated era at a time t_1 is[11]

$$\mathcal{N}(q) \equiv \ln\left(\frac{a(t_1)}{a(t_q)}\right) = \mathcal{N}_0 + \ln\left(\frac{a(t_0)\,H(t_0)}{q}\right) + \ln\left(\frac{H(t_q)}{H(t_1)}\right) , \quad (10.1.39)$$

where

$$\mathcal{N}_0 \equiv \ln\left(\frac{a(t_1)H(t_1)}{a(t_0)H(t_0)}\right) . \quad (10.1.40)$$

and as usual t_0 is the present time. In particular, we cannot observe any fluctuation unless q is large enough to have entered the horizon by the present time, which requires that the present physical wavelength $a(t_0)/q$ be less than the present horizon distance $\approx 1/H(t_0)$, so the maximum number of e-foldings before the beginning of the radiation-dominated era that can

---

[11]A. R. Liddle and S. M. Leach, *Phys. Rev. D* **68**, 103503 (2003) [astro-ph/0305263].

ever be observed is

$$\mathcal{N}_{\max} = \mathcal{N}_0 + \ln \left(\frac{H(t)}{H(t_1)} \right)_{\max} , \qquad (10.1.41)$$

in which the second term on the right is the maximum value of $\ln[H(t)/H(t_1)]$ for t within the last $\mathcal{N}_{\max}$ e-foldings of inflation. In general we expect energy to be lost during inflation and in the reheating phase at the end of inflation, so $H(t_{\text{exit}}) > H(t_1)$ and hence $\mathcal{N}_{\max} > \mathcal{N}_0$, but for slow roll inflation it may not be a bad approximation to neglect this energy loss, in which case the second term in Eq.(10.1.41) can be neglected, and

$$\mathcal{N}_{\max} \simeq \mathcal{N}_0 . \qquad (10.1.42)$$

The reason that we have chosen to write the formula for $\mathcal{N}(q)$ as in Eq. (10.1.39) is that we have already calculated $\mathcal{N}_0$ in Section 4.1; it is

$$\mathcal{N}_0 = \ln \left(\frac{\rho_1^{1/4}}{0.037 \, h \, \text{eV}} \right) , \qquad (10.1.43)$$

where ρ_1 is the energy density at the beginning of the radiation-dominated era. For instance, if we take $h = 0.7$ and $\rho_1 = G^{-2} = [1.2 \times 10^{19} \text{ GeV}]^4$, then $\mathcal{N}_0 \simeq 68$, and according to Eq.(10.1.42) we can only explore the final 68 e-foldings of inflation. We will re-evaluate this bound in Section 10.3, with a better estimate of ρ_1.

There is one form of the potential for which the constant $\mathcal{R}_q^o$ can be calculated analytically, with no further approximations.[12] It is the exponential potential

$$V(\varphi) = g \, e^{-\lambda \varphi} \qquad (10.1.44)$$

(with g and λ arbitrary real constants) which we have already considered in Section 4.2. Of course, the potential cannot have this form for all φ, or inflation would never end, but as remarked in the previous paragraph, in the calculation of $\mathcal{R}_q^o$ it is only relevant that the potential should take this form for values of the field near the value it takes at horizon crossing.

The solution of Eqs. (10.1.6) and (10.1.7) for the exponential potential is

$$\bar{\varphi}(t) = \frac{1}{\lambda} \ln \left(\frac{8\pi \, G g \epsilon^2 t^2}{3 - \epsilon} \right) , \qquad (10.1.45)$$

---

[12]L. F. Abbott and M. B. Wise, *Nucl. Phys.* **B 244**, 541 (1984); F. Lucchin and S. Matarrese, *Phys. Rev. D* **32**, 1316 (1985); *Phys. Lett. B* **164**, 282 (1985); D. H. Lyth and E. D. Stewart, *Phys. Lett. B* **274**, 168 (1992).

and

$$H = 1/\epsilon t , \tag{10.1.46}$$

where ϵ is the positive dimensionless quantity

$$\epsilon \equiv -\frac{\dot{H}}{H^2} = \frac{\lambda^2}{16\pi G} . \tag{10.1.47}$$

Inflation with this potential is often called power-law inflation, because $a \propto t^{1/\epsilon}$. It is convenient for the moment to normalize the co-moving coordinates so that

$$a = t^{1/\epsilon} . \tag{10.1.48}$$

Note that $a\dot{\varphi}^2/H \propto t^{(1-\epsilon)/\epsilon}$, so our previous assumption that $a\dot{\varphi}^2/H$ vanishes for $a(t) \to 0$ is satisfied if $\epsilon < 1$, as we will assume.

For this potential it is convenient to take the constant t_* in the definition (10.1.36) of conformal time as $t_* = \infty$, in which case the conformal time is negative

$$\tau = -\left(\frac{\epsilon}{1-\epsilon}\right) t^{-(1-\epsilon)/\epsilon} . \tag{10.1.49}$$

As t and $a(t)$ run from zero to infinity, τ runs from $-\infty$ to zero. In terms of τ, Eq. (10.1.35) now reads

$$\frac{d^2 \mathcal{R}_q}{d\tau^2} - \frac{2}{(1-\epsilon)\tau} \frac{d\mathcal{R}_q}{d\tau} + q^2 \mathcal{R}_q = 0 , \tag{10.1.50}$$

whose solutions are proportional to τ^ν times a Hankel function $H_\nu^{(1)}(-q\tau)$ or $H_\nu^{(2)}(-q\tau)$, where

$$\nu = \frac{1}{2}\left(1 + \frac{2}{1-\epsilon}\right) = \frac{3}{2} + \frac{\epsilon}{1-\epsilon} . \tag{10.1.51}$$

For large real x the first Hankel function has the asymptotic behavior

$$H_\nu^{(1)}(x) \to \sqrt{\frac{2}{\pi x}} \exp\left(ix - i\nu\pi/2 - i\pi/4\right) ,$$

and $H_\nu^{(2)}(x) = H_\nu^{(1)*}(x)$, so the initial condition (10.1.38) picks out the solution $\propto \tau^\nu H_\nu^{(1)}(-q\tau)$, and fixes its normalization so that

$$\mathcal{R}_q(t) = -\frac{\lambda\sqrt{\pi}}{4(2\pi)^{3/2}\epsilon}\left(\frac{\epsilon}{1-\epsilon}\right)^{-1/(1-\epsilon)}$$

$$\times \exp\left(\frac{i\pi\nu}{2} + \frac{i\pi}{4}\right)(-\tau)^\nu H_\nu^{(1)}(-q\tau) . \tag{10.1.52}$$

For small real x

$$H_\nu^{(1)}(x) \to \frac{-i\Gamma(\nu)}{\pi} \left(\frac{x}{2}\right)^{-\nu}$$

so outside the horizon, in the limit $q/a \ll H$ where $-q\tau \ll 1$, the quantity $\mathcal{R}_q$ approaches a constant $\mathcal{R}_q^o$, given by Eq. (10.1.52) as

$$\mathcal{R}_q^o = i\frac{\lambda\,\Gamma(\nu)}{8\sqrt{2}\,\pi^2\epsilon} \left(\frac{\epsilon}{1-\epsilon}\right)^{-1/(1-\epsilon)} \exp\left(\frac{i\pi\nu}{2} + \frac{i\pi}{4}\right) \left(\frac{q}{2}\right)^{-\nu}. \qquad (10.1.53)$$

For purposes of comparison with results for other potentials, it is convenient to rewrite this in terms of quantities at the time t_q of horizon crossing, defined by

$$q/a(t_q) = H(t_q) . \qquad (10.1.54)$$

Solving this gives $t_q = (\epsilon q)^{\epsilon/(1-\epsilon)}$ and $H(t_q) = q^{-\epsilon/(1-\epsilon)}\epsilon^{-1/(1-\epsilon)}$. It is also convenient to use Eq. (10.1.47) to express λ in terms of ϵ and G. Then Eq. (10.1.53) can be written

$$\zeta_q^o = \mathcal{R}_q^o = i\sqrt{16\pi\,G}q^{-3/2}H(t_q)\frac{\Gamma(\nu)2^{\nu-3/2}}{4\pi^2\sqrt{\epsilon}}(1-\epsilon)^{1/(1-\epsilon)} \exp\left(\frac{i\pi\nu}{2} + \frac{i\pi}{4}\right).$$

$$(10.1.55)$$

Note that this formula does not depend on the convention we have chosen in Eq. (10.1.48) for the constant factor in $a(t)$.

To have a sufficient number of e-foldings in inflation, it is necessary for the potential to be fairly flat, so we are chiefly interested in the case where ϵ is small. For $\epsilon \ll 1$, we use $\Gamma(3/2) = \sqrt{\pi}/2$, and write Eq. (10.1.55) as

$$\zeta_q^o = \mathcal{R}_q^o = -i\sqrt{16\pi\,G}q^{-3/2}H(t_q)\frac{1}{8\pi^{3/2}\sqrt{\epsilon}} . \qquad (10.1.56)$$

The most important point is that for a nearly flat potential, for which $H(t_q)$ is nearly q-independent, $\mathcal{R}_q^o$ is nearly proportional to $q^{-3/2}$. This result is not limited to potentials of exponential form, as long as the potential is fairly flat for fields near $\bar\varphi(t_q)$.

The approximate $q^{-3/2}$ dependence of $\mathcal{R}_q^o$ on q is well supported by observation. We saw in Section 8.1 that the power spectral function $P(k)$ that governs the distribution of dark matter has a dependence on the physical wave number k that for small k is given by a factor $k^4|\mathcal{R}_{ka_0}^o|^2$, so any potential that is fairly flat for fields near $\bar\varphi(t_q)$ gives a spectrum close to the Harrison–Zel'dovich spectrum, $P(k) \propto k$ for small k. Also, as we saw in Section 6.3, the spectral function $\mathcal{P}_\phi(q)$ for the Newtonian gravitational potential is proportional for long wavelengths to $|\mathcal{R}_q^o|^2$, which as shown in Section 2.6 implies that (aside from the integrated Sachs–Wolfe effect and the effects of reionization) the quantity $\ell(\ell+1)C_\ell$ becomes independent of ℓ for small ℓ. And we saw in Sections 7.2 and 7.4 that the near proportionality of $\mathcal{R}_q^o$ to $q^{-3/2}$ gives results for the correlations of temperature and polarization fluctuations in the cosmic microwave background at larger values of ℓ in good agreement with observation.

To be more precise, Eq. (10.1.53) shows that for the exponential potential, the scalar spectral index defined in Section 7.2 by $|\mathcal{R}_q^o|^2 \propto q^{-4+n_S}$ has the constant value

$$n_S = 4 - 2\nu = 1 - \frac{2\epsilon}{1-\epsilon} \,.$$

The third-year **WMAP** result $n_S = 0.958 \pm 0.016$ quoted in Section 7.2 thus shows that for the exponential potential, $\epsilon = 0.021 \pm 0.008$. This is small enough so that we can use the approximate formula (10.1.56). Writing $|\mathcal{R}_q^o|^2 \simeq |N|^2 q^{-3}$, we see that for the exponential potential

$$|N|^2 \simeq \frac{GH_{\text{exit}}^2}{4\pi^2\epsilon} \,,$$

where H_{exit} is the expansion rate at horizon exit, now ignoring its weak dependence on q. The third-year **WMAP** result $|N|^2 = (1.93 \pm 0.12) \times 10^{-10}$ quoted in Section 7.2 thus shows that for the exponential potential, $H_{\text{exit}} \simeq 2\pi|N|\sqrt{\epsilon/G} \simeq 1.5 \times 10^{14}$ GeV, corresponding to an energy density $3H_{\text{exit}}^2/8\pi G \simeq [2.6 \times 10^{16}\text{ GeV}]^4$.

Of course, the exponential potential is just one special case, with the special property that $V''/V = (V'/V)^2$, so there is no reason to expect these results to apply in detail for other potentials. In Section 10.3 we will see what to expect for general potentials, under the slow-roll approximation. But within the slow-roll approximation, the results for general potentials are similar in order of magnitude to those we have already found for the exponential potential.

We are now in a position to consider the plausibility of the assumption that the action has the simple form (10.1.1).[13] With a single real scalar field χ, there is no loss of generality in taking the action in the form (10.1.1) if

---

[13]There is a large literature on other possibilities. Here is a partial list: C. Armendariz-Picon, T. Damour, and V. Mukhanov, *Phys. Lett. B* **458**, 209 (1999) [hep-th/9904075]; J. Martin and R. H. Brandenberger, *Phys. Rev. D* **63**, 1235012 (2001) [hep-th/0005209]; R. H. Brandenberger and J. Martin, *Mod. Phys. Lett. A* **16**, 999 (2001) [astro-ph/0005432]; J. C. Niemeyer, *Phys. Rev. D* **63**, 123502 (2001) [astro-ph/0005533]; J. C. Niemeyer and R. Parentani, *Phys. Rev. D* **64**, 101301 (2001) [astro-ph/0101451]; A. Kempf and J. C. Niemeyer, *Phys. Rev. D* **64**, 103501 (2001) [astro-ph/0103225]; R. Easther, B. R. Greene, W. H. Kinney, and G. Shiu, *Phys. Rev. D* **64**, 103502 (2001) [hep-th/0104102];

we assume that the action contains at most two spacetime derivatives. The most general action in this case is

$$I_\chi = \int d^4x \sqrt{-\text{Det}g} \left[-\frac{1}{2}K(\chi)g^{\mu\nu}\frac{\partial\chi}{\partial x^\mu}\frac{\partial\chi}{\partial x^\nu} - U(\chi) \right],$$

where K and U are real functions of χ that are arbitrary, except that unitarity requires $K \geq 0$. This can be put in the form (10.1.1) by introducing $\varphi = \int d\chi \, K^{1/2}(\chi)$.

But why should the terms in the action contain no more than two spacetime derivatives? This is only one of a number of similar questions: Why should the action of gravitation contain only the Einstein term, proportional to $\int d^4x\sqrt{-\text{Det}g}R$, and not other generally covariant terms with more than two derivatives of the metric? And why should the action of the standard electroweak model or quantum chromodynamics contain only renormalizable terms?

We do not know the answers to any of these questions with any certainty, but we have at least a plausible possible answer. Any of these additional terms in the action would involve operators of higher dimensionality, but dimensional analysis requires that such terms must be accompanied by coefficients containing additional negative powers of some fundamental mass. If that mass is large enough enough, then the additional terms in the action are suppressed under ordinary conditions. For the success of general relativity, it is only necessary that the length $1/M$ be sub-macroscopic, but unless we impose a condition of baryon and lepton conservation, the fundamental mass appearing in the action of quarks and leptons has to be at least of order 10^{16} GeV, in order to suppress proton decay below experimental bounds.

The inflaton field is considered to take values of the order of the Planck mass, so there is nothing to suppress arbitrary powers of the scalar field in the inflaton action, which is why we take the potential $V(\varphi)$ as an arbitrary function. But spacetime derivatives of the inflaton field yield factors of order q/a, so each additional spacetime derivative would introduce a factor of order q/aM. At horizon exit, q/a equals H, which we have seen is of order 10^{14} GeV, while we would expect M to be much larger, somewhere

---

Phys. Rev. D **66**, 023518 (2002); *Phys. Rev. D* **67**, 063508 (2003) [hep-th/0110226]; R. H. Brandenberger, S. E. Joras, and J. Martin, *Phys. Rev. D* **66**, 083514 (2002) [hep-th/0112122]; N. Kaloper, M. Kleban, A. Lawrence, and S. Shenker, *Phys. Rev. D* **66**, 123510 (2002) [hep-th/0201158]; N. Kaloper, M. Kleban, A. Lawrence, S. Shenker, and L. Susskind, *J. High Energy Phys.* **0211**, 037 (2002) [hep-th/0209231]; U. H. Danielsson, *Phys. Rev. D* **66**, 023511 (2002); L. Bergström and U. H. Danielsson, *J. High Energy Phys.* **12**, 38 (2002); C. P. Burgess, J. M. Cline, F. Lemieux, and R. Holman, *J. High Energy Phys.* **2**, 48 (2003); J. Martin and R. Brandenberger, *Phys. Rev. D* **68**, 063513 (2003); J. Martin and C. Ringeval, *Phys. Rev. D* **69**, 083515 (2004); T. Okamoto and E. A. Lim, *Phys. Rev. D* **69**, 083519 (2004). R. de Putter and E. V. Linder, 0705.0400.

in the range from $|V|^{1/4} \approx 10^{16}$ GeV to $1/\sqrt{G} \simeq 10^{19}$ GeV. Hence it is plausible that at and after horizon exit, *and for some time before horizon exit*, terms with more than the minimum number of spacetime derivatives are suppressed. This is all we need to justify the calculations of this section, at least as a good first approximation, as long as we stick to a single inflaton field. The possibility of more than one inflaton field will be discussed in Section 10.4.

10.2 Tensor fluctuations during inflation

The fluctuations in the tensor field $D_{ij}(\mathbf{x}, t)$ during inflation can be treated in much the same way as the scalar fluctuations considered in the previous section. Since the tensor anisotropic inertia π_{ij}^T vanishes for scalar field theories, the field equation (5.1.53) for the tensor modes takes the simple form

$$\nabla^2 D_{ij} - a^2 \ddot{D}_{ij} - 3a\dot{a}\dot{D}_{ij} = 0 . \qquad (10.2.1)$$

We recall also that D_{ij} satisfies conditions that eliminate any vector or scalar contributions:

$$D_{ij} = D_{ji} , \qquad D_{ii} = 0 , \qquad \partial_i D_{ij} = 0 . \qquad (10.2.2)$$

The plane wave solutions have the form $e_{ij} \mathcal{D}_q(t) e^{i\mathbf{q}\cdot\mathbf{x}}$, where $\mathcal{D}_q(t)$ satisfies the differential equation

$$\ddot{\mathcal{D}}_q + 3H\dot{\mathcal{D}}_q + (q^2/a^2)\mathcal{D}_q = 0 \qquad (10.2.3)$$

and e_{ij} is a time-independent polarization tensor satisfying the conditions

$$e_{ij} = e_{ji} , \qquad e_{ii} = 0 , \qquad q_i e_{ij} = 0 . \qquad (10.2.4)$$

We recall from Section 5.2 that for a given unit vector $\hat{q}$, there are two independent polarization tensors satisfying these conditions. For $\hat{q}$ in the three-direction, these can be chosen to have components

$$e_{11} = -e_{22} = 1/\sqrt{2} , \quad e_{12} = e_{21} = \pm i/\sqrt{2} , \quad e_{3i} = e_{i3} = 0 . \quad (10.2.5)$$

For $\hat{q}$ in any other direction we define e_{ij} to be the tensor obtained by applying to (10.2.5) the standard rotation (7.4.12) that takes the three-axis into the direction of $\hat{q}$. The polarization tensors constructed in this way are called $e_{ij}(\hat{q}, \pm 2)$, because Eq. (10.2.5) describes a wave of helicity ± 2.

At early times, when $q/a \gg H$, Eq. (10.2.3) has WKB solutions of the form

$$\mathcal{D}_q(t) \to h(t) \exp\left(-iq \int_{t_*}^t \frac{dt'}{a(t')}\right) , \qquad (10.2.6)$$

where $h(t)$ varies much more slowly than the argument of the exponential, and t_* is arbitrary. Then the terms in Eq. (10.2.3) of second order in q/a cancel, while the terms of first order give $2\dot{h} + 2Hh = 0$, so that $h \propto a^{-1}$. From now on we will define $\mathcal{D}_q(t)$ as the solution of Eq. (10.2.3) which is normalized so that, for $a(t) \to 0$,

$$\mathcal{D}_q(t) \to \frac{\sqrt{16\pi G}}{(2\pi)^{3/2}\sqrt{2q}\,a(t)} \exp\left(-iq \int_{t_*}^{t} \frac{dt'}{a(t')}\right). \tag{10.2.7}$$

Since $\mathcal{D}_q(t)$ and $\mathcal{D}_q^*(t)$ are a complete set of solutions of Eq. (10.2.3), the most general real tensor field satisfying the conditions (10.2.1) and (10.2.2) takes the form

$$D_{ij}(\mathbf{x}, t) = \sum_{\lambda=\pm2} \int d^3q \left[\mathcal{D}_q(t)e^{i\mathbf{q}\cdot\mathbf{x}}\beta(\mathbf{q}, \lambda)e_{ij}(\hat{q}, \lambda)\right.$$
$$\left. + \mathcal{D}_q^*(t)e^{-i\mathbf{q}\cdot\mathbf{x}}\beta^*(\mathbf{q}, \lambda)e_{ij}^*(\hat{q}, \lambda)\right] \tag{10.2.8}$$

(That is, instead of characterizing two independent solutions $\mathcal{D}_{1q}(t)$ and $\mathcal{D}_{2q}(t)$ of Eq. (10.2.3) by their behavior at late times, as we did in Section 5.2, we now take them as $\mathcal{D}_q(t)$ and $\mathcal{D}_q^*(t)$, characterized by their behavior at early times.) With $\mathcal{D}_q(t)$ normalized to satisfy Eq. (10.2.7), the canonical commutation relations require that

$$[\beta(\mathbf{q}, \lambda), \beta(\mathbf{q}', \lambda')] = 0, \quad [\beta(\mathbf{q}, \lambda), \beta^*(\mathbf{q}', \lambda')] = \delta^3(\mathbf{q} - \mathbf{q}')\delta_{\lambda\lambda'}. \tag{10.2.9}$$

Thus $\beta(\mathbf{q}, \lambda)$ and $\beta^*(\mathbf{q}, \lambda)$ can be interpreted as the annihilation and creation operators for a graviton of helicity λ.

As in the case of scalar perturbations, we assume that during inflation the universe is in a quantum state $|0\rangle$ satisfying the vacuum condition $\beta(\mathbf{q}, \lambda)|0\rangle = 0$. Then expectation values of products of Ds are Gaussian, with pairings

$$\langle 0|D_{ij}(\mathbf{x}, t)\, D_{kl}(\mathbf{y}, t)|0\rangle = \int d^3q\, e^{i\mathbf{q}\cdot(\mathbf{x}-\mathbf{y})}\left|\mathcal{D}_q(t)\right|^2 \Pi_{ij,kl}(\hat{q}) \tag{10.2.10}$$

where $\Pi_{ij,kl}(\hat{q})$ is the helicity sum given by Eq. (5.2.25):

$$\Pi_{ij,k\ell}(\hat{q}) \equiv \sum_\lambda e_{ij}(\hat{q}, \lambda)\, e_{k\ell}^*(\hat{q}, \lambda)$$
$$= \delta_{ik}\delta_{j\ell} + \delta_{i\ell}\delta_{jk} - \delta_{ij}\delta_{k\ell} + \delta_{ij}\hat{q}_k\hat{q}_\ell + \delta_{k\ell}\hat{q}_i\hat{q}_j - \delta_{ik}\hat{q}_j\hat{q}_\ell - \delta_{i\ell}\hat{q}_j\hat{q}_k$$
$$- \delta_{jk}\hat{q}_i\hat{q}_\ell - \delta_{j\ell}\hat{q}_i\hat{q}_k + \hat{q}_i\hat{q}_j\hat{q}_k\hat{q}_\ell. \tag{10.2.11}$$

For any given potential, the function $\mathcal{D}_q(t)$ is to be calculated by integrating the differential equation (10.2.3) with the initial condition (10.2.7).

Outside the horizon, for $q/a \ll H$, the solution becomes a constant $\mathcal{D}_q^0$, which provides an initial condition for the gravitational wave when it re-enters the horizon. In carrying this out, it is again useful to replace ordinary time as the independent variable with conformal time

$$\tau \equiv -\int_t^\infty \frac{dt'}{a(t')} \,, \qquad (10.2.12)$$

and take $t_* = \infty$ in Eq. (10.2.7). Eq. (10.2.3) then becomes

$$\frac{d^2\mathcal{D}_q}{d\tau^2} + 2Ha\frac{d\mathcal{D}_q}{d\tau} + q^2\mathcal{D}_q = 0 \,. \qquad (10.2.13)$$

Note that

$$Ha = \frac{da}{dt} = \frac{1}{a}\frac{da}{d\tau} \,.$$

Thus Eq. (10.2.13) is the same as the Mukhanov–Sasaki equation (10.1.35) except that $z \equiv a\dot{\bar{\varphi}}/H$ is replaced with a.

The initial condition (10.2.7) for $q/aH \gg 1$ is independent of the details of the potential, while for $q/aH \ll 1$ the tensor amplitude $\mathcal{D}_q(t)$ simply approaches a constant $\mathcal{D}_q^0$. Hence, just as for the scalar amplitude $\mathcal{R}_q^0$, the tensor amplitude $\mathcal{D}_q^0$ outside the horizon can only depend on the behavior of the potential at values of $\bar{\varphi}(t)$ at around the time of horizon exit, when q/aH is of order unity. Thus the measurements of the tensor amplitude after horizon re-entry can only tell us about the last $\mathcal{N}$ e-foldings of inflation, with $\mathcal{N}$ bounded by Eq. (10.1.40), just as for scalar modes.

As for scalar modes, it is useful to consider as a test case the one potential for which $\mathcal{D}_q^0$ can be calculated analytically without relying on the slow-roll approximation, the exponential potential (10.1.44). For this potential $\dot{\bar{\varphi}}/H$ is the constant $2\epsilon/\lambda$, so Eq. (10.1.37) gives $z \propto a$, and in this case Eq. (10.2.13) is precisely the same as the Mukhanov–Sasaki equation. Also, the initial condition (10.2.7) is the same as the initial condition (10.1.38), except for a factor $-2\epsilon\sqrt{16\pi G}/\lambda = -2\sqrt{\epsilon}$. Hence for the exponential potential:

$$\mathcal{D}_q(t)/\mathcal{R}_q(t) = -2\epsilon\sqrt{16\pi G}/\lambda = -2\sqrt{\epsilon} \,. \qquad (10.2.14)$$

Of course, the values $\mathcal{D}_q^0$ and $\mathcal{R}_q^0$ outside the horizon then also have this ratio. This is usually expressed in terms of a scalar tensor ratio, conventionally defined as

$$r_q \equiv 4|\mathcal{D}_q^0/\mathcal{R}_q^0|^2 \,. \qquad (10.2.15)$$

We see that for the exponential potential, r_q has the wavelength-independent value

$$r = 16\epsilon \,. \qquad (10.2.16)$$

We saw in the previous section that for the exponential potential the third-year WMAP results give $\epsilon = 0.021 \pm 0.008$, so that $r = 0.34 \pm 0.13$. This is almost incompatible with the upper bound $r < 0.3$ on r set (for this value of ϵ) by the third-year WMAP results,[1] so a potential $V(\varphi)$ that is exponential around the value that φ takes at the time of horizon exit is almost ruled out. (Of course, we already knew that the potential could not be exponential over the whole range of φ, for then inflation would never end, but we are here not relying on any assumption about the form of the potential except for the values of φ taken around the time of horizon exit.) To analyze tensor as well as scalar perturbations for more general potentials, we need to invoke the slow-roll approximation, to which we turn in the next section.

10.3 Fluctuations during inflation: The slow-roll approximation

It is not possible to calculate the scalar and tensor perturbations $\mathcal{R}_q(t)$ and $\mathcal{D}_q(t)$ analytically for general potentials. However, the need (discussed in Chapter 4) for a substantial number of e-foldings of expansion during inflation suggests that H should have been slowly varying during an era long compared with $1/H$. We will therefore now assume that H varies little throughout a "slow-roll" era, during which q/aH goes from much less to much greater than unity. We saw in Sections 10.1 and 10.2 that the asymptotic values $\mathcal{R}_q^o$ and $\mathcal{D}_q^o$ depend only on the evolution of the fields around the time of horizon crossing, when $q/a \approx H$, so it will not be necessary for us to assume that the slow-roll era extends back to the beginning of the expansion, or forward to the end of inflation.

We will work with the Mukhanov–Sasaki equation (10.1.35):[1]

$$\frac{d^2\mathcal{R}_q}{d\tau^2} + \frac{2}{z}\frac{dz}{d\tau}\frac{d\mathcal{R}_q}{d\tau} + q^2\mathcal{R}_q = 0 , \qquad (10.3.1)$$

where τ is the conformal time, and

$$z \equiv \frac{a\dot{\varphi}}{H} . \qquad (10.3.2)$$

---

[1] See Figure 14 of D. Spergel *et al.*, *Astrophys. J. Suppl. Ser.* **170**, 377 (2007) [astro-ph/0603449].
[1] This is usually given in the equivalent form

$$\frac{d^2u_q}{d\tau^2} + \left[q^2 - \frac{1}{z}\frac{d^2z}{d\tau^2}\right]u_q = 0 ,$$

where $u_q \equiv z\mathcal{R}_q$. It is easier to work with it in the form (10.3.1), if only because in this way we only need to calculate the first derivative of z.

Recalling once again that $\dot{H} = -4\pi G \dot{\varphi}^2$, we can write

$$\frac{1}{z}\frac{dz}{d\tau} = aH(1 + \delta + \epsilon) \tag{10.3.3}$$

where for a general potential

$$\epsilon \equiv -\dot{H}/H^2 , \qquad \delta \equiv \ddot{H}/2H\dot{H} . \tag{10.3.4}$$

We also need a formula for aH in terms of τ. For this purpose, we note that

$$\frac{d}{d\tau}\left(\frac{1}{aH}\right) = -1 + \epsilon . \tag{10.3.5}$$

So far, everything is exact. The slow-roll approximation requires that ϵ and δ are small during the era of horizon crossing, which has the consequence that ϵ varies little during this era, because

$$\dot{\epsilon} = 2\epsilon \left(\epsilon + \delta\right) H . \tag{10.3.6}$$

We will also assume that δ varies little.[2] (Of course, ϵ and δ cannot be strictly constant except for an exponential potential, for which $H \propto 1/t$ and therefore $\delta = -\epsilon$.) Integrating Eq. (10.3.5) then gives, for a suitable choice of an additive constant in τ,

$$aH = -\frac{1}{(1 - \epsilon)\tau} . \tag{10.3.7}$$

As in the case of the exponential potential, τ is negative, and its magnitude goes from $-\tau \gg 1/q$ early in inflation when $q/aH \gg 1$, to $-\tau \ll 1/q$ for $q/aH \ll 1$. Using Eqs. (10.3.3) and (10.3.7) in Eq. (10.3.1) then gives, to first order in ϵ and δ,

$$\frac{d^2\mathcal{R}_q}{d\tau^2} - \frac{2(1 + \delta + 2\epsilon)}{\tau}\frac{d\mathcal{R}_q}{d\tau} + q^2\mathcal{R}_q = 0 , \tag{10.3.8}$$

The general solution of this equation for constant δ and ϵ is a linear combination of $\tau^\nu H_\nu^{(1)}(-q\tau)$ and $\tau^\nu H_\nu^{(2)}(-q\tau)$, where now

$$\nu = \frac{3}{2} + 2\epsilon + \delta . \tag{10.3.9}$$

---

[2]This is the case for a power-law potential, under the same condition $|\varphi| \gg 1/\sqrt{4\pi G}$ that was found necessary in Section 4.2 to justify the slow-roll approximation. For the consequences of dropping this assumption, see S. Dodelson and E. D. Stewart, *Phys. Rev. D* **65**, 101301 (2002) [astro-ph/0109354]; E. D. Stewart, *Phys. Rev. D* **65**, 103508 (2002) [astro-ph/0110322].

This agrees to first order in ϵ with the result (10.1.51) for the exponential potential, for which $\delta = -\epsilon$.

Now, we are not assuming that the slow-roll approximation applies all the way back to the beginning of the expansion, but fortunately this is not necessary. Eq. (10.1.38) should still apply by the beginning of the slow-roll era, when q/aH is still very large, so this initial condition fixes the solution during the whole of the slow-roll era as

$$\mathcal{R}_q(t) = -\frac{\sqrt{-\pi\tau}}{2(2\pi)^{3/2}\,z(\tau)}e^{i\pi\nu/2+i\pi/4}H_\nu^{(1)}(-q\tau) \,. \qquad (10.3.10)$$

(Note that Eqs. (10.3.3) and (10.3.7) give $z(\tau) \propto \tau^{-\nu+1/2}$ during the slow-roll era.) Then late in the slow-roll era, when $q/aH \ll 1$, Eq. (10.3.10) has the asymptotic value

$$\mathcal{R}_q^o = i\frac{\sqrt{-\tau}\,\Gamma(\nu)}{2\sqrt{\pi}\,(2\pi)^{3/2}\,z(t)}e^{i\pi\nu/2+i\pi/4}\left(\frac{-q\tau}{2}\right)^{-\nu} \,, \qquad (10.3.11)$$

which is constant because Eqs. (10.3.3) and (10.3.7) give $z \propto \tau^{-\nu+1/2}$. Thus $\mathcal{R}_q^o$ has the q-dependence[3]

$$\mathcal{R}_q^o \propto q^{-\nu} = q^{-3/2-2\epsilon-\delta} \,. \qquad (10.3.12)$$

This may be regarded as a generalization of the result $\mathcal{R}_q^o \propto q^{-3/2-\epsilon}$ that we found for the exponential potential, to the case where $\delta \neq -\epsilon$.

Because $\mathcal{R}_q^o$ is time-independent, it can be calculated by setting t in Eq. (10.3.11) to any convenient value. We shall evaluate it at the time t_q of horizon crossing, defined as in Sections 10.1 and 10.2 by

$$q/a(t_q) = H(t_q) \,. \qquad (10.3.13)$$

(But note that Eq. (10.3.10) with $t = t_q$ does *not* give the correct value of $\mathcal{R}_q^o$.) Ignoring corrections of order ϵ or δ except in exponents, Eq. (10.3.7) gives

$$\tau(t_q) = -\frac{1}{(1-\epsilon)q} \simeq -\frac{1}{q} \,. \qquad (10.3.14)$$

Also, $\dot{H} = -4\pi G\dot{\varphi}^2$, so

$$\dot{\varphi}(t_q) = \pm\sqrt{-\dot{H}(t_q)}/\sqrt{4\pi G} = \pm H(t_q)\sqrt{\epsilon(t_q)}/\sqrt{4\pi G} \,.$$

[3]E. D. Stewart and D. H. Lyth, *Phys. Lett.* B **302**, 171 (1993). This calculation was carried to the next order of the slow-roll approximation by A. R. Liddle and M. S. Turner, *Phys. Rev.* D **50**, 758 (1994). For a review, see J. E. Lidsey, A. R. Liddle, E. W. Kolb, E. J. Copeland, T. Barreiro, and M. Abney, *Rev. Mod. Phys.* **69**, 373 (1997).

and so

$$z(t_q) = \frac{\pm\sqrt{\epsilon(t_q)}\,q}{H(t_q)\sqrt{4\pi G}} \ . \tag{10.3.15}$$

The slow-roll approximation thus gives[4]

$$\mathcal{R}_q^o = \mp i \frac{\sqrt{16\pi G}\,q^{-3/2}H(t_q)}{8\pi^{3/2}\sqrt{\epsilon(t_q)}} \ . \tag{10.3.16}$$

This is the same as the result (10.1.56) for an exponential potential, except that the factor $1/\sqrt{\epsilon(t_q)}$ now contributes to the q-dependence of $\mathcal{R}_q^o$. (To check that the q dependence of $H(t_q)$ and $\epsilon(t_q)$ gives the extra factor $q^{-2\epsilon-\delta}$, we differentiate Eq. (10.3.13) with respect to q, and find

$$\frac{dt_q}{dq} = \frac{1}{a(t_q)\left(H^2(t_q) + \dot{H}(t_q)\right)} \ .$$

Then

$$\frac{q}{H(t_q)}\frac{d\,H(t_q)}{dq} = \frac{\dot{H}(t_q)}{H^2(t_q) + \dot{H}(t_q)} = -\frac{\epsilon(t_q)}{1 - \epsilon(t_q)}$$

and Eq. (10.3.6) gives

$$\frac{q}{\epsilon(t_q)}\frac{d\,\epsilon(t_q)}{dq} = \frac{2(\epsilon(t_q) + \delta(t_q))}{1 - \epsilon(t_q)} \ .$$

Hence, replacing the denominators $1 - \epsilon$ with unity, $H(t_q) \propto q^{-\epsilon}$ and $\epsilon(t_q) \propto q^{2\epsilon+2\delta}$, and so $H(t_q)/\sqrt{\epsilon(t_q)} \propto q^{3/2-\nu}$, which gives Eq. (10.3.16) the q-dependence (10.3.12).)

Now let us apply the slow-roll approximation to the tensor modes. In general, the tensor wave equation (10.2.3) can be written as

$$\frac{d^2\mathcal{D}_q}{d\tau^2} + 2aH\frac{d\mathcal{D}_q}{d\tau} + q^2\mathcal{D}_q = 0 \ . \tag{10.3.17}$$

During the slow-roll era, we can use Eq. (10.3.7) to put this in the form

$$\frac{d^2\mathcal{D}_q}{d\tau^2} - \frac{2}{(1-\epsilon)\tau}\frac{d\mathcal{D}_q}{d\tau} + q^2\mathcal{D}_q = 0 \ . \tag{10.3.18}$$

---

[4]S. W. Hawking, *Phys. Lett.* **B 115**, 295 (1982); A. A. Starobinsky, *Phys. Lett.* **B 117**, 175 (1982); A. Guth and S.-Y. Pi, *Phys. Rev. Lett.* **49**, 1110 (1982); J. M. Bardeen, P. J. Steinhardt and M. S. Turner, *Phys. Rev. D* **28**, 679 1983); D. H. Lyth, *Phys. Lett.* **B 147**, 403 (1984); **B 150**, 465 (1985); *Phys. Rev. D* **31**, 1792 (1985).

The general solution for constant ϵ is a linear combination of $\tau^\mu H_\mu^{(1)}(-q\tau)$ and $\tau^\mu H_\mu^{(2)}(-q\tau)$, with

$$\mu = \frac{3}{2} + \frac{\epsilon}{1-\epsilon} . \tag{10.3.19}$$

The solution during the slow-roll era that satisfies the initial condition (10.2.7) is

$$\mathcal{D}_q(t) = \frac{\sqrt{16\pi\, G}}{(2\pi)^{3/2}\sqrt{2qa(t)}} \sqrt{\frac{-q\tau\pi}{2}} \exp(i\mu\pi/2 + i\pi/4)\, H_\mu^{(1)}(-q\tau) . \tag{10.3.20}$$

(Note that Eq. (10.3.7) gives $(\tau/a)da/d\tau = aH\tau = -1/(1-\epsilon)$, so $a \propto \tau^{-1/(1-\epsilon)}$ and therefore $\sqrt{-\tau}/a \propto \tau^\mu$.) The asymptotic solution for $q/a \ll H$ is then

$$\mathcal{D}_q^o = -i\frac{\sqrt{16\pi\, G}\,\Gamma(\mu)\sqrt{-\tau}}{2\sqrt{\pi}\,(2\pi)^{3/2}a(t)} \exp(i\mu\pi/2 + i\pi/4) \left(\frac{-q\tau}{2}\right)^{-\mu} . \tag{10.3.21}$$

Thus $\mathcal{D}_q^o$ has the q-dependence[3]

$$\mathcal{D}_q^o \propto q^{-\mu} \simeq q^{-3/2-\epsilon} . \tag{10.3.22}$$

In contrast with the case of the exponential potential, the asymptotic q-dependence of the tensor modes is in general different from that of the scalar modes.

We can give a more convenient expression for $\mathcal{D}_q^o$ by setting $t = t_q$ in Eq. (10.3.21). Then using Eq. (10.3.14) and $a(t_q) = q/H(t_q)$ and taking $\epsilon \to 0$ everywhere but in the q-dependence of $H(t_q)$, Eq. (10.3.21) gives[5]

$$\mathcal{D}_q^o = i\frac{\sqrt{16\pi\, GH(t_q)}}{4\pi^{3/2}q^{3/2}} . \tag{10.3.23}$$

It is conventional to write the q-dependence of the squared magnitudes of the tensor and scalar amplitudes outside the horizon as

$$|\mathcal{D}_q^o|^2 \propto q^{-3+n_T(q)} , \qquad |\mathcal{R}_q^o|^2 \propto q^{-4+n_S(q)} . \tag{10.3.24}$$

Then in the slow-roll approximation, Eqs. (10.3.12) and (10.3.22) give

$$n_T(q) = -2\epsilon(t_q) , \qquad n_S(q) = 1 - 4\epsilon(t_q) - 2\delta(t_q) . \tag{10.3.25}$$

(As a check, recall that for the exponential potential $\delta = -\epsilon$, so Eq. (10.3.25) gives $n_S = 1 - 2\epsilon$, in agreement with the result of Section 10.1.) Also,

---

[5] A. A. Starobinsky, *JETP Lett.* **30**, 683 (1979).

comparison of Eqs. (10.3.16) and (10.3.23) yields the relation

$$r(q) = 16\epsilon(t_q) = -8n_T(q) , \qquad (10.3.26)$$

where, with the conventional definition of r, $r(q) \equiv 4|\mathcal{D}_q^o/\mathcal{R}_q^o|^2$. This relation among measurable quantities is known in the literature as the *slow-roll consistency condition*. For any potential other than the exponential potential the scalar/tensor ratio r depends on q.

For inflation with a single inflaton field, the relation $\dot{H} = -4\pi G \dot{\varphi}^2$ tells us that $\epsilon(t)$ is always positive, but $\delta(t)$ can have either sign, so in general Eq. (10.3.25) gives $n_T(q) < 0$, while $n_S(q)$ can be greater or less than unity. Nevertheless, experience with many models shows[6] that physically plausible potentials that are not finely tuned tend to have $n_S(q)$ less than unity, and even less than 0.98. But for slow-roll inflation, ϵ and δ are small, so $n_S(q)$ cannot be very much less than unity. Thus the general picture of slow-roll inflation received some support from the third-year WMAP result quoted in Section 7.2, that $n_S = 0.958 \pm 0.016$.

As we saw in Chapter 7, the quantities $q^3|\mathcal{D}_q^o|^2$ and $q^3|\mathcal{R}_q^o|^2$ provide a measure of the contribution of tensor and scalar fluctuations to the multipole coefficients $C_{TT,\ell}$ in the angular distribution of the cosmic microwave background temperature. From Eq. (10.3.26) we can see that the tensor modes are likely to contribute much less to the $C_{TT,\ell}$ than the scalar modes. Also, Eq. (10.3.16) and the fact that anisotropies in the cosmic microwave background temperature are small but not too small to be observed indicates that the Hubble constant during the slow-roll era must be small compared with the Planck mass $1/\sqrt{G}$, but not too small. In the slow-roll limit, where $\nu \simeq 3/2$, we can write $|\mathcal{R}_q^o|^2 = |N|^2 q^{-3}$, and Eq. (10.3.16) shows that

$$|N|^2 = \frac{16\pi G H_{\text{exit}}^2}{64\pi^3 |\epsilon_{\text{exit}}|} = \frac{(8\pi G)^2 \bar{\rho}_{\text{exit}}}{96\pi^3 |\epsilon_{\text{exit}}|} ,$$

where the subscript "exit" denotes the time of horizon exit, and in accordance with the slow-roll approximation we here ignore the weak dependence of this time on q. As we saw in Section 7.2, the factor $|N|^2$ has the value $(1.93 \pm 0.12) \times 10^{-10}$, so

$$\frac{\bar{\rho}_{\text{exit}}}{|\epsilon_{\text{exit}}|} = [(6.70 \pm 0.10) \times 10^{16} \text{ GeV}]^4 . \qquad (10.3.27)$$

---

[6]M. B. Hoffman and M. S. Turner, *Phys. Rev D* **64**, 023506 (2001); W. H. Kinney, *Phys. Rev. D* **66**, 083508 (2002); H. V. Peiris *et al.*, *Astrophys. J. Suppl. Ser.* **1148**, 213 (2003); G. Efstathiou and K. J. Mack, *J. Cosm. Astropart. Phys.* **05**, 008 (2005); L. A. Boyle, P. J. Steinhardt, and N. Turok, *Phys. Rev. Lett.* **96**, 111301 (2006).

Measurements of cosmic microwave background anisotropies have so far been sensitive only to the spectral index of the scalar rather than the tensor modes, and therefore have yielded information only on $\delta + 2\epsilon$, not ϵ, but unless there is a cancelation between δ and 2ϵ, these measurements suggest that $|\epsilon|$ is probably not much greater than a few percent. If for instance we take $|\epsilon|_{\text{exit}} = 0.05$, then $\bar{\rho}_{\text{exit}} \simeq [3.2 \times 10^{16} \text{ GeV}]^4$. In any case, we now see that in inflationary theories, the smallness of cosmic fluctuations before horizon re-entry is simply a reflection of the fact that, for reasons that are still mysterious, the energy scale defined by the energy density of the universe at horizon exit is a few orders of magnitude less than the Planck energy scale, $(8\pi G)^{-1/2} = 2.4 \times 10^{18} \text{ GeV}$.

This mystery is strongly reminiscent of another mystery encountered in elementary particle physics: the unification energy scale, where the three coupling constants of the electroweak and strong interactions all come together,[7] is about $2 \times 10^{16} \text{ GeV}$, also a few orders of magnitude less than the Planck energy scale $(8\pi G)^{-1/2}$. Perhaps they are the same mystery.

The measured values of n_S and $|N|$ and the observational upper limit on the tensor/scalar ratio r already allow us to put useful constraints on the inflaton potential. We saw in the previous section that this data is close to ruling out any potential $V(\bar{\varphi})$ with an exponential dependence on $\bar{\varphi}$ for the values that $\bar{\varphi}(t)$ takes around the time of horizon exit. To go further, it is useful first to express ϵ and δ in terms of the potential. Using the general relation (4.2.3)

$$\dot{H} = -4\pi G \dot{\bar{\varphi}}^2 ,$$

and the slow-roll formula (4.2.8)

$$\dot{\varphi} = -\frac{V'(\varphi)}{3H} = -\frac{V'(\varphi)}{\sqrt{24\pi G V(\varphi)}} ,$$

gives

$$\epsilon(t) = \frac{1}{16\pi G} \left(\frac{V'(\bar{\varphi}(t))}{V(\bar{\varphi}(t))} \right)^2 . \tag{10.3.28}$$

Also, the time-derivative of Eq. (4.2.3) gives $\ddot{H} = -8\pi G \dot{\bar{\varphi}}\ddot{\bar{\varphi}}$, and using Eq. (4.2.12) then gives

$$\delta(t) = \frac{1}{16\pi G} \left(\frac{V'^2(\bar{\varphi}(t))}{V^2(\bar{\varphi}(t))} - \frac{2V''(\bar{\varphi}(t))}{V(\bar{\varphi}(t))} \right) . \tag{10.3.29}$$

---

[7] See e.g. QTF, Vol. II, Sec. 28.2.

For instance, for a power-law potential $V(\varphi) \propto \varphi^\alpha$, we have

$$\epsilon(t) = \frac{\alpha^2}{16\pi G\bar{\varphi}^2(t)}, \qquad \delta(t) = \frac{2\alpha - \alpha^2}{16\pi G\bar{\varphi}^2(t)}, \qquad (10.3.30)$$

so $\delta(t) = (2/\alpha - 1)\epsilon(t)$, and therefore

$$n_S(q) = 1 - \frac{(\alpha + 2)r(q)}{8\alpha}. \qquad (10.3.31)$$

The experimental bound on n_S depends on the value assumed for r, so observations define an allowed region in the n_S–r plane.[8] At present, the straight line (10.3.31) intersects the (68% confidence level) allowed region for all positive α, even when WMAP three year data is combined with data from the CBI and VSA microwave backgrounds, or from the Sloan or 2dF sky surveys. However, low values of α are favored, and even a modest shrinking of the allowed area would rule out high values of α.

To go further, we need to say something about the value of the scalar field at horizon exit. For this purpose, we can make use of the relation (4.2.14), which gives the number $\Delta\mathcal{N}$ of e-foldings when the scalar field goes from φ_1 to φ_2, under the assumption that the slow-roll approximation holds over this period, as

$$\Delta\mathcal{N} = -\int_{\varphi_1}^{\varphi_2} \left(\frac{8\pi G V(\varphi)}{V'(\varphi)} \right) d\varphi. \qquad (10.3.32)$$

If $|V'/V| \simeq \sqrt{16\pi G\epsilon}$ is essentially constant over the range of φ from φ_1 to φ_2, then the number of e-foldings associated with this change in φ is

$$\Delta\mathcal{N} = \Delta\varphi\sqrt{4\pi G/\epsilon} \qquad (10.3.33)$$

Lyth[9] has used this relation in the case where φ_1 and φ_2 are the field values at horizon exit for wave numbers corresponding to $\ell \simeq 1$ and $\ell \simeq 100$, for which $\Delta\mathcal{N} = \ln 100 = 4.6$, to show that if ϵ is large enough to give a detectable tensor mode, then the scalar field must change by an amount that is at least as large as the Planck scale $1/\sqrt{4\pi G}$.

If we make the strong assumption that the slow roll approximation holds over the whole era from horizon exit to the end of inflation, but do not now assume that ϵ is necessarily constant through this era, then for a power-law

[8]See Figure 14 of D. Spergel *et al.*, *Astrophys. J. Suppl. Ser.* **170**, 377 (2007) [astro-ph/0603449].

[9]D. H. Lyth, *Phys. Rev. Lett.* **78**, 1861 (1997) [hep-ph/9606387]. Also see G. Efstathiou and K. J. Mack, *J. Cosm. & Astropart. Phys.* **05**, 008 (2005) [astro-ph/0503360]; R. Easther, W. H. Kinney, and B. A. Powell, *J. Cosm. Astropart. Phys.* **08**, 004 (2006) [astro-ph/0601276].

potential $V(\varphi) \propto \varphi^{\alpha}$, Eq. (10.3.32) gives the number of e-foldings from horizon exit for a wave number q to the end of inflation as

$$\mathcal{N}(q) = \frac{4\pi G}{\alpha} \left[\bar{\varphi}^2(t_q) - \bar{\varphi}^2(t_{\text{end}}) \right],$$ (10.3.34)

where t_q and t_{end} are the times of horizon exit and the end of inflation, respectively. If we further assume that $|\bar{\varphi}(t_{\text{end}})| \ll |\bar{\varphi}(t_q)|$, then

$$\bar{\varphi}^2(t_q) \simeq \frac{\alpha \mathcal{N}(q)}{4\pi G},$$ (10.3.35)

so Eq. (10.3.30) gives

$$\epsilon(t_q) \simeq \frac{\alpha}{4\mathcal{N}(q)}, \qquad \delta(t_q) \simeq \frac{2 - \alpha}{4\mathcal{N}(q)},$$

and therefore[10]

$$n_S(q) \simeq 1 - \frac{\alpha + 2}{2\mathcal{N}(q)}, \qquad r(q) = \frac{4\alpha}{\mathcal{N}(q)}.$$ (10.3.36)

We noted in Section 10.1 that if the energy density at the beginning of the radiation-dominated era is the Planck density G^{-2}, then $\mathcal{N}(q) \simeq 68$ for wave numbers that are just coming into the horizon at the present, and correspondingly less for larger wave numbers; for instance, for wave numbers corresponding to $\ell = 100$, $\mathcal{N}$ would be less by an amount $\ln 100 = 4.6$. To derive a better estimate of $\mathcal{N}$, we can use Eq. (10.3.27). Under the risky assumption that the energy density ρ_1 at the beginning of the radiation-dominated era is the same as at horizon exit, and taking $\epsilon = O(.02)$, we have $\rho_1 \simeq [2.5 \times 10^{16} \text{ GeV}]^{1/4}$, so Eq. (10.1.43) with $h = 0.7$ shows that for a wave number that just enters the horizon at the present, $\mathcal{N} \simeq 62$, while for the wave number corresponding to $\ell \simeq 100$, $\mathcal{N} \simeq 57$. Taking $\mathcal{N} = 60$, for a quadratic potential Eq. (10.3.36) gives $n_S = .97$ and $r = 0.13$, which is consistent with the WMAP third-year results, while for a quartic potential Eq. (10.3.35) gives $n_S = 0.95$ and $r = 0.26$, which is barely outside the range allowed by WMAP.[11] Any $\alpha > 4$ is ruled out. But this conclusion is contingent on the assumption of a slow-roll inflation from the time of horizon exit until the end of inflation, with $V(\bar{\varphi}(t)) \propto \bar{\varphi}^{\alpha}(t)$ over this whole period.

---

[10]D. H. Lyth and A. Riotto, *Phys. Rep.* **314**, 1 (1999).
[11]D. Spergel, ref. 8.

10.4 Multifield inflation

Observations of the cosmic microwave background and large scale structure indicate that the primordial scalar fluctuations outside the horizon are

- nearly Gaussian,

- adiabatic,

- nearly scale invariant, in the sense that $\mathcal{R}_q^o$ is nearly proportional to $q^{-3/2}$,

- weak, in the sense that $q^{3/2}\mathcal{R}_q^o \ll 1$. (We saw in Section 7.2 that $q^{3/2}\mathcal{R}_q^o$ is of order 10^{-5}.)

We have seen in Sections 10.1 and 10.3 that these properties of primordial scalar fluctuations follow under the assumptions that

1. *The energy density during inflation receives appreciable contributions from just a single real "inflaton" scalar field.* This implies that the fluctuations are adiabatic during inflation, in which case they remain so thereafter.

2. *During the era of horizon exit (say, for q/aH falling from 10 to 0.1), H is sufficiently small so that q/a is less than whatever fundamental scale (such as the grand unification scale or the Planck scale) characterizes the theory, not only during this era but for some time before it.* (For the case considered in Section 10.1, we estimated that $H \approx 10^{14}$ GeV, which is probably small enough.) This implies that during this era the scalar field is described by a simple effective action, involving no more than two spacetime derivatives. It follows that for some time before the era of horizon exit the inflaton behaves like a free field, so that the fluctuations are Gaussian.

3. *For observed fluctuations, in the era of horizon exit inflation is "slow-roll," in the sense that $|\dot{H}|/H^2 \ll 1$ and $|\ddot{H}/H\dot{H}| \ll 1$.* Together with assumptions 1 and 2, this implies that the fluctuations are nearly scale invariant.

But there is no particular reason to believe that the energy density during inflation is dominated by a *single* scalar field, so we are naturally led to consider the case of several inflaton fields $\varphi^n(x)$. We shall show that the same properties of primordial scalar fluctuations follow if we make assumptions 2 and 3, but replace assumption 1 with

1*. *The energy in all the scalar fields is converted at the end of inflation into ordinary matter and radiation in local thermal equilibrium, in which all chemical potentials vanish.* (Baryon and lepton number would then have to be generated later, as discussed in Section 3.3.)

Whatever the number of scalar fields, under assumption 2, the effective action of the scalar fields is dominated by terms with a minimum number of spacetime derivatives during and after the era of horizon exit and for some time before it. For arbitrary numbers of scalar fields, the most general such action takes the form

$$I_\varphi = \int d^4x \sqrt{-\mathrm{Detg}} \left[-\frac{1}{2} g^{\mu\nu} \gamma_{nm}(\varphi) \frac{\partial \varphi^n}{\partial x^\mu} \frac{\partial \varphi^m}{\partial x^\nu} - V(\varphi) \right], \qquad (10.4.1)$$

where $V(\varphi)$ is an arbitrary real potential, repeated scalar field indices are summed, and $\gamma_{nm}(\varphi)$ is an arbitrary real symmetric positive-definite matrix, which we shall call the *field metric*. (This matrix must be positive-definite to give the right sign to commutators of fields and their time derivatives.) The energy-momentum tensor, which serves as the source of the gravitational field, is derived as described in Appendix B from the action (10.4.1), and takes the form

$$T_{\mu\nu} = g_{\mu\nu} \left[-\frac{1}{2} g^{\rho\sigma} \gamma_{nm}(\varphi) \frac{\partial \varphi^n}{\partial x^\rho} \frac{\partial \varphi^m}{\partial x^\sigma} - V(\varphi) \right] + \gamma_{nm}(\varphi) \frac{\partial \varphi^n}{\partial x^\mu} \frac{\partial \varphi^m}{\partial x^\nu}. \quad (10.4.2)$$

The scalar field equations are derived from the principle that I_φ must be stationary with respect to infinitesimal variations in the the scalar fields, and take the Euler–Lagrange form

$$\frac{\partial}{\partial x^\mu} \left(\sqrt{-\mathrm{Detg}}\, g^{\mu\nu} \gamma_{nm}(\varphi) \frac{\partial \varphi^n}{\partial x^\nu} \right) = \sqrt{-\mathrm{Detg}}$$

$$\times \left(\frac{1}{2} g^{\mu\nu} \frac{\partial \gamma_{lm}(\varphi)}{\partial \varphi^n} \frac{\partial \varphi^l}{\partial x^\mu} \frac{\partial \varphi^m}{\partial x^\nu} + \frac{\partial V(\varphi)}{\partial \varphi^n} \right). \qquad (10.4.3)$$

We take each scalar field $\varphi^n(x)$ as an unperturbed term $\bar{\varphi}^n(t)$ that depends only on time, plus a small perturbation $\delta\varphi^n(\mathbf{x}, t)$:

$$\varphi^n(\mathbf{x}, t) = \bar{\varphi}^n(t) + \delta\varphi^n(\mathbf{x}, t). \qquad (10.4.4)$$

Similarly, as in Chapters 5–8, the metric is assumed to be given by the unperturbed Robertson–Walker metric $\bar{g}_{\mu\nu}(t)$ (with $K = 0$) plus a small perturbation $h_{\mu\nu}(\mathbf{x}, t)$

$$g_{\mu\nu}(\mathbf{x}, t) = \bar{g}_{\mu\nu}(t) + h_{\mu\nu}(\mathbf{x}, t). \qquad (10.4.5)$$

The unperturbed energy-momentum tensor is of the perfect fluid form (5.1.35), with unperturbed energy density, pressure, and velocity

$$\bar{\rho} = \frac{1}{2}\gamma_{nm}(\bar{\varphi})\dot{\bar{\varphi}}^n\dot{\bar{\varphi}}^m + V(\bar{\varphi}) \,, \tag{10.4.6}$$

$$\bar{p} = \frac{1}{2}\gamma_{nm}(\bar{\varphi})\dot{\bar{\varphi}}^n\dot{\bar{\varphi}}^m - V(\bar{\varphi}) \,, \tag{10.4.7}$$

$$\bar{u}^0 = 1 \,, \quad \bar{u}^i = 0 \,. \tag{10.4.8}$$

The scalar field equation (10.4.3) for the unperturbed fields is

$$\ddot{\bar{\varphi}}^n + \gamma_{ml}^n(\bar{\varphi})\dot{\bar{\varphi}}^m\dot{\bar{\varphi}}^l + 3H\dot{\bar{\varphi}}^n + \gamma^{nm}(\bar{\varphi})\frac{\partial V(\bar{\varphi})}{\partial \bar{\varphi}^m} = 0 \,, \tag{10.4.9}$$

where γ^{nm} is the reciprocal of the matrix γ_{nm}, γ_{ml}^n is the affine connection in field space:

$$\gamma_{ml}^n(\bar{\varphi}) = \frac{1}{2}\gamma^{nk}(\bar{\varphi})\left(\frac{\partial\gamma_{km}(\bar{\varphi})}{\partial\bar{\varphi}^l} + \frac{\partial\gamma_{kl}(\bar{\varphi})}{\partial\bar{\varphi}^m} - \frac{\partial\gamma_{ml}(\bar{\varphi})}{\partial\bar{\varphi}^k}\right) \,, \tag{10.4.10}$$

and H is the expansion rate $H \equiv \dot{a}/a = \sqrt{8\pi G\bar{\rho}/3}$. The reader may check that Eq. (10.4.9) guarantees that the energy-conservation equation $\dot{\bar{\rho}} = -3H(\bar{\rho} + \bar{p})$ is satisfied by the quantities (10.4.6) and (10.4.7). From Eqs. (10.4.6) and (10.4.9) we find the convenient formula

$$\dot{H} = -4\pi G\gamma_{nm}(\bar{\varphi})\dot{\bar{\varphi}}^n\dot{\bar{\varphi}}^m \,. \tag{10.4.11}$$

For more than one scalar field $T_{\mu\nu}$ is not of the perfect fluid form (5.1.35) to all orders in perturbations, but by comparing the first-order terms in Eq. (10.4.2) with Eqs. (5.1.39)–(5.1.41), we see that to first order the anisotropic inertia vanishes, and the perturbations to the energy density, pressure, and velocity potential are

$$\delta\rho = \gamma_{nm}(\bar{\varphi})\dot{\bar{\varphi}}^n\delta\dot{\varphi}^m + \frac{1}{2}\dot{\bar{\varphi}}^n\dot{\bar{\varphi}}^m\frac{\partial\gamma_{nm}(\bar{\varphi})}{\partial\bar{\varphi}^k}\delta\varphi^k$$
$$+ \frac{\partial V(\bar{\varphi})}{\partial\bar{\varphi}^n}\delta\varphi^n + \frac{1}{2}h_{00}\gamma_{nm}(\bar{\varphi})\dot{\bar{\varphi}}^n\dot{\bar{\varphi}}^m \,, \tag{10.4.12}$$

$$\delta p = \gamma_{nm}(\bar{\varphi})\dot{\bar{\varphi}}^n\delta\dot{\varphi}^m + \frac{1}{2}\dot{\bar{\varphi}}^n\dot{\bar{\varphi}}^m\frac{\partial\gamma_{nm}(\bar{\varphi})}{\partial\bar{\varphi}^k}\delta\varphi^k$$
$$- \frac{\partial V(\bar{\varphi})}{\partial\bar{\varphi}^n}\delta\varphi^n + \frac{1}{2}h_{00}\gamma_{nm}(\bar{\varphi})\dot{\bar{\varphi}}^n\dot{\bar{\varphi}}^m \,, \tag{10.4.13}$$

$$\delta u = -\frac{\gamma_{nm}(\bar{\varphi})\dot{\bar{\varphi}}^n\delta\varphi^m}{\gamma_{kl}(\bar{\varphi})\dot{\bar{\varphi}}^k\dot{\bar{\varphi}}^l} \,. \tag{10.4.14}$$

The reader can easily check that these formulas reduce to Eqs. (10.1.9)–(10.1.11) in the single-field case, with $\gamma_{11} = 1$.

Since there is no first-order anisotropic inertia, in Newtonian gauge we have $\Phi = \Psi$, so $h_{00} = -2\Psi$, and the Einstein field equation Eq. (5.3.21) takes the form

$$\dot{\Psi} + H\Psi = 4\pi G \gamma_{nm}(\bar{\varphi})\dot{\bar{\varphi}}^n \delta\varphi^m \, , \qquad (10.4.15)$$

The terms of first order in the field equation Eq. (10.4.3) are much simplified if we now adopt a notation that reveals the transformation of quantities under redefinitions $\varphi^n \to \varphi'^n(\varphi)$ of the scalar fields. Under such transformations, quantities like $\dot{\bar{\varphi}}^n$ and $\delta\varphi^n$ transform as contravariant vectors, in the sense that

$$\delta\varphi'^n = \frac{\partial\bar{\varphi}'^n}{\partial\bar{\varphi}^m}\delta\varphi^m \, , \qquad \dot{\bar{\varphi}}'^n = \frac{\partial\bar{\varphi}'^n}{\partial\bar{\varphi}^m}\dot{\bar{\varphi}}^m \qquad (10.4.16)$$

For any vector v^n that transforms in this way, we can define a rate of change that is also a vector:

$$\frac{D}{Dt}v^n \equiv \frac{\partial}{\partial t}v^n + \gamma_{lm}^n(\bar{\varphi})\,\dot{\bar{\varphi}}^l v^m \, . \qquad (10.4.17)$$

With this notation, the first-order terms in the field equation (10.4.3) give

$$\frac{D^2}{Dt^2}\delta\varphi^n + 3H\frac{D}{Dt}\delta\varphi^n + \gamma^{nm}(\bar{\varphi})\frac{\partial^2 V(\bar{\varphi})}{\partial\bar{\varphi}^m\,\partial\bar{\varphi}^l}\delta\varphi^l - \left(\frac{\nabla^2}{a^2}\right)\delta\varphi^n$$

$$= -2\gamma^{nm}(\bar{\varphi})\,\Psi\,\frac{\partial V(\bar{\varphi})}{\partial\bar{\varphi}^m} + 4\dot{\Psi}\,\dot{\bar{\varphi}}^n + \gamma^n{}_{lmk}(\bar{\varphi})\,\dot{\bar{\varphi}}^l\dot{\bar{\varphi}}^m\delta\varphi^k \, , \quad (10.4.18)$$

where $\gamma^n{}_{lmk}(\bar{\varphi})$ is the Riemann–Christoffel tensor in field space:

$$\gamma^n{}_{lmk}(\bar{\varphi}) \equiv \frac{\partial\gamma_{ml}^n(\bar{\varphi})}{\partial\bar{\varphi}^k} - \frac{\partial\gamma_{mk}^n(\bar{\varphi})}{\partial\bar{\varphi}^l} + \gamma_{lm}^r(\bar{\varphi})\gamma_{kr}^n(\bar{\varphi}) - \gamma_{lk}^r(\bar{\varphi})\gamma_{mr}^n(\bar{\varphi}) \, . \quad (10.4.19)$$

(The scalar fields can be redefined to make $\gamma_{nm} = \delta_{nm}$ if and only if $\gamma^n{}_{lmk} = 0$. We are not assuming that this is the case.) Also, the constraint (5.3.26) is here

$$\left(\dot{H} - \frac{\nabla^2}{a^2}\right)\Psi = 4\pi G\gamma_{nm}(\bar{\varphi})\left(-\dot{\bar{\varphi}}^n\frac{D}{Dt}\delta\varphi^m + \delta\varphi^m\frac{D}{Dt}\dot{\bar{\varphi}}^n\right) . \qquad (10.4.20)$$

The solutions are written as superpositions of plane waves

$$\delta\varphi^n(\mathbf{x}, t) =$$
$$\sum_N \int d^3q \left[\delta\varphi_{Nq}^n(t)e^{i\mathbf{q}\cdot\mathbf{x}}\alpha(\mathbf{q}, N) + \delta\varphi_{Nq}^{n*}(t)e^{-i\mathbf{q}\cdot\mathbf{x}}\alpha^*(\mathbf{q}, N)\right] \qquad (10.4.21)$$

$$\Psi(\mathbf{x}, t) =$$

$$\sum_N \int d^3q \left[\Psi_{Nq}(t) e^{i\mathbf{q}\cdot\mathbf{x}} \alpha(\mathbf{q}, N) + \Psi^*_{Nq}(t) e^{-i\mathbf{q}\cdot\mathbf{x}} \alpha^*(\mathbf{q}, N) \right]. \quad (10.4.22)$$

Here N labels different solutions of the coupled equations (10.4.15)–(10.4.17), with ∇^2 replaced with $-q^2$:

$$\dot{\Psi}_{Nq} + H\Psi_{Nq} = 4\pi G \gamma_{nm}(\bar{\varphi}) \dot{\bar{\varphi}}^n \delta\varphi^m_{Nq}, \quad (10.4.23)$$

$$\frac{D^2}{Dt^2} \delta\varphi^n_q + 3H \frac{D}{Dt} \delta\varphi^n_q + \gamma^{nm}(\bar{\varphi}) \frac{\partial^2 V(\bar{\varphi})}{\partial \bar{\varphi}^m \partial \bar{\varphi}^l} \delta\varphi^l_q + \left(\frac{q^2}{a^2}\right) \delta\varphi^n_q$$

$$= -2\gamma^{nm}(\bar{\varphi}) \Psi_q \frac{\partial V(\bar{\varphi})}{\partial \bar{\varphi}^m} + 4\dot{\Psi}_q \dot{\bar{\varphi}}^n + \gamma^n{}_{lmk}(\bar{\varphi}) \dot{\bar{\varphi}}^l \dot{\bar{\varphi}}^m \delta\varphi^k_q \quad (10.4.24)$$

$$\left(\dot{H} + \frac{q^2}{a^2}\right) \Psi_q = 4\pi G \gamma_{nm}(\bar{\varphi}) \left(-\dot{\bar{\varphi}}^n \frac{D}{Dt} \delta\varphi^m_q + \delta\varphi^m_q \frac{D}{Dt} \dot{\bar{\varphi}}^n \right). \quad (10.4.25)$$

There is one second-order equation for each scalar field, and one first-order equation for Ψ, so with one constraint on first derivatives the number of independent solutions equals twice the number of scalar fields. Since $(\delta\varphi^n_{Nq}, \Psi_{Nq})$ and $(\delta\varphi^{n*}_{Nq}, \Psi^*_{Nq})$ are all independent solutions, the index N takes as many values as the index n.

To find the initial conditions for Eqs. (10.4.23)–(10.4.25), and to find the commutation relations for the operators $\alpha(\mathbf{q})$ and $\alpha^*(\mathbf{q})$ in Eqs. (10.4.21) and (19.4.22), we note that for some time before the era of horizon exit we will have $q/a \gg H$, and q^2/a^2 much greater than any element of $\partial^2 V / \partial \bar{\varphi}^n \partial \bar{\varphi}^m$. Hence the solutions up to the beginning of the era of horizon exit take the WKB form

$$\delta\varphi^n_{Nq}(t) \to f^n_{Nq}(t) \exp\left(-iq \int_{t_1}^t \frac{dt'}{a(t')}\right),$$

$$\Psi_{Nq}(t) \to g_{Nq}(t) \exp\left(-iq \int_{t_1}^t \frac{dt'}{a(t')}\right), \quad (10.4.26)$$

where $f^n_{Nq}(t)$ and $g_{Nq}(t)$ vary much more slowly than the argument of the exponential, and t_1 is arbitrary. Eqs. (10.4.23) and (10.4.25) are both satisfied to leading order in q/a if we take

$$g_{Nq} = \frac{4i\pi G a}{q} \gamma_{nm}(\bar{\varphi}) \dot{\bar{\varphi}}^n f^m_{Nq}. \quad (10.4.27)$$

The terms in Eq. (10.4.24) of first order in q/a then give

$$\frac{D}{Dt} f^n_{Nq} + H f^n_{Nq} = 0. \quad (10.4.28)$$

To solve this, we note that, because $\gamma^{nm}\left(\bar{\varphi}(t)\right)$ is positive-definite, it can be written in terms of a set of vielbein vectors $e_N^n(t)$ (with N running over as many values as n), as

$$\gamma^{nm}\left(\bar{\varphi}(t)\right) = \sum_N e_N^n(t) e_N^m(t) . \qquad (10.4.29)$$

These vielbeins can be defined to satisfy the equation of parallel transport[1]

$$\frac{D}{Dt} e_{Nq}^n = 0 , \qquad (10.4.30)$$

so the solution of Eq. (10.4.28) is

$$f_{Nq}^n(t) \propto a^{-1}(t) e_N^n(t) . \qquad (10.4.31)$$

For reasons that will soon become apparent, we shall normalize these solutions so that, for $q/a \gg H$,

$$f_{Nq}^n = (2\pi)^{-3/2}(2q)^{-1/2}a^{-1}e_N^n .$$

With this normalization, at the beginning of the era of horizon exit we have[2]

$$\delta\varphi_{Nq}^n(t) = \frac{1}{(2\pi)^{3/2}a(t)\sqrt{2q}} e_N^n(t) \exp\left(-iq\int_{t_1}^t \frac{dt'}{a(t')}\right) , \qquad (10.4.32)$$

$$\Psi_{Nq}(t) = \frac{4i\pi\, G\gamma_{nm}(\bar{\varphi})e_N^m(t)\dot{\bar{\varphi}}^n(t)}{(2\pi)^{3/2}\sqrt{2q^3}} \exp\left(-iq\int_{t_1}^t \frac{dt'}{a(t')}\right) . \qquad (10.4.33)$$

For times early enough so that $q/a \gg H$ the commutation relations for φ^n can be obtained from the action (10.4.1) with Ψ neglected, and therefore

---

[1] We define the vectors $e_{Nq}^n(t)$ to satisfy the first-order differential equation (10.4.30), and, as an initial condition, to satisfy Eq. (10.4.29) at some initial time $t = t_1$. From Eq. (10.4.30) and the definition (10.4.10) it follows that for all times

$$\dot{D}_m^n = \left[-\gamma_{lk}^n D_m^k + \gamma_{lm}^k D_k^n\right]\dot{\bar{\varphi}}^l ,$$

where

$$D_m^n \equiv \sum_N e_N^n e_N^k \gamma_{km} .$$

This differential equation for D_m^n has a solution $D_m^n = \delta_m^n$, and our initial condition tells us that $D_m^n = \delta_m^n$ at $t = t_1$, so this is the solution for all times. It follows that the vectors $e_{Nq}^n(t)$ defined in this way satisfy the condition (10.4.29) for vielbeins for all times.

[2] A result equivalent to Eq. (10.4.32) is given in Eq. (4.4) of H.-C. Lee, M. Sasaki, E. D. Stewart, T. Tanaka, and S. Yokoyama, *J. Cosm. & Astropart. Phys.* **0510**, 004 (2005) [astro-ph/0506262], using a "δN" formalism due to M. Sasaki and E. D. Stewart, Prog. Theor. Phys. **95**, 71 (1996) [astro-ph/9507001]. But their paper does not reach the conclusion (10.4.41) found in this section.

take the form:

$$[\varphi^n(\mathbf{x}, t), \varphi^m(\mathbf{y}, t)] = 0 \,,$$

$$[\varphi^n(\mathbf{x}, t), \dot{\varphi}^m(\mathbf{y}, t)] = ia^{-3}(t)\, \gamma^{nm}\!\left(\bar{\varphi}(t)\right)\delta^3(\mathbf{x} - \mathbf{y}) \,,$$

With $\delta\varphi^n$ normalized as in Eq. (10.4.32), this implies that the time-independent operator coefficients in Eqs. (10.4.21) and (10.4.22) satisfy the commutation relations

$$\Big[\alpha(\mathbf{q}, N)\,,\, \alpha(\mathbf{q}', N')\Big] = 0 \,, \quad \Big[\alpha(\mathbf{q}, N)\,,\, \alpha^*(\mathbf{q}', N')\Big] = \delta^3(\mathbf{q} - \mathbf{q}')\delta_{NN'} \,.$$

$$(10.4.34)$$

Assuming that there is enough time before horizon exit for the state of the world to decay into the Bunch–Davies vacuum $|0\rangle$, with $\alpha(\mathbf{q}, N)|0\rangle = 0$, it follows then from Eqs. (10.4.21) and (10.4.22) that the observed perturbations will be *Gaussian*, just as in the single-field case.

According to assumption 3, during the era when q/a drops from being somewhat larger to somewhat smaller than H — say, from $10H$ to $0.1H$, the scalar fields are rolling slowly down the potential hill. We assume that $V(\varphi)$ satisfies whatever flatness conditions are necessary to allow us to drop all terms in Eqs. (10.4.23)–(10.4.25) proportional to $\dot{\bar{\varphi}}^n_N$ or $\ddot{\bar{\varphi}}^n_N$ and to ignore the second derivative of the potential in Eq. (10.4.24). Then during the era of horizon exit Eq. (10.4.24) is approximately

$$\delta\ddot{\varphi}^n_{Nq} + 3H\delta\dot{\varphi}^n_{Nq} + \left(\frac{q^2}{a^2}\right)\delta\varphi^n_{Nq} = 0 \,. \qquad (10.4.35)$$

H is roughly constant during this era,[3] so the independent solutions of Eq. (10.4.35) are proportional to $(1 + iq\tau)\exp(-iq\tau)$ and its complex conjugate, where τ is again the conformal time

$$\tau = \int_\infty^t \frac{dt'}{a(t')} \simeq -\frac{1}{Ha(t)} \,.$$

The scalar field perturbations at the beginning of the era of horizon exit are given by Eq. (10.4.32), so during this era we have

$$\delta\varphi^n_{Nq} \simeq \frac{1}{(2\pi)^{3/2}\sqrt{2q}}\left(\frac{1}{a} + \frac{iH}{q}\right)e^{+iq/aH}e^n_N \,. \qquad (10.4.36)$$

---

[3]The fractional change in H during the era of horizon exit is $|\dot{H}/H| \times \ln(100)/H$, which is small if $|\dot{H}|/H^2 \ll 1/\ln(100) = 0.22$.

By the end of the era of horizon exit we have $q/a \ll H$, and the scalar field perturbations approach the quantities

$$\delta\varphi^n_{Nq} \to \frac{iH}{(2\pi)^{3/2}\sqrt{2q^3}}e^n_N .$$

(10.4.37)

Since H and e^n_N are slowly varying, they can be evaluated at the time of horizon exit, and depend weakly on q.

At some time after the era of horizon exit the slow-roll conditions must become violated, if only in order that the energy in the inflaton fields can eventually be converted into ordinary matter and radiation. The potential term on the left-hand side of Eq. (10.4.24) is then no longer negligible, and things get complicated. But once q/a becomes much less than H, the subsequent evolution of the scalar fields during inflation cannot depend on q, so until horizon re-entry all scalar field perturbations $\delta\varphi^n_{Nq}$ will have the same wave number dependence, close to $q^{-3/2}$, as given by Eq. (10.4.37) at the end of the era of horizon exit. The same applies to the scalar metric perturbations; when a becomes sufficiently large so that $q^2/a^2 \ll |\dot{H}|$, Eq. (10.4.25) gives Ψ_{Nq} approximately proportional to $q^{-3/2}$.

In general, it is not easy to see what these results imply for the perturbations observed in the cosmic microwave background or large scale structure. However, there is one case where an important conclusion can be reached. If according to assumption 1* the energy in all the scalar fields is converted at the end of inflation into ordinary matter and radiation in local thermal equilibrium, and if at this time all conserved quantities like electric charge have zero density, then as remarked in Section 5.4, the perturbations become adiabatic, with $\mathcal{R}_q$ taking a constant value $\mathcal{R}^o_q$ until horizon re-entry. For small fluctuations, $\mathcal{R}^o_q$ will be some linear combination of the perturbations $\delta\varphi^n_{Nq}$ and Ψ_{Nq} at the end of the era of horizon exit. We do not know the coefficients in this linear combination, which in general will depend on the shape of the potential experienced as the field evolves until the end of inflation, as well as on the mechanism of energy transfer to matter and radiation. But we can be sure that these coefficients are independent of wave number, because once the era of horizon exit is over the perturbations are far outside the horizon. Hence we can conclude that in this case $\mathcal{R}^o_q$ will have the same wave number dependence as $\delta\varphi^n_{Nq}$ and Ψ_{Nq} at the end of the era of horizon exit, which for slow-roll inflation will be close to $q^{-3/2}$.

The ubiquity of the $q^{-3/2}$ wavelength dependence can be understood on very general grounds. For negligible spatial curvature, nothing should depend on how the co-moving coordinate vector $\mathbf{x}$ is normalized, so suppose we change its scale by a transformation $\mathbf{x} \to \lambda\mathbf{x}$, with λ constant. To keep $\mathbf{q} \cdot \mathbf{x}$ unchanged, we must then change the scale of co-moving wave numbers by a transformation $\mathbf{q} \rightarrow \lambda^{-1}\mathbf{q}$. Consider any perturbation $Z(\mathbf{x}, t)$ that, like $\delta\varphi^n(\mathbf{x}, t)$ and $\Psi(\mathbf{x}, t)$, is given by a Fourier integral

$$Z(\mathbf{x}, t) = \sum_N \int d^3q \left[e^{i\mathbf{q}\cdot\mathbf{x}} a(\mathbf{q}, N) z_{Nq}(t) + e^{-i\mathbf{q}\cdot\mathbf{x}} a^*(\mathbf{q}, N) z^*_{Nq}(t) \right],$$

where N labels the various solutions of the field equations, and $a(\mathbf{q}, N)$ and $a^*(\mathbf{q}, N)$ are annihilation and creation operators satisfying the commutation relations

$$\left[a(\mathbf{q}, N), a^*(\mathbf{q}', N') \right] = \delta_{NN'} \delta^3(\mathbf{q} - \mathbf{q}').$$

Under the transformation $\mathbf{q} \rightarrow \lambda^{-1}\mathbf{q}$, the delta function in this commutation relation transforms as $\delta^3(\mathbf{q} - \mathbf{q}') \rightarrow \lambda^3 \delta^3(\mathbf{q} - \mathbf{q}')$, so we must have $a(\mathbf{q}, N) \rightarrow \lambda^{3/2} a(\mathbf{q}, N)$. Also, of course $d^3q \rightarrow \lambda^{-3} d^3q$. Hence in order for $Z(\mathbf{x}, t)$ to be unaffected by this change of scale, we must have $z_{Nq}(t) \rightarrow \lambda^{3/2} z_{Nq}(t)$. This condition is satisfied if $z_{Nq}(t)$ has a q-dependence $\propto q^{-3/2}$, *and* if it does not depend on the scale of $a(t)$. Of course, to keep the physical coordinate vector $\mathbf{x}a(t)$ independent of the normalization chosen for the co-moving coordinates, $a(t)$ has the scale transformation $a(t) \rightarrow \lambda^{-1}a(t)$, so $q/a(t)$ is scale-invariant, and if $z_{Nq}(t)$ depends on the scale of the function $a(t)$ then we cannot conclude that it is proportional to $q^{-3/2}$. But if $z_{Nq}(t)$, like $\delta\varphi^n_{Nq}(t)$ and $\Psi_{Nq}(t)$, takes a nearly time-independent value z^o_{Nq} after horizon exit, then outside the horizon it will not depend strongly on $a(t)$. It could still depend on $H(t_q)$, $\dot{H}(t_q)$, etc., where t_q is the time of horizon exit, defined by the scale-invariant condition $q/a(t_q) = H(t_q)$, but in the limit of very slow roll inflation H, $\dot{H}$, etc. depend only weakly on time, so z^o_{Nq} can depend only weakly on the scale of the function a, and the scale-invariance of $Z(\mathbf{x}, t)$ requires that z^o_{Nq} be nearly proportional to $q^{-3/2}$.

The intensity of observed adiabatic fluctuations is related to the quantity $\mathcal{R}$ defined by Eq. (5.4.1). Using Eqs. (10.4.14), (10.4.11), (10.4.21), and (10.4.22), during inflation this is

$$\begin{aligned}
\mathcal{R}(\mathbf{x}, t) &\equiv -\Psi(\mathbf{x}, t) + H(t)\,\delta u(\mathbf{x}, t) \\
&= -\Psi(\mathbf{x}, t) + \frac{4\pi\, GH(t)}{\dot{H}(t)} \gamma_{nm}\big(\bar{\varphi}(t)\big) \dot{\bar{\varphi}}^n(t) \delta\varphi^m(\mathbf{x}, t) \\
&= \sum_N \int d^3q \left[e^{i\mathbf{q}\cdot\mathbf{x}} a(\mathbf{q}, N) \mathcal{R}_{Nq}(t) + e^{-i\mathbf{q}\cdot\mathbf{x}} a^*(\mathbf{q}, N) \mathcal{R}^*_{Nq}(t) \right],
\end{aligned}$$

$$(10.4.38)$$

where

$$\mathcal{R}_{Nq}(t) = -\Psi_{Nq}(t) + \frac{4\pi\, GH(t)}{\dot{H}(t)} \gamma_{nm}\big(\bar{\varphi}(t)\big) \dot{\bar{\varphi}}^n(t) \delta\varphi^m_{Nq}(t). \qquad (10.4.39)$$

Because of the factor $\dot{H}$ in the denominator, at the end of the era of horizon exit the second term in Eq. (10.4.39) dominates over the term $-\Psi_{Nq}$, and Eq. (10.4.37) then gives at this time

$$\mathcal{R}_{Nq} \rightarrow i \frac{4\pi G H^2}{(2\pi)^{3/2}\sqrt{2q^3}\,\dot{H}} \gamma_{nm}(\bar{\varphi})\dot{\bar{\varphi}}^n e_N^m \;. \tag{10.4.40}$$

The definition (10.4.29) of the vielbeins and the formula (10.4.11) then lead immediately to a sum rule

$$\left(\sum_N |\mathcal{R}_{Nq}|^2 \right)^{1/2} = \frac{\sqrt{G}H^2}{2\pi\,q^{3/2}\sqrt{|\dot{H}|}} \;. \tag{10.4.41}$$

That is, the root-mean-square value of the quantities $|\mathcal{R}_{Nq}|$ at the end of the era of horizon exit is the same as the value of $|\mathcal{R}_q|$ outside the horizon in the single-field case, given by Eq. (10.3.16). But Eqs. (10.4.38) and (10.4.34) tell us that the correlation function of $\mathcal{R}$ in the Bunch–Davies vacuum is

$$\int d^4x\, e^{-i\mathbf{q}\cdot(\mathbf{x}-\mathbf{y})} \langle \mathcal{R}(\mathbf{x}, t)\, \mathcal{R}(\mathbf{y}, t) \rangle = (2\pi)^3 \sum_N |\mathcal{R}_{Nq}|^2 \tag{10.4.42}$$

So from Eq. (10.4.41) it follows that *the correlation function $\langle \mathcal{R}(\mathbf{x}, t)\, \mathcal{R}(\mathbf{y}, t) \rangle$ is the same at the end of the era of horizon exit as in the single field case.* With more than one scalar field the correlation function (10.4.42) is not in general time-independent outside the horizon, but it is plausible that the value of $|\mathcal{R}_q|$ during a period of thermal equilibrium after inflation will not be orders of magnitude different from Eq. (10.4.41) at the end of the era of horizon exit. Thus the observed strength and spectral shape of anisotropies in the cosmic microwave background suggests a value of H at horizon exit of order 10^{14} GeV, as in the single field case.

One can derive stronger results in the case where all but one of the eigenvalues of the matrix $\gamma^{nm}(\bar{\varphi})\,\partial^2 V(\bar{\varphi})/\partial\bar{\varphi}^m\partial\bar{\varphi}^l$ in Eq. (10.4.23) are large and positive. In this case, the unperturbed scalar fields roll along the direction of the eigenvector for the small eigenvalue, and the only significant perturbations lie in that direction. The problem then reduces to the single field case; we have an essentially adiabatic perturbation, with the amplitude $\mathcal{R}_q^o$ given in the slow-roll approximation by Eq. (10.3.16), and the slope $n_S(q)$ given in this approximation by Eq. (10.3.25), where ϵ and δ are to be calculated in terms of the time-derivatives of the expansion rate by Eq. (10.3.4). But there is no known reason why the potential should have the properties needed to justify these results. On the other hand, tensor perturbations during inflation are governed by Eq. (10.3.17) however many scalar fields there are, and the only role played by the scalar fields is to contribute to the

Hubble rate H. Thus for slow-roll inflation with any number of scalar fields, the tensor amplitude $\mathcal{D}_q^o$ is given by Eq. (10.3.23), and the slope parameter $n_T(q)$ is given by Eq. (10.3.25), just as in the single field case.

The fact that observations of cosmic microwave background anisotropies and large scale structure indicate that scalar fluctuations outside the horizon are adiabatic and Gaussian, with $\mathcal{R}_q^o$ approximately proportional to $q^{-3/2}$, and with $q^{3/2}\mathcal{R}_q^o \ll 1$, evidently is consistent with a very large class of models of inflation. This is encouraging, because it supports the general idea of slow-roll inflation, but also disappointing, because it shows that these observations so far do not really tell us anything specific about the details of inflation. With further improvements in experimental precision, we can look forward to a more decisive test of theories of the early universe.

Appendix A

Some Useful Numbers

Numerical Constants

$$\pi = 3.1415927 \qquad 1'' = 4.84814 \times 10^{-6} \text{ radians}$$
$$e = 2.7182818 \qquad \ln 10 = 2.3025851$$
$$\gamma = 0.5772157 \qquad \zeta(3) = 1.2020569$$

Physical Constants[1]

Speed of light in vacuum	$c \equiv 2.99792458 \times 10^{10} \text{ cm sec}^{-1}$
Planck constant	$h = 6.6260693(11) \times 10^{-27} \text{ erg-sec}$
Reduced Planck constant	$\hbar \equiv h/2\pi = 1.05457168(18) \times 10^{-27} \text{ erg sec}$
	$= 6.58211915(56) \times 10^{-22} \text{ MeV sec}$
Electronic charge (unrat.)	$e = 4.80320441(41) \times 10^{-10} \text{ esu}$
Electron volt	$1 \text{ eV} = 1.60217653(14) \times 10^{-12} \text{ erg}$
	$\hbar c = 197.326968(17) \times 10^{-13} \text{ MeV cm}$
Fine structure constant	$\alpha \equiv e^2/\hbar c = 1/137.03599911(46)$
Electron mass	$m_e = 9.1093826(16) \times 10^{-28} \text{ g}$
	$m_e c^2 = 0.510998918(44) \text{ MeV}$
Rydberg energy	$hcR \equiv m_e e^4/2\hbar^2 = 13.6056923(12) \text{ eV}$
Thomson cross section	$\sigma_T = 8\pi e^4/3m_e^2 c^4 = 0.665245873(13) \times 10^{-24} \text{ cm}^2$
Proton mass	$m_p = 1.67262171(29) \times 10^{-24} \text{ g}$
	$m_p c^2 = 938.272029(80) \text{ MeV}$
Neutron mass	$m_n c^2 = 939.565360(81) \text{ MeV}$
Deuteron mass	$m_d c^2 = 1875.61282(16) \text{ MeV}$
Atomic mass unit	$m(C^{12})/12 = 1.66053886(28) \times 10^{-24} \text{ g}$
	$m(C^{12})c^2/12 = 931.494043(80) \text{ MeV}$
Avogadro's number	$N_A = 6.0221415(10) \times 10^{23}/\text{mole}$
Boltzmann constant	$k_B = 1.3806505(24) \times 10^{-16} \text{ erg/K}$
	$= 8.617343(15) \times 10^{-5} \text{ eV/K}$
Radiation energy constant	$a_B = \frac{8\pi^5 k_B^4}{15h^3 c^3} = 7.56577(5) \times 10^{-15} \text{ erg cm}^{-3} K^{-4}$

---

[1] From *Review of Particle Physics*, S. Eidelman *et al.* (Particle Data Group), *Phys. Lett.* **B 592**, 1 (2004).

509

Weak coupling constant $G_{wk} = 1.16637(1) \times 10^{-5} \text{ GeV}^{-2}$
Gravitational constant $G = 6.6742(10) \times 10^{-8} \text{ dyn cm}^2 \text{ g}^{-2}$
Planck energy $\sqrt{\hbar c/G} = 1.22090(9) \times 10^{19} \text{ GeV}$

Astronomical Constants[2]

Julian year   1 year $\equiv$ 365.25 days = 3.1557600×10^7 sec
Light year   1 light (Julian) year = $9.460730472 \times 10^{17}$ cm
Mean earth-sun distance   1 A.U. = $1.4959787066 \times 10^{13}$ cm
Parsec   1 pc $\equiv$ $648000/\pi$ A.U. = 3.0856776×10^{18} cm
                            = 3.2615638 light (Julian) year
Solar mass $M_{\odot} = 1.9891 \times 10^{33}$ g
Solar luminosity $L_{\odot} = 3.845(8) \times 10^{33} \text{ erg sec}^{-1}$
Apparent luminosity for apparent magnitude m
$$\ell = 2.52 \times 10^{-5} \text{ erg cm}^{-2} \text{ sec}^{-1} \times 10^{-2m/5}$$
Absolute luminosity for absolute magnitude M
$$\mathcal{L} = 3.02 \times 10^{35} \text{ erg sec}^{-1} \times 10^{-2M/5}$$
For a Hubble constant $H_0 = h \times 100 \text{ km sec}^{-1} \text{ Mpc}^{-1}$:
Hubble time $H_0^{-1} = 3.0857\, h^{-1} \times 10^{17} \text{ sec} = 9.778\, h^{-1} \times 10^9$ years
Hubble distance $c/H_0 = 2997.92458\, h^{-1}$ Mpc
Critical density $\rho_{\text{crit}} \equiv \frac{3H_0^2}{8\pi G} = 1.878\, h^2 \times 10^{-29} \text{ g cm}^{-3}$
                            $= [0.00300 \text{ eV}]^4\, h^2$

---

[2]From *Allen's Astrophysical Quantities*, ed. A. N. Cox (AIP Press, New York, 2000).

510

Appendix B

Review of General Relativity

In this appendix we offer a brief introduction to the General Theory of Relativity, Einstein's theory of gravitation. This appendix is not a substitute for a thorough treatment of the theory, but it outlines the parts of the theory that are used in this book, and it serves to establish our notation.

1 The Equivalence Principle

General Relativity is based on the Principle of the Equivalence of Gravitation and Inertia, or the Equivalence Principle for short. The Equivalence Principle is a generalization of the familiar observation that, because of the equality of gravitational and inertial mass, freely falling observers do not feel the effects of gravitation. According to the Equivalence Principle, at any spacetime point in an arbitrary gravitational field there is a "locally inertial" coordinate system in which the effects of gravitation are absent in a sufficiently small spacetime neighborhood of that point. This Principle allows us to write the equations governing any sufficiently small physical system in a gravitational field if we know the equations governing it in the absence of gravitation: it is only necessary to write the equations in a form which is generally covariant — that is, whose form is independent of the spacetime coordinates used — and which reduce to the correct equations in the absence of gravitation. Such equations will be true in the presence of a gravitational field, because general covariance guarantees that they are true in any set of coordinates if they are true in any other set of coordinates, and the Equivalence Principle tells us that there *is* a set of coordinates in which the equations are true — the set of coordinates that is locally inertial at the spacetime location of the system in question. In general there will be more than one set of generally covariant equations that reduce to the correct equations in the absence of gravitation, but the differences between these equations always involve terms with extra spacetime derivatives, which become negligible if we restrict ourselves to a spacetime region that is small compared with the scale of distances and times over which the gravitational and other fields vary appreciably.

2 The metric: Ticking clocks

As an example of this procedure, consider the equation that governs the rate at which clocks tick in a gravitational field. Special Relativity tells us that if a clock ticks once in every time interval dT when at rest in the absence of a gravitational field, then the separation $d\xi^\alpha$ between the spacetime locations of successive ticks when the clock is moving in the absence of a gravitational field is governed by the relation

$$\eta_{\alpha\beta}\, d\xi^\alpha d\xi^\beta = -dT^2 \,. \tag{B.1}$$

(Here ξ^1, ξ^2, and ξ^3 are the Cartesian space coordinates, using units of length in which the speed of light c is unity; $\xi^0 \equiv t$; $\eta_{\alpha\beta}$ is the Minkowski metric, the diagonal matrix with $\eta_{11} = \eta_{22} = \eta_{33} = 1$ and $\eta_{00} = -1$; and repeated indices are summed.) The correct equation governing the ticking of the clock in a general gravitational field is then

$$g_{\mu\nu}(x)dx^\mu dx^\nu = -dT^2 \,, \tag{B.2}$$

where $g_{\mu\nu}(x)$ is the *metric*, a field defined by the two properties that, first, a transformation to a coordinate system x'^μ changes the metric to

$$g'_{\rho\sigma}(x') = g_{\mu\nu}(x)\,\frac{\partial x^\mu}{\partial x'^\rho}\frac{\partial x^\nu}{\partial x'^\sigma} \,, \tag{B.3}$$

and, second, in coordinates that are locally inertial and Cartesian at a point x, the metric at x is $\eta_{\alpha\beta}$ and its first derivatives at x vanish. (In Eq. (B.3), x^μ and x'^μ are the coordinates of the same physical point in two different coordinate systems.) Eq. (B.2) is generally covariant, because the coordinate differentials have the obvious transformation property

$$dx'^\rho = \frac{\partial x'^\rho}{\partial x^\mu}dx^\mu \,, \tag{B.4}$$

so that

$$g'_{\rho\sigma}(x')\, dx'^\rho dx'^\sigma = g_{\mu\nu}(x)\,\frac{\partial x^\mu}{\partial x'^\rho}\frac{\partial x^\nu}{\partial x'^\sigma}\frac{\partial x'^\rho}{\partial x^\kappa}dx^\kappa\frac{\partial x'^\sigma}{\partial x^\lambda}dx^\lambda = g_{\mu\nu}(x)\, dx^\mu dx^\nu \,.$$

To repeat our general argument, Eq. (B.2) is true, because in locally inertial Cartesian coordinates it reduces to the equation (B.1) that describes clocks in the absence of gravitation, and its general covariance means that if it is true in one set of coordinates then it is true in any other set of coordinates. In the same way, the spacetime separation dx^μ of the ends of a small ruler whose length is dL when measured at rest in the absence of gravitation will in general be given by

$$g_{\mu\nu}(x)\, dx^\mu dx^\nu = +dL^2 \,.$$

Likewise, the differences dx^μ between the spacetime coordinates of two successive positions along a ray of light are governed by the equation

$$g_{\mu\nu}(x)\, dx^\mu dx^\nu = 0 \,.$$

3 Tensors, vectors, scalars

Quantities that transform as in (B.3) and (B.4) are known as *covariant tensors* and *contravariant vectors*, respectively. In general, contravariant and covariant quantities are labeled with upper and lower indices, respectively, and for each such index there is a factor in the transformation rule of $\partial x'/\partial x$ or $\partial x/\partial x'$, respectively. It is also possible to have mixed quantities, with some upper and some lower indices. For instance, there is a mixed tensor δ^μ_ν, defined in any coordinate system by

$$\delta^\mu_\nu \equiv \begin{cases} 1 & \mu = \nu \\ 0 & \mu \neq \nu \end{cases} . \tag{B.5}$$

Even though its components are the same in all coordinate systems this is a tensor because

$$\delta^\mu_\nu \frac{\partial x'^\rho}{\partial x^\mu} \frac{\partial x^\nu}{\partial x'^\sigma} = \frac{\partial x'^\rho}{\partial x^\mu} \frac{\partial x^\mu}{\partial x'^\sigma} = \delta^\rho_\sigma .$$

There is also a contravariant tensor $g^{\mu\nu}$, defined as the inverse of the metric

$$g^{\mu\lambda} g_{\lambda\nu} = \delta^\mu_\nu . \tag{B.6}$$

To see that this is a tensor, just note that

$$\left(g^{\rho\sigma}(x) \frac{\partial x'^\mu}{\partial x^\rho} \frac{\partial x'^\lambda}{\partial x^\sigma} \right) g'_{\lambda\nu}(x') = g^{\rho\sigma}(x) \frac{\partial x'^\mu}{\partial x^\rho} \frac{\partial x'^\lambda}{\partial x^\sigma} \frac{\partial x^\eta}{\partial x'^\lambda} \frac{\partial x^\tau}{\partial x'^\nu} g_{\eta\tau}(x)$$

$$= g^{\rho\sigma}(x) \frac{\partial x'^\mu}{\partial x^\rho} \delta^\eta_\sigma \frac{\partial x^\tau}{\partial x'^\nu} g_{\eta\tau}(x) = \delta^\rho_\tau \frac{\partial x'^\mu}{\partial x^\rho} \frac{\partial x^\tau}{\partial x'^\nu} = \delta^\mu_\nu .$$

Thus the quantity in parenthesis in the first line is the reciprocal of the transformed metric

$$g^{\rho\sigma}(x) \frac{\partial x'^\mu}{\partial x^\rho} \frac{\partial x'^\lambda}{\partial x^\sigma} = g'^{\mu\lambda}(x')$$

verifying that the reciprocal of the metric is a contravariant tensor.

A *scalar* $s(x)$ is a quantity whose value at a physical spacetime point is not changed by a coordinate transformation; that is, using a set of coordinates x'^μ it is

$$s'(x') = s(x) \tag{B.7}$$

The derivative $v_\mu \equiv \partial s/\partial x^\mu$ of a scalar $s(x)$ is a *covariant vector*:

$$v'_\rho \equiv \frac{\partial s'}{\partial x'^\rho} = \frac{\partial s}{\partial x^\mu} \frac{\partial x^\mu}{\partial x'^\rho} = v_\mu \frac{\partial x^\mu}{\partial x'^\rho}. \tag{B.8}$$

Scalars and vectors may be regarded as tensors with no indices or one index, respectively. We can make tensors out of other tensors by taking the direct product; for instance, if $A^{\mu}{}_{\nu}$ and $B_{\rho\sigma}$ are tensors, then so is the direct product $C^{\mu}{}_{\nu\rho\sigma} \equiv A^{\mu}{}_{\nu}B_{\rho\sigma}$. We can also make tensors with fewer indices out of other tensors by contracting upper and lower indices, as Eqs. (B.2) and (B.6). As a special case, we often lower (or raise) an index on a tensor by taking the direct product of the tensor with the metric (or its inverse) and then contracting an upper (or lower) index on the tensor with an index on the metric (or its inverse). For instance, if $A^{\mu\nu}$ is a tensor, then so is $A_{\rho}{}^{\nu} \equiv g_{\mu\rho}A^{\mu\nu}$, while if $B_{\rho}{}^{\nu}$ is a tensor then so is $B^{\sigma\nu} = g^{\sigma\rho}B_{\rho}{}^{\nu}$. Note that raising and lowering the same index just gives back the original tensor; for instance $g^{\sigma\rho}\left(g_{\mu\rho}A^{\mu\nu}\right) = A^{\sigma\nu}$. Any equation that states the equality of two tensors of the same type or that a tensor of any type vanishes is generally covariant.

4 The affine connection: Falling bodies

But not everything is a tensor. For instance, Eq. (B.4) tells us that the first derivative of the coordinate x^{μ} of a particle with respect to some scalar quantity u that parameterizes position along the particle's trajectory (such as the time on some fixed clock) is a vector

$$\frac{dx'^{\rho}}{du} = \frac{\partial x'^{\rho}}{\partial x^{\mu}}\frac{dx^{\mu}}{du} , \tag{B.9}$$

but the second derivative is *not* a vector

$$\frac{d^2 x'^{\rho}}{du^2} = \frac{d}{du}\left(\frac{\partial x'^{\rho}}{\partial x^{\mu}}\frac{dx^{\mu}}{du}\right)$$

$$= \frac{\partial x'^{\rho}}{\partial x^{\mu}}\frac{d^2 x^{\mu}}{du^2} + \frac{\partial^2 x'^{\rho}}{\partial x^{\mu}\partial x^{\nu}}\frac{dx^{\mu}}{du}\frac{dx^{\nu}}{du} . \tag{B.10}$$

This means that the correct generalization of the equation $d^2\xi^{\alpha}/du^2 = 0$ for the motion of a particle in the absence of gravitation is not $d^2 x^{\mu}/du^2 = 0$, because this equation is not generally covariant. Instead, to cancel the second term on the right-hand side of Eq. (B.10) we must introduce a quantity $\Gamma^{\lambda}_{\mu\nu}(x)$ defined by the transformation property

$$\Gamma'^{\tau}_{\sigma\rho} = \frac{\partial x'^{\tau}}{\partial x^{\lambda}}\frac{\partial x^{\mu}}{\partial x'^{\sigma}}\frac{\partial x^{\nu}}{\partial x'^{\rho}}\Gamma^{\lambda}_{\mu\nu} - \frac{\partial^2 x'^{\tau}}{\partial x^{\mu}\partial x^{\nu}}\frac{\partial x^{\mu}}{\partial x'^{\sigma}}\frac{\partial x^{\nu}}{\partial x'^{\rho}} , \tag{B.11}$$

and the proviso that $\Gamma^{\lambda}_{\mu\nu}(x)$ vanishes in a coordinate system that is locally inertial and Cartesian at x. The correct equation of motion for a particle that is freely falling in a gravitational field is then

$$\frac{d^2 x^\lambda}{du^2} + \Gamma^\lambda_{\mu\nu} \frac{dx^\mu}{du} \frac{dx^\nu}{du} = 0 , \tag{B.12}$$

because this *is* generally covariant, and reduces to the correct equation of motion $d^2 \xi^\alpha / du^2 = 0$ in the absence of gravitation. The field $\Gamma^\lambda_{\mu\nu}$ is known as the *affine connection*, and of course is not a tensor.

There is a simple formula for the affine connection in terms of the metric

$$\Gamma^\lambda_{\mu\nu} = \frac{1}{2} g^{\lambda\rho} \left[\frac{\partial g_{\rho\mu}}{\partial x^\nu} + \frac{\partial g_{\rho\nu}}{\partial x^\mu} - \frac{\partial g_{\mu\nu}}{\partial x^\rho} \right] . \tag{B.13}$$

It is straightforward to check that this is generally covariant, and it is true in a locally inertial Cartesian coordinate system because in such a system both sides vanish, so it is correct for general gravitational fields and coordinate systems.

A trajectory that satisfies Eq. (B.12) is called a *spacetime geodesic*, because on such a trajectory the integral

$$\int_{u_1}^{u_2} \sqrt{g_{\mu\nu}(x(u)) \frac{dx^\mu(u)}{du} \frac{dx^\nu(u)}{du}} \, du$$

is stationary under variations that leave $x^\mu(u)$ fixed at the endpoints u_1 and u_2. Often instead of specifying the metric we specify the *line element* $ds^2 = g_{\mu\nu} dx^\mu dx^\nu$ for arbitrary differentials dx^μ.

The equation of motion (B.12) is not valid for just any choice of the parameter u. To see this, note that Eq. (B.12) implies a conservation law

$$\frac{d}{du} \left[g_{\mu\nu} \frac{dx^\mu}{du} \frac{dx^\nu}{du} \right] = \frac{\partial g_{\mu\nu}}{\partial x^\lambda} \frac{dx^\mu}{du} \frac{dx^\nu}{du} \frac{dx^\lambda}{du} + g_{\mu\nu} \frac{d^2 x^\mu}{du^2} \frac{dx^\nu}{du} + g_{\mu\nu} \frac{dx^\mu}{du} \frac{d^2 x^\nu}{du^2}$$

$$= \left(\frac{\partial g_{\mu\nu}}{\partial x^\lambda} - g_{\mu\kappa} \Gamma^\kappa_{\nu\lambda} - g_{\nu\kappa} \Gamma^\kappa_{\mu\lambda} \right) \frac{dx^\mu}{du} \frac{dx^\nu}{du} \frac{dx^\lambda}{du}$$

$$= 0 . \tag{B.14}$$

It follows that u must be a linear function of the *proper time* τ, defined by

$$d\tau \equiv \sqrt{-g_{\mu\nu} dx^\mu dx^\nu} , \tag{B.15}$$

which according to Eq. (B.2) is the time told by a clock that falls freely along with the particle. The only exception to this conclusion is for massless particles like photons, whose spacetime trajectory satisfies the same equation of motion (B.12) as for a massive particle, but for which the conserved quantity

$d\tau/du$ vanishes. For a massless particle or a ray of light we need to choose the parameter u as the time told by some other freely falling clock.

In the case of non-zero mass m, it is convenient to take the affine parameter u as $u = \tau/m$, for then we can define the energy-momentum four-vector as

$$p^\mu = m\frac{dx^\mu}{d\tau} = \frac{dx^\mu}{du} \; ,$$

and, using Eq. (B.15),

$$g_{\mu\nu}p^\mu p^\nu = -m^2 \; .$$

For massless particles we have $g_{\mu\nu}(dx^\mu/du)(dx^\nu/du) = 0$ however u is normalized, so we can simply suppose that it is normalized in such a way that the energy-momentum four-vector is $p^\mu = dx^\mu/du$.

5 Gravitational time dilation

These results allow us to derive one of the most important consequences of the Equivalence Principle. For a slowly moving particle, dx^i/du is much less than dx^0/du, so Eq. (B.12) becomes

$$\frac{d^2x^i}{du^2} + \Gamma^i_{00}\frac{dx^0}{du}\frac{dx^0}{du} = 0 \; . \tag{B.16}$$

For a weak gravitational field, the metric $g_{\mu\nu}$ is nearly equal to the Minkowski metric $\eta_{\mu\nu}$, so

$$g_{\mu\nu} = \eta_{\mu\nu} + h_{\mu\nu} \; , \tag{B.17}$$

with the components of $h_{\mu\nu}$ much less than unity. We then can take $u = \tau \simeq x^0 \equiv t$, so the equation of motion of a freely falling slowly moving particle in a weak gravitational field is

$$\frac{d^2x^i}{dt^2} = -\Gamma^i_{00} \; , \tag{B.18}$$

where i runs over the values 1, 2, 3, labeling spatial directions in a Cartesian coordinate system. The affine connection for a weak gravitational field is

$$\Gamma^\lambda_{\mu\nu} \simeq \frac{\eta^{\lambda\rho}}{2}\left[\frac{\partial h_{\rho\mu}}{\partial x^\nu} + \frac{\partial h_{\rho\nu}}{\partial x^\mu} - \frac{\partial h_{\mu\nu}}{\partial x^\rho}\right] \; . \tag{B.19}$$

In particular, for a weak time-independent gravitational field we have

$$\Gamma^i_{00} \simeq -\frac{1}{2}\frac{\partial h_{00}}{\partial x^i} \; . \tag{B.20}$$

Eqs. (B.18) and (B.20) allow us to identify $-h_{00}/2$ as the Newtonian gravitational potential ϕ, so in a weak static gravitational field we have

$$g_{00} \simeq -1 - 2\phi . \tag{B.21}$$

Now consider a clock at rest in such a field. According to Eq. (B.2), if the time between ticks in the absence of a gravitational field is dT, then in the presence of the field it is dt, where

$$(-1 - 2\phi)dt^2 \simeq -dT^2 .$$

Hence the time between ticks is no longer dT, but rather

$$dt \simeq (1 - \phi)dT . \tag{B.22}$$

In the negative gravitational potential at the surface of a star clocks therefore tick more slowly than in interstellar space, or in the much weaker gravitational potential at the surface of the earth. This could not be observed on the star's surface, since all physical processes would be slowed there by the same factor, but it is observed at a distance, by measuring the decrease in the frequency of photons emitted from atomic transitions on the star's surface. The gravitational time dilation is measured most accurately by observing the shift of spectral lines as photons rise or fall in the earth's gravitational field.

6 Covariant derivatives

Although the spacetime derivative of a scalar field is a vector, the derivative of a vector or a tensor field is in general not a tensor. For instance, a contravariant vector field v^μ has the transformation property

$$v'^\rho = \frac{\partial x'^\rho}{\partial x^\mu} v^\mu , \tag{B.23}$$

so that

$$\frac{\partial v'^\rho}{\partial x'^\sigma} = \frac{\partial x'^\rho}{\partial x^\mu} \frac{\partial x^\nu}{\partial x'^\sigma} \frac{\partial v^\mu}{\partial x^\nu} + \frac{\partial^2 x'^\rho}{\partial x^\mu \partial x^\nu} \frac{\partial x^\nu}{\partial x'^\sigma} v^\mu . \tag{B.24}$$

To construct a tensor we must add a term that cancels the second term in this transformation law. In this way we are led to introduce a *covariant derivative*

$$v^\mu_{;\nu} \equiv \frac{\partial v^\mu}{\partial x^\nu} + \Gamma^\mu_{\nu\lambda} v^\lambda , \tag{B.25}$$

which does transform as a mixed tensor

$$v'^\rho_{;\sigma} = \frac{\partial x'^\rho}{\partial x^\mu} \frac{\partial x^\nu}{\partial x'^\sigma} v^\mu_{;\nu} . \tag{B.26}$$

Likewise, the covariant derivative of a covariant vector is defined by

$$v_{\nu\,;\mu} \equiv \frac{\partial v_\nu}{\partial x^\mu} - \Gamma^\lambda_{\mu\nu} v_\lambda \,, \tag{B.27}$$

and is a covariant tensor. More generally, the covariant derivative of a tensor with any number of upper and/or lower indices is given by similar formulas, with a $+\Gamma$ for every upper index and a $-\Gamma$ for every lower index. It is easy to check that the covariant derivatives $g_{\mu\nu\,;\lambda}$ of the metric tensor all vanish, as they must, since in a locally inertial Cartesian frame the covariant derivative is an ordinary derivative and the first derivatives of the metric vanish, so that the covariant derivatives vanish, and a tensor $g_{\mu\nu\,;\lambda}$ that vanishes in one coordinate system must vanish in all coordinate systems.

7 Effects of gravitation: The Maxwell equations

Given the equations that govern some set of fields in the absence of gravitation, we can find the equations that apply (at least in sufficiently small regions) in a gravitational field by replacing all Minkowski metrics η with metric tensors and all derivatives with covariant derivatives. As mentioned earlier, the procedure does not give a unique result, since there are tensors formed from second and higher derivatives of the metric that vanish in the absence of gravitation, but the effect of including such tensors in the generally covariant field equations would be negligible in a sufficiently small spacetime region.

For instance, in Cartesian coordinates in the absence of gravitation electric and magnetic fields are governed by Maxwell's equations

$$\partial_\alpha F^{\alpha\beta} = -J^\beta \,, \tag{B.28}$$

$$\partial_\alpha F_{\beta\gamma} + \partial_\beta F_{\gamma\alpha} + \partial_\gamma F_{\alpha\beta} = 0 \,, \tag{B.29}$$

where $F^{\alpha\beta}$ is the electromagnetic field strength tensor (with $F^{01} = E_1$, $F^{23} = B_1$, etc.), J^α is the electric current four-vector (with J^0 the charge density and J^1, J^2, J^3 the electric current density), and

$$F_{\alpha\beta} \equiv \eta_{\alpha\gamma} \eta_{\beta\delta} F^{\gamma\delta} \,. \tag{B.30}$$

Hence in the presence of gravitation the field equations are

$$F^{\nu\mu}{}_{;\nu} = -J^\mu \tag{B.31}$$

$$F_{\mu\nu;\lambda} + F_{\nu\lambda;\mu} + F_{\lambda\mu;\nu} = 0 \,, \tag{B.32}$$

with

$$F_{\mu\nu} \equiv g_{\mu\lambda}g_{\nu\kappa}F^{\lambda\kappa} . \tag{B.33}$$

We use the same letter of the alphabet for tensors like $F_{\mu\nu}$ and $F^{\mu\nu}$ that are related by raising and lowering indices by contraction with the metric, because they represent the same physical quantity.

8 Currents and conservation laws

For a system of particles labeled by an index n, with spacetime coordinates $x_n^\mu(u)$ and electric charges e_n, the electric current four-vector is given by

$$
\begin{aligned}
J^\mu(x) &\equiv \frac{1}{\sqrt{-\mathrm{Det}g(x)}} \int du \sum_n e_n \, \delta^4\Big(x_n(u) - x\Big) \frac{dx^\mu(u)}{du} \\
&= \frac{1}{\sqrt{-\mathrm{Det}g(x)}} \sum_n e_n \, \delta^3\Big(\mathbf{x}_n(t) - \mathbf{x}\Big) \frac{dx^\mu(t)}{dt} .
\end{aligned}
\tag{B.34}
$$

Here $\delta^4(z)$ is a fictitious function with an infinitely narrow and infinitely high peak at $z = 0$, normalized so that, for any smooth function $f(z)$,

$$\int d^4x f(y)\, \delta^4(y - x) = f(x) ,$$

and $\delta^3(\mathbf{z})$ is the same in three dimensions. In particular, the integral of $\sqrt{-\mathrm{Det}g}\, J^0$ over a finite three-dimensional volume equals the total electric charge within that volume. (In cosmology we would be more interested in the baryon current, with the baryon number of the nth particle or the nth galaxy appearing instead of e_n.) The factor $1/\sqrt{-\mathrm{Det}g(x)}$ is needed here, because the four-dimensional delta function is not a scalar. We can see this by noting that, under a transformation from coordinates x^μ to coordinates x'^μ, the differential spacetime volume element is changed to

$$d^4x' = \left| \frac{\partial x'}{\partial x} \right| d^4x \tag{B.35}$$

where $|\partial x'/\partial x|$ is the Jacobian of the coordinate transformation — that is, the determinant of the matrix whose components are $\partial x'^\mu/\partial x^\nu$. The Jacobian can be expressed in terms of the determinants of the metrics; by taking the determinant of Eq. (B.3), we find

$$\mathrm{Det}\, g' = \left| \frac{\partial x'}{\partial x} \right|^{-2} \mathrm{Det}\, g . \tag{B.36}$$

Hence it is the spacetime volume $d^4x\sqrt{-\text{Det}g}$ rather than d^4x that transforms as a scalar. (A minus sign is inserted in front of the determinant of the metric, because in physical spacetimes this determinant is negative.) From the defining equation of the delta function

$$f(y) = \int d^4x f(x)\, \delta^4(x - y)$$

$$= \int \left(d^4x\, \sqrt{-\text{Det}g(x)} \right) f(x) \left(\delta^4(x - y)/\sqrt{-\text{Det}g(x)} \right),$$

we see that it is the ratio $\delta^4(x_n - x)/\sqrt{-\text{Det}\, g(x)}$ appearing in the current (B.34) rather than the delta function itself that transforms as a scalar. This current satisfies the conservation law

$$\partial_\mu \left(\sqrt{-\text{Det}g(x)}\, J^\mu(x) \right) = \int du \frac{d}{du} \sum_n e_n\, \delta^4(x_n(u) - x) = 0 , \qquad \text{(B.37)}$$

provided x is not at the value of any $x_n(u)$ at either endpoint of the integral. This is the same as the generally covariant conservation law

$$0 = J^\mu_{;\mu} \equiv \frac{\partial}{\partial x^\mu} J^\mu + \Gamma^\mu_{\nu\mu} J^\nu \qquad \text{(B.38)}$$

because

$$\Gamma^\nu_{\mu\nu} = \frac{1}{2} g^{\nu\lambda} \partial_\mu g_{\nu\lambda} = \frac{1}{2} \partial_\mu \ln \left(-\text{Det}g \right) .$$

This is the correct conservation condition, because in the absence of gravitation there is a current that in Cartesian coordinate systems satisfies the conservation law $\partial_\mu J^\mu = 0$, and therefore in general coordinates in a gravitational field must satisfy the generally covariant generalization (B.38).

9 The energy-momentum tensor

Likewise, in the absence of gravitation any set of particles and/or fields will have a symmetric energy-momentum tensor $T^{\alpha\beta}$, which is conserved in the sense that

$$\frac{\partial T^{\alpha\beta}}{\partial x^\beta} = 0 . \qquad \text{(B.39)}$$

Just as J^β is the β component of the current of electric charge, we can think of $T^{\alpha\beta}$ as the β component of the current of p^α. In the presence of a gravitational field the conservation law becomes

$$T^{\mu\nu}_{;\nu} \equiv \frac{\partial T^{\mu\nu}}{\partial x^\nu} + \Gamma^\mu_{\kappa\nu} T^{\kappa\nu} + \Gamma^\nu_{\kappa\nu} T^{\mu\kappa} = 0 . \qquad \text{(B.40)}$$

The final Γ term here is a geometric effect, similar to what we found for conserved currents, but the other Γ term represents the exchange of energy and momentum between gravitation and the other fields.

For instance, for an ideal gas of particles that move freely except for gravitational forces and perhaps for collisions that are localized in space, the energy-momentum tensor is given by replacing e_n in Eq. (B.34) with the energy-momentum four-vector $p_n^\mu = E_n\, dx_n^\mu/dt$, whose spatial components are the components of the three-momentum, and $p_n^0 = E_n$:

$$T^{\mu\nu}(x) = \frac{1}{\sqrt{-\mathrm{Det}g(x)}} \int du \sum_n \delta^4\Big(x_n(u) - x\Big) \frac{dx_n^\mu(u)}{du} p_n^\nu(u)$$

$$= \frac{1}{\sqrt{-\mathrm{Det}g(x)}} \sum_n \delta^3\Big(\mathbf{x}_n(t) - \mathbf{x}\Big) p_n^\mu(t) p_n^\nu(t)/E_n(t) . \quad \text{(B.41)}$$

In particular, T^{00} is the energy density. Direct calculation using the equation of motion (B.12) shows that this satisfies the covariant conservation law (B.40).

10 Perfect and imperfect fluids

A perfect fluid is defined as a medium for which at every point there is a locally inertial Cartesian frame of reference, moving with the fluid, in which the fluid appears the same in all directions. In such a locally inertial co-moving frame the components of the energy momentum tensor must take the form

$$T^{ij} = \delta_{ij}p , \quad T^{i0} = T_{0i} = 0 , \quad T^{00} = \rho ,$$

where i and j run over the three Cartesian coordinate directions $1, 2, 3$. (This is because non-zero value of T^{i0} and any term in T^{ij} other than one proportional to δ_{ij} would select out special directions in space, such as the direction of T^{i0}, or of one of the non-degenerate eigenvectors of T^{ij}.) The coefficients p and ρ are known as the pressure and energy density, respectively. Then in a locally inertial Cartesian frame with an arbitrary velocity, the energy-momentum tensor takes the form

$$T^{\alpha\beta} = p\,\eta^{\alpha\beta} + (p+\rho)u^\alpha u^\beta , \quad \text{(B.42)}$$

where ρ and p are defined to be the same as in the co-moving inertial frame, and u^α is defined by the conditions that it transforms as a four-vector under Lorentz transformations, and that in the locally Cartesian co-moving inertial frame it has components $u^0 = 1$ and $u^i = 0$. This four-vector, known as the velocity vector, is normalized so that, in any inertial frame, $\eta_{\alpha\beta}u^{\alpha}u^{\beta} = -1$. It follows that in a general gravitational field the energy-momentum tensor of a perfect fluid is

$$T^{\mu\nu} = p\,g^{\mu\nu} + (p+\rho)u^{\mu}u^{\nu}, \quad g_{\mu\nu}u^{\mu}u^{\nu} = -1 \tag{B.43}$$

where p and ρ are defined by the condition that they are equal to the coefficients in the energy-momentum tensor in a locally co-moving inertial coordinate system, so that they are scalars, and u^{μ} is defined by the conditions that it transforms as a four-vector under general coordinate transformations and has the components $u^0 = 1$ and $u^i = 0$ in the locally co-moving Cartesian inertial frame. This formula for $T^{\mu\nu}$ is correct because it is generally covariant and it is true in locally inertial Cartesian coordinate systems. The equations of relativistic hydrodynamics in a gravitational field are derived by imposing the conservation condition (B.40) on this tensor. In addition, if the pressure depends on the density n of some conserved quantity such as baryon number as well as on the energy density ρ, then we need the equation of conservation, which in locally inertial Cartesian frames reads

$$\frac{\partial}{\partial x^{\alpha}}\left(n\,u^{\alpha}\right) = 0. \tag{B.44}$$

Thus in a general coordinate system in an arbitrary gravitational field, we have

$$\left(n\,u^{\mu}\right)_{;\mu} = 0. \tag{B.45}$$

For an imperfect fluid, there is a small correction $\Delta T^{\alpha\beta}$ to formula (B.42) for the energy-momentum tensor in locally inertial Cartesian coordinate system:

$$T^{\alpha\beta} = p\eta^{\alpha\beta} + (p+\rho)u^{\alpha}u^{\beta} + \Delta T^{\alpha\beta} \tag{B.46}$$

and a small correction ΔN^{α} to whatever current may be conserved

$$\frac{\partial}{\partial x^{\alpha}}\left(n\,u^{\alpha} + \Delta N^{\alpha}\right) = 0. \tag{B.47}$$

The scalar ρ is defined as the energy density observed in a co-moving frame in which $u^i = 0$, so that in this frame $\Delta T^{00} \equiv 0$. This implies that in all locally inertial Cartesian frames $u_{\alpha}u_{\beta}\Delta T^{\alpha\beta} = 0$, since this quantity is a scalar and vanishes in a co-moving frame. The scalar n can be defined as the value of the conserved density observed in such a co-moving frame, so by the same reasoning in all locally inertial Cartesian frames we have $u_{\alpha}\Delta N^{\alpha} = 0$. The pressure can be defined as whatever function of ρ and perhaps n gives the pressure in a static homogeneous fluid. (A different definition is used in Chapter 5 *et seq.*) But the definition of the velocity four-vector u^α remains somewhat ambiguous. We could define u^i to be the velocity of particle transport,[1] in which case in a co-moving frame with $u^i = 0$ we would have $\Delta N^i = 0$ as well as $\Delta N^0 = 0$, so that in general locally inertial Cartesian frames $\Delta N^\alpha = 0$. Instead we will define u^i to be the velocity of energy transport,[2] so that in co-moving frames we also have $T^{i0} = \Delta T^{i0} = 0$, and so $u_\beta \Delta T^{\alpha\beta} = \Delta T^{\alpha 0} = 0$ in this frame, which implies that $u_\beta \Delta T^{\alpha\beta} = 0$ in all locally inertial Cartesian frames, but in general $\Delta N^\alpha \neq 0$. With this definition of velocity, the second law of thermodynamics together with the conditions $u_\beta \Delta T^{\alpha\beta} = 0$ and $u_\alpha \Delta N^\alpha = 0$ gives[3]

$$\Delta T_{\alpha\beta} = -\eta \left(\frac{\partial u_\alpha}{\partial x^\beta} + \frac{\partial u_\beta}{\partial x^\alpha} + u_\beta u^\gamma \frac{\partial u_\alpha}{\partial x^\gamma} + u_\alpha u^\gamma \frac{\partial u_\beta}{\partial x^\gamma} \right)$$
$$- (\zeta - \frac{2}{3}\eta) \frac{\partial u^\gamma}{\partial x^\gamma} \left(\eta_{\alpha\beta} + u_\alpha u_\beta \right), \tag{B.48}$$

$$\Delta N_\alpha = -\chi \left(\frac{nT}{\rho + p} \right)^2 \left[\frac{\partial}{\partial x^\alpha} \left(\frac{\mu}{T} \right) + u_\alpha u^\beta \frac{\partial}{\partial x^\beta} \left(\frac{\mu}{T} \right) \right]. \tag{B.49}$$

Here η, ζ, and χ are the positive coefficients of shear viscosity, bulk viscosity, and heat conduction, respectively, and μ is the chemical potential associated with the conserved quantum number, defined by the condition that the entropy density is $(p + \rho - \mu n)/T$. It is then an immediate consequence of the Equivalence Principle that in general coordinate systems in arbitrary gravitational fields that vary little over a mean free path or mean free time,

$$\Delta T_{\mu\nu} = -\eta \left(u_{\mu;\nu} + u_{\nu;\mu} + u_\nu u^\kappa u_{\mu;\kappa} + u_\mu u^\kappa u_{\nu;\kappa} \right)$$
$$- (\zeta - \frac{2}{3}\eta) u^\kappa{}_{;\kappa} \left(g_{\mu\nu} + u_\mu u_\nu \right), \tag{B.50}$$

$$\Delta N_\mu = -\chi \left(\frac{nT}{\rho + p} \right)^2 \left[\frac{\partial}{\partial x^\mu} \left(\frac{\mu}{T} \right) + u_\mu u^\nu \frac{\partial}{\partial x^\nu} \left(\frac{\mu}{T} \right) \right]. \tag{B.51}$$

---

[1] This is the option adopted by C. Eckart, *Phys. Rev.* **58**, 919 (1940), and also in Secs. 2.11 and 15.10 of G&C.

[2] This is the definition used by L. D. Landau and E. M. Lifschitz, *Fluid Mechanics*, trans. by J. B. Sykes and W. H. Reid (Pergamon Press, London, 1959), Section 127. We will adopt this definition of velocity in this book, because it imposes the maximum possible constraint on the energy-momentum tensor at the cost of putting less of a constraint on the current of conserved quantities, and in cosmology we frequently have to do with situations in which there are either no non-zero conserved quantities at all, such as in the early universe before cosmological leptogenesis or baryongenesis, or no conserved quantities that are large enough to seriously affect the relation between pressure and density, as in the radiation-dominated era at temperatures above about 10^4 K.

[3] Landau & Lifschitz, *op. cit.*

11 The action principle

There is a general algorithm for deriving the energy-momentum tensor for systems that may be more complicated than an ideal gas or perfect fluid, provided only that they are governed by an action principle. According to the action principle, the differential equations governing the behavior of particles and fields can be expressed as the statement that the "matter action," a functional I_m of the fields and particle trajectories, is stationary with respect to infinitesimal variations of the fields and particles. The Equivalence Principle tells us to include the metric in the matter action in such a way that I_m is invariant under general coordinate transformations. Then the change in the action when we make an infinitesimal change $\delta g_{\mu\nu}$ in the metric (leaving all other dynamical variables unchanged) must be of the form

$$\delta I_m = \frac{1}{2} \int d^4x \sqrt{-\text{Det}\, g(x)}\; T^{\mu\nu}(x)\, \delta g_{\mu\nu}(x)\,, \tag{B.52}$$

where $T^{\mu\nu}(x)$ is a symmetric tensor, which we identify as the energy-momentum tensor.

For instance, the action for a gas of charged particles with masses m_n, charges e_n, and trajectories $x_n^\mu(u)$ interacting with electromagnetic fields is taken as

$$I_m = -\frac{1}{4} \int d^4x \sqrt{-\text{Det}\, g}\; F_{\mu\nu} F_{\rho\sigma} g^{\mu\rho} g^{\nu\sigma}$$

$$- \sum_n m_n \int du \left[-g_{\mu\nu}(x_n(u)) \frac{dx_n^\mu(u)}{du} \frac{dx_n^\nu(u)}{du} \right]^{1/2}$$

$$+ \sum_n e_n \int du \frac{dx_n^\mu(u)}{du} A_\mu(u)$$

in which the homogeneous Maxwell equations (B.29) are enforced by writing the field strength tensor in terms of a vector potential A_μ as $F_{\mu\nu} = \partial_\mu A_\nu - \partial_\nu A_\mu$. The reader can verify that the conditions for the matter action to be stationary with respect to arbitrary small variations in $A_\mu(x)$ and $x_n^\mu(u)$ (arbitrary, except that they vanish for $x^\mu \to \infty$ or $u \to \pm\infty$) are the inhomogeneous Maxwell equations (B.28) together with the equations of motion of charged particles in a combined gravitational and electromagnetic fields

$$m_n \left[\frac{d^2 x_n^\mu}{d\tau_n^2} + \Gamma^\mu_{\nu\rho} \frac{dx_n^\nu}{d\tau_n} \frac{dx_n^\rho}{d\tau_n} \right] = e_n \frac{dx_n^\nu}{d\tau_n} F^\mu{}_\nu(x_n)\,,$$

where τ_n is the invariant proper time along the nth particle trajectory, defined by

$$\frac{d\tau_n}{du} \equiv \left[-g_{\mu\nu}(x_n(u)) \frac{dx_n^\mu(u)}{du} \frac{dx_n^\nu(u)}{du} \right]^{1/2} .$$

Eq. (B.52) gives an energy-momentum tensor

$$T^{\mu\nu}(x) = [-\mathrm{Det}g(x)]^{-1/2} \sum_n m_n \int d\tau_n \frac{dx_n^\mu}{d\tau_n} \frac{dx_n^\nu}{d\tau_n} \delta^4(x - x_n)$$

$$+ F^{\rho\mu}(x) F_\rho{}^\nu(x) - \frac{1}{4} g^{\mu\nu}(x) F^{\rho\sigma}(x) F_{\rho\sigma}(x) ,$$

which is the same as Eq. (B.41), with extra terms representing the energy and momentum in the electromagnetic field. (In the derivation we use the relations $\delta\mathrm{Det}g = \mathrm{Det}\, g\, g^{\mu\nu}\, \delta g_{\mu\nu}$ and $\delta g^{\mu\rho} = -g^{\mu\nu}\delta g_{\nu\sigma} g^{\sigma\rho}$.)

The justification of the identification of $T^{\mu\nu}$ in Eq. (B.52) is that this tensor is conserved, in the sense of Eq. (B.40). To show this, note that in general our assumption that I_m is a scalar tells us that it is unchanged if we simultaneously make the replacements

$$d^4x \to d^4x' , \qquad \frac{\partial}{\partial x^\mu} \to \frac{\partial}{\partial x'^\mu} , \tag{B.53}$$

$$x_n^\mu(u) \to x'^\mu_n(u) , \qquad A_\mu(x) \to A'_\mu(x') = \frac{\partial x^\nu}{\partial x'^\mu} A_\nu(x) , \tag{B.54}$$

$$g_{\mu\nu}(x) \to g'_{\mu\nu}(x') = \frac{\partial x^\rho}{\partial x'^\mu} \frac{\partial x^\sigma}{\partial x'^\nu} g_{\rho\sigma}(x) , \tag{B.55}$$

and likewise for any other fields entering in the action. But the coordinate x'^μ (unlike $x'^\mu_n(u)$) is just a variable of integration, so we can change x'^μ back to x^μ everywhere without changing I_m. It follows that I_m is unchanged by the replacements

$$x_n^\mu(u) \to x'^\mu_n(u) , \tag{B.56}$$

$$A_\mu(x) \to A'_\mu(x) = \frac{\partial x^\nu}{\partial x'^\mu} A_\nu(x) - [A'_\mu(x') - A'_\mu(x)] \tag{B.57}$$

$$g_{\mu\nu}(x) \to g'_{\mu\nu}(x) = \frac{\partial x^\rho}{\partial x'^\mu} \frac{\partial x^\sigma}{\partial x'^\nu} g_{\rho\sigma}(x) - [g'_{\mu\nu}(x') - g'_{\mu\nu}(x)] , \tag{B.58}$$

with x^μ and $\partial/\partial x^\mu$ now left unchanged. (This combination of a coordinate transformation and a relabeling of coordinates is sometimes called a *gauge transformation*.) For a general infinitesimal coordinate transformation we have $x^\mu \to x^\mu + \epsilon^\mu(x)$, with $\epsilon^\mu(x)$ arbitrary infinitesimal functions of x.

Then the transformations (B.56)–(B.58) become

$$x_n^\mu(u) \to x_n^\mu(u) + \epsilon^\mu(x_n) , \tag{B.59}$$

$$A_\mu(x) \to A_\mu(x) - \frac{\partial \epsilon^\nu(x)}{\partial x^\mu} A_\nu(x) - \frac{\partial A_\mu(x)}{\partial x^\nu} \epsilon^\nu(x) , \tag{B.60}$$

$$g_{\mu\nu}(x) \to g_{\mu\nu}(x) - \frac{\partial \epsilon^\rho(x)}{\partial x^\mu} g_{\rho\nu}(x) - \frac{\partial \epsilon^\sigma(x)}{\partial x^\nu} g_{\mu\sigma}(x)$$

$$- \frac{\partial g_{\mu\nu}(x)}{\partial x^\rho} \epsilon^\rho(x) . \tag{B.61}$$

Now, as long as the equations of motion of particles and the field equations for "matter" (including electromagnetic) fields are satisfied, the action is unaffected by any infinitesimal changes in particle trajectories and matter fields. On the other hand, unlike the total action of matter plus gravitation, the matter action is not stationary under variations in the metric, even when the field equations are satisfied. For a general infinitesimal coordinate transformation, using Eq. (B.61) in Eq. (B.52) lets us write the condition that I_m is a scalar as

$$0 = \delta I_m = \int d^4x \sqrt{-\mathrm{Det}g(x)}\; T^{\mu\nu}(x) \left[-\frac{\partial \epsilon^\rho(x)}{\partial x^\mu} g_{\rho\nu}(x) - \frac{\partial \epsilon^\rho(x)}{\partial x^\nu} g_{\mu\rho}(x) \right.$$

$$\left. - \frac{\partial g_{\mu\nu}(x)}{\partial x^\rho} \epsilon^\rho(x) \right] , \tag{B.62}$$

Integrating by parts and setting the coefficient of $\epsilon^\rho(x)$ equal to zero then yields the conservation condition (B.40).

12 Scalar field theory

We will frequently encounter cosmological models involving a scalar field $\varphi(x)$, with an action[4]

$$I_\varphi = - \int d^4x \sqrt{-\mathrm{Det}g} \left[\frac{1}{2} g^{\mu\nu} \frac{\partial \varphi}{\partial x^\mu} \frac{\partial \varphi}{\partial x^\nu} + V(\varphi) \right] , \tag{B.63}$$

where $V(\varphi)$ is a function known as the potential. The field equations for φ in a gravitational field are given by the condition that this be stationary with respect to variations in φ:

$$\frac{1}{\sqrt{-\mathrm{Det}g}} \frac{\partial}{\partial x^\mu} \left[\sqrt{-\mathrm{Det}g}\; g^{\mu\nu} \frac{\partial \varphi}{\partial x^\nu} \right] = \frac{\partial V(\varphi)}{\partial \varphi} . \tag{B.64}$$

---

[4]In general, the coefficient of the first term in square brackets might depend on φ, but such a field-dependent coefficient can always be eliminated by a redefinition of the scalar field. This simplification is not generally possible, however, with more than one scalar field, the case discussed in Section 10.4.

The energy-momentum tensor for the scalar field is found by varying the metric and comparing with Eq. (B.40):

$$T_\varphi^{\mu\nu} = -g^{\mu\nu}\left[\frac{1}{2}g^{\rho\sigma}\frac{\partial\varphi}{\partial x^\rho}\frac{\partial\varphi}{\partial x^\sigma} + V(\varphi)\right] + g^{\mu\rho}g^{\nu\sigma}\frac{\partial\varphi}{\partial x^\rho}\frac{\partial\varphi}{\partial x^\sigma}. \qquad \text{(B.65)}$$

This has the same form as the energy-momentum tensor (B.43) for a perfect fluid, with energy density, pressure, and velocity four-vector given by[5]

$$\rho = -\frac{1}{2}g^{\mu\nu}\frac{\partial\varphi}{\partial x^\mu}\frac{\partial\varphi}{\partial x^\nu} + V(\varphi) \qquad \text{(B.66)}$$

$$p = -\frac{1}{2}g^{\mu\nu}\frac{\partial\varphi}{\partial x^\mu}\frac{\partial\varphi}{\partial x^\nu} - V(\varphi) \qquad \text{(B.67)}$$

$$u^\mu = -\left[-g^{\rho\sigma}\frac{\partial\varphi}{\partial x^\rho}\frac{\partial\varphi}{\partial x^\sigma}\right]^{-1/2}g^{\mu\tau}\frac{\partial\varphi}{\partial x^\tau} \qquad \text{(B.68)}$$

(The sign of u^μ does not affect the energy-momentum tensor, so it cannot be found by comparing Eqs. (B.65) and (B.43). It is chosen here so that u^0 should have the value $u^0 = +1$ in the case considered in Chapter 4 — a spacetime with $g^{00} = -1$ and a scalar field that does not depend on position and increases with time — provided the square root is understood to be positive.) The reader can check that $T_\varphi^{\mu\nu}$ is conserved in the sense of Eq. (B.40) as a consequence of the field equation (B.64) for φ.

13 Parallel transport

A body carried along an orbit $x^\mu = x^\mu(t)$ may be characterized by one or more t-dependent vectors or tensors. If these vectors or tensors do not change at time t in a frame of reference that is locally inertial at $x^\mu(t)$, then

----

[5]This result for pressure is different from that given (without explanation) by E. W. Kolb and M. S. Turner, *The Early Universe* (Addison–Wesley, Redwood City, CA, 1990), Eq. (8.21), according to whom the pressure of a scalar field in a Robertson–Walker metric is

$$p = \dot\varphi^2/2 - V(\varphi) - (\nabla\varphi)^2/6a^2 ,$$

instead of the result for a Robertson–Walker metric given by Eq. (B.67)

$$p = \dot\varphi^2/2 - V(\varphi) - (\nabla\varphi)^2/2a^2 .$$

The Kolb–Turner result is obtained if we define the pressure as the value of $T^i{}_i/3$ measured by an observer in a locally inertial coordinate system moving with four-velocity v_μ

$$p \equiv \frac{1}{3}\left(g_{\mu\rho} + v_\mu v_\rho\right)T^{\rho\sigma}(\delta^\mu_\sigma + v_\sigma v^\mu) ,$$

but take the velocity v^μ to have components $v^0 = 1$, $v^i = 0$. It seems more natural in using this formula for pressure to take v^μ as the velocity u^μ given by Eq. (B.68), for this is the velocity that appears in the perfect fluid formula (B.43) for the energy-momentum tensor. This choice of v^μ leads to Eq. (B.67).

in an arbitrary frame of reference they will undergo a change known as *parallel transport*. For instance, a contravariant vector $v^\mu(t)$ will have a rate of change

$$\frac{dv^\mu(t)}{dt} = -\Gamma^\mu_{\nu\lambda}\Big(x(t)\Big)v^\nu(t)\frac{dx^\lambda(t)}{dt} \, . \tag{B.69}$$

To check that this is valid, we note that it is assumed to be true at time t in frames of reference in which the affine connection $\Gamma^\mu_{\nu\lambda}\Big(x(t)\Big)$ vanishes, and it is generally covariant, and it is therefore true in all frames if it is true in such locally inertial frames. In particular, Eq. (B.12) shows that the momentum $p^\mu = dx^\mu/du$ satisfies the equation of parallel transport. Similarly, any tensor carried along an orbit that does not change in a frame that is locally inertial at a point along the orbit will have a rate of change in general frames at that point given by a sum of terms like the right-hand side of Eq. (B.69), with a $-\Gamma$ for every contravariant index and a $+\Gamma$ for every covariant index. For instance, for a covariant tensor $J_{\mu\nu}$, the equation of parallel transport is

$$\frac{dJ_{\mu\nu}(t)}{dt} = +\Gamma^\rho_{\mu\lambda}\Big(x(t)\Big)J_{\rho\nu}(t)\frac{dx^\lambda(t)}{dt} + \Gamma^\rho_{\nu\lambda}\Big(x(t)\Big)J_{\mu\rho}(t)\frac{dx^\lambda(t)}{dt} \, .$$

14 The gravitational field equations

It remains only to give the equations that govern the gravitational field itself. These must satisfy two requirements: they must be generally covariant, and for weak slowly changing gravitational fields they must yield the Poisson equation for the Newtonian potential ϕ in Eq. (B.21)

$$\nabla^2\phi = 4\pi G\, T^{00} \, , \tag{B.70}$$

where G is Newton's constant. If we limit ourselves to partial differential equations that (like the Poisson equation) have just two spacetime derivatives of the metric, then the field equations are unique:

$$R_{\mu\nu} - \frac{1}{2}g_{\mu\nu}g^{\lambda\kappa}R_{\lambda\kappa} = -8\pi\, GT_{\mu\nu} \, , \tag{B.71}$$

where $T_{\mu\nu}$ is the energy-momentum tensor with lowered indices

$$T_{\mu\nu} \equiv g_{\mu\lambda}\, g_{\nu\kappa}\, T^{\lambda\kappa} \tag{B.72}$$

and $R_{\mu\nu}$ is the *Ricci tensor*:

$$R_{\mu\nu} \equiv \frac{\partial\Gamma^\lambda_{\mu\lambda}}{\partial x^\nu} - \frac{\partial\Gamma^\lambda_{\mu\nu}}{\partial x^\lambda} + \Gamma^\kappa_{\mu\lambda}\Gamma^\lambda_{\nu\kappa} - \Gamma^\kappa_{\mu\nu}\Gamma^\lambda_{\lambda\kappa} \, . \tag{B.73}$$

For weak static fields the time–time component of this equation is the same as the Poisson equation (B.70), provided g_{00} is related to ϕ by Eq. (B.21). The tensor appearing on the left-hand side of Eq. (B.71) satisfies a set of differential *Bianchi identities*:

$$\left[g^{\lambda \nu} \left(R_{\mu\nu} - \frac{1}{2} g_{\mu\nu} g^{\rho\kappa} R_{\rho\kappa} \right) \right]_{;\lambda} = 0 \,. \tag{B.74}$$

This is why this is the linear combination of the tensors $R_{\mu\nu}$ and $g_{\mu\nu} g^{\lambda\kappa} R_{\lambda\kappa}$ that appears on the left-hand side of the field equations; otherwise the field equations would not be consistent with the energy-momentum conservation law (B.40). The field equations could have been derived more easily by including a gravitational term in the action

$$I_g = -\frac{1}{16\pi G} \int d^4x \sqrt{-\mathrm{Det}g(x)} \; g^{\lambda\kappa}(x) \, R_{\lambda\kappa}(x) \,. \tag{B.75}$$

The Bianchi identities (B.74) can be derived from the fact that I_g is automatically invariant under general coordinate transformations of the metric, and the field equations (B.71) can be derived from the condition that the total action $I_g + I_m$ be stationary with respect to arbitrary variations of the metric.

If we allow terms in the gravitational field equation with fewer than two spacetime derivatives, then it is possible to include a term on the left-hand side of the field equation (B.71) proportional to $g_{\mu\nu}$. This is the so-called *cosmological constant* term, discussed in Section 1.5. It can be regarded as a "vacuum-energy" correction to $T_{\mu\nu}$. Aside from a cosmological constant, the only other possible modification of the left-hand side of Eq. (B.71) would involve terms with *more* than two spacetime derivatives. (They can be derived, for instance, by including terms in the integrand of the gravitational action proportional to $\sqrt{-\mathrm{Det}g} \, R^{\mu\nu} R_{\mu\nu}$ or $\sqrt{-\mathrm{Det}g} \, (R^{\mu}{}_{\mu})^2$.) Dimensional analysis tells us that such terms would have coefficients whose dimensionality (relative to that of the factor $1/16\pi G$ in Eq. (B.75)) is a positive power of length. The experimental success of General Relativity shows that this length is much smaller than the scale of the solar system, so the effect of such terms would be completely negligible at cosmological distance scales.

Appendix C

Energy Transfer between Radiation and Electrons

One often needs to know the rate at which a photon will lose or gain energy when passing through an ionized gas. For instance, we need this information in calculating the Sunyaev–Zel'dovich effect, discussed in Section 2.5, and in understanding the preservation of thermal equilibrium between matter and radiation, treated in Section 2.2. This appendix will first derive formulas for the mean and the mean square change in energy of a photon in striking a single electron, and will then use these results to derive the rate of change of the photon energy distribution function in passing through an ionized gas. For the purposes of this section, we take $\hbar = c = 1$.

Suppose that an electron traveling in the three-direction with momentum four-vector

$$p = (0, 0, p_e, E_e) , \qquad E_e \equiv \sqrt{p_e^2 + m_e^2} , \tag{C.1}$$

is struck by a photon with energy ω moving along a direction with polar and azimuthal angles η and ϕ, giving the photon a new energy ω' and a new direction with polar and azimuthal angles η' and ϕ'. That is, the initial four-momentum q and final four-momentum q' of the photon take the forms

$$q = \left(\sin \eta \cos \phi, \ \sin \eta \sin \phi, \ \cos \eta, \ 1 \right) \omega , \tag{C.2}$$

$$q' = \left(\sin \eta' \cos \phi', \ \sin \eta' \sin \phi', \ \cos \eta', 1 \right) \omega' . \tag{C.3}$$

To calculate the cross section for this scattering event, we perform a Lorentz transformation to a frame of reference in which the electron is initially at rest, in which case the cross section takes a simple and well-known form. In the electron rest-frame, the initial and final photon four-momenta are Lq and Lq', where $L^\mu{}_\nu$ is the Lorentz transformation

$$L = \begin{pmatrix} 1 & 0 & 0 & 0 \\ 0 & 1 & 0 & 0 \\ 0 & 0 & \gamma & -\beta\gamma \\ 0 & 0 & -\beta\gamma & \gamma \end{pmatrix} , \tag{C.4}$$

where $\beta \equiv p_e/E_e$ is the electron velocity, and $\gamma \equiv (1 - \beta^2)^{-1/2}$. This gives the initial and final photon four-momenta in the electron rest frame as

$$Lq = \left(\sin \alpha \cos \phi, \ \sin \alpha \sin \phi, \ \cos \alpha, 1 \right) k , \tag{C.5}$$

$$Lq' = \left(\sin \alpha' \cos \phi', \ \sin \alpha' \sin \phi', \ \cos \alpha', 1 \right) k' . \tag{C.6}$$

where the initial and final photon energies in the electron rest frame are

$$k = (-\beta\gamma\cos\eta + \gamma)\omega, \qquad k' = (-\beta\gamma\cos\eta' + \gamma)\omega' \qquad (\text{C.7})$$

while the initial and final polar angles of the photon's velocity in this frame are given by

$$\cos\alpha = \frac{\cos\eta - \beta}{1 - \beta\cos\eta}, \qquad \cos\alpha' = \frac{\cos\eta' - \beta}{1 - \beta\cos\eta'}, \qquad (\text{C.8})$$

and there is no change in the azimuthal angles. We also recall that the conservation of energy and momentum in the electron rest frame gives the final photon energy in this frame as[1]

$$k' = \frac{k}{1 + (k/m_e)(1 - \cos\theta)}, \qquad (\text{C.9})$$

where θ is the scattering angle in the electron rest frame

$$\cos\theta \equiv \widehat{Lq} \cdot \widehat{Lq'} = \cos\alpha\cos\alpha' + \cos(\phi - \phi')\sin\alpha\sin\alpha'. \qquad (\text{C.10})$$

The fractional change in the photon energy in the original frame of reference can then be expressed in terms of quantities in the electron rest frame by

$$\frac{\omega' - \omega}{\omega} = \frac{1}{1 + \beta\cos\alpha}\left(\frac{1 + \beta\cos\alpha'}{1 + (k/m_e)(1 - \cos\theta)} - \beta\cos\alpha - 1\right). \qquad (\text{C.11})$$

In most cases of interest k/m_e and β are much less than unity, so we will keep only terms of first order in k/m_e and zeroth order in β together with terms of (for the moment) arbitrary order in β and zeroth order in k/m_e:

$$\frac{\omega' - \omega}{\omega} \simeq -(k/m_e)(1 - \cos\theta) + \frac{\beta(\cos\alpha' - \cos\alpha)}{1 + \beta\cos\alpha}. \qquad (\text{C.12})$$

For an electron at rest and for photon energy $k \ll m_e$, the cross-section differential (summed over final spins and polarizations, and averaged over initial spins and polarizations) is[2]

$$d\sigma = \frac{3\sigma_T}{16\pi}\left(1 + \cos^2\theta\right)d(\cos\alpha')\,d\phi', \qquad (\text{C.13})$$

---

[1]QTF, Vol. I, Eq. (8.7.14).
[2]QTF, Vol. I, Eq. (8.7.42).

where σ_T is the Thomson cross section $e^4/6\pi m_e^2$. The cross-section differential is itself Lorentz invariant, though the variables on which it depends are not, so the average photon energy change per collision is

$$\langle \omega' - \omega \rangle = \frac{1}{\sigma_T} \int (\omega' - \omega) \, d\sigma$$

$$= \frac{3}{16\pi} \int_{-1}^{1} d(\cos\alpha') \int_{0}^{2\pi} d\phi' \, (\omega' - \omega) \, (1 + \cos^2\theta)$$

$$= -\frac{\beta\omega\cos\alpha}{1 + \beta\cos\alpha} - \frac{k\omega}{m_e}. \qquad (C.14)$$

We must average this over the electron direction of motion, or equivalently over α. In calculating this average we must keep in mind that it is $\cos\eta$ rather than $\cos\alpha$ whose distribution function in the interval from -1 to $+1$ is flat. Also, the transition probability is proportional not only to $d\sigma$ but also to the relative speed[3]

$$u = \frac{|p \cdot q|}{E_e\omega} = 1 - \beta\cos\eta = \frac{1 - \beta^2}{1 + \beta\cos\alpha}. \qquad (C.15)$$

Hence the average of the energy change over the relative direction of the initial electron and photon is

$$\langle\langle \omega' - \omega \rangle\rangle = \int_{-1}^{+1} \langle \omega' - \omega \rangle u \, d\cos\eta \Big/ \int_{-1}^{+1} u \, d\cos\eta$$

$$= \frac{1}{2} \int_{-1}^{+1} \frac{(1 - \beta^2)^2}{(1 + \beta\cos\alpha)^3} \left[-\frac{\beta\omega\cos\alpha}{1 + \beta\cos\alpha} - \frac{k\omega}{m_e} \right] d\cos\alpha .$$

$$(C.16)$$

The terms of first order in β make no contribution to the integral over α, so the leading terms in the fractional energy transfer are those of second order in β or first order in ω/m_e:

$$\langle\langle \omega' - \omega \rangle\rangle/\omega \simeq \frac{4}{3}\beta^2 - \frac{k}{m_e} \simeq \frac{4}{3}\beta^2 - \frac{\omega}{m_e}. \qquad (C.17)$$

Since the mean photon fractional energy change contains terms of order β^2 and ω/m_e, we will be interested in any terms in the average squared fractional energy change of the same order. Inspection of Eq. (C.12) shows that to this order, we have

$$(\omega' - \omega)^2/\omega^2 \simeq \beta^2(\cos\alpha' - \cos\alpha)^2 \qquad (C.18)$$

---

[3] QTF, Vol. I, Eq. (3.4.17).

To this order we can neglect the difference between u and unity and between $d\cos\eta$ and $d\cos\alpha$, so that

$$\langle\langle(\omega'-\omega)^2\rangle\rangle \simeq \frac{3\beta^2\omega^2}{32\pi} \int_{-1}^{+1} d\cos\alpha \int_{-1}^{+1} d\cos\alpha'$$

$$\times \int_0^{2\pi} d\phi' (\cos\alpha' - \cos\alpha)^2 (1+\cos^2\theta)$$

$$= \frac{2}{3}\beta^2\omega^2 . \tag{C.19}$$

All this is for an electron of a fixed speed β. If the number of electrons with speed between β and $\beta + d\beta$ is given by the Maxwell–Boltzmann distribution with electron temperature T_e, and is hence proportional to $\beta^2 \exp(-m_e\beta^2/2k_BT_e)\,d\beta$, the average value of β^2 is $3k_BT_e/m_e$, and Eqs. (C.17) and (C.19) become

$$\langle\langle\omega'-\omega\rangle\rangle \simeq \frac{4k_BT_e}{m_e}\omega - \frac{\omega^2}{m_e} , \tag{C.20}$$

and

$$\langle\langle(\omega'-\omega)^2\rangle\rangle = \frac{2k_BT_e}{m_e}\omega^2 \tag{C.21}$$

Now suppose that photons with a number $N(\omega)d\omega$ with energies between ω and $\omega + d\omega$ interact with a gas of non-relativistic electrons. For photon energies $\omega \ll m_e$, the rate of change of the distribution function is

$$\dot{n}(\omega)d\omega = n_e \int d\omega' \, n(\omega') \, R(\omega' \to \omega)\, d\omega \, (1+N(\omega))$$

$$-n_e \, n(\omega) \, d\omega \int d\omega' \, R(\omega \to \omega')\, (1+N(\omega')) . \tag{C.22}$$

Here n_e is the number density of electrons; $R(\omega \to \omega')d\omega'$ is the average of $u\,d\sigma$ over initial electron velocities for collisions in which a photon of initial energy ω is given an energy between ω' and $\omega' + d\omega'$; $N(\omega) = (2\pi\hbar)^3 n(\omega)d\omega/8\pi\omega^2 d\omega$ is the number of photons per quantum state of energy ω (the denominator is $8\pi\omega^2 d\omega$ instead of $4\pi\omega^2 d\omega$ because of the two polarization states of photons); and the factors $1+N(\omega)$ and $1+N(\omega')$ are included to take account of the stimulated emission of photons into states that are already occupied. The first term in Eq. (C.22) gives the increase in $n(\omega)$ due to scattering of photons of any initial energy ω' into energy ω, while the second term gives the decrease in $n(\omega)$ due to scattering of photons of energy ω into any final energy ω'. We saw that collisions of photons of energy $\omega \ll m_e$ with a non-relativistic gas of electrons at temperature T typically change photon energies by only small fractional amounts, of order ω/m_e or $k_B T/m_e$, so the rate constants $R(\omega' \to \omega)$ and $R(\omega \to \omega')$ are sharply peaked around $\omega' \simeq \omega$. It is therefore convenient to change the variable of integration from ω' to $\Delta \equiv \omega - \omega'$ in the first term and to $\Delta \equiv \omega' - \omega$ in the second term of Eq. (C.22). Also canceling the unintegrated differential $d\omega$ and multiplying with $(2\pi\hbar)^3/8\pi\omega^2$, this formula now reads

$$\dot{N}(\omega) = \frac{n_e}{\omega^2} \int_{-\infty}^{\infty} d\Delta \, (\omega - \Delta)^2 \, N(\omega - \Delta) \, R(\omega - \Delta \to \omega)\,(1 + N(\omega))$$

$$- n_e N(\omega) \int d\Delta \, R(\omega \to \omega + \Delta)\,(1 + N(\omega + \Delta)) \,. \tag{C.23}$$

Since the rate coefficients $R(\omega - \Delta \to \omega)$ and $R(\omega \to \omega + \Delta)$ are sharply peaked around $\Delta = 0$, we can expand the other factors in Eq. (C.23) (and the difference between $R(\omega - \Delta \to \omega)$ and $R(\omega \to \omega + \Delta)$) to second order in Δ:

$$\dot{N}(\omega) = -\frac{n_e}{\omega^2} \int_{-\infty}^{\infty} d\Delta \, \Delta \left(1 + N(\omega)\right) \frac{\partial}{\partial\omega}\left[\omega^2 N(\omega)\, R(\omega \to \omega + \Delta)\right]$$

$$+ \frac{n_e}{2\omega^2} \int_{-\infty}^{\infty} d\Delta \, \Delta^2 \left(1 + N(\omega)\right) \frac{\partial^2}{\partial\omega^2}\left[\omega^2 N(\omega)\, R(\omega \to \omega + \Delta)\right]$$

$$- n_e N(\omega) \int_{-\infty}^{\infty} d\Delta \, \Delta \, \frac{\partial N(\omega)}{\partial\omega}\, R(\omega \to \omega + \Delta)$$

$$- \frac{n_e}{2} N(\omega) \int_{-\infty}^{\infty} d\Delta \, \Delta^2 \, \frac{\partial^2 N(\omega)}{\partial\omega^2}\, R(\omega \to \omega + \Delta) \,.$$

This can be rewritten as

$$\dot{N}(\omega) = -\frac{n_e}{\omega^2} \int_{-\infty}^{\infty} d\Delta \, \Delta \, \frac{\partial}{\partial\omega}\left[\omega^2 N(\omega)\, R(\omega \to \omega + \Delta)\left(1 + N(\omega)\right)\right]$$

$$+ \frac{n_e}{2\omega^2} \int_{-\infty}^{\infty} d\Delta \, \Delta^2 \frac{\partial^2}{\partial\omega^2}\left[\omega^2 N(\omega)\, R(\omega \to \omega + \Delta)\left(1 + N(\omega)\right)\right]$$

$$- \frac{n_e}{\omega^2} \int_{-\infty}^{\infty} d\Delta \, \Delta^2 \frac{\partial}{\partial\omega}\left[\omega^2 N(\omega)\, R(\omega \to \omega + \Delta)\, \frac{\partial N(\omega)}{\partial\omega}\right] \,.$$

Inverting the order of integration and differentiation, this becomes

$$\dot{N}(\omega) = -\frac{n_e \sigma_T}{\omega^2} \left[\frac{\partial}{\partial\omega}\left(\omega^2 N(\omega)\left(1 + N(\omega)\right)\langle\langle\omega' - \omega\rangle\rangle\right)\right.$$

$$- \frac{1}{2}\frac{\partial^2}{\partial\omega^2}\left(\omega^2 N(\omega)\left(1 + N(\omega)\right)\langle\langle(\omega' - \omega)^2\rangle\rangle\right)$$

$$\left.+ \frac{\partial}{\partial\omega}\left(\omega^2 N(\omega)\,\langle\langle(\omega' - \omega)^2\rangle\rangle\,\frac{\partial N(\omega)}{\partial\omega}\right)\right] \,, \tag{C.24}$$

in which we use

$$\int_{-\infty}^{\infty} d\Delta \, \Delta \, R(\omega \to \omega + \Delta) = \sigma_T \langle\langle \omega' - \omega \rangle\rangle \,,$$

$$\times \int_{-\infty}^{\infty} d\Delta \, \Delta^2 \, R(\omega \to \omega + \Delta) = \sigma_T \langle\langle (\omega' - \omega)^2 \rangle\rangle \,. \qquad \text{(C.25)}$$

Inserting the values (C.20) and (C.21) of $\langle\langle \omega' - \omega \rangle\rangle$ and $\langle\langle (\omega' - \omega)^2 \rangle\rangle$ gives the *Kompaneets equation*:[4]

$$\dot{N}(\omega) = \frac{n_e \sigma_T k_B T_e}{m_e \omega^2} \frac{\partial}{\partial \omega} \left[\omega^4 \frac{\partial N(\omega)}{\partial \omega} \right] + \frac{n_e \sigma_T}{m_e \omega^2} \frac{\partial}{\partial \omega} \left[\omega^4 N(\omega) \left(1 + N(\omega) \right) \right] .$$

$$\text{(C.26)}$$

As a check, we may note that if the photon distribution function is given by the Planck formula *with photon temperature equal to the electron temperature*, so that $N(\omega) = [\exp(\omega/k_B T_e) - 1]^{-1}$, then the two terms in Eq. (C.26) cancel, giving no change in $N(\omega)$, as of course must be the case for photons in equilibrium with electrons. As a further check, we note that (as long as $\omega^4 N(\omega)$ and $\omega^4 \, \partial N/\partial \omega$ both vanish at $\omega = 0$ and $\omega \to \infty$) the total number density of photons $\int_0^\infty 4\pi \omega^2 \, N(\omega) \, d\omega$ does not change with time, as could also be seen by integrating Eq. (C.22) over ω, and interchanging ω and ω' in the second term.

---

[4]A. Kompaneets, *Zh. Exper. Teor. Fiziki* **312+**, 876 (1956).

Appendix D

The Ergodic Theorem

In cosmology we often have to deal with position-dependent variables like temperature or density or scalar fields whose fluctuations are governed by some sort of probability distribution. In this appendix we will consider a general real random variable $\varphi(x)$ depending on a D-dimensional Euclidean coordinate x. The generalization to several random variables will be obvious.

We will assume that the distribution function giving the probabilities of various functional forms for $\varphi(x)$ is *homogeneous*, in the sense that the average of any product of φs with different arguments depends only on the differences of the arguments. That is, for arbitrary z,

$$\langle \varphi(x_1)\varphi(x_2)\cdots\varphi(x_n)\rangle = \langle\varphi(x_1+z)\varphi(x_2+z)\cdots\varphi(x_n+z)\rangle . \quad \text{(D.1)}$$

In cases where x is a time coordinate, such distributions are said to be *stationary*.

We will further assume that the φs at distant arguments are uncorrelated. To put this formally, we assume that for $|u| \to \infty$,

$$\langle\varphi(x_1+u)\varphi(x_2+u)\cdots\varphi(y_1-u)\varphi(y_2-u)\cdots\rangle$$

$$\to \langle\varphi(x_1+u)\varphi(x_2+u)\cdots\rangle\,\langle\varphi(y_1-u)\varphi(y_2-u)\cdots\rangle$$

$$= \langle\varphi(x_1)\varphi(x_2)\cdots\rangle\,\langle\varphi(y_1)\varphi(y_2)\cdots\rangle , \quad \text{(D.2)}$$

with the final expression justified by Eq. (D.1).

Under these conditions, we can prove an important result known as the *Ergodic Theorem*: If the limit in Eq. (D.2) is approached sufficiently rapidly, then the root mean square difference between any product $\varphi(x_1+z)\varphi(x_2+z)\cdots$, averaged over a range R of z values around an arbitrary point z_0, and the ensemble average of the same product, vanishes as $R^{-D/2}$ for large R. That is, if we define

$$\Delta_R^2(x_1, x_2, \ldots) \equiv \left\langle\left(\int d^D z\, N_R(z)\, \varphi(x_1+z)\varphi(x_2+z)\cdots \right.\right.$$

$$\left.\left. -\langle\varphi(x_1)\varphi(x_2)\cdots\rangle\right)^2\right\rangle , \quad \text{(D.3)}$$

where[1]

$$N_R(z) \equiv (\sqrt{\pi}R)^{-D} \exp\left(-|z - z_0|^2/R^2\right), \tag{D.4}$$

then for $R \to \infty$,

$$\Delta_R \to O(R^{-D/2}). \tag{D.5}$$

To prove this theorem, we first use the condition $\int N_R(z)\,d^Dz = 1$ to rewrite Eq. (D.3) as

$$\Delta_R^2(x_1, x_2, \ldots) = \left\langle\left(\left[\int d^Dz\, N_R(z)\left[\varphi(x_1 + z)\varphi(x_2 + z)\cdots\right.\right.\right.\right.$$
$$\left.\left.\left.\left. -\langle\varphi(x_1)\varphi(x_2)\cdots\rangle\right]\right)^2\right\rangle,$$

Expanding the square and again using Eq. (D.1) and the normalization condition for N_R, and introducing new integration variables $u \equiv (z - w)/2$ and $v \equiv (z + w)/2$, we have

$$\Delta_R^2 = \int d^Dz\, N_R(z) \int d^Dw\, N_R(w)$$
$$\times \left[\left\langle\varphi(x_1 + z)\varphi(x_2 + z)\cdots\varphi(x_1 + w)\varphi(x_2 + w)\cdots\right\rangle\right.$$
$$\left. -\left\langle\varphi(x_1)\varphi(x_2)\cdots\right\rangle^2\right]$$
$$= \left(\frac{2}{\pi R^2}\right)^D \int d^Dv\, \exp\left(-2|v - z_0|^2/R^2\right) \int d^Du\, \exp\left(-2|u|^2/R^2\right)$$
$$\times \left[\left\langle\varphi(x_1 + u)\varphi(x_2 + u)\cdots\varphi(x_1 - u)\varphi(x_2 - u)\cdots\right\rangle\right.$$
$$\left. -\left\langle\varphi(x_1)\varphi(x_2)\cdots\right\rangle^2\right]$$
$$= \left(\frac{2}{\pi R^2}\right)^{D/2} \int d^Du\, \exp\left(-2|u|^2/R^2\right)$$

---

[1] The specific form (D.4) for the function N_R is not essential. It is only important that the function be constant for $|z - z_0|^2 \ll R^2$, vanish rapidly for $|z - z_0|^2 \gg R^2$, and be normalized so that $\int N_R(z)\,d^Dz = 1$.

$$\times \left[\left\langle \varphi(x_1 + u)\varphi(x_2 + u) \cdots \varphi(x_1 - u)\varphi(x_2 - u) \cdots \right\rangle \right.$$

$$\left. - \left\langle \varphi(x_1)\varphi(x_2) \cdots \right\rangle^2 \right].$$

Assuming that the limit in Eq. (D.2) is approached sufficiently rapidly, the u integral would converge even without the factor $\exp\left(-2|u|^2/R^2\right)$, so for $R \to \infty$ we can take this factor as unity, and find our final expression

$$\Delta_R^2 \to \left(\frac{2}{\pi R^2}\right)^{D/2} \int d^D u$$

$$\times \left[\left\langle \varphi(x_1 + u)\varphi(x_2 + u) \cdots \varphi(x_1 - u)\varphi(x_2 - u) \cdots \right\rangle \right.$$

$$\left. - \left\langle \varphi(x_1)\varphi(x_2) \cdots \right\rangle^2 \right], \qquad\qquad (D.6)$$

thus confirming Eq. (D.5).

Appendix E

Gaussian Distributions

Consider a random variable $\varphi(x)$ depending on a D-dimensional coordinate x. We will define it to have zero average value, by subtracting from φ any non-zero average it may have. The distribution function governing φ is said to be *Gaussian* if the average of a product of an even number of φs is the sum over all ways of pairing φs with each other of a product of the average values of the pairs:

$$\langle \varphi(x_1)\varphi(x_2)\cdots\rangle = \sum_{\text{pairings}} \prod_{\text{pairs}} \langle \varphi\varphi\rangle , \qquad (E.1)$$

(with the sum over pairings not distinguishing those which interchange coordinates in a pair, or which merely interchange pairs), while the average of the product of any odd number of φs vanishes. For instance

$$\langle \varphi(x_1)\varphi(x_2)\varphi(x_3)\varphi(x_4)\rangle = \langle \varphi(x_1)\varphi(x_2)\rangle\langle \varphi(x_3)\varphi(x_4)\rangle$$
$$+ \langle \varphi(x_1)\varphi(x_3)\rangle\langle \varphi(x_2)\varphi(x_4)\rangle + \langle \varphi(x_1)\varphi(x_4)\rangle\langle \varphi(x_2)\varphi(x_3)\rangle . \quad (E.2)$$

(Of course, there is no way of pairing all of an odd number of φs, which is why for Gaussian distributions the average of any odd number of φs must vanish.) More generally, in the average of a product of $2n$ factors of φ, each of $(2n)!$ permutations of the coordinates defines a pairing (with the first permuted φ paired with the second, the third with the fourth, and so on), but $2^n n!$ of these differ only by permutations of the two coordinates within a pair or permutations of the n pairs, so in general this average contains $(2n)!/2^n n!$ terms.

If $\varphi(x)$ is governed by a Gaussian distribution, then so is any linear functional of $\varphi(x)$. For instance, consider a set of linear functionals of the form

$$F_i[\varphi] \equiv \int d^D x \, f_i(x) \, \varphi(x) . \qquad (E.3)$$

By multiplying Eq. (E.2) by $f_1(x_1)f_2(x_2)f_3(x_3)f_4(x_4)$ and integrating, we find

$$\langle F_1 F_2 F_3 F_4\rangle = \langle F_1 F_2\rangle\langle F_3 F_4\rangle + \langle F_1 F_3\rangle\langle F_2 F_4\rangle + \langle F_1 F_4\rangle\langle F_2 F_3\rangle . \qquad (E.4)$$

Such distributions are called Gaussian because of the form taken by the probability distribution of general linear functionals $F[\varphi] = \int d^D x f(x)\varphi(x)$ of φ. In $\langle F^{2n}[\varphi]\rangle$, each of the $(2n)!/2^n n!$ terms makes the same contribution

$\langle F^2[\varphi]\rangle^n$, so
$$\langle F^{2n}[\varphi]\rangle = \langle F^2[\varphi]\rangle^n (2n)!/2^n n! \ . \tag{E.5}$$

This implies that the probability $P(F)\,dF$ that the functional is between F and $F + dF$ is the Gaussian function

$$P(F)\,dF = \frac{1}{\sqrt{2\pi\,\langle F^2\rangle}} \exp\left(-\frac{F^2}{2\langle F^2\rangle}\right) dF \ . \tag{E.6}$$

Distributions of this sort arise commonly when $\varphi(x)$ is the sum of a large number of independently fluctuating terms. The *central limit theorem* tells us that in this case the distribution of the sum is Gaussian. In the application that most concerns us here, the fluctuations in the temperature of the cosmic microwave background are believed to be nearly Gaussian because they arise (as discussed in Chapter 10) from the quantum fluctuations of one or more nearly free quantum fields.

Appendix F

Newtonian Cosmology

During the era when the energy density of the universe is dominated by cold dark matter, the behavior of perturbations can be adequately treated by the methods of Newtonian mechanics. This has applications both for our introductory study of anisotropies in the cosmic microwave background in Section 2.6, and for the treatment of the large scale structure of matter in Section 8.1.

The equations of non-relativistic hydrodynamics and Newtonian gravitational theory for a fluid with mass density ρ, velocity $\mathbf{v}$, zero pressure, and gravitational potential ϕ are the equation of continuity

$$\frac{\partial \rho}{\partial t} + \nabla \cdot \left(\mathbf{v} \rho \right) = 0 , \tag{F.1}$$

the Euler equation

$$\frac{\partial \mathbf{v}}{\partial t} + \left(\mathbf{v} \cdot \nabla \right) \mathbf{v} = -\nabla \phi , \tag{F.2}$$

and the Poisson equation

$$\nabla^2 \phi = 4\pi G \rho . \tag{F.3}$$

These equations have an unperturbed solution (distinguished by bars) of the form

$$\bar{\rho} = \rho_0 (a_0/a)^3 , \quad \bar{\mathbf{v}} = H\mathbf{X} , \quad \bar{\phi} = 2\pi G \bar{\rho} \mathbf{X}^2 /3 , \tag{F.4}$$

where $H \equiv \dot{a}/a$; a_0 and ρ_0 are constants; and $a(t)$ satisfies the equation

$$\dot{a}^2 + K = 8\pi G \bar{\rho} a^2 /3 , \tag{F.5}$$

with K a constant. (We use $\mathbf{X}$ to denote the ordinary Euclidean coordinate vector, to distinguish it from the co-moving coordinate vector $\mathbf{x}$.) This of course corresponds to a cosmological theory with a Robertson–Walker metric, and indeed we have already encountered this solution in Section 1.5, as an alternative approach to the derivation of the Friedmann equation (F.5). In particular, the solution of the equation $d\mathbf{X}/dt = \bar{\mathbf{v}} = H\mathbf{X}$ for the motion of a co-moving object is $\mathbf{X}(t) = [a(t)/a(t_0)]\mathbf{X}(t_0)$, in agreement with Eq. (1.5.22). The co-moving coordinate is thus $\mathbf{x} = \mathbf{X}/a$.

To this unperturbed solution for ρ, $\mathbf{v}$, and ϕ we now add small perturbations $\delta\rho$, $\delta\mathbf{v}$, and $\delta\phi$. The terms in Eqs. (F.1)–(F.3) of first order in these perturbations are

$$\frac{\partial \delta\rho}{\partial t} + 3H\delta\rho + H\mathbf{X} \cdot \nabla\delta\rho + \bar\rho\nabla \cdot \delta\mathbf{v} = 0 \,, \tag{F.6}$$

$$\frac{\partial \delta\mathbf{v}}{\partial t} + H\mathbf{X} \cdot \nabla\delta\mathbf{v} + H\delta\mathbf{v} = -\nabla\delta\phi \,, \tag{F.7}$$

$$\nabla^2\delta\phi = 4\pi G\delta\rho \,. \tag{F.8}$$

Eqs. (F.6)–(F.8) do not appear translation-invariant, but the underlying spatial homogeneity of this problem can be restored by writing the equations in terms of the co-moving coordinate $\mathbf{X}/a$. This is brought out most conveniently if we write the perturbations as Fourier transforms in the co-moving coordinate:

$$\delta\rho(\mathbf{X}, t) = \int d^3q \, \exp\left(\frac{i\mathbf{q} \cdot \mathbf{X}}{a(t)}\right) \delta\rho_{\mathbf{q}}(t) \,, \tag{F.9}$$

and likewise for $\delta\mathbf{v}$ and $\delta\phi$. The partial differential equations (F.6)–(F.8) then become the ordinary differential equations

$$\frac{d\delta\rho_{\mathbf{q}}}{dt} + 3H\delta\rho_{\mathbf{q}} + ia^{-1}\bar\rho\,\mathbf{q} \cdot \delta\mathbf{v}_{\mathbf{q}} = 0 \,, \tag{F.10}$$

$$\frac{d\delta\mathbf{v}_{\mathbf{q}}}{dt} + H\delta\mathbf{v}_{\mathbf{q}} = -ia^{-1}\mathbf{q}\delta\phi_{\mathbf{q}} \,, \tag{F.11}$$

$$q^2\delta\phi_{\mathbf{q}} = -4\pi Ga^2\delta\rho_{\mathbf{q}} \,. \tag{F.12}$$

The solutions of Eqs. (F.10)–(F.12) can be classified according to the transformation properties of the dependent variables under three-dimensional rotations:

Vector modes: In these modes, all scalars vanish: not only $\delta\rho_{\mathbf{q}}$ and $\delta\phi_{\mathbf{q}}$, but also $\mathbf{q} \cdot \delta\mathbf{v}_{\mathbf{q}}$. Then Eqs. (F.10) and (F.12) are automatically satisfied, while Eq. (F.11) becomes

$$\frac{d\delta\mathbf{v}_{\mathbf{q}}}{dt} + H\delta\mathbf{v}_{\mathbf{q}} = 0 \,, \tag{F.13}$$

whose solution is simply $\delta\mathbf{v}_{\mathbf{q}} \propto 1/a$. Because the vector modes simply decay, they are widely ignored.

Scalar Modes: In these modes, the velocity perturbation $\delta\mathbf{v}(\mathbf{X}, t)$ can be expressed as the gradient (with respect to $\mathbf{x} = \mathbf{X}/a$) of a scalar potential perturbation $\delta u(\mathbf{X}, t)$. For the Fourier transforms, this means that

$$\delta\mathbf{v}_{\mathbf{q}} = i\mathbf{q}\,\delta u_{\mathbf{q}} \,. \tag{F.14}$$

Then, using Eq. (F.12) to eliminate $\delta\phi_\mathbf{q}$, Eq. (F.11) becomes

$$\frac{d\delta u_\mathbf{q}}{dt} + H\delta u_\mathbf{q} = -a^{-1}\delta\phi_\mathbf{q} = \frac{4\pi Ga}{q^2}\delta\rho_\mathbf{q}\,, \qquad (F.15)$$

while Eq. (F.10) gives

$$\frac{d\delta\rho_\mathbf{q}}{dt} + 3H\delta\rho_\mathbf{q} - a^{-1}\bar{\rho}\,\mathbf{q}^2\delta u_\mathbf{q} = 0\,. \qquad (F.16)$$

For $K = 0$ we have $4\pi G\bar{\rho} = 3H^2/2$. Using $\bar{\rho} \propto a^{-3}$ and the definition of H, we can eliminate $\delta u_\mathbf{q}$ from these coupled equations and write them as a second-order differential equation for the fractional density perturbation

$$\frac{d}{dt}\left(a^2\frac{d}{dt}\left(\frac{\delta\rho_\mathbf{q}}{\bar{\rho}}\right)\right) - \frac{3}{2}H^2a^2\left(\frac{\delta\rho_\mathbf{q}}{\bar{\rho}}\right) = 0\,, \qquad (F.17)$$

or, recalling that for $K = 0$ we have $a \propto t^{2/3}$ and $H = 2/3t$,

$$\frac{d}{dt}\left(t^{4/3}\frac{d}{dt}\left(\frac{\delta\rho_\mathbf{q}}{\bar{\rho}}\right)\right) - \frac{2}{3}t^{-2/3}\left(\frac{\delta\rho_\mathbf{q}}{\bar{\rho}}\right) = 0\,. \qquad (F.18)$$

The general solution of this equation is a linear combination of the powers $t^{2/3}$ and t^{-1}. It is reasonable to suppose that by the time of last scattering only the leading mode with $\delta\rho_\mathbf{q}/\bar{\rho} \propto t^{2/3} \propto a$ will have survived. Since the mean density $\bar{\rho}(t)$ is proportional to $a^{-3}(t)$, we have $\delta\rho_\mathbf{q}(t) \propto a^{-2}(t)$. Eq. (F.12) then shows that $\delta\phi_\mathbf{q}$ is time independent, a result used in Section 2.6.

To verify that these are the most general solutions of Eqs. (F.10)–(F.12), we need only count equations and solutions. With $\delta\phi_\mathbf{q}(t)$ eliminated by use of Eq. (F.12), Eqs. (F.10) and (F.11) are a set of $1 + 3 = 4$ coupled first-order differential equations, so they have four linearly independent solutions. We have found two independent vector modes (corresponding to the two directions perpendicular to $\mathbf{q}$) and two independent scalar modes (with $\delta\rho_\mathbf{q}/\bar{\rho}$ proportional to $t^{2/3}$ or t^{-1}), so these are indeed the most general solutions.

The cases of non-zero pressure or $K \neq 0$ are treated in detail in Section 15.9 of G&C.

Appendix G

Photon Polarization

The polarization of photons is of great interest in cosmology, not only because it can be observed in the cosmic microwave background, but also because it affects the anisotropic inertia terms appearing in the gravitational field equations for both scalar and tensor perturbations. This appendix will review the description of the spin state of an individual photon in terms of photon polarization vectors and the related density matrix, and the somewhat unusual parallel transport equation that governs the time dependence of photon polarization vectors and density matrices in gravitational fields.

Let us first recall how we describe the polarization of a photon in the absence of gravitational fields. The most general pure state of a single photon is a linear superposition $\alpha_+ \Psi_+ + \alpha_- \Psi_-$, where $\Psi_\pm$ are states of helicity ∓ 1 (that is, eigenstates of the component of angular momentum in the direction $\hat{n}$ from which the photon is coming[1] with eigenvalues $\sigma = \pm \hbar$, and normalization $(\Psi_\sigma, \Psi_{\sigma'}) = \delta_{\sigma\sigma'}$) and $\alpha_\pm$ are complex numbers satisfying the normalization condition $|\alpha_+|^2 + |\alpha_-|^2 = 1$. We represent such a state by a polarization vector $\mathbf{e} = \alpha_+ \mathbf{e}_+ + \alpha_- \mathbf{e}_-$ with $\mathbf{e}^* \cdot \mathbf{e} = 1$, where $\mathbf{e}_\pm$ is the polarization vector for photons of helicity ∓ 1. For instance, for a photon that is seen coming from the direction[2]

$$\hat{n} = (\sin\theta \, \cos\phi, \, \sin\theta \, \sin\phi, \, \cos\theta) , \tag{G.1}$$

we can take

$$\mathbf{e}_\pm = (\hat{\theta} \pm i\hat{\phi})/\sqrt{2} , \tag{G.2}$$

where $\hat{\theta}$ and $\hat{\phi}$ are orthogonal unit vectors in the plane perpendicular to $\hat{n}$:

$$\hat{\theta} \equiv (\cos\theta \, \cos\phi, \, \cos\theta \, \sin\phi, \, -\sin\theta) , \quad \hat{\phi} \equiv (-\sin\phi, \, \cos\phi , 0) . \tag{G.3}$$

A linearly polarized photon has $|\alpha_+| = |\alpha_-|$, and hence a polarization vector that is real up to an unimportant over-all phase factor. The opposite case is circular polarization, for which either $|\alpha_+|$ or $|\alpha_-|$ vanishes. Between these

---

[1]This is the opposite of the direction of the photon's motion, so the helicity, which is defined as the component of angular momentum along the direction of the photon's motion, is the negative of the component along the direction from which it is coming. This is why we use a label $\pm$ to indicate photons of helicity ∓ 1.

[2]For correlations between polarizations in a small patch of sky, one can choose the three-axis to be in the direction $\hat{n}$ of this patch, in which case the polarization vectors for helicity ∓ 1 could be chosen simply as $(1, \pm i, 0)/\sqrt{2}$.

two extremes is elliptic polarization, for which $|\alpha_+|$ and $|\alpha_-|$ are unequal, but neither vanishes.

An individual photon will in general have probabilities P_m of being in any one of various pure states $\Psi_m = \alpha_{m+}\Psi_+ + \alpha_{m-}\Psi_-$, so according to the rules of quantum mechanics, the probability of finding it in a particular state $\Psi = \alpha_+\Psi_+ + \alpha_-\Psi_-$ represented by a polarization vector $\mathbf{e} = \alpha_+\mathbf{e}_+ + \alpha_-\mathbf{e}_-$, will be

$$P(\mathbf{e}) = \sum_m P_m |(\Psi_m, \Psi)|^2 = \sum_m P_m |\alpha_{m+}^*\alpha_+ + \alpha_{m-}^*\alpha_-|^2 = \mathbf{e}^{*i}\mathbf{e}^j N^{ij} \quad \text{(G.4)}$$

where N^{ij} is the *density matrix* for that photon

$$N^{ij} \equiv \sum_m P_m e_m^i e_m^{j*} . \quad \text{(G.5)}$$

(Repeated indices i, j, etc. are summed over the coordinate indices 1,2,3.) Because the probabilities are real and positive, the matrix N^{ij} is hermitian and positive:

$$N^{ij*} = N^{ji} , \quad N^{ij}\xi^i\xi^{j*} \geq 0 \text{ for all } \xi^i , \quad \text{(G.6)}$$

and because $\sum_m P_m = 1$ and $e_m^i e_m^{i*} = 1$ (not summed over m), this matrix has unit trace

$$N^{ii} = 1 . \quad \text{(G.7)}$$

Of course, the photon polarization vectors are all orthogonal to the photon's direction of motion $-\hat{n}$, so also

$$N^{ij} \hat{n}^i = N^{ij}\hat{n}^j = 0 , \quad \text{(G.8)}$$

The scattering of light by non-relativistic electrons does not produce circular polarization, and therefore we expect that all microwave background photons are linearly polarized, in which case N^{ij} is real.

So far, we have defined the polarization vector e^i in the absence of gravitation as a unit three-vector, transverse to the photon's direction of motion. We can if we like define it in a gravitational field as the spatial part of a four-component object e^μ that happens to have $e^0 = 0$, and that satisfies

$$p_i e^i = p_\mu e^\mu = 0 , \quad g_{ij}e^i e^{j*} = g_{\mu\nu}e^\mu e^{\nu*} = 1 , \quad \text{(G.9)}$$

where $p^\mu = g^{\mu\nu} p_\nu$ is the photon four-momentum.[3] But since this object obviously does not transform as a four-vector there is no reason to think that its time dependence would be given by the equation (B.69) of parallel transport. Indeed, it cannot evolve by parallel transport, for if it did then the condition $e^0 = 0$ would not be preserved along the photon trajectory $x^i = x^i(t)$ unless it so happened that $\Gamma_{ij}^0(\mathbf{x}(t), t)$ vanishes, which is not usually the case. Instead, we expect that in a locally inertial frame of reference in which the affine connection $\Gamma_{\nu\lambda}^\mu \big(x(t) \big)$ vanishes, the polarization vector of a photon will be time-independent only up to a gauge transformation $e^\mu \to e^\mu + \alpha p^\mu$, so that gauge-invariant quantities like the field strength $p^\mu e^\nu - p^\nu e^\mu$ will be time-independent. Then in a general frame of reference, a photon polarization vector will undergo parallel transport up to a gauge transformation:

$$\frac{de^\mu(t)}{dt} = -\Gamma_{\nu\lambda}^\mu \big(x(t) \big) e^\nu(t) \frac{dx^\lambda(t)}{dt} + \alpha(t) \frac{dx^\mu(t)}{dt} . \tag{G.10}$$

The gauge transformation parameter $\alpha(t)$ can then be determined from the condition that $e^0 = 0$, so that $de^0/dt = 0$:

$$\alpha(t) = \Gamma_{\nu\lambda}^0 \big(x(t) \big) e^\nu(t) \frac{dx^\lambda(t)}{dt} . \tag{G.11}$$

With $e^0 = 0$, Eq. (G.10) then reads

$$\frac{de^j(t)}{dt} = \left[-\Gamma_{i\lambda}^j \big(x(t) \big) + \Gamma_{i\lambda}^0 \big(x(t) \big) \frac{dx^j(t)}{dt} \right] e^i(t) \frac{dx^\lambda(t)}{dt} . \tag{G.12}$$

Note that the conditions (G.9) are preserved with time, for $p_\mu e^\mu$ and $g_{\mu\nu} e^\mu e^\nu$ are preserved by ordinary parallel transport because they are scalars, and are preserved under gauge transformations because $p_\mu p^\mu = 0$ and $p_\mu e^\mu = 0$.

In a gravitational field we continue to define the statistical matrix N^{ij} by Eq. (G.5), and it continues to satisfy the hermiticity and positivity conditions (G.6), but now instead of (G.7) and (G.8) we have

$$g_{ij} N^{ij} = 1 , \quad p_i N^{ij} = p_j N^{ij} = 0 . \tag{G.13}$$

---

[3] We can construct e^μ in a gravitational field in a general coordinate system x^μ from the polarization vector e_{flat}^i of the photon in a locally inertial frame with coordinates ξ^α, as

$$e^\mu = \frac{\partial x^\mu}{\partial \xi^i} e_{\text{flat}}^i + \beta p^\mu ,$$

with β adjusted to make $e^0 = 0$.

It follows from Eq. (G.12) that the statistical matrix of a photon moving in a gravitational field satisfies the parallel transport equation

$$\frac{dN^{ij}(t)}{dt} = \left[-\Gamma^i_{k\lambda}\big(x(t)\big) + \Gamma^0_{k\lambda}\big(x(t)\big)\frac{dx^i(t)}{dt}\right] N^{kj}(t)\frac{dx^\lambda(t)}{dt}$$
$$+ \left[-\Gamma^j_{k\lambda}\big(x(t)\big) + \Gamma^0_{k\lambda}\big(x(t)\big)\frac{dx^j(t)}{dt}\right] N^{ik}(t)\frac{dx^\lambda(t)}{dt}.$$

$$(G.14)$$

Appendix H

The Relativistic Boltzmann Equation

In this appendix we derive the Boltzmann equations for neutrinos and photons, that govern the evolution of the distribution of these relativistic particles in phase space. This will provide the basis for our account in Section 6.1 of the widely used numerical calculations of the evolution of scalar perturbations. In addition, these results will be needed in Section 6.6, where we evaluate the damping of tensor modes due to the anisotropic inertia of free-streaming neutrinos, and in Section 7.4, where we calculate the polarization of the microwave background produced by the scattering of an anisotropic distribution of photons by non-relativistic electrons. This appendix will first consider the simpler case of neutrinos, for which scattering may be neglected, and then turn to the more complicated case of photons, for which scattering plays an essential role. In an elementary application of the results for photons, we will derive a formula for the rate of damping of acoustic waves in a medium of photons and charged particles, used in Section 6.4.

Throughout this appendix we shall adopt a coordinate system for which

$$g_{00} = -1, \qquad g_{i0} = 0, \tag{H.1}$$

while $g_{ij}(\mathbf{x}, t)$ is unconstrained. In the linear approximation, this form of the metric is automatic for tensor modes, and follows if we work in synchronous gauge for scalar modes. However, we will not specialize to the case of linear perturbations until later. With this metric, the only non-vanishing components of the affine connection are

$$\Gamma_{ij}^k = \frac{1}{2} g^{kl} \left(\frac{\partial g_{li}}{\partial x^j} + \frac{\partial g_{lj}}{\partial x^i} - \frac{\partial g_{ij}}{\partial x^l} \right), \tag{H.2}$$

$$\Gamma_{i0}^j = \frac{1}{2} g^{jk} \dot{g}_{ki}, \qquad \Gamma_{ij}^0 = \frac{1}{2} \dot{g}_{ij}. \tag{H.3}$$

1 Neutrinos

It will be convenient to work with a neutrino distribution function $n_\nu(\mathbf{x}, \mathbf{p}, t)$, defined by

$$n_\nu(\mathbf{x}, \mathbf{p}, t) \equiv \sum_r \left(\prod_{i=1}^3 \delta(x^i - x_r^i(t)) \right) \left(\prod_{i=1}^3 \delta(p_i - p_{ri}(t)) \right), \tag{H.4}$$

with r labeling trajectories of individual neutrinos (or antineutrinos). This expression has the defining property of a number density, that the integral of n_ν over any volume of phase space equals the number of neutrinos in that volume. According to Section 4 of Appendix B, the momentum variable in Eq. (H.4) is defined by $p_{ri} = g_{ij} p_r^j$, where p^μ is the momentum four-vector

$$p_r^\mu \equiv dx_r^\mu / du_r \ . \tag{H.5}$$

where u_r is a suitably normalized affine parameter for which the spacetime trajectory satisfies Eq. (B.12):

$$\frac{d^2 x_r^\lambda}{du_r^2} + \Gamma_{\mu\nu}^\lambda(x_r) \frac{dx_r^\mu}{du_r} \frac{dx_r^\nu}{du_r} = 0 \ . \tag{H.6}$$

Then between collisions the rate of change of the momentum is simply given by

$$\dot{p}_{ri} = \frac{1}{2p_r^0} p_r^j p_r^k \left(\frac{\partial g_{jk}}{\partial x^i} \right)_{\mathbf{x}=\mathbf{x}_r} \ , \tag{H.7}$$

while the rate of change of the coordinate is

$$\dot{x}_r^i = p_r^i / p_r^0 \ . \tag{H.8}$$

It follows then directly from Eqs. (H.4), (H.7), and (H.8) that in the absence of collisions, n_ν satisfies a Boltzmann equation

$$\frac{\partial n_\nu}{\partial t} + \frac{\partial n_\nu}{\partial x^i} \frac{p^i}{p^0} + \frac{\partial n_\nu}{\partial p_i} \frac{p^j p^k}{2p^0} \frac{\partial g_{jk}}{\partial x^i} = 0 \ . \tag{H.9}$$

It should be understood that p^i and p^0 are expressed here in terms of the independent variable p_i by $p^i = g^{ij}(\mathbf{x}, t) p_j$ and $p^0 = \left(g^{ij}(\mathbf{x}, t) p_i p_j \right)^{1/2}$, so they depend on position and time as well as on p_i.

We now specialize to the case of a small perturbation. The spatial metric is then of the form

$$g_{ij}(\mathbf{x}, t) = a^2(t) \delta_{ij} + \delta g_{ij}(\mathbf{x}, t) \ , \tag{H.10}$$

with δg_{ij} small. With only a small perturbation to the metric, the neutrino distribution function never gets very different from its equilibrium form, so we write

$$n_\nu(\mathbf{x}, t) = \bar{n}_\nu \left(a(t) \sqrt{g^{ij}(\mathbf{x}, t) p_i p_j} \right) + \delta n_\nu(\mathbf{x}, t) \ , \tag{H.11}$$

where

$$\bar{n}_\nu(p) \equiv \frac{1}{(2\pi)^3} \left[\exp \left(p/k_B a(t) \bar{T}(t) \right) + 1 \right]^{-1} \ , \tag{H.12}$$

and δn_ν is a small perturbation. (Note that the factor $a(t)$ in the argument of $\bar{n}$ in Eq. (H.11) is canceled by the factor $a(t)$ multiplying $\bar{T}(t)$ in Eq. (H.12); this factor has been included in Eqs. (H.11) and (H.12) because at times of interest, when $\bar{T} < 10^{10}$K, we have $\bar{T}(t) \propto 1/a(t)$, so that Eq. (H.12) defines $\bar{n}_\nu$ as a time-independent function of its argument.) The first term in Eq. (H.11) is the neutrino distribution we would expect according to the Principle of Equivalence in a perturbed gravitational field, if the distribution in locally inertial frames were just the equilibrium distribution $\bar{n}_\nu(p)$; the second term δn_ν thus represents the departure of the neutrino distribution from its equilibrium form.

We can derive an initial condition for δn_ν by noting that, at a time t_1 corresponding to a temperature $\bar{T}_1 \approx 10^9$ K, the neutrino scattering rate had already dropped well below the expansion rate, but there had not yet been time for the perturbations in the metric to distort the neutrino distribution away from *local* thermal equilibrium, but with a perturbed temperature. Thus as a convenient initial condition we may take

$$\delta n_\nu(\mathbf{x}, \mathbf{p}, t_1) = -p\bar{n}'_\nu(p)\left[\frac{\delta T(\mathbf{x}, t_1)}{\bar{T}(t_1)} + \frac{\hat{p}_k \cdot \delta u_k(\mathbf{x}, t_1)}{a(t_1)}\right], \qquad \text{(H.13)}$$

where $\bar{T}(t_1) \approx 10^9$ K. The second term in square brackets represents the Doppler shift due to a possible neutrino streaming velocity $\delta\mathbf{u}$, analogous to that given for microwave background photons by Eqs. (2.4.5) and (2.4.6). In local thermal equilibrium this is the same velocity perturbation as for baryons and photons. (Note that the metric position vector is ax^k, so the metric velocity vector is $au^k = a^{-1}u_k$.)

To first order in metric and density perturbations, Eq. (H.9) reads

$$0 = \frac{\partial \delta n_\nu(\mathbf{x}, \mathbf{p}, t)}{\partial t} + \frac{\partial \delta n_\nu(\mathbf{x}, \mathbf{p}, t)}{\partial x^i}\frac{p_i}{a(t)p}$$

$$+ \frac{\bar{n}'_\nu(p)}{2p}\frac{\partial}{\partial t}\left(a^2(t)\delta g^{ij}(\mathbf{x}, t)\right)p_i p_j$$

$$+ \frac{a(t)\bar{n}'_\nu(p)}{2p^2}\frac{\partial \delta g^{ij}(\mathbf{x}, t)}{\partial x^k}p_i p_j p_k$$

$$+ \bar{n}'_\nu(p)\frac{p_i p_j p_k}{2a^3(t)p^2}\frac{\partial \delta g_{ij}(\mathbf{x}, t)}{\partial x^k},$$

where here $p \equiv \sqrt{p_i p_i}$. (The third from last and penultimate terms on the right-hand side arise from the dependence of the argument of $\bar{n}_\nu$ in Eq. (H.11) on t and $\mathbf{x}$ through the combination $a^2(t)g^{ij}(\mathbf{x}, t_1)p_i p_j$.) To first order, we have $g^{ij} = a^{-2}\delta_{ij} - a^{-4}\delta g_{ij}$, so the penultimate and last terms on the right-hand side cancel, leaving us with the much simpler result

$$\frac{\partial \delta n_\nu(\mathbf{x}, \mathbf{p}, t)}{\partial t} + \frac{\hat{p}_i}{a(t)} \frac{\partial \delta n_\nu(\mathbf{x}, \mathbf{p}, t)}{\partial x^i} = \frac{p\bar{n}'(p)}{2} \hat{p}_i \hat{p}_j \frac{\partial}{\partial t} \left(a^{-2}(t) \delta g_{ij}(\mathbf{x}, t) \right) ,$$

(H.14)

where again $p \equiv \sqrt{p_k p_k}$, and as usual we write $\hat{p}_i \equiv p_i/p$.

2 Photons

The Boltzmann equation for photons is considerably more complicated than for neutrinos, because of the necessity of taking photon scattering into account. Since scattering can change the polarization of photons, we can no longer write a separate Boltzmann equation for each helicity state of photons, as we did for neutrinos. Instead, we now define a *number density matrix*:

$$n_\gamma^{ij}(\mathbf{x}, \mathbf{p}, t) \equiv \sum_r \left(\prod_{k=1}^3 \delta\left(x^k - x_r^k(t) \right) \right) \left(\prod_{k=1}^3 \delta\left(p_k - p_{rk}(t) \right) \right) N_r^{ij}(t) ,$$

(H.15)

with r here labeling trajectories of individual photons, and $N_r^{ij}(t)$ the polarization density matrix of the rth photon. As discussed in Appendix G, if an individual photon can have any one of several polarization vectors e_m^i, with probabilities P_m, then it has a polarization density matrix

$$N^{ij} \equiv \sum_m P_m e_m^i e_m^{j*} .$$

(H.16)

If we observe whether the polarization of a photon with polarization density matrix N^{ij} is in a particular direction e^i rather than in an orthogonal direction, we find a probability $g_{ik}g_{jl}e^{*i}e^j N^{kl}$. Recall that for a photon of three-momentum p_i in a general three-metric g_{ij}, the photon polarization vectors are defined so that

$$p_i e^i = 0 , \qquad g_{ij} e^i e^{j*} = 1$$

(H.17)

so the polarization density matrix of photon r satisfies

$$p_{ri}(t) N_r^{ij}(t) = p_{ri}(t) N_r^{ji}(t) = 0 , \qquad g_{ij}\left(\mathbf{x}_r(t), t \right) N_r^{ij}(t) = 1 ,$$

(H.18)

and the number density matrix correspondingly satisfies

$$p_i n_\gamma^{ij}(\mathbf{x}, \mathbf{p}, t) = p_i n_\gamma^{ji}(\mathbf{x}, \mathbf{p}, t) = 0 , \qquad g_{ij}(\mathbf{x}, t) n_\gamma^{ij}(\mathbf{x}, \mathbf{p}, t) = n_\gamma(\mathbf{x}, \mathbf{p}, t) ,$$

(H.19)

where $n_\gamma(\mathbf{x}, \mathbf{p}, t)$ is the phase space number density of photons, defined just as in Eq. (H.4). By their definition, both $N_r^{ij}(t)$ and $n_\gamma^{ij}(\mathbf{x}, \mathbf{p}, t)$ are Hermitian matrices.

As discussed in Appendix G, between collisions the time dependence of the polarization vectors is given by a combination of parallel transport with a gauge transformation that keeps the time component of $e_r^{j\mu}(t)$ vanishing:

$$\frac{de_r^j(t)}{dt} = \left[-\Gamma_{i\lambda}^j\left(\mathbf{x}_r(t), t\right) + \Gamma_{i\lambda}^0\left(\mathbf{x}_r(t), t\right) \frac{dx_r^j(t)}{dt} \right] e_r^i(t) \frac{dx_r^\lambda(t)}{dt} , \quad \text{(H.20)}$$

and consequently the time dependence of the density matrix of the rth photon is

$$\frac{dN_r^{ij}(t)}{dt} = \left[-\Gamma_{k\lambda}^i\left(\mathbf{x}_r(t), t\right) + \Gamma_{k\lambda}^0\left(\mathbf{x}_r(t), t\right) \frac{dx_r^i(t)}{dt} \right] N_r^{kj}(t) \frac{dx_r^\lambda(t)}{dt}$$

$$+ \left[-\Gamma_{k\lambda}^j\left(\mathbf{x}_r(t), t\right) + \Gamma_{k\lambda}^0\left(\mathbf{x}_r(t), t\right) \frac{dx_r^j(t)}{dt} \right] N_r^{ik}(t) \frac{dx_r^\lambda(t)}{dt} . \quad \text{(H.21)}$$

The time dependence of the variables $p_{ri}(t)$ and $x_r^i(t)$ is given by the same equations (H.7) and (H.8) as for neutrinos. It follows then directly from Eqs. (H.15), (H.21), (H.7) and (H.8) that $n_\gamma^{ij}(\mathbf{x}, \mathbf{p}, t)$ satisfies a Boltzmann equation:

$$\frac{\partial n_\gamma^{ij}}{\partial t} + \frac{\partial n_\gamma^{ij}}{\partial x^k} \frac{p^k}{p^0} + \frac{\partial n_\gamma^{ij}}{\partial p_k} \frac{p^l p^m}{2p^0} \frac{\partial g_{lm}}{\partial x^k}$$

$$+ \left(\Gamma_{k\lambda}^i - \frac{p^i}{p^0} \Gamma_{k\lambda}^0 \right) \frac{p^\lambda}{p^0} n_\gamma^{kj}$$

$$+ \left(\Gamma_{k\lambda}^j - \frac{p^j}{p^0} \Gamma_{k\lambda}^0 \right) \frac{p^\lambda}{p^0} n_\gamma^{ik} = C^{ij} , \quad \text{(H.22)}$$

where C^{ij} is a term representing the effect of photon scattering. (As a reminder, we note that in Eq. (H.22) p^i and p^0 are functions of $\mathbf{x}$, t, and the p_j, given by $p^i \equiv g^{ij} p_j$ and $p^0 = [g^{ij} p_i p_j]^{1/2}$.)

To evaluate the collision term C^{ij} in Eq. (H.22), let us first consider the case of flat spacetime, where $g_{ij} = \delta_{ij}$. Since the Thomson scattering of an unpolarized or linearly polarized photon can only produce linearly polarized photons, we will limit ourselves here to the case of linear polarization, for which polarization vectors can be taken to be real, and the polarization density matrix and number density matrices are real and symmetric. We will first consider the case of electrons at rest, and later (in dealing with scalar perturbations) take up the case of a non-zero plasma velocity.

The scattering of photons of momentum $\mathbf{p}$ into some other direction causes a decrease in $n_\gamma^{ij}(\mathbf{x}, \mathbf{p}, t)$ which is simply given by a term in C^{ij}:

$$C_-^{ij}(\mathbf{x}, \mathbf{p}, t) = -\omega_c(t) n_\gamma^{ij}(\mathbf{x}, \mathbf{p}, t) , \qquad (\text{H}.23)$$

where $\omega_c(t)$ is the total collision rate. (We hold here to the convention of writing n_γ^{ij} and C^{ij} with upper indices, though of course in flat spacetime there is no distinction between contravariant and covariant spatial tensors.) There is also an increase in $n_\gamma^{ij}(\mathbf{x}, \mathbf{p}, t)$ caused by the scattering of photons with some initial momentum $\mathbf{p}_1$ into momentum $\mathbf{p}$, with energy conservation requiring (assuming that $|\mathbf{p}| \ll m_e$) that $|\mathbf{p}_1| = |\mathbf{p}|$. The Klein–Nishina formula[1] tells us that when a photon with momentum p_{1i} and real polarization vector e_1^i is scattered by an electron at rest, the probability of finding a photon in the final state with real polarization vector e^i is proportional to $(e \cdot e_1)^2$, with no dependence on the initial photon momentum p_{1i} or the final photon momentum p_i, except of course that e_1^i and e^i must be orthogonal to p_{1i} and p_i, respectively. If the initial photon can have various polarizations e_{1n}^i with probabilities P_n, then correspondingly the probability of finding the photon in the final state with polarization e^i is proportional to $e^i e^j N_1^{ij}$, where $N_1^{ij} = \sum_n P_n e_{1n}^i e_{1n}^j$. As already mentioned, this probability must equal $e^i e^j N^{ij}$, where N^{ij} is the polarization density matrix of the photon in the final state. Together with the conditions that $p_i N^{ij} = 0$ and $N^{ii} = 1$, this tells us that the scattered photon has polarization density matrix

$$N^{ij}(\hat{p}) = S^{-1}(\hat{p}) \left[N_1^{ij} - \hat{p}_i \hat{p}_k N_1^{kj} - \hat{p}_j \hat{p}_k N_1^{ik} + \hat{p}_i \hat{p}_j \hat{p}_k \hat{p}_l N_1^{kl} \right], \qquad (\text{H}.24)$$

where

$$S(\hat{p}) \equiv 1 - \hat{p}_i \hat{p}_j N_1^{ij} . \qquad (\text{H}.25)$$

The rate of increase in $n_\gamma^{ij}(\mathbf{x}, \mathbf{p}, t)$ due to scattering of photons from an arbitrary direction $\hat{p}_1$ into direction $\hat{p}$ is then

$$C_+^{ij}(\mathbf{x}, \mathbf{p}, t) = n_e \int d^2\hat{p}_1 \frac{d^2\sigma}{d^2\hat{p}}$$
$$\times S^{-1}(\hat{p}) \Big[n_\gamma^{ij}(\mathbf{x}, |\mathbf{p}|\hat{p}_1, t) - \hat{p}_i \hat{p}_k \, n_\gamma^{kj}(\mathbf{x}, |\mathbf{p}|\hat{p}_1, t)$$
$$- \hat{p}_j \hat{p}_k \, n_\gamma^{ik}(\mathbf{x}, |\mathbf{p}|\hat{p}_1, t) + \hat{p}_i \hat{p}_j \hat{p}_k \hat{p}_l \, n_\gamma^{kl}(\mathbf{x}, |\mathbf{p}|\hat{p}_1, t) \Big] . \qquad (\text{H}.26)$$

where n_e is the electron density, and $d^2\sigma/d^2\hat{p}$ is the differential scattering cross section. Summing $(e \cdot e_1)^2$ over final polarizations gives a differential cross section proportional to $S(\hat{p})$, and $\int d^2\hat{p}\, S(\hat{p}) = 8\pi/3$, so equating

---

[1] See, e.g., QTF, Vol. I, Sec. 8.7.

$\int d^2\hat{p} \, (d^2\sigma/d^2\hat{p})$ to the Thomson scattering cross section σ_T, we have

$$\frac{d^2\sigma}{d^2\hat{p}} = \frac{3\sigma_T}{8\pi} S(\hat{p}) \,. \tag{H.27}$$

Then Eqs. (H.23), (H.26), and (H.27) give a net change in $n_\gamma^{ij}(\mathbf{x}, \mathbf{p}, t)$ equal to

$$C^{ij}(\mathbf{x}, \mathbf{p}, t) = C_-^{ij}(\mathbf{x}, \mathbf{p}, t) + C_+^{ij}(\mathbf{x}, \mathbf{p}, t)$$

$$= -\omega_c(t) n_\gamma^{ij}(\mathbf{x}, \mathbf{p}, t) + \frac{3\omega_c(t)}{8\pi} \int d^2\hat{p}_1$$

$$\times \Big[n_\gamma^{ij}(\mathbf{x}, |\mathbf{p}|\hat{p}_1, t) - \hat{p}_i\hat{p}_k \, n_\gamma^{kj}(\mathbf{x}, |\mathbf{p}|\hat{p}_1, t) - \hat{p}_j\hat{p}_k \, n_\gamma^{ik}(\mathbf{x}, |\mathbf{p}|\hat{p}_1, t)$$

$$+ \hat{p}_i\hat{p}_j\hat{p}_k\hat{p}_l \, n_\gamma^{kl}(\mathbf{x}, |\mathbf{p}|\hat{p}_1, t) \Big] \,, \tag{H.28}$$

in which we use $\omega_c(t) = n_e(t)\sigma_T$.

As a check, it is easy to see that the collision term (H.28) vanishes in local thermal equilibrium. In equilibrium photons are unpolarized, so $n_\gamma^{ij} \propto \delta_{ij} - \hat{p}_i\hat{p}_j$, and their momentum distribution is homogeneous and isotropic, so the coefficient of proportionality f depends only on $|\mathbf{p}|$ and t:

$$n_{\text{eq}}^{ij}(\mathbf{x}, \mathbf{p}, t) = f(|\mathbf{p}|, t)\Big[\delta_{ij} - \hat{p}_i\hat{p}_j\Big] \,.$$

Then

$$\int d^2\hat{p}_1 n_{\text{eq}}^{ij}(\mathbf{x}, |\mathbf{p}|\hat{p}_1, t) = \frac{8\pi}{3} f(|\mathbf{p}|, t)\delta_{ij} \,.$$

Using this in Eq. (H.28) shows that in equilibrium the two terms in C^{ij} cancel, as of course they must.

To find the result for an arbitrary three-metric g_{ij}, we must simply write Eq. (H.28) in a form that is invariant under general three-dimensional coordinate transformations, and reduces to Eq. (H.28) in the case of a flat three-metric with $g_{ij} = \delta_{ij}$. In this way, we find

$$C^{ij}(\mathbf{x}, \mathbf{p}, t) = -\omega_c(t) n_\gamma^{ij}(\mathbf{x}, \mathbf{p}, t)$$

$$+ \frac{3\omega_c(t)}{8\pi} \int \frac{d^3 p_1 \sqrt{\text{Det}g(\mathbf{x}, t)}}{p^{02}(\mathbf{x}, \mathbf{p}, t)} \delta\Big(p^0(\mathbf{x}, \mathbf{p}, t) - p^0(\mathbf{x}, \mathbf{p}_1, t)\Big)$$

$$\times \Big[n_\gamma^{ij}(\mathbf{x}, \mathbf{p}_1, t) - \frac{g^{ik}(\mathbf{x}, t) p_k p_l}{p^{02}(\mathbf{x}, \mathbf{p}, t)} n_\gamma^{lj}(\mathbf{x}, \mathbf{p}_1, t)$$

$$- \frac{g^{jk}(\mathbf{x}, t) p_k p_l}{p^{02}(\mathbf{x}, \mathbf{p}, t)} n_\gamma^{il}(\mathbf{x}, \mathbf{p}_1, t)$$

$$+ \frac{g^{ik}(\mathbf{x}, t) g^{jl}(\mathbf{x}, t) p_k p_l p_m p_n}{p^{04}(\mathbf{x}, \mathbf{p}, t)} n_\gamma^{mn}(\mathbf{x}, \mathbf{p}_1, t) \Big] \,, \tag{H.29}$$

where

$$p^0(\mathbf{x}, \mathbf{p}, t) \equiv \sqrt{g^{ij}(\mathbf{x}, t)p_i p_j} \, . \tag{H.30}$$

We now again specialize to the case of a small perturbation, writing the metric as in Eq. (H.10), and putting the photon number density matrix in a form analogous to Eq. (H.11):

$$n_\gamma^{ij}(\mathbf{x}, \mathbf{p}, t) = \frac{1}{2}\bar{n}_\gamma\left(a(t)p^0(\mathbf{x}, \mathbf{p}, t)\right)\left[g^{ij}(\mathbf{x}, t) - \frac{g^{ik}(\mathbf{x}, t)g^{jl}(\mathbf{x}, t)p_k p_l}{p^{02}(\mathbf{x}, \mathbf{p}, t)}\right]$$

$$+ \, \delta n_\gamma^{ij}(\mathbf{x}, \mathbf{p}, t) \, , \tag{H.31}$$

where $\bar{n}_\gamma(p)$ is the equilibrium phase space number density, a time-independent function of its argument

$$\bar{n}_\gamma(p) \equiv \frac{1}{(2\pi)^3}\left[\exp\left(p/k_B a(t)T(t)\right) - 1\right]^{-1}, \tag{H.32}$$

and δn^{ij} is a small perturbation. Fortunately, to first order in perturbations the Boltzmann equation (H.22) is greatly simplified by the fact that the quantity in square brackets in Eq. (H.31) satisfies a collisionless Boltzmann equation

$$0 = \left(\frac{\partial}{\partial t} + \frac{p^k}{p^0}\frac{\partial}{\partial x^k} + \frac{p^l p^m}{2p^0}\frac{\partial g_{lm}}{\partial x^k}\frac{\partial}{\partial p^k}\right)\left[g^{ij} - \frac{p^i p^j}{p^{02}}\right]$$

$$+ \frac{p^\lambda}{p^0}\left(\Gamma^i_{k\lambda} - \frac{p^i}{p^0}\Gamma^0_{k\lambda}\right)\left[g^{kj} - \frac{p^k p^j}{p^{02}}\right]$$

$$+ \frac{p^\lambda}{p^0}\left(\Gamma^j_{k\lambda} - \frac{p^j}{p^0}\Gamma^0_{k\lambda}\right)\left[g^{ik} - \frac{p^i p^k}{p^{02}}\right] . \tag{H.33}$$

(This can be proved directly, or more easily by noting that if we set $p_i = p_{ri}(t)$ and $x^i = x_r^i(t)$, then the quantity in square brackets is the sum over two orthogonal polarizations of the quantity $e_r^i(t)e_r^j(t)$, and therefore satisfies Eq. (H.33) as a consequence of Eq. (H.20). Since Eq. (H.33) holds in this sense for any photon trajectory, it is necessary for it to hold for arbitrary x^i and p_i.) Acting on the factor $\bar{n}_\gamma(ap^0)$, the derivative operators in Eq. (H.22) give to first order

$$\left(\frac{\partial}{\partial t} + \frac{p^k}{p^0}\frac{\partial}{\partial x^k} + \frac{p^l p^m}{2p^0}\frac{\partial g_{lm}}{\partial x^k}\frac{\partial}{\partial p_k}\right)\bar{n}_\gamma(ap^0) = -\frac{1}{2}\sqrt{p_i p_i}$$

$$\times \bar{n}_\gamma'(\sqrt{p_i p_i})\,\hat{p}_k \hat{p}_l\frac{\partial}{\partial t}\left(a^{-2}\delta g_{kl}\right), \tag{H.34}$$

with the effect of the position and momentum derivative terms in parentheses on the left-hand side canceling. There is another simplification provided by the fact that the the first term in Eq. (H.31) does not contribute to the collision term, since even in a gravitational field collisions by themselves do not alter equilibrium distribution functions. Keeping only terms of first order in perturbations, the Boltzmann equation (H.22) then becomes

$$
\frac{\partial \, \delta n_\gamma^{ij}(\mathbf{x}, \mathbf{p}, t)}{\partial t} + \frac{\hat{p}_k}{a(t)} \frac{\partial \, \delta n_\gamma^{ij}(\mathbf{x}, \mathbf{p}, t)}{\partial x^k} + \frac{2\dot{a}(t)}{a(t)} \, \delta n_\gamma^{ij}(\mathbf{x}, \mathbf{p}, t)
$$

$$
- \frac{1}{4a^2(t)} p \bar{n}_\gamma'(p) \hat{p}_k \hat{p}_l \frac{\partial}{\partial t} \left(a^{-2} \delta g_{kl}(\mathbf{x}, t) \right) \left(\delta_{ij} - \hat{p}_i \hat{p}_j \right)
$$

$$
= -\omega_c(t) \, \delta n_\gamma^{ij}(\mathbf{x}, \mathbf{p}, t) + \frac{3\omega_c(t)}{8\pi} \int d^2 \hat{p}_1
$$

$$
\times \left[\delta n_\gamma^{ij}(\mathbf{x}, p\hat{p}_1, t) - \hat{p}_i \hat{p}_k \, \delta n_\gamma^{kj}(\mathbf{x}, p\hat{p}_1, t) - \hat{p}_j \hat{p}_k \, \delta n_\gamma^{ik}(\mathbf{x}, p\hat{p}_1, t) \right.
$$

$$
\left. + \hat{p}_i \hat{p}_j \hat{p}_k \hat{p}_l \, \delta n_\gamma^{kl}(\mathbf{x}, p\hat{p}_1, t) \right], \tag{H.35}
$$

where again $p \equiv \sqrt{p_i p_i}$, and $\hat{p} \equiv \mathbf{p}/p$. As a check, note that Eq. (H.35) is consistent with the condition $p_i \delta n^{ij} = 0$. As a further check, note that the first-order perturbation to the photon phase space number density $g_{ij} n_\gamma^{ij}$ is $a^2 \delta n^{ii}$, and according to the trace of Eq. (H.35), the Boltzmann equation for $a^2 \delta n_\gamma^{ii}$ is the same as the Boltzmann equation (H.14) for the perturbation to the neutrino phase space density, aside from the presence of the collision term and the appearance of $\bar{n}_\gamma'$ instead of $\bar{n}_\nu'$.

For tensor modes, Eq. (H.35) can be used as it stands, but in dealing with scalar (or vector) modes, we need to take up a complication that has been ignored until now: in general the plasma has a small velocity $\delta \mathbf{u}(\mathbf{x}, t)$ (which in Section 6.1 *et seq.* is denoted $\delta \mathbf{u}_B(\mathbf{x}, t)$). Since $\delta \mathbf{u}$ is itself a first-order perturbation, in calculating its effect on the collision term C^{ij}, to first order we can ignore any perturbations to the gravitational field or the photon number density matrix. The effect of the plasma velocity is to shift the energy $|\mathbf{p}_1|$ of the incident photon that when scattered yields a final photon with momentum $\mathbf{p}$. To calculate this effect, let's first consider the case of flat spacetime. Since photon energy is conserved in the rest frame of the plasma, in which the plasma metric four-velocity is simply $v^\mu = (0, 0, 0, 1)$, we have $(p_1 - p) \cdot v = 0$; evaluating this scalar in the "lab" frame in which the plasma velocity four-vector is $v^\mu = [\delta \mathbf{v}, \sqrt{1 - (\delta \mathbf{v})^2}]$ we have

$$
0 = v_\mu (p_1 - p)^\mu = \delta \mathbf{v} \cdot (\mathbf{p}_1 - \mathbf{p}) - \sqrt{1 - (\delta \mathbf{v})^2} (p_1^0 - p^0) \,,
$$

or, to first order in velocity,

$$|\mathbf{p_1}| = |\mathbf{p}|\left[1 + (\hat{p}_1 - \hat{p}) \cdot \delta\mathbf{v}\right].$$

As already noted, the flat-space velocity components v^k are related to the velocity components in a Robertson–Walker three-metric $g_{ij} = a^2\delta_{ij}$ by $v^k = a u^k = a^{-1}u_k$, so

$$|\mathbf{p_1}| = |\mathbf{p}|\left[1 + \frac{(\hat{p}_1 - \hat{p})_k \delta u_k}{a}\right]. \tag{H.36}$$

As a result, the two terms in the collision term (H.28) do not cancel even when we set the three-metric g_{ij} equal to $a^2\delta_{ij}$ and set the photon number density matrix equal to the form it would have in equilibrium with this three-metric:

$$n^{ij}_{\gamma,\text{eq}}(\mathbf{x}, \mathbf{p}, t) = \frac{1}{2a^2(t)}\bar{n}_\gamma(|\mathbf{p}|)\left[\delta_{ij} - \hat{p}_i\hat{p}_j\right].$$

Since in this case these two terms in C^{ij} *would* cancel if $|\mathbf{p_1}|$ were equal to $|\mathbf{p}|$, we find a new term in $C^{ij}(\mathbf{x}, \mathbf{p}, t)$, equal to the difference between the term $-\omega_c(t)n^{ij}_{\gamma,\text{eq}}(\mathbf{x}, \mathbf{p}, t)$ and the same term with $|\mathbf{p}|$ replaced with $|\mathbf{p_1}|$, averaged over $\hat{p}_1$. The linearized Boltzmann equation (H.35) now becomes

$$\frac{\partial\,\delta n^{ij}_\gamma(\mathbf{x}, \mathbf{p}, t)}{\partial t} + \frac{\hat{p}_k}{a(t)}\frac{\partial\,\delta n^{ij}_\gamma(\mathbf{x}, \mathbf{p}, t)}{\partial x^k} + \frac{2\dot{a}(t)}{a(t)}\delta n^{ij}_\gamma(\mathbf{x}, \mathbf{p}, t)$$

$$-\frac{1}{4a^2(t)}p\bar{n}'_\gamma(p)\hat{p}_k\hat{p}_l\frac{\partial}{\partial t}\left(a^{-2}(t)\delta g_{kl}(\mathbf{x}, t)\right)\left(\delta_{ij} - \hat{p}_i\hat{p}_j\right)$$

$$= -\omega_c(t)\,\delta n^{ij}_\gamma(\mathbf{x}, \mathbf{p}, t) + \frac{3\omega_c(t)}{8\pi}\int d^2\hat{p}_1$$

$$\times\left[\delta n^{ij}_\gamma(\mathbf{x}, p\hat{p}_1, t) - \hat{p}_i\hat{p}_k\,\delta n^{kj}_\gamma(\mathbf{x}, p\hat{p}_1, t) - \hat{p}_j\hat{p}_k\,\delta n^{ik}_\gamma(\mathbf{x}, p\hat{p}_1, t)\right.$$

$$\left.+\hat{p}_i\hat{p}_j\hat{p}_k\hat{p}_l\,\delta n^{kl}_\gamma(\mathbf{x}, p\hat{p}_1, t)\right]$$

$$-\frac{\omega_c(t)}{2a^3(t)}\left(p_k\delta u_k(\mathbf{x}, t)\right)\bar{n}'_\gamma(p)\left[\delta_{ij} - \hat{p}_i\hat{p}_j\right]. \tag{H.37}$$

(The term in Eq. (H.36) proportional to $\hat{p}_{1k}\delta u_k$ gives no contribution to the integral over $\hat{p}_1$.)

*   *   *

In an important application of Eq. (H.37), we may derive the decay rate of acoustic waves in a homogeneous and time-independent plasma in flat spacetime (with $a = 1$ and ω_c time-independent) when the collision rate is much larger than the sound frequency. (This result will be used in Section 6.4.) We seek a solution of the Boltzmann equation (H.37) of the form

$$\delta n_\gamma^{ij}(\mathbf{x}, \mathbf{p}, t) = e^{i\mathbf{k}\cdot\mathbf{x}} e^{-i\omega t} \delta n_\gamma^{ij}(\mathbf{p}) , \quad \delta u_j(\mathbf{x}, t) = e^{i\mathbf{k}\cdot\mathbf{x}} e^{-i\omega t} \delta u_j . \quad \text{(H.38)}$$

Then with $a = 1$ and ω_c constant, Eq. (H.37) becomes

$$\left[\omega_c - i\omega + i\hat{p} \cdot \mathbf{k} \right] \delta n_\gamma^{ij}(\mathbf{p})$$

$$= \frac{3\omega_c}{8\pi} \int d^2\hat{p}_1$$

$$\times \left[\delta n_\gamma^{ij}(p\hat{p}_1) - \hat{p}_i\hat{p}_k \, \delta n_\gamma^{kj}(p\hat{p}_1) - \hat{p}_j\hat{p}_k \, \delta n_\gamma^{ik}(p\hat{p}_1) + \hat{p}_i\hat{p}_j\hat{p}_k\hat{p}_l \, \delta n_\gamma^{kl}(p\hat{p}_1) \right]$$

$$- \frac{\omega_c}{2} \left(\mathbf{p} \cdot \delta\mathbf{u} \right) \bar{n}_\gamma'(p) \left[\delta_{ij} - \hat{p}_i\hat{p}_j \right] . \quad \text{(H.39)}$$

We now need a formula for the plasma velocity δu_j. For this purpose, we note that in flat space the first-order photon and baryonic plasma contributions to the energy-momentum tensor have

$$\delta T_{\gamma j}^i = \int d^3 p \, a^2 n_\gamma^{kk}(\mathbf{p}) \, |\mathbf{p}|\hat{p}_i\hat{p}_j , \quad \delta T_{\gamma j}^0 = \int d^3 p \, a^2 n_\gamma^{kk}(\mathbf{p}) \, |\mathbf{p}|\hat{p}_j , \quad \text{(H.40)}$$

and

$$\delta T_{Bj}^i = 0 , \quad \delta T_{Bj}^0 = \bar{\rho}_B \delta u_j , \quad \text{(H.41)}$$

so the equation of momentum conservation may be written

$$\omega \, \bar{\rho}_B \delta u_j = \int d^3 p \, a^2 \delta n_\gamma^{kk}(\mathbf{p}) \, p_j \left(\mathbf{k} \cdot \hat{p} - \omega \right) . \quad \text{(H.42)}$$

To evaluate the right-hand side, multiply Eq. (H.39) with p_j and integrate over $\mathbf{p}$; this gives

$$i \int d^3 p \, (\mathbf{k} \cdot \hat{p} - \omega) \, p_j \, \delta n_\gamma^{kk}(\mathbf{p}) = -\omega_c \int d^3 p \, p_j \delta n_\gamma^{kk}(\mathbf{p}) + \frac{4\omega_c}{3} \delta u_j \bar{\rho}_\gamma ,$$

where $\bar{\rho}_\gamma = \int d^3 p \, p \, \bar{n}_\gamma(p) = -(1/4) \int d^3 p \, p^2 \, \bar{n}_\gamma'(p)$. Using this in Eq. (H.42) gives our formula for $\delta\mathbf{u}$:

$$\left(\omega \, \bar{\rho}_B + \frac{4i\omega_c}{3} \bar{\rho}_\gamma \right) \delta u_j = i\omega_c \int d^3 p \, p_j \delta n_\gamma^{kk}(\mathbf{p}) . \quad \text{(H.43)}$$

For scalar modes, $\mathbf{k}$ is the only parameter in the problem with a sense of direction, so we can express the integrals appearing in Eqs. (H.39) and

(H.43) as

$$\int d^3p |\mathbf{p}| \, \delta n_\gamma^{ij}(\mathbf{p}) = \bar{\rho}_\gamma \left[X \delta_{ij} + Y \hat{k}_i \hat{k}_j \right], \tag{H.44}$$

$$\int d^3p \, \delta n_\gamma^{ii}(\mathbf{p}) \, p_j = \bar{\rho}_\gamma \, Z \, \hat{k}_j, \tag{H.45}$$

where X, Y, and Z are here functions of $k \equiv |\mathbf{k}|$ and ω. Eq. (H.43) then lets us express $\delta \mathbf{u}$ in terms of Z:

$$\delta \mathbf{u} = \frac{3Z}{4[1 - i\omega t_c R]} \, \hat{k}, \tag{H.46}$$

where $R \equiv 3\bar{\rho}_B / 4\bar{\rho}_\gamma$, and $t_c \equiv 1/\omega_c$ is the mean time between collisions. Using Eqs. (H.44) and (H.46) in Eq. (H.39) then yields a formula for the perturbed intensity:

$$4\pi \int \delta n_r^{ij}(p\hat{p}) p^3 \, dp = \frac{3\bar{\rho}_\gamma}{2[1 - i\omega t_c + i\hat{p} \cdot \mathbf{k} t_c]} \left[\left(X + \frac{Z(\hat{p} \cdot \hat{k})}{1 - i\omega t_c R} \right) \left(\delta_{ij} - \hat{p}_i \hat{p}_j \right) \right.$$

$$\left. + Y \left(\hat{k}_i - \hat{p}_i(\hat{p} \cdot \hat{k}) \right) \left(\hat{k}_j - \hat{p}_j(\hat{p} \cdot \hat{k}) \right) \right]. \tag{H.47}$$

By inserting this back in the definitions (H.44) and (H.45), we find three homogeneous linear relations among X, Y, and Z. For these to be consistent, the determinant of the coefficients must vanish, which yields a relation between ω and $k = |\mathbf{k}|$.

The resulting dispersion relation is quite complicated, but it becomes much simpler in the case of small mean free time t_c. As an *ansatz*, we can try taking Y/X of first order in t_c, while Z/X is of order unity, leaving it for later to check whether this leads to a consistent solution. To second order in t_c, the terms in Eq. (H.44) proportional to δ_{ij} and $\hat{k}_i \hat{k}_j$ and the coefficient of $\hat{k}_i$ in Eq. (H.45) yield the homogeneous linear relations

$$0 = \left(i\omega t_c - \omega^2 t_c^2 - \frac{2}{5} k^2 t_c^2 \right) X + \frac{1}{10} (1 + i\omega t_c) Y$$

$$+ \frac{2}{5} \left(-ikt_c + \omega k t_c^2 (2 + R) \right) Z \tag{H.48}$$

$$0 = \frac{1}{5} k^2 t_c^2 X + \left(-\frac{3}{10} + \frac{7i}{10} \omega t_c \right) Y$$

$$+ \left(\frac{i}{5} k t_c - \frac{(2+R)}{5} \omega k t_c^2 \right) Z \tag{H.49}$$

$$0 = \left(-ikt_c + 2\omega k t_c^2\right) X - \frac{1}{5}ikt_c\, Y$$

$$+ \left(i\omega t_c(1+R) - \omega^2 t_c^2(1+R+R^2) - \frac{3}{5}k^2 t_c^2\right) Z\,. \quad \text{(H.50)}$$

(As anticipated, these relations are consistent with the assumption that Y is of order t_c relative to X and Z.) To first order in t_c, the vanishing of the determinant of the coefficients in these relations yields the condition

$$0 = 15(1+R)\omega^2 - 5k^2 + i\omega^3 t_c\left[-5 - 5R + 15R^2\right] + 7i\omega k^2 t_c\,.$$

This has the solution, again to first order in t_c:

$$\omega = \pm\frac{k}{\sqrt{3(1+R)}} - i\Gamma\,, \quad \text{(H.51)}$$

where Γ is the decay rate

$$\Gamma = \frac{k^2 t_c}{6(1+R)}\left\{\frac{16}{15} + \frac{R^2}{1+R}\right\}\,. \quad \text{(H.52)}$$

This is the same as the formula (6.4.25) for the damping rate given in Section 6.4, originally derived by Kaiser.[2] As discussed there, it is equivalent to the formulas (6.4.24) for the coefficients of shear viscosity and heat conduction of a non-relativistic plasma in which momentum and energy are transported by photons.

---

[2]N. Kaiser, *Mon. Not. Roy. Astron. Soc.* **202**, 1169 (1983).

Glossary of Symbols

(These are symbols used in more than one section. Numbers following the symbol give the section in which the symbol is first used.)

a	1.1	scale factor in Robertson–Walker metric
a_B	2.1	radiation energy constant
$a_{\ell m}$	2.6	temperature partial wave amplitude (same as $a_{T,\ell m}$)
A	5.1	one of scalar perturbations to metric
B	3.3	baryon number
	5.1	one of scalar perturbations to metric
	6.1	subscript denoting baryonic plasma
C_i	5.1	one of vector perturbations to metric
C_ℓ	2.6	temperature multipole coefficient (same as $C_{TT,\ell}$)
d	1.2	proper distance
d_A	1.4	angular diameter distance
d_H	2.6	horizon distance
d_L	1.4	luminosity distance
D	6.1	subscript denoting cold dark matter
$D^{(\ell)}$	7.4	representative of the rotation group for spin ℓ
D_{ij}	5.1	tensor perturbation to metric
$\mathcal{D}_q$	5.2	gravitational wave amplitude
$\mathcal{D}_q^o$	6.6	constant gravitational wave amplitude outside horizon
e_{ij}	5.2	graviton polarization tensor
E	5.1	one of scalar perturbations to metric
EQ	6.2	subscript indicating the time of matter–radiation equality
F	5.1	one of scalar perturbations to metric
	6.6	function appearing in tensor anisotropic inertia
	7.1	form factor in scalar temperature fluctuation
g	3.1	multiplicity of states
$g_{\mu\nu}$	1.1	spacetime metric
$\tilde{g}_{ij}$	1.1	spatial part of Robertson–Walker metric
G	1.5	Newton's gravitational constant
	7.1	form factor in scalar temperature fluctuation
G_i	5.1	one of vector perturbations to metric
h	1.3	Hubble constant in units of 100 km sec^{-1} Mpc^{-1}
	2.4	Planck's constant $= 2\pi\hbar$
$h_{\mu\nu}$	5.1	perturbation to the metric
H	2.2	expansion rate $\dot{a}(t)/a(t)$
H_0	1.2	Hubble constant $\dot{a}(t_0)/a(t_0)$

$H_\nu^{()}$ 10.1 Hankel functions

I 1.12 action

ISW 2.6 contribution of integrated Sachs–Wolfe effect

j_ℓ 2.6 spherical Bessel function

J 6.1 dimensionless neutrino intensity

J^μ 1.1 current four-vector

J_{ij} 6.1 dimensionless photon intensity matrix

k 8.1 physical wave number at present $= q/a_0$

k_B 2.2 Boltzmann constant

K 1.1 curvature constant in Robertson–Walker metric

  6.6 function appearing in tensor anisotropic inertia

ℓ 1.3 apparent luminosity

  2.6 multipole order

L 1.3 absolute luminosity

  2.1 subscript indicating time of last scattering

  3.3 lepton number

m 1.3 apparent magnitude

  1.5 particle mass

M 1.3 absolute magnitude

  1.9 total mass

Mpc 1.3 million parsecs

n 1.10 number density

n^{ij} 6.1 photon number density matrix

n_S 7.2 slope parameter in $\mathcal{R}_q^o$

N 7.2 coefficient of $q^{-3/2}$ in $\mathcal{R}_q^o$

$\mathcal{N}$ 3.1 effective number of particle types

  4.1 number of e-foldings in inflation

p 1.1 pressure

pc 1.3 parsec

P_ℓ 2.4 Legendre polynomials

P 8.1 power spectral function of density fluctuations

$\mathcal{P}_\phi$ 2.6 power spectral function of gravitational potential

$\mathbf{q}$ 2.6 co-moving wave number vector

q_0 1.4 deceleration parameter

r 1.1 radial coordinate in Robertson–Walker metric

  7.3 tensor/scalar ratio

R 2.6 3/4 the ratio of baryon to photon energy density

$R_{\mu\nu}$ 1.5 Ricci tensor

$\mathcal{R}$ 5.4 curvature perturbation

$\mathcal{R}^o$ 6.2 constant curvature perturbation outside horizon

s 1.1 proper length

  3.1 entropy density

S	7.1	superscript indicating scalar mode
$S_{\mu\nu}$	1.5	source term in Einstein field equation
$S(\hat{n})$	7.4	standard rotation from three-axis to direction $\hat{n}$
$\mathcal{S}$	6.5	a scalar transfer function
SW	2.6	contribution of Sachs–Wolfe effect
t	1.1	time
t_0	1.5	present age of universe
t_1	6.1	initial time in solution of Boltzmann equation
T	2.1	temperature
	7.1	superscript indicating tensor mode
$T_{\mu\nu}$	1.1	energy-momentum tensor
$\mathcal{T}$	6.5	the main scalar transfer function
u	1.1	affine parameter
	5.1	velocity potential
u_μ	5.1	velocity four-vector
$\mathcal{U}$	6.6	a tensor transfer function
v_s	2.6	sound speed
V	1.12	scalar field potential
w	1.1	ratio of pressure to density
$\mathbf{x}$	1.1	co-moving spatial coordinate
X	2.3	fractional ionization
$\mathbf{X}$	1.9	Newtonian spatial coordinate
y	6.3	ratio of matter and radiation densities$= a/a_{\mathrm{EQ}}$
Y_ℓ^m	2.6	ordinary spherical harmonics
$\mathcal{Y}_\ell^m$	7.4	spin $+2$ spherical harmonics
z	1.2	redshift
	10.1	quantity $a\dot{\bar{\varphi}}/H$ in Mukhanov–Sasaki equation
α	5.2	stochastic parameter of scalar modes
	9.1	angle between gravitational lens and true source positions
β	5.2	stochastic parameter of tensor modes
	6.4	ratio of baryon and total matter densities
	9.1	angle between gravitational lens and apparent source positions
β_E	9.1	angular radius of Einstein ring
γ	6.1	subscript denoting photons
	6.4	Euler constant
Γ	6.4	acoustic decay rate
$\Gamma^\mu_{\nu\kappa}$	1.1	affine connection
$\tilde{\Gamma}^i_{jk}$	1.1	spatial affine connection for Robertson–Walker metric
δ_{ij}	1.1	Kronecker delta
δ_{Xq}	6.2	$\delta\rho_{Xq}/(\bar{\rho}_X + \bar{p}_X)$ (for $X = B, D, \gamma, \nu$)
Δ	5.3	change due to gauge transformation

	6.5	a scalar transfer function
$\Delta_T^{(S)}$	6.1	scalar photon temperature amplitude
$\Delta_P^{(S)}$	6.1	scalar photon polarization amplitude
$\Delta_T^{(T)}$	6.6	tensor photon temperature amplitude
$\Delta_P^{(T)}$	6.6	tensor photon polarization amplitude
ϵ	10.1	$-\dot{H}/H^2$
ϵ_0, ϵ^S	5.3	gauge transformation parameters
ζ	5.4	perturbation conserved outside horizon
η	3.2	baryon/photon ratio
$\eta_{\mu\nu}$	1.5	Minkowski spacetime metric
θ	1.1	polar angle (except in Sec. 9.5)
κ	6.2	dimensionless rescaled wave number
μ	3.1	chemical potential
ν	1.2	frequency
	6.1	subscript denoting neutrinos & antineutrinos
Ξ	6.6	a tensor transfer function
π_{ij}	1.5	total anisotropic inertia tensor
π^S	1.5	scalar anisotropic inertia
π_i^V	1.5	vector anisotropic inertia
π_{ij}^T	1.5	tensor anisotropic inertia
Π	6.1	a scalar source function
ρ	1.1	energy density
σ	1.10	cross section
	2.2	entropy per baryon
	8.1	standard deviation of density perturbations
τ	1.1	proper time
	1.10	optical depth
	5.3	parameter for transformation between synchronous gauges
	10.1	conformal time
ϕ	1.1	azimuthal angle (except in Sec. 9.5)
	2.6	gravitational potential
φ	1.12	scalar field
Φ	5.3	a scalar metric perturbation in Newtonian gauge
	6.1	a scalar source function
ψ	5.3	a scalar metric perturbation in synchronous gauge
Ψ	5.3	a scalar metric perturbation in Newtonian gauge
	6.6	tensor source function
Ω_K	1.5	$-K/a_0^2 H_0^2$
Ω_M	1.5	ratio of matter density to critical density
Ω_R	1.5	ratio of radiation density to critical density
Ω_Λ	1.5	ratio of vacuum density to critical density

Assorted Problems

1. Consider a universe described by a Robertson–Walker metric with $K = +1$. Give a transformation of co-moving space coordinates that leaves the metric unchanged, and that takes the point $\mathbf{x} = (0, 0, r)$ into a point $\mathbf{x} = (0, 0, r')$, with no change in the time. (Hint: Consider the three-dimensional space as the surface of a four-dimensional ball, construct this transformation as a rotation in four dimensions, and then express it in terms of the Robertson–Walker coordinates.) Also give the corresponding transformation for $K = -1$.

2. Suppose that the total pressure and energy density of the universe are related by $p = -\rho + \rho^2/\rho_1$, where ρ_1 is a constant. Assume zero spatial curvature. Calculate ρ as a function of the Robertson–Walker scale factor a, taking $a = a_1$ when $\rho = \infty$. Calculate a and ρ as functions of time, taking $t = 0$ as the time when $\rho = \infty$. Calculate the age of the universe and the deceleration parameter q_0 for a given present density ρ_0.

3. Consider the empty cosmology, with $\Omega_M = \Omega_R = \Omega_\Lambda = 0$. Calculate the luminosity distance and angular diameter distance as functions of redshift in this cosmological model. What is the age of the universe as a function of the present Hubble constant?

4. Suppose that astronomers measure the age of a galaxy with redshift $z = 2.5$. How old would this galaxy have to be (at the time the light from it was emitted) in order to rule out the hypothesis that $\Omega_M = 1$ with negligible vacuum and radiation energy density. Use $H_0 = 70$ km/sec/Mpc.

5. Suppose that $\Omega_M = 0.25$ and $\Omega_V = 0.75$, with Ω_R negligible. What is the redshift at which the expansion of the universe stopped decelerating and began to accelerate?

6. Suppose that the gravitational potential energy of any pair of galaxies with separation r decreases as r^{-n} instead of r^{-1}. What combination of the mass of a virialized cluster of galaxies and the Hubble constant could be calculated from measurements of angular separations and velocity dispersions of its individual galaxies?

7. Suppose that the fluctuations in the temperature of the cosmic microwave background are governed by a Gaussian isotropic

translation-invariant probability distribution. Calculate the quantity

$$\left\langle \left(C_\ell^{\text{obs}} - C_\ell \right)^3 \right\rangle / C_\ell^3$$

as a function of ℓ.

8. Suppose that there is some new massless elementary particle that effectively decouples from everything else at a temperature 10^{16} K. Suppose that the particles with mass less (and in fact much less) than 1 TeV are those of the minimal Standard Model: three generations of quarks, leptons, antiquarks, and antileptons with helicity $\pm 1/2$; the $W^\pm$ and Z^0 with helicity ± 1 and 0; the photon and eight massless gluons with helicity ± 1; and a single neutral boson of helicity zero; together with their antiparticles (where distinct from the particles). What is the ratio at present of the temperature of the new massless particles and the temperature of photons?

9. Suppose that the binding energy of deuterium were 5% less. Estimate the effect that this would have on the abundance of He^4 produced in the early universe.

10. Suppose that instead of $B - L$, only the combination $3B - L$ of baryon number and lepton number were conserved, along with the usual conserved quantities of the Standard Model. Taking account of the particles of the minimal (non-supersymmetric) Standard Model, what is the density of baryon number and lepton number in thermal equilibrium for a given density of $3B - L$?

11. Suppose that the Robertson–Walker scale factor increases during inflation by a factor e^{50}; that the scale factor at the beginning of the radiation-dominated era was the same as at the end of inflation; that the energy density at the end of inflation was $[2 \times 10^{16} \text{GeV}]^4$; and that the energy density at the beginning of the radiation-dominated era was $[2 \times 10^{15} \text{GeV}]^4$. Over what range of angles would you expect the cosmic microwave background now to have a nearly uniform temperature?

12. Use the "slow-roll" approximation to calculate the inflaton field $\varphi(t)$ as a function of time for a potential $V(\varphi) = g\,\varphi^n$, where g and n are positive constants.

13. Find the two independent solutions of the field equations and conservation equations for scalar perturbations far outside the horizon in Newtonian gauge in the case of a perfect fluid having $p = w\rho$, with

w constant. Use the results to calculate $\mathcal{R} \equiv -\Psi + H\,\delta u$ for each solution.

14. Consider the equations for scalar fluctuations in Newtonian gauge in a universe containing only cold dark matter. Find the two solutions of the coupled equations for Φ_q and δu_q. (You can normalize these solutions any way you like, but keep it simple.) For each solution, calculate $\mathcal{R}_q$, $\delta\rho_q$, and ζ_q. Do not make any assumption about the magnitude of the wave number.

15. Consider a vector field

$$V^\mu(\mathbf{x}, t) = \bar{V}^\mu(t) + \delta V^\mu(\mathbf{x}, t) ,$$

where the unperturbed space components $\bar{V}^i$ are zero, and the unperturbed time component $\bar{V}^0$ is some non-zero function only of time. What is the change in δV^0 and δV^i induced by a general gauge transformation?

16. Re-do the analysis of Section 5.2, but for vector fluctuations.

17. Suppose that the differential scattering cross section of neutrinos on all other particles were a constant, independent of direction or energy. (It isn't.) For a given total neutrino collision rate $\omega_c(t)$, derive the Boltzmann equation that would be satisfied by the neutrino phase space distribution function $n_\nu(\mathbf{x}, \mathbf{p}, t)$ in a spacetime metric with $g_{00} = -1$, $g_{i0} = 0$, and $g_{ij}(\mathbf{x}, t)$ arbitrary. For the case where $g_{ij}(\mathbf{x}, t) = a^2(t)\delta_{ij} + \delta g_{ij}(\mathbf{x}, t)$, derive the Boltzmann equation for the phase space density perturbation $\delta n_\nu(\mathbf{x}, t)$ defined by Eq. (6.1.39), to first order in δg_{ij} and δn_ν.

18. Suppose that H_0 were doubled, and Ω_B and Ω_M reduced by a factor four. What would be the qualitative effect on the plot of $\ell^2 C_\ell$ vs. ℓ for $\ell \gg 1$?

19. Assume that $C_{TT,\ell}^S$ is dominated by a co-moving wave number at last scattering equal to ℓ/r_L, where r_L is the co-moving radial coordinate of the surface of last scattering. Also, neglect the Doppler effect, and use $\mathcal{S} = 5$ and $\mathcal{T} = 1$ for the transfer functions. With these assumptions, give an approximate formula for the ratio of the heights of the first and second acoustic peaks in $C_{TT,\ell}^S$.

20. What value of the multipole order ℓ corresponds to a co-moving wavelength equal to the diameter of a co-moving sphere that encloses the total mass $10^{15}\,M_\odot$ (including cold dark matter) of a cluster of galaxies?

21. Define a mean square matter density fluctuation as

$$\sigma_R^2 \equiv \left\langle \left(\frac{1}{(\sqrt{\pi} R)^3} \int d^3x \, \delta_M(\mathbf{x}, t) \, e^{-\mathbf{x}^2/R^2} \right)^2 \right\rangle$$

where the integral extends over all space. At what time does σ_R^2 become equal to 0.1 for R equal to the diameter of a co-moving sphere that encloses the total mass $10^{12} \, M_\odot$ (including cold dark matter) of a large galaxy? Assume a Harrison–Zel'dovich primordial spectrum with $|N| = 2 \times 10^{-5}$.

22. Derive a formula for the multipole coefficients in the correlation between fluctuations in the cosmic microwave background temperature in a direction $\hat{n}$ and the cosmic mass density fluctuation at a redshift z and direction $\hat{n}'$. Use any approximation that you think is reasonable, but make it clear what approximations you are using.

23. What is the effect of a general infinitesimal rotation on the polarization tensor $e_{ij}(\hat{n}, \pm 2)$ of a gravitational wave travelling in a direction $\hat{n}$ along the 1-axis?

24. Suppose that the probability distribution of tensor fluctuations respected the symmetries of translation and rotation invariance, but not invariance under space inversion. What would that imply for the average $\langle \beta(\mathbf{q}, \lambda) \beta^*(\mathbf{q}', \lambda') \rangle$, where $\beta(\mathbf{q}, \lambda)$ is the stochastic parameter for wave number $\mathbf{q}$ and helicity λ? Give a formula for the multipole coefficient $C_{EB,\ell}^T$ in this case.

25. Suppose that light passing a galaxy is deflected toward the center of the galaxy by an angle A/b^n, where A and n are constants, and b is the distance of closest approach of the light to the galaxy center. Find the lens equation, i.e. a relation between the observed angle β between the lines of sight from the earth to the lensing galaxy and a more distant source of light and the angle α that there would be between these lines of sight if light were not deflected by the galaxy. Give a formula for the apparent luminosity of the image of the source in terms of β.

26. Define multipole coefficients $C_{\gamma E,\ell}$ by

$$\langle a_{\gamma,\ell m} \, a_{E,\ell'm'}^* \rangle = \delta_{\ell\ell'} \, \delta_{mm'} \, C_{\gamma E,\ell} ,$$

where $a_{\gamma,\ell m}$ and $a_{E,\ell m}$ are the coefficients in the partial wave expansion of the shear and the Stokes parameters, defined in Sections 9.5 and 7.4. Give a formula for the correlation functions $\langle \gamma_1(\hat{n}) \, Q(\hat{n}') \rangle$,

$\langle \gamma_2(\hat{n})\, Q(\hat{n}') \rangle$, $\langle \gamma_1(\hat{n})\, U(\hat{n}') \rangle$, $\langle \gamma_2(\hat{n})\, U(\hat{n}') \rangle$ of the shear and Stokes parameters in terms of these multiipole coefficients.

27. Assume that inflation is driven by a single inflation field φ, with a potential $V(\varphi) = g \exp(-\lambda \varphi^2)$. What are the conditions on λ and φ to allow the use of the slow-roll approximation? Assuming these conditions to be satisfied, give values for the scalar and tensor slope parameters n_S and n_T. By how many e-foldings does the universe expand when φ changes by a factor 100?

28. In multi-field inflation, what is the dependence of scalar field perturbations on wave number outside the horizon when you take account of a slight variation of the potential during the era of horizon exit?

Author Index

Where page numbers are in italics, they refer to figures, footnotes or references.

Wright, E. L. *147*
Wu, K. K. S. *1*
Wyse, R. F. *126*

Yadav, J. *1*
Yamaguchi, M. *328*
Yamashita, K. *85*
Yanagida, T. *179*
Yao, W.-M. *157*
Yee, H. K. C. *68*
Yock, P. *443*
Yokayama, J. *328*
Yokoyama, S. *502*
York, D. G. *415*
Yoshii, Y. *85, 172, 446*

Yoshimura, M. *176*
Yu, J. T. *135, 257, 294, 418*

Zakharov, A. F. *442*
Zaldarriaga, M. viii, *77, 82, 257, 267,
334, 346, 366, 370, 374, 376, 377,
379, 391, 397*
Zaritsky, D. *23, 46*
Zatsepin, G. T. *107*
Zee, A. *176, 246*
Zel'dovich, Ya B. *116, 132, 133, 135,
142, 160, 207, 257, 294,* 414, *418*
Zinn-Justin, J. *325, 348*
Zioutas, K. *199*
Zlatev, I. *90, 96*
Zwicky, F. *66, 68*

Subject Index